THE ROUTLEDGE HANDBOOK OF PLACEMAKING

This Handbook is the first to explore the emergent field of 'placemaking' in terms of the recent research, teaching and learning, and practice agenda for the next few years. Offering valuable theoretical and practical insights from the leading scholars and practitioners in the field, it provides cutting-edge interdisciplinary research on the placemaking sector.

Placemaking has seen a paradigmatic shift in urban design, planning, and policy to engage the community voice. This Handbook examines the development of placemaking, its emerging theories, and its future directions. The book is structured in seven distinct sections curated by experts in the areas concerned. Section One provides a glimpse at the history and key theories of placemaking and its interpretations by different community sectors. Section Two studies the transformative potential of placemaking practice through case studies on different places, methodologies, and theoretical frameworks. It also reveals placemaking's potential to nurture a holistic community engagement, social justice, and human-centric urban environments. Section Three looks at the politics of placemaking to consider who is included and who is excluded from its practice and if the concept of placemaking needs to be reconstructed. Section Four deals with the scales and scopes of art-based placemaking, moving from the city to the neighborhood and further to the individual practice. It juxtaposes the voice of the practitioner and professional alongside that of the researcher and academic. Section Five tackles the socio-economic and environmental placemaking issues deemed pertinent to emerge more sustainable placemaking practices. Section Six emphasizes placemaking's intersection with urban design and planning sectors and incudes case studies of generative planning practice. The final seventh section draws on the expertise of placemakers, researchers, and evaluators to present the key questions today, new methods and approaches to evaluation of placemaking in related fields, and notions for the future of evaluation practices. Each section opens with an introduction to help the reader navigate the text. This organization of the book considers the sectors that operate alongside the core placemaking practice.

This seminal Handbook offers a timely contribution and international perspectives for the growing field of placemaking. It will be of interest to academics and students of placemaking, urban design, urban planning and policy, architecture, geography, cultural studies, and the arts.

Cara Courage is a placemaking, arts, activism, and museums academic-practitioner, and Head of Tate Exchange, Tate. Cara is author of *Arts in Place: The Arts, the Urban and Social Practice* (Routledge, 2017), and the co-editor of *Creative Placemaking and Beyond* (Routledge, 2018).

Tom Borrup is an international consultant and author of *The Power of Culture in City Planning* and *The Creative Community Builders' Handbook.* He is Senior Lecturer and Director of Graduate Studies for the University of Minnesota's Master of Professional Studies in Arts and Cultural Leadership.

Maria Rosario Jackson's expertise is comprehensive community revitalization, systems change, dynamics of race and ethnicity, and roles of arts and culture in communities. She is Institute Professor at Arizona State University and also has a long career in strategic planning, research and evaluation with philanthropy, government and nonprofit organizations.

Kylie Legge is the CEO and founder of place data analytics company Place Score and place-making consultancy Place Partners. Kylie is a passionate advocate for human-centred design in cities and is the author of *Doing it Differently* and *Future City Solutions*.

Anita McKeown is an interdisciplinary artist, curator, scholar, and educator with research interests in Sustainable Creative Placemaking and Open Source Culture and Technology. She is the Co-Director of SMARTlab Skelligs, research lab in South Kerry and SMARTlab's NAISC Skellig Kerry Diaspora Network Fellow.

Louise Platt is a senior lecturer in Festival and Events at Manchester Metropolitan University. Her research interests predominantly lie in the role of festivity in places. She is on the executive committee of the Leisure Studies Association and the editorial board of *Leisure Studies* journal.

Jason Schupbach is a nerd, the Dean of the Westphal College of Media Arts and Design at Drexel University, and a nationally recognized expert in the role that arts and design play in improving communities. He was the federal liaison to the design community in his role as Director of Design and Creative Placemaking Programs for the National Endowment for the Arts.

THE ROUTLEDGE HANDBOOK OF PLACEMAKING

Edited by Cara Courage

WITH

TOM BORRUP
MARIA ROSARIO JACKSON
KYLIE LEGGE
ANITA MCKEOWN
LOUISE PLATT
JASON SCHUPBACH

Routledge
Taylor & Francis Group
LONDON AND NEW YORK

First published 2021
by Routledge
2 Park Square, Milton Park, Abingdon, Oxon OX14 4RN

and by Routledge
52 Vanderbilt Avenue, New York, NY 10017

Routledge is an imprint of the Taylor & Francis Group, an informa business

British Library Cataloguing-in-Publication Data
A catalogue record for this book is available from the British Library

Library of Congress Cataloging-in-Publication Data
A catalog record has been requested for this book

ISBN: 978-0-367-22051-8 (hbk)

ISBN:978-0-367-68409-9(pbk)

ISBN: 978-0-429-27048-2 (ebk)

Typeset in Bembo
by Deanta Global Publishing Services, Chennai, India

First show up, in a while people will come to know you, keep showing up, and in a while, you will have a community.

*Carol Bebelle, Founder of Ashe Cultural Center in New Orleans**

* see Cook, this volume.

CONTENTS

CONTENTS CURATED BY TOPICS

Art and artists

Climate and environmental ecology

Community of practice

COVID-19

Creative placemaking

Digital and technology

Equitable and intersectional placemaking

Evaluation

Festivals

Funding

Governance and stewardship

Indigenous placemaking and place

LGBTQI+

Migrant and displaced peoples

Narrative

Non-human

Place branding

Place economics

Place knowledge

Placemaking and community development

Placemaking definition

Placemaking futures

Placemaking history

Placemaking practice

Planning and generative planning

Policy

Productivity and innovation

Protest and resistance

Regeneration, development, and gentrification

Rural placemaking

Settler colonialism

Social practice placemaking

Spatial access

Sustainable Development Goals

Tactical urbanism

Teaching and learning

Tourism

Urban design

Wellbeing and healing

FIGURES

ABBREVIATIONS

ACF	Asia Culture Forum
ADD	*Art du Déplacement*
AHRC	Arts and Humanities Research Council (UK)
AICP	American Institute of Certified Planners
API	Asian and Pacific Islander
AQAL	All-Quadrants/All-Levels
AUD	Australian Dollar
BID	Business Improvement District
BSCG	Berkeley Street Community Garden
CABE	Commission for Architecture and the Built Environment
CAC	Citizen Advisory Committee
CBD	Central Business District
CBP&RC	Committee for Better Parks and Recreation in Chinatown
CBPAR	Community Based Participatory Action Research
CCDC	Chinatown Community Development Center
CCF	Continuous Cover Forest
CDC	Center for Disease Control
CDC	Community Development Corporation
CDI	Community Development Investments
CEO	Chief Executive Officer
CETA	Comprehensive Employment and Training Act of 1973
CFC	chlorofluorocarbon
CIP	*Centro Internacional de la Papa*
CLT	Community Land Trust
COVID-19	Coronavirus Disease 2019
CRA	Community Redevelopment Agency
CRC	Chinatown Resource Center
CSA	Community Supported Agriculture
CURS	Centre for Urban and Regional Studies
DE	Developmental Evaluation
DOT	US Department of Transportation

DRC	Design Review Committee
EA	Embedded Artist
EDC	Economic Development Corporation
ELNA	East Lawrence Neighborhood Association
EP	English Partnerships
EPA	Environmental Protection Agency (US)
EZ	Empowerment Zone
FFIG	Fruit Futures Initiative Gary
FHA	Federal Housing Administration
FOI	Freedom of Information
FOVC	Fear of violent crime
GCM	General Circulation Models
GDP	Gross Domestic Product
GNMP	Global NGO Master's Program
GRG	Great Rivers Greenway
GSD	Harvard Graduate School of Design
HCI	Human–Computer Interaction
HMRI	Housing Market Renewal Initiative
HOLC	Home Owners' Loan Corporation
HUD	US Department of Housing and Urban Development
iD+Pi	Indigenous Design and Planning Institute
IP	Indigenous planning
ioby	in our back yards
IoT	Internet of Things
IPCC	Intergovernmental Panel on Climate Change
JACCC	Japanese American Cultural and Community Center
JACS-AI	American Community Services-Asian Involvement
KCIA	National Intelligence Service
LAPD	Los Angeles Police Department
LGBTQ	Lesbian, gay, bisexual, transgender, and questioning (or queer)
LGBTQ+	Lesbian, gay, bisexual, transgender, and questioning/queer; the 'plus' is intended as an all-encompassing representation of sexual orientations and gender identities.
LISC	Local Initiatives Support Corporation
LLB	Bachelor of Laws
LT-PRO	Little Tokyo Peoples' Rights Organization
LTRP	Little Tokyo Redevelopment Project
LTSC	Little Tokyo Service Center
MAPC	Metropolitan Area Planning Council
MAV	Municipal Association of Victoria
MCC	Manchester City Council
MCM	Midcentury modern
MCST	Ministry of Culture, Sports and Tourism
MDPAG	Manchester Disable Peoples Access Group
MoMA	Museum of Modern Art (New York)
NACEDA	National Alliance of Community and Economic Development Associations
NCRR	Nikkei for Civil Rights and Redress
NEA	National Endowment for the Arts

NGO	Non-governmental organization
NSW	New South Wales
NYC	New York City
OHCHR	Office of the United Nations High Commissioner for Human Rights
pCr	PermaCultural Resilience
PEARL	Pendle Enterprise and Regeneration (Brierfield Mill) Ltd
PEC	People's Emergency Center
PIE	Psychologically Informed Environments
PPP	People, Process, Place [framework]
PPS	Project for Public Spaces
PRADS	Parks Related Anti-Displacement Strategies
QMem	Queer Memorials: International Comparative Perspectives on Sexual Diversity and Social Inclusivity
RFP	Requests for Proposals
RTPI	Royal Town Planning Institute
SDG	Sustainable Development Goal
SIAP	Social Impact of the Arts Project
SRO	Single Room Occupancy
STEAM	Science, Technology, Engineering Arts, Math(s)
T4A	Transportation for America
TIF	Tax Increment Financing
TPAR	Transformative Participatory Action Research
TRIP	Transportation Research and Improvement Project
TSAC	Times Square Advertising Coalition
UCL	University College London
UCLA	University of California, Los Angeles
UCLG CISDP	United Cities and Local Governments Committee on Social Inclusion, Participatory Democracy and Human Rights
UK	United Kingdom
UN	United Nations
UNCRPD	United Nations Convention on the Rights of Persons with Disabilities
UNEP	United Nations Environment programme
UNESCO	United Nations Educational, Scientific and Cultural Organization
URC	Urban Regeneration Companies
US	United States of America
USDAC	US Department of Arts and Culture
USP	Unique selling point
VALI	Validating Arts and Livability Indicators
WHO	World Health Organization
WHRCF	World Human Rights Cities Forum
WMO	World Metrological Organization
WPA/FAP	Works Progress Administration Federal Art Project
WWF	World Wildlife Fund
YMCA	Young Men's Christian Association
YWCA	Young Women's Christian Association

EDITORS

Cara Courage, Handbook Editor and Section Editor, is a placemaking, and arts, urban and museums academic and practitioner, and Head of Tate Exchange, Tate's platform dedicated to socially engaged art. Cara is author of *Arts in Place: The Arts, the Urban and Social Practice* (Routledge, 2017), the co-editor of *Creative Placemaking and Beyond* (Routledge, 2018), with Anita McKeown, and instigated this Handbook, convening its curatorial panel. Cara has a 20-plus-year career in the arts largely working with museums and galleries, arts in the public realm, and public engagement with the built environment, as a consultant and project manager for public and private initiatives and her own projects, alongside a research practice on the same. At Tate Exchange, Cara has developed its specialism in particular in social practice arts, social justice, climate emergency, and creative aging through its curation and programming and across international practice and research networks, in the museum and arts sectors. Cara speaks internationally on topics covering the twenty-first-century museum, the civic and activist museum, socially engaged art in community and museum settings and arts, and urban design, placemaking, and planning.

Section editors

Tom Borrup is an international consultant, speaker, and lecturer addressing cultural planning, creative placemaking, and cultural district planning. As founder of Creative Community Builders, he consults with cities, foundations, and nonprofits to develop synergy between arts, economic development, urban planning and design. His 2006 book *The Creative Community Builders' Handbook*, remains a leading text in the field. Tom's new book, *The Power of Culture in City Planning: Living Together Through the 21st Century* is expected in Fall 2020, published by Routledge. Tom earned his PhD in Leadership and Change from Antioch University researching the role of social and organizational networks in the planning and management of cultural districts. Tom has an MA in Communications and Public Policy from Goddard College and was a 2001–2002 Fellow in the Knight Program in Community Building at the University of Miami School of Architecture. He serves as Director of Graduate Studies and Senior Lecturer for the University of Minnesota's Master of Professional Studies in Arts and Cultural Leadership. He is also a Visiting Professor at the University of International Business and Economics in Beijing, China and teaches Cultural Planning for Drexel University's Arts Administration Graduate Program.

Maria Rosario Jackson is an expert in comprehensive community revitalization, urban poverty, systems change, dynamics of race and ethnicity and roles of arts and culture in communities. She is an Institute Professor in the Herberger Institute for Design and the Arts and the Watts College of Public Service and Community Solutions at Arizona State University where she also leads the Studio for Creativity, Place and Equitable Communities. Maria is senior advisor to the Kresge Foundation working with the Strategic Research Learning and Evaluation and the Arts and Culture Programs. President Obama appointed Maria to the National Council on the Arts in 2012. Maria co-chairs the County of Los Angeles Cultural Equity and Inclusion Initiative and serves on the boards of Smithsonian Center for Folklife and Cultural Heritage and The Music Center (Los Angeles County) among other organizations. Previously, Maria worked at the Urban Institute, a national public policy research organization in Washington DC for almost 20 years. Maria earned a PhD in Urban Planning from the University of California, Los Angeles and a Master of Public Administration from University of Southern California.

Kylie Legge is a leading voice in the evolving profession of placemaking. Kylie is an architecture graduate, planner, author, facilitator, curator, entrepreneur. She is the CEO and founder of place data analytics company Place Score and placemaking consultancy Place Partners. Kylie is a passionate advocate for human-centred design in cities, the relationship between people and their urban environments and how we can work better collaboratively to create the kinds of places people want to spend time in. Her commitment to understanding the trends and external factors influencing decision-making ensures that projects she works on are fit for the future. Kylie is the author of the Urban Trends book series which includes two highly regarded titles, Doing it Differently, and Future City Solutions.

Anita McKeown is an award-winning itinerant artist, curator and researcher working at the intersection of art, equitable spatial planning, and technology. Using a range of tactics – situated arts practice, publication, and education – Anita works to facilitate the co-creation of locally scaled interventions that are context-responsive and ecologically sensitive. Working across artforms, digitally and analogue, the projects arise from extended relationships with people and place to contribute to and encourage systemic self-organization and resilience.

Louise Platt is an interdisciplinary researcher and Senior Lecturer in Festival Management at Manchester Metropolitan University. Louise is a fellow of the Institute of Place Management and member of the Executive Committee of the Leisure Studies Association. Her research focus is on placemaking and festivity with a particular focus on processional forms, and experiences of festivals and leisure spaces. Louise's work predominantly draws on cultural geography, dance/performance theory, and post-structural philosophy to elucidate a more fluid understanding of place and community using festivity as a lens. She teaches festival studies at both undergraduate and postgraduate levels and supports PhD students on festival related topics.

Jason Schupbach is the Dean of the Westphal College of Media Arts and Design at Drexel University. Jason was previously the director of Design and Creative Placemaking programs for the National Endowment for the Arts, where he oversaw all design and creative placemaking grant-making and partnerships, including *Our Town* and *Design Art Works* grants, the Mayor's Institute on City Design, the Citizens' Institute on Rural Design, and the NEA's Federal agency collaborations. Previous to his current position, Jason served Governor Patrick of Massachusetts as the Creative Economy director, tasked with growing creative and tech businesses in the state. He formerly was the director of ArtistLink, a Ford Foundation–funded initiative to stabilize and revitalize communities through the creation of affordable space and innovative environments for creatives. Jason has also worked for the Mayor of Chicago and New York City's Department of Cultural Affairs. He has written extensively on the role of arts and design in making better communities, and his writing has been featured as a Best Idea of the Day by the Aspen Institute.

CONTRIBUTORS

Maribel Alvarez holds the Jim Griffith Chair in Public Folklore at the Southwest Center, University of Arizona, where she also is Associate Dean for Community Engagement in the College of Social & Behavioral Sciences. Maribel is the founder of the Southwest Folklife Alliance, a nonprofit affiliated with the University of Arizona, which produces programs connecting artisanal economies, foodways, and traditional arts to community planning and neighbourhood-based economic development throughout the US–Mexico border corridor. In 2018 the American Folklore Society awarded Maribel the prestigious Americo Paredes Prize for 'excellence in integrating scholarship and engagement with the people and communities one studies.'

Husam AlWaer is an urbanist with a multi-disciplinary background in Architecture, Urban Planning, and Sustainability, and an award-winning author and curator of events, focusing on issues of placemaking and sustainable urban design practice and their social impacts. Husam is currently a professorial Reader position in Sustainable Urban Design at University of Dundee. He has published over 45 peer-reviewed international journal papers, professional reports, and books relating to the broad topic of Urban Design, Spatial Planning, and Sustainability, including his recent published book *Site and Composition: Design Strategies in Architecture and Urbanism* with Bandyopadhyay and Aldallal (Routledge, 2016), and a new international edited book with Barbara Illsley on *Rethinking Masterplanning: Creating Better Places* (ICE Publisher, 2017).

Jeff Baldwin is a Professor of Geography in the Geography, Environment, and Planning Department at Sonoma State University. Jeff's research projects have explored various relationships between social systems and environmental communities. His explorations on the implications of recognizing the agency of non-humans for environmental ethics focuses upon the promise of and barriers to beaver recolonization in the increasingly drought-prone American West. Throughout, Jeff has argued that non-human beings have their own agency, their own projects, and that we could increase human wellbeing by cultivating partnerships with our very active biospheric co-inhabitants.

Neil Brenner is Professor of Urban Theory at the Harvard Graduate School of Design (GSD). Neil's writing and teaching focus on the theoretical, conceptual, methodological, and cartographic dimensions of urban questions. His work builds upon, and seeks to extend, the fields of critical urban and regional studies, comparative geopolitical economy, and radical sociospatial theory.

Sarah Calderon is the Managing Director of ArtPlace America. Previously, Sarah Calderon was the Executive Director of Casita Maria Center for Arts & Education (Bronx, NY) from 2008 to 2015. During her tenure, she has overseen the opening of a new, 90,000-square-foot facility for the Center's arts and education programming and developed partnerships with organizations ranging from Lincoln Center to the NYC Housing Authority. Before joining Casita, Sarah founded and ran Stickball Printmedia Arts in East Harlem, a printmaking and digital arts organization for youth.

Roy Chan is an urban planner who practices place-based oral history to inform how we plan our neighbourhoods in culturally inclusive ways. As the Community Planning Manager at the Chinatown Community Development Center, Roy leads a Creative Placemaking initiative that brings local artists, community organizers, and residents together to preserve and activate cultural spaces in San Francisco Chinatown. Roy previously served as Co-Executive Director at the Oakland Asian Cultural Center, where he continues to direct the Oakland Chinatown Oral History Project. He currently serves as a Cultural Affairs Commissioner for the City of Oakland.

Eliza Charley is a Strategic Consultant at CoDesign Studio in Melbourne, Australia. She is a writer, actor, and business strategist who has worked in industries as varied as urban development, data analytics, corporate consulting, interior design, and film and television. Eliza has a passion for storytelling and understanding how people find their sense of belonging in life. Wherever Eliza is planted, she works towards delivering projects that promote people's sense of belonging whether that be through content creation, strategic business practices, or developing technology.

Samantha Choudhury has over 15 years' experience in Australia and the USA as a passionate urbanist whose interest lies in creating prosperous places that matter and that people can connect to. She has worked on a range of urban scales including city-wide urban renewal, place planning and activation, co-designing with stakeholders and community, providing urban design advice, facilitating urban development and providing strategic research for all sectors including retail, housing, mixed use, and commercial environments. Across the globe, Samantha has used her urban planning and international development expertise to deliver places that make people prosper.

Kim Cook's work in Creative Placemaking began in 1994 when she started to question why Oakland was so culturally rich and yet so asset poor. Kim spent 10 years working with teens in after-school and juvenile detention programs where there was always a struggle for adequate financial support even as the youth flourished creatively. This question of access to resources led Kim to spend time cultivating and applying her knowledge in nonprofit finance, cross-sector collaboration, and large civic initiatives. Her work is place based, contextualized in collaboration with local residents, and runs the spectrum from social practice to spectacle.

Ian Cooper originally trained as an architect. He has spent his working life as an independent research consultant in private practice as a partner in Eclipse Research Consultants (1984–2017). Ian is an associate of Cambridge Architectural Research Ltd, a director of Cambridge Energy and of the not-for-profit Sustainable Development Foundation. He is an Honourable Research Fellow of the Geddes Institute of Urban Research at the University of Dundee and a tutor on the Integrated Design for the Built Environment Masters course for the Institute of Sustainable Leadership at the University of Cambridge.

Jonathan Jae-an Crisman is an artist and urban scholar whose work considers the intersections between culture, politics, and place. He is Assistant Professor of Public & Applied Humanities at the University of Arizona.

Clara Crivellaro is an Interaction Design researcher at Newcastle University's Open Lab. Clara's work focuses on the design of novel socio-technical processes to support social innovation, collaborative service transformation, and social activism. Clara collaborates with diverse groups, organizations, and institutions to explore this, including local authorities, social justice charities, grassroots and urban social movements.

Sherry Dobbin served as Director of Times Square Arts and Creative Director for Times Square Alliance from 2012 to 2016, and is now Partner at Futurecity UK, a London-based global placemaking and public art commissioning agency, where she consults for developers, cities, business improvement districts to establish permanent and programmatic cultural sustainability for place identity. Sherry has worked across all artforms for 35 years, led cultural organizations, and curated for public realm in four continents, is widely published, and the Chair of the Urban Art Forum UK for the Urban Land Institute; a FRSA of the Royal Society of Arts; and sits on Board of Directors organizations ranging from focus on architecture; cultural and gender diversity; moving image in public space, and performing arts in public and digital platforms.

Lisa Eckenwiler, PhD, is Professor of Philosophy in the Department of Philosophy at George Mason University, where she teaches courses in bioethics and global health ethics and co-directs the Global Health Fellows program. Lisa is author of *Long-term Care, Globalization and Justice* (Johns Hopkins University Press, 2012) and lead editor of *The Ethics of Bioethics: Mapping the Moral Landscape* (Johns Hopkins University Press, 2007) and co-founder of the Resisting Borders network.

Tim Edensor is Professor of Human Geography at Manchester Metropolitan University and a Research Fellow in Geography at Melbourne University. He is the author of *Tourists at the Taj* (1998), *National Identity, Popular Culture and Everyday Life* (2002), *Industrial Ruins: Space, Aesthetics and Materiality* (2005), *From Light to Dark: Daylight, Illumination and Gloom* (2017), and *Stone: Stories of Urban Materiality* (2020). He is also the editor of *Geographies of Rhythm* (2010) and co-editor of *The Routledge Handbook of Place* (2020), *Rethinking Darkness: Cultures, Histories, Practices* (2020), and *Weather: Spaces, Mobilities and Affects* (2020).

Claire Edwards is Director of the Institute for Social Science in the 21st Century (ISS21) and Lecturer in the School of Applied Social Studies, University College Cork. Having formerly worked for the Disability Rights Commission in the UK, Claire's research and teaching interests focus on sociological and geographical understandings of disability, and on dynamics of socio-spatial in/justice in the lives of people with disabilities. Claire has conducted research on disabled people's interactions with the criminal justice system, disability organizations' engagement in urban regeneration initiatives, and Universal Design in children's play spaces, and has published widely on these areas.

Samantha Edwards-Vandenhoek is a design researcher, digital media artist and senior lecturer in the School of Design, Swinburne University of Technology, Melbourne. Samantha's current research seeks to understand how situated art practices informed by participatory design processes can transform the experience of place in remote Aboriginal communities in Australia. More broadly, Samantha's research explores the role of place-based design and placemaking in community development, leading to social and economic resilience. Samantha holds a BA (Hons) in Aboriginal Prehistory and a PhD in Media and Communication Studies.

Catherine Fennell is an urban anthropologist and associate professor of Anthropology at Columbia University, with a joint appointment in the Center for the Study of Ethnicity and Race at Columbia University. Catherine's work examines how the social and material legacies of twentieth-century urbanism shape the politics of social difference, collective obligation, and utopian imagination in the United States. Her first book, *Last Project Standing*, won the 2016 Book Prize from the Association of Political and Legal Anthropology. Catherine is at work on new research concerning the aftermaths of private homes in the late industrial urban Midwest.

Cathy Fitzgerald, PhD by Creative Practice, National College of Art and Design, Dublin, Ireland, is an Irish-based New Zealander and an independent ecosocial art practitioner, researcher and educator. Hollywood Forest, 'the little wood that could,' that she lives with in rural Ireland – a monoculture plantation that is becoming a permanent forest – inspires her ongoing (since 2008) ecosocial art practice. In 2019, Cathy founded *Haumea Online* and teaches ecoliteracy to creative professionals from across the world.

Philip Graus, FAIA, MPIA, is an architect and urban planner with experience in practice and research. He was previously a director of Cox Architecture. He is currently director, Western City at the Greater Sydney Commission and Chair, North Sydney Design Excellence Panel.

Shin Gyonggu taught linguistics at Jeonnam National University in Gwangju, South Korea, retiring in 2013. He was named Best Teacher in 2006 and served as Language Center Director and International Affairs Dean for four years each. He was a Fulbright scholar in Indiana University Bloomington in 1995. Presently he is co-founder and Director of the Gwangju International Center and Senior Advisor to Gwangju City for Human Rights.

Jamie Hand is Director of Research Strategies at ArtPlace America, where since 2014 she has designed and led cross-sector knowledge and network building. Prior to ArtPlace, Jamie worked at the National Endowment for the Arts, where she launched the *Our Town* grant program, oversaw the Mayors' Institute on City Design and the Citizens' Institute on Rural Design, and advised on several interagency initiatives. Jamie comes to the US national creative placemaking field from previous work in landscape infrastructure, parks and public space, community-engaged design, and public art. Jamie is co-editor of *Gateway: Visions for an Urban National Park* (Princeton Architectural Press, 2011) and board chair of ioby ('in our back yards'), a nonprofit civic tech organization that supports local neighborhood leaders to grow and implement great ideas through crowdfunding, training, and community organizing.

Lucinda Hartley is CoFounder and Chief Growth Officer, Neighbourlytics. Urban designer turned entrepreneur, Lucinda uses big data to measure the quality of life and wellbeing of neighbourhoods. She is a co-founder of Neighbourlytics, a social analytics platform which harnesses digital data to provide real-time insights into place performance. Lucinda is also an Adjunct Senior Fellow at the University of Melbourne.

Jen Hughes currently serves as director of design and creative placemaking at the National Endowment for the Arts (NEA), where she oversees two grant portfolios, as well as leadership initiatives that include the Mayors' Institute on City Design and the Citizens' Institute on Rural Design. Since joining the NEA in 2011, Jen has fulfilled multiple roles and has managed federal, philanthropic, and local relationships to strategically integrate arts, culture, and design into comprehensive community development plans. Prior to her work at the NEA, Jen was an urban planner for the District of Columbia and has held multiple positions in the private sector.

Theodore (Ted) S. Jojola, PhD, is a Distinguished Professor and Regents' Professor in the Community & Regional Planning Program, School of Architecture + Planning, University of New Mexico (UNM). Currently, Ted is the founder and Director of the Indigenous Design + Planning Institute. Ted is an enrolled member of the Pueblo of Isleta.

Andrew J. Jones, Associate, James Lima Planning + Development, is an economic and policy researcher primarily focused on innovation, downtown revitalization, and parks and open space. At JLP+D, Andrew has advised on the economic and fiscal impacts of downtown university projects, led stakeholder engagement for workforce development strategies, evaluated governance models for downtowns and urban places, and conducted asset and cluster-based economic analysis to inform district-scale master plans. Andrew is passionate about cultivating inclusive entrepreneurial ecosystems through planning and place-based economic development strategies.

Jason F. Kovacs is an Associate Professor in the Department of Urban Administration at the University of Seoul. Jason obtained his PhD in Planning from the University of Waterloo. Prior to moving to South Korea in 2017, Jason briefly taught at the University of Toronto and before that for several years at Nipissing University. Jason's current research interests include arts-led revitalization and cultural and creative tourism. He has published in the past on an array of topics including, among others, cultural planning, public art, and heritage conservation.

Xander Lenc is a PhD candidate in geography at UC Berkeley, where he studies landscapes of incarceration in California. Xander's work is supported by the National Science Foundation and the Global Urban Humanities Initiative. Thank you to Britt Young, Jason Luger, Angela Marino, Julia Sizek, Susan Moffat, and C. Greig Crysler for your invaluable guidance and commentary on this project.

James F. Lima, President, James Lima Planning + Development, has extensive experience in the planning and implementation of urban revitalization projects throughout North America. James' real estate and economic advisory firm, James Lima Planning + Development, helps public and private sector clients create more vibrant, equitable, and resilient places. JLP+D provides planning, policy, real estate, and economic advisory services for downtown and waterfront revitalization, institutional real estate value creation, great placemaking, and shaping impactful public policy. Previously, James served as a NYC affordable housing official and later was appointed by then-NYC Mayor Michael Bloomberg as founding President of the public corporation charged with planning, redeveloping, and operating historic Governors Island in New York Harbor for public use.

Jeremy Liu is a Senior Fellow at PolicyLink where he guides a national initiative to integrate arts, culture, and creative placemaking into policy change and equitable development. The initiative includes: field building through artists and creative strategists partnering with community development organizations and community developers and policy strategists supporting equity-focused arts and culture organizations, research, and creative projects. Jeremy co-authored *Creative Change: Arts, Culture, and Equitable Development*, and co-edited the Federal Reserve Bank of San Francisco Community Development Innovation Review Issue Transforming Community Development through Arts and Culture.

Dave Loewenstein is a muralist, printmaker, and community organizer based in Lawrence, Kansas. His community-based murals can be found across the United States, and in Northern Ireland, South Korea, and Brazil. Loewenstein's prints, which focus on social justice issues, are exhibited internationally and are in the permanent collections of the Museum of Modern Art in New York, Yale University, and the Center for the Study of Political Graphics in Los Angeles. Dave is the co-author of *Kansas Murals: A Traveler's Guide*; and is the subject of *Called to Walls*, a feature length documentary that premiered in 2016. In 2014, Dave was named one of the founding Cultural Agents for the US Department of Arts and Culture.

Graham Marshall is a built environment expert, with grounded regeneration experience in the private, public, and research sectors. With a desire to spend a lifetime creating and improving places for all communities, Graham has been in the vanguard of placemaking throughout his career. He has contributed to national, regional, and local design policy; was a director of the governments pilot Urban Regeneration Company Liverpool Vision; currently a design advisor to Department for Communities Northern Ireland; Built Environment Expert with Design Council CABE; Expert Advisor with High Streets Task Force; and an active member of several regional design review panels and CABE national panel.

Harriet McKindlay is a Strategic Operations Lead at CoDesign Studio in Melbourne, Australia. Harriet is an Urban Planner with a passion for human-centred placemaking and locally led approaches to solve complex urban issues. She believes a multidisciplinary approach is key for achieving desired place outcomes and creating places that prosper. Harriet holds a Bachelor of Environments (Urban Design & Planning) and a Master of Urban Planning, both from the University of Melbourne. She also completed a six-month study exchange program, during her Masters, at the University of Copenhagen.

Lyndsey Ogle is a PhD Candidate in Performance Studies at the University of California, Berkeley (MA: NYU BFA: Northern Illinois University). Lyndsey is also an interdisciplinary artist and curator exploring the intersections of cultural discourse, narratives of 'the social,' and technologies of self through performance, public engagement, and online content.

Adelina Ong completed her PhD at the Royal Central School of Speech and Drama, University of London, UK in 2018. Her thesis, 'Compassionate mobilities,' proposes a theory for negotiated living inspired by parkour, *Art du Déplacement*, breakin' (breakdancing), and graffiti. Adelina has published in *Theatre, Dance and Performance Training Journal* and *Research in Drama Education: The Journal of Applied Theatre and Performance* (RiDE). She has recently co-edited a special issue of *RiDE* 'On Access' with Colette Conroy and Dirk Rodricks which has also been published as an edited collection under Routledge.

Hayun Park is a PhD candidate in the Department of Urban Administration at the University of Seoul. She obtained a MA in International Law and Human Rights with a thesis entitled 'Realizing the right to the city in urban redevelopment in Seoul.' Hayun has since worked on several city government projects and grassroots initiatives in South Korea, being tasked with obtaining community input on planning projects and strengthening community capacity for sustainable management. Her current doctoral research builds upon her work experiences and is focused on participatory planning and community capacity building in Seoul.

Sean Peacock is a PhD researcher based at Open Lab, Newcastle University, UK. Sean is researching how digital technologies can support young people self-organizing in educational and youth work settings to realise better futures for their cities. His professional background is as an urban planner, working in London and North East England before returning to university. Through his research, Sean has developed expertise in designing digital methods to amplify youth voice in placemaking and environmental issues.

Aare Puussaar is a research software engineer at Newcastle University's Urban Observatory in the UK. Aare's research is exploring opportunities for developing data-driven tools and processes to support civic participation and advocacy. He has a background in computer engineering and before joining the university he worked as a data scientist studying spatial mobility. Aare has more than a decade of experience in designing and developing software for spatial data, location-based services, environmental justice, and placemaking.

Paul Graham Raven researches the narrative rhetorics of sociotechnical and climate imaginaries for Lund University, Sweden, where he will start a Marie Skłodowska-Curie Postdoctoral Fellowship from October 2020. His doctoral thesis proposed a novel model of sociotechnical change based on social practice theory, and a narrative prototyping methodology for infrastructure foresight. He's also a writer and critic of science fiction, an occasional journalist and essayist, a collaborator with designers and artists, and a lapsed consulting critical futurist. He currently lives in Malmö with a cat, some guitars, and sufficient books to constitute an insurance-invalidating fire hazard.

Morag Rose is an artist, activist, academic, and anarchoflaneuse. Morag worked in the voluntary sector before completing her PhD 'Women walking Manchester: desire lines through the original modern city' (University of Sheffield 2018). In 2006 Morag founded Manchester psychogeographical collective The LRM (Loiterers Resistance Movement) and has performed, presented, and exhibited widely. Morag is a lecturer in Human Geography at The University of Liverpool and a member of the Walking Artists Network. Her research interests include protecting and promoting public space, accessible architecture, radical histories, creative mischief-making, Americana music, and the geographies of *Doctor Who*.

Victor Rubin is Senior Fellow at PolicyLink, a nonprofit institute advancing equitable policy change in the US. He leads the research about ArtPlace America's Community Development Investments initiative and was guest editor of the special issue on arts and culture of the *Community Development Innovation Review* (2019). Victor has been an advisor to the American Planning Association, the American Institute of Architects, and other organizations. Victor served as Director of the HUD Office of University Partnerships and was Adjunct Associate Professor in City and Regional Planning at the University of California, Berkeley, the department where he earned his MCP and PhD.

Patricia Moore Shaffer, PhD, is the Deputy Director, Research & Analysis, at the National Endowment for the Arts. She has led over 50 program evaluations ranging from national evaluation studies conducted for federal agencies including the Library of Congress and NASA to smaller-scale research, policy, and evaluation studies for state agencies, nonprofit organizations, institutions of higher education, and school districts. Prior to her federal career, Patricia held education positions in several art museums in the US and Canada and served on two local arts councils in Georgia and Virginia.

Trude Schjelderup Iversen is a philosopher, curator, and critic. Trude is previous director of UKS, Young Artist Society (2001–5), PhD Candidate in art theory, University of Oslo, curatorial resident and lecturer in contemporary art theory at Center for Curatorial Studies, Bard College, New York (2008–9). Trude has contributed to numerous periodicals, books, journals, and magazines including *Lights On* (Astrup Fearley Museum), *Capital it Fails us Now* (Simon Sheikh, ed.), *Art and its Institutions* (Nina Montmann, ed.), and is editor of *Materialets tale* (2019) and the anthologys *The New Administration of Aesthetics* (Torpedo Press 2007) with Tone Hansen. She works as a senior curator at KORO, Public Art Norway.

Michaela P. Shirley, MCRP, is Program Specialist for the Indigenous Design + Planning Institute. She is a PhD student in American Studies at UNM and is Diné (Navajo) from Kin Dah Lichii, Arizona.

Gail Skelly is a PhD candidate in the Human Geography Department, School of Science and the Environment, at Manchester Metropolitan University. Her research is inspired by 20 years practice working in the Greater Manchester arts sector, and concerns the impact of community light festivals in Northern England. Gail was Creative Director of Ordsall Community Arts, Salford, for 14 years, where she produced an annual lantern festival, *Lighting the Legend*, within a small urban community. Gail's work is motivated by investigating the cycles of making and creativity involved in the production of community light festivals and how they contribute to a deep sense of place.

Nigel Smith has 20 years' experience working with communities through placemaking. While his background in architecture has given him 'street-cred,' it is liveability and productivity that energize his current work for local governments, not-for-profits, academia, and private clients. Nigel teaches urban design at RMIT University, Melbourne, and is a guest lecturer and panel member at conferences and workshops. Innovative work on major urban developments for government and global property companies saw him become an originator of the placemaking movement in Australia. Nigel lives in Melbourne's CBD and loves being part of a diverse and renowned neighbourhood.

Erik Takeshita has been named as a Senior Fellow to ArtPlace America, supported by the Bush Foundation. His thought partnership will come from more than 20 years of culturally rooted community development experience. Erik joined the Bush Foundation as Community Creativity Portfolio Director in August of 2015. From 2008 to 2015 he led a breadth of work at the Local Initiatives Support Corporation (LISC), including launching a nationwide Creative Placemaking initiative.

Daniel Tucker works as an artist, writer, and organizer developing documentaries, publications, exhibitions, and events inspired by his interest in social movements and the people and places from which they emerge. Since 2014 Daniel has been an Assistant Professor and founding Graduate Program Director in Socially Engaged Art at Moore College of Art & Design in Philadelphia. He founded AREA Chicago and Never The Same with Rebecca Zorach, and has developed projects in association with the A Blade of Grass, Common Field, Creative Time, Pew Center for Art & Heritage, Leeway Foundation, and University of California Institute for Research in the Arts.

Rosanna Vitiello is a cultural strategist, designer, and researcher, whose work encourages people to see their cities through new eyes to open urban potential. Her work explores identity and meaning among places and communities, drawing on shared narratives as an untapped design tool. Rosanna works worldwide across scales, from consensus-building at government level to drawing out unheard stories with local people at street level. Rosanna is co-founder of creative studio The Place Bureau and collaborative platform PlaceLabs. She is a Research Associate at Central Saint Martins and holds an MA in Design for Public Space from Elisava, Barcelona.

Frances Whitehead is a civic practice artist bringing the methods, mindsets, and strategies of contemporary art practice to the process of shaping the future city. Connecting emerging art practices to discourses of climate change, post-humanism, counter-extinction, and sustainability, Frances works as a Public Artist, expanding the role of artists in society and within multiple ecologies, asking, *What do Artists Know?* Frances has worked professionally as an artist since the mid-1980s and has worked collaboratively as ARTetal Studio since 2001. She is Professor of Sculpture at the School of the Art Institute of Chicago.

Ian Wight, PhD, FCIP, GTB, is a Senior Scholar, City Planning, University of Manitoba, Canada, currently enjoying some 're-firement' in his native Scotland. His scholarship of integration and application features integral approaches to understanding the nexus of place, placemaking, and planning. He is an advocate of planning as placemaking, as wellbeing by design. Ian's current action research centres on the related 'evolving professionalism beyond the status quo,' in the context of contemplating the education of the agents of the next enlightenment.

Marcus Willcocks is a practising designer and action researcher, who works with social, spatial, and collaborative forms of design. He collaborates with widely diverse communities, industry disciplines, authorities, and decision-makers, towards more equitable, creative public spaces, and more playfulness in how we respond to tricky problems. Marcus is a Research Fellow at Central Saint Martins, University of the Arts London, with the Design Against Crime Research Centre, and the Socially Responsive Design Innovation unit. He is Senior Urban Designer with active travel charity, Sustrans, and is one of the Design Council's network of Built Environment Experts.

Martin Zebracki is an Associate Professor of Critical Human Geography at the University of Leeds. His research straddles the areas of public art, sexuality, digital culture, and social inclusivity and has been widely published in journals including *Progress in Human Geography*, *Urban Studies*, *Citizenship Studies*, and *Social & Cultural Geography*. Martin is joint editor of the Routledge anthologies *Public Art Encounters* (2018) and *The Everyday Practice of Public Art* (2016) and is Editorial Board Member of *Public Art Dialogue*. He is Principal Investigator of the AHRC-funded project *Queer Memorials: International Comparative Perspectives on Sexual Diversity and Social Inclusivity*.

PREFACE

Cara Courage

Curating *The Routledge Handbook of Placemaking*

The Routledge Handbook of Placemaking sets the placemaking research, teaching and learning, and practice agenda for the sector's next era. The Handbook and its curated sectional approach was conceived by Dr Cara Courage, its Editor, who has convened a cohort of placemaking experts from across research and practice to co-curate the Handbook sections with contributors from across the breadth of the placemaking sector, including its exemplar theorists and researchers, and, as placemaking is, at first, an applied practice, practitioners from its professional fields.

The cover image, of the street mural on 16 St, Washington DC, leading to the White House, was chosen for two reasons. Firstly, to literally put on the cover of the Handbook the political stance taken within. Secondly, as representative of placemaking's inherent political imperative. The mural was painted overnight in June 2020 in permanent street paint by city workers and volunteers, as so many murals and placemaking street art are, and was commissioned by Mayor Muriel Bowser, to be in place before the weekend's Black Lives Matter demonstrations. The street-scale BLACK LIVES MATTER words take up two blocks north of Lafayette Square, the site of police charging protestors to clear the way for Trump's photo-op at St. John's Church, a site the mural also reaches. The mural was offered as a show of solidarity, but later, Black Lives Matter DC denounced the street mural as 'performative and a distraction from her [Bowser's] active counter organizing to our demands to decrease the police budget and invest in the community. Black Lives Matter means Defund the police' (Ebert, 2020). In this one image, we have an art practice that has become synonymous with placemaking, undertaken by a familiar cohort from across municipal, public, and commercial life, and since, rendered in site-specific words in further cities across the United States. We also have contested public realm and citizen-administration relations, contested politics, racial injustice, and the right to the city and to protest (and indeed, such street murals have been the site since of racist acts of removal and painting over.) This is the politics of placemaking – and it is the job of placemakers to work in the service of our communities and, for those white placemakers, to offer our allyship at all times, platforming, amplifying, and making space for our people-of-colour colleagues and communities (natch, any and all protected charcataerisics.) It should never be forgotten that the street mural was intended as monument to lives lost from the Black community across America, and the world. Having this image on the front cover of this book is an act of remembrance and of resistence and a reminder of who we serve as placemakers and of our duty to 'do better.'

The Handbook comes at a pivotal time for the placemaking sector. From its roots in the new urbanism movement of the 1960s, placemaking is in the academic and professional spotlight: as part of the paradigmatic shift in urban design, planning, and policy to include the user and *in situ* community voice; in the arts sector's increasing concern for, and operation in, place-based community development and politics; and the surge of culturally-led urban regeneration. Placemaking as a term is gaining currency outside of the professional built environment sibling sectors and is a growing field of academic research, often practice-based. A library of key texts has formed – from Jane Jacobs' *The Death and Life of Great American Cities* (1961) to the National Endowment for the Arts White Paper, *Creative Placemaking* (2010) for example – and the sector is seeing increasing practitioner and researcher publishing across academic, professional, and popular media. Thus, now is the time to bring seminal thinking and new writing into one volume, to both designate the field to date and to signpost placemaking's substantive new era. Placemaking being a transdisciplinary practice, chapters speak to themes outside of the section they are placed in, each taking the purview of the section as their start, middle, and end point, however. Dominant, normative, placemaking practice can be accused of being located in its 'capital P' placemaking practice formation, and many of the contributors and projects mentioned are located in the northern, western, hemisphere, and particularly the UK and US; however, in urging a move from a centralized and nearing-homogeneous practice, the Handbook centres contributions from indigenous researchers, practitioners, and communities, and holds a global and intersectional position, as well as the non-human. Each section is comprised of chapters from a cohort of placemaking thought leaders, researchers, professional and creative practitioners, defining and presenting established and emerging placemaking theory, research, and practice within the context of its global, intersectional, and multidisciplinary field. Placemaking examples feature prominently, of all scales, sites, and communities of practice, and of international scope. Importantly, at a time when placemaking is gaining popular currency, it presents some of the best of extant placemaking writers alongside newly commissioned writing, to first galvanize and second, extend, critical thinking on placemaking theory and practice.

How the Handbook is curated

The Handbook is presented in seven curated topic sections, representing the breadth of the placemaking operational sector: Section 1, History and Theory of Placemaking, edited by Jason Schupbach; Section 2, Practices of Placemaking, edited by Dr Tom Borrup; Section 3, Problematizing Placemaking, edited by Dr Louise Platt; Section 4, Art, Artists, and Placemaking, edited by Dr Cara Courage; Section 5, Placemaking, Environment, and Sustaining Ecologies, edited by Dr Anita McKeown; Section 6, Placemaking, Urban Design, and Planning, edited by Kylie Legge; and Section 7, Researching and Evaluating Placemaking, edited by Prof Maria Rosario Jackson. To help the reader navigate the Handbook and its intersections from their bespoke purview and speaking directly to the inter-sectorial nature of placemaking practice, the contents have been further curated by (topline) topics, such as equitable and intersectional placemaking, practice, protest, arts and artists (as found in the Handbook ebook), and at the end of each chapter is a select list of some of those complementary chapters in the Handbook, outside of the section in which the chapter is sited, suggested as further reading.

Section 1, History and Theory of Placemaking, is edited by Jason Schupbach, the Dean of the Westphal College of Media Arts and Design at Drexel University, and previously Director of Design and Creative Placemaking Programs for the National Endowment for the Arts. It is a critical appraisal of the history and key theories of placemaking, and its interpretations by different community sectors, with a specific focus on creative placemaking, and examination of the USA's federal government's and foundation's role in creative placemaking.

Section 2, Practices of Placemaking, is edited by Dr Tom Borrup, founder of Creative Community Builders. The section queries the transformative potential of placemaking practice through an assortment of case studies drawn from diverse geographies, methodologies, and theoretical positions and deconstructs the practice of placemaking to uncover its generative potential for holistic community conversation, social justice, and the creation of human-centred urban environments.

Section 3, Problematizing Placemaking, is edited by Dr Louise Platt, interdisciplinary researcher and Senior Lecturer in Festival Management at Manchester Metropolitan University. The chapters presented in this section address debates around the politics and ethics of placemaking, who is included or excluded from placemaking practice either by choice or through structural inequalities, and brings together a range of work that address issues faced by those marginalised in the urban experience, as well as considering whether placemaking as a concept needs to be retheorised or reconceptualised according to different contexts.

Section 4, Art, Artists and Placemaking, is edited by Dr Cara Courage, the Handbook Editor. This section presents and examines diverse scales, scopes, and locales of arts-led placemaking, moving from the city region, to the neighbourhood, to the intimate of one-to-many artist enquiry. As a field within placemaking that is practitioner and end-user led, this section presents the practitioner and professional voice alongside that of the researcher and academic.

Section 5, Placemaking, Environment, and Sustaining Ecologies, is edited by Dr Anita McKeown, itinerant artist, curator, and researcher working at the intersection of art, equitable spatial planning, and technology. This section addresses aspects of socio-economic and environmentally equitable placemaking concerns, the contributors providing a theoretical, historical, and philosophical underpinning that raises issues pertinent to placemaking practices that will need to be addressed for a more sustainable more-than-human process rather than simply humans making geo-locations.

Section 6, Placemaking, Urban Design, and Planning, is edited by Kylie Legge, founding Director of Place Partners. This section focuses on placemaking's inter-relation with urban design and planning sectors and includes case studies of generative planning practice, and its contributors are all global planning and policy experts, from across urban design and governmental sectors.

Section 7, Researching and Evaluating Placemaking, is edited by Professor Prof Maria Rosario Jackson, Senior advisor to the Arts and Culture Program, The Kresge Foundation, Institute Professor at Arizona State University, Herberger Institute for Design and the Arts, and the College of Public Service and Community Solutions. This section includes the perspectives of placemakers, researchers, and evaluators and features writings on key questions evaluators are facing, new methods and approaches to evaluation of placemaking and related fields, and thoughts about the future of evaluation practices.

References

Ebert, G. (2020) 'A Bold Black Lives Matter Statement Transforms a Street Leading to the White House in Washington D.C.' in *Colossal*, 5 June 2020 [online] Available: https://www.thisiscolossal.com/2020/06/black-lives-matter-washington-dc/. [Accessed: 9 June 2020].

Jacobs, J. (1961) *The Death and Life of Great American Cities*. New York: Random House.
Markusen A. and Gadwa, A. (2010) *Creative Placemaking*, National Endowment for the Arts [online] Available: https://www.arts.gov/publications/creative-placemaking. [Accessed: 20 April 2020].

List of topics (A–Z):

Art and Artists
Climate and environmental ecology
Community of practice
COVID-19
Creative placemaking
Digital and technology
Equitable and intersectional placemaking
Evaluation
Festivals
Funding
Governance and stewardship
Indigenous placemaking and place
LGBTQI+
Migrant and displaced peoples
Narrative
Non-human
Place branding
Place economics
Place knowledge
Placemaking and community development
Placemaking definition
Placemaking futures
Placemaking history
Placemaking practice
Planning and generative planning
Policy
Productivity and innovation
Protest and resistance
Regeneration, development and gentrification
Rural placemaking
Settler colonialism
Social practice placemaking
Spatial access
Sustainable Development Goals
Tactical urbanism
Teaching and learning
Tourism
Urban design
Wellbeing and healing

ACKNOWLEDGEMENTS

Thanks are due to the Section Editors for their inspired co-curation and to them and to all the contributors for their collective dedication to this publication, its workload an addition to that of their professional roles and projects, and undertaken without payment and through individual life-changing events; to the team at Routledge that have seen this project through from its initial conception to the published version you are reading now; and to all the placemakers in the world for creating, nurturing, and sustaining this practice, that participate in research and inform theory, and that make placemaking the radical practice it can truly be.

Thank you to my placemaking muse AM, and to SL and AH at the Dames Road Studio and CM at Uppstigen 124 for support in the making of this Handbook.

Thank you to Nadia Aziz for the most generous permission to use her image for the Handbook cover. At her request, in lieu of fee, a donation of £250 is given to the Arab American Institute Foundation.

Both UK and American English feature in this Handbook, recognizing the voice of contributors and direct quotes from other text, artists, and project participants.

Citation as: Courage et al., 2021.

1

INTRODUCTION

What really matters: moving placemaking into a new epoch

Cara Courage

This Handbook was initiated at a time when the ten-year anniversary of the Markusen and Gadwa (2010) National Endowment for the Arts Creative Placemaking White Paper was on the near horizon; was being written when the nomenclature of climate change was changing to climate and ecological emergency; and its final editing took place during the first months of the COVID-19 pandemic and the beginnings of the global Black Lives Matter and race justice protests after the (at the time of writing, accused) murder of George Floyd while in Minneapolis police custody. The Markusen and Gadwa White Paper became an era-defining moment for placemaking, and as we look from 2020 to 2030 and beyond, undoubtedly COVID-19 and racial and climate justice will engender change in placemaking practice, and these concerns are foregrounded in this Handbook: indeed, as this book calls for looking again at current placemaking narratives and demands new ones, it signals the time for placemaking to evolve, as is only correct for a people- and place-responsive practice with a radical potential at its core.

COVID-19 has put what really matters in our public realm into sharp relief, and in sum, what matters is the human. The pandemic is anti-place: in particular, it is counter to the particularly urban design of collective occupation, and has created a fear of human proximity and taken from us our familiar collective social experiences and sites of serendipitous encounter (Sennett, 2012, p. 38; Soja, 1997, p. 20; Watson, 2006, p. 6) in the public realm. Our common response of a physical distancing from others is contra to the human desire for interaction and contra to public-realm and built-environment design: these places are designed to be animated. As fiscal economies suffer worldwide, so too we find ourselves in something akin to a place-based social or cultural recession. There is an emotional toll in seeing from a distance but feeling ever close the places we have a connection with, that we have made and shaped in our own lifetimes, lost to us, even if, we hope at the time of writing, temporarily. Singularly and cumulatively, the virus does not discriminate – but the pandemic, within our structural and social systems of oppression, does, cleaving further open structures of intersectional discrimination and vulnerability, and, in large part, the burden of social distancing and care falling upon the marginalised and the lowest paid. Those that kept our places working for us but who were previously consigned to its margins – the street cleaners, the bus drivers, the delivery drivers, amongst many others – are not 'unskilled workers' but 'key workers'; public life has not shrunk for them, their worth far in excess of their remuneration. What has placemaking ever looked like for those that have been overlooked in public realm life?

If we ever took it for granted, we now appreciate the full value of human-centred public space when it has been taken away from us – not least, the right to protest in it, and remark has to be given to those places mentioned in this Handbook that have now become sites of Black Lives Matter protest and of police brutality. We see people resist the term 'social distancing,' in favour of 'physical distancing,' as they take their social life online and animate their neighbourliness on the front step when clapping and banging pots and pans for the pandemic frontline workers in the UK for example, and as they give time to street-level community support networks. Our spatial perspective has pivoted to the hyperlocal of place, where grocery shopping is a walk away and people buy from suppliers in the local economy. Outdoor exercise is within a close-by boundary, and people are using parks and discovering walks previously unknown or out of reach to them in their previous day-to-day routines: a 'relocalism,' if you will. If the public realm has changed, our need for social ritual hasn't. Its site may have moved, however: rainbow posters in windows, balconies as galleries and concert venues, rooftops becoming distanced gym classes. Our distanced sociability affirms to ourselves and to others that we are still here, that the public realm is still there, and that our community is relevant and must be seen and heard.

What is placemaking?

Ask those balcony singers, rooftop exercisers, or street cleaners what placemaking is however, and its highly probable that you will be met with a blank look. Despite the intense and pervasive placemaking activity of the past ten years, and indeed the years before it, there is still a need to explain what placemaking is to those within our sibling sectors as much as to those outside them. In all my research and practice, I have not come across one single placemaking definition that is used by all (and indeed, different definitions appear throughout this Handbook), and to a degree, definitions are avoided. One could be left wondering if placemaking is even 'a thing,' if it is so amorphous or undefinable. There are several reasons why a placemaking practitioner or organisation would want to avoid a definition: naming what you do too tightly may limit funding and commercial opportunities, and for the individual practitioner, one can define one's own work and position in and vis-à-vis this field. Conversely, naming what you do as placemaking, in your own definition, is a self-fulfilling prophecy – 'I make place, therefore I am a placemaker' covers a wide range of practices in opposition to each other from ideological purviews (see below.) Placemaking may be so new as sector, when placed next to architecture or urban design for example, that it could also be fair to say that we are still defining the field – and indeed, that is a function of this Handbook. But equally, could the term placemaking have become meaningless as a result of its disparate, diluted, and obscured use? While not having an accepted definition of placemaking can be used to one's advantage, it is far from certain that this is to the sector's advantage. Just as with the terms 'green' or 'eco,' and their use to the point of losing any meaning, the placemaking term is being applied to almost any project in place, and those projects – urban farmers' markets, an outdoor cinema, a pop-up park – are all looking familiar no matter where in the world you may find them, despite their relative merit.

For me, what differentiates placemaking from other built environment sectors, and should be central to any understanding or definition of it, is that placemaking is an approach and a set of tools that puts the community front and centre of deciding how their place looks and how it functions. There is a community imperative in placemaking. As artist Jeanne van Heeswijk said, 'The community is the expert in being the community' (van Heeswijk, 2012): in placemaking, the community, however defined in the particular context, is recognised and valued as the expert in being the community. The moment you take the community out of placemaking as both spearheading and equal stakeholder in its process, the process is no longer placemaking and the radical potential of this place-based process is completely lost. Placemaking, when done well, has an agency of relative expertism (Courage, 2017), joining equitably community (however self-defined), architects, urban designers, artists, policymakers, planners, developers, Mayors and city administrations, educators, housing departments... and uses the existing assets of a place to their best effect and facilitates creative patterns of activities and connections – cultural, economic, social, environmental – that define a place and support its ongoing evolution. Placemaking represents a paradigm shift in thinking about planning and urban design, from a primary focus on buildings and macro urban form to a focus on public space and human activity – what happens in these spaces, why, how, and with and by whom, and not: this is all the stuff of placemaking. There is a twofold need for the processes of placemaking. First, it demands all those involved to work across sectors and out of silos, and often with art practice, especially that of community and social practice or socially engaged art. These particular art practices, as collaborative and transdisciplinary, are best placed to lead by example here, and indeed to break some of our sibling sectors' fear of the trial-and-error of process. Second, there is a need for architects, urban designers, and planners to pay ever more attention to local knowledges and desires in order to give depth to the meaning of their place designs, in a generative, open source, and open-ended process without, or with an unknown, built environment output. Again, artists have a vital role here, driving and incubating the conversations through community-based and explorative and testing methodologies.

In *this* placemaking, people have their love of place confirmed, renewed, valued; their place attachment activates as place stewardship; which leads to increased social cohesion and wellbeing; which in turn results in the genuine formation of the vibrant, liveable places that administrations, planners, and developers the world over are working to variously create or secure (Courage, 2017). When projects are done with integrity and hold the expertism of the community paramount, the community are active producers, not consumers, of the public realm. But like anything that happens in the public realm, placemaking is of contested power and politics. The very real situation for many at the community end of placemaking – or, at least, what is branded as placemaking – is of wholescale social cleansing, communities evicted and dispersed by developers, artists brought in to place wash (Pritchard, 2019). The reader may not recognise in this book what has become normative placemaking practice. This is not bad thing. This Handbook deliberately (re)situates into placemaking discourse non-normative, subaltern (Louai, 2012), and diverse knowledges.

Placemaking as a community of practice?

Placemakers have a concern with community as an outward entity, but we turn the lens inwards now, to the consideration of placemaking as a community of practice. Succinctly, a community of practice is a group of people who 'share a concern or passion for something they do and learn how to do it as they interact regularly' (Wenger, 2006, p. 1). The community can be any group engaged in a common intentional process of knowledge sharing and creation (ibid.), developing unique perspectives on their shared concern and a body of common language and approaches (Wenger et al., 2002, p. 5). Activities undertaken by a community of practice may include: problem-solving, seeking experience, discussing developments, documentation projects, visits, mapping knowledge and identifying gaps (Wenger, 2006, p. 2), pondering common issues, exploring ideas, acting as sounding boards, creating tools and standards, and developing tacit shared knowledge (Wenger et al., 2002, pp. 4–5). Participation is essential to the community of practice: it is through participation that identity and mutual recognition and practices develop, and connection, meaning, negotiation, and action occur (Handley et al., 2006, p. 643). The outcome of this is a practice based on a 'craft intimacy,' 'close interactions around shared problems and sense of commonality' (Wenger et al., 2002, pp. 120–2). Communities of practice are comprised of three dimensions: the *domain*, its joint enterprise, what it is about; the *community*, its mutual engagement, how it functions; and the *practice*, its shared repertoire, what capability it produces (Wenger, 1998, p. 2; Wenger, 1998, p. 72ff, in Fuller, 2007, p. 21). Some members will participate as they care about the domain and want to see it developed; others because of the value of having a community to interact with as peers; and others to learn about the practice and develop craft (Wenger et al., 2002, p. 44).

The domain of placemaking, on first look, is self-evident, but in practice is contested and negotiated. Whereas one can identify as a placemaker, the intent and outcomes of one's placemaking may be wholly different and in opposition – think here, the difference in practice between top-down and bottom-up placemaking (see, Placemaking Typology, Courage, 2017, pp. 72–76). However, as a relatively new practice, negotiating the domain is both the first task of the community and its ongoing task, critical as it is to community development (Wenger et al., 2002, p. 45). With placemaking, a global and intentional community, as understood above, and a body of knowledge is iteratively and generatively forming, comprised of a shared repertoire of resources – experiences, tools, language, problem-solving, for example (Wenger, 2006, p. 2). The activities undertaken by a community of practice are akin to those undertaken by placemakers through their practice, and, via increasing formal and informal knowledge exchange, these activities are being turned inwards to the placemaking community itself; a craft intimacy is emerging that could situate placemakers from different locales in a common frame of practice. This however is a moot point for the sector. Wenger et al. (2002) present seven principles for cultivating a community of practice – design for evolution, open to dialogue between inside and outside perspectives, invite different levels of participation, develop both public and private community spaces, focus on value, combine familiarity and excitement, and create a rhythm for the community (ibid., p. 451) – and state that it requires a coordinator and that those in the community take on leadership roles (ibid., p. 5) to connect those at the core of the community to those on its periphery through 'build[ing] benches' (ibid., p. 57). While placemaking encourages an evolving design and intra- and inter-sectorial dialogue for example, this is not uniform across sector, locale, or site (see below), and whereas one can identify a coordinator at a project level, as well as accommodate differing participation levels and tenures, is this possible or correct for the sector as a whole? Furthermore, the 'top-down' and the 'bottom-up' dialogue is given much noise in placemaking, but is this dialogue meaningful? One cannot deliberately cultivate a community of practice, but 'elicit and foster participation' (ibid., p. 13). Is it more prudent then to consider a number of coalescing communities of practices as a sector ecology?

This latter point may be borne out by the following consideration. The value of a community of practice is both short- and long-term: help with immediate problems, reduction in time searching for information or solutions, problem-solving enhanced with a wider perspective, community-supported risk-taking, benchmarking of expertise, and sustained professional and practice development (ibid., p. 15). Here again, we can recognise a project-level practice of placemaking, if not a sector one, that is developed enough to ascribe community-of-practice naming to. However, for a community of practice, value may not be apparent at the start; it may not be apparent to those outside the community; it will change over time and form a systemic body of knowledge (ibid., p. 59). The surfacing of value and its sustained (re)creation is key to the success of the community of practice (ibid.). A lack of value recognition and the constant re-creation of value to those outside of the practice will be something that many placemakers recognise, if only anecdotally.

If placemaking is a community of practice, it is a distributed (ibid., p. 118) one – across intent as much as cultures and geography – and this poses a challenge to it emerging and evolving as a community of practice, as each subcommunity has its own habitus (Bourdieu, 1997) which can lead to miscommunication or misunderstanding (Wenger et al., 2002, p. 18) when joined as a whole. Wenger points to designing distributed communities to overcome issues of connections and visibility, geographical distance, hierarchies, isolation, and lack of encounter opportunity (ibid., p. 116). However, one cannot avoid the tension that may arise from distributed communities' crossing of boundaries, diverse affiliations, and competing priorities (ibid., p. 117). The concept of the micropublic (Amin, 2008) or the microcommunity (Kester, 2011) may prove useful to apply here, as both concepts work with difference in a group through relational means, accepting disagreement as a positive force in collective endeavour; as too might an ecology of culture thinking, with dynamic aspects within a larger ecosystem, feedback loops, emergent behaviour, and interdependence and self-organisation. Furthermore, as a community of practice grows in size and duration, the capacity to know all may diminish, and those at its beginning, and as an operational core, may feel antagonised by newcomers. This risks any ensuing multiple sites or trajectories of practice (Hughes et al., 2007, p. 5) not being acknowledged and the reproduction of continuity rather than the fostering of transformation (Hager, in Fuller, 2007, p. 22). Arnstein's *Ladder of Participation* (1969) may prove useful here, understanding as it does both progression through, and differentiated status of, participation – a more nuanced position than the community of practice 'participation or not' purview.

The next placemaking epoch

Over the last ten years, placemaking has evolved through its collaborative and iterative approach, gathering new constituents of practitioners and professions along the way and becoming increasingly cognizant of its political implications and agency. To become a community of practice, an ally to all citizens, and to respond to times of crisis, there is much to evaluate across the placemaking sector. This is alongside an asking of the place-based, and ergo placemaking, 'new normal' of the COVID-19 pandemic what value does public space offer in a lockdown for those of us placemakers privileged enough to have time during the pandemic to reflect? We are also asking: what is ahead of us and what do we want to keep from this time and take forward to our 'next normal'?

The next ten years pose challenges for society that have place and placemaking implications. While placemaking alone cannot change the conditions that produce and maintain structural inequalities – from climate crisis, ageing populations, poor mental health, to systematic inequality, social segregation, and the place-based aspects of these – as placemakers we have to ask ourselves if we are fit for purpose to serve our communities and places and co-create real and integrative solutions to the most pressing concerns of contemporary and future living. The ideal of placemaking offers a practical, proactive, and integrated approach to place: does its reality match that? Furthermore, in the next ten years, the placemaking and regeneration projects of recent years will be reaching a point of maturity that opens them up to a lived-in, experiential evaluation. How do we ensure these evaluations don't remain in the realm of a performative participation, being side-lined and silenced, and instead are listened to and generatively embraced?

As placemakers, we know that social infrastructure is key to social action and social capital, and that a community's place attachment is key in this, and it is these factors that in turn ensure strong social support networks. We are adept at seeing challenges from the purview of possibility and in connecting people. It is this asset, of connecting and networking people, that will come to the fore as we move into our 'next normal': joining in collaboration all local stakeholders and the communities of place, to deliberate on what we don't want to return to in our lived and situated experience, as much as what we want to innovate, and to translate this into action.

An imperative of placemaking going forward is a public and enacted commitment to intersectionality: a place that is unwelcome for one is unwelcoming for all and serves only to perpetuate divisive social, political and material relations. A reliance on certain metric evaluations of place can obscure the experiences of those marginalised in the public realm and excluded from its discourse and decision-making. Who is allowed a place in our places, who is heard there, and who sits at the decision-making table, determines what our places can be, and an emancipatory and intersectional placemaking is key to achieving place justice. Achieving such a place is not a matter of engagement and empowerment. As placemakers, we do not empower people; people hold an intrinsic power, and it is our job to 'create' platform and share resources for people to enact their power. To quote co-director of Fun Palaces, Stella Duffy OBE: 'We all have power. It is the system that denies some of us use of it' (Duffy, 2019). It is within our gift as placemakers to change our sector systems and to use our knowledge of working with communities though change to assist mutual support networks to (further) develop, maintain, and grow as the keystone in creating equitable and inclusive places. As placemakers we also have a responsibility to help amplify people's voices – not to speak for them, but to use our professional privilege beside and with them – and interject this into the systems of place to put people at the centre of the decisions government and the private sector are making. This is a co-created engagement that is honest, transparent, and values-based and that respects and is in the service of the communities we work with.

The COVID-19 pandemic and Black Lives Matter have not just brought into sharp focus what really matters about place, but also that the stories, myths, and imaginaries we have built our economy and cities around are unfit to confront climate breakdown or intersectional social, racial, or economic inequity. We need new stories, and we need to listen to diverse narratives and knowledges. At the time of writing, we have no concrete way of knowing how the pandemic or Black Lives Matter will change the world and our places therein. What we do know however is that the material conditions and the psychosocial trauma many are now collectively experiencing for the first time has been the long-lived experience of indigenous people, migrants, and people and communities of colour – our responsibility, as placemakers, is not to extract from this lived experience, but to listen deeply and to learn from it and apply that learning forward, as well the white placemakers of us deeply listen to and learn from our people pf colour colleagues, and support emerging placemaking talent from marginalised demographics. Our recovery from the pandemic and our ability to adapt to current climate change and to mitigate what we can of climate emergency requires all of the work we have done before as placemakers, but also a radical transformation. Mental health, civic voice, community building, and equitable societies rely on inclusive, active, public space, where social infrastructure supports social integration. To get to this place, literally and metaphorically, placemaking needs to be the radical practice proposed above, a community-based process, and not an imposed top-down strategy or instrumentalized solution. Placemaking needs to work for social and environmental justice – this is what really matters and should be what shapes the next era of placemaking: to create places that heal, rather than harm.

References

Amin, A. (2008). 'Collective culture and urban pubic life', *City*, 12(1), pp. 5–24[online]. Available at: https://www.tandfonline.com/doi/abs/10.1080/13604810801933495 (Accessed: 11 April 2020).

Arnstein, S.R. (1969). 'A ladder of citizen participation', *Journal of the American Planning Association*, 35(4), pp. 216–224 [online]. Available at: https://www.participatorymethods.org/sites/participatorymethods.org/files/Arnstein%20ladder%201969.pdf (Accessed: 11 April 2020).

Bourdieu, P. (1977). *Outline of a Theory of Practice*. Cambridge: Cambridge University Press.
Courage, C. (2017). *Arts in Place: The Arts, the Urban, and Social Practice*. Abingdon: Routledge.
Duffy, S. (2019). 'Here's why we need to stop 'empowering' people', *The* Visionary Arts Foundation [online]. Available at: https://visionaryarts.org.uk/stella-duffy-heres-why-we-need-to-stop-empowering-people/ (Accessed: 21 April 2020).
Fuller, A. (2007). 'Critiquing theories of learning and communities of practice', in Hughes, J., Jewson, N. and Unwin, L. (eds.) *Communities of Practice Critical Perspectives*. Abingdon: Routledge.
Handley, K., Sturdy, A., Fincham, R. and Clark, T. (2006). 'Within and beyond communities of practice: Making sense of learning through participation, identity and practice', *Journal of Management Studies*, 43(3).
Hughes, J., Jewson, N. and Unwin, L. (2007). 'Introduction. Communities of Practice: A contested concept in flux', in Hughes, J., Jewson, N. and Unwin, L. (eds.) *Communities of Practice Critical Perspectives*. Abingdon: Routledge.
Kester, G.H. (2011). *The One and the Many: Contemporary Collaborative Art in a Global Context*. Durham: Duke University Press.
Louai, E.H. (2012). 'Retracing the concept of the subaltern from Gramsci to Spivak: Historical developments and new applications', *African Journal of History and Culture*, 4(1), pp. 4–8 [online]. Available at: http://www.academicjournals.org/AJHC (Accessed: 21 April 2020).
Markusen A. and Gadwa, A. (2010). *Creative Placemaking* [online]: National Endowment for the Arts. Available at: https://www.arts.gov/publications/creative-placemaking (Accessed: 20 April 2020).
Pritchard, S. (2019). 'Place guarding: Activist art against gentrification', in Courage, C. and McKeown, A. (eds.) *Creative Placemaking: Research, Theory and Practice*. Abingdon: Routledge.
Sennett, R. (2012). *Together: The Rituals, Pleasures and Politics of Cooperation*. London: Allen Lane.
Soja, E.W. (1997/2005/2011) 'Six discourses on the postmetropolis' in Westwood, S. and Williams, J. (eds.) *Imagining Cities: Scripts, Signs, Memory*. London: Routledge.
van Heeswijk, J. (2012). *Public Art and Self-Organisation*, London (conference) [online]. Available at: http://ixia-info.com/events/next-events/public-art-and-self-organisation-london/ (Accessed: 15 January 2016).
Watson, S. (2006). *City Publics: The (Dis)enchantments of Urban Encounters*. Abingdon: Routledge.
Wenger, E. (1998). 'Communities of practice: Learning as a social system', *Systems Thinker*, 9(5), pp. 2–2 [online]. Available at: https://thesystemsthinker.com/communities-of-practice-learning-as-a-social-system/ (Accessed: 8 April 2020).
Wenger, E. (2006). *Communities of Practice: A Brief Introduction* [online]. Available at: https://scholarsbank.uoregon.edu/xmlui/bitstream/handle/1794/11736/A%20brief%20introduction%20to%20CoP.pdf?sequence=1&isAllowed=y (Accessed: 8 April 2020).
Wenger, E., McDermott, R. and Snyder, W.M. (2002). *Cultivating Communities of Practice*. Boston: Harvard Business School Press.

Further reading in this volume

Preface: Placemaking in the age of COVID-19 and protest
Jason Schupbach
Chapter 4: A future of creative placemaking
Sarah Calderon and Erik Takeshita
Preface: 'Disastrous forces, accidental actions, and grassroots responses'
Tom Borrup
Preface: The problem with placemaking
Louise Platt
Chapter 14: Experts in their own tomorrows: placemaking for participatory climate futures
Paul Graham Raven
Chapter 20: Displacemaking 2015 and 2020
Catherine Fennell and Daniel Tucker
Preface: Towards developing equitable economies: the concept of *Oikos* in placemaking
Anita McKeown
Chapter 35: Planning governance – lessons for the integration of placemaking
Nigel Smith

Chapter 39: *Translating Outcomes*: laying the groundwork for interdisciplinary evaluation of creative placemaking
Jamie Hand
Chapter 40: Transforming community development through arts and culture: a developmental approach to documentation and research
Victor Rubin
Chapter 43: A theory of change for creative placemaking: the experience of the National Endowment for the Arts' *Our Town* program: an interview with Patricia Moore Shaffer, PhD
Maria Rosario Jackson
Chapter 44: Creative Placemaking and comprehensive community development: rethinking neighborhood change and evaluation
Maria Rosario Jackson

SECTION 1

History and theory of placemaking

Section Editor: Jason Schupbach

PREFACE

Placemaking in the age of COVID-19 and protest

Jason Schupbach

Two months ago, I was thinking that this preface should outline for you how this opening chapter explores the interconnected policies and procedures that make up the history of placemaking, with a large focus on the role of creatives in the placemaking process. How the term 'placemaking' is itself challenged, and how the concepts of doing no harm, building communities from asset-based strategies, equitable development, and placemaking that emerges from repression and expression of cultures are outlined. That was before COVID-19. Before the world was turned upside down economically and physically. Before the tears, before the fear and the defiance, before worldwide protests against racial and economic injustice... just *before*. Before was when we could connect physically with minimal worry. Before was when public space could be safe and a place to meet, buy and sell, and celebrate. Before was when cultural anchors were places of group celebration, spiritual movement, and economic success. Before was when economic, environmental, and racial injustices were already bad, and now they are horrible.

We must not give up hope and longing for 'before.' The worst thing that could happen in response to this crisis is if it becomes an excuse for privileged placemaking 'experts' to experiment with new or broken unjust urban policies. As Alissa Walker wrote recently:

> If the coronavirus has made anything clear, it's that cities cannot be fixed if we do not insist on dismantling the racial, economic and environmental inequities that have made the pandemic deadlier for low-income and nonwhite residents. Yet many prominent urbanists have simply tweaked the language from their January 2020 tweets and fed them back into the propaganda machine to crank out COVID-tagged content, perpetuating the delusion that all cities need are denser neighborhoods, more parks, and open streets to magically become 'fairer.'
>
> *(Walker, 2020)*

We can and will do better, we must. As the chapters in this section outline, there is much to be learned from the history of placemaking efforts that is relevant to the work we all have ahead of us to rebuild a more just society.

What we learn from this chapter

The intention of this chapter is to lay some groundwork for the reader in the historical policies and procedures which make up the scope of placemaking, with a large focus on the 'creative placemaking' movement of the last ten years. Following chapters challenge these ideas and policies, as do chapters here.

James Lima and Andrew Jones' chapter describes an American history of placemaking from an economic development policy perspective. Lima describes the many different forms the movement has taken, and the good and bad outcomes of those policies, and expresses a perspective that many in the current urban planning field in the States take when viewing this history.

Both the chapter from Jennifer Hughes and that from Sarah Calderon and Erik Takeshita describe in detail a partnership between the federal government and foundations to respond to the Great Recession – what some called the 'creative placemaking' movement. Because the Great Recession in the States was primarily a housing crisis in which homeowners' houses became worth less than their mortgages, Americans' mobility for employment plummeted. Hence a response was needed that helped people where they lived. A series of 'place-based' initiatives were launched by the Federal government, foundations, and NGOs in response to these conditions. These writings focus on a slice of this larger 'place-based' movement called 'creative placemaking,' which is generally assumed to be the role that creatives or a local culture of a place have in its development. Just over a decade old, the policies and funding that supported creative placemaking, and the movement's accomplishments and failures are described in these essays in detail. These essays also work to try and describe what needs to happen next.

An enormous amount of knowledge was generated in the past ten years about the ways that humans can use their creativity and cultural expressions to build the equitable communities that they want and deserve. Chapters by Jeremy Liu and Kim Cook describe these processes in detail, using examples of where people build place despite harsh physical conditions (the Nevada desert were Burning Man occurs) and/or great injustice lives. Lui presents a history of repressed cultures and the places they built in response to that repression, suggesting that there is much to be learned from these histories in our new work ahead. Cook similarly gives examples from her work in New Orleans and with Burning Man to describe projects, and the processes behind them. Those looking for answers about 'what works' in placemaking practice should spend time with these essays.

What's next?

Obviously, the authors in this chapter do not represent the wide scope of people whose voices must be heard as we rebuild. They do not outline in full the 400 years of racial injustice and violence that has occurred in the United States that influences almost all urban policy in the country. We offer these essays as a starting point for a conversation. It will be critical going forward to embrace the multitude of voices necessary to build communities of justice, and to not repeat the mistakes of the past. People make place, and they are demanding change. Place still matters, and how we build it together matters more than ever. The last ten years of placemaking work grew from a crisis – the Great Recession – and a new one must take shape to respond to this crisis. Let us remember that we have been through trauma and must reform our society through healing-centered practices that work to solve the longstanding issues we suffer from. We must collectively work together to pull the best practices from the history of placemaking and build a new collective response, a response that values and respects the humans who live in a place. What a complex puzzle we have to solve, and what opportunity to unleash creative thought into that process!

Reference

Walker, A. (2020) [online]. Available at: www.Curbed.com (20 May 2020).

Further reading in this volume

Preface: The radical potential of placemaking
Cara Courage
Chapter 1: Introduction: what really matters – moving placemaking into a new epoch
Cara Courage
Preface: 'Disastrous forces, accidental actions, and grassroots responses'
Tom Borrup
Chapter 7: Conflict and memory: human rights and placemaking in the City of Gwangju
Shin Gyonggu
Chapter 19: Placemaking in the ecology of the human habitat
Graham Marshall
Chapter 24: Artists, creativity, and the heart of city planning
Tom Borrup
Preface: The only thing constant is change
Kylie Legge
Chapter 43: A theory of change for creative placemaking: the experience of the National Endowment for the Arts' *Our Town* program: an interview with Patricia Moore Shaffer, PhD
Maria Rosario Jackson

2
PLACEMAKING AS AN ECONOMIC ENGINE FOR ALL

James F. Lima and Andrew J. Jones

Introduction

As recently as a few decades ago, America's cities were being drained of residents and businesses. Yet, in 2020, demand for proximity to urban centers was approaching one of the highest points in recent memory, before the onset of the COVID-19 pandemic. The proliferation of desirable urban 'places,' or high-amenity districts that are designed, activated, and managed in a way that cultivates a connection between people and the physical space around them, is among the most compelling explanations for this extraordinary turnaround. Cities, and the open space assets that typically anchor them, attract concentrations of workers, firms, residents, students, visitors, and capital investment, generating value for and enhancing the economic competitiveness of their host communities. As a result, cities and governments, developers and corporations, and anchor institutions and nonprofits alike are increasingly turning to placemaking as an effective strategy to drive economic growth, productivity, and innovation. While cities will always face challenges and crises that threaten to reverse the social and economic progress they have made, placemaking will have an important role to play in inclusive strategies to achieve more resilient communities and economies.

How we got here

Beginning in the early 1900s in the United States, more than six million African Americans migrated northward to escape persecution wrought by the 'Jim Crow' laws that governed much of the American South, which harshly enforced racial segregation and denied non-white individuals basic civil rights. They largely settled in the industrial cities of the Northeast and Upper Midwest, which had been predominantly white until that point (Boustan, 2010). These white populations expressed discomfort with the influx of people of color and employed a number of strategies to maintain a similar degree of segregation experienced in the South, such as housing discrimination and exclusionary zoning. The practices then became codified, setting in motion a reconfiguration of American socioeconomic geography that would result in people of color living in city centers and whites living in suburban rings on the periphery of metro areas. It is no coincidence that the cities who witnessed the largest in-migration of nonwhites between 1940 and 1970 also lost the greatest share of their white population (Boustan, 2010).

The segregation of and 'divestment' from city centers were intensified by federal housing and transportation policy. In 1934, the Federal Housing Administration (FHA) was established to insure and regulate the terms of mortgages for single-family homes and multifamily development projects. However, the FHA's Underwriting Manual only permitted these benefits to be extended to geographies deemed 'best' or 'still desirable' by the Home Owners' Loan Corporation (HOLC), which almost exclusively consisted of white, exurban communities. This federal housing policy encouraged residents and real estate capital to leave cities, ushering in an era of racially charged suburbanization that would degrade urban neighborhoods throughout the twentieth century (Massey and Denton, 2003).

In 1956, construction began on the Interstate Highway System, a 42,500-mile network of roadways that would link cities across the United States (Mohl, 2001). The building of these roads throughout the 1950s not only razed many urban neighborhoods but enabled centrally located manufacturing companies to defect to greenfield sites in the suburbs and in Sun Belt cities. This decentralization shifted hundreds of thousands of middle-class (and white collar) jobs out of the hubs of the Northeast and Upper Midwest, exacerbating urban exodus and leaving behind an 'underclass' of residents lacking economic opportunity (Sugrue, 2005). Between 1947 and 1972, the central cities of the 33 most populous metropolitan areas lost approximately 880,000 manufacturing jobs and 867,000 retail-wholesale jobs (Wilson, 2012).

The outcome of these mid-twentieth-century structural changes in the American society and economy left many cities in disarray. Between 1950 and 1990, 18 of the nation's 25 largest cities suffered a population loss (Lewyn, 1996). And while in the 1950s to 1980s, many American cities sought ways to combine available federal, State and local funding through 'urban renewal' efforts focused on high-density housing, these often large-scale, heavy-handed approaches to rebuilding the inner city often cleared out entire neighborhoods and only sometimes succeeded in rebuilding suitable replacement housing, new industry, and neighborhood amenities. Meanwhile, as cities lost residents and property values declined, local tax bases shrank and cities found themselves unable to fund critical municipal services such as public education and police protection, maintain vital infrastructure, or provide support to their increasingly distressed populations (Dreier, 1993).

Rebuilding the strength of urban settlements through innovative, multi-pronged investment strategies

To remediate blight and revitalize inner cities across the country, the federal government changed course, launching two ambitious initiatives in the 1990s: the HOPE VI and Empowerment Zone programs. HOPE VI distributed billions of dollars in block grants to fund the development of transit-accessible, mixed-income housing in dense, pedestrian-friendly districts. For every $1 of HOPE IV funds contributed to a given project, $1.30 in private sector funding was also deployed (US Department of Housing and Urban Development, 2016). The Empowerment Zone (EZ) Program offered grants, low-cost loans, and sizable tax incentives for businesses hiring locally and making capital investments in designated economically distressed communities (US Government Accountability Office, 2010). Both of these programs were innovative in their use of public–private partnerships to finance economic development, setting a precedent for future urban revitalization initiatives.

Local stakeholders, both public and private, sought to build on the momentum of these federally driven investments and their understanding of the intrinsic economic and social value of density. They began forming entities that could funnel capital into catalytic projects in central business districts and brownfield sites along post-industrial waterfronts. Some of these endeavors took the form of economic development corporations (EDCs), quasi-public non-profits charged with making strategic investments to enhance the vitality and competitiveness of a city or district. EDCs are typically governed by boards comprised of local business leaders, property owners, and public officials (Strom, 2008). Another option that became increasingly popular throughout the 1990s and 2000s was the business improvement district (BID), a defined geographic area within which business and property owners are required to pay an additional tax in order to fund public infrastructure projects (streetscape improvements, public spaces, etc.). BIDs have similarly mixed boards that ensure public, private, and nonprofit stakeholders are engaged in the development and activation of district assets (Strom, 2008).

Through this often-careful choreography between private and public funds and entities, the resurgent capital flowing back into cities was highly effective at transforming urban spaces into more vibrant, amenity-rich places that attracted businesses and residents, particularly those with a college degree. Between 2000 and 2010, America's largest metro areas saw a collective population growth rate of more than 13 per cent within two miles of city hall, a proxy for 'downtown' (US Census Bureau, 2012). Over roughly the same period, more college-educated professionals moved downtown than to the suburbs in 39 of the 50 largest US metro areas (US Census Bureau, 2012). Businesses and jobs soon followed. Between 2007 and 2011, jobs grew faster in city centers than in suburbs for the first time in decades (Cortright, 2015).

However, the renewed appeal of cities has had unintended consequences, including the rising cost of living, increasingly strained transit infrastructure, and shrinking amounts of open space per capita. What's more, the benefits of this urban revival are not equally distributed. Traditionally marginalized communities are disproportionately bearing the brunt of the downsides yielded by this latest wave of urbanization (Hyra, 2014). Placemaking as a strategy is well-positioned to help address these complex issues, creating and sustaining concentrations of economic activity in an equitable and inclusive way (Vey, 2018). It accomplishes this in part through the design and management of inviting urban spaces that are open to all.

How 'place' drives productivity and shifts the geography of innovation

Critics have pointed to urban density as the cause of a range of social and economic issues, from public health crises to high housing prices. Yet, evidence continues to suggest that density generates far greater benefits than it does harm, cultivating dynamic and diverse communities that are exceptionally productive, sustainable, and livable. Dense places are also often resilient and highly adaptable ones, capable of reinventing themselves and evolving in response to new challenges and circumstances.

The economic value of density in particular has been well documented. America's densest cities drive a disproportionate share of the country's job growth (Abel et al., 2011). Regions with dense populations also lead the country in firm birth rates (Armington and Acs, 2002). While density increasingly seems like a prerequisite for a community's economic success, there is not a deterministic relationship between the two. Economic performance varies greatly between American metro areas with comparable population sizes and densities.

Quality of place offers a compelling answer as to what distinguishes certain high-performing metros from their similarly dense but less competitive counterparts. As Jennifer Vey, Director of the Brookings Institution's Bass Center for Transformative Placemaking argues, 'density absent of investments in placemaking may yield few, if any, benefits at all… a relatively compact but poorly designed neighborhood can discourage social interaction, make walking more dangerous, and worsen congestion and localized pollution' (Vey et al., 2019). Indeed, many of density's key economic benefits rely on components of placemaking to come to fruition. Research has found that density drives economic growth by encouraging frequent interactions between workers and firms that allow them to share inputs and collaborate to develop ideas (Abel et al., 2011). If a city or district lacks public spaces and other 'social infrastructure' to facilitate these productive connections, however, it could fail to maximize the potential value of its density. Firm agglomeration has been strongly associated with highly productive and competitive markets (Vey et al., 2019). Yet, if a community that contains a significant number of firms is detached from potential consumers through strict zoning and use segregation or limited transit options, the capacity for such a market to develop could be stunted. In fact, this is one of the reasons why startup companies are abandoning suburbs and cities with poor transit infrastructure for those with fast and abundant options (Credit, 2018). Moreover, the majority of high-growth firms that choose to relocate to dense cities with vibrant, accessible communities are doing so because of the critical mass of young, educated workers they contain (Welch and Anderson, 2017). As of 2014, two-thirds of America's 25- to 34-year-olds with a bachelor's degree live in the country's 51 largest metro areas. Since 2000, this demographic's population has grown twice as fast in city centers than those in other parts of the metro area. These trends even held true for metro areas that experienced a net loss in population such as Cleveland and New Orleans (Cortright, 2014).

This development is a reversal of the geographic and demographic patterns that persisted for the latter half of the twentieth century. As recently as a few decades ago, America's most high-powered companies wanted to be situated in 'nerdistans,' or suburban havens with sprawling corporate campuses (Florida and Hathaway, 2018). These corporate giants were initially drawn to low-density locations by the sizable communities of educated, white-collar workers that had developed there, a product of the urban exodus discussed earlier (Mozingo, 2011). At the time, most Americans considered the suburban lifestyle to be highly desirable. The General Social Survey, which seeks to measure the American population's subjective sense of wellbeing, consistently found that Americans are happier in low-density communities and small cities (Okulicz-Kozaryn and Valente, 2018).

The 'millennial' generation was the first to buck this trend. This cohort, which consists of those entering adulthood between 2000 and 2014, reported that they were most happy in places with a population of more than 250,000 people and least happy in places with fewer than 8,000 (Okulicz-Kozaryn and Valente, 2018). Multiple arguments have been offered to explain this phenomenon. Some posit that millennials' student-loan debt burdens prevent them from pursuing the traditional route of purchasing a home and a car, forcing them to live in places where they can rent and ride public transit (Winters and Tabit, 2019). Yet, this theory implies that upon attaining financial stability, millennials will exit the city, which has largely not been the case (Lee et al., 2019). Others argue that the millennial penchant for an urban lifestyle is a counterreaction to the 'separatist geography' of the suburbs, where they had little ability to walk or bike places on their own (Wyckoff et al., 2015). Recent analysis found that proximity to consumption, entertainment, and cultural amenities was more strongly associated with millennial urban in-migration than any other potential explanatory variable (Lee et al., 2019). Subsequent research has confirmed millennials' prioritization of quality places containing these elements. Two-thirds of college-educated millennials reported that they will decide where to live first before looking for a job (Wyckoff et al., 2015). The economic importance of this place-driven demographic cannot be overstated. In the digital age, human capital has supplanted physical capital as the primary input for economic growth. Since the early 1980s, the American 'knowledge' sector has exploded, adding 1.9 million employees per year on average. Over the same period, other sectors have grown at much slower rates, averaging 100,000–250,000 new jobs per year (Welch and Anderson, 2017). The impact of this shift has never been more significant than in cities. A one percentage point increase in a metropolitan area's proportion of residents with a college degree is associated with a two per cent increase in its GDP per capita (Abel and Gabe, 2011).

Cities have two potential strategies for cultivating a base of skilled workers. The first is to become better at educating their current population through investments in public schools and community college systems. However, this option yields no benefits if these newly educated workers defect to a more appealing place (Cortright, 2014). The more effective alternative in the long term is a placemaking play to become an attractive place for mobile, well-educated workers. While there are talented workers at all age levels, those below the age of 35 are statistically the most likely to move, more than twice as likely as their older counterparts, making them the talent base that is 'up for grabs' (Cortright, 2014). This population strongly prefers communities that are bustling with activity, full of places to socialize, provide plenty of housing and transit options, and contain a diverse mix of residents. The demographic shift and population preferences have implications for the locational choice of businesses. Cities that are successful in leveraging placemaking and developing a talent pipeline using such desirable, high-amenity districts tend to subsequently see an influx of new firms. Access to young, high-quality labor is a critical competitive factor for dynamic firms who seek to grow. As a result, it has become a primary motivating factor driving firms' location decisions (Cortright, 2014). Businesses tend to follow the residential preferences of workers. Just as the decentralization of the population in the mid-twentieth century caused businesses to move to dispersed locations to be closer to potential workers and customers, the re-urbanization of American skilled labor is bringing them back to central business districts (Glaeser and Kahn, 2001). Between 2010 and 2015, nearly 500 firms have relocated to or opened a new office in a central business district, including dozens of Fortune 500 companies. The majority of these companies were moving from suburban locations. Nearly two-thirds of them were in 'knowledge' sectors (information, finance or professional, scientific, and technical services). These companies overwhelmingly cited the ability to use the vibrant surrounding neighborhood as a selling point to attract and retain top talent as a main reason for their move (Anderson, 2015).

Not only is a critical mass of young, educated workers a key factor in luring major companies to locate in urban centers, it is also a significant input for catalyzing entrepreneurial activity. The number of firms, especially those in high-tech industries, rises faster in places with an abundant supply of young workers (Ouimet and Zarutskie, 2014). Rates of new-firm birth are of great economic consequence to cities and regions. A 10 per cent increase in a metropolitan area's startup rate raises overall wages and employment by up to 2 per cent over the following decade (Lee, 2016). What's more, high rates of entrepreneurship are also associated with more research and development activity (Fazio et al., 2016).

Whether through incumbent relocation, startup creation or both, the introduction of new, innovative firms in the knowledge sector yields positive spillover effects for cities and regions. Economist Enrico Moretti found that for each new high-tech job in a city, five additional jobs are ultimately created in the local service sector. These jobs consist of both skilled occupations and unskilled ones (Moretti, 2013). Moretti argues that the cities who harbor this explosive job creation potential are those with a solid base of human capital that enables them to keep attracting employers in key industrial clusters. Those who are unable to cultivate such a workforce are susceptible to being on the wrong side of what Moretti dubs a 'great divergence' in the economic fates of American cities (Moretti, 2013). Cities recognize this critical juncture and are pursuing policies to seed local 'innovation ecosystems' that attract talented workers, proliferate high-growth firms, and generate the associated economic gains. Innovation ecosystems have been defined in a number of ways but are most simply described as a collection of stakeholders, assets, and their interactions within an urban environment that result in new technologies.

'Innovation districts' are cities' attempts at engineering these ecosystems through planning and placemaking. They are distinct geographic areas where leading-edge institutions cluster and connect with startups and the organizations that support them to develop new ideas, products, and enterprises (Katz and Wagner, 2014).

However, whether these collaborations occur and the extent to which they are productive is determined by place. If an innovation district's urban fabric does not promote interactions between individuals, firms, and institutions that spawn, sharpen, and advance ideas, it will likely fail to nurture an ecosystem. By contrast, dense, walkable, and highly connected places foster the open culture of innovation by creating spaces that facilitate idea generation and exchange, the most fundamental input for innovation. Successful innovation districts require the development of public spaces that encourage networking, mixed uses to create a 'buzzing' atmosphere, and a variety of options that connect people to the district and within it. These place-based ingredients are so integral to the success of an innovation district that a recent handbook for 'auditing' your city's 'innovation district' includes a section on evaluating a district's 'quality of place.' All of an innovation district's built environment assets should aim to create an 'experience of proximity' in which highly visible activity engages, excites, and inspires (Vey et al., 2018).

The majority of the United States' innovation districts are located in well-established technology hubs such as Seattle and Cambridge, who have transformed neighborhoods such as South Lake Union and Kendall Square into vibrant, thriving spaces that are now home to some of the country's most innovative and valuable companies (Katz and Wagner, 2014). In recent years, cities across the country have pursued this strategy and enjoyed great success. Chattanooga, once dubbed the 'dirtiest city in America' during the fallout of deindustrialization, has had an economic renaissance having redeveloped its riverfront into a high-amenity, mixed-use district and introduced high-speed public broadband, among other initiatives (digital connectivity in particular is poised to become an increasingly critical component of urban places in the twenty-first century). These investments in quality of life paid dividends for Chattanooga, contributing to new residential development downtown and luring young entrepreneurs as well as a new Volkswagen assembly plant to the city, bringing a significant number of new jobs and revenue (Storring and Benz, 2018).

Place-oriented development: how parks and open space enhance real estate value

Parks and open spaces are typically the centerpiece of any economically competitive 'place.' They offer a number of benefits to cities and their residents, both quantifiable and not. Previously, parks were seen as nice to have but nonessential investments that improve quality of life, contributing to the health of the population, supporting local biomes, and providing people with a place to gather, regardless of their ability to pay (National Recreation and Park Association, 2018). While there is a sizable body of research providing evidence that parks can also be economic assets with the potential for significant real estate and economic value creation, city officials have only recently started to seriously incorporate parks investments into their overall economic development plans and growth strategies.

The primary impetus for this reevaluation of open space's economic potential has been the rise of park 'mega-projects' that have transformed once sleepy or undesirable districts into must-see cultural attractions. While high-profile projects in major American cities such as New York's High Line and Chicago's Millennium Park are among the most well-known examples of this, a number of other, smaller cities have undertaken large-scale initiatives to create public spaces that achieve similar results and become city-defining landmarks. This raises the question of how cities are financing and subsequently justifying the capital expenditures and maintenance costs associated with such ambitious plans. There are two primary mechanisms through which parks directly generate economic value and revenue for local governments.

The first is the 'proximity premium,' or the notion that people are willing to pay more for real estate, primarily residential, within a reasonable distance of a park (100–2,600 feet depending on a variety of factors), which in turn raises assessed property values. This concept is predicated on the 'hedonic' model that economists use to try and estimate the impact of particular amenities on the prices of various assets (Harnik and Welle, 2009). The rationale is that cities make investments in a park or public space and proceed to capture a portion of proximity premiums through increases in the assessed value of properties within a certain radius of the park. They then leverage this revenue stream to service bond payments (if the project was debt financed) or directly fund site maintenance, improvements, and even programming in some cases (Crompton, 2005). Yet, not all parks or open spaces provide the same economic benefits and there are several factors that influence the scale of the proximity premium conferred to individual property owners and city governments alike. The bulk of the premium, approximately 75 per cent, is found within a travel distance of 600 feet of the park. Properties adjacent to parks see a premium that is approximately 22 per cent higher than those 2,600 feet away (Miller, 2001). What's more, the type of open space and its relative location to other amenities within the city are also key determinants of premium size.

'Natural parks' or those that dedicate at least 50 per cent of their land area to natural habitat preservation have been found to deliver the largest proximity premium. Specialty parks (those with a single, primary use such as boating or golf) and standard urban parks also have a significant positive impact on property values. While there are perceived negative externalities associated with having a home adjacent to or nearby to each of these types of parks (noise, foot traffic, etc.) studies have found that none of these are enough to significantly offset premiums (Lutzenhiser and Netusil, 2001). Open space siting is also consequential. The characteristics and demographics of a park's surrounding neighborhoods have a measurable impact on proximity premiums. Residents of high-density neighborhoods that are close to a city's central business district (CBD) place a higher premium on open space proximity than their suburban counterparts. In one study, neighborhoods that were twice as dense also saw proximity premiums that were three times higher than the city's average (Anderson and West, 2006). Seeing as many placemaking initiatives occur in dense neighborhoods that are close to CBDs, the inclusion of a park or open space component could greatly enhance the project's economic impact. Evidence from cities across the United States attests to the findings of this research. In 2001, Indianapolis designated five neighborhoods near the city's core as cultural districts and proposed the development of a trail to link all of the districts' assets and provide a venue for public art. Between 2008 and 2014 (the period of the trail's construction), the total assessed value of properties within 500 feet of these new public spaces increased by 148 per cent (Majors and Burow, 2019). Dallas' park system enhanced the value of existing real estate within a 750-foot radius by $119 million aggregately. Downtown parks, including recent Dallas investments such as Klyde Warren Park and Katy Trail, drove a sizable portion of this premium (HR&A Advisors, 2016).

Parks are both a value-adding amenity and a critical piece of loss-mitigation infrastructure protecting economic, cultural, and civic assets from climate risk. For example, it is now an imperative for parks commissioners to focus on the range of existential threats from climate change, from flooding to urban heat island effects. With respect to sea level rise and other flood risk due to climate change, public waterfront parks are increasingly seen as the first line of defense. Unlike previous strategies for mitigating flooding such as levies, which have created a sharp urban edge or divided communities, parks are being reimagined in cities as a new form of resilience infrastructure with an overlay of 'social infrastructure' informed by community engagement. Plans for the 10-mile 'BIG U' integrated flood protection system for Lower Manhattan, with which our firm has been associated since its inception, provide more intensive and varied community programming, and stronger connections to the water's edge and upland communities, while insulating one of New York City's most flood-vulnerable and economically and socially diverse string of neighborhoods from sea-level rise, storm surges, and other impacts of climate change (Rebuild by Design, 2019). If executed true to its vision, the BIG U can become exemplary of the notion of 'hedonistic sustainability' or the idea that communities can become more climate resilient and improve the quality of life for residents simultaneously (Ingels, 2012).

The second primary mechanism through which parks and open spaces create economic value for cities is induced new development. Real estate developers are increasingly recognizing the opportunities for asset value enhancement that parks provide as well as the increased foot traffic parks generate for commercial operations. They are shifting the geographic focus of their development activity to park-proximate neighborhoods and increasingly are willing to fund some or all of the cost of building new parks (Norris and Singh, 2018). Since 2000, Pittsburgh has invested approximately \$130 million in the city's riverfront parks. This has helped to catalyze \$2.6 billion in private riverfront development activity, a 20:1 return on investment ratio (Riverlife, 2015). Even smaller cities like Greenville, South Carolina are seeing outsized returns on open space investments. The city's \$13.5 million investment in Falls Park downtown has yielded nearly \$600 million in nearby development between 2004 to 2015 (White, 2015). Developers are also beginning to invest in publicly accessible open spaces themselves. As municipal budgets have tightened, the public resources available for open-space creation, operations, programming, and improvements have flatlined or even diminished many cities. This disinvestment restricted neighborhoods' access to open space and denied many communities the benefits they bring about (National Recreation and Park Association, 2018). Developers have helped resolve this funding gap by directly investing in open-space projects or making financial contributions to park stewardship intermediaries such as conservancies or business improvement districts. Not only do these investments help to transform underused public assets into vibrant community spaces, they also help provide economic returns to the developers as well (Norris and Singh, 2018). While many cities and neighborhoods welcome the infusion of capital and long-term partnership that developers offer, others are wary of the private sector encroaching on the public realm.

Pre-emptive efforts to keep places open and accessible

The notion that a series of place-based investments will transform a neighborhood into a vibrant district that can attract young, educated workers, bring new business activity, and increase property values raises the specter of gentrification for city officials and community members alike. Such concern is not unfounded. However, placemaking is an economic development strategy that has a resolute focus on inclusion. Placemaking seeks to 'shape the public realm in order to maximize shared value' while 'paying particular attention to the physical, cultural, and social identities that define a place and support its ongoing evolution' (PPS, 2007). Yet, policymakers are often unaware of how they can preemptively address concerns regarding gentrification and incorporate principles of equity into placemaking project design and implementation processes. This section surveys the most promising policies cities are pursuing to anticipate and prevent place and open-space-driven gentrification, dubbed 'Parks Related Anti-Displacement Strategies' or 'PRADS' by a recent study (Rigolon and Nemeth, 2019). As renewed urban challenges underscore the inequities that exist in American communities, it has become all the more imperative that the voices of marginalized residents are amplified and that their needs are prioritized through inclusive placemaking initiatives.

Proactive, community-oriented housing policies are one strategy to quell displacement pressures and preserve community character while facilitating growth. Community Land Trusts (CLTs) have emerged as a promising solution to offset unintended negative impacts of revitalization and investment on neighborhood housing markets. CLTs are nonprofit organizations that provide affordable housing in perpetuity through their ownership of land in a given community, leasing it to low-income families. This model seeks to provide long-term affordability and stability in communities by offering housing options that are not subject to real estate speculation and keep monthly costs at an affordable rate due to the lack of profit incentives. What's more, CLTs create opportunities for wealth building by including a 'resale formula' clause in the contract. This grants families exiting the trust a payout that is a portion of the increased value of the property they leased (the remainder is kept by the trust to help preserve long-term affordability) (Choi et al., 2017). Cities have also recently begun utilizing tax increment financing (TIF) districts to fund affordable housing projects. Traditionally, a city designates a group of contiguous parcels as a TIF district and subsequently earmarks property tax revenue from increases in their assessed value to finance a public infrastructure project nearby, whether directly or by servicing bond payments. In addition to leveraging these funds for typical TIF projects, policymakers are now also designating TIF funds for the development, rehabilitation, and preservation of affordable housing (Dye and Merriman, 2006). In 2017 alone, Chicago's TIF districts generated $660 million in funds, a substantial base of funds that the city can use to address its growing housing shortage (Office of the Cook County Clerk, 2018). If a city were to establish a TIF district that encapsulates all of the parcels within the radius of a park, open space, or placemaking investment, it could capture a portion of the proximity premium and use these funds to maintain affordability for the neighborhood's existing residents.

Preserving access to places also means taking measures to ensure that they feel welcome to people from all walks of life. A major component of this is placing community input at the center of planning processes. Washington, DC recently launched an ambitious plan to build the district's first elevated park atop Anacostia's 11th Street Bridge, linking the traditionally low-income, African American neighborhood to Washington's Navy Yard, one of the city's fastest-growing areas. To address concerns regarding gentrification, Mayor Muriel Bowser and other leaders engaged the Anacostia community in a year-long process to draft an 'equitable development plan' that would use their feedback to drive the conceptualization and design of the park. This led to a number of parallel initiatives to create and preserve Anacostia's affordable housing, celebrate the neighborhood's culture and heritage, and mandate that the construction and operation of the park employs as many community members as possible (Bernard and Kratz, 2018). Governance and stewardship organizations are another critical component that helps protect and promote community voice in decisions concerning open-space assets. For example, while there is an evolving set of models for the improvement, programming, and maintenance of cities' parks ranging from park conservancies to community development corporations (CDCs), they vary in the extent to which they address equity considerations through their institutional structure. Our firm's 2018 study with The Trust for Public Land found that the park alliance model is most conducive to community involvement. This equity orientation is due to its flat and representative structure that allows for shared decision-making authority between the city and the community on all aspects of park stewardship (Trust for Public Land and JLP+D, 2018).

In order to sustain inclusivity throughout the course of a placemaking initiative, planners, policymakers, and other stakeholders must take dedicated measures to engage and listen to the community at every step of the process. Failure to do so risks denying these exciting new urban districts the asset that has been critical to the economic success of American cities in the twenty-first century: a diverse population. America's fastest growing metropolitan areas from 2000 to 2014, which accounted for more than half of the country's new jobs, were also among the country's most ethnically heterogeneous (DuPuis et al., 2017). Diverse communities are also one of the last harbingers of upward mobility in the United States. Economist Raj Chetty and his team recently found that integrated neighborhoods (mixed-income, mixed-race) give American children of all sorts the best odds of attaining a higher standard of living than their parents (Chetty et al., 2014). Placemaking in its broadest sense strives to create such communities, as an impactful response seeking to better connect communities and organize economic, social, cultural, and environmental hubs of activity and benefit. In this way, placemaking can play an important role in undoing the economic and social harms inflicted by decades of segregation and suburbanization while setting cities up for sustainable, resilient, and long-term growth that can be shared by all.

Conclusion

Through targeted, partnership-driven investments in place, cities across the United States have achieved a positive turnaround of their social and economic fortunes, transforming previously empty or underutilized swaths of land into amenity-rich districts anchored by inviting, value-generating open spaces. These revived and well-connected urban destinations have become a magnet for young, talented workers and firms that leverage their built environments to facilitate and accelerate economic and innovation activity. This experience offers lessons for how cities can recover from future crises and resolve persistent challenges through placemaking initiatives that prioritize inclusion, social connectivity, and ingenuity.

References

Abel, J., Dey, I. and Gabe, T. (2011). 'Productivity and the density of human capital', *Journal of Regional Science*, 52(4), pp. 562–586.

Anderson, G. (2015). *Core Values: Why American Companies are Moving Downtown*. Washington, DC: Smart Growth America.

Anderson, S. and West, S. (2006). 'Open space, residential property values, and spatial context', *Regional Science and Urban Economics*, 36(6), pp. 773–789.

Armington, C. and Acs, Z. (2002). 'The determinants of regional variation in new firm formation', *Regional Studies*, 36(1), pp. 33–45.

Bernard, R. and Kratz, S. (2018). *11th Street Bridge Park's Equitable Development Plans*. Washington, DC: Building Bridges Across the River, pp. 4–31.

Boustan, L. (2010). 'Was postwar suburbanization "white flight"? Evidence from the black migration', *Quarterly Journal of Economics*, 125(1), pp. 417–443.

Chetty, R., Hendren, N., Kline, P. and Saez, E. (2014). '"Where is the land of opportunity?" The geography of intergenerational mobility in the United States', *The Quarterly Journal of Economics*, 129(4), pp. 1553–1623.

Choi, M., Van Zandt, S. and Matarrita-Cascante, D. (2017). 'Can community land trusts slow gentrification?' *Journal of Urban Affairs*, 40(3), pp. 394–411.

Cortright, J. (2014). 'The young and restless and the nation's cities', in City Report. Portland, OR: City Observatory, pp. 1–30.

Cortright, J. (2015). 'Surging city center job growth', in City Report. Portland, OR: City Observatory.

Credit, K. (2018). 'Transitive properties: A spatial econometric analysis of new business creation around transit', *Spatial Economic Analysis*, 14(1), pp. 26–52.

Crompton, J. (2005). 'The impact of parks on property values: Empirical evidence from the past two decades in the United States', *Managing Leisure*, 10(4), pp. 203–218.

Dreier, P. (1993). 'America's urban crisis: Symptoms, causes, solutions', *North Carolina Law Review*, 71(5), pp. 1352–1401.
DuPuis, N., Stahl, E. and Rainwater, B. (2017). 'The future of equity in cities', National League of Cities, pp. 23–26.
Dye, R. and Merriman, D. (2006). *Tax Increment Financing: A Tool for Local Economic Development*. Cambridge, MA: Lincoln Institute of Land Policy, pp. 1–7.
Fazio, C., Guzman, J., Murray, F. and Stern, S. (2016). *A New View of the Skew: A Quantitative Assessment of the Quality of American Entrepreneurship*. Cambridge, MA: MIT Innovation Initiative, pp. 1–3.
Florida, R. and Hathaway, I. (2018). *How the Geography of Startups and Innovation is Changing*. Cambridge, MA: Harvard Business Review.
Glaeser, E. and Kahn, M. (2001). 'Decentralized employment and the transformation of the American city', *Brookings-Wharton Papers on Urban Affairs*, 2001(1), pp. 1–63.
Harnik, P. and Welle, B. (2009). *Measuring the Economic Value of a City Park System*. Washington, DC: Trust for Public Land, pp. 2–15.
HR&A Advisors. (2016). *Economic Value and Benchmarking Study of the Dallas Park System*. New York: HR&A Advisors, pp. 25–30. https://www.hraadvisors.com/portfolio/dallas-parks/
Hyra, D. (2014). 'The back-to-the-city movement: Neighbourhood redevelopment and processes of political and cultural displacement', *Urban Studies*, 52(10), pp. 1753–1773.
Ingels, B. (2012). *Hedonistic Sustainability*. New York: TEDxEast. https://www.archdaily.com/203599/tedx-hedonistic-sustainability-bjarke-ingels
Katz, B. and Wagner, J. (2014). *The Rise of Innovation Districts: A New Geography of Innovation in America. Innovation Districts*. Washington, DC: Brookings Institution, pp. 1–21.
Lee, Y. (2016). 'Entrepreneurship, small businesses and economic growth in cities' *Journal of Economic Geography*, pp. 311–343.
Lee, Y., Lee, B. and Shubho, M. (2019). 'Urban revival by millennials? Intraurban net migration patterns of young adults, 1980–2010', *Journal of Regional Science*, 59(3), pp. 538–566.
Lewyn, M. (1996). 'The urban crisis: Made in Washington', *Journal of Law and Policy*, 4(2), pp. 513–550.
Lutzenhiser, M. and Netusil, N. (2001). 'The effect of open spaces on a home's sale price', *Contemporary Economic Policy*, 19(3), pp. 291–298.
Majors, J. and Burow, S. (2019). *Assessment of the Impact of the Indianapolis Cultural Trail: A Legacy of Gene and Marilyn Glick*. Indianapolis, IN: IU Public Policy Institute, pp. 1–40.
Massey, D. and Denton, N. (2003). *American Apartheid*. Cambridge, MA: Harvard University Press, pp. 53–57.
Miller, A. (2001). *Valuing Open Space: Land Economics and Neighborhood Parks. Master of Architecture*. Cambridge, MA: Massachusetts Institute of Technology.
Mohl, R. (2001). 'Urban expressways and the racial restructuring of postwar American cities', *Jahrbuch für Wirtschaftsgeschichte/Economic History Yearbook*, 42(2), 89–104.
Moretti, E. (2013). *The New Geography of Jobs*. Boston: Houghton Mifflin Harcourt, pp. 45–121.
Mozingo, L. (2011) *Pastoral Capitalism: A History of Suburban Corporate Landscapes*. Cambridge, MA: MIT Press, pp. 23–43.
National Recreation and Park Association. (2018). *Revitalizing Inner-City Parks: New Funding Options Can Address the Needs of Underserved Urban Communities*. Ashburn, VA: National Recreation and Park Association, pp. 1–4. https://www.nrpa.org/contentassets/f768428a39aa4035ae55b2aaff372617/urban-parks.pdf
Norris, M. and Singh, R. (2018). *The Case for Open Space: Why the Real Estate Industry Should Invest in Parks and Open Spaces*. Washington, DC: Urban Land Institute, pp. 10–21.
Office of the Cook County Clerk. (2018). *Cook County TIFs to Bring in a Record $1 Billion According to Clerk Orr: Transit TIF revenue doubled*. Chicago, IL: Cook County.
Okulicz-Kozaryn, A. and Valente, R. (2018). 'No urban malaise for Millennials', *Regional Studies*, 53(2), pp. 195–205.
Ouimet, P. and Zarutskie, R. (2014). 'Who works for startups? The relation between firm age, employee age, and growth', *Journal of Financial Economics*, 112(3), pp. 386–407.
PPS (Project for Public Spaces). (2007). *What is Placemaking* [Online]. Available at: https://www.pps.org/article/what-is-placemaking (Accessed: 8 June 2020).
Rebuild by Design. (2019). *The Big 'U': Promoting Resilience Post-Sandy Through Innovative Planning, Design, & Programming*. New York: Rebuild by Design, pp. 1–17. http://www.rebuildbydesign.org/data/files/504.pdf

Rigolon, A. and Németh, J. (2019). 'Green gentrification or 'just green enough': Do park location, size and function affect whether a place gentrifies or not?', *Urban Studies*, 57(2), pp. 402–420.
Riverlife. (2015). *Three Rivers Park: Economic Impact Analysis*. Pittsburgh, PA: Riverlife, pp. 6–8.
Storring, N. and Benz, C. (2018). *Chattanooga Innovation District, Tennessee. Opportunities for Transformative Placemaking*. New York: Project for Public Spaces, pp. 3–5.
Strom, E. (2008). 'Rethinking the politics of downtown development', *Journal of Urban Affairs*, 30(1), pp. 37–61.
Sugrue, T. (2005). *The Origins of the Urban Crisis*. Princeton: Princeton University Press, pp. 126–152.
Trust for Public Land and JLP+D. (2018). *Building Bridges: A Community-Based Stewardship Study for an Equitable East River Park*. New York: Trust for Public Land, pp. 15–22. http://www.rebuildbydesign.org/our-work/research/building-bridges-a-community-based-stewardship-study-for-an-equitable-east-river-park
US Census Bureau. (2012). *Patterns of Metropolitan and Micropolitan Population Change: 2000 to 2010*. Washington, DC: US Census Bureau, pp. 29–31.
US Department of Housing and Urban Development. (2016). *HOPE VI Data Compilation and Analysis*. Washington, DC: US Department of Housing and Urban Development, pp. 11–15.
US Government Accountability Office. (2010). *Revitalization Programs: Empowerment Zones, Enterprise Communities, and Renewal Communities*. Washington, DC: US Government Accountability Office, pp. 1–4.
Vey, J. (2018). *Why We Need to Invest in Transformative Placemaking. Innovation Districts*. Washington, DC: Brookings Institution.
Vey, J., Hachadorian, J., Andes, S., Wagner, J. and Storring, N. (2018). *Assessing Your Innovation District: A How-to Guide. Innovation Districts*. Washington, DC: Brookings Institution, pp. 54–72.
Vey, J., Shearer, C. and Kim, J. (2019). *Where Jobs Are Concentrating and Why It Matters to Cities and Regions*. Washington, DC: Brookings Institution, pp. 4–26.
Welch, C. and Anderson, L. (2017). *Place Matters: The Role of Placemaking in Economic Development*. Washington, DC: International Economic Development Council, pp. 6–8.
White, K. (2015). *Falls Park on the Reedy*. Cambridge, MA: Rudy Bruner Award for Urban Excellence, pp. 148–149.
Wilson, W. (2012). *Truly Disadvantaged*. Chicago: University of Chicago Press.
Winters, J. and Tabit, P. (2019). *Rural Brain Drain: Examining Millennial Migration Patterns and Student Loan Debt. Consumer & Community Context*. Washington, DC: Federal Reserve Board of Governors, pp. 7–14.
Wyckoff, M., Neumann, B., Pape, G. and Schindler, K. (2015). *Placemaking as an Economic Development Tool: A Placemaking Guidebook*. Michigan: MSU Land Policy Institute, pp. 75–85.

Further reading in this volume

Chapter 15: Un/safety as placemaking: disabled people's socio-spatial negotiation of fear of violent crime
Claire Edwards
Chapter 22: Embedded Artist Project: Epistemic Disobedience + Place
Frances Whitehead
Chapter 25: 'If you can make it there, you can make it anywhere…': cultural placemaking at the heart of cities
Sherry Dobbin
Chapter 32: *The Hollywood Forest Story*: Placemaking for the Symbiocene
Cathy Fitzgerald
Chapter 34: Reconnecting *cité* and *ville*
Philip Graus
Chapter 35: Planning governance – lessons for the integration of placemaking
Nigel Smith
Chapter 36: Facilitator skills for effective collaborative placemaking
Husam AlWaer and Ian Cooper
Chapter 37: *The Neighbourhood Project*: a case study on community-led placemaking by CoDesign Studio
Lucinda Hartley, Eliza Charley, Sama Choudhury, and Harriet McKindlay
Chapter 38: Public seating: a small but important place in the city
Kylie Legge

Chapter 42: Creative Placemaking and Placekeeping evaluation challenges from the practitioner perspective: an interview with Roy Chan
Maria Rosario Jackson
Chapter 45: How the city speaks to us and how we speak back: rewriting the relationship between people and place
Rosanna Vitiello and Marcus Willcocks

3

AN ANNOTATED HISTORY OF CREATIVE PLACEMAKING AT THE FEDERAL LEVEL

Jen Hughes

Introduction

This chapter offers a history of creative placemaking from the perspective of and the programming led by the National Endowment for the Arts (NEA), a federal government agency that funds arts, culture, and design in the United States. In 2010, the NEA released a White Paper called *Creative Placemaking* (Markusen and Gadwa, 2010). It documented two decades of American creative placemaking, drawing on case studies and research that demonstrated the ways that 'creative initiatives were animating places and sparking economic development' (ibid., p. 3). The White Paper laid the groundwork for the establishment of a creative placemaking grant program at the NEA in 2011, called *Our Town:* a collaboration of philanthropic foundations to fund creative placemaking, ArtPlace America, and federal interagency collaborations that integrated arts and culture as a key strategy for advancing economic and community development. Various creative placemaking initiatives led by the NEA supported the evolution of the field over the past decade, with an emphasis and focus on partnership across sectors. Some initiatives and programming focused on making creative placemaking legible to non-arts sectors, while others such as the *Our Town* grant program explicitly piloted investments in local creative placemaking partnerships and projects. As a federal agency, the NEA has played a unique role in advancing creative placemaking practice in communities as diverse as the country itself, supporting the context-specific work of rural, tribal, suburban, and urban places.

Responding to the Great Recession

Creative placemaking was born out of a response to the Great Recession, and a recognition that communities faced a myriad of challenges that were inextricably linked to one another and could not be solved or addressed in isolation. The Great Recession presented significant new challenges to American cities, beginning in 2007 and lasting well into 2009. Homeowners lost a significant percentage of their net worth during the recession, resulting from the bursting of a 'housing bubble' and the global financial crisis. The Great Recession bankrupted government coffers and left residents unable to afford their mortgages, while housing prices simultaneously declined. Offering an innovative approach to economic development, arts and culture were perceived to hold great promise for helping communities recover from the economic hardship of the Great Recession. With artists and cultural assets present in every corner of the United States, creative placemaking offered a fresh approach for elected officials and local leaders to drive economic development by capitalizing on the unique assets of their place.

The NEA's support of creative placemaking was seeded through a research endeavour that demonstrated the ways that creative initiatives were animating places and sparking economic development. Authored by Ann Markusen and Anne Gadwa Nicodemus, the 2010 NEA White Paper 'Creative Placemaking' served as a foundation for informing the programmatic elements of the NEA's creative placemaking grant program, offering communities tangible arts and cultural case studies to mimic or learn from. The White Paper also defined creative placemaking:

> In creative placemaking, partners from public, private, non-profit, and community sectors strategically shape the physical and social character of a neighborhood, town, city, or region around arts and cultural activities. Creative placemaking animates public and private spaces, rejuvenates structures and streetscapes, improves local business viability and public safety, and brings diverse people together to celebrate, inspire, and be inspired.
>
> *(Markusen and Gadwa, 2010)*

The compelling case made by the White Paper was that 'creative placemaking generates economic returns in multiple ways' (Markusen and Gadwa, 2010). The connection between the role of the arts and economic development was a key driver in the early adoption of creative placemaking by a range of local mayors and other community leaders who were looking for new, and arguably low-cost ways, to revitalize their communities. Many of the projects profiled in the White Paper resulted in transformative physical impacts, such as the design of a new public space, artist live–work space, and public art initiatives. Economic impacts, such as increased business revenue, new local jobs, a growing tax base, were lauded as indicators of progress in community revitalization. Creative placemaking promised to ultimately demonstrate tangible economic impact and a visible, physical transformation. In the urgency of the post–Great Recession, communities were desperate to identify swift recovery and new ways to invest in their local ecosystems. However, as the field evolved, the US moved beyond the Great Recession, and as inequity rose in many urban areas; the arts and cultural response shifted towards a more comprehensive approach to community development; one that was also rooted in social equity and systemic impact.

Seeding creative placemaking with federal and philanthropic funding

While artists, cultural organizations, and designers have worked to improve their communities for centuries, federal funding, resources and attention to explicitly supporting creative placemaking began in 2010. Under the leadership of the NEA Chairman Rocco Landesman, the terminology of creative placemaking was born and translated into a national federal funding program called *Our Town*.

While a new federal grant program at the NEA was one way to incentivize local communities to integrate arts and culture into their economic development strategies, Chairman Landesman recognized the limitation of NEA dollars and sought out collaborative federal partnerships that could advance the integration of arts, culture, and design into a holistic approach to supporting quality of life in American communities. In an effort to respond to the Great Recession, the Obama Administration sought to advance place-based strategies that were more comprehensive and less siloed, creating an opportunity for the NEA to set the community development table with arts and culture as a key contributor to driving local, positive impact. With federal agency collaborations such as the Partnership for Sustainable Communities underway, the NEA began to introduce creative placemaking not as a panacea, but as a complementary approach to driving holistic community development via arts and culture. The Partnership for Sustainable Communities was an unprecedented federal collaboration among the US Department of Housing and Urban Development (HUD), the US Department of Transportation (DOT), and the US Environmental Protection Agency (EPA). The main objective of this collaboration was to empower federal agencies to break out of their silos and to jointly solve interrelated challenges in communities recognizing that jobs, transportation options, housing, education, and environment are inextricably interconnected, and impact the quality of life of individuals (The White House, 2010). Federal agencies moved towards addressing community challenges through the lens of place and moved away from a technocratic one-size-fits-all approach. NEA's creative placemaking program, *Our Town*, was intended to complement this holistic and place-based approach by acknowledging the potential for arts and culture to play a role in responding to local conditions and in helping to solve complex community challenges. *Our Town* aligned with other place-based federal funding programs that were focused on improving the quality of life and livability in American communities.

Another approach to seeding creative placemaking was to corral additional funding commitments from the philanthropic community. NEA leadership convened the heads of several large philanthropic organizations to birth ArtPlace America. As a foundation collaborative, ArtPlace was envisioned as an entity to advance the creative placemaking field through investment in local projects and research. Uninhibited by federal funding regulations and processes, ArtPlace could serve to advance the field in complementary ways to federal investment programs and often fund projects and initiatives in a way that was not hamstrung by federal regulations that governed the work of the NEA. Unlike the NEA, ArtPlace could make investments in the construction of cultural facilities and public space, and even make direct grants to artists and organizations other than non-profit 501c3 organizations. Over the past decade, the collaborative partnership between the NEA and ArtPlace proved to be critical in advancing and establishing a solid field of creative placemaking practitioners and practice. Via a thoughtful approach to comprehensive community development, ArtPlace has led comprehensive community development sector research, funded over 285 local projects, piloted a community development investment program, and convened practitioners, demonstrating the value of arts and culture in driving 'equitable, healthy, and sustainable communities' (ArtPlace America, 2020). ArtPlace presented a unique opportunity for the NEA to collaborate with philanthropy to ultimately enhance creative placemaking practice in communities all across the country. This collaboration enabled creative placemaking to have staying power beyond the federal government, advancing research, field building, and experimentation that responded to an ever-evolving field of practice.

Our Town: NEA funding for local pilot projects

While artists, designers, and cultural organizations had been improving places for centuries, looking to them as equal partners driving positive community and economic development was in fact a radical shift in government policy. This shift was incentivized through NEA funding for one- to two-year creative placemaking projects. In 2010, the NEA first issued grant guidelines for creative placemaking under a pilot initiative (and the precursor to *Our Town*) via a program called the 'Mayors' Institute on City Design 25th Anniversary Initiative.' This pilot program received applications from cities, as a prerequisite for applying required that the mayor had participated in the NEA's 'Mayors' Institute on City Design Program' (www.micd.org). Ultimately,

21 grants were awarded, including projects such as the public space design of the Main Terrain park in Chattanooga that focused on public health outcomes, a collaborative public art installation by Mary Miss in Indianapolis to illuminate knowledge about a local waterway, and the architectural design of Culture Shed in Hudson Yards, New York. Ultimately, this pilot round of grants set the stage for a more robust creative placemaking grant program. In 2011, the *Our Town* program was officially launched as the agency's signature creative placemaking grant program with funding that was specifically allocated via the federal congressional budgeting process.

The *Our Town* program, named after the famous American play by Thornton Wilder, had some unique requirements that delineated it from other existing NEA grant programs. First, all applications required a partnership between a local government or federally recognized tribal government and a non-profit 501c3 organization. One of the two primary partners had to demonstrate an arts, culture, or design mission. Second, the application required a letter of support from the highest-ranking official in the community, such as a mayor, county judge, tribal leader, or town manager. The idea was to elevate the role of arts and culture within communities by incentivizing partnerships between the local government and cultural organizations, with a political leader to help champion the initiative locally. As a result of the grant application opportunity, unexpected local partners began to unite around a vision for the community's future with arts and culture taking center stage.

Another first was the introduction of livability, a new outcome area for the agency via the *Our Town* program. Livability in the *Our Town* grant guidelines was articulated in the following way:

> Livability: American communities are strengthened through the arts. The anticipated long-term results for livability projects are measurable community benefits, such as growth in overall levels of social and civic engagement; arts- or design-focused changes in policies, laws, and/or regulations; job and/or revenue growth for the community; and changes in in-and-out migration patterns.
>
> *(National Endowment for the Arts, 2011)*

The grant guidelines specified that funding amounts were available, 'ranging from $25,000 to $250,000, for creative placemaking projects that contribute toward the livability of communities and help transform them into lively, beautiful, and sustainable places with the arts at their core.' While other grant programs at the NEA were capped at $100,000, *Our Town* offered a larger grant award amount; and in turn also awarded some requests for funding in full.

The NEA's process for selecting grantees occurs via a peer review process. Grant review panelists representing diverse perspectives, geographies, and artistic disciplines are invited to review, score, and comment on applications to the NEA; and ultimately provide recommendations to the agency on those worthy of receiving American tax-payer funding. For *Our Town*, applications are reviewed in panels divided by similar community size, tapping the expertise of artistic practitioners that understand the unique geographic contexts of the places applying to the program. Grant review panelists also represent the diversity of the creative placemaking field, tapping the expertise of community development practitioners, urban planners, social practice artists, local government agencies, designers, and more. The result is an incredibly diverse body of projects that are funded via the *Our Town* portfolio each year, representing a wide range of creative placemaking project approaches, unique partnerships, and artistic disciplines. By spring 2020 NEA has supported over 636 creative placemaking projects and invested over $49.3 million in rural, tribal, suburban, and urban communities throughout the nation.

From Ajo, Arizona, to the Cheyenne Rivers Sioux Reservation in South Dakota, the *Our Town* program extended the reach of the agency by inviting new applicants, partners, and project activities into NEA's funding portfolio. The *Our Town* program dramatically expanded the number of new and first-time applicants to the NEA. With a grant program that explicitly required local government or tribal government partnership, new entities sought to obtain NEA grant funding to support their local arts initiatives. While the program has remained competitive, only funding about 20–25 per cent of applicants each year, the program's application call has helped to catalyze new relationships in places between the arts and non-arts sectors, as unexpected local partners pull together an *Our Town* application. Those leading creative placemaking projects are often driven by a coalition of diverse local actors; including artists, economic developers, housing advocates, public safety officials, community activists, farmers, religious and business leaders. Oftentimes, 10–15 local organizations came together and committed to partnering on an *Our Town* project proposal. The groundbreaking nature of these new local partnerships cannot be overemphasized. Many communities reported to the NEA that multiple stakeholders gathered around tables for the first time to talk about the future of their place and expressed that they should have been working together for years. The unique projects that emerged from these conversations are a true testament to the ingenuity of artists partnering with community entities to bring about positive change to a place. Under the banner of creative placemaking, new applicants to the agency shifted their view of arts and culture as not just a nice-to-have or an add-on, but rather as a central component of their community's future vision and success. For example, in Austin, Texas, Forklift Danceworks collaborated with the City of Austin Parks and Recreation Department Aquatics Division and several East Austin neighborhoods in an effort to organize the community around saving their local pools. Via a two-year-long artistic engagement, choreographers worked with recreation department staff and local residents to produce a series of performances that sought to strengthen the citizens' sense of ownership over public pools and reenergize the community to save the valued assets of public pools. This innovative partnership and pilot project demonstrated the value of artists and cultural events to unite community and inform local government budgets and planning processes. As a result, other city departments have attempted to partner with Forklift and other artists to establish a new approach to community engagement around city issues.

Making creative placemaking projects legible

While a new federal grant program was an important contribution to incentivizing local creative placemaking projects, the emerging field was in need of concrete case studies that could relate to the diverse rural, suburban, tribal, and urban contexts that exist across the country. Under the leadership of NEA Design and Creative Placemaking Director, Jason Schupbach, the *Our Town* program made two critical investments that helped to grow creative placemaking practice and make the work accessible and legible to local communities. The first was an investment in developing a microsite called 'Exploring *Our Town*' on the NEA website. GO Collaborative, led by Lynn Osgood, was the selected contractor to develop the microsite and engaged a team of professionals from the fields of art, design, planning, programming, and writing. Released in the fall of 2014, 'Exploring *Our Town*' profiled case studies from NEA grant investments, and offered a suite of additional project insights that enabled communities to learn ways that they might approach a creative placemaking project. The in-depth case studies were a first attempt to make the work happening in this burgeoning field legible and relevant to various community contexts. For many, 'Exploring *Our Town*' revealed that creative placemaking was not a field limited to urban cities, but rather was thriving in tribal, rural, and suburban contexts. 'Exploring *Our Town*' was an early robust tool for effectively communicating creative placemaking and enabled new connections with other sectors and federal agencies that were investing in place-based programming and grant support.

In December 2016, the NEA published a book, *How to Do Creative Placemaking*, to round out case studies produced in the online resource of 'Exploring *Our Town*.' The book reflects on where the field has been and where it has the potential to head. The book features a diverse set of chapter authors, and encapsulates perspectives from creative placemaking practitioners, a housing authority leader, a mayor, an economic developer, and various artists. It was a piece intended to capture the important moment of creative placemaking as a compendium of key thinkers and thought leaders that embodied a range of sectors. The publication release culminated in a one-day convening, 'Creative Placemaking: The Role of Arts in Community Development,' in December 2016 hosted by the Wilson Center, in partnership with the NEA, the Kresge Foundation, ArtPlace America, and Partners for Livability Communities (Wilson Center, 2016). The convening, the 'Exploring *Our Town*' web resource, and *How to Do Creative Placemaking* publication helped to build momentum and make the work more legible to non-arts sectors.

Investing in knowledge-building and network organizations

Another key strategy for building knowledge to advance the creative placemaking field was to establish a new program category of funding within *Our Town*, called 'Knowledge Building.' This new program category was first funded in fiscal year 2015 and had an explicit call to 'build and disseminate creative placemaking knowledge more broadly' (National Endowment for the Arts, 2015). One- to two-year projects led by either arts service or design organizations or economic and community development organizations were incentivized to partner on advancing the field of creative placemaking through a range of activities such as 'mentorships, training opportunities and convenings, technical assistance, research linked back to practice, technology projects, and other projects appropriate to the organizations' internal system of learning' (National Endowment for the Art, 2015). In short, this new program category helped to seed and further cement new ways of embedding creative placemaking practice and approaches in non-arts networks; including a diverse assortment of community development organizations from Smart Growth America's Transportation for America (T4A) to National Alliance of Community and Economic Development Associations (NACEDA). A requirement for closing out a grant award included the delivery of a final publication, conference agendas, reflections of learning, media, or other material that documented the activities that took place. The intention was to also ensure that the final deliverables were disseminated widely throughout the organization's network.

The pilot round of knowledge-building grantees seeded several resources that helped to make legible creative placemaking for local leaders in the non-arts sectors. For example, The Trust for Public Land, in partnership with City Parks Alliance, hosted a colloquium for park administrators and arts and cultural organizations to facilitate peer learning, culminating in *The Field Guide for Parks and Creative Placemaking* (Clarke, 2017). Similarly, Springboard for the Arts partnered with the International Downtown Association to develop a toolkit for downtown managers, helping to guide the integration of arts, culture, and design into the development of downtowns (Springboard for the Arts, 2017). The toolkit included thoughtful approaches on how to hire and contract with artists, with nuts-and-bolts tools including sample artist contracts, how to issue a request for proposals, and how artists can help address the mission and goals of downtown managers. Even arts network organizations, such as OPERA America, embarked on an interesting pivot as they sought to pilot civic-practice learning workshops for opera companies, ultimately piloting a new grant program that helped the art form of opera's 'authentic creative assets address public priorities and community needs' (Barto, 2018).

The NEA '*Our Town* Knowledge Building' program funded a total of 44 projects over the course of the program's five-year run which concluded in the last round of awards made in

2019. From an American Institute of Certified Planners (AICP) accredited creative placemaking curriculum for the American Planning Association members, to a peer exchange program called the 'Creative Counties Placemaking Challenge' for the National Association of Counties members, the knowledge-building program helped to seed a multitude of new publications, programs, and learning tools. A key component of the program was to ensure that new resources were created to assist local leaders and communities to advance creative placemaking work, and that these resources were disseminated widely to ultimately influence and inform local practice. This program enabled experimentation in forging new partnerships for network organizations between arts and community development sectors and helped to seed new streams of creative placemaking work including research, publications, conferences, training institutes, mentorship programs, and more. Network organizations were empowered to translate creative placemaking into language that resonated best with their member base and continued to make this work more legible to local practitioners that were undertaking local projects. The NEA's investment in 'Knowledge Building' grantees was often accelerated and enabled by strategic and significant investment made by The Kresge Foundation Arts and Culture Program's National Networks grants. Unlike the one- to two-year project-based grants made by the NEA, The Kresge Foundation made significant, longer-term commitments to national network organizations that built out creative placemaking programs through staff positions and other transformational organization investments. In the fall of 2019, NEA and The Kresge Foundation brought together the cohort of 'Knowledge Building' grantees and National Network grantees to further connect those working to advance and strengthen the field of creative placemaking.

Accelerating community capacity to support local work

While the NEA '*Our Town* Knowledge Building' grant program resulted in new tools and resources for the field, the NEA recognized a need to also directly support its local project grantees through non-financial technical assistance in the form of mentors, coaches, peer exchanges, and creative placemaking consultancies. In an emerging and evolving field requiring complex partnership, and goals of community change, *Our Town* grantees were unique among the NEA's portfolio and often necessitated additional support beyond grant funding. Other federal agencies have long supported technical assistance as a complement to their grant awards to ensure that grantees were well-equipped to put federal dollars to good use, harnessing best-practice approaches and broader national learning.

In its first iteration of technical assistance, the NEA issued a program solicitation and scope of work to ultimately select a cooperator to launch and run a 'Pilot Creative Placemaking Technical Assistance Program.' Conceptually, the program was designed to serve select *Our Town* grantees in the execution of their grant award and project activities, providing an opportunity to harness rich learning from a select group of grantees for the agency to consider scaling wraparound support that goes beyond grant awards. The cooperator, Local Initiatives Support Corporation (LISC) in partnership with PolicyLink, was selected via a competitive process and, in February 2016, formally launching the program. The Kresge Foundation brought additional funding to the table to enable LISC and PolicyLink to make the program more robust by serving nine Kresge grantees with the same programmatic approach, in addition to the *Our Town* grantees. Grantees were invited to apply to receive technical assistance and were ultimately selected via a competitive process, ensuring that the cohort represented the range of communities and organizations that interface with the *Our Town* program. For example, it was critical to have the grantees represent rural, urban, suburban, and tribal communities. Similarly, the pilot program sought to serve a range of lead grantee organizations that spanned municipal, non-profit, arts, non-arts organization typologies. This type of grantee cohort enabled a rich body of learning and was intended to be more broadly representative of the diverse grantees and applicants served by the NEA *Our Town* program.

Over a two-year period, LISC and PolicyLink delivered intensive technical assistance to a cohort of 16 grantees, including both NEA *Our Town* and Kresge grantees, to help advance their project work. Via mentorships, deployment of consultants to work directly with grantees, webinars, and peer learning opportunities, technical assistance ultimately benefitted the grantees, but also the funders in identifying gaps and needs in the field that went beyond project funding. For the NEA, this was a groundbreaking venture and seemingly the first time that the agency engaged in providing technical assistance directly to grantees to help them succeed in the implementation of grant funding. The pilot program enabled deeper learning for NEA program staff and an opportunity to think more explicitly about wraparound support that can ensure the effective use of grant funding, with the goal of ultimately seeding long-term outcomes that positively benefit local residents in communities.

LISC also delivered a public webinar series that ran from Fall 2018 through Spring 2019 and was open to the broader public on key topical areas that resonated consistently with the grantees that received deep hands-on technical assistance (Local Initiatives Support Corporation, 2019). The webinars had a far reach, capturing the attention of over 1,300 unique viewers. Nuts-and-bolts learning on how to commission an artist, run a call for entry, or issue a request for proposals was paired with creative community engagement strategies to introduce viewers to a wide range of topics, ultimately expanding the imagination on what creative placemaking can do for communities. Webinars featured a mix of experts on particular subjects such as partnership, as well as artists or practitioners on the ground who have executed projects. Grounding each webinar in specific case studies was critical to spark inspiration and share both successes and pitfalls. An accompanying workbook was launched in spring 2020 as a compendium to the webinars, with various templates, prompting questions, and curriculum to guide the development and execution of creative placemaking projects in local communities.

Experimentation in approaches to technical assistance continued in the summer of 2019 when the NEA and LISC hosted the first-ever 'Local Leaders' Institute on Creative Placemaking.' While the technical assistance delivery to date primarily focused on supporting grantees and their partners, the NEA was interested to establish an approach to serve the roughly 80 per cent of *Our Town* applicants who are unsuccessful at obtaining a grant award. A significant percentage of *Our Town* applicants are first-time applicants to the agency, and likely embarking on creative placemaking endeavors for the first time. In the spring of 2019, rejected *Our Town* applicants were invited to apply to attend the inaugural 'Local Leaders' Institute on Creative Placemaking.' The Institute was conceived by iterating on successful programs such as the 'Mayors' Institute on City Design,' with the same notion of bringing together teams of local leaders with resource team members that had expertise in creative placemaking to share. Ultimately, six local teams were selected, representing mid-size and rural communities, and they met in Washington, DC for two-days in July 2019. The community teams each consisted of a local government representative and arts/cultural representative, which is the partnership required to establish eligibility to apply to the *Our Town* program. Over the course of two days, Institute participants presented case study projects rooted in their community, and exchanged ideas, challenges, and opportunities with peers and the resource team. Ultimately, the investment in technical assistance over the course of the pilot program and 'Local Leaders' Institute' enabled an expanded strategic direction for the *Our Town* program and served as a new approach for the NEA as a federal agency.

Reflecting on a decade of federal investment in creative placemaking

The first decade of federal investment in creative placemaking was both exciting and experimental, seeding new relationships, approaches, pilot programs, local projects, and resources to support an evolving field. While the early years of *Our Town* focused on livability and economic development outcomes, there was a noticeable evolution among the *Our Town* grantees that were less interested in driving economic development, but rather sought to advance social change and civic engagement on the local level. Similarly, there was a shift in the grantee portfolio that moved from a focus on the artistic output of a project (such as a piece of public art, a newly designed public space, a cultural plan) to valuing and elevating the artistic process (such as artist-led community engagement). The artistic process demonstrated a compelling way to drive long-term change in a place and seed new local partnerships that proved transformational.

In Spring 2019, approaching the 10-year anniversary of the *Our Town* program, the NEA attempted to codify a more expanded theory of change for *Our Town*, and the way in which it supports the ecosystem of creative placemaking. Via a collaborative contract with 2M and Metris Consulting, the NEA commissioned the development of a theory of change, logic models, research case studies, and a program evaluation framework that thoroughly examined the *Our Town* grant portfolio (National Endowment for the Arts, 2019). This collaborative research work resulted in a substantial rewrite of the *Our Town* grant guidelines in 2019 which articulated that successful creative placemaking projects 'ultimately lay the groundwork for systemic changes that sustain the integration of arts, culture, and design into local strategies for strengthening communities' (National Endowment for the Arts, 2019). The theory of change acknowledged that artists, culture bearers, and designers are uniquely positioned to collaborate with cross-sector partners in the following ways: by envisioning, imagining new possibilities for a community or place – a new future, a new way of overcoming a challenge or a new approach to problem-solving; connecting, bringing together communities, people, places, and economic opportunity via physical spaces or new relationships; illuminating, bringing new attention to, or elevating key community assets and issues, voices of residents, local history, or cultural infrastructure; energizing, injecting new or additional energy, resources, activity, people, or enthusiasm into a place, community issue, or local economy (National Endowment for the Arts, 2019).

The *Our Town* program also shifted its outcome of livability to 'strengthening communities: providing opportunities for the arts to be integrated into the fabric of community life.' Project outcomes were defined in an expanded way in 2019, enabling communities to select outcomes most relevant to their place: economic change – economic improvements of individuals, institutions, or the community including local business growth, job creation/labor force participation, professional development/training, prevention of displacement, in-migration, and tourism; physical change – physical improvements that occur to the built and natural environment including beautification and/or enhancement of physical environment, new construction, and redevelopment (including arts, culture, and public space); social change – improvements to social relationships, civic engagement and community empowerment, and/or amplifying community identity including civic engagement, collective efficacy, social capital, social cohesion, and community attachment; systems change – improvements to community capacity to sustain the integration of arts, culture, and design into strategies for advancing local economic, physical, and/or social outcomes including, for example, establishment of new and lasting cross-sector partnerships, shifts in institutional structure, practices, or policies, replication or scaling of innovative project models, and the establishment of training programs or dissemination of informational resources to support the creative placemaking field (National Endowment for the Arts, 2019). The expanded outcomes and theory of change for the *Our Town* program offers a broader reflection of the big tent of creative placemaking and can offer guideposts for states, regions, or local funders that seek to establish programs that center arts in community development. Via the NEA's creative placemaking investments, communities have been empowered to chart the course of their projects and identify the outcomes most desirable to them. The evolution of the *Our Town* grant portfolio illuminates the ways in which local leaders have sought to engage arts, culture, and design in achieving goals that stretch beyond economic development, such as social change to drive equity.

Collaborations between the arts and non-arts sectors have prototyped new ways of seeding systems change within our society. It is for this reason that creative placemaking holds particular promise in the face of an unknown future. A global pandemic, historic rates of unemployment, social unrest, and climate change are upending our previous ways of life. As we seek to heal and reimagine a way forward, the vision of artists will be more necessary than ever. The old ways of doing and being will require new creative approaches and cross-sector partnerships to tackle systemic issues. The field of creative placemaking has been exercising its muscles for this moment to rise to the challenges that lie ahead: to ultimately strengthen communities.

References

ArtPlace America. (2020). [online]. Available at: http://www.artplaceamerica.org/introduction (Accessed: 31 May 2020).

Barton, L. (2018). *The National Opera America Center: Civic Practice* [online]. Available at: https://www.operaamerica.org/content/about/CivicPractice/files/intro.pdf (Accessed: 1 June 2020).

Clarke, M. (2017). *Field Guide for Creative Placemaking* [online]. Available at: https://www.tpl.org/sites/default/files/files_upload/FINAL_FieldGuide_Layout_sm_0.pdf (Accessed: 1 June 2020).

Local Initiatives Support Corporation. (2019). *How to do Creative Placemaking Webinar Series* [online]. Available at: https://www.lisc.org/our-resources/resource/how-do-creative-placemaking-webinar (Accessed: 30 May 2020).

Markusen, A. and Gadwa, A. (2010). *Creative Placemaking* [online]. Available at: https://www.arts.gov/publications/creative-placemaking (Accessed: 1 June 2020).

National Endowment for the Arts. (2011–2019). *Grant Guidelines* [online] Available at: https://www.arts.gov/grants-organizations/our-town/grant-program-description (Accessed: 1 June 2020).

National Endowment for the Arts. (2019). *Our Town: A Theory of Change and Logic Model for the National Endowment for the Arts' Creative Placemaking Grants Program* [online]. Available at: https://www.arts.gov/sites/default/files/Our-Town-Theory-of-Change.pdf (Accessed: 1 June 2020).

Springboard for the Arts. (2017). *Guide for Business Districts to Work with Local Artists* [online]. Available at: https://springboardexchange.org/guide-for-business-districts/ (Accessed: 1 June 2020).

The White House. (2010). *Sustainable Communities Partnership Fact Sheet* [online]. Available at: https://obamawhitehouse.archives.gov/sites/default/files/uploads/SCP-Fact-Sheet.pdf (Accessed: 30 May 2020).

Wilson Center. (2016). *Creative Placemaking: The Role of Arts in Community Development* [online]. Available at: https://www.wilsoncenter.org/event/creative-placemaking-the-role-arts-community-development (Accessed: 30 May 2020).

Further reading in this volume

Chapter 10: From moon village to mural village: The consequences of creative placemaking in Ihwa-dong, Seoul
Jason F. Kovacs and Hayun Park
Chapter 11: *Free State Boulevard* and the story of the East 9th Street Placekeepers
Dave Lowenstein
Chapter 20: Displacemaking 2015 and 2020
Catherine Fennell and Daniel Tucker
Chapter 24: Artists, creativity, and the heart of city planning
Tom Borrup

Chapter 26: Sculpturing sound in space: on *The Circle and the Square* (2016) by Suzanne Lacy
Trude Schjelderup Iversen
Chapter 41: Rituals of regard: on festivals, folks, and findings of social impact
Maribel Alvarez
Chapter 44: Creative Placemaking and comprehensive community development: rethinking neighborhood change and evaluation
Maria Rosario Jackson
Chapter 42: Creative Placemaking and Placekeeping evaluation challenges from the practitioner perspective: an interview with Roy Chan
Maria Rosario Jackson
Chapter 43: A theory of change for creative placemaking: the experience of the National Endowment for the Arts' *Our Town* program: an interview with Patricia Moore Shaffer, PhD
Maria Rosario Jackson

4
A FUTURE OF CREATIVE PLACEMAKING

Sarah Calderon and Erik Takeshita

Introduction

Employing arts and cultural strategies to address social challenges is not a new idea, particularly during times of crisis. Crises such as the massive influx of new arrivals to cities in the early twentieth century, the Great Depression, the economic challenges of the 1970s, or the Great Recession of 2008 disrupt the status quo. During these times, some believe that the thinking and approaches that created the problem will not solve it and there is an openness to new ideas and ways of thinking. These are ideal times to harness the power of arts and culture.

In the early twentieth century, settlement houses integrated arts and culture to help welcome new arrivals to cities across the United States. Thousands of artists were employed during the Great Depression by the Works Progress Administration Federal Art Project (WPA/FAP) and later through the Comprehensive Employment and Training (CETA) Act of 1973 (Burnham and Durland, 2016). As noted in the chapter by Jen Hughes in this volume, in response to the Great Recession of 2008, the National Endowment for the Arts created the *Our Town* program and helped spearhead the creation of the public–private partnership, ArtPlace America. ArtPlace America was developed as a place-based community and economic development response to the loss of mobility as housing markets crashed and jobs were lost. Over the last 10 years, ArtPlace has strengthened the creative placemaking field – supporting and embedding arts and culture strategies that are working across community planning and development sectors to create equitable, healthy, and sustainable communities. Similar to previous crises, the COVID-19 crisis creates an opportunity to advance creative placemaking as a critical tool, particularly as community wellbeing becomes an ever more urgent need and focus. During the pandemic, the world turned to artists, designers, and culture bearers for solace and comfort. While many are all too aware of the inequities in our society, for others the multiple current crises put the stark realities of inequity into the spotlight, reinforcing the need to envision a different future and the importance of our relationships and the realities of our interconnectedness.

The future of creative placemaking is intrinsically linked to the uncertain future of our world, country, and community at-large. The COVID-19 pandemic is not just a public health crisis that impacted people's physical and mental health. It has an impact on our economy and will impact everything from education to incarceration and from immigration to food security. There are fewer dollars available and more demands. There will need to be new ideas and ways of working. In addition, there will be other crises and disruptions again in the future. The cornerstones of creative placemaking must be further strengthened creating a field that can be nimble, collaborative, collective, participatory, and responsive for our futures.

The future

Creative placemaking is, at its core, about unleashing the unlimited power of arts and culture to advance community wellbeing. Over the past 10 years it has been demonstrated that arts and culture can contribute to positive outcomes in other community development sectors ranging from public health to economic development, and from immigration to food and agriculture. In the future, artists and culture bearers can be leaders and allies in the problem-solving, envisioning, and advancing of the collaborative practices necessary to create more equitable, healthy, and sustainable communities. Creative placemaking can and should flourish in the future by supporting communities to: (1) imagine the future and remember the past; (2) advance equity; (3) help people build stronger relationships; and (4) support cross-sector collaboration and impact. More on the promise of each of these aspects of creative placemaking and a few current examples are provided below.

Imagining and remembering

Society will continue to struggle with a multitude of 'wicked' problems – a problem that is difficult or impossible to solve because of incomplete, contradictory, and changing requirements that are often difficult to recognize. Everything from climate change, to ending disparities and poverty, to terrorism. These wicked problems are abstract and systemic issues that have profound impacts on people's everyday lives. Critical issues like these need future-imagining – allowing folks to 'remove the cognitive, emotional, and temporal distance between us and the future' and 'enact and interact with potential crises as if that future is happening now' (Zaidi, 2018). Future imagining helps open up individuals' abilities to creatively problem-solve, think critically, evoke empathy, and provide their immediate attention to an issue. Interventions that allow communities to future imagine might allow them to innovate proactively instead of reactively. Arts and cultural strategies employed in community development structures can help us do just this – imagine our future. In the future, creative placemaking will help connect people to these large-scale issues and provide them with tools to help them imagine things that do not currently exist.

The Harrison Center for the Arts' PreEnactment Theater in Indianapolis, Indiana is a good example. To help people get a glimpse into a future, artists, and community organizers stage 'PreEnactments' that include everything from set designers building temporary facades on vacant lots and sprucing up abandoned buildings to actors modeling inclusive and equitable ways of living. It is a way for the artists and the community to both envision and experience the neighborhood they want for their future. A neighborhood that *ought* to be – one that is more just, equitable, and economically vibrant.

As important as envisioning the future may be, it is essential to also remember the past. When it comes to issues of climate change or food security, there are ancient ways of being, including Native American traditions, that can inform contemporary solutions. There is wisdom in their traditions for all of us on how our actions may impact the planet – the land, water, plants, and animals – and how decisions made today will impact future generations. In the future there may be more efforts like the work being done to reclaim Indigenous Maskoke land and the establishment of an ecovillage in rural Weogufka, Alabama. This work is a great example of how old ways of being can help develop contemporary solutions. The collective is committed to embracing the role of protecting and reviving traditional relationships to the earth while revitalizing language and culture. In 2018, Ekvnv Yefolecvlke became official land owner of 577 acres of Alabama woods. Colonization and capitalism are antithetical to their culture and are therefore not seen in their ecovillage – where only the resources that the language describes are found. In addition, initiatives that introduce regenerative agriculture, aquaponics, and livestock farming also aim to improve the holistic health of the community while their communal, shared wealth structure, creates a self-sufficiency leading to a more sustainable community (Thomas, 2010).

Advancing equity

Perhaps one of the most wicked problems facing the United States now and in the future is addressing issues of equity. This country was built on 400 years of inequality. Historic inequities, racism, and trauma have led to disparities for black people, indigenous people, and people of color in nearly every category imaginable from income to incarceration, from health to wealth, and from morbidity to mortality (Stanford Center on Poverty and Inequality, 2017). While creative placemaking can help advance equity in a number of ways, it will not magically address inequity in the United States. While not an exhaustive list, a few ways creative placemaking can help address equity in the future include helping communities heal from racism and trauma, helping people in the community better understand other people's experiences, and being a driver for jobs, sustainability, and community wealth-building. As a backdrop for this work, the arts and culture fields must continue to move beyond the Western canon and recognize artistic and cultural contributions of artists and culture bearers that are black, indigenous, and people of color. Simultaneously, arts and cultural organizations must examine whom they are benefitting and how, moving beyond audience development and toward becoming more inclusive civic and cultural spaces that serve as anchor institutions whose mission includes building equitable, sustainable, and healthy local communities.

An example of how creative placemaking can advance equity by helping people who have experienced historic racism and trauma heal while also, simultaneously, helping others see inequities, is Clemmons Family Farm. Clemmons Family Farm is a beautiful, rare African American–owned farm that serves as a platform to push for racial equity and empower a growing network of Vermont artists of African descent by celebrating the arts and cultural heritage of the African diaspora in Vermont, a state that is only 1.2 per cent black. They also 'create opportunities for healthy dialogue around the identity and cultures of all people (people of different races, ethnicities, nationalities, religions, etc.) for a stronger and more supportive multicultural community' (Clemmons Family Farm, n.d.). To advance equity in the future, communities must have a sustainable workforce pipeline and be able to build community wealth. The work of the Sweetwater Foundation in Chicago demonstrates how this can be done using creative placemaking. They are building a regenerative community by lifting up the local assets and creativity of community members and helping them connect to the uniqueness of their place. The arts, education, aquaponics, carpentry, and farming that Sweetwater engages in within their community values an essential economy – one that liberates the community from a single bottom line and looks to a triple bottom line that recognizes people, planet, and profit and is both emergent and flexible for the future.

Building relationships

Human beings are social creatures. Relationships are critical to our physical and psychosocial wellbeing. Arts and culture are a valuable tool in helping us build stronger relationships, both with people who are like us and with people who are different from us. These relationships are critical to building healthier, more equitable, and sustainable communities in the future.

Before folks can build relationships with people who are different from themselves, they must be comfortable with themselves and with others who are like them. Many black people, indigenous people, and people of color have been traumatized by racism and marginalization. As we move forward, it is critical that people are provided the time and space to grieve and heal. People then need to feel safe and to have a sense of belonging with other people who have similar experiences. An example of this is the work artists and culture bearers in Natchez, Mississippi, did to host a series of events to help the community own and reframe their cultural history. This was part of a broader HEAL Community Natchez effort to use community arts and culture to build toward equitable community development through a lens of health and wellbeing. To achieve this longer-term vision for the overall community, it was vital for there to be time and space for members of the African American community to take ownership over the reframing of their own cultural history as a way for them to start to address the past before they could imagine a shared future with and for the broader community. To build more cohesive communities in the future, arts and culture can be employed to build relationships with people across differences. 'Dear Tamaqua,' organized via the Tamaqua Community Art Center, is a good example of how this can be done. 'Dear Tamaqua' (Tamaqua Area.com, 2015) invited people to write letters, sing songs, or submit drawings about their community. They got input from playgrounds and libraries but also from bars by offering opportunities for patrons to participate by writing or drawing on coasters. Making it easy and fun meant more people got involved, especially those who might otherwise not, and helped these folks 'see themselves' in the end product, building a greater sense of ownership and pride in the outcomes.

As these examples show, arts and culture are critical assets to a community in terms of helping to build the relationships needed as human beings. A community's culture, relationships, and organizations help determine that community's ability to recognize, respond, and recover from challenges. Our future as a country will need this. Arts and cultural organizations are critical anchors in our neighborhoods and communities. They often serve as gathering places for artists, culture bearers, and other community leaders. They are community lifelines. They are places for people to learn, practice, and hone their creative expression, through social events, food, health, technology, and housing activities and they are also places for connections for students to each other.

Fostering cross-sector collaboration

The future, particularly one that may have fewer resources, demands holistic cross-sector solutions to the challenges facing people and communities. Human beings do not live in silos; an event that impacts public health inevitably affects the economy, food security, education, and public safety, and more. Creative placemaking is, by its very nature, cross-disciplinary and cross-sector work. The creativity of artists and culture bearers is a renewable resource that exists in every community; it is like wind or solar energy – while it is there, mechanisms are needed to harness this power (Springboard for the Arts, n.d.). Artists and culture bearers working in creative placemaking can contribute by both helping other sectors achieve their desired outcomes and fostering innovation across sectors.

As outlined by Jamie Hand elsewhere in this volume, ArtPlace America has worked collaboratively over the past seven years with a range of partners from different sectors to develop knowledge and resources about the ways that arts and culture can contribute to community outcomes in the areas of agriculture and food, environment and energy, workforce development, economic development, public safety, housing, transportation, immigration, youth, and public health. This work has highlighted examples of arts and culture strategies that have successfully been employed toward community outcomes across sectors or community development silos. A great example is the Zuni Youth Enrichment Project in New Mexico (see also Jojola and Shirley, and Rubin, volume). H'on A:Wan Park is a 2.5-acre complex 'designed to cultivate a sense of belonging among youth that is rooted in awareness of and pride in Zuni traditional culture [to counter] the effects of intergenerational trauma, including poverty, obesity, diabetes, substance abuse and youth suicide' (Sonke et al., 2019). H'on A:Wan is more than just a park project or a youth development effort; it is a community hub made to build a healthier, more equitable and sustainable community with arts and culture at its core. In the future, artists and culture bearers can help ask cross-sector questions, thread the needles between silos, and help envision a new reality. The Fargo Project serves as a good example of how an artist can help foster cooperation across municipal departments, deepen engagement with the community, and achieve several outcomes within the same project. Artist Jackie Brookner worked with a variety of departments within the City of Fargo and engaged hundreds of community members to help reimagine and transform a barren stormwater basin into a vibrant social, cultural, and ecological hub. The result was an amenity that many residents feel a sense of attachment to that still achieves the required stormwater management objectives. In addition to the new park and stormwater management system, the Fargo project helped foster new relationships across departments and new ways of involving residents' voices (Westlake, 2018.)

Conclusion

Society is changing, and as the future becomes increasingly uncertain, communities need a multitude of tools and multiple supporting structures to realize the full potential of creative placemaking to help develop healthy, equitable, and sustainable communities. There are promising signs. The field has a growing cadre of practitioners using the moniker of creative placemaking, and shared values and standards of practice are emerging (Waller, 2019). There is a growing body of knowledge that is being co-created alongside movement leaders, engaging in culturally appropriate research methods, and helping to codify and validate practice-based evidence. Practitioners, funders, and policymakers have been identified, supported, connected, and engaged. Public agencies and private philanthropy have used creative placemaking as a frame for their investments and grantmaking. Leaders in the field are rising from the grassroots as well as from the grass-tops. There are leaders at the local, regional, and national levels, not only from the arts, culture, and design sectors, but also from other sectors of community planning and development.

Yet, there is more to be done. For example, for creative placemaking to realize its full potential in the future, it must include support for more leaders from black communities, indigenous communities, and communities of color and other historically underrepresented communities. If one compares the creative placemaking field to the ages and stages of humans, it may be a 'tween' – somewhere between a child and a teenager. The creative placemaking field is moderately strong (The Bridgespan Group, 2018) – able to do some things on its own, with still more room for growth, pushing boundaries but still in need of support, not able to be left alone at home just yet, and capable of building on current strengths.

As described earlier, there are four specific reasons why it is important to have a robust creative placemaking field in the future. It starts with being able to imagine a future that is better than the past while also remembering ancient wisdom and ways that have worked, and still do today. The future demands advancing equity in the United States. As social cohesion becomes a key outcome of community development across all sectors, relationships must be built both within communities and across difference. Finally, as the silos of community development erode and scarce resources force collaboration, cross-sector work will rise. This is hard work made even more difficult when you are attempting to strengthen a field that is at its core an intersection of two previously established or 'original' fields – arts and culture and community planning and development. Change must occur in each of the 'original' fields at the same time as we are evolving a new, distinct, yet related, field – creative placemaking. Creative placemaking was officially established as a federal policy in response to a societal crisis. The rise of the field post-disruption has demonstrated that arts and cultural strategies can allow community development to be responsive and nimble, and are crucial. The future needs an even stronger creative placemaking field with substantial networks that operate and collaborate effectively; creative placemaking work that allows for place-based contexts and different approaches; and decision-making power and significant resources that flow directly into the hands of the folks who live, work, and play in a place.

References

Burnham, L.F. and Durland, S. (2016). 'Looking for CETA', *Public Art Review*, 54[online]. Available at: https://forecastpublicart.org/looking-for-ceta/(Accessed: 2 June 2020).

Clemmons Family Farm. (n.d.). [online]. Available at: http://www.clemmonsfamilyfarm.org/our-vision.html (Accessed: 2 June 2020).

Sonke, J., Golden, T., Francois, S., Hand, J., Chandra, A., Clemmons, L., Fakunle, D., Jackson, M.R., Magsamen, S., Rubin, V., Sams, K. and Springs, S. (2019). *Creating Healthy Communities through Cross-Sector Collaboration* [White paper]. University of Florida Center for Arts in Medicine/ArtPlace America [online]. Available at: https://arts.ufl.edu/site/assets/files/174533/uf_chc_whitepaper_2019.pdf (Accessed: 2 June 2020).

Springboard for the Arts. (n.d.). Creative People Power [online]. Available at: https://springboardforthearts.org/creative-people-power/ (Accessed: 2 June 2020).

Stanford Center on Poverty and Inequality. (2017). 'State of the Union, 2018', in Pathways Special Issue [online]. Available at: https://inequality.stanford.edu/sites/default/files/Pathways_SOTU_2017.pdf (Accessed: 2 June 2020).

Tamaqua Area.com. (2015). *Over 1,400 See Tamaqua in a New Light*, August 6, 2015 [online]. Available at: https://tamaquaarea.wordpress.com/2015/08/06/over-1400-see-tamaqua-in-a-beautiful-new-light/ (Accessed: 2 June 2020).

The Bridgespan Group. (2018). *Creative Placemaking Field Study* [online]. Available at: https://www.artplaceamerica.org/blog/creative-placemaking-field-study (Accessed: 2 June 2019).

Thomas, K.E. (2010). 'Building a post-colonial community starts with vocabulary', in *NextCity*, 15 May 2020 [online]. Available at: https://nextcity.org/daily/entry/building-a-post-colonial-community-starts-with-vocabulary (Accessed: 2 June 2020).

Waller, M. (2019). *How Do We Know It's Creative Placemaking? Let's Talk About Shared Values*, 20 November 2019 [online]. Available at: https://www.artplaceamerica.org/blog/lets-talk-about-our-shared-values (Accessed: 2 June 2020).

Westlake, S. (2018). *Artists Improve Fargo's Storm Basins*, 18 October 2018 [online]. Available at: https://www.artplaceamerica.org/blog/artists-improve-fargo-storm-basins (Accessed: 2 June 2020).

WPA Federal Art Project: United States History (n.d.). *Encyclopedia Britannica* [online]. Available at: https://www.britannica.com/topic/WPA-Federal-Art-Project (Accessed: 5 May 2020).

Zaidi, L. (2018). *Experiencing the Future for Changemaking*, 20 November 2020 [online]. Available at: https://medium.com/predict/experiencing-the-future-for-changemaking-5cea7e2f4e10 (Accessed: 2 June 2020).

Further reading in this volume

Chapter 1: Introduction: what really matters – moving placemaking into a new epoch
Cara Courage
Chapter 5: Making places for survival: looking to a creative placemaking past for a guide to the future
Jeremy Liu

Chapter 9: From the dust of bad stars: disaster, resilience, and placemaking in Little Tokyo
Jonathan Jae-an Crisman
Chapter 11: *Free State Boulevard* and the story of the East 9th Street Placekeepers
Dave Lowenstein
Chapter 12: Public transformation: affect and mobility in rural America
Lyndsey Ogle
Chapter 14: Experts in their own tomorrows: placemaking for participatory climate futures
Paul Graham Raven
Chapter 26: Sculpturing sound in space: on *The Circle and the Square* (2016) by Suzanne Lacy
Trude Schjelderup Iversen
Chapter 28: Integral placemaking: A poiesis of sophrosynes?
Ian Wight
Chapter 30: Ecological selves as citizens and governance as ethical placemaking
Lisa Eckenwiler
Chapter 3: Seven generations: a role for artists in Zuni PlaceKnowing
Theodore S. Jojola and Michaela P. Shirley
Chapter 39: *Translating Outcomes*: laying the groundwork for interdisciplinary evaluation of creative placemaking
Jamie Hand
Chapter 40: Transforming community development through arts and culture: a developmental approach to documentation and research
Victor Rubin
Chapter 42: Creative Placemaking and Placekeeping evaluation challenges from the practitioner perspective: an interview with Roy Chan
Maria Rosario Jackson
Chapter 43: A theory of change for creative placemaking: the experience of the National Endowment for the Arts' *Our Town* program: an interview with Patricia Moore Shaffer, PhD
Maria Rosario Jackson
Chapter 45: How the city speaks to us and how we speak back: rewriting the relationship between people and place
Rosanna Vitiello and Marcus Willcocks

5

MAKING PLACES FOR SURVIVAL

Looking to a creative placemaking past for a guide to the future

Jeremy Liu

Author's note: this chapter begins not with a Forward, but a Backward – a perspective on where I have been, on a journey from community development to creative placemaking and back again. This is my practitioner story that picks up, along the way, three inspiring examples of how communities of color have always relied upon creativity to resist and grow. These are three, among many, examples I have encountered in my own search for the future of creative placemaking grounded in equity, justice, and resilience.

Backward?

Bent over, pulling weeds from the dirt, beads of sweat worked their way down my face, stinging my eyes. In Boston, the summer of 2000 was sweltering. The community garden plots farmed by Chinese seniors at the Berkeley Street Community Garden (BSCG), sheltered under meter after square meter of lush bitter melon and its fuzzy vines, offered a welcome respite. These cool, green 'rooms' also shielded their occupants from a mainstream Boston society that did not know what to make of the rambunctious, ad hoc, and sprawling reach of these bountiful gardens, nor the gardeners who tended them. These garden plots manifested food sovereignty and economic self-sufficiency while simultaneously expressing a cultural and social identity constantly under threat of erasure by policies, actions, and neglect.

I was the Director of Community Programs for the Asian Community Development Corporation. Volunteering at the BSCG was a way for me to get to know the neighborhoods of Chinatown and the South End in Boston. Working alongside our community youth leaders and Chinese American seniors to pull weeds was definitely in the job description. The BSCG was riotously verdant and layered with purpose and meaning: flowers and ornamentals interspersed, plot-by-plot, with bitter melon and bok choy. Well-tended, polite gardens competed for attention – largely losing – against intensively cultivated micro-farms. The clash of Western and Eastern, expressed via overlapping (sometimes even clashing) aesthetic, gardening, and agricultural traditions, and even social norms, was as much a feature of the space as the daily practices of the garden.

It was eye-opening to meet these seniors and learn about how they squatted on the land after the cancelling of the infamous 'Inner Belt' highway project in 1972, which became the first step towards the permanent protection of the BSCG land by a land trust. They not only helped save the land, but became ongoing stewards, organized and intently focused on protecting the rights of the Chinese American community to have a place to grow food for their families and neighbors. The Chinese American seniors who tended the plots were doing much more than gardening, however. These seniors were carrying on a traditional agricultural practice that manifested food sovereignty and economic self-sufficiency while simultaneously expressing a cultural and social identity constantly under threat of erasure.

The more I literally dug into the soil of the BSCG, the more I uncovered the true culture of Chinatown – a culture that extends far deeper than the stereotypes most tourists know. This culture is expressed through economic self-sufficiency, creative resilience, and active resistance against erasure, homogenization, and disenfranchisement. I came to regard the BSCG as the most authentic example of Chinese culture in Chinatown. I also came to see it as the most vibrant manifestation of democracy in a city where most have little sway over the forms and uses of spaces. This enraging fact was even more true for communities of color like Chinatown, which had suffered successive waves of trauma from eminent domain displacement, ongoing marginalization, and underlying racism embedded in mainstream land use policies, such as the urban renewal land takings for the Interstate 93 and 90 highways in the 1950s and 1960s, and the rezoning of all adult entertainment uses from around the city into this one neighborhood in 1974. In these gardens, in this nexus of social, cultural and political agency, held by these Chinese and Chinatown seniors, I found creative placemaking. The work of the Asian community development corporation (CDC) became supporting the self-organization of and advocacy by the gardeners. We commissioned an artist to create a pathway, a fence and other infrastructure to both showcase the garden and protect the rights of individuals to grow and express themselves in each plot.

Traditional Asian enclaves in the United States have names such as 'Chinatown,' 'Little Tokyo,' and 'Little Saigon,' and they emerged from an era where communities of color were segregated into defined areas by redlining and other policies. Although these place names seem representative of diversity, they actually serve for many outsiders as little more than place markers of generic exoticism as exemplified by the common refrain that a visit to Chinatown is the cheapest ticket to the Far East. My experiences in the BSCG garden led me to wonder how Asian American communities across the United States can transcend historical and ongoing marginalization. How can labels such as 'Chinatown' serve as beacons of ethnic pride and possibility when they originate out of a racist and discriminatory legacy? How can Chinatown be politically, economically and culturally valued – –as much by the people rooted there and for what the place means to them, as it is for those who visit as a destination?

Names do not a place make, and the difference is vast between merely naming and doing the authentic work of building up a community where the voice, power, and agency of people are at the heart of its physical manifestation. Into this chasm, creative placemaking and community development can either fall – and fail – together, or together build a bridge that crosses the greatest divides of racial, economic, and social inequity that all societies face.

People making places

Over my decade of work in Boston Chinatown, I began to seek out inspiration and explore the legacies of creative, community responses to oppression, disenfranchisement, discrimination, racism. The mainstream use of the term creative placemaking fails to define and advance a community development practice because it does not encompass and recognize the historic and contemporary legacies of placemaking that have harmed communities, such as Robert Moses' remaking of the very fabric of New York City in 1900s, nor recognize the way communities have utilized placemaking as an act of resistance against oppression. When it is most authentic, creative placemaking isn't superficial or even a new strategy for community or economic development; it is an essential part of community self-determination. The three cases that follow are some of the most inspiring to me as they represent this definition of creative placemaking among communities of color that hold different sorts of promise and resonance for the future.

Among all of them, survival is a central driver – survival in the face of slavery, White Supremacy, oppression, and exclusion. Likewise, they each represent a community's expression of this survival, and each happens in place, where the community and the geography are inextricably linked. In picking these cases, I asked myself: What are the revolutionary, radical, and resistance moments in US history that involved taking 'ownership' of a place, that transformed that place into what we desire, need, express, and used that place as a platform or stepping stone from which to further liberation? The risk in casting a net across history so broadly for these cases is that their social and political contexts are so different, and thus, while there are lessons to be drawn from each, their selection does not mean that their situations are equal or even comparable. Rather, I am connecting these past cases to present projects and efforts and to highlight how advocacy or policy change happened as a result in the 'Forward!' part of this chapter.

Maroon settlements of the Great Dismal Swamp

Maroon settlements – renegade communities of escaped slaves – have been documented in the Americas and the Caribbean as far back as the 1500s. The largest known Maroon settlement in what was to become the United States was in the Great Dismal Swamp – an almost impossibly remote 2,000-square-mile marshland rife with dense underbrush, snakes, alligators, and insects. The swamp stretches from southern Virginia to northeast North Carolina along the Savannah River. Documentation of life in the Great Dismal Swamp comes both from firsthand accounts, which are relatively scant, and archeological research.

Some estimates surmise that thousands of escaped slaves, along with Native Americans and some Whites, took refuge in the area between the 1600s through the end of the Civil War. It is believed that the earliest inhabitants of the Great Dismal Swamp were indigenous Americans seeking refuge from European settlers (Youssef, 2017). Escaped slaves began to settle in the swamp in the 1700s. Some became acquainted with the swamp when they were put to work dredging a canal connecting the Chesapeake Bay to the Albermarle Sound. After the canal was completed, companies of slaves were sent to the swamp for months at a time to produce shingles and planks from the cypress and juniper trees that grew there. 'The wilderness offered them a modicum of freedom not found on the plantation; they fished, hunted, and worked at their own pace, the requirement being that each person produce a given number of shingles' (Diouf, 2014). As their work took them deeper and deeper into the swamp, some of these slaves seized the opportunity to take permanent refuge there.

Some of the Maroons sustained themselves by working in timber camps, where advantageous arrangements could be made between the slaves also working in the camps (the slaves were credited a fixed amount of money for every thousand shingles they produced and Maroons helped them increase their shingle output) or directly with the White lumbermen. Still others worked as entrepreneurs, trading their shingles for money or provisions (Diouf, 2014). Life in the outer swamp was arduous – pay, food, and clothing were all meager, and the Maroons were at constant risk of capture by bounty hunters. At the same time, they enjoyed a freedom and self-determination denied their brothers and sisters on the plantations. Diouf cites a Maroon who escaped for Canada and fondly recalled moving freely in his bark canoe, hunting wild hogs and cows, chasing bears away, and enjoying singing sessions with the community's preacher (Diouf, 2014).

Other Maroons lived much more remotely in the inner swap, some for generations, keeping the locations of their settlements secret from anyone with whom they came in contact. That contact was limited: Diouf 'cautiously concludes' these Maroons lived in small, family-oriented groups (Diouf, 2014). Some Maroons never saw a White person and it is believed some stayed in the swamp for years after Emancipation because they had not heard the news. Maroon housing was responsive to its environment and designed to avoid detection. In the Great Dismal Swamp, this typically meant small shanties on stilts or braced against trees and constructed from materials in the immediate environs. Given the dearth of tools, construction required supreme ingenuity. Yet the shelters could be large and elaborate, containing furniture, stoves, and wooden pipes. These abodes were often built beneath sturdy trap doors that kept them camouflaged. Maroons in other locations constructed equally complex yet hidden abodes in caves and tree trunks. While the swamp's 'Borderland Maroons' often traded, stole, or were given necessary supplies, the 'Hinterland Maroons' had to be entirely self-reliant. They farmed, growing crops such as corn and sunflowers, despite the extreme inhospitality of their environment. They also fished, hunted, made baskets, and traded. They married and created families. One advantage to living so remotely was that the threat of capture was minimized. While militias frequently attacked more accessible Maroon settlements, history shows no report of a 'hunt' in the inner Great Dismal Swamp (Diouf, 2014).

Despite the extreme conditions, the Great Dismal Swamp and other settlements afforded the Maroons an autonomy and a feeling of security denied them in slavery. 'I felt safer among the alligators than among white men,' observed one Maroon (Youssef, 2017). Historical archaeologist Dan Sayers characterizes the swamp as a 'landscape of power,' noting that the settlement's existence shifts the narrative of slaves solely as victims and shows that they were 'people who were resilient, and constructed their lives and constructed the places in which they lived' (Shapiro-Perl, 2014). The Great Dismal Swamp offers an intrinsically place-based narrative of self-determination, resistance, and intentional community-making, albeit under the most tyrannical of conditions.

These communities of resistance and resilience presage more contemporary sites of Black self-determination in place, including the Black Panthers' free breakfast and education programs in Oakland, California, and the Village of Arts and Humanities in Philadelphia, Pennsylvania. The vernacular creativity of an alienated people manifests in these places that also operate on the symbolic level: known, expressive, and influential far beyond their geographic boundaries. Like the Maroon Settlements of the Great Dismal Swamp, these two more contemporary cases are 'rare examples of people undermining inequalities and oppressions inherent to capitalistic modes of production and social worlds by forging and perpetuating a novel social world outside the capitalistic world.' The 10 generations of Maroons of the Great Dismal Swamp 'stand among the few who have successfully undercut the brutal and racist world… They were successful because they accurately critiqued the racialized capitalistic world within which they were imbedded' (Sayers, 2016).

Casitas of South Bronx and New York City

In the first half of the twentieth century, the South Bronx provided stable neighborhoods for largely European immigrants. As the century progressed, the area became home to increasing numbers of Puerto Rican, West Indian, and African-American residents, many fleeing destabilization in their own countries, or steered there due to illegal redlining practices, and seeking job opportunities for unskilled laborers in the post–World War II boom. (Redlining is the concerted and systematic discrimination by federal, state, local policies, and private business against residents of specific neighborhoods. In practice, it impacts primarily African American and other racial and ethnic populations. It continues to this day in various guises, including in emergency response times, availability of ride-sharing services, and the targeting of predatory lending practices.) Between World War I and World War II, corporate and government forces wrought extensive destruction on the South Bronx community. From the 1930s through the 1960s, under the direction of city planner Robert Moses, New York created one of the nation's densest concentrations of public housing in the South Bronx. Moses also spearheaded the construction of four highways and three bridges through the community.

These projects dislocated tens of thousands of residents and, when completed, segregated races and concentrated poverty (four housing projects were targeted to low-income Latinos and African Americans, while one was targeted to middle-income Whites) while also disrupting the collegial character of the neighborhood. Where small businesses and front stoops once lent themselves to friendly exchanges, imposing towers now kept people isolated. Another Moses-led project, Co-op City – a collection of 35 high-rise buildings constructed in the North Bronx – 'sucked' thousands of middle-class residents from the South and West Bronx during the 1960s and 1970s, while Moses' expressways facilitated the migration of the borough's more affluent residents to other parts of the city or the suburbs.

Meanwhile, corporations began to automate their manufacturing facilities or move them to the American South or overseas, in search of ever-cheaper labor. Between 1947 and 1976, New York City lost 500,000 manufacturing jobs (McLaughlin, 2019). Although trade jobs were theoretically a viable alternative for those without a college degree, many unions in the higher-paying trades denied membership to non-White laborers. While civil service provided a career path for many, unions in the more lucrative uniformed professions such as firefighters and police exercised similar racist practices. By the 1960s, one-quarter of the South Bronx's residents received welfare. A decade later, that number increased to 40 per cent (McLaughlin, 2019). This systematic racism hollowed out a stable economic base and set the stage for an influx of heroin, marijuana, and crack cocaine. The sale and use of these drugs, coupled with new laws imposing harsh sentences for drug crimes, led to dual epidemics of AIDS and mass incarceration, further decimating the community (McLaughlin, 2019). The impact on the physical environment was stark. Seven census tracts in The Bronx lost more than 97 per cent of their buildings to fire and abandonment between 1970 and 1980; 44 tracts lost more than half their buildings (Flood, 2019). Some of the fires were due to arson, often on behalf of landlords losing money on their investments. Severe cuts to the fire department exacerbated the problem.

Amid this landscape of segregation and disinvestment, many of the South Bronx's sizeable Puerto Rican population – 30 per cent of the total population in the 1980s – began reclaiming their community through the creation of community gardens and the establishment of *casitas de madera*, one-story houses constructed on vacant city-owned land in a style that evoked the colorful vernacular architecture of Puerto Rico's working poor. (*Casitas* also began to appear in East Harlem and the Lower East Side, two other Puerto Rican enclaves.) Collectively built and often constructed from salvaged materials, *casitas* are often fully outfitted, with comfortable furniture, televisions, and full kitchens. Electricity is often tapped from a nearby lamppost or secured through arrangement with the superintendent of a neighboring building. Water comes from a rain barrel or nearby spigot. The *casitas* are painted in vibrant colors reminiscent of the Caribbean and purposefully decorated with found materials. Notes Joseph Sciorra: 'These are not haphazard uses of ephemera reflecting a 'culture of poverty' but like the *casita* itself, are instead a deliberate and conscious manifestation of deeply felt values, beliefs and needs in culturally specific and meaningful ways' (Sciorra, 1966). The *casita* is typically surrounded by gardens and a non-vegetated yard known as a *batey*, which may feature benches, domino tables, chicken coops, gardens, barbecue pits, a stage, and a patio for dancing (Sciorra, 1994).

Casita activities are plentiful, their role in the life of the community myriad, from sheltering the homeless to serving as social clubs and cultural centers (Sciorra, 1994). *Rincon Criollo* ('Creole Corner' or 'Downhome Corner') is possibly the first and one of the best-known *casitas*. Established in 1974 by José (Chema) Soto, the original structure was built on a formerly garbage-filled city lot, cleared by Soto and his friends. Soto and community members built a larger structure on the site following a fire. *Rincon Criollo* hosted renowned Puerto Rican performers such as salsa band leader Andy Montanez and television personality Iris Chaco and was known as a hub for traditional music and dance forms, such as the *bomba* and *plena* (Sciorra, 1994). The *casitas* became safe, family-friendly places for daily gatherings and community celebrations in the midst of a scarred neighborhood where crime was a daily reality. Children were introduced to traditional dance, drumming and other arts in concerted attempts to keep them from succumbing to street life. Members – admitted not through application but via introduction and disproportionately, but not exclusively, male – watch sports, play dominoes, and tend gardens. *Casitas* also became neighborhood centers – places to discuss and organize around local concerns. *Villa Puerto Rico*, for instance, regularly offers assistance with voter registration, public benefits registration, and housing concerns. 'More than a sentimental backdrop for the garden, the casita is a workshop where craftsmen carve drums and speckled carnival masks and where local children learn dance steps to rhythms that first came to this hemisphere aboard slave ships' (Gonzalez, 1990).

Despite the vibrant colors and the safe, convivial environment, Sciorra notes that the *casitas* have their roots in a culture of oppression, based as they are on the vernacular architecture that emerged from colonization of the homeland by tobacco and sugar farmers:

> For those who have lived the *casita*, pieced together its walls and inhabited its space, the moribundly nostalgic *jibaro* iconography mass produced in the service of tobacco and alcohol companies as well as the islands tourist industry invades their memories. But for many of the former *casita* inhabitants I spoke with in New York City, the building recounts a history of toil, suffering and struggle that stands in marked opposition to such saccharine imagery.
>
> *(Sciorra, 1966)*

By the late 1970s, the *casitas* became targets of government harassment, such as in the form of fines for improper trash disposal after garbage-strewn lots had been cleared or the attempted collection of back-taxes for 'continuous use' of city-owned property. In 1978, then-Mayor Ed Koch established the 'GreenThumb' program to regulate the unofficial use of city-owned land for community gardens. Leases for the gardens specified that no illegal structures were permitted, and the Koch administration began annulling leases for properties that housed *casitas*. When Jane Weissman took over the 'GreenThumb' program in 1984, she informally reversed that practice. In 1991, 'GreenThumb' developed a standardized permitted structure. The city continued to demolish *casitas* that do not possess 'GreenThumb' leases. Despite a movement to have the *casitas* declared historic landmarks, the remaining structures remain unprotected. *Rincon Criollo* was forced to move one block north in 2007 when the city reclaimed the property where it sat for a housing development.

Despite encroaching development and the aging of its founders (Soto died in 2015), a younger generation keeps the *casita* tradition alive. 'We congregate on Sundays. It's all about community,' noted Ivan Carrero, garden coordinator at the United We Stand garden. 'We've been here for eons' (Kensinger, 2019). Observes Sciorra: 'The vibrant, life-affirming culture of New York *casitas* is a counter voice questioning political negligence and economic tyranny that have left so much destruction in their wake' (Sciorra, 1966). The Peralta Hacienda Historical Park in the Fruitvale neighborhood of Oakland, California and the Sweet Water Foundation in Chicago, Illinois carry on this legacy of raising a counter voice to the dominant economic and political mainstream. Each brings together the literal and iconic presence of a home or house structure situated amidst a garden that represents not only a connection to the past, but a path forward towards liberation just as 'the casitas were places where the community built their structures in order to establish their notion of an ideal world' (Goldin, 2013).

Indian Canyon in the Unceded Ohlone Lands of California

By the mid-1880s, after nearly 250 years of contact with colonizers, the indigenous (Native American) population of the North American continent had been decimated through government-mandated genocide and forced relocation. Due to different colonial legacies, the atrocities committed against Native populations manifested themselves in distinct ways between the eastern and western parts of the continent. In the West, the enslavement and indentured servitude of tribes is a central feature of the founding of California, through the Mission system under Spanish occupation and even through the late 1880s, through the 1850 Act for the Government and Protection of Indians 'known to critics as the Indian Slave Act, which allowed whites to basically kidnap Indians and force them to work against their will' (Markus, 2013). The overlapping impositions of Spanish missions and Mexican and American rule disrupted, and then brought to the brink of utter destruction, the social, cultural, and ecological systems that had sustained thriving and vibrant indigenous populations for over 10,000 years.

Between 1769 and 1900, the Native American population in the area of California declined by 95 per cent. By 1990, California had the second-largest Native American population of any state, with 242,000 Native American residents. While most of these residents were Native Americans indigenous to California, many came to California from other states to seek employment in the metropolitan areas of the state, resulting in relatively large and diverse Urban Indian populations. There are over 100 reservations, many of them in remote rural areas, and in addition to the groups recognized by the federal government, with rights to government services and sovereign status, there are dozens of tribes attempting to gain federal recognition. Among these tribes seeking recognition are the Ohlone. Like many of the Native American populations in California, the Ohlone are not one tribe, but more of an 'extended family' of geographically bound villages with similar languages and cultural practices. Ohlone lands extended from the Carquinez Straits in the northern portion of the San Francisco Bay Area, south along the coast to Big Sur (Clinger, 2014).

There are innumerous stories and histories of resistance and struggle against oppression among California's Native American population. Near the present-day city of Hollister and in the Unceded Ohlone Lands of California sits Indian Canyon. The story of how it came to serve as ceremonial grounds for all indigenous people seeking traditional lands encompasses an inspiring intergenerational struggle to resist encroachment by the dominant White society. As the only land continuously held by the Ohlone people – the indigenous inhabitants of the San Francisco and Monterey Bay Areas – Indian Canyon is a fundamental example of placemaking as a continuous rejuvenation of community at the nexus of people and place. In the 1880s, Indian Canyon served as a safe haven for Native people who were escaping the oppression of the nearby Spanish mission at San Juan Bautista. The canyon is just a mile long, but it is vibrant with oak trees, streams, and a waterfall, making it an attractive haven then, and now (Chitnis, 2015).

The contemporary story of Indian Canyon starts with The Indian Allotment Act of 1887, also known as 'The Dawes Act,' a federal act that created a legal mechanism to break up tribal collective land stewardship into individual household ownership. The act was created under the guise of allowing Native American households to become self-sufficient as farmers, and allocating adult males 160 acres, adult females, 80 acres, and children 40 acres, each. However, the federal government's true motivation was to enable White settlers to acquire lands that were formerly recognized as tribal territory. By some accounts, between 1887 and 1905, over 90 million acres of Indian-owned land disappeared under a White wave of policy and capitalism. Just over 100 years later, in 1988, Ann Marie Sayers successfully used the very same Dawes Act to take back control of her family's land from the Federal Bureau of Land Management. Sayers used the Act to reclaim land that had been in her family for centuries in Indian Canyon (Trust Patent number 04-88-0047). It was the very same Act that Sayers' great-grandfather used in 1911 to establish claim to the land in Indian Canyon as 'an individual trust allotment' of almost 160 acres. 'My brother and I inherited the land that my mother owned who inherited it from her grandfather, my great-grandfather, Sebastian Garcia,' says Sayers (Clinger, 2014). From 1980 to 1988, Sayers developed an argument that she could fulfill the stringent Dawes Act requirements: generate enough revenue enough to live on, reside on the property exclusive of a home elsewhere, and conduct grazing without irrigation. According to Sayers:

> I worked with the Soil Conservation Service to see what we could do that this land could support. That is when we came up with West African pygmy goats. Then I had to build a home which is the log cabin I now live in... This allowed me to establish claim to the land adjacent to the original trust allotment. In 1988 I got the trust patent title for this property which consists of 123 acres.
>
> *(Clinger, 2014)*

'The canyon is alive through the power of ceremonies,' Sayers says (Clinger, 2014). The canyon has a large arbor, seven sweat lodges, and 30–40 more sites where individuals can come for their 'hamblechiya' or vision/nature quest where they can commune with nature without being disturbed by anyone. These facilities are offered by Sayers to indigenous communities from around the world for their ceremonies, storytelling gatherings, dances, and cultural events. For nearly three decades, Sayers and her family have produced a storytelling festival where people have the opportunity to hear Native stories told by Native people. Since 2015 they have hosted a Sacred Prayer Run from Mission San Juan Bautista to Indian Canyon, a 19-mile relay retracing the steps many Ohlone took to flee the oppression of the Mission system (Indian Canyon Life, n.d.).

Other expressive practices for making the invisible visible include an 'Imaginary Burdens Basket' at the Canyon's entrance for visitors, not to mention the overall subversion of the perverse Dawes Act policy, using Pygmy Goats no less. The legacy of Indian Canyon can be seen in more recent efforts to make manifest the intangible culture of threatened places, such as the Mauna Kea protest village that blocks access to the mountain by the construction crews of the controversial Thirty Meter Telescope. The village, in an expression of Native Hawaiian culture, encompasses all elements of life to sustain the *kia'i*, or protectors, including places of ceremony and worship, living quarters, and an onsite school. The creative application of land tenure strategies to preserving and expanding the indigenous stewardship of Indian Canyon are echoed in the Sogorea Te' Land Trust, an urban indigenous women-led community organization founded in 2012 that facilitates the return of Chochenyo and Karkin Ohlone lands in the San Francisco Bay Area to Indigenous stewardship. One particularly creative approach used by Trust is the Shuumi Land Tax that remixes the Chochenyo Ohlone word for 'a gift' with a not-subtle-at-all statement about the obligation all non-Ohlone residents to be part of supporting indigenous communities to reclaim land that was stolen from them. Describing the invitation that the Shuumi Land Tax and the Trust itself represents, co-founder, Corrina Gould asks:

> How does this land trust become not just my dream, but it a dream of people in the Bay Area that really want to see something different? A way for us to humanize ourselves, a way for us to be together on land. We need a whole community to envision this, to dream this out with us, beyond the parameters that we've all been given.
>
> *(Sogorea Te' Land Trust, n.d.)*

Forward!

During the 1800s, the phrase 'a Chinaman's chance' was used to describe futile situations and desperate circumstances. Today, does 'a Chinatown's chance' represent the dead end of a marginalized population or the thriving hope of ethnic, place-based neighborhoods that nurture generations of community growth and opportunity? Since my time in Boston Chinatown, I have operated under the assumption that officially designated, ethnic-affiliated neighborhoods do indeed have a nurturing effect on community growth and opportunity. I believe they play a symbolic and strategic role in the struggle for social justice and equality among communities of color. The rise of formally designated cultural districts as a cultural policy strategy suggest that the reclaiming of place identity politics is underway, as in the officially designated Mission District of San Francisco and in the work to establish the Black Cultural Zone in Oakland.

The Thai community of Los Angeles worked for years to establish 'Thai Town' in the Hollywood neighborhood. Historic Filipino Town in Los Angeles, the three remaining Japantowns in the United States, and many other communities are working to restore, enhance, or revive their identities as positive monikers through creative placemaking. For a community to be officially recognized means that, at some level, there is an acknowledgment of their existence and a practical justification of their needs. By merely requiring those in power to say names like Chinatown aloud, we make mainstream society acknowledge differences. The irony, I would suggest, is that the name Chinatown isn't ethnic enough to give pause to those outside our community. Visitors won't stumble over its pronunciation, and that missing pronunciation pause is transformative; it has the power to transport someone from comfort to confusion, from the United States to somewhere abroad, from complacency to complexity. It is this amplification of difference that can lead to respect for diversity, to acknowledgment of a community's contributions to the greater good, and to recognition of community needs that the mainstream cannot otherwise imagine.

For this reason, the Asian CDC created other creative placemaking projects that strengthened the community through resident-led planning and planning, engagement, and organizing around neighborhood advocacy, and housing development, and expressed Chinatown's identity beyond its geographic boundaries. Working with artist Mike Blockstein and 10 community youth leaders, we combined youth organizing, digital storytelling, and a community master-plan process into 'A Chinatown Banquet,' an interactive project designed to generate empathy among policy and political leaders of Boston and others outside of Chinatown for the neighborhood's priorities. We revived the legacy of movie theaters as intergenerational and intercultural social spaces in Chinatown through an annual week-long, free, outdoor film festival called 'Films at the Gate.' We reclaimed parts of Chinatown that had been seized through urban renewal and eminent domain by asserting a form of site control – moral site control that used history and art to catalyze organizing, community planning, and political advocacy to successfully influence a land-use decision. The Hudson Street for Chinatown campaign, which reclaimed Parcel 24 from the Massachusetts Highway Department at nominal cost, became a significant expansion of housing, open space, and cultural space for the community in the One Greenway development, including the Pao Arts Center. We supported community youth, leaders, and advocates to operate a pop-up library for several years to recreate and reimagine the Chinatown Branch Library that had been shuttered and never replaced decades earlier. A new Chinatown Branch Library is now being planned as part of the Asian CDC's next development. Using a virtual world gaming engine, we created a game called *Participatory Chinatown* that used a digital manifestation of Boston Chinatown, archetypal neighborhood stakeholder avatars, and community quests to generate intra-neighborhood empathy for complex and contentious community planning processes.

Reflecting on these examples, I realize that my creative practice as a community developer was channeling legacies of placemaking that persist from the Maroon Settlements, Casitas, and Indian Canyon. We were building up a vibrant and thriving community, in plain sight of those who sought to disempower and oppress us, where Chinatown could grow and maintain an authentic identity, economy and society. We were supporting expressive placemaking that projected this identity through forms both ancient and contemporary at the Berkeley Street Community Garden on land previously deemed unwanted. Finding ways for policy to serve our needs, to the point of subverting the mainstream or status quo in our own unique ways, led to the rare successful reclaiming of land taken by urban renewal at Parcel 24. Around the country in rural and urban places, in communities of all kinds, these legacies persist.

Creative placemaking as a defined field of practice arrives just as the challenges to community development threaten to swamp past gains and hinder sustained equitable development; seemingly intractable and deepening economic inequity, unforeseen and unforeseeable disruptive changes to the systems of work, housing, and health, and the fraying of social cohesion across the country call for new approaches. We can and must look to 'landscapes of power' – and resistance – in communities around the country and in our past to shape creative placemaking's growth.

References

Chitnis, R. (2015). 'In the land of my ancestors: Native woman stands her ground in Ohlone Territory', in *Truthout*, 12 October 2015 [online]. Available at: https://truthout.org/articles/in-the-land-of-my-ancestors-native-woman-stands-her-ground-in-ohlone-territory/ (Accessed: 1 June 2020).

Clinger, J. (ed.) (2014). *Ohlone Elders and Youth Speak: Restoring a California Legacy* [online]. Available at: http://communityworkswest.org/wp-content/uploads/2016/09/Ebook-Final.pdf (Accessed: 1 June 2020).

Diouf, S.A. (2014). *Slavery's Exiles: The Story of the American Maroons*. New York: New York University Press.

Flood, J. (2019). 'Why the Bronx burned', in *New York Post*, 16 May 2010 [online]. Available at: https://nypost.com/2010/05/16/why-the-bronx-burned/ (Accessed: 1 June 2020).

Goldin, E. (2013). 'Casitas: Places of community power in the south Bronx', in *The Surreal Estate: Perspectives on Tenant Organizing from the Urban Homesteading Assistance Board*, 21 February 2013 [online]. (Accessed: 14 November 2019).

Gonzalez, D. (1990). 'Las Casitas': Oases or Illegal Shacks?', in *The New York Times*, 20 September 1990 [online]. Available at: https://www.nytimes.com/1990/09/20/nyregion/las-casitas-oases-or-illegal-shacks.html (Accessed: 1 June 2020).

Indian Canyon Life. (n.d.). Indian Canyon Life [online]. Available: https://indiancanyonlife.org/ (Accessed: 1 June 2020).

Kensinger, N. (2019). 'Inside the casitas of the south bronx's community gardens,' in *Curbed*, 1 October 2015 [online]. Available at: https://ny.curbed.com/2015/10/1/9915402/inside-the-casitas-of-the-south-bronxs-community-gardens (Accessed: 1 June 2020).

Markus, B.P. (2013). 'The 'golden state's' brutal past through native eyes', in *Truthout*, 16 November 2013 [online]. Available at: https://truthout.org/articles/the-golden-states-brutal-past-through-native-eyes/ (Accessed: 1 June 2020).
McLaughlin, C. (2019). *South Bronx Battles: Stories of Resistance, Resilience, and Renewal*. Berkeley, CA: University of California Press.
Sayers, D.O. (2016). *Desolate Place for a Defiant People: The Archaeology of Maroons, Indigenous Americans, and Enslaved Laborers in the Great Dismal Swamp*. Gainesville, FL: University Press of Florida.
Sciorra, J. (1966). 'Return to the future: Puerto Rican vernacular architecture in New York City', in King A.D. (ed.) *Re-presenting the City: Ethnicity, Capital, and Culture in the 21st-Century Metropolis*. New York: New York University Press, pp. 60–92.
Sciorra, J. (1994). 'We're not just here to plant. We have a culture. An ethnography of the south bronx casita rincón criollo', in *New York Folklore*, XX (3–4), pp. 19–41.
Shapiro-Perl, N. (2014). Landscape of power: Freedom and slavery in the great dismal swamp [online]. Available at: https://vimeo.com/134317981 (Accessed: 1 June 2020).
Smith, S.L. (2013). 'Freedom for California's Indians', in *The New York Times*, April 29, 2013.
Sogorea Te' Land Trust Our Vision. (n.d.). in *Sogorea Te' Land Trust* [online]. Available at: https://sogoreate-landtrust.org/our-vision/ (Accessed: 1 June 2020).
Youssef, S. (2017). 'The great dismal swamp,' in 99% Invisible, episode 271, August 15, 2017 [online]. Available at: https://99percentinvisible.org/episode/great-dismal-swamp/ (Accessed: 1 June 2020).

Further reading in this volume

Preface: 'Disastrous forces, accidental actions, and grassroots responses'
Tom Borrup
Chapter 7: Conflict and memory: human rights and placemaking in the city of Gwangju
Shin Gyonggu
Chapter 8: Queer placemaking, settler colonial time, and the desert imaginary in Palm Springs
Xander Lenc
Chapter 9: From the dust of bad stars: disaster, resilience, and placemaking in Little Tokyo
Jonathan Jae-an Crisman
Chapter 10: From moon village to mural village: the consequences of creative placemaking in Ihwa-dong, Seoul
Jason F. Kovacs and Hayun Park
Chapter 18: '*Homomonument* sounds like a poem': queer placemaking 30 years on: a conversational dialogue with Thijs Bartels, author of *Dancing on the Homomonument* (2003)
Martin Zebracki
Chapter 21: Placemaking through parkour and *Art du Déplacement* (ADD) as a Singaporean applied performance practitioner in London
Adelina Ong
Chapter 28: Integral placemaking: A poiesis of sophrosynes?
Ian Wight
Chapter 31: Seven generations: a role for artists in Zuni PlaceKnowing
Theodore S. Jojola and Michaela P. Shirley
Chapter 35: Planning governance: lessons for the integration of placemaking
Nigel Smith
Chapter 41: Rituals of regard: on festivals, folks, and findings of social impact
Maribel Alvarez
Chapter 42: Creative Placemaking and Placekeeping evaluation challenges from the practitioner perspective: an interview with Roy Chan
Maria Rosario Jackson
Chapter 43: A theory of change for creative placemaking: the experience of the National Endowment for the Arts' *Our Town* program: an interview with Patricia Moore Shaffer, PhD
Maria Rosario Jackson

6

LISTEN, CONNECT, ACT

Kim Cook

Context matters

In New Orleans there is an expression, 'I know this dirt.' It speaks to time, and passage of time, and fertility, and soil, and longevity, and labor, and love of home. *I know this dirt.* It summons the idea of connecting in community and looking deeply at the soul of a place. It tells us to listen and learn the stories of the people. When we develop the tools of listening to the storytelling and language and history that is derived from among the people in the places where we undertake our work, we root that work deeply and generate possibilities that can be timeless. We cannot franchise creative placemaking – context matters. It is not formulaic in its realization. We can commit to clear qualities of listening and engagement and truly foster cross-sector collaborations that lead to lasting partnerships and coalitions that deepen and positively change both the cultural and economic future of a place. Creative placemaking, when done well, is inherently idiosyncratic; it responds to and reflects the cultural assets centered in a location and those who reside there. It is with this perspective that true creative placemaking can occur.

Humans

In the end, when we 'make' places we are talking about the places where humans live, work, play, learn, and come together (or not) as a community. In its best form, creative placemaking will add to the vitality of a community, through engagement, collaboration, participation, and collective decision-making. Then the economic and other benefits follow from these processes. Very often, creative placemaking has garnered attention as a method for advancing economic development goals – in some ways serving as a response to the elimination of the value attributed to the arts and culture as a core element of civic life. Creative placemaking added utility to the conversation about the value of the arts – when the intrinsic-value conversation could not sway policy or move capital, economic benefit could serve to do so. Yet, the very heart of creative placemaking sits in the efforts of humans who see possibility in unlikely places, who forge partnerships with unusual suspects, and who move mountains to find the capital needed to realize their aspirations for community. Sometimes those mountains are moved one shovel of dirt or can of paint at a time. The work, and the beauty, of creative placemaking is best realized through human interaction, through conscious listening, and by valuing the contributions of residents in cities and rural settings who have deep wisdom of their places, filled with memory, culture, and opportunity to advance civic goals.

Exemplars of the work

Remarkable examples of this approach are everywhere. Here are a few to consider: Carol Bebelle, Founder of Ashe Cultural Center in New Orleans; Artist Yazmany Arboleda; the Network of Ensemble Theaters' 'Microfest'; Ghana Think Tank; Maud Le Floch's Polau des Artes in France; Dr Mindy Fullilove; Joy Mboya, Founder of GoDown Center in Kenya; and Karin Lekberg's work at Subtopia in Sweden.

What these people and projects share is a profound ability to be present, to demonstrate respect for the people who inhabit the places they live and work in, and a practice of responsive listening. Carol Bebelle is known to say: 'First show up, in a while people will come to know you, keep showing up, and in a while, you will have a community' (Bebelle, 2015). Her work over more than 20 years – at first the Efforts of Grace, and then translated into the Ashe Cultural Center in Central City New Orleans – served as the catalyst for a revitalization of the O.C. Haley corridor. Carol Bebelle's work led to the creation of a multi-disciplinary exhibition and learning space, 29 residential units, a theater and community black box space, a credit union, and a legacy of emphasizing culture, community, and economic development (Asante and Lopez, 2017; Worthy, 2019). Artist Yazmany Arboleda is deeply engaged in working directly with people and expressing their stories. While Yazmany's work is often ephemeral he is committed to the role of listening and stimulating people to collaborate within the spaces and places where they live. This is transformative work both for participants and for witnesses of his public art projects and place activation (Bric. 2019; TEDx, 2014). Network of Ensemble Theaters' *Microfest USA* project took a deep look at the communities of New Orleans, Detroit, and Appalachia, culminating in a summit in Honolulu. With each of their explorations they made tangible efforts to include the voices of the people who lived in those locations while also examining the role that local theater plays in advancing community and economic development strategies. In her paper, '*Microfest USA*: A synthesis of learning about art, place, and culture,' Pam Korza states: 'Artists and culturally specific organizations play a catalytic role in revitalizing public spaces and neighborhoods and providing civic as well as cultural gathering spaces' (Korza, 2013). The *Microfest USA* project points to engaging with and documenting work that is locally grounded and vitally important. Ghana Think Tank's innovative work proposes that we subvert the notion that the answers reside in the Western nations (Ghana Think Tank, n.d.); Maud Le Floch's work at Polau des Artes Urbaines asks developers to embed artists in the building process as site-specific contributors to community engagement that influence the build decisions (Le POLAU, n.d.); Dr Mindy Fullilove points to reconnecting cities to their past while planning for their future and carefully listening to those who reside there (Fulliove, 2013); Joy Mboya's work in Nairobi took raw space and generated a community center that serves thousands of artists, entrepreneurs, and community members at the GoDown Arts Center (Engage Talk, 2018; SmartMonkey, 2013); while Karin Lekberg's work at Subtopia, located in the Botkyrka municipality outside of Stockholm has generated a creative cluster that positions the migrant community as core contributors to the future of Sweden (Subtopia, n.d.).

Examples of process

As someone who has been doing this work for 25 years, I often get asked 'how' to do the work, and what I have learned. To examine some specific process choices, I will reference work from my own direct experience in New Orleans, Louisiana and more recently in Washoe County, Nevada and Mountain View, California.

Arts Council New Orleans

When I moved to New Orleans to take the role of President/CEO of the Arts Council in early 2013, the city was rapidly advancing a narrative of itself as a new tech hub, a place for entrepreneurs, a destination in the midst of a post-Katrina renaissance. This narrative, while it had merit, was also in contrast to the other stories I was learning as I listened to people who lived there. I heard about New Orleans as the center of mass incarceration for the world; as the place where 30,000 blighted properties remained and community development organizations were struggling with the challenges of funding based in affordable housing legislation that did not allow for renovation or blight remediation as strategies for helping neighborhoods; a city where thousands of youth between 15 and 24 years were out of school and out of work; and the influx of new and enthusiastic transplants was seen as much as an intrusion as an infusion of talent. Initially the strategy for me was to listen, ask who else people thought I should talk to, and listen some more. I believed that a small agency, like the Arts Council, might be able to partner with larger entities and leverage resources that would aid the larger population. This resulted in two initiatives as well as some changes to our grant-making processes.

The first initiative was a launch of an art and technology festival, *LUNA Fête*, centered on project mapping. The hoped for outcomes were many, including that we might: raise awareness with regard to urban lighting strategies as a mechanism for increased safety with lighting as art as well as a safety solution; build a rapport between long-term residents and younger newcomers through the creation of a large-scale, family-friendly spectacle, in the streets – that was free. This paradigm would be familiar to New Orleans while also introducing technology that the younger entrepreneurs would identify with, and thus could allow for connectivity across these two factions within New Orleans; generate enthusiasm for these technologies and foster new learning and tools capabilities for New Orleans artists; and create partners across the New Orleans tourism, arts, and technology sectors. Launched in 2014, *LUNA Fête* is the longest-running and largest project-mapping festival in the United States and has advanced the technical skills and opportunities for 200 artists and 60 youth since its inception. In 2019 *LUNA Fête* was attended by more than 100,000 people and has established a firm place in the crowded New Orleans festival calendar. It continues to be a cross-sector collaboration among several partners.

The second initiative, *Youth Solutions*, began with a working group of 12 people who sat together and met for a year. Comprised of architects, public health professionals and researchers, a social worker, an artist, graphic designers, a community activist, along with myself and one other staff member of the Arts Council, we came together once a month, broke bread, checked in on each other's worlds and wellbeing, before our conversation about how to address youth trauma alongside the trauma to the physical place of New Orleans would commence. We talked amongst ourselves, and we also reached out to have conversations with partners city-wide. Consequently, *Youth Solutions* was created as a living laboratory to: examine the linkages between what social science social/emotional/wellness indicators show is effective and what artists naturally do; build a handbook for artists such that they might be better able to articulate their work in the vocabulary of social science; test out the impact that design education could have for local youth who might be empowered to propose and enact changes in their community; use the built environment and work with architects and artists to create a sense of agency for the youth through changing the world around them; and create toolkits and documentation that could be shared broadly for the field. After a year of developing the project, *Youth Solutions* was funded by both the National

Endowment for the Arts *Our Town* grant and ArtPlace America. The program took hold in the summer of 2015 and resulted in many young people participating in the creation of public art interventions throughout Central City in New Orleans. Perhaps more importantly, *Youth Solutions* created an environment that evolved the practice of youth empowerment and public art for the Arts Council that continues to this day – work that is grounded in a clear understanding of how to articulate and communicate impact when artists work with youth. In terms of our grant-making, we made substantial revisions to the funding categories so that we could support and celebrate the street parading and traditional cultural communities of New Orleans.

Partners and *Burning Man Project*

When I joined *Burning Man Project* as the Director, Art and Civic Engagement, I was invited to collaborate with officials at the County offices in Washoe County, Nevada. We secured funding through the NEA *Our Town* grant, for a 200-mile ArTrail which we developed with specific sensitivity to the role storytelling, memory, and placekeeping could take as a part of a placemaking effort. The county hosted over five community story circles, learning about the history of miners, ranchers, indigenous people, and residents over chili dinners, pancake breakfasts, and coffee. In fact, recording and archiving these stories became an outcome from the process that has ignited and inspired Washoe County leaders. The *Burning Man* team and the Washoe County staff worked together to make sure that all of our encounters reflected the intent of our project. Thus, when we held a large stakeholder meeting with county, state, and local organizations represented around the table, we took more than a full hour at the top of the meeting asking each person to bring something with them that they connected to their relationship to Washoe County. This time of storytelling in the circle was established with transparency about our intent to create an 'us' before we presented our plans to the group. We did not want to stand up in front of a room with chairs in a row and state our case and then defend it. We wanted to explicitly foster a sense of unity through story and a common feeling for the place that was home. This gentle opening then set the stage for the presentation of the plan for the trail and we avoided steering into the pitfalls of public hearings that are so often fraught with challenging discourse. Once we became an 'us' we were able to use that unity to nurture a feeling of common cause while also providing legitimate methods for stakeholders to participate and influence the overall project.

Another project emerged in 2018 as the *Burning Man Project* team began to explore creative placemaking with Google. In this partnership we endeavored to ensure that even with a large corporate partner we could find ways to influence art selection and community input processes that would be authentic and participatory. *Burning Man* is at heart a culture of engagement and not one of production. While we were eager to see the art of *Burning Man* translate into more permanent public settings, we were equally interested in an approach that was values-aligned with the *Burning Man* ethos. Understanding that ultimately the decision-making would lay within the Google leadership, we also knew that our partners at Google were invested in creating avenues for local residents to influence that outcome. Plus, in our meetings with City staff members we heard a common desire between Mountain View and Google to create a public space that was inviting, where people would gather, and linger, and feel welcome. There was a sense of urgency that this not become a 'workplace amenity' but rather a central plaza that drew residents in and reflected their interests.

To undertake the engagement process, we developed several input systems, the first, a storytelling evening at a local cafe centered on stories of 'where you went when you were a kid and what you loved about it.' Again, as with the Washoe County stakeholder group, transparent communication was key; we shared that we didn't want to create a situation where they told us what they really wanted was a neon banana and then we end up picking a ceramic frog and they wonder why we asked in the first place. We shared that while literal desires might not be met, design principles and aesthetic preferences could be garnered through this evening of sharing. We started the evening off and soon people in the room were eagerly talking about where they played, how they hid and discovered and engaged in adventures as children. The room had people from early twenties to early seventies and the experience quickly turned into an evening of laughter and good feelings (it turns out that asking people to remember where they played as children is a good mood machine). We were able to glean such factors as verticality for wayfinding (represented in their youth by water towers and street lights), ways to be in public while also having privacy (hiding in juniper bushes, hanging out at the wall at the basketball court), and finding portals to new spaces (tunnels and following creek beds). These were drawn onto small poster boards by my team and brought into the room for the selection committee. Similarly, we hosted two Human Centered Design workshops for community residents, asking them to consider how we might create places to gather, places to feel ownership, places to linger, and through a series of drawing exercises and maquette-building with found objects we were able to get at impulses for play, free spiritedness, and interactivity. Additionally, we hosted online commenting opportunities which both boosted the visibility of artists who submitted and gave us a broader input reach (comments were not broadcast). We also created and taught a seminar on archetype and identity for Google employees to get at underlying symbols that would be meaningful for those who worked there. In the course of the last two years we've been able to facilitate the commissioning of 15 artists for Google with $2.588 million in artist fees, all grounded in the principles and design influences of the community at large.

Black Rock City

Burning Man, which takes place each year in the Nevada desert, is, itself, a creative placemaking engine. Known as Black Rock City, with over 32 years of innovation, improvisation, play, and co-creation it has evolved into a laboratory where learning by doing, collective effort, and year-over-year iteration has led to a deep body of practice and knowledge that translates well into creative placemaking. In her 2019 article for the *New York Times*, titled 'A Nobel-winning economist goes to Burning Man,' Emily Badger quotes Paul Romer as saying about *Burning Man* and Black Rock City:

> I picture an economist showing up… and saying: 'Oh, look! This is the miracle of the invisible hand. All of this stuff happens by self-interest, and it just magically appears.' And there's this huge amount of planning that actually is what's required beneath it to make the order emerge.
>
> *(Badger, 2019)*

In fact, it does take planning, but it also takes innovation and practice. The results of this integrated approach are what *Burning Man* has to offer the field of creative placemaking. Most certainly the practice of building a temporary city in the desert results in a huge body of knowledge, but I would argue it is the approach to year over year, learning by doing, and then iterating that is an even more crucial element in what *Burning Man* can offer others. Creating a prototyping mindset and then improving on that initial effort, engaging with community members in a whole-hearted collective build effort, and generating practices that improve year over year have essential qualities and approaches to offer those who work in more permanent environments. Too often the world of city planning and building becomes hampered by risk avoidance, long-term and slow-moving capital flows, and competing interests. *Burning Man* offers cities and placemakers a potential toolkit for subverting those overly defined methodologies and testing out new practices. One example of a city where community members can directly impact the built environment is in Bologna, Italy, where the local government has created a micro-contract system for citizens to make improvements to their neighborhoods. This sort of risk-taking and participatory action within a permanent city is at the heart of the temporary metropolis known as Black Rock City where *Burning Man* takes place.

Thinking about culture and creative placemaking in a post–COVID-19 environment

When we are fortunate enough to all arrive at a post–COVID-19 environment we will find ourselves at risk of cultural destitution that is profoundly deeper than what we witnessed in the 2008–2009 recession. As I consider this, I am reminded of the paper, 'Cultivating 'natural' cultural districts,' co-authored by Mark J. Stern and Susan C. Seifert, created in partnership with The Reinvestment Fund. Their work comes to mind because they reference the idea of clusters of cultural assets, both formal and informal arts production, and while not called out specifically within their paper, I would suggest that a cultural cluster must also include migrants, restaurants, food trucks, and the social spaces that anchor our communities. These cultural generators will play a critical role in revitalizing our cities after the pandemic. As stated by Stern and Seifert:

> While the arts are commerce, they revitalize cities not through their bottom-line but through their social role. The arts build ties that bind – neighbor-to-neighbor and community-to-community. It is these social networks that translate cultural vitality into economic dynamism.
>
> *(Stern and Seifert, 2007)*

After an extended period of social isolation – where even our most casual social encounters with the small-business owners who built the dry cleaners, the shoe repair, the beauty salon, and others who make up our communities, were denied us, we will want those relationships to survive and those businesses to thrive. Success will require substantial capital flows to the street level. These small-business owners are the backbone of our economic health, our service providers, and our connective tissue. Along with artists, they have the capacity to bring back a vibrant society, if we invest wisely. Bringing both cultural sector activists and small-business owners together and insisting that the investment in our cities be aimed at restoring our 'third places' will be essential. This will be the essence of creative placemaking when we re-emerge to rebuild our economies and our communities. We need, fundamentally, to connect. One positive result of the global pandemic is that we are reminded of our shared humanity and vulnerability; to translate that into action for our future will mean thoughtful and responsive investments at the neighborhood level. Creative placemakers will be well poised to support hyper-local efforts that can generate a global revitalization through banding together and insisting on responsible capital deployment at the grassroots level. Otherwise we will suffer the tearing apart of the rich and vibrant fabric that makes each place unique and allows us to enjoy the many incredible benefits of living in diverse and dynamic communities.

References

Asante-Muhammad, D. and Lopez, J. (2017). *Partner Spotlight: Ashé Cultural Arts Center Provides Equitable Futures to New Orleans Residents* [online]. Available at: https://prosperitynow.org/blog/partner-spotlight-ashe-cultural-arts-center-provides-equitable-futures-new-orleans-residents (Accessed: 1 March 2020).

Badger, E. (2019). 'A nobel-winning economist goes to burning man', in *New York Times*, 9 September 2019 [online]. Available at: https://www.nytimes.com/2019/09/05/upshot/paul-romer-burning-man-nobel-economist.html (Accessed: 7 March 2020).

Bebelle, C. (2015). Community Leader. (Personal Communication, April 2015).

Bric, T.V. (2019). *Yazmany Arboleda Left the Gallery World to Create Art with the Public in Public | BK Stories* [online]. Available at: https://www.youtube.com/watch?v=sJXsI5FSkiE (Accessed: 3 March 2020).

Engage Talk. (2018). *Vision: The Bigger Picture_Joy Mboya* [online]. Available at: https://www.youtube.com/watch?v=X36kgCZgFMQ (Accessed 7 March 2020).

Fulliove, M.T. (2013). *Urban Alchemy: Restoring Joy in America's Sorted-Out Cities.* New York: New Village Press.

Ghana Think Tank. (n.d.). *GhanaThinkTank Developing the First World* [online]. Available at: http://www.ghanathinktank.org/about (Accessed: 4 March 2020).

Gorenflo, N. (2015). 'Bologna celebrates one year of a bold experiment in urban commoning', in Krytyka Polityczna and European Cultural Foundation (eds.) *Build the City: Perspectives on Commons and Culture.* Amsterdam, The Netherlands: European Cultural Foundation, pp. 147–151 [online]. Available at: https://www.culturalfoundation.eu/library/build-the-city-book (Accessed: 1 June 2020).

Korza, P. (2013). *MicroFest: USA | A Synthesis of Learning About Art, Culture, and Place* [online]. Available at: https://www.ensembletheaters.net/sites/default/files/files/NETSynthesis_FINAL.pdf (Accessed: 4 March 2020).

Le POLAU. (n.d.) *Polau Arts Urbanisme* [online]. Available at: http://polau.org/le-polau/structure/ (Accessed: 5 March 2020).

SmartMonkey, T.V. (2013). *Joy Mboya, Godown Arts Centre: The Railway Property Development and Making a Cultural Precinct* [online]. Available at: https://www.youtube.com/watch?v=w7uOtYrqEmA (Accessed: 7 March 2020).

Stern, M.J. and Seifert, S.C. (2007). 'Cultivating 'natural' cultural districts', in *'Natural' Cultural Districts: A Three-City Study* 1 [online]. Available at: http://repository.upenn.edu/siap_cultural_districts/1 (Accessed: 1 June 2020).

Subtopia. (n.d.) *Subtopia: A Center for Art, Culture and Social Engagement* [online]. Available at: https://www.subtopia.se/about/ (Accessed: 7 March 2020).

TEDx. (2014). *10,000 Reasons to Believe in the Power of Art in Public Space: Yazmany Arboleda at TEDxUNC* [online]. Available at: https://www.youtube.com/watch?v=rkSvSkV_KCM (Accessed: 3 March 2020).

Worthy, T.C. (2019). 'City council honors Ashé cultural arts center co-founder Carol Bebelle,' in *Uptown Messenger* [online]. Available at: https://uptownmessenger.com/2019/12/city-council-honors-ashe-cultural-arts-center-co-founder-carol-bebelle/ (Accessed: 1 March 2020).

Further reading in this volume

Chapter 11: *Free State Boulevard* and the story of the East 9th Street Placekeepers
Dave Lowenstein
Chapter 12: Public transformation: affect and mobility in rural America
Lyndsey Ogle
Chapter 14: Experts in their own tomorrows: placemaking for participatory climate futures
Paul Graham Raven
Chapter 16: More than a mural: participatory placemaking on Gija Country
Samantha Edwards-Vandenhoek
Chapter 17: 'I am not a satnav': Affective placemaking and conflict in 'the ginnel that roared'
Morag Rose
Chapter 18: '*Homomonument* sounds like a poem': queer placemaking 30 years on: a conversational dialogue with Thijs Bartels, author of *Dancing on the Homomonument* (2003)
Martin Zebracki
Chapter 22: Embedded Artist Project: Epistemic Disobedience + Place

Frances Whitehead
Chapter 23: Routing out place identity through the vernacular production practices of a community light festival
Gail Skelly and Tim Edensor
Chapter 24: Artists, creativity, and the heart of city planning
Tom Borrup
Chapter 25: 'If you can make it there, you can make it anywhere…': cultural placemaking at the heart of cities
Sherry Dobbin
Chapter 31: Seven generations: a role for artists in Zuni PlaceKnowing
Theodore S. Jojola and Michaela P Shirley
Chapter 37: *The Neighbourhood Project*: a case study on community-led placemaking by CoDesign Studio
Lucinda Hartley, Eliza Charley, Sama Choudhury, and Harriet McKindlay
Chapter 41: Rituals of regard: on festivals, folks, and findings of social impact
Maribel Alvarez
Chapter 45: How the city speaks to us and how we speak back: rewriting the relationship between people and place
Rosanna Vitiello and Marcus Willcocks

SECTION 2

Practices of placemaking

Section Editor: Tom Borrup

PREFACE

'Disastrous forces, accidental actions, and grassroots responses'

Tom Borrup

I write this essay during a tragic and stressful week in my upper Midwest American city of Minneapolis: the quote above, that gives the title to this preface from Jonathan Jae-an Crisman (this volume), speaks to this time as well as to the chapters herein. During the peak of COVID-19, we saw the brutal murder of an African American man, George Floyd, by a police officer while three other cops looked on. Caught on video, as so many such horrible events recently, this was fed live on Facebook. Outrage prompted immediate protests. Multiple demonstrations were large and loud and, yes, sometimes unlawful – as in, blocking traffic or defying curfews. Burning the police precinct where the officers worked provided symbolic cause for celebration by some. Protests continued across the city, while at very late hours of the night, widespread arson took place. Most believe it was instigated by White supremacists. Over three nights major vandalism, looting, and fires consumed several areas of the city. I witnessed a large part of my own neighborhood as it was stripped and burned. Across an area recently designated by the city as a cultural corridor, we lost low-income housing under construction, a library, post office, nonprofit agencies, and over 100 small businesses. The vast majority of these businesses were locally owned by People of Color and migrants who for decades had toiled to build what had become a vibrant and beautiful multi-cultural, multi-ethnic community. While thriving, this community was still economically marginal before. Together with COVID-19, violence virtually destroyed this place that had been made. The day after the third and most devastating night of fires, with no guarantee there wouldn't be a fourth and while protests continued, hundreds of residents were joined by hundreds more volunteers from across the city in clean-up efforts. Those businesses still standing were boarded and people distributed food and other necessities to residents in need. At the same time, a couple hundred more residents gathered in a nearby park to self-organize to protect themselves and their neighbors against possible future violence.

All these actions and reactions are part of placemaking: a process that never ends. Placemaking happens in fits and starts, in good times and bad. It moves forward and is set back. Placemaking might involve a festival or a mural or it may involve spontaneous organizing, and rebuilding. And at its best, it is a process most centrally about people. When my neighborhood burned, people showed up and organized because they cared about each other and the community they had built. This section explores a number of different ways people understand and relate to place – the meanings and purpose of land and the built environment, how meaning that connects people comes to be, and how conflict is frequently part of the process. If placemaking is really about people, it must involve striving for equity and justice. Unfortunately, too often it has been associated with gentrification and real estate profiteering. Such opposing values and forces inevitably play out through conflict, and in this section, we read stories of conflicts that are big and small, brief and protracted.

As we know from writers in other parts of this book, the very terms placemaking and creative placemaking are loaded. Places exist, they change, but they are not made. Dakota and Ojibwe people who lived with the land for thousands of years in and around what is now called Minneapolis, like Indigenous peoples elsewhere, were forcefully and deceptively dispossessed of their lands. In relationships they maintained with the land, the idea of *making* a place makes no sense. People come from the land, live with the land, and are responsible for caring for it on behalf of future generations. '"Land" refers not just to the materiality of land, but also its "spiritual, emotional, and intellectual aspects",' wrote McCoy et al. (2016, p. 9). The very idea of *making* a place considers land as an object to be possessed, exploited, and altered in conformance with the desires of those who see themselves possessing it. Indigenous ways of understanding and living with land were disrupted across the world beginning in the sixteenth century when Western Europeans began to travel and by hook or crook conquer and settle many places. As we will read in chapters that follow, such invasions can come in the form of armed soldiers and even well-meaning artists. Invaders don't always understand or care that they are trampling on other ways of living in place or the lives and histories of those in that place. For different people, land represents places to settle, to own, to profit from, to build a life on, to hold, to commemorate, or to be responsible to care for. That people hold different relationships is sometimes misunderstood, confused, or conflated.

Placemaking, creative placemaking, city planning, and academic research carry forward elements of settler colonialism. Wrote McCoy et al. (2016):

> Though some may attempt to dismiss discussions of settler colonialism as overly concerned with the past, settler colonialism is important to analyze because it relies upon assumptions about other cultures that are alive and well in the most powerful societies in the contemporary world.
>
> *(p. 2)*

In remarkable work on research practices as an extension of colonialism and imperialism, Linda Tuhiwai Smith (1999) wrote:

> And, in each place, after figures such as Columbus and Cook had long departed, there came a vast array of military personnel, imperial administrators, priests, explorers, missionaries, colonial officials, artists, entrepreneurs and settlers, who cut a devastating swathe, and left a permanent wound, on the societies and communities who occupied the lands named and claimed under imperialism.
>
> *(p. 21)*

Explorers, missionaries, and military paved the way for settlers, administrators, artists, and entrepreneurs to carry on the work – sometimes in the guise as placemakers.

As Wolfe (2006) wrote: 'Settler colonizers come to stay: invasion is a structure not an event' (p. 388). Structures or embedded approaches are not episodic but find their way into related practices and ongoing work, including placemaking, city planning, and even environmental work. Paperson (2014) described how settler colonialism continues to live within and inform contemporary environmental education that often builds on the false concept of 'terra nullius, the colonial fiction of "empty land" or "and not legally belonging to anyone"' (p. 117). He extended this to how such practices as 'greening the ghetto can mask a neoliberal curriculum of whitening the ghetto with "better-educated," ecologically "responsible," global citizens' (p. 121). In the same way, the legacy of settler colonialism is found in city planning. Libby Porter, in her important book, *Unlearning the Colonial Cultures of Planning*, found that within planning practice, 'the *cultural* traits of colonialism endure as "a variety" of colonial representations and encounters [that] both precede and succeed periods of actual possession and rule' (2016, p. 16). Planning, Porter wrote, 'arises as an activity and set of practices from a locatable and cultural world-view' (p. 44). The very process of spatial ordering reflects this particular view that she described as a 'colonial process of producing space for certain ends, to favour certain people (their cultural lifeways and economic systems)' (p. 46). This makes conflict not only common but sometimes necessary for communities seeking justice for historical wrongs. While never simple or linear, places evolve through various processes including periodic conflict. Different ways of understanding place and relationships to place contribute to many conflicts. Conflict and the struggle for shared understanding that result from different relationships to land and to place underpin the chapters in this section. Conflict is not always something welcomed or understood. It comes in many forms that are sometimes subtle or shrouded in policies, sometimes violent or aggressive. Actions are big and small and can involve love, anger, or compassion. Placemaking work can cause conflict and can take place in the aftermath of conflict, in a cycle that repeats. Recognizing conflict as part of placemaking – like death as part of life – is important both to address it constructively and to find ways to rebuild in its aftermath.

This section begins with conflict that is both horrific and hopeful. In his chapter, 'Conflict and memory: human rights and placemaking in the City of Gwangju,' Shin Gyonggu chronicles the May 18 Democracy Movement in South Korea to illustrate how a violent conflict brought about powerful meaning of place. Since 1980, an area of the city as well as structures both old and new have become important sites of memory and meaning for both older and younger generations. In 2017, I visited Gwangju, where stories of students fighting and dying, and the level of pride in what they achieved, were incredibly moving. Equally impressive were multiple efforts over recent decades to honor key sites and people, to create a central space within the city, along with institutions and programs that build on an identity for the city around human rights and culture.

Some of the efforts Shin writes about involved struggles among different players within the city over saving or demolishing buildings. These might be considered 'usual' conflicts in other places such as Palm Springs, California as we read in Xander Lenc's layered chapter, 'Queer Placemaking, settler colonial time, and the desert imaginary in Palm Springs.' Lenc combines ways of understanding the desert and its homoerotic (his)stories and the complicated and sometimes uncomfortable coexistence among Indigenous people and White 'settlers.' Choices that are based in both aesthetics and understanding of land put mostly gay, White architectural preservationists in conflict with Native people and their land and enterprises – all set against the backdrop of this desert resort community, that itself is an outgrowth of Hollywood imagery and, more recently, heavily populated by wealthy, gay men.

In the next chapter, 'From the dust of bad stars: disaster, resilience, and placemaking in Little Tokyo,' Jonathan Jae-an Crisman describes placemaking as a 'continual negotiation between disastrous forces, accidental actions, and grassroots responses to those things.' He takes the reader on a journey with Los Angeles' Little Tokyo neighborhood from its evolution during the early twentieth century to the time its residents were forcefully removed to concentration camps during World War II. After returning to find themselves dispossessed, Japanese Americans painstakingly rebuilt their community. Crisman focuses on how contemporary property rights – the very system that took land from Native peoples – has helped return Little Tokyo as a centralized and spiritual home. Against the current threat of land speculation and forces of gentrification, this community hopes these same property rights will help them retain many important places of meaning and a sense of community life.

In 'From moon village to mural village,' Jason F. Kovacs and Hayun Park describe unintended consequences resulting from efforts at creative placemaking in Ihwa-dong, a working-class neighborhood in Seoul, South Korea. The authors reflect on the city's efforts to uplift the neighborhood through art, substituting it for more-needed investment in new housing and infrastructure. Artists not from the community transform the neighborhood into a mural village that attracts visitors, while media icons and social media make it a heavily visited destination. This chapter brings in the phenomenon of 'over tourism' that has afflicted several European cities in recent years. Residents of Ihwa-dong, tired of the high volume of tourists and disrespectful behaviors, push back and deface some of the murals. The conflict brings about a new level of communication and engagement among residents and with the project's lead artist who now has a studio in the neighborhood.

Another grand scheme to remake a working-class neighborhood also ignored grassroots involvement from its initiation. In his chapter, '*Free State Boulevard* and the story of the East 9th Street Placekeepers,' Dave Lowenstein recounts steps by a local art center, the city government, and national funders in this small city of Lawrence, Kansas, a liberal university town in a conservative state. While there are similarities to Ihwa-dong, the shoe was on the other foot in Lawrence. Lowenstein, an artist living in the neighborhood, joined with other community members to resist the effort to 'revitalize' the community through art. The art center thought its affiliation with the city gave it permission to remake the image and physical characteristics of the neighborhood without community involvement. Neighbors organized, and with favorable political changes in city government, the project was halted. Later, efforts were reimagined and relaunched with involvement of neighborhood leaders to move forward using a different approach.

As largely a rural area, Kansas is one of the states in the US identified in contemporary political terms as a 'red' state – in other words a more conservative state. The division between conservative and progressive states, known as 'blue' states, is explored by Lyndsey Ogle in her chapter, 'Public transformation: affect and mobility in rural America.' She follows the work of rural Minnesota-based artist, Ashley Hanson, and her tour of rural communities across the United States to collect and document stories that connect people spanning this so-called red–blue divide. Hanson attempts to disrupt the one-dimensional narrative of rural communities and to see them as hybrid, mobile, and affectively linked both with each other and with urban areas.

Finally, and as a hopeful nod to the future, Sean Peacock, Aare Puussaar, and Clara Crivellaro describe their work with school children in Newcastle, UK, who re-envision, inform, and reform placemaking. Their chapter, 'Sensing our streets: involving children in making people-centred smart cities,' examines ways young people see place both through their own eyes and through technology, using sensing devices. While conflicts in this case are not open or result in radical changes, both the young people and the technology push the boundaries of placemaking practice and bring in new ways of understanding place.

Collectively these authors address how conflicting ideas of place and ways of approaching placemaking play out in a variety of contexts. Their stories capture relatively brief moments in the time of those places. Resilience and resolution will come while new areas of misunderstanding inevitably emerge. Similarly, people in my Minneapolis neighbourhood will pick up the pieces, build on their relationships that are now even stronger, and make the community anew.

References

McCoy, K., Tuck, E. and McKenzie, M. (2016) *Land Education: Rethinking Pedagogies of Place from Indigenous, Postcolonial, and Decolonizing Perspectives*. London: Routledge.
Paperson, L. (2014) 'A ghetto land pedagogy: An antidote for settler environmentalism' in *Environmental Education Research*, 20(1), pp. 115–130.
Porter, L. (2016) *Unlearning the Colonial Cultures of Planning*. Ashgate Publishing, Surrey, UK: Ashgate.
Smith, L. (1999) *Decolonizing Methodologies*. London: Zed Books, Ltd.
Wolfe, P. (2006) 'Settler colonialism and the elimination of the native', *Journal of Genocide Research*, 8(4), pp. 387–409.

Further reading in this volume

Preface: The radical potential of placemaking
Cara Courage
Chapter 1: Introduction: what really matters – moving placemaking into a new epoch
Cara Courage
Preface: Placemaking in the age of COVID-19 and protest
Jason Schupbach
Chapter 2: Placemaking as an economic engine for all
James F. Lima and Andrew J. Jones
Chapter 3: An annotated history of creative placemaking at the federal level
Jen Hughes
Chapter 11: Free State Boulevard and the story of the East 9th Street Placekeepers
Dave Lowenstein
Preface: The problem with placemaking
Louise Platt
Chapter 12: Public transformation: affect and mobility in rural America
Lyndsey Ogle
Chapter 14: Experts in their own tomorrows: placemaking for participatory climate futures
Paul Graham Raven
Chapter 17: 'I am not a satnav': Affective placemaking and conflict in 'the ginnel that roared'
Morag Rose
Chapter 23: Routing out place identity through the vernacular production practices of a community light festival
Gail Skelly and Tim Edensor
Chapter 25: 'If you can make it there, you can make it anywhere…': cultural placemaking at the heart of cities
Sherry Dobbin
Chapter 26: Sculpturing sound in space: on *The Circle and the Square* (2016) by Suzanne Lacy
Trude Schjelderup Iversen
Chapter 27: Is 'tactical urbanism' an alternative to neoliberal urbanism?: reflections on an exhibition at the MoMA
Neil Brenner
Chapter 35: Planning governance – lessons for the integration of placemaking
Nigel Smith
Chapter 41: Rituals of regard: on festivals, folks, and findings of social impact
Maribel Alvarez
Chapter 42: Creative Placemaking and Placekeeping evaluation challenges from the practitioner perspective: an interview with Roy Chan
Maria Rosario Jackson

7
CONFLICT AND MEMORY
Human rights and placemaking in the city of Gwangju

Shin Gyonggu

Places become meaningful because they hold memories and generate new ones. Shared or collective memories, especially of tumultuous events, leave a lasting and powerful impact on both people and places. Few places hold as much collective meaning for the people of South Korea as key sites of conflict during the 1980s in what is known as the May 18 Democratization Movement – pivotal events that led to the current democracy of South Korea and the democratic way of containment of the COVID-19 pandemic. This chapter tells the story of this conflict and how its sites of memory have become important for the people of Gwangju, the entire country, and the world. In the placemaking that has emerged since those violent days the identity of the city has aligned with human rights foremost, and with culture second. Both have been honored as the city builds on those important values and recognizes them in the physical landscape and in many activities that both commemorate and move the city and the nation forward. Some opposition to progressive democracy remains while the placemaking work of the past four decades continues. Yet, as I will point out, it is incomplete.

The May 18 Democratization Movement

In the modern history of South Korean democracy, there were many tragic incidents of bloodshed. In 1948 under the US military government, nearly 30,000, or one-tenth of the population of Jeju Island, were killed. In the Korean war from 1950 to 1953 about 2.5 million of the 30 million population were annihilated. In April 1960, 186 citizens were killed while protesting against the authoritarian government. In May 1961, the democratically elected one-year-old government was toppled by General Park Jeonghui. He continued to rule the country with absolute power, ruthlessly oppressing democracy movements till 26 October 1979, when he was assassinated by his KCIA (National Intelligence Service) chief. This was followed by the coup of the so-called new military group and the declaration of martial law on 12 December the same year. In the absence of the strong man that Park was, demonstrations continued demanding democracy until 17 May 1980, when the military regime expanded martial law to the whole nation and stationed paratroopers in all major university campuses including Jeonnam (or, Chonnam) National University in the southern city of Gwangju. On 18 May, university students confronted paratroopers who were preventing them from getting onto the campus, while the whole nation was silent. This confrontation led to the so-called Gwangju Riot, which continued for nine days till 27 May, resulting in the death of 165 citizens, 84 confirmed missing, 376 deaths from injuries, and more than 360 people missing, though this number was not confirmed officially. The development of the situation before and after the May 18 Movement is well elaborated in the books of *Kwangju Diary: Beyond Death, Beyond the Darkness of the Age* (Lee, 1999) and *South Korean Democracy: Legacy of the Gwangju Uprising* (Katsiaficas and Na, 2006).

Character of Gwangju City

The May 18 Movement was the beginning of the long process of the democratization of South Korea and the transformation of the heart of the city of Gwangju as it focused on important sites of memory. This occurred not only because of the size and importance of the sacrifice but also other factors including the sociopolitical character of the city and its commitment to human rights. Historically the city, along with the surrounding region, has been well-known for its spirit of resistance to invasions from outside Korea. The 1.5 million population of Gwangju and 3.5 million people of the surrounding province have usually voted in favor of candidates from liberal parties in free elections since the country's independence. The people naturally supported the pro-democracy movement during and after the May 18 uprising. Without the unconditional support of the citizens, such a large-scale resistance to the military oppression would not have been possible.

Citizen army leaders' sacrifice

The leading group of the May 18 Movement, with about 150 citizens, stayed at the Provincial Hall on the last day of the uprising despite the ultimatum of the paratroopers. Consequently 17 young people were killed. The power of the legend of the uprising would have been much weaker if they had surrendered or vacated that last stronghold. Bradley Martin of the *Baltimore Sun*, the last foreign reporter who stayed in Gwangju, interviewed the citizen army spokesperson Yun Sang-won. Yun said to the reporter: 'I am staying here not to kill but to put my blood in the hand of the dictators to be accountable.' The reporter later recollected: 'I was struck by the look in his eyes – he seemed clearly aware that his own death was imminent, yet he never lost his gentle quality and kindness' (Martin, 2016). Their blood tainted the military regime as a group of criminals who killed their own people, igniting continuous resistance from inside and outside Korea, resulting in constitutional change in 1987 and the prosecution of the generals in 1997.

Solidarity: domestic and international

International support was also critical, without which the military government would have silenced the opposition with uttermost violence. National networks of students, laborers, activists, and religious leaders were also crucial. No other uprisings have ever been given a higher level of attention than the May 18 Democratization Movement. In this regard, Gwangju citizens are also indebted to many people outside Gwangju. The sacrifice of young activists and the attention of the people in and outside the city created and spread numerous legends, and the May 18 Memorial area came to be a place of historical meaning, overcoming the oppression of the military rulers and the distortion of the truth by the conservative governments.

The presence of the media, especially the international media, amplified the impact inside and outside of Korea, with the tragic scenes of massacre. The military regime continued to condemn the movement as a riot, but the younger generation and activist groups shared information through widely available copying machines and personal communication. Lee Jae-eui, one of the survivors of the uprising, wrote a book comprehensively describing the uprising based on his personal experiences and those of other survivors, discrediting the distortion of the military government in 1985. The Justice and Peace Commission of the Catholic Church also published photo records of the massacre in 1985. People outside Gwangju felt indebted to the victims of Gwangju, and their dedication kept the spirit of resistance alive despite the harsh military oppression. This was the first movement to initiate a large-scale grass-roots resistance to the authoritarian military government in South Korea.

A student movement for democracy through culture

After the May 18 Movement, most activist students carried out conscientization training: the process of developing a critical awareness of the social reality through reflection and action on and off of campus. And their action achieved political change, getting rid of the dictatorship supported by US imperialism. To do so, they engaged in activities of rediscovering traditional culture such as native vocabulary, clothes, and traditional farmers' dance and music. They created their own people's songs in place of the protest songs of civil rights movements of the 1960s and 1970s in the US. The period of the 1980s and 1990s in South Korea is like the hippies and the anti-war movement period of the US in both political and cultural aspects. The students made the most of the power of music and dance to consolidate their will to fight for democracy and justice. This cultural conscientization can be considered the beginning of the Korean cultural wave, which began to bloom in the mid-1990s and to spread outside Korea from 2000 till now. Young producers of cultural products wanted to share their dreams of democracy, justice, and humanism through their movies, dramas, and books based on their reflections and actions in student life. Their products began to be consumed by the people who were also exposed to the social issues directly and indirectly through the smell of tear gas and scenes of atrocities perpetrated by the military and authoritarian government. The growth of Korean culture coincided with the development of its economy and democracy.

All things combined, Gwangju is given credit as one of the most important movements for the current democracy and culture of South Korea. The movement was initially named Gwangju Riot in May 1980 by the military regime, but it was quickly renamed Gwangju Incident even by the military government. In 1988 the National Parliament officially renamed it the May 18 Democratization Movement. Since 1990, a series of special acts by the National Parliament restored the honor of the victims of the Gwangju uprising to the level of the soldiers who have died in battles, and the Ministry of Patriots and Veterans Affairs now manages the May 18 National Cemetery.

The deaths during the May 18 Movement were the indelible crime and the weakest point of the conservative party. On the other hand, it is the most important movement galvanizing the support of the public for the progressive government. That is why ultra-right-wing groups continue to defame it with false allegations asserting that 600 North Korean soldiers carried out the killing in 1980. The positive aspect of such distortion seems to be the increasing attention of the public to the May 18 Movement. The current progressive Korean government launched the May 18 Investigation Commission in December 2019. In addition, some parliament members are preparing a law prohibiting the groundless defamation of the May 18 Movement. Throughout this historical development and confrontation, Gwangju itself has spontaneously become a place of significance along with its many places of important meaning.

Places of struggle: 1980–1998

Many places have spontaneously evolved to be historically meaningful during the struggle for democracy in Gwangju since May 18, 1980. They are the record of blood and devotion of activists and angry citizens against the military regime, providing inspiration to visitors and the upcoming generations. The main gate of Jeonnam National University was the starting point of the uprising. The university campus continued to be a place of conflict between students and riot police throughout the 1980s. However, the Provincial Hall became the symbol of the resistance by the citizen army who stayed there until it was retaken by the soldiers. Geumnam-ro Street lies to the front of the Provincial Hall, and it naturally continued to be the major battleground between the demonstrators and the paratroopers during the May 18 Uprising and with the riot police through much of the 1980s. The street also includes other meaningful places such as the YWCA, YMCA, Catholic Center, and Jeon-il Building.

As the starting point of the May 18 Uprising, though many people feel sorry about the condition of the university gate, it was one of the symbols at the time of the conflict, changed through renovation. However, there are many other places of memory on the campus. The May 18 mural painted by students at the tenth commemoration of the May 18 Movement in 1990 is one of them. Covering one side of the four-story building, with a width of 10m and a height of 16m, it describes the citizen army armed with guns in a military jeep and women cooking for the demonstrators. It continues to keep students and visitors exposed to the essence of the spirit of the movement. This mural survived due to the failure of the university administration in procuring replacement funding from the Ministry of Education several times.

The old main building of the 1980s has been transformed into a memorial museum, which describes the May 18 Movement and the history of the democratization of South Korea. It also houses the May 18 Research Institute, which was established in 1996 to carry out academic research on the democratization movement. The Institute organizes an annual conference and publishes the *Journal of Democracy and Human Rights* on the themes of democracy, human rights, peace, and the May 18 democratization movement. The Institute has an education function with a non-governmental organization (NGO) course, which began in 2002, operating master's and doctoral degree programs. There are two more memorial halls in the campus: one in memory of the leader of the citizen army Yun Sang-won, who refused to surrender and was killed at the dawn of the May 27 in 1980; another in memory of Kim Namju, one of the most widely known radical anti-government poets in late-twentieth-century Korea, who considered himself a fighter rather than a poet.

Connecting Provincial Hall, Democracy Plaza, the fountain, and Geumnam-ro Street

At 2pm on May 21, 1980 there was a mass shooting of the demonstrators. This ignited a citywide uprising. Citizens began to arm themselves with rifles from the military reserve corps weapon storages. Martial law army forces retreated along with the police forces. Citizens continued to have mass rallies around the fountain in front of the Provincial Hall demanding democracy and an apology for the atrocities committed by the martial law army forces. On the morning of 27 May, around 150 citizens stayed, disregarding the warning of the martial law army to vacate the hall. On 26 May, they advised young students and women to go back home. Students did not share their identity with each other in fear of spies planted by the Defense Security Command. They confronted the soldiers to keep their honor and the spirit of the movement in the full understanding that they could not kill the soldiers who were fully armed with advanced armaments. Officially 17 citizens were killed including the citizen army commander Mr Yun Sang-won, who is now one of the most well-known legends of the Uprising. The survivors were imprisoned and tortured. Without the sacrifice of the brave young people, the May 18 Movement would not be commemorated as it is now. In addition, there were no crimes such as shoplifting or bank break-ins. There was no panic buying during the movement, either.

In 2005, at the site of the former Provincial Hall, the central government began to build the Asia Culture Center to reward Gwangju for its sacrifice. It was completed in 2015. In front of the Asia Culture Center is the Democracy Plaza created around the fountain. The main street of Gwangju, Geumnam-ro stretches to the north. It was the battle ground of the demonstrators against the paratroopers for nine days in May 1980. The violent confrontation of the citizens continued against the riot police throughout the 1980s. The street was full of tear gas fumes. Demonstrators shouted slogans such as 'Down with the dictators,' 'Bring back democracy,' and 'Recover the honor of the May 18 victims.' Many young people were arrested, imprisoned, tortured, and even enlisted into the army. The street was a place of anger and violence. With democracy growing, however, it has become a place of commemoration and celebration. An international visitor Taryn Assaf (2013) correctly described the current mood of the street.

A commemoration continues every year on May 18 to honor the uprising and its historical place in the lineage of Korean democratization. A variety of festivals and cultural events fill the historic Geunnam-ro – the street upon which the battles between citizens and soldiers were fought – for the purpose of educating people about the uprising under the theme of remembrance. A variety of stalls line the street, some promoting various political causes and charities, some offering the opportunity to make Gwangju-themed wood-block art, and others dressed in sketches of political satire. Cheers and songs can be heard in the distance from groups rallying for workers' rights; indeed, the feeling is light and jovial. Smiles can be seen on the faces of most – and the energy is emblematic of a spirit that was first ignited in May of 1980.

Other key buildings restored

The Jeon-il Building and the Catholic Center are the only buildings that have been restored with their original appearance intact. Most other buildings, including the Provincial Hall and the university main gate, have been drastically transformed or have disappeared, losing their original shapes. Professional architects originally diagnosed that the Jeon-il building was too old to be kept and did not have aesthetic value worthy of renovation. The city government, which bought the building, originally decided to tear it down to rebuild. However, most of the citizens wanted to save it as part of the May 18 heritage. In addition, 245 bullet holes fired from helicopters were discovered in 2017. The building has been renovated to remain as one of the significant historic buildings to witness the atrocities of the military groups on their own people.

Both the Gwangju YMCA and the YWCA are two of the most important buildings as they helped to develop the modern civil society movement in Gwangju. The Gwangju YMCA building was the place where protest leaders frequently held indoor rallies. This is also where the citizen army had firearms training. Throughout the 1980s, numerous anti-military government gatherings were held here. In the 1980, the Gwangju YWCA building behind the Jeon-il Building was another important place for the activists. Here they produced the *Fighters Newsletter*. There were frequent meetings of the Citizens' Settlement Committee to mediate between the demonstrators and the martial law army. It was also a stronghold of the citizen army and one of the targets of the martial law army at the dawn of 27 May. However, the Gwangju YWCA moved from the historical building to a bigger space with the financial support of the West German government. The new building lost historical significance, but it housed many civil society organizations throughout the 1980s. There are civic groups requesting the restoration of the original site of the YWCA.

Establishing the May 18 National Cemetery

The May 18 National Cemetery is one of the three most important symbols of the Gwangju Uprising along with the Provincial Hall and the Main Gate of the Jeonnam National University. The old cemetery was commonly called 'Mangwol-dong Cemetery.' In 1930, families and relatives carried the bodies of the victims in carts and trucks and buried them here without any formal ceremonies. Since 1980, many students and laborers from all over the country have visited this place to pay tribute to the victims and to strengthen the spirit of resistance to the then military dictators. This site also became a must-see site for international reporters when visiting Gwangju. In order to remove the evidence of their crimes, the military government paid compensation to the bereaved families to move the graves, but most families refused to take the money. At the request of the Gwangju citizen groups, the government began a new cemetery project in 1994, and finished the May 18 Cemetery in 1997 with 120 tombs transferred from the old cemetery. Now there are 680 people buried here, including those who died of injuries afterwards and the 82 missing. There are 160 more applications that failed to be acknowledged as missing by the government. The families complain that the government criteria are too strict since they cannot find evidence of missing victims because the military buried many bodies without any trace. The cemetery was promoted as the National Cemetery in 2002. Since 2002, the May 18 commemoration ceremony has begun to be hosted by the President or the Prime Minister. The cemetery has an exhibition hall, a shrine with photos of the victims, and a small theater showing a video of the May 18 Uprising, and an educational experience hall, which serves as an educational venue for young visitors. The new national cemetery is also exerting its own influence by officially commemorating the legacies of the people who took part in the May 18 Movement. The two cemeteries play complementary roles as places of memory and of inspiration. A total of 605,900 people, including 8,517 foreigners, visited the cemetery in 2019.

Places of commemoration and promotion: 1998 to present

While the level of democratization has become mature, some citizen groups of Gwangju began to expand the issue of democratization to promote Gwangju as a human rights city. On the other hand, some citizen groups tried to soften the image of the city from a radical city to a city of culture since it also has a strong tradition in art and culture. The City Hall started the Gwangju Art Biennale in 1995 and the Gwangju Design Biennale in 2005. It also established the Gwangju Design Center and Gwangju Culture Foundation in 2010. President Roh (2003–08) started the construction of the Asia Culture Center (ACC) in 2005 and the renovation of the former Provincial Hall. The ACC opened in 2015 with an investment equivalent to nearly 700 million US dollars. The central government is still investing nearly 100 million dollars in the ACC each year. In 2006, the Ministry of Culture also started an annual conference to promote the ACC and Gwangju as a city of culture: Asia Culture Forum (ACF). The government designated Gwangju as the first East Asia Culture City along with Quanzhou in China and Yokohama in Japan. Gwangju also joined the UNESCO Creative City Network in Media in 2014. In spite of the lower investment in democracy and human rights than in culture, citizen groups were successful in creating meaningful places to promote the spirit of Gwangju, such as the May 18

Memorial Foundation and the May 18 Archives. The presence of these two institutes continues to make possible the activities and projects to promote the spirit of the May 18 Movement.

Establishing a human rights identity

The May 18 Memorial Foundation is an interesting example of a human rights institution created mainly by the May 18 Movement participants in 1994 with the aim of discovering the truth about the massacre, bringing about the prosecution of the military dictators, and recovering the honor of the victims. The energy of the movement would have been much weaker without this institution. It is now active in promoting the spirit of the May 18 Movement in diverse ways including education, events, and volunteer activities in and outside Korea. It is developing May 18 Movement teaching materials for students in collaboration with teachers. It is also fighting against the ultra-right-wing groups who are distorting the facts of the May 18 Movement. The Foundation managed to get a leader of such a group sentenced to a two-year prison term in February 2020. There are international projects such as the Gwangju Prize for Human Rights from 2000 and the Gwangju Asia Forum from 1999, both of which are well-known among human rights activists in Asia. The Global NGO Master's Program (GNMP) trains future human rights professionals in collaboration with Jeonnam National University. It has continued to widen the scope of influence of the May 18 Movement in the UN as well as been shown at the UN conference in New York in 2017 and the UNOHCHR conference in Geneva in 2019. The Foundation organized a virtual international commemoration of the May 18 Movement in 2020, replacing the physical event. The contribution of the Foundation has been made possible by government funding and with the support of citizens. It is also a good example showing the importance of an institutional approach to commemorating and promoting any movement. The foundation has developed a wide-ranging network with Asian human rights activists through the Gwangju Asia Forum, which started in 1999. It invites around 300 participants from more than 20 countries to wrestle with diverse human rights issues. Since 1999, it has also supported human rights activists with the Gwangju Prize for Human Rights. Some of the past winners of the award include Xanana Gusmão, the first president of the East Timor, and Basil Fernando, the head of the Asian Human Rights Commission, among many others. Indonesia's Bedjo Untung, founder of the 1965 Murder Victims Research Foundation, is the recipient of Gwangju Prize for Human Rights, with prize money of $50,000, in 2020.

May 18 Democratization Movement Archives

The May 18 Archives is another important institution of the movement which was initiated in 2010 by a group of people including a long-time researcher on the May 18 Movement This initiative was implemented with the funding from City Hall, which also purchased the former Catholic Center for more than $10 million in order to establish the May 18 Archives. The Archives is listed in the UNESCO World Heritage list of human rights documents. One of the obstacles to approval by the UNESCO was the strong opposition of right-wing groups, who staged a demonstration at the UNESCO headquarters in Paris. In spite of this objection, UNESCO approved the May 18 Democratization Movement Archives being listed in the World Human Rights Documentary Heritage in May 2011. The approval boosted not only the value of the May 18 Movement but also the morale of Gwangju citizens against groundless attacks by the ultra-right wing who want to undermine the legitimacy of the progressive party, since the massacre committed by the military junta in the May 18 Movement criminalizes the past military government and the right-wing parties. In this regard, the World Heritage listing of the

May 18 documents is one of the most important steps in handling the challenges from the right-wing groups. From 2012, most of the Korean high-school history textbooks began to include the fact that the May 18 Movement Archives are listed as part of UNESCO world heritage.

2011 World Human Rights Cities Forum

The City of Gwangju started the annual World Human Rights Cities Forum (WHRCF) in 2011. It is not a physical space but an annual event. However, it has been functioning as an effective space for dialogue among local and international participants. Now it is becoming an important space of cooperation to strengthen ties among human rights cities and NGO activists around the world. It has also contributed to exposing the city of Gwangju globally. In October 2020, the forum is being organized with the involvement of 40 organizations – international, national, and local –such as UNESCO, UN OHCHR, UCLG CISDP, Raoul Wallenberg Institute, Korean Ministry of Education, and Korean National Human Rights Commission, among others. In 2019, it attracted 2,000 Koreans and 250 international participants from 131 cities in 48 countries. Gwangju civil society activists are also benefitting from this event by expanding the scope of their understanding on human rights and related issues while widening their international network. It is a training venue for local people regarding the localization of world issues such as the Sustainable Development Goals and the New Urban Agenda. The Forum has been an effective device for the promotion of human rights cities globally with the Korean government at the center of influencing the UN Human Rights Council to adopt resolutions on local government and human rights. The attraction of the issue of human rights in the context of Gwangju is significant if the WHRCF is compared with the ACF, which started five years before WHRCF. ACF was organized by only seven local organizations without the presence of international or regional organizations such as UNESCO and OHCHR. The ACF participation is also limited in numbers and areas. It had 250 local participants in 2019, while WHRCF has 250 international participants. More than half of the participants came to Gwangju at their own expense. Without the historical legacy of Gwangju, such strong international attention would not be possible.

Concluding remarks

The May 18 Democracy Movement and the dedication of human rights activists have transformed Gwangju into a human rights-based city with numerous public and cultural spaces with rich historical meaning. They are the product of the tears and blood of the activists and victims of the uprising with the support and participation of common citizens. Citizens also managed to restore the Provincial Hall and save the Jeon-il building from destruction. Citizen initiatives supported by both national and local governments have created additional spaces for the promotion of the May 18 spirit such as the May 18 Archives and the May 18 National Cemetery. Here in Gwangju, the younger generation visits these spaces with their friends, parents, or teachers and grows by being exposed to the important value of democracy, human rights, and culture. The spirit of the May 18 Movement can be further promoted for the future of democracy, human rights, and culture by more professional management, including more spaces for people to gather for both formal and informal meetings and conversation.

References

Assaf, T. (2013). 'A sign of struggle: The Gwangju uprising and the politics of symbolism' in Korea Bridge [online]. Available at: http://koreabridge.net/post/sign-struggle-gwangju-uprising-and-politics-symbolism-isc-korea (Access: 2 June 2020).

Katsiaficas, G. and Na, K. (eds.) (2006). *South Korean Democracy: Legacy of the Gwangju Uprising*. London: Routledge.
Lee, Jae-eui. (1999). *Kwangju Diary: Beyond Death, Beyond the Darkness of the Age*. University of California. Los Angeles. Second edition in English translated by SEOL Kap-Su and Nick Mamatas.
Martin, B. (2016). 'A special meeting between Gwangju spokesman's father and a reporter', in *Hankyoreh Shinmun* [online]. Available at: http://english.hani.co.kr/arti/english_edition/e_international/745553.html (Accessed: 1 June 2020).

Further reading in this volume

Chapter 4: A future of creative placemaking
Sarah Calderon and Erik Takeshita
Chapter 5: Making places for survival: looking to a creative placemaking past for a guide to the future
Jeremy Liu
Chapter 9: From the dust of bad stars: disaster, resilience, and placemaking in Little Tokyo
Jonathan Jae-an Crisman
Chapter 16: More than a mural: participatory placemaking on Gija Country
Samantha Edwards-Vandenhoek
Chapter 17: 'I am not a satnav': Affective placemaking and conflict in 'the ginnel that roared'
Morag Rose
Chapter 18: '*Homomonument* sounds like a poem': queer placemaking 30 years on: a conversational dialogue with Thijs Bartels, author of *Dancing on the Homomonument* (2003)
Martin Zebracki
Chapter 22: Embedded Artist Project: Epistemic Disobedience + Place
Frances Whitehead
Chapter 31: Seven generations: a role for artists in Zuni PlaceKnowing
Theodore S. Jojola and Michaela P Shirley
Chapter 32: *The Hollywood Forest Story*: Placemaking for the Symbiocene
Cathy Fitzgerald
Chapter 42: Creative Placemaking and Placekeeping evaluation challenges from the practitioner perspective: an interview with Roy Chan
Maria Rosario Jackson
Chapter 45: How the city speaks to us and how we speak back: rewriting the relationship between people and place
Rosanna Vitiello and Marcus Willcocks

8

QUEER PLACEMAKING, SETTLER COLONIAL TIME, AND THE DESERT IMAGINARY IN PALM SPRINGS

Xander Lenc

The first structure most tourists notice as they drive East into the California desert city of Palm Springs is a cross-stich of I-beams forming an enormous triangular canopy looming towards the road. Though only 55 years old, this former Tramway Gas Station is listed under the National Register of Historic Places and serves as the city's visitor center. Inside, three themes emerge. Half of the room is filled with historical pictures of 1950s buildings and Hollywood celebrity entourages, architectural history books, and other midcentury memorabilia. A splash of rainbows sits adjacent to this section, with colorful souvenirs offered with LGBTQ+ tourists in mind; I pick up a free copy of *The Gay Yellow Pages for Palm Springs and Desert Cities* with nearly 200 pages. Finally, in a separate booth across the entryway, a representative of the Agua Caliente Indian Reservation is prepared with pamphlets describing outdoor tourism opportunities on tribal land. If tourist development operates through a neoliberal planned placemaking agenda that draws from material and imaginary features of a space (Lew, 2017), what is the relationship between the three aspects of the area's character on display at the gateway to the city?

In the extensive literature on architecture, non-normative sexuality and gender, and settler-colonialism, researchers tend to address only two of these three at a time. For example, Henry Urbach (1996) has written of the importance of closets as an architectural feature of (rather than a simple offhand metaphor for) homosexuality; Scott Lauria Morgensen (2010) has drawn on Jasbir Puar's work to describe settler homonationalism, in which queer movements 'naturalize settlement'; and Janet McGay, Anoma Pieris, and Emily Potter (2011) have argued that architecture helps navigate Australian settler placemaking. This chapter uses the case study of the built urban landscape, queerness, and settler ideology in Palm Springs to explore how time functions in American placemaking. The city is widely celebrated for its high concentration of modern (and especially *midcentury* modern, hereafter 'MCM') homes, stores, and government buildings, many of which sit on federally recognized tribal land. While several excellent monographs have been written about the ongoing colonial project of building Palm Springs (e.g. Przeklasa, 2011; Kray, 2009) and the unique role played by gay men in revitalizing interest in 'desert modernism' (LoCascio, 2013), these two stories are rarely told together. I argue that attending to the *uses* of different temporal registers helps integrate different historiographies of space and real estate in

California's Colorado Desert. The first part of this chapter argues that queer colonial eyes have interpreted desert landscapes as objects of temporal disjuncture and liberatory possibility. The second part introduces two different historiographic approaches to Palm Springs history and contends that they rely on different temporal modes of memorialization. I conclude by arguing that the desert's temporality affects *both* narrativizations, providing the background noise to placemaking in the city.

Ruins, settler-colonialism, and the erotics of the desert

In *The Necessity for Ruins,* landscape theorist J.B. Jackson (1980) describes a shift in American memorialization practices in the mid-twentieth century. Prior to this transformation, history is treated as a sequence of explicitly political events, and the present social order is the continuation or reenactment of a foundational covenant between people and their leaders: a constitution, a treaty, etc. Today (especially in the American West), the covenant model has largely been replaced by an *evolutionary* one that treats the past as a remote spacetime, one that is unstructured, depoliticized, and most of all, *gone*. These two temporal modes correspond with two different political-religious attitudes, which produce distinct practices of memorialization. The earlier form honors canonical and sacrosanct figures and produces monumental statues or parades that honor specific features of the covenant (names, dates, etc.). By contrast, the newer evolutionary approach honors an amorphous, romanticized, almost prehistorical golden age through vernacular reenactments that restore as much of the 'original' landscape as possible: 'There is no lesson to learn, no covenant to honor; we are charmed into a state of innocence and become part of the environment. History ceases to exist' (p. 102). Unlike the earlier 'Latin' model of history, the narrative structure of evolutionary memorialization relies on a period of ruination for stability: 'Ruins provide an incentive for restoration, and for a return to origins' (ibid.).

Jackson doesn't theorize the role these landscape temporalities play in narrativizing American settler-colonialism, but his description of evolutionary historical registers is helpful for understanding the role of the landscape in understanding the doxic background noise of settler placemaking; in other words, its *common sense*. Two closely related notions of common sense have emerged in Native American Studies in recent years. Anti-Indian common sense, as described by Nick Estes (2019), is a modification of Antonio Gramsci's theory of hegemony; unlike Gramscian common sense, it does not rely on the internalization of common sense among the governed/colonized to operate, and 'originates from the continued assertion of patriarchal White sovereignty over Native lands and lives' (p. 50). Estes urges readers to disrupt anti-Indian common sense by reframing cities, towns, and urban spaces as colonial settlements and border towns (p. 50). Similarly, Mark Rifkin's theorization of settler common sense signifies 'the ways in which the legal and political structures that enable non-native access to Indigenous territories come to be lived as given, as simply the unmarked, generic conditions of possibility for occupancy, association, history, and personhood' (Rifkin, 2014, p. xvi). This is a common theme in Native American Studies, but what distinguishes Rifkin's approach is his insistence on wedding queer theories of time (especially in Elizabeth Freeman's (2010) notions of *chrononormativity* and *temporal drag*) to the question of landedness and settler-colonialism. He argues that the nineteenth-century American writers he studies (Hawthorne, Thoreau, and Melville) deployed a queer critique of the 'nuclear normality' and state violence of 'the politics of placemaking,' but that these 'queerings take shape' through an unexamined conceptual architecture that entrenches settler sovereignty: 'The impression of anachronism that surrounds *Indianness*, then, helps orient and provide momentum for the feeling of givenness that marks nonnative's relation to place' (Rifkin, 2014, p. 31; emphasis in original).

Jackson's schema is *inadvertently* helpful for thinking about the different settler colonial narrativizations of the deserts of the American and Mediterranean worlds, and especially how the denigration of landscapes and their indigenous inhabitants that Tracy Voyles (2015) has termed 'wastelanding' occurs in the desert. Kuletz (2016) is correct in diagnosing European and US American writings on the desert as 'wasteland discourse' that dismisses arid lands as marginal, lifeless, and unproductive (pp. 3–4), but this has not always been the case. Though ancient Europeans often saw deserts as dangerous, they were not considered ruined, abnormal, or defective landscapes (Davis, 2016, p. 35). This was also true of many early Christians, including the third-century 'Desert Fathers' who traveled to the Scetis Desert (Wadi El Natrun) to find ascetic (and, some argue, homoerotic) religious experiences (Schroeder, 2009). This began to shift after the rise of Islam; during the Crusades, Christians often blamed 'destructive Arab invasions' for desertification (Davis, 2016, p. 42). In the modern period, Diana K. Davis (2016) has shown how the 'dessicationist' narrative of the desert as a once-fertile space that has grown lifeless due to misuse has helped articulate and mobilize French colonial intervention in North Africa (pp. 81–116). This North African setting was a site of counterhegemonic (but nonetheless colonial) queer life for some French men. Thus, we see scenes such as André Gide's moment of queer genesis with an Arab boy on the sand dunes of Tunisia on a journey to meet Oscar Wilde in Algiers (Dollimore, 1987). Historian Robert Aldrich writes of François Augiéras, a French mentee of Gide who traveled through Algeria:

> Only in a venue such as North Africa, a romanticized and mythologized North Africa, could a man find true happiness... Augiéras expresses ... the sense of being receptive to the beauties of the region, the joys of solitary contemplation, the excitement of market-places, cafes, and brothels, and pleasures of sex. All come together in physical and poetic enjoyment... in the 'Oriental' desert.
>
> *(Aldrich, 2008, p. 212)*

Here the queer colonial gaze involves the visual consumption of both desert landscapes *and* colonized male bodies. Arab culture is linked to the past, a fossilized space shielded from modernity but made accessible to the colonial citizen, who may pass to and from the dominated periphery and sample its products and subjects at will. This mythologized North Africa is utopian to Augiéras not because it represents the future, but because it represents the past. 'Old' World deserts were seen as inherently peopled and full of history; their existence is explained in terms of human activity. For an emerging White *queer* common sense, arid colonies were exotic spaces where the flesh of the colonized and the desert landscape itself are made available through empire. For an emerging *imperial* common sense, deserts are ruins, and ruins have a history; perhaps one that demands restoration to an Edenic past, even if that requires genocide.

In the 'New' World, settler travelogues have extolled the perceived virtues of Californian deserts since Mary Austin published *The Land of Little Rain* in 1903, but the use of the desert as a symbol of vacuity, timelessness, and meaninglessness persists. For Jean Baudrillard (1988), to take one prominent example, Southwestern deserts are not only diagnostic of American culture, but are also unlike Old World deserts: 'Nothing is more alien to American deserts than symbiosis (loose-fitting clothing, slow rhythms, oases) such as you find in native desert cultures. Here, everything human is artificial' (p. 66). He struggles to recognize extant native desert cultures in America, referring to spaces like Monument Valley as 'the mausoleum of the Indians' and suggesting that 'The extermination of the Indians put an end to the natural cosmological rhythm of these landscapes' (p. 70). In reality, Monument Valley is managed by very-much-alive Indians of the Navajo Nation. The desert's supposed lack of history also makes it a *non-erotic* space: 'I speak of the American deserts and of the cities which are not cities. … Everywhere: Los Angeles or Twenty-Nine [sic] Palms, Las Vegas or Borrego Springs. No desire: the desert' (p. 123).

Despite, or perhaps because of, the de-eroticization of the American desert by heterosexual writers like Baudrillard, arid lands have assumed the role of a space of (largely White) escape and refuge. In fact, the existential flight into the desert is a widespread trope in settler queer media. This is especially true of queer travel narratives, including *The Price of Salt*, *Transamerica*, *Thelma and Louise*, *Desert Hearts*, *The Adventures of Priscilla Queen of the Desert*, and *My Own Private Idaho*. The desert is treated as a space of refuge in emptiness, of queer embrace of the death drive, of the timelessness and negation of history that enables self-becoming in a hostile world. Unlike earlier depictions, these portrayals treat the desert as unpeopled and devoid of colonial history. When indigenous people do appear, as in the dreamlike appearance and disappearance of a group of aboriginal people in *Priscilla*, they are marginal to the narrative, placed in a phantasmagoric relationship to history. More than just a literary or movie trope, this spatial imaginary is integral to American settler queer life: thus we see the 'utopian queers' who gather in Nevada's Black Rock Desert for Burning Man (Hornik, 2019), Foucault's supposedly-pivotal epiphanies during a 1975 LSD trip in California's Death Valley (Wade, 2019), or as I argue here, Palm Springs.

Palm Springs

There are two stories about landscape that are commonly employed by Palms Springs place-makers: one describes a struggle for tribal sovereignty through the nineteenth and twentieth centuries, and the other describes a movement to restore and memorialize architecture. In what follows I will consider their respective temporal mechanics before asking how the figure of the desert might help reconcile their tensions.

A story of tribal restoration

In 1862 the thirty-seventh US Congress handed over large swaths of the Western United States to railway barons in the form of financing, right-of-way assurances, and land grants. In many areas this land was allocated in alternating plots, so within a decade the General Land Office's platting maps of much of the West resembled a checkerboard, with allocated odd-numbered squares braided with unallocated even-numbered squares (Ainsworth, 1965). In the years following the Civil War, the federal government periodically allocated some even-numbered sections to the Iviatim people, more commonly known to settlers as the Cahuilla (probably from the Spanish *kawiya*, or 'master'; Pritzker, 2000, p. 118). The Iviatim had lived in the Sec-he ('boiling water') region for over 5,000 years, with strong ties to the Agua Caliente Hot Spring and a nearby petroglyph-covered canyon inhabited by a powerful and malevolent being named Tahquitz (Hough, 2004). In 1876, President Grant formally recognized the Agua Caliente Band of Cahuilla Indians (now one of many recognized Cahuilla tribes; hereafter 'Agua'), establishing a reservation that included Section 14 of the checkerboard (including the spring) and part of Section 22 (including Tahquitz Canyon; Agua Caliente Band of Cahuilla Indians, 2020). A larger allocation the following year left Agua with tens of thousands of acres of checkerboarded land, but this didn't always translate into political sovereignty. In 1891 Congress ostensibly clarified and secured Agua's title even as it (like the 1886 Dawes Act) expanded federal control over Indian life through suppression of Cahuilla language and religion, formal restrictions on agriculture, and proletarianization of Cahuilla people in the cattle, peat, and asbestos industries (Pritzker, 2000, p. 119).

In the 1920s, the beneficiaries of a burgeoning film industry in Los Angeles nurtured a tourism economy in the young and unincorporated town of Palm Springs, but Agua was hamstrung with federal bureaucracy that managed tribal planning and leasing. The White-controlled Palm Springs Chamber of Commerce had de facto authority over development until the town's incorporation in 1932 (Przeklasa, 2011, pp. 3–4), but even thereafter, city leaders maintained White supremacy by ensuring a White 'guardianship' program over reservation land, a 'negro removal' program targeting Agua's relative willingness to rent to Black tenants, and doggedly opposed proposed federal housing projects that would have served the lower-income Black, Latino, and Native American construction workers who were building modernism in town (Kray, 2009, p. 175). Manuel Shvartzberg Carrió (2019) has shown how the Alexander Construction Company bypassed the 'problem' of a nonwhite proletariat in town by collaborating with US Steel to automate the construction process, and has suggested that the unique role of the modernist architect and modernism's modular possibilities were useful to those who sought to undermine Cahuilla sovereignty. Thanks to tribal leaders and organizations such as Vyola Ortner (Ortner and duPont, 2012) and the Mission Indian Federation (Przeklasa, 2011) Agua's members secured the ability to issue long-term leases in 1959, and White conservatorship was eliminated in 1968 following a state investigation that charged Palm Springs leaders with conducting a 'city-engineered holocaust' (Kray, 2004). Since then, Agua has made major developments on Section 14: they have built a wildly successful casino, and as of writing are building both a minor-league hockey arena (Powers, 2020) and a Cahuilla cultural museum at the site of the original spring. 'It's where our creation story is based,' says tribal chairman Jeff Grubbe. 'It's one of the most important pieces of land we have, so for us to have the ability to start over from scratch and build something that reflects us as a people – our history and culture and traditions and how we got here – is very special to us' (quoted in de Crinis, 2019).

The Tribal Revival narrative of Palm Springs, which is present in public-facing Agua material and a handful of excellent historical monographs, stands out in part because, *contra* Jackson, it relies on *both* of the historical modalities described in *The Necessity of Ruins*. As a 'Latin' view of history, it presents the current social order of Palm Springs as the outcome of a highly-politicized nineteenth-century *legal compact* that must be defended; specific dates, documents, and leaders are named and honored, and the original promises of the compact guide a continuous, if bumpy, progression of history into the present. But as an 'evolutionary' view of history, it identifies creation itself as the origin of the social order, one that has fallen into disrepair but can now return as Agua 'start[s] over from scratch.' They also may imply different political projects; while the former defends what territorial claims were ultimately secured through settler colonial legal institutions and practices such as checkerboarding and the Bureau of Indian Affairs, the latter suggests that those precise institutions and practices are illegitimate impositions of a political-legal system. Meredith Alberta Palmer (2020) has argued that 'settler sovereign landscapes' are constructed not only through racialized dispossession of land, but also through maintaining the hegemony of the concept of land as an object that can be owned. Given that the US Supreme Court has restricted the expansion of Indian sovereignty into areas with 'distinctly non-Indian character' (Palmer, 2020) the evolutionary modality of the Tribal Revival narrative hints at a more radical break from the politics of federal recognition.

Crucially, *both* temporal modalities are issued in response to settler ideology and attending to them helps explain how settler common sense is physically manifested in Palm Springs. For instance, the role of landscape in developing settler common sense becomes clearer when compared to anti-Black racism in Palm Springs. It is easy for a pedestrian in Downtown Palm Springs to enter and exit the reservation several times without ever knowing that they have done so. The same cannot be said of the border of the historically Black Crossley Tract, which was explicitly cordoned off from the neighboring golf course by a row of enormous tamarisk trees from the early 1960s until ongoing protests led to their removal in 2018 (Kennedy, 2018). Though anti-Blackness is a forceful part of Palm Springs placemaking with intimate ties to attacks on indigenous sovereignty, it operates differently on the landscape: anti-Black landscapes were explicit and written through partitioning, while landscapes of anti-Indian common sense were written through erasure and a revanchist denial of the checkerboard.

A story of architectural restoration

The second narrativization of Palm Springs begins during the expansion of the desert tourism industry in the 1920s but is more attentive to architecture than to land. In 1924 construction began on the Oasis Hotel, the first of thousands of modern-style buildings to come in subsequent decades (Weznell, 2014). Southern California's film industry was surprisingly resilient during the depression, and Palm Springs' connection to Hollywood wealth allowed it to grow through the 1930s as other resort towns sputtered (Leet, 2004, p. 45). Where the modernism narrative acknowledges the checkerboard, it carries a different significance. For example, in a book on Richard Neutra's Miller House, Stephen Leet writes that 'The temporary but beneficial consequence of this development pattern was that the land to the east of Miller's property [the reservation] remained a pristine desert landscape, largely developed until the mid-1950s' (p. 62). Neutra insisted that the surrounding desert must be consumed in a comfortable and modern setting. This was couched in unambiguously racialized and temporal terms: 'There is nothing Indian, Spanish, or Mediterranean in the Colorado desert… this new love for nature can not [sic] find its right architectural expression in the imitation of the rusticity of the pioneer or the native' (quoted in Leet, 2004, p. 131).

Modernism wasn't the only architectural style in town – Palm Springs followed the wave of Spanish Colonial Revival architecture in the 20s and 30s (Hart, 2017) – but by the midcentury it became best known for its distinctive style of 'desert' modernism (Chavkin, 2016). These private houses were often second homes designed for pleasure rather than professionalism. As a result, the landscape of Palm Springs lies on the extreme end of modernism's spectrum of whimsy, with flamboyant butterfly roofs and ubiquitous swimming pools and wet bars. In the mid-50s, the father-and-son real estate development team George and Robert Alexander worked with younger architects like William Krisel to mass-produce desert modernism through modular design and assembly-line construction; by the end of the decade, Palm Springs was a major destination for the White middle class (Lagdameo, 2020).

Midcentury Palm Springs was a space of bellicose compulsory heterosexuality, but this was not heterosexuality modeled on the nuclear family. During the 'Rat Pack' era, celebrities such as Frank Sinatra used their svelte Palm Springs properties as getaway homes with mistresses (Goolsby, 2015), and as the town grew it became a major Spring Break destination. The 1963 film *Palm Springs Weekend* depicts the vain and supposedly comic attempts by the Palm Springs Police to control the wild parties, fights, drag races, and sexual liaisons that erupt amongst the visiting college students at the modernist Riviera Hotel and what appears to be an Alexander home. The narrator in a trailer for the film describes Palm Springs as a natural ecology for migrating heterosexual college students:

> In the springtime the swallows fly to Capistrano, bees head for the flowers, and kids here in Southern California head for Palm Springs! It seems to be a kind of primitive, instinctual urge… the boys come for girls, the girls come for boys, and when they come together… man, it's wild!
>
> *(Movieclips Classic Trailers, 2014)*

Spring Break grew increasingly raucous until 1986, when a riot broke downtown and a mob of men began sexually assaulting women in the street. In response, then-mayor Sonny Bono banned thong bikinis, poolside drinking after 11pm, and traffic along Palm Canyon Drive during spring break to curb cruising. The spring breakers left town, and for the most part have not returned since (Kelman, 2018).

Their departure accelerated an ongoing decline in Palm Springs tourism. After the tumult of the late 1960s and early 1970s – during which Palm Springs witnessed the escalation of the Vietnam War and the death of nearly the entire Alexander family in a 1965 plane crash – the goofy optimism of Palm Springs' landscape and alleged lifestyle fell out of style (LoCascio, 2013, pp. 31–32). Many modern buildings were left vacant or remodeled in Spanish Colonial or Neo-Eclectic styles, and the town's occupants and visitors both became more geriatric in character; for 20 years the city was considered (as one journalist wrote) 'a character-free, climate-controlled elephants' graveyard' (Anderson, 1998). (The Modernism Revival narrative's strangest theme is its ageism, a remarkable habit in a city with a median age 18 years higher than the state average.) But the elderly was not the only marginalized group that began to cluster in Palm Springs; for reasons that are not entirely clear, the city became a major destination for lesbians and gay men of means in the 1970s and 1980s. The Dinah Shore Golf Tournament evolved into a lesbian gathering, and the Warm Sands neighborhood south of Downtown grew increasingly popular with gay men who began to gentrify the area by the early 1990s (Conrad, 2018). Many of them joined the ranks of the gay 'keepers of culture' (Fellows, 2005), becoming committed architectural preservationists and devotees of MCMism through activism and preservation-minded homeownership (LoCascio, 2013).

When I speak to colleagues about Palm Springs, they often suggest that these preservationists are attracted to the archeological remnants of a hyperbolically heteronormative era, either as bids for a form of acceptance denied to them in childhood or as defiant gestures of queer reclamation. As I have previously written (Lenc, 2018), I found very little evidence for either of these suggestions during my visits. On the contrary, colleagues chafed at the idea that their homes were 'museums' of the midcentury and denied that nostalgia fueled their attraction to modernism. 'I love the 40s, 50s, and 60s. But I don't want to *live with* them,' said one respondent who lives in a restored MCM home. Instead, what attracted him was the alleged *timelessness* of the era's design. When pressed about what timelessness means, his answer was comparable to that of most others I spoke to:

> If you look at antiques now, they look out of style with today's world... Charles chairs [a midcentury chair model] are like sculpture; nobody would ever find it dated or old... There may be people who have nostalgia, maybe its subconscious and they saw it in a magazine when they were 16 or 18, but let's be honest: gays are just more design-focused than straight guys.

One preservation activist insisted that 'This is not a re-enactment! I hadn't been exposed to modernism as a child, my mother preferred early American furniture.' Time held paradoxical significances for him: he fell in love with MCM antiques and architecture because it seemed so 'new,' but also because they seemed so 'timeless'; he said that 'It would be so fun to go back in time when all these buildings were new, when excitement was high' before saying, 'I have no connection to that period at all.' These turbulent temporal flows and alternating avowals/disavowals of the midcentury – which as everyone noted, was a conservative and homophobic period in Palm Springs and elsewhere – were present in the testimonies of every gay or modern homeowner or preservationist I spoke to.

While the Tribal Restoration Narrative is usually marginal to the Modernism Restoration Narrative, the recent demolition of the Spa Resort Hotel (a celebrated MCM landmark on the reservation; Descant, 2014), to build the aforementioned cultural museum, forced these homeowners and activists to face the reality of Section 14. Their responses shared a common disappointment but varied in their temporal descriptions of tribal land and life. Several argued that the tribe's decision to demolish the hotel was 'shortsighted'; one respondent said that there was 'no future' for MCM homes on tribal land and lamented the temporal ambiguity of the reservation's newest non-modernist structures:

> I don't think they're worried about any consistent style that has anything to do with Palm Springs. They're not interested in the midcentury [...] I don't even know what a Cahuilla motif is. I think they lived in temporary structures. I don't know if there was a style, a Cahuilla style. There's no style that I can see. Adobe maybe? A grass roof?

It's worth mentioning that in addition to the structures, including MCM homes, that Cahuilla people have built since checkerboarding, there is a long tradition of pre-colonial Cahuilla architectural practices that includes dome-shaped shelters constructed with brush, rectangular thatch houses, granaries, sweat houses, and ceremonial lodges (Pritzker, 2000, p. 119).

Others were more diplomatic. One respondent emphasized the ancientness of Agua's ties to the land, reframing the evolutionary component of the Tribal Restoration Narrative in liberal multi-cultural terms:

> Really, this is their land. They have a different vision of what is historic and what should be preserved based on thousands of years of their own culture. I would think that going forward as more of these issues that both sides learn to live together, learn not to mistrust one another. The [Spa Resort] Hotel was on the most sacred spot, it's the heart of the whole region for them.

Unlike the Tribal Restoration narrative, which uses tightly wound strands of both the 'Latin' *and* 'evolutionary' modes of landscape history that Jackson treats as distinct, the Architectural Restoration narrative is more thoroughly evolutionary. While preservationists could in theory erect monuments to Richard Neutra (and the names of a few modernists like the Alexanders are admittedly present in the Hollywood-style 'Walk of Stars' downtown), instead they tend to deemphasize politicized political compacts as key historical structuring moments. Instead, there is an emphasis on a golden era in the past, the ruins of which stimulated a movement to restore this historical landscape in its 'original' form. But while Jackson understands this view of history as merely unstructured outside of the structuring device of the ruins, what we see in Palm Springs is a knot of temporal contradictions: timelessness flows into novelty flows into nostalgia even as they negate one another. The knottedness only becomes clearer when we try to consider preexisting theories of queer narrative and time. The exodus of lesbians and gay men from Los Angeles to Palm Springs doesn't fit the pattern of what Halberstam (2005) identifies as the 'metronormative' coming-out narrative in which queer subjects migrate from the country to self-actualize in a liberal metropolis; most of the men I spoke to have been out for years or even decades in large cities (especially San Francisco and Los Angeles) *before* moving to the desert. Nor do we see Tongson's (2011) 'queer suburban imaginaries' or Muñoz's (2009) 'queer futurity,' which endeavor to explain queer of color responses to and negotiations with White supremacist landscapes. Instead, we see a variety of seemingly contradictory temporal modalities used to articulate the value of MCM architecture and design, sexual community, and settler presence.

Conclusion: desert time in Palm Springs

While the differences between the Architectural and Tribal restoration narratives are easy to spot – Alexander homes as democratizing architecture vs. the Alexanders as racial capitalists, the tragic demolition of the Spa Resort Hotel vs. the triumphant establishment of a tribal museum, settler common sense vs. indigenous sovereignty – it is easy to miss their common dependence on preexisting attitudes towards the desert itself. The desert is an inconsistent signifier in American settler colonial discourse: it simultaneously represents the space of the ancient other *and* a space of emptiness and timelessness. Settlers seek the desert as an otherwise and elsewhere, embodying erotic lifeways that may or may not be 'normal' but do not belong in the urban core *or* the atomic family home, from Frank Sinatra use of Palm Springs as a site for extramarital affairs to LGBTQ+ people (and gay men in particular) who seek to build 'a different kind of Eden' in the Coachella Valley. But in a settler colonial context, these flights to elsewhere also leverage the figure of the *empty* desert to stage a break with history and the creation of alternative sexual possibility. This figure is present as the settler common sense of midcentury developers enables them to deny an indigenous presence *even on a reservation*. The temporal logic of modern architecture provided a convenient grammar for this historical erasure, as when Richard Neutra 'likened the desert setting with its rocks and mountains to the landscape of the moon and conceived of the [Kaufmann] house as a gem-like pavilion in a small, lush oasis in the midst of a grand but relatively barren place' (Hines, 1994). The turbulent temporal knots found in MCM preservationist discourse were already present in queer desert discourse that deploys timelessness, ancientness, and newness where convenient for settler purposes. Reconciling the Tribal and Architectural Revival narratives in Palm Springs will require more than 'dialogue' between settlers and tribal members: if placemakers are to dismantle the settler common sense that holds them apart, they must question assumptions about the history and use of the desert as a marginal, ruinous, and ahistorical wasteland.

References

Agua Caliente Band of Cahuilla Indians. (2020). *Cultural History* [online]. Vision Agua Caliente. Available at: https://www.aguacaliente.org/content/History%20and%20Culture (Accessed 12 May 2020).

Ainsworth, E. (1965). *Golden Checkerboard*. Palm Desert, CA: Desert-Southwest.

Aldrich, R. (2008). *Colonialism and Homosexuality*. New York: Routledge.

Anderson, K. (1998). 'Desert cool', in *New Yorker*, pp. 128–137.

Baudrillard, J. (1988). *America*. London: Verso.

Chavkin, D. (2016). *Unseen Midcentury Desert Modern*. Layton, UT: Gibbs Smith.

Conrad, T. (2018). 'Gay community has shaped and influenced Palm Springs for decades', in *Desert Sun*, 6 April 2018 [online]. Available at: https://www.desertsun.com (Accessed: 13 May 2020).

Davis, D.K. (2016). *The Arid Lands: History, Power, Knowledge*. Cambridge, MA: MIT Press.

de Crinis, M. (2019). 'A spring awakening: Ancient waters and tribal visionaries pump new life into downtown Palm Springs', in *Palm Springs Life*, 6 March 2019 [online]. Available at: https://www.palmspringslife.com (Accessed: 18 May 2020).

Descant, S. (2014). 'Spa tear-down catches PS off guard', in *Lansing State Journal*, 4 September 2014 [online]. Available at: https://www.lansingstatejournal.com (Accessed: 19 May 2020).

Dollimore, J. (1987). Different desires: Subjectivity and transgression in Wilde and Gide', in *Textual Practice*, 1(1), pp. 48–67.

Estes, N. (2019). 'Anti-Indian common sense: Border town violence and resistance in Mni Luzahan', in Dorries, H., Henry, R., Hugill, D., McCreary, T. and Tomiak, J. (eds.) *Settler City Limits: Indigenous Resurgence and Colonial Violence in the Urban Prairie West*. East Lansing, MI: Michigan State University Press, pp. 44–69.

Fellows, W. (2005). A *Passion to Preserve: Gay Men As Keepers of Culture*. Madison, WI: University of Wisconsin Press.

Freeman, E. (2010). *Time Binds: Queer Temporalities, Queer Histories*. Durham, NC: Duke University Press.

Goolsby, D. (2015). 'Barbara Sinatra among women who ran with the Rat Pack', in *Desert Sun*, 19 October 2015 [online]. Available at: https://www.desertsun.com (Accessed: 16 May 2020).

Halberstam, J.J. (2005). In *a Queer Time and Place: Transgender Bodies, Subcultural Lives*. New York: NYU Press.

Hart, L.M. (2017). 'Thrill of the old', in *Palm Springs Life*, 2 April 2017 [online]. Available at: https://www.palmspringslife.com (Accessed: 15 May 2020).

Hines, T.S. (1994). *Richard Neutra and the Search for Modern Architecture: A Biography and History*. Berkeley, CA: University of California Press.

Hornik, S. (2019). 'Utopian queers of burning man', in *Los Angeles Blade: America's LGBT News Source*, 22 August 2019 [online]. Available at: https://www.losangelesblade.com (Accessed: 13 May 2020).

Hough, S.E. (2004). 'Macroscope: Writing on the walls', in *American Scientist*, 92(4), pp. 302–304.

Jackson, J.B. (1980). *The Necessity for Ruins, and Other Topics*. Amherst, CA: University of Massachusetts Press.

Kennedy, C.S. (2018). 'Palm springs starts removal of controversial tamarisk trees between neighborhood, golf course', in *Desert Sun*, 22 May 2018 [online]. Available at: https://www.desertsun.com (Accessed: 17 May 2020).

Kelman, B. (2018) 'In 1986, a spring break riot changed palm springs, in *Desert Sun* [online]. Available at: https://www.desertsun.com (Accessed: 17 My 2020).

Kray, R.M. (2004). 'The path to paradise', in *Pacific Historical Review*, 73(1), pp. 85–126.

Kray, R. (2009). *Second-class citizenship at a first-class resort: Race and public policy in Palm Springs*. Doctoral Dissertation, University of California, Irvine.

Kuletz, V.L. (2016). *The Tainted Desert: Environmental and Social Ruin in the American West*. New York: Routledge.

Lagdameo, J.B. (2020). 'What are Alexander Homes, and why are they still so beloved?' in *Dwell*, 10 January 2020 [online]. Available at: https://www.dwell.com (Accessed: 16 May 2020).

Leet, S. (2004). *Richard Neutra's Miller House*. New York: Princeton Architectural Press.

Lenc, X. (2018). 'Gay desert modern: Sexuality, architecture and indigeneity in Palm Springs, California', in *Journal of Urban Cultural Studies*, 5(3), pp. 391–401.

Lew, A.A. (2017). 'Tourism planning and place making: Place-making or placemaking?' in *Tourism Geographies*, 19(3), pp. 448–466.

LoCascio, J.P. (2013). *A different kind of Eden: Gay men, modernism, and the rebirth of Palm Springs* (Order No. 1546742). Available from ProQuest Dissertations & Theses Global. (1459216958).

McGaw, J., Pieris, A., and Potter, E. (2011). 'Indigenous place-making in the city: Dispossessions, occupations and implications for cultural architecture', in *Architectural Theory Review*, 16(3), pp. 296–311.

Morgensen, S.L. (2010). 'Settler homonationalism: Theorizing settler colonialism within queer modernities', in *A Journal of Lesbian and Gay Studies*, 16(2), pp. 105–131.

Movieclips Classic Trailers. (2014). Palm Springs Weekend (1963) Official Trailer: Troy Donahue, Connie Stevens Movie HD [Video]. Available: https://www.youtube.com/watch?v=JuoWM66AsRQ (Accessed: 17 May 2020).

Muñoz, J.E. (2009). *Cruising Utopia: The Then and There of Queer Futurity*. New York: NYU Press.

Ortner, V., and Pont, D. (2012). *You Can't Eat Dirt: Leading America's First All-Women Tribal Council and How We Changed Palm Springs*. Palm Springs, CA: Fan Palm Research Project.

Palmer, M.A. (2020). 'Rendering settler sovereign landscapes: Race and property in the Empire State', in *EPD: Society and Space*, 38(5), pp. 793–810.

Powers, S. (2020). 'NHL Seattle names new president for AHL affiliate in Palm Springs' in *The Desert Sun*, 28 April 2020 [online]. Available at: https://www.desertsun.com (Accessed: 14 May 2020).

Pritzker, B. (2000). *A Native American Encyclopedia: History, Culture, and Peoples*. Oxford: Oxford University Press.

Przeklasa, T.R., Jr. (2011). *The band, the bureau, and the business interests: The Mission Indian Federation and the fight for the Agua Caliente Indian Reservation* (Order No. 1495933). Available from ProQuest. (873577681).

Rifkin, M. (2014). *Settler Common Sense: Queerness and Everyday Colonialism in the American Renaissance*. Minneapolis, MN: University of Minnesota Press.

Schroeder C.T. (2009). 'Queer eye for the ascetic guy? Homoeroticism, children, and the making of Monks in late antique Egypt' in *Journal of the American Academy of Religion*. American Academy of Religion, 77(2), 333–347.

Svartzberg-Carrió, M. (2019). 'Palm Springs and the nomos of modernity', in Kockelkorn, A. and Zschocke, N. (eds.) Productive Universals—Specific Situations: Critical Engagements in Art, *Journal of Architecture and Urbanism*. Berlin: Sternberg Press, pp. 163–208.

Tongson, K. (2011). *Relocations: Queer Suburban Imaginaries.* New York: NYU Press.
Urbach, H. (1996). 'Closets, clothes, disclosure', in *Assemblage*, 30, pp. 63–73.
Voyles, T.B. (2015). *Wastelanding: Legacies of Uranium Mining in Navajo Country.* Minneapolis, MN: University of Minnesota Press.
Wade, S. (2019). *Foucault in California: [A True Story--Wherein the Great French Philosopher Drops Acid in the Valley of Death].* Berkeley, CA: Heyday.
Wenzell, N. (2014). 'Palm Springs history: The Oasis hotel', in *Desert Sun* [online]. Available at: https://www.desertsun.com (Accessed: 21 May 2020).

Further reading in this volume

Chapter 4: A future of creative placemaking
Sarah Calderon and Erik Takeshita
Chapter 6: Listen, connect, act
Kim Cook
Preface: 'Disastrous forces, accidental actions, and grassroots responses'
Tom Borrup
Chapter 10: From moon village to mural village: the consequences of creative placemaking in Ihwa-dong, Seoul
Jason F. Kovacs and Hayun Park
Chapter 12: Public transformation: affect and mobility in rural America
Lyndsey Ogle
Chapter 14: Experts in their own tomorrows: placemaking for participatory climate futures
Paul Graham Raven
Chapter 18: '*Homomonument* sounds like a poem': queer placemaking 30 years on: a conversational dialogue with Thijs Bartels, author of *Dancing on the Homomonument* (2003)
Martin Zebracki
Chapter 21: Placemaking through parkour and *Art du Déplacement* (ADD) as a Singaporean applied performance practitioner in London
Adelina Ong
Chapter 22: Embedded Artist Project: Epistemic Disobedience + Place
Frances Whitehead
Chapter 25: 'If you can make it there, you can make it anywhere…': cultural placemaking at the heart of cities
Sherry Dobbin
Chapter 30: Ecological selves as citizens and governance as ethical placemaking
Lisa Eckenwiler
Chapter 31: Seven generations: a role for artists in Zuni PlaceKnowing
Theodore S. Jojola and Michaela P Shirley
Chapter 32: *The Hollywood Forest Story*: Placemaking for the Symbiocene
Cathy Fitzgerald
Chapter 45: How the city speaks to us and how we speak back: rewriting the relationship between people and place
Rosanna Vitiello and Marcus Willcocks

9

FROM THE DUST OF BAD STARS

Disaster, resilience, and placemaking in Little Tokyo

Jonathan Jae-an Crisman

Introduction: Disaster, resilience, placemaking

How is a place made? This chapter argues that places – and especially ethnic and immigrant places – are made through continual negotiation between disastrous forces and grassroots responses to those forces. Certain cultures of participation, collectivity, and futurity ensure that a place and its community will be resilient in the face of disaster, while communities lacking these cultures – what some have lumped under the term 'social capital' – often succumb to external pressures too great for a brittle place to endure (Putnam, 2000; Chetty et al., 2014; Chetty et al., 2016). This quality turns out to be far more important for a place's resilience than the often more-visible technological and infrastructural attempts to prefigure disaster. As the case of Little Tokyo in Los Angeles demonstrates, while disaster is not something to celebrate, it is also the bad seed that has in part made these places what they are today.

In this chapter, after providing a short introductory history of development and disaster in Little Tokyo, I will share three stories of contingent placemaking in this historic Japanese American community, micro histories which demonstrate the link between things falling apart and community coming together. First, shared elements between the development of the Japanese American Cultural and Community Center (JACCC) and the Little Tokyo Service Center (LTSC) demonstrate how the coincidental timing of community development and urban destruction created avenues for building local power. Second, the disastrous decline of Los Angeles' economy during the 1990s enabled Little Tokyo to recentralize as the cultural and spiritual home for Japanese Americans throughout the Southland. Third, a land seizure, of the historic Union Church in Little Tokyo, gone wrong ultimately led to its rebirth as one of the most important sites for Asian American art and culture throughout the US.

In sum, these narratives point toward a theory of urban placemaking that is likely to make any urban planner or designer uneasy: many of the most critical urban developments that make place and define community are contingent phenomena. The urban designer, understood for a long time now to lack the 'master planner' capacity valorized in narratives of modernity, might even lack what little agency is left within urban plans. But this reality ought to be seen as liberatory; rather than toil over plans that might remain on paper or, worse, cause harm, urban planners and designers are invited to enter into a participatory and dynamic exchange between the vagaries of life and the resilience of communities to make place and sustain the future.

Little Tokyo: A history of development and disaster

Little Tokyo, a historically Japanese American neighborhood of about 6,000 residents and workers in Los Angeles, is a community forged through concerted responses to crisis and disaster. From redlining and discrimination, to the wholesale uprooting of the community during the forced incarceration of Japanese Americans during World War II; from urban renewal and redevelopment to the gentrification of today, Little Tokyo is a neighborhood that should not still exist. It is a beautiful aberration to the inexorable processes of urbanization under neoliberal capitalism, contingent upon the ad hoc responses to unexpected challenges (Brenner et al., 2009; Brenner, 2013). Urban scholar Dana Cuff has termed this quality of the city, its never-ending process of destruction and rebirth, as being 'provisional' (2001). Who could have predicted Executive Order 9066? Or that Little Tokyo would end up today sandwiched between some of the highest real estate values in Los Angeles? Yet Little Tokyo abides, standing as an example that other communities, desperate from the impending threat of gentrification, look to for inspiration and survival tactics.

From the Latin for a 'bad star,' the word 'disaster' originated from astrological explanations as to why bad things happen without apparent reason. Chalk it up to troubled alignments between the planets and stars. We now know, of course, that there is no such thing as disaster from the stars: bad things happen because people will them to (Cazdyn, 2007; Stabile, 2007). Poorly designed buildings, mismanaged food supplies, the unexpected consequences of new technologies, greed and caprice from those in power: Katrina and Haiti taught us that these are the reasons why disaster strikes (Virilio, 2007; Smith, 2007; Frye, 2007). We are the bad stars. In our contemporary moment of hyperrational management and regulation, it takes an alignment of bad stars for multiple redundant systems to fail – or, as some might argue, for them to succeed beyond their wildest expectations – for disaster to truly befall us (Cazdyn, 2011). Disaster has been befalling Little Tokyo since its beginnings.

So how is a place made? People, embodying the weight of gender, race, class, and history, interact with space to give it meaning and life (Lefebvre, 1991; Tuan, 1977; Massey, 1994; Massey, 2005; Lipsitz, 2007; Rios et al., 2012; Rios and Lachapelle, 2015; Kaplan, 2018). In the case of Little Tokyo, it begins with a single immigrant entrepreneur, a former sailor, Hamanosuke 'Charlie Hama' Shigeta. Shigeta opened the Kame Restaurant on East First Street in 1885, acting as a beacon that drew in additional Japanese immigrants who opened businesses, started families, created temples, churches, and other cultural institutions. This immigrant urbanism flourished even through the Great Depression, where second-generation Nisei took their business savvy to their parents' flagging shops and restaurants, launching Nisei Week in 1934. The week-long festival unified the community, bringing a new generation of customers to Little Tokyo's shops, along with a distinctly Japanese American culture – neither Japanese nor American, but a hybridized third culture – which married, for example, traditional Japanese odori street dances with American beauty pageant queens. Thus, the place of Little Tokyo was made through daily life, commerce, and culture, growing to a thriving community of some 30,000 Japanese Americans who lived, worked, and communed in the several city blocks just to the east of Los Angeles's downtown.

That is, until February 19, 1942, just two months after the United States entered World War II, when President Roosevelt signed Executive Order 9066 calling for the forced relocation of Japanese Americans to incarceration camps in remote areas (the degree to which internment disrupted Little Tokyo is difficult to convey in this short section, and further reading is recommended; see: Reeves, 2015; Inada, 2000). This virtually erased Little Tokyo from the map as its entire population was incarcerated in faraway places, forcing leases to be broken, places of worship to see their congregations dwindle down to zero, and business to close their doors. (While Little Tokyo was a thriving place, it did so in the face of redlining, racial covenants, and other forms of discrimination which ensured that Japanese Americans could not own land, but instead were forced to rent, making their tenure tenuous.) What happens when a place is unmade? In its place emerged Brownsville, a hotbed of Black life and culture as African Americans from Central Avenue just to the south of Little Tokyo moved up and took over the plentiful leases offered by desperate landowners. Destruction begets creation. Eventually, after 1945, when most Japanese Americans were released from the camps, the community of Little Tokyo slowly re-created their home using the shuttered temples and churches as hubs for rebuilding. A handful of sympathetic, white clergy remained as stewards of the cultural buildings which held belongings and the right-of-use for those returning from Manzanar, Heart Mountain, and the other camps that dotted remote locales in the American West.

Little Tokyo was never the same. All places change and evolve over time, but the wholesale removal of the community from its place in the city meant that change here was disastrous, sudden, and permanent. As Little Tokyo was rebuilt, it became known as a spiritual home for Japanese Americans throughout Southern California, a site for religious and cultural institutions, and a place where Japanese commerce was centered. But its residential population was unable to return to leases long given away and, instead, Japanese Americans dispersed throughout the Southland. Its geographic footprint in Los Angeles shrank, diminished to only a few city blocks. Adding salt to this recent wound, even some of these blocks were seized by the City through eminent domain to build a new police headquarters, a jail, and a handful of other administrative buildings.

But a new generation of Japanese Americans, the *Sansei*, came of age during the 1960s as the student and civil rights movements swept the nation. Consciousness building about community and race gave these young artists and activists tools to confront the powers which had sent their parents to concentration camps and had stolen their land (Ishizuka, 2016; Jeung et al., 2019). Moreover, the experience of internment shaped attitudes toward power, government, and solidarity: Little Tokyo had to stick together, it had to build a culture of organizing and activism, and its community were not content to accept things as they came, or to allow outside powers dictate what was to happen to their place in the city. So began a new history of resistance, resilience, and contingent placemaking in the face of challenges to Little Tokyo's place and sense of belonging in the city.

The destruction of urban renewal begets community organizing

The Sun Building, once located in the heart of Little Tokyo, was a squat, three-story building containing spaces for community organizations, artists, and low-income residents. In interviews with community leaders and artists from Little Tokyo, the Sun Building came up frequently as a key site where local arts and culture flourished. But it also stood out as a symbol for the predecessor of what we now call gentrification: urban redevelopment of the 1960s and 1970s. Despite it being an integral part of community life and despite numerous protests from local activists, it was demolished in 1976 to make way for the New Otani Hotel and Weller Court, high-end properties designed to cater to tourists from Japan. These developments were key pieces in the Little Tokyo Redevelopment Project (LTRP), a long-term urban renewal plan enacted in 1970 as a partnership between the City of Los Angeles, its semi-autonomous Community Redevelopment Agency (CRA), and local and international businesses (Blount et al., 2014; Marks, 2004; Suga, 2004).

The destruction of the Sun Building happened to coincide with several other events. First, there was a new generation of young, Japanese American artists and activists in Little Tokyo who were coming of age amidst the various social movements of the 1960s and 1970s. This generation of *Sanseis* were more likely to speak up and act out against perceived injustices in comparison to past generations. They started numerous activist organizations centered around Little Tokyo including Nikkei for Civil Rights and Redress (NCRR) which advocated for reparations from Japanese American incarceration during WWII, *Gidra* which was a radical newspaper covering issues around race, class, and politics, and Japanese American Community Services-Asian Involvement (JACS-AI) which provided social services. Activists affiliated with the multi-generational Little Tokyo Peoples' Rights Organization (LT-PRO) were some of the key actors in protesting the Sun Building's demolition through marches, sign drops, and petitioning.

Second, what would ultimately become JACCC was founded in 1971 as part of LTRP. Its team began to raise funds from local and international donors, corporations, and public sources, creating a vision for the organization and its building, plaza, and theater which would ultimately open in 1980. By this time, Little Tokyo activists worked to ensure that the organization would not only be a bridge for introducing American audiences to culture from Japan but also a site for sharing the distinct Japanese American culture in Little Tokyo which had flourished for a century. Furthermore, activists ensured that the organization and its properties would support not only high culture, but also local arts and community organizations. Thus, its name evolved from the Japanese Cultural Center to the Japanese American Cultural and Community Center – though, to this day, the Japanese transliteration of the Center's name remains the 'Japan–US Culture Center,' perhaps as a means to placate Japanese funders.

The third element of this story is perhaps the most important, even as it is the most accidental. Bill Watanabe, a local community leader and activist, had applied for and received a community development block grant from the City of Los Angeles which he had hoped to use to provide rent subsidies to the numerous tenants struggling after their eviction from the Sun Building. But after the funds were awarded, Watanabe discovered that they were only permitted to be used for direct services, not the rent subsidies he had hoped for. Watanabe and several of the small service organizations that had popped up throughout Little Tokyo were forced to join forces to use the money, and LTSC was born in 1979. Furthermore, these activists rallied to secure rent subsidies from the developing JACCC and its new building. The organization would live up to its new name, providing low-rent spaces for numerous arts and community organizations who once had a home in the Sun Building.

While the eviction and demolition of the Sun Building is marked with sorrow by many of Little Tokyo's leaders, it is also the flashpoint for new beginnings in the neighborhood. The complicated history of the Little Tokyo Development Project is hardly one to be lauded without qualification, yet the many events that unfolded under its development program led to the establishment of JACCC, which then acted as a new home for the very tenants that it displaced – an outcome, it should be noted, that only came with hard-earned resistance from a galvanized community and its activist political leaders. Furthermore, LTSC went on to become an institutional home for many of these young activists and community leaders, eventually moving beyond its origin as an accidental service provider to a multi-dimensional organization. Its activities now include a community development corporation that has built affordable housing and community spaces, advocacy for Little Tokyo's interests in local policy debates, and a support system for small businesses and culture bearers in the neighborhood. This giant in Little Tokyo's history and development may not have been established were it not for the accidental misreading of a grant's restricted fund uses.

A damaged urban economy enables community land ownership

Beyond the establishment of JACCC, LTSC, and rental spaces for dozens of arts and community organizations, these organizations have also served to migrate significant blocks of land out of the speculative real estate market and into the non-profit sector. This happy byproduct of a contingent history has ensured that current pressures of gentrification are at least slowed to a degree where Little Tokyo has not vanished from the map like so many immigrant neighborhoods across the US. This process began decades earlier, sparked through efforts by community leaders to ensure that something like the forced removal and incarceration of Japanese Americans would never happen again, especially after facing ongoing issues around racial covenants and discrimination, and the City of Los Angeles's use of eminent domain to build the LAPD headquarters.

Coincident with LTRP in the 1970s, many key religious institutions in the neighborhood made the decision to begin capital campaigns, acquire land, and build new buildings. Some institutions, including Higashi Honganji Buddhist Temple and Centenary United Methodist Church, had been forced out of Little Tokyo in the 1910s and 1920s, and this was a return to a shared place in the city. Others, including Hompa Hongwanji Buddhist Temple and Union Church of Los Angeles, were concerned about the LTRP's plans and wanted to get ahead of eminent domain by relocating to new buildings outside of the LTRP map. Construction was completed on the various new architectural insertions into the landscape of Little Tokyo for each of these four institutions in 1976, 1995, 1969, and 1976, respectively. Here, too, we see activism and community organizing in response to unexpected and even imagined threats, placemaking contingent on response to threat, disaster, and discrimination.

In the 1990s, Los Angeles's economy was in the doldrums. The exodus of important aerospace and defense industries disproportionately impacted the local economy. Other factors included racial unrest and police violence symbolized by Rodney King and the uprisings of 1992; natural disasters including the 1994 Northridge earthquake, fires, and landslides; and the apotheosis of suburbanization as core areas continued to be subject to white flight with new immigrants moving in and wealthier, whiter residents moving to the city's edges (Davis, 1990; Klein, 2008). Los Angeles's downtown, Little Tokyo included, were areas that many people avoided, and many of the small businesses that formed the backbone of Little Tokyo either left for other areas or were shuttered entirely. Community leaders knew that more needed to be done. This process of re-centralizing Japanese American cultural life in Little Tokyo continued: LTSC established its community development arm in 1994 and began plans for low-income housing, cultural facilities, and more.

Thus, over the course of decades, Little Tokyo has gradually returned as the dominant site for Japanese American culture and symbolic identity – despite the fact that it has very little residential stock and is ethnically diverse (residents are only about a quarter Japanese American; Painter et al., 2016). This process has happened through fits and starts as institutions and organizations undertook their own steps to secure a place in what some community leaders have referred to as 'the mother ship.' In aggregate, the recentralization of Japanese American culture in Little Tokyo has ensured its long-term viability within both the Japanese American community and the broader public imaginary as a place with a marked Japanese American identity. Many other ethnic and immigrant neighborhoods cease to function as subsequent generations assimilate not only culturally but also geographically as they move to other areas, in contrast to initial immigrants who are forced into enclaves and are dependent on mutual aid and familial connections (Ellis and Wright, 2005; Li, 2006; Toji and Umemoto, 2003). Furthermore, the fact that these institutions have operated as cultural, non-profit, or public entities has removed their land from the speculative real estate market in a spatial pattern geographically distributed throughout Little Tokyo, inoculating it from the worst of gentrification pressures – an impressive accomplishment, given its location amidst some of the highest value land in LA.

An earthquake sparks community cultural development

One structure that deserves special attention is the old Union Church building. As the US National Park Service describes this historic site,

> The Japanese Union Church of Los Angeles was established February 7, 1918 through the merger of three congregations, the Los Angeles Presbyterian Church (est. 1905), the Los Angeles Congregational Church (est. 1908), and the Japanese Bethlehem Congregational Church of Los Angeles (est. by 1911).
>
> *(Waugh et al., 1988)*

The church completed its shared home in 1923 and was active until World War II when the entire congregation was incarcerated. The building was used for several years as a community center for the new Black community which had moved into Little Tokyo. After the return of Japanese Americans to Little Tokyo in 1945, the building was rechristened as the Evergreen Hotel, and it was used as a resettlement and rebuilding center. It finally returned to its original purpose in 1949, after the community center found a new home and the Evergreen Hotel ceased operations.

Later, as LTRP gained steam through the 1960s and officially began in 1970, church members were concerned about their location in an area that was slated for redevelopment. They preemptively relocated, selling their property to the City of Los Angeles, which in turn leased it to the CRA. The church ultimately found a new parcel of land a few blocks away, and they moved into their new facility upon its completion in 1976, completing the sale of their previous site to the city in 1978. While Union Church undertook these activities to preempt the expected eviction through eminent domain by LTRP, the imagined threat would never even materialize: the old Union Church building sat abandoned for decades. So placemaking can also be contingent even to imagined threats and unfulfilled expectations. But the formation of a new home for Union Church is only part of the story.

Among the many issues facing LA in the 1990s was the 1994 Northridge earthquake. The old Union Church building was condemned as unsafe following the earthquake, though this had little impact as it had been sitting vacant for 20 years. Nevertheless, this jolt compelled the City to figure out what to do with the property, ultimately leasing it to LTSC with its newly instituted community development corporation. The agreement was for 44 years at 1 dollar per year with the understanding that LTSC would renovate and seismically retrofit the building – work that ended up costing $3.4 million at the time. The rechristened Union Center for the Arts was envisioned as a community center for the arts, and the first tenants who were all signed in 1995 have remained to this day: the historic Asian American media arts organization Visual Communications, the Asian American studio arts and exhibition center LA Art Core, and the oldest Asian American theater group in the US, East West Players. The site has also become home to the oldest Asian American open mic night, 'Tuesday Night Cafe,' which has become an important venue for activism and organizing by youth.

The Union Center for the Arts is yet another example of space that has been held in public and non-profit trust, out of the speculative real estate market, acting as a buffer against rapid increases in rents and property costs. But, moreover, its role as an internationally recognized home for Japanese American and Asian American arts and culture has cemented Little Tokyo's identity in the public imaginary as one fundamentally interwoven with these ethnic and immigrant urbanisms. East West Players, in particular, has international acclaim and yet it was itinerant, located in various sites elsewhere in Los Angeles, belying its historico-geographical connection to Japanese American and Asian American identity. The creation of the Union Center for the Arts located this identity within Little Tokyo, contributing to the making of place. And this would not have occurred but for several disasters and grassroots responses to these events: the threat of imminent domain by the CRA, which forced the relocation of Union Church, which volunteered to sell their property to the City of Los Angeles; the long-term abandonment of the property, leaving this fenced-up blight in the eyes and minds of community members and, especially, LTSC; and the Northridge earthquake itself compelled the City to act, giving actors in Little Tokyo a window of opportunity to reclaim the property and transform it into one that could benefit the neighborhood.

Conclusion: Sustaining Little Tokyo

History delicately balances between multiple forces, inexorably moving forward as places are made, destroyed, and rebirthed in its wake, as Walter Benjamin described in his *Ten Theses on the Philosophy of History* (1940). The outcome of this process is one determined based on a calculus between the vagaries of life – contingent on disaster, serendipity, chance, and other unpredictable external factors – and the culture of the community based in that place. These stories of Little Tokyo and its continued life, even as so many immigrant urbanisms of its kind have faded away, is hardly one where disaster was absent. Rather, bad stars shone especially bright on the neighborhood, but a distinct culture based on the specific place of Little Tokyo and its experiences in coming together to meet these challenges created an active and communal resilience capable of withstanding these unexpected forces. This form of urban resilience, based in specific, place-based cultural praxis, is one worth further examination. Too often, resilience – especially in a world which is rapidly heating and has risk and disaster on the mind – defaults to top-down, infrastructural interventions that can have the unintended consequence of disrupting the very cultural practices that give a place resilience. Walls, levees, and transportation networks stand in for a disappeared population.

These stories of Little Tokyo also point beyond mere survival in the face of disaster. At other times, the unexpected serendipity of coincidental timing and the fortunate misreading of scripts – combined with a participatory, communal, and creative culture – led to moments of placemaking that have injected new life into the community. The histories of placemaking are, at times, intentional and determined. But, more often than not, after a plan has been envisioned and drawn, its execution goes awry. Unexpected factors, human error, and any number of other forces lead to placemaking contingent on the unpredictable. But communities that have built up a culture of resilience, participation, engagement, and creativity can respond to these contingent moments to turn the proverbial lemon into lemonade.

One additional theme that has repeatedly emerged amidst this process is the relationship between placemaking and property rights. Little Tokyo demonstrates the importance of this invisible and often overlooked social construction, which undergirds nearly everything about urban development. Because of the not-entirely intentional stockpiling of land in Little Tokyo into cultural, non-profit, institutional, and public uses, the neighborhood has been effectively inoculated against the worst pressures of gentrification. The slicing and dicing of property rights into religious institutions, non-profit housing, and cultural organizations has guaranteed that a certain amount of square footage will always remain outside of the speculative real estate market. Moreover, its spatial distribution across Little Tokyo has also acted to suppress some of the most rapid forms of speculation that occur as parcels are aggregated, joined, and expanded, with the spread of a 'hip, saleable vibe' from parcel to parcel, as if the buildings on them were on fire.

The theory of urban placemaking built out of these elements – as one contingent on disaster and the culture of resilience embodied by its community, combined with the underlying factor of property rights – is manifesting again today in Little Tokyo (Crisman and Kim, 2019; Kim, 2012; and Kim, 2011). In response to the recognition that the three last large parcels of land in the neighborhood are publicly owned and slated for development, the community has come together to envision the future it desires and to stake a claim over its place in the city under the auspices of a coalition named Sustainable Little Tokyo. This action was, again, the communal response to a bad star: LA Metro razed the entirety of one of the last remaining blocks of Little Tokyo to make way for construction on its Regional Connector project. This block, along with two others, is publicly owned and subject to the pressures of community organizing and activism, unlike most of the privately held land that occupies urban spaces today. If history serves as a guide, these factors point toward a sustained yet entirely unknowable future for Little Tokyo and its place in the city.

Acknowledgment

The author would like to thank Annette Kim for her generous advising during this project, and to Scott Oshima and the whole Little Tokyo community for its welcoming spirit.

References

Brenner, N. (2013). 'Open city or the right to the city?, *TOPOS: The International Review of Landscape Architecture and Urban Design*, 85, pp. 42–45.

Brenner, N., Marcuse, P. and Mayer, M. (2009). 'Cities for people, not for profit: An introduction', *City*, 13(2–3), pp. 173–181.

Casey Blount, Wendy Ip, Ikuo Nakano, Elaine Ng. 2014. *Redevelopment Agencies in California: History, Benefits, Excesses, and Closure.* Washington DC: US Department of Housing and Urban Development. Accessed online at https://www.huduser.gov/portal/publications/Redevelopment_WhitePaper.pdf.

Cazdyn, E. (2007). 'Disaster, crisis, revolution', *South Atlantic Quarterly*, 106(4), pp. 647–662.

Cazdyn, E. (2011). 'Semiology of a disaster or, toward a non-moralizing materialism', *Scapegoat: Architecture/Landscape/Political Economy*, 2, pp. 32–34.

Chetty, R., Hendren, N. and Katz, L.F. (2016). 'The Effects of Exposure to Better Neighborhoods on Children', *The American Economic Review*, 106(4), pp. 855–902.

Chetty, R., Hendren, N., Kline, P. and Saez, E. (2014). 'Where is the land of opportunity? The geography of intergenerational mobility in the United States', *The Quarterly Journal of Economics*, 129(4), pp. 1553–1623.

Crisman, J. and Kim, A. (2019). 'Property outlaws in the Southland: The potential and limits of guerrilla urbanism in the cases of arts gentrification in boyle heights and street vending decriminalization in Los Angeles', *Urban Design International*, 24(3), pp. 159–170.

Cuff, D. (2001). *The Provisional City: Los Angeles Stories of Architecture and Urbanism.* Cambridge, MA, MIT Press.

Davis, M. (1990). *City of Quartz: Excavating the Future in Los Angeles.* New York: Verso.

Ellis, M. and Wright, R. (2005). 'Assimilation and differences between the settlement patterns of individual immigrants and immigrant households', *Proceedings of the National Academy of Sciences of the United States of America*, 102(43), pp. 15325–15330.

Frye, I.S. (2007). 'When Disaster Is a Bureaucrat', *South Atlantic Quarterly*, 106(4), pp. 709–726.

Gary Painter, Jung Hyun Choi, Vincent Reina, Derek Hung, Jacob Denney and Jovanna Rosen. 2016. *Little Tokyo Community Assessment.* Los Angeles: Sol Price Center for Social Innovation, University of Southern California. Accessed online at https://socialinnovation.usc.edu/files/2013/01/Little-Tokyo-Community-Assessment-093016-Final_with-description.pdf.

Inada, L.F. (2000). *Only What We Could Carry: The Japanese American Internment Experience.* Berkeley, CA: Heyday Books.

Ishizuka, K. (2016). *Serve the People: Making Asian America in the Long Sixties.* London: Verso.

Jeung, R., Umemoto, K., Dong, H., Mar, E., Tsuchitani, L.H., Pan, A. (eds.) (2019.) *Mountain Movers: Student Activism and the Emergence of Asian American Studies*. Los Angeles: UCLA Asian American Studies Center.
Kaplan, D.H. (2018). *Navigating Ethnicity: Segregation, Placemaking, and Difference*. Lanham, MD: Rowman & Littlefield.
Kim, A.M. (2011). 'Introduction: Real rights to the city—Cases of property rights changes towards equity in Eastern Asia', *Urban Studies*, 48(3), pp. 459–469.
Kim, A.M. (2012). 'The Mixed-Use Sidewalk: Vending and Property Rights in Public Space', *Journal of the American Planning Association*, 78(3), pp. 225–238.
Klein, N.M. (2008). *The History of Forgetting: Los Angeles and the Erasure of Memory*. New York: Verso.
Lefebvre, H. (1991). *The Production of Space*. Oxford: Blackwell.
Li, W. (2006). *From Urban Enclave to Ethnic Suburb: New Asian Communities in Pacific Rim Countries*. Honolulu: University of Hawai'i Press.
Lipsitz, G. (2007). 'The racialization of space and the spatialization of race: Theorizing the hidden architecture of landscape', *Landscape Journal*, 26(1), pp. 10–23.
Marks, M.A. (2004). 'Shifting ground: The rise and fall of the Los Angeles community redevelopment agency', *Southern California Quarterly*, 86(3), pp. 241–290.
Massey, D.B. (1994). *Space, Place, and Gender*. Minneapolis, MN: University of Minnesota Press.
Massey, D.B. (2005). *For Space*. Thousand Oaks: SAGE.
Putnam, R. (2000). *Bowling Alone: The Collapse and Revival of American Community*. New York: Simon & Schuster.
Reeves, R. (2015)., *Infamy: The Shocking Story of the Japanese American Internment in World War II*. New York: Holt.
Rios, M. and Lachapelle, P. (2015). 'Community Development and Democratic Practice: Pas de Deux or Distinct and Different?' *Community Development*, 46(3), pp. 190–197.
Rios, M., Vazquez, L. and Miranda, L. (2012). *Diálogos: Placemaking in Latino Communities*. Milton Park and Abingdon: Routledge.
Smith, N. (2007). 'Disastrous Accumulation', *South Atlantic Quarterly*, 106(4), pp. 769–787.
Stabile, C.A. (2007). 'No shelter from the storm', *South Atlantic Quarterly*, 106(4), pp. 683–708.
Suga, M.S. (2004). 'Little Tokyo reconsidered: Transformation of Japanese American community through the early redevelopment projects', *Japanese Journal of American Studies*, 15, pp. 237–255.
Toji, D. and Umemoto, K. (2003). 'The paradox of dispersal: Ethnic continuity & community development among Japanese Americans in little Tokyo', *AAPI Nexus: Policy, Practice, and Community*, 1(1), pp. 21–46.
Tuan, Y. (1977). *Space and Place: The Perspective of Experience*. Minneapolis, MN: University of Minnesota Press.
Virilio, P. (2007). *The Original Accident*. Cambridge, MA: Polity.
Waugh, I.A., Yamato, A. and Okamura, R.Y. (1988). *Japanese Americans in California, Five Views: An Ethnic Historic Site Survey for California*. Sacramento, CA: California Department of Parks and Recreation, Office of Historic Preservation.

Further reading in this volume

Chapter 2: Placemaking as an economic engine for all
James F. Lima and Andrew J. Jones
Chapter 5: Making places for survival: looking to a creative placemaking past for a guide to the future
Jeremy Liu
Chapter 10: From moon village to mural village: the consequences of creative placemaking in Ihwa-dong, Seoul
Jason F. Kovacs and Hayun Park
Chapter 11: *Free State Boulevard* and the story of the East 9th Street Placekeepers
Dave Lowenstein
Chapter 14: Experts in their own tomorrows: placemaking for participatory climate futures
Paul Graham Raven
Chapter 17:- 'I am not a satnav': Affective placemaking and conflict in 'the ginnel that roared'
Morag Rose
Chapter 18: '*Homomonument* sounds like a poem': queer placemaking 30 years on: a conversational dialogue with Thijs Bartels, author of *Dancing on the Homomonument* (2003)

Martin Zebracki
Preface: The radical potential of placemaking
Cara Courage
Chapter 21: Placemaking through parkour and *Art du Déplacement* (ADD) as a Singaporean applied performance practitioner in London
Adelina Ong
Chapter 23: Routing out place identity through the vernacular production practices of a community light festival
Gail Skelly and Tim Edensor
Chapter 25: 'If you can make it there, you can make it anywhere…': cultural placemaking at the heart of cities
Sherry Dobbin
Preface: The only thing constant is change
Kylie Legge
Chapter 39: Translating Outcomes: Laying the groundwork for interdisciplinary evaluation of creative placemaking
Jamie Hand
Chapter 41: Rituals of regard: on festivals, folks, and findings of social impact
Maribel Alvarez
Chapter 42: Creative Placemaking and Placekeeping evaluation challenges from the practitioner perspective: an interview with Roy Chan
Maria Rosario Jackson

10

FROM MOON VILLAGE TO MURAL VILLAGE

The consequences of creative placemaking in Ihwa-dong, Seoul

Jason F. Kovacs and Hayun Park

Introduction

That creative placemaking is nearly always framed in a favourable light should not be surprising given that the various activities that may comprise it are meant to enhance the built environment and qualities of place and, with that, improve people's quality of life. This chapter, however, addresses a notable case in which creative placemaking did the exact opposite for a large segment of the affected population. In 2006, the small residential community of Ihwa-dong, Seoul became the focus of arts-led revitalization. Situated adjacent to a portion of the city's historic city walls, this community of late 1950s-era housing stock was transformed through the work of artists into a mural village under the auspices of a government-funded public art project. Ihwa Mural Village, as it is now commonly known, soon became a tourist magnet, bringing with it some investment and further beautification, but also crowds of visitors and, with that, all of the problems that are associated with 'overtourism.'

A decade after the residential community's transformation, two of the most iconic murals were destroyed, not by outsiders, but by local residents angered at the unequal benefits accruing from tourism as well as the negative impact that excessive visitor numbers was having on everyday life. The story of Ihwa Mural Village has been the focus of many studies in the Korean language, but remains largely unknown outside South Korea. This chapter will first briefly address interpretations of creative placemaking. The chapter will then examine the transformation of Ihwa-dong into a mural village and its subsequent rise as a popular destination for domestic and international tourists. The chapter will then address the problem of overtourism as well as other issues that contributed to the 2016 vandalism incident. Finally, the chapter will conclude with some of the lessons learned.

Creative placemaking

Creative placemaking has received a growing amount of attention over the past decade by scholars, planners, and cultural policy experts in many countries around the world. Although interpretations vary, the notion of creative placemaking according to Gadwa Nicodemus (2013,

p. 213) 'emphasize[s] art-centred initiatives with place-based physical, economic, and/or social outcomes.' In contrast to traditional placemaking projects such as public space design and façade improvements, creative placemaking is distinguished by the role played by arts and culture as well as the centrality of community participation and public–private engagement. As Markusen and Gadwa (2010, p. 3) explain in their foundational White Paper on the subject, 'In creative placemaking, partners from public, private, non-profit, and community sectors strategically shape the physical and social character of a neighborhood, town, city, or region around arts and cultural activities.' They go on to note that this activity 'animates public and private spaces, rejuvenates structures and streetscapes, improves local business viability and public safety, and brings people together to celebrate, inspire, and be inspired.' Examples of such creative placemaking include activities such as outdoor concerts and community festivals, and projects including the creation of live–work spaces for artists, the display of public art, and the beautification of neighbourhoods with mural art.

Some experts however argue that, rather than being centred on arts and culture or on the revitalization of the built environment through planning, creative placemaking is more about an ongoing process, one that animates space, enhances the identity and unique qualities of place, connects people with place and other people, and promotes capacity building for local organizations and the civic sector, amongst other outcomes (Borrup, 2016). As Webb (2013, p. 35) asserts, 'Current models of creative placemaking are tethered to the built environment and urban revitalization.' Webb advocates for an expanded framework in which creative placemaking 'is guided by civic engagement activities that foster cultural stewardship,' 'spurs systematic social change,' and 'articulates a shared aesthetic of belonging' (p. 46). The interpretation of creative placemaking that the authors of this chapter use builds upon both of these general interpretations. That is, the arts are understood to have an important role in differentiating creative placemaking from traditional placemaking activities that also transform the built environment, but the process also ideally engages in a meaningful way those members of the public whose quality of life will be affected by the creative placemaking initiative.

While much of the literature on creative placemaking supports the process without acknowledging the potential drawbacks, there are some exceptions. For example, Frenette (2017, p. 341) states that creative placemaking differs from creative class-centred initiatives in its 'stronger emphasis on equity' amongst other things, but he nonetheless acknowledges in his study of creative placemaking in the United States that inequality is a 'persistent problem for creative placemaking stakeholders' (p. 338). In particular, gentrification and the failure of creative placemaking initiatives in expanding opportunities for marginalized groups have been identified as key concerns (Frenette, 2017; Markusen and Gadwa Nicodemus, 2014; Zitcer, 2018). This case study adds overtourism to the list of potential negative outcomes. Thus, while tourism development may certainly bring with it many economic benefits and be seen as a potentially positive outcome of creative placemaking, it is important that proponents of creative placemaking also recognize the potential negative consequences that excessive tourist numbers to a transformed place may pose to residents' quality of life and plan accordingly.

A community transformed: Ihwa-dong, Seoul

Ihwa-dong (hereafter Ihwa Village, or simply Ihwa) was an inconspicuous residential community noted more for its ageing infrastructure and housing than anything else prior to its transformation into a mural district. Characterized by a higher elevation and steep terrain, the community was simply one of many so-called 'moon villages' or *daldongnae* in the city. A moon village is a low-income, hillside residential area, which owing to its higher elevation was traditionally thought to offer a better view of the night sky. The term is also sometimes applied strictly to urban hillside shantytowns, which Ihwa Village *was* prior to its redevelopment in the late 1950s. Beginning in 1958, the slum-like environment of Ihwa, which had become home to many refugees and migrants from the countryside during and in the years following the Korean War (1950–53), was transformed by the National Housing Corporation (now Korea Land and Housing Corporation). To deal with the poor housing environment, the makeshift homes in the area were removed and replaced by 57 two-storey residential structures. While these houses served their purpose well for decades in providing impoverished families with a proper roof over their heads, a half-century later, residents were calling for more modern housing to replace their deteriorating homes. Although a proposal for the redevelopment of Ihwa Village was in discussion for some time, it never materialized due to the lack of profitability of new housing structures with building height restrictions imposed because of the area's location next to Seoul's historic city walls. In its place came the Ministry of Culture, Sports and Tourism's (MCST) 2006 *Art in City* public art initiative, which was intended to ameliorate the urban environment of this and other residential neighbourhoods in decline, albeit through the more modest means of 'rehabilitation-type regeneration' rather than demolition-centred redevelopment (Kim and Park 2016).

It is worth noting that the MCST's *Art in City* initiative predates the creative placemaking term that arose in the United States by several years. Nonetheless, the objectives of this South Korean initiative, which also went by the lengthy official title 'Public Art Enterprise for Neighborhood Improvement of Neglected Regions' fit well with explanations that have since been given on creative placemaking. First, the public art project was intended to tackle social polarization and the unequal presence of art and culture within cities, which was often confined to higher-income neighbourhoods. Second, the initiative was meant to address the right that all citizens have to live in pleasant urban environments; art would contribute to enhancing the quality of place in the targeted residential communities. Finally, *Art in City* was intended to foster a new model of public art policy, one that actively involved resident participation. Altogether 11 urban residential areas in South Korea were selected for the 2006 initiative, which was funded through a lottery commission fund. In Seoul, the initiative involved the participation of district-level government, including Jongno-gu in the case of Ihwa Village.

Unlike the 10 other residential areas, which were selected through an open competition, Ihwa Village was specially chosen by the program's Public Art Program Committee. The committee's choice tied into the area's character as a post-Korean War residential area in need of rehabilitation, which fit perfectly with the type of place the public art initiative was meant to effect change in. Moreover, according to the lead artist who oversaw the community's transformation, Ihwa's location and 'countryside atmosphere' near to a trendy university street was also a consideration; the arts-led revitalization project would create a place ideal for post-secondary students nearby to visit, linking a modern area of the city with a more traditional residential environment.

The old residential area of narrow streets and steep stairwells was transformed from unremarkable moon village to popular mural village over the course of a few months in late 2006 through the work of 68 artists from outside the community. Altogether 70 murals and other artworks were completed. Significantly, although an attempt had been made to involve local residents as called for in the objectives of the *Art in City* initiative, their involvement in cooperative painting remained limited to a handful of murals. The limited involvement of residents in the transformation of Ihwa-dong likely owed to the hurried nature of the project, which was tied to several delays culminating with the resignation of the original project director in the summer. The lead artist subsequently took over the position to oversee the project, which needed to be completed by year's end. Not surprisingly, a survey conducted soon after the project's completion indicated that a large number of residents felt dissatisfied with their level of involvement in the community's transformation (Korea Art Management Service, 2007). Nonetheless, most viewed the transformation of Ihwa into a mural village favourably, particularly given that most believed it to be a temporary measure to improve the district until the post-war-era housing could be replaced with modern apartments and infrastructure.

Tourism, complaints, and community responses

Overtourism, which the World Tourism Organization defines as 'the impact of tourism on a destination, or parts thereof that excessively influences perceived quality of life of citizens and/or quality of visitors' experiences in a negative way' (UNWTO, 2018), has been a growing issue in Ihwa Village ever since its transformation in 2006. One of the first clear signs of this problem dates back to 2010 when one artist insisted that their 'angel wings' mural be removed after hearing about growing complaints by local residents of the excessive number of visitors engaged in taking photos of themselves in front of the artwork to post online. Like many other tourism hotspots in Seoul, social media has served as an important catalyst in boosting visitor numbers at Ihwa Mural Village (Jang and Park 2020; Oh, 2020). In particular, this first notable incident of a growing public backlash against the murals was tied to the use of the village as a set for episodes of a popular Korean TV show, which involved the visit of a celebrity entertainer whose photo in front of the angel wings mural led to an influx of young visitors aiming to take their own photos there. For the next few years, Ihwa Village underwent an intense phase of 'tourism gentrification' (Cocola-Gant, 2018) as the residential area was transformed into a tourist destination. In particular, several outsiders, including artists, began to move into Ihwa Village. A number of homes were converted into studios as well as bistros, cafes, galleries, and private museums. In addition, some of the original murals that were already in a state of decay or had been previously erased were replaced in 2013 through the work of professional artists. Their work was supported through maintenance funds from the MCST. Notably, Ihwa Mural Village's popular painted 'flower staircase' was redone in more durable tile and a new staircase mural depicting koi fish was completed that same year, also by the original lead artist, who has since moved his studio into the community.

Through these interventions, Ihwa Mural Village soon became the focus not only of domestic tourists, but also of international visitors, particularly from China and to a lesser extent Japan. Media reports captured the growing complaints about noise pollution, litter, and even public urination. However, despite the promotion of a 'Silent Tourism Campaign' meant to remind visitors that they were in a residential environment, the issue of overtourism continued to remain a problem, eventually boiling over with the 2016 mural vandalism incident which saw the destruction of the lead artist's staircase murals. The vandalized artworks were accompanied by graffiti adjacent to them that were directed towards government officials and tourists alike. Some messages, such as 'we are against urban rehabilitation and the murals,' were aimed at the government's seeming reluctance to deal with the ageing housing and infrastructure while others, such as 'this is a residential area, not a tourist site,' were clearly aimed at local visitors and domestic tourists.

Several researchers have acknowledged the problem that overtourism has posed for the wellbeing of Ihwa's residents. For example, Woo et al. (2017) conducted surveys with residents in both Ihwa Village and in Seoul's even more popular tourist destination of Bukchon Historic District. They maintain that the 'touristification' of both residential areas has 'negatively affect[ed] residents' community life, economic life, and health [and] safety' (p. 417). In particular, they as well as others have identified several key problems in these two saturated tourist sites, including: noise, rubbish, and graffiti; the lack of parking facilities and public facilities such as restrooms; increased anxiety amongst residents with privacy and safety-related concerns being an issue; a deterioration of normal community life with the influx of visitors; and concerns about inequality in tourism-generated revenue. Our own research on Ihwa Village, particularly on government and community-led efforts to deal with the problem since the incident, has revealed that apart from the issues outlined above, a key matter that has been overlooked ties into problems in governance and management. Specifically, some of the key problems of this creative placemaking project, which have exacerbated the problem of overtourism, include poor communication with stakeholders on the part of government, as well as confusion on the part of citizens with regard to who their concerns should be directed to, particularly given that receipt of many of their formal complaints over the years was purportedly never acknowledged. Our interviews also revealed the role of a proposed and since abandoned land-use plan that would have excluded a large segment of the estimated 134 households in the area (Oh and Hwang, 2018) from potentially benefiting economically from tourism in sparking the 2016 incident. Specifically, the unveiled land-use plan a few months prior to the incident infuriated many residents since it designated certain areas of Ihwa Village as residential only and others as mixed commercial-residential spaces, including those areas 'newcomers' had invested in during the intense phase of tourism gentrification in the early 2010s.

What is perhaps more notable than the 2016 vandalism incident and the reasons behind it are the responses that have since emerged by government and especially by the community itself. The vandalism that led to the destruction of Ihwa Village's iconic 'flower' and 'koi fish' staircases as well as graffiti adjacent to them drew much media attention to the longstanding concerns of the area's residents. Government responses since have included the appointment of a community planner tasked with encouraging conflict resolution, and the establishment of a new local decision-making committee ('united community group') meant to act as a bridge between residents and government. It remains to be seen what if anything will materialize from these developments, particularly since some residents already claim that the local committee is merely symbolic; that its suggestions to government will never be given serious consideration. On the other hand, the government has recently invested more into improving the built environment with additions being made, such as handrails along stretches of some of the area's steep streets.

Given what creative placemaking is ideally about, a more impressive development since the 2016 incident involves the former lead artist's effort to engage with community members, particularly with those residents who have always felt excluded from the public art project. Specifically, the artist has acknowledged the problems of the original project, admitting that much of the original art lacked meaning since it was introduced by artists parachuted in during the rushed transformation phase in 2006 and to a lesser extent again in 2013 with the creation of additional murals. In response, he has promoted meetings with residents, including with some who originally supported the vandalism of his popular staircase murals. Although the artist has a complicated relationship with the community, having been the project director during Ihwa's transformation, at the time of writing he was acting as an intermediary, encouraging community dialogue, listening to residents' ideas and concerns, and making an effort to relay this information to other stakeholders, all through a non-official, voluntary capacity. In addition, he was also volunteering his time responding to requests by residents for advice on a range of local matters. Thus, the current effort by the artist to engage the community fits well with explanations of socially engaged art and the role that artists may play as intermediaries between different stakeholders, including residents, fellow artists, or government employees. Artists, as Frennette (2017, p. 340) states, are often 'well-practiced at communicating with many publics and adjusting to the expectations of various parties.' Similarly, Redaelli (2018) asserts that creative placemaking brings artists into the centre of a community where they can act as researcher in helping to celebrate the local history and culture of place, collaborator in adding layers of meaning, and facilitator in helping to create a common vision by facilitating conversation amongst people.

While the future of Ihwa Village's murals remains uncertain, at the time of writing most of the graffiti has been removed, some murals have been repainted, and there is talk of repairing the staircase murals and others, albeit with input from the community. With regard to tourism, the destruction of some of the murals did have a negative impact on visitor numbers. However, there were still some 325,000 visitors on average per month according to an estimate for 2016/17, and many international tourists were still to be seen during our visits to Ihwa Village in the autumn of 2019. In fact, Ihwa Village was still being frequented by some visitors in late March 2020 in spite of the coronavirus pandemic situation.

Conclusion

This case study has revealed the potential failure of creative placemaking in enhancing livability and quality of life, particularly when there is inadequate community engagement or consideration of the potential negative effects of such initiatives should the affected site become too popular with visitors. Thus, this chapter contributes not only to the literature on creative placemaking (Courage and McKeown, 2019) but also to the specialized works on murals tourism (Skinner and Jolliffe, 2017) and the growing body of studies on overtourism (Capocchi et al., 2019; Dodds and Butler, 2019; Milano et al., 2019; Pechlaner et al., 2020), which has to date been largely confined to addressing the issue in popular tourist destinations in Europe. In terms of creative placemaking, several lessons can be highlighted. First, community members must be actively involved in the process of creative placemaking and should be clearly informed what the project is about, including what the longer-term timeline is for the project. Despite community engagement being an objective of the *Art in City* initiative, participation was lacking in the Ihwa case from the get-go, and residents were unaware of the long-term nature of the project, hence their initial indifference both to the project and to the early budding interest towards the area by outsiders. Second, since creative placemaking is ideally about highlighting the stories of place and revealing the unique qualities of place, this should be reflected in the nature of the project. In the case of Ihwa Mural Village, the mural art was almost exclusively made by outsiders and had little meaning or connection to the history of the place, an issue that is now likely to be corrected. Third, responsibility for the long-term management of such projects should be clearly defined from the onset of the process and understood by community stakeholders. Much of the difficulties that boiled over with the 2016 mural vandalism incident can be attributed to poor communication between government and residents. Fourth, the potential consequences of creative placemaking, including social and economic impacts, should be considered and planned for accordingly. In the case of Ihwa Village, tourism was not factored in as a potential consequence of the area's transformation nor was the question of who benefits adequately considered, especially during the development of a land-use plan to deal with the area's popularity as a tourist destination.

On a final note, the process of creative placemaking should ideally emerge organically within a community, rather than being assigned to a place by outsiders. In the case of Ihwa Mural Village, the idea to transform the marginalized residential area through mural art originated outside the community. Moreover, most of the beneficiaries of the area's transformation were not the original residents but rather outsiders who moved into the community during its 'tourism gentrification' phase. Ultimately, all of these issues, as well as the problem of overtourism, contributed to the shocking incident in which several individuals, with the apparent approval of many fellow residents, attempted to destroy the tourism base of their own community. However, it would seem that conciliation efforts since the 2016 incident, including current attempts by the former lead artist and now day-time resident of Ihwa Village to engage community members in visioning the future of the mural village, are serving to correct some of the problems in how the *Art in City* project was originally implemented in Ihwa-dong. Consequently, current efforts, it would seem, are bringing the ongoing rehabilitation process in Ihwa closer to the ideals set forth by leading experts and proponents of creative placemaking.

References

Borrup, T. (2016). 'Creative placemaking: Arts and culture as a partner in community revitalization', in Boyle-Clapp, D., Brown, M. and Gard-Ewell, M. (eds.) *Fundamentals of Arts Management*. 6th edn. Amherst, MA: University of Massachusetts Arts Extension Service, pp. 50–69.

Capocchi, A., Vallone, C., Pierotti, M. and Amaduzzi, A. (2019). 'Overtourism: A literature review to assess implications and future perspectives', *Sustainability*, 11(12), 3303. https://doi.org/10.3390/su11123303.

Cocola-Gant, A. (2018). 'Tourism gentrification', in Lees, L. and Phillips, M. (eds.) *Handbook of Gentrification Studies*. Cheltenham and Northhampton: Edward Elgar Publishing, pp. 281–292.

Courage, C. and McKeown, A. (eds.) (2019). *Creative Placemaking: Research, Theory and Practice*. London: Routledge.

Dodds, R. and Butler, R.W. (eds.) (2019). *Overtourism: Issues, Realities and Solutions*. Berlin: De Gruyter.

Frenette, A. (2017). 'The rise of creative placemaking: Cross-sector collaboration as cultural policy in the Unites States', *The Journal of Arts Management, Law, and Society*, 47(5), pp. 333–345.

Gadwa Nicodemus, A. (2013). 'Fuzzy vibrancy: Creative placemaking as ascendant US cultural policy', *Cultural Trends*, 22(3–4), pp. 213–222.

Jang, H. and Park, M. (2020). 'Social media, media and urban transformation in the context of overtourism', *International Journal of Tourism Cities*, 6(1), pp. 233–260.

Kim, M. and Park, J. (2016). 'The effect of village regeneration on settlement and residential satisfaction', *Journal of Asian Architecture and Building Engineering*, 15(3), pp. 519–526.

Korea Arts Management Service. (2007). *2006 Evaluation Report of Public Art Project for Improving the Living Environment of Disadvantaged Areas* (in Korean), https://m.gokams.or.kr

Markusen, A. and Gadwa, A. (2010). *Creative Placemaking*, Washington, DC: National Endowment for the Arts.

Markusen, A. and Gadwa Nicodemus, A. (2014). 'Creative placemaking: how to do it well', *Community Development Investment Review*, 2, pp. 35–42.

Milano, C., Cheer, J.M. and Novelli, M. (eds.) (2019). *Overtourism: Excesses, Discontents and Measures in Travel and Tourism*. Boston: CABI.

Oh, D. and Hwang, J. (2018). 'A comparative study on Ehwa Mural Village Seoul and Chemainus Mural Project, Vancouver Island', *Journal of the Korean Regional Development Association*, 30(5), 207–234 (in Korean).

Oh, Y. (2020). 'From concrete walls to digital walls: Transmedia construction of place myth in Ihwa Mural Village, South Korea', *Media, Culture & Society* [OnlineFirst]. https://doi.org/10.1177%2F0163443720916410.

Pechlaner, H., Innerhofer, E. and Eschbamer, G. (eds.) (2020). *Overtourism: Tourism Management and Solutions*. New York: Routledge.

Redaelli, E. (2018). 'Creative placemaking and theories of art: Analyzing a place-based NEA policy in Portland, OR', *Cities*, 72, pp. 403–410.

Skinner, J. and Jolliffe, L. (eds.) (2017). *Murals and Tourism: Heritage, Politics and Identity*. New York: Routledge.

UNWTO (United Nations World Tourism Organisation). (2018). '*Overtourism*'? *Understanding and Managing Urban Tourism Growth beyond Perceptions*. Executive Summary. Madrid, Spain: UNWTO.

Webb, D. (2013). 'Placemaking and social equity: Expanding the framework of creative placemaking', *Artivate: A Journal of Entrepreneurship in the Arts*, 3(1), pp. 35–48.

Woo, E., Kim, Y. and Nam, J. (2017). 'Impacts of touristification on residents life from a qualitative approach: Focusing on Bukchon Hanok Village and Ihwa Mural Village', *Journal of Tourism and Leisure Research*, 29(11), pp. 417–436 (in Korean).

Zitcer, A. (2018). 'Making up creative placemaking', *Journal of Planning Education and Research*, 40(3), pp. 278–288.

Further reading in this volume

Chapter 2: Placemaking as an economic engine for all
James F. Lima and Andrew J. Jones
Chapter 5: Making places for survival: looking to a creative placemaking past for a guide to the future
Jeremy Liu
Chapter 6: Listen, connect, act
Kim Cook
Chapter 11: *Free State Boulevard* and the story of the East 9th Street Placekeepers
Dave Lowenstein
Chapter 16: More than a mural: participatory placemaking on Gija Country
Samantha Edwards-Vandenhoek
Chapter 17: 'I am not a satnav': Affective placemaking and conflict in 'the ginnel that roared'
Morag Rose
Chapter 18: '*Homomonument* sounds like a poem': queer placemaking 30 years on: a conversational dialogue with Thijs Bartels, author of *Dancing on the Homomonument* (2003)
Martin Zebracki
Preface: The radical potential of placemaking
Cara Courage
Chapter 20: Displacemaking 2015 and 2020
Catherine Fennell and Daniel Tucker
Chapter 22: Embedded Artist Project: Epistemic Disobedience + Place
Frances Whitehead
Chapter 23: Routing out place identity through the vernacular production practices of a community light festival
Gail Skelly and Tim Edensor
Chapter 24: Artists, creativity, and the heart of city planning
Tom Borrup
Chapter 25: 'If you can make it there, you can make it anywhere…': cultural placemaking at the heart of cities
Sherry Dobbin
Chapter 26: Sculpturing sound in space: on *The Circle and the Square* (2016) by Suzanne Lacy
Trude Schjelderup Iversen
Chapter 31: Seven generations: a role for artists in Zuni PlaceKnowing
Theodore S. Jojola and Michaela P Shirley
Chapter 35: Planning governance: lessons for the integration of placemaking
Nigel Smith
Chapter 40: Transforming community development through arts and culture: a developmental approach to documentation and research
Victor Rubin
Chapter 42: Creative Placemaking and Placekeeping evaluation challenges from the practitioner perspective: an interview with Roy Chan
Maria Rosario Jackson

11

FREE STATE BOULEVARD AND THE STORY OF THE EAST 9TH STREET PLACEKEEPERS

Dave Loewenstein

I am a 30-year resident of East Lawrence, Kansas, a community-based artist and neighborhood organizer. I was drawn to this place right after I left (I should say quit) graduate school in search of a way of being an artist that aligned better with the work I was doing as an organizer and activist. My neighborhood of East Lawrence is the oldest in the city. It has been and continues to be a working-class neighborhood full of radical history and commitment to social justice, but it didn't start that way. Many Lawrencians would like to forget that the land the City now occupies was taken from the Kanza and Delaware people as part of an early placemaking project that concluded with the Kansas–Nebraska Act of 1854. That tragic history is too often obscured by the wild pride residents have for the Bloody Kansas years leading up to the Civil War, when transplants from Massachusetts fought to ensure that Kansas would enter the Union as a 'Free State.'

Adjacent to present-day downtown, East Lawrence gradually slopes down to the railroad and old factory buildings that run along the Kansas River. It's still full of small single-family homes and backyard gardens. It's where Langston Hughes went to church as a young boy, where Civil Rights marches began and ended and more recently where a massive creative placemaking project funded by ArtPlace was proposed to revitalize us. That project, known as *Free State Boulevard* is the subject of this chapter. As a firsthand witness to it, I was a participant in fighting it and, in the end, one of the people who reimagined it as a more just and equitable endeavor.

The panel

It was a big deal, for me at least, to be invited to present on a panel with the Chairman of the National Endowment for the Arts about Creative Placemaking. The panel, hosted by the Spencer Museum of Art, included professors from Haskell Indian Nations University and the University of Kansas, a noted architect, and the chair of the Nebraska Arts Council (the Kansas Arts Commission had been defunded by our governor, so there was no chairperson to attend.) The auditorium was full.

It was early 2012, and there had yet to be an official placemaking project in town, so I think most folks didn't fully understand the term (and I still wonder if I do), and because of that, most of the panelists talked around the idea, instead commenting on the disconnect between nature and culture, the overuse of cell phones, and the need for strong arts education. As a community-based muralist, I was familiar with and often a part of placemaking projects. But this was new. Never had the Spencer Museum or any other arts organization in Lawrence taken notice of this work in such a formal way. Chairman Landesman's presence and the knowledge of big new funding opportunities through NEA's 'Our Town' grants, and ArtPlace changed that. And that new attention made me a little nervous. I felt that not so comfortable feeling of something scary approaching, like when the barometric pressure drops before a storm, and it came out in my comments, where I ended by asking, 'I wonder if maybe we should adapt the phrase Creative Placemaking, to creative place-sustaining, place-keeping, or place-enhancing, because in some cases the places don't need to be made, they just need to be recognized and cared for' (Lowenstein, 2012).

When I think back, there had been signs that something was brewing. There was new real estate speculation happening along East 9th Street (a real estate agent cold-called me to ask if I wanted to sell my studio, but wouldn't say who they were working for), a hotel and new farm-to-table restaurant were being built on what had been a beloved community green space, and there was lots of talk from the chamber of commerce about how East Lawrence artists were an engine for economic development. What I didn't know then was that Chairman Landesman's visit coincided with a yet-to-be revealed effort by our local Arts Center to mount a massive placemaking project in East Lawrence.

And then it hit; news broke that the Lawrence Arts Center was proposing a $4.5 million project funded by ArtPlace (a privately funded placemaking organization initiated in part by Rocco Landesman in 2011) and the City called 'Free the Radicals' that would 'revitalize' a seven-block stretch of East 9th Street, comprised of small single-family homes and a few local businesses (including my studio), into a kind of hip outdoor arts and culture corridor. I couldn't believe it, even though the proposed project cut right through the middle of our neighborhood, the East Lawrence Neighborhood Association (I was and still am a member of the ELNA board), had not been consulted on the development of the proposal at all. That was not a good sign.

East Lawrence

The eastside comprises about six neighborhoods east of downtown. East Lawrence, the City's original townsite now adjacent to downtown, is one of them. For more than a century, it was known as the Bottoms (closer to the river and railroad), where working-class people lived and few others visited. It's always been a working-class neighborhood with a progressive spirit. East Lawrence's proximity to downtown has always made it a place of contention. For a long time, it was thought of by people who didn't live there as a place to ignore and to go around. This was true in the 1970s when plans were hatched to build a four-lane bypass, the Haskell Loop, that would go straight through East Lawrence into downtown. The fight to stop it (and it *was* stopped) gave rise to new neighborhood activism and led to the formation of our neighborhood association. A similar road project was proposed 20 years later, this time called the Eastern Parkway, which was also defeated in large part by the work of neighborhood activists. Both projects were pushed by City Commissioner Bob Schumm, who will reenter this story later. In more recent years, the relatively cheap property and preserved vernacular and old factory architecture has ignited a flurry of plans to flip East Lawrence into the next hip place for young white professionals to move. Through these challenges, East Lawrencians learned how to stand up for their values and fight back against developers and their cheerleaders in City Hall. This would prove to be crucial in the struggle that would follow.

The proposal

'Free the Radicals' made it to the final round but did not win a grant. ArtPlace's two main criticisms were that the project did not fall within an already existing Cultural District and that the City did not have a specific Arts and Culture staff position. So, guess what happened? The Arts Center with the support of City Commissioner Bob Schumm fast-tracked the designation of a Cultural District that conveniently circumscribed East Lawrence, and the hiring of a full-time Director of Arts and Culture. Within a year both of these objectives were achieved, and the Arts Center made its second proposal to ArtPlace, this time titling it *Free State Boulevard*. It worked. In April of 2013, the headline on the front page of the newspaper read, 'Massive undertaking aims to transform Lawrence from arty to ArtPlace.'

Here's how the Arts Center's new winning proposal began:

> Free State Boulevard: From the Studios to the Streets, led by the Lawrence Arts Center, the City of Lawrence, and a Creative Team, will revitalize six blocks of 9th Street that link the Warehouse Arts Area and Downtown Lawrence, creating multi-modal paths, upgraded amenities, and a new model of urban infrastructure that will enable local artists to engage our community in ways inspired by the revolutionary and counter-culture spirit of Lawrence and our favorite iconoclasts John Brown, Langston Hughes, and William S. Burroughs.

And under Economic Impact, the proposal said:

> Free State Boulevard will encourage new investment along 9th street, including investments in real estate development, small businesses, and original art. Specifically, Free State Boulevard will connect major 'bookend' development investments, each emblematic of the diverse character of the Cultural District.

This was not the first time residents here had been faced with a development project that aimed to erase their history in the name of progress. But the way the Arts Center was exploiting Lawrence's Civil War and 'Iconoclast' history for a project the real aim of which was to speed gentrification, was alarming.

Early organizing

Dread. That was my initial feeling upon reading the proposal. I knew that not consulting with the neighborhood on a project this big could only mean they thought they could push it through without us. The potential impacts, intentional or not, were clear – bars, restaurants, lofts, and sky-rocketing property values, which would push many long-time residents out of the neighborhood. In a word, gentrification with placemaking as a catalyst. Trying to understand what we were up against led me to Roberto Bedoya's 2013 essay, 'Placemaking and the politics of belonging and dis-belonging,' where he says,

> The blind love of Creative Placemaking that is tied to the allure of speculation culture and its economic thinking of 'build it and they will come' is suffocating and unethical, and supports a politics of dis-belonging employed to manufacture a 'place.'

That was it. The question was what to do.

It was a strange sort of upside down for me. Since I was a community-based artist whose studio was at ground zero of *Free State Boulevard*, friends and neighbors asked me what role I had in the project or how I participated in drafting the proposal. None, I had to tell them, I heard about it just like they did, through the media. I had spent more than 20 years in this neighborhood doing exactly the kind of work this proposal claimed to support and no one talked with me or asked me to collaborate. So, no consulting with the neighborhood or with one of the only local artists who actually had experience doing this type of work. That couldn't be good. But I am an organizer and my first impulse was not to complain, it was to plan. What was the strategy and what could the tactics be to give us a chance at gaining power within the project, or if that failed, building public support to stop it? First, I thought, people needed to know what was actually being proposed. Second, we needed to demand transparency in documents and meetings. And third, we needed to provide a different vision for the project, alternatives not just critiques.

A group of neighborhood friends and supporters began meeting at my studio to talk about the implications of the project, what we were feeling and what we wanted to do. I was asked to facilitate the meetings and led them in a way that was similar to the way I organized community-art projects. We got to know each other. We used a story-circle format where each person got a chance to share without interruption. One of us took notes and shared them back with the group after the meeting. We discussed action items and got people to commit to certain tasks. And, we agreed to try to limit speculation and personal attacks. Our first action items included: beginning formal dialogue with the City and Arts Center; writing letters to the newspaper; sharing public documents with the neighborhood association; demanding representation on all committees; and requesting an access to the full ArtPlace proposal including the budget.

The Watergate moment

This last action item was one of the first chinks in the armor of *Free State Boulevard*. The Arts Center refused to provide their full proposal, claiming since they were a private non-profit they didn't have to. We pushed. We knew that the project was dependent on \$3.5–4 million in public funds and that most people would want to know how that money was going to be spent. And this is the only real thriller part of this story. Out of the blue, we received a copy of the full proposal, anonymously. Someone on the inside wanted us to know what was in it (we never found out who). Reading it over it was just as we had feared, but seeing it on the page was still surprising.

In the income category were big cash donations by developers at the east and west ends of 9th Street (the 'bookend' developments), and among other private donors was by-then-former NEA Chairman, Rocco Landesman who contributed \$25,000. Also surprising was that a design firm, el dorado (who was then working for one of the 'bookend' developers), had already been named in the Arts Center's application to run the project even though at that very moment there was a public search, led by the City, to choose a design team. It was no surprise when out of the seven candidates, el dorado was chosen. There was never such a search for the lead artists. They came with el dorado, a Canadian duo, known best for working on arts-based city infrastructure projects. Local or regional artists were never considered. With no response from repeated outreach to the Arts Center about these issues, we decided it was time to go to the press and to do a Freedom of Information (FOI) request, since most of the emails about the project included City staff and or commissioners in conversation with the Arts Center and ArtPlace.

It doesn't matter if you're right if they have the votes

The newspaper picked up on our concern about the full proposal and for the first time asked critical questions about the transparency of the project. We also received hundreds of emails through our FOI request, many of which exposed a cavalier and dismissive attitude about East Lawrence and especially ELNA. But we were still losing. The City Commission which had initiated a Citizen Advisory Committee (CAC) for the project, stacked it with supporters and only due to our protest allowed more than one person from East Lawrence. I was one of them. In our meetings, the votes were consistently 13 for moving forward with the plan and 2 against. I recognized that our token representation on the CAC committee gave the Arts Center cover and didn't really help us at all. There had to be another way.

If one or two seats at the table weren't going to work, we were going to have to build our own table, so that's what we did. It was called 'Imagine East 9th Street,' a community event to engage in conversation about the ArtPlace proposal and to envision alternatives to it. We modeled it after the community Imagining we had held earlier in the year at the kick-off for the new US Department of Arts and Culture (USDAC), a performative (not real) government agency dreamed up by Adam Horowitz and Arlene Goldbard, whose motto was, 'Together, We Create' (Horowitz and Goldbard, 2019). Changing the frame around the ArtPlace proposal proved to be critical. A big crowd gathered in an old barbed wire factory along the river to share food, memories, and dreams for the neighborhood. We also were able to build trust and share a set of facts about the project that many were unaware of. The media was also there and reported on the event. We still didn't have the votes, but we were building support and arming people with the information they needed to ask tough questions.

If I can't dance, it's not my revolution

In meetings and City Hall, it had been difficult to describe the value of this place when the frame for the discussion was always centered on economic development and vibrancy, whatever the hell that is code talk for. What I realized thanks to a certain fox from the book *The Little Prince* was that what was most important was invisible. It was our memories and aspirations that were woven into the social fabric and sparked by the bricks and trees and porches and sounds and smells of the neighborhood: things that were apparent to those of us who knew the place but invisible to those who didn't.

This couldn't have been made clearer than when the project's lead artists, the duo from Canada, came to engage the community in what they called a charette (first mistake.) As a part of it, they presented a 3D model of East Ninth Street made from what looked to me like sugar cubes. It was pure white and only included built structures: no trees, no people, no life at all. It was as if the street was a blank slate available for others to project their desires onto without concern for who lived there or what meanings and stories were embedded within it. And that was our next revelation. We had to bring the street to life, sharing our collective internal visions for everyone to see. Luckily, many of them were artists we had organized in the past and we knew just what to do. It was going to be fun!

How could we tell our stories in a way that would make people stop and listen? I thought they needed to be larger than life and in a form that would be inviting. I had just seen the video fountains in Chicago in front of the Art Institute depicting regular people, who because of their scale and presentation were completely captivating doing simple things like smiling or blinking, and I thought, that's it! I shared these thoughts with my filmmaker friend Nicholas Ward and we came up with the idea for *Facing East*. We did short interviews shot in my studio and gave our subjects prompts that never appear in the film (you only hear their answers). Nicholas edited them down to create a beautiful and funny 12-minute film that celebrates and worries for East Lawrence. We showed it outside on the front of the studio during the monthly artwalk to an audience of neighbors and passersby. Those beautiful images of our friend's giant faces sharing stories of their love for their place and concerns about its future was one of the biggest reasons I kept on fighting.

One of our other big concerns was that people in Lawrence did not have a good grounding in what placemaking actually meant, and that other communities across the country were also facing problems with projects like ours. A piecemeal approach of putting up flyers and social media posts didn't seem like it was enough, so we created the 'East 9th St. Placekeepers' website (Placekeepers, 2013). It would be a serious space for information and also a place that collected all of the creative ways in which people had been responding to placemaking projects across the country. We also decided it wouldn't be a place for us to editorialize, beyond choosing the material we put up. We would let folks read for themselves. It included all of the media stories and public documents related to *Free State Boulevard*, and also many articles about placemaking from around the country.

One of the 'media' sources appearing on 'Placekeepers' was the intrepid neighborhood muckraker, Biff Beluga. He was a regular on *Live TV Live*, a short-segment local news program inspired by the *Daily Show*, created by Nicholas and another friend of ours, Amber Hansen. Biff, a send-up of a 1940s newscaster mixed with John Waters and John Stewart, could say things we couldn't and in a way that made you laugh even if you disagreed. His riffs on carpetbaggers, backstabbers, and gentrifiers were cutting and hilarious. Even if we lost this fight, we were gonna have blast going down.

The surprise(s)

By this time, it was April of 2015. In spite of our efforts, *Free State Boulevard* was sailing through City Commission and was nearing the point of no return. And then, three things happened that we never could have planned or expected. All along, one of the biggest obstacles to the project was that we didn't have the votes in City Commission. The mayor, Jeremy Farmer, and long-time commissioner Bob Schumm, were two of the project's biggest cheerleaders and the other commissioners were either nominally for the project or on the fence. But elections change things and that city commission election changed everything. The two incumbent candidates, Terry Riordan and Schumm whose campaign focused on how East 9th Street could become like the bucolic scene depicted in a famous painting by Georges Seurat, both lost. A third decided not to run again, and in their place came three new commissioners, two of which we knew had serious doubts about *Free State Boulevard*.

And then, Mayor Farmer was arrested for embezzling from a local food pantry and promptly resigned. In his place, a new commissioner, Lisa Larsen, was appointed, and she had serious reservations about *Free State Boulevard*. So, all of a sudden, we maybe did have the votes to affect some change. But turning around this behemoth of a project with all of its momentum was not going to be easy, unless... Unless – and this was the case – the Work Plan already approved by the previous Commission was found at the last minute to be on an official truck route, meaning all of the street-narrowing and multi-modal paths would have to be scrapped. el dorado would have to start over after nearly two years, while the new sitting Commission was no longer an ally.

The end and the beginning

For the next year, el dorado presented altered versions of their plan, none of which met with Commission approval. They presented to the Historic Resources Commission and endured four-hour-long meetings with residents. Skins began to wear thin while the Placekeepers and Biff Beluga kept reporting. By the time July of 2015 came around, the project came up for a final vote to approve funding allocation in the budget for the City portion of the project. There weren't enough votes to pass it and the issue was tabled, to be brought up again 'at future date.' With no funds in the coming year's budget, *Free State Boulevard* as it had been conceived was effectively dead.

In 2016 and 2017 there were a few attempts to revive the project, but none of them got anywhere, and I got to thinking that if we accepted the circumstances as they were, what would a good ArtPlace project for East Lawrence look like? In January of 2016, the Arts Center CEO, who had almost singlehandedly pushed the project and fought against those who questioned it, announced her retirement, but the organization still held the funds from ArtPlace, and so what began as an exercise, turned into a plan. I had been so demonized by some at the Arts Center that the thought of reengaging with them about the project seemed ridiculous, but I also knew I would always wonder what might have happened if I didn't at least try. I made the call, in this case to the interim CEO, Cindy Maude. I asked if she'd be willing to meet. She said yes.

Role reversal

Paradigm shifts are hard on the brain and heart. To see what was once the ground as the figure or what was once the disaster as the opportunity takes, simultaneously, concentration and letting go. I had spent years fighting alongside my neighbors to contain and or stop *Free State Boulevard* that to imagine now switching roles and trying to imagine our own vision of an ArtPlace project was disorienting and risky. At the same time, I realized I probably could do it. I knew the neighborhood. I was a community-based artist and many folks trusted me. So, I went to work drafting an outline that I called 'Rebuilding East 9th Street Together.' To help me and give me confidence I called on Arlene Goldbard of the USDAC and cultural activist Julia Cole from Kansas City. They were instrumental in guiding my thinking and encouraged me to go for it. It took about a month to get it into shape. Here's how the intro read:

> Rebuilding East 9th St. Together
>
> How to move forward with the East 9th Street project may seem fraught with difficulty due to its contentious past, but what if we viewed what happened as an event that allowed us to see circumstances anew, and act to change them for the better? Like rebuilding after a storm, or adjusting to an unexpected tragedy, we could see the East 9th St. project not only as a failure or loss, but as an opportunity to create new ties, fortify important relationships, and develop more sustainable and equitable planning processes.

Artists, like scientists, often learn most from experiments or projects that don't go as planned. It is in those moments when they reflect most deeply and honestly, and open themselves to solutions previously unseen. This also happens in the aftermath of some natural disasters as Rebecca Solnit writes in her book *A Paradise Built in Hell* (Solnit, 2010). Communities are often at their most creative, collaborative, and empathetic at times of disaster, she says. Why? Because in order to recover and plan for the future, they need each other. And this can lead to new ways of working together and new solutions to formerly intractable problems.

We are in a similar situation with the East 9th Street project. New circumstances provide an opportunity to reevaluate and rebuild – the street, mutual trust, and a measure of shared agency and responsibility. The alternative would be to ignore or try to forget what happened, hiding the wounds but not trying to heal them. Instead we could reimagine this perceived failure as a platform to explore the fundamental issues that bind and challenge us, and then act to address them with new strategies led by artists and neighbors. The process of this work would likely be healing in and of itself. It also might reveal new pathways to more equitable and sustainable solutions to the questions East 9th Street has raised.

The new plan kept some of the original structure, giving 15 artists or artist teams grants, while removing the biggest expense, the three $100,000 integrated art commissions. The new plan also added or adapted many key points, including: grantees did not have to be traditional artists, but did need to show some connection to the neighborhood; the projects no longer had to happen on East 9th Street; and jurors, two from Lawrence and one from out of town, were chosen and paid to evaluate the proposals They could be located or occur anywhere in East Lawrence. We also added two new categories for participation: Neighborhood Specialists, who had special knowledge or skills that could be useful to the selected Project Artists; and a Youth Corps, which funded five projects designed and executed by young people.

I added an updated budget and timeline and shared the whole thing with Cindy when we met. She was encouraged and agreed to partner. Next, we brought it to another ELNA board member, Josh Davis, who had been a strong advocate for the original project, understanding that there was going to be resistance if it were just me. The three of us then presented our plan to ELNA, the City and the Arts Center board. Eventually they all approved, and my job was done. I knew that as the main designer of the plan, I would not be able to apply to be a project artist or the project manager. That was the bargain I made going into it.

The Arts Center led a search for a project manager (I was on the search committee), and were lucky to get Amanda Enfield, an arts administrator who had worked for years with Van Go another highly regarded arts non-profit in town. In the spring of 2019, the 15 artist projects were selected along with the 5 neighborhood specialists and youth corps projects. All of the projects were set to be completed by June of 2020 but have been pushed back due to the pandemic. And although the new plan didn't require projects to exist on East 9th Street, the City followed through and funded a basic street repair of those seven blocks. Go to rebuildingeastninth.com to find out more.

I'd like to say that this resolution healed our neighborhood and that we all buried the hatchet, but that's not true. Although the rebuilding project has been well received, the anger and resentment born of the years of conflict remains to some degree. And, we also didn't stop the developers from continuing to buy property and add to their fantasy of neo-Soho in Kansas. Gentrification continues, it's just not bolstered by a shortsighted placemaking project. What I feel today is that we did what neighbors do – we looked out for each other. We stood up even when there appeared to be no chance that we would prevail, and at least part of the time we had fun doing it. We made our place, East Lawrence, a little better by strengthening our bonds, articulating and sharing our stories, and providing opportunities for our neighbors to celebrate together. We were good placekeepers.

Coda

I can't finish this chapter without talking about what it means to be writing this in the midst of the COVID-19 pandemic. The idea of creative placemaking and its focus on vibrancy and density feel pretty distant, almost nostalgic today. Most of my work has been cancelled or postponed, and like many of my peers in the community-based art world, I applied for unemployment for the first time in my life. But there's something else. East Lawrence and our neighborhood association ELNA has come together during a difficult time once again. We have set up a mutual aid network to ensure that everyone has what they need, including the opportunity to just talk. And there's been some interesting new art that's gone up on the street too – hopeful signs that our spirit is strong.

References

Bedoya, R. (2013). 'Placemaking and the politics of belonging and dis-belonging', *Grantmakers in the Arts READER*, 24(1). (the table of contents does not show page numbers https://www.giarts.org/reader-24-1)

Horowitz, A. and Goldbard, A. (2019). The U.S. Department of Arts and Culture (USDAC) [online]. Available at: https://usdac.us/ (Accessed: 2 June 2020).

Lowenstein, D. (2012). Creative Placemaking Conference Presentation. Lawrence, KS: Spencer Museum of Art.

Placekeepers. [online]. Available at: https://eastninth.net/ (Accessed: 2 June 2020).

Solnit, R. (2010). *A Paradise Built in Hell: The Extraordinary Communities that Arise in Disaster.* London: Penguin Books.

Further reading in this volume

Chapter 1: Introduction: what really matters – moving placemaking into a new epoch
Cara Courage
Chapter 2: Placemaking as an economic engine for all
James F. Lima and Andrew J. Jones
Chapter 3: An annotated history of creative placemaking at the federal level
Jen Hughes
Chapter 4: A future of creative placemaking
Sarah Calderon and Erik Takeshita
Chapter 5: Making places for survival: looking to a creative placemaking past for a guide to the future
Jeremy Liu
Chapter 10: From moon village to mural village: the consequences of creative placemaking in Ihwa-dong, Seoul
Jason F. Kovacs and Hayun Park
Chapter 14: Experts in their own tomorrows: placemaking for participatory climate futures
Paul Graham Raven
Preface: The radical potential of placemaking
Cara Courage
Chapter 20: Displacemaking 2015 and 2020
Catherine Fennell and Daniel Tucker
Chapter 22: Embedded Artist Project: Epistemic Disobedience + Place
Frances Whitehead
Chapter 24: Artists, creativity, and the heart of city planning
Tom Borrup
Chapter 26: Sculpturing sound in space: on *The Circle and the Square* (2016) by Suzanne Lacy
Trude Schjelderup Iversen
Chapter 31: Seven generations: a role for artists in Zuni PlaceKnowing
Theodore S. Jojola and Michaela P Shirley
Chapter 35: Planning governance:– lessons for the integration of placemaking
Nigel Smith
Chapter 39: Translating Outcomes: Laying the groundwork for interdisciplinary evaluation of creative placemaking
Jamie Hand
Chapter 41: Rituals of regard: on festivals, folks, and findings of social impact
Maribel Alvarez
Chapter 42: Creative Placemaking and Placekeeping evaluation challenges from the practitioner perspective: an interview with Roy Chan
Maria Rosario Jackson

12
PUBLIC TRANSFORMATION
Affect and mobility in Rural America

Lyndsey Ogle

We live in a moment of profound political division in the United States, exacerbated by the monolithic categorization of supposedly antithetical communities. Labels like blue and red, liberal and conservative, urban and rural obscure the nuance of place and its communities in order to frame anyone outside the immediate homophilic sphere as 'Other' – something to be feared and rejected. These polarizing narratives go far beyond party affiliation, pitting, for example, the highly educated against those with less schooling, long-standing community groups against community colonizers, the beneficiaries of technological change against those disenfranchised by it, advocates for diversity against openness and those who privilege stability and security (Galston, 2017). These tensions circle around issues of policy but are fueled by an emotional resonance that drives us to cluster into affective communities, siloed by geography and media, where our biases are reinforced, political leanings intensified (Kintz, 1997; Bishop and Cushing, 2008). Here our view of the opposition shifts from incomprehensible to intolerable.

Drawing on an examination of the iterative and experimental artwork, *Department of Transformation*, conceived by Minnesota-based theatre artist Ashley Hanson, this chapter explores how creative placemaking, as a performative practice of encounter, addresses political polarization in the United States within the context of the rural–urban divide. Challenging a singular narrative of the 'American rural,' which limits the agency and visibility of rural artists and residents, this chapter traces the trajectory of Hanson's project to present a picture of 'rural America' that is hybrid, mobile, and affectively linked. It concludes by looking at the multidisciplinary exhibition that was held as part of Hanson's project and asks how the story of the place can be reperformed and recirculated within an artworld frame without erasing its multiplicity.

Creative placemaking and conversation

> I wanted more, more conversations, just more.
>
> *(Ashley Hanson)*

National conversations about creative placemaking have taken up the topic of community polarization with limited success. As an increasingly interdisciplinary field, the discourse around placemaking is often stymied by divergent disciplinary priorities and the contradictory implications of 'fuzzy concepts' (Markusen, 2003), or terms that not only hold different meanings within different fields but signal very different institutional or social power dynamics. For example, scholarly journals, academic conferences, and museum symposia often bring together stakeholders ranging from artists, city planners, development leaders, cultural organizers, and social interest non-profits to debate the role of creative placemaking in addressing issues like gentrification, the climate crisis, race and class division, and exploitive or exclusionary legislation. These conversations, however, can fall apart when the focus shifts to desired outcomes, including metrics and deliverables, or the parameters for success or failure. Further still, debates about policy struggle to transcend geographically situated knowledges. When focusing on primarily urban issues, creative placemaking discourse fails to consider *how* or *why* proposed interventions might not work in rural contexts, or why what might work in one 'American rural' might not work in them all.

Minnesota-based theatre artist Ashley Hanson is regularly involved in national conversations about creative placemaking and is often one of the only people in the room with her eyes toward the rural. After attending the 'Next Generation Rural Creative Placemaking Summit' in October 2016, an event which brought together rural artists and organizers from around the country, Hanson began to note that creative placemaking discussions – be they situated in a rural or urban context – rely on affective narratives and histories *as much or more* than quantitative data when trying to connect local issues to broader discourses about inequality and reform.

These affective narratives or 'narratives *as felt*' can be understood through what sociologist Arlie Russel Hoschield (2016) calls the 'deep story':

> A deep story is a feels-as-if story—it's the stories feelings tell, in the language of symbols. It removes judgment. It removes fact. It tells us how things feel. Such a story permits those on both sides of the political spectrum to stand back and explore the subjective prism through which the party on the other side sees the world.
>
> *(Chapter 9, para. 1)*

Hanson saw untapped potential in sharing stories and connecting disparate artists from around the country but resisted the formal structure of the conference, where often the most fruitful dialogue takes place on the fringes of formalized conversation – in hallways, hotel rooms, and late-night drinking sessions. She knew more rural artists had to be out there, ones without the means or access to congregate in a central forum, and so she began to think about what it might look like to bring the conversation to them instead.

The concept of creative placemaking has evolved into an umbrella term bringing together projects concerned with the material or socioeconomic conditions of space and place alongside practices elsewhere labeled 'social practice,' 'socially engaged,' or 'dialogic' art. Though geographically situated, these projects also examine the experience of encounter fostered *through* or *as* artistic practice. Creative placemaking then, if thought of as *intentional* collaboration with and in communities, is particularly equipped to engage with and unpack the affective narratives that drive partisan divide, especially when situated in communities that are ambivalent or even antagonistic to the field or idea of art practice. Such works not only challenge the *assumed* progressive politics of place-based work or the conservative agendas of local community agents but are forced to reconcile the goals of outcome-focused institutions with process-oriented or iterative art practice.

Most importantly, however, these works take up the 'thorny issue' of 'how, when, and with whom' one should enter into dialogue. By engaging with the 'opposition,' we acknowledge what and whom we do not know. We confront our assumptions about the beliefs and priorities of those we consider to be de facto members of our community, and we acknowledge who or what we want to admit into an art world context. In this way, writer and curator Monika Szewczyk (2009) argues, conversation is always political *and* aesthetic:

> [I]f as an art, conversation is the creation of worlds, we could say that to choose to have the conversation with someone is to admit them into the field where worlds are constructed. And this ultimately runs the risk of redefining not only 'the other' but us as well. Art and conversation share this space of invention, yet only conversation comes with a precondition of plurality that might totally undo the creative agent.
>
> *(p. 2)*

By engaging with oppositional communities through conversation *as* aesthetic form (rather than simply framing 'the opposition' as a subject to resist), this work allows both practitioners and practice to be transformed.

For Hanson, devising artistic works through conversation is nothing new. With over a decade working in applied theatre and arts programing, Hanson has developed a practice that draws on a wide range of ethnographic methodologies, rooted in oral history traditions and feminist activism. As Co-Founder of the theatre company PlaceBased Productions, Hanson partners with rural agencies to develop community-led, site-specific musicals. Rather than simply creating a 'a vague aesthetic of progressive uplift' (Davis, 2013) in conservative communities, Hanson uses multidisciplinary art practices and open-ended dialogue to address issues of policy and the long-standing roots of community conflict.

Though PlaceBased Productions works exclusively with rural communities, as of 2016, Hanson was a self-described 'nomadic artist,' splitting time between her home-bases in Minneapolis, Minnesota and Boulder, Colorado. With investments in both rural and urban contexts, she had a unique vantage point from which to observe the changing discourse around American politics leading up to and following United States presidential election. Mobilized by what she saw as a profound disconnect between rural and urban communities, Hanson launched the *Department of Public Transformation* in the winter of 2017 to not only better understand the role of art in the construction of a national narrative of the rural within the United States, but to also confront essentialist depictions of place that exacerbate political polarization and the rural–urban divide.

Complicating the rural–urban binary

Recognizing that in the United States a national understanding of the rural is largely shaped by images and narratives created by and for urban audiences (Robinson, 2016), the rural comes to be known in terms of what it is not – as 'the other' of the urban, or 'what is left' when the specificities of the urban have been taken into account (Ratcliffe et al., 2016). The US Census Bureau, for example, has defined the rural as everything that is not urban (Ratcliffe et al., 2016). As urban centers have spread into sprawling metros and suburbs, the census has adjusted its definition of the urban to take into account 'urban clusters' of at least 2,500 residents linked together by spatial 'hops' and 'jumps,' without adjusting the criteria of the rural. The result is a convoluted metric for making the geographic distinction in which a single county or tract can be classified as both urban and rural. This ambiguity reinforces a rural–urban binary that queer theorist Scott Herring (2010) contends 'is as much context-specific, phantasmatic, performative, subjective' – and, he emphasizes – '*standardizing*, as it is geographically verifiable' (p. 8). While it is not generally thought to be harmful, in the way that binaries of race or gender are often discussed (Dymitrow and Brauer, 2017), the rural–urban binary does establish a power hierarchy. It is maintained through 'structures of intense feeling' or affective narratives that demand rural individuals or groups experience themselves in relation to the 'dominant spatial performatives of the 'urban'' (Herring, 2010, p. 13), narratives that preclude any commonality of culture or experience, and disavow any aesthetic expression outside the mandates of hegemonic culture.

The rural–urban binary not only marks the two geographies as antithetical but as mutually constitutive. Thus, the prevailing narrative of the rural as a white, patriarchal, heteronormative, conservative, working-class monoculture functions, in part, as the antithesis of the progressive, multicultural urban. The urban is the site of individuality, culture, and progress, while the rural is 'a bastion of the "mass," undifferentiated, unhip people and perspectives' (Johnson, 2008, p. 5). The urban is constantly evolving. The rural is static. The urban is future. The rural is past.

These narratives construct a 'moral geography' (Massey, 2007) in which the rural is determined to be either a success or failure based on its use-value in the urban marketplace. The 'rural idyll' (Little and Austin, 1996) or the notion of rural as a 'culturally laden' place (Darby, 2000, p. 54) functions as a 'theatre of consumption' (Leiss et al., 1986) for urban tourists or migrants to reconnect with nature and in turn, find their true, authentic selves. Alternatively, the rural is described as a site of 'deprivation' (Woodward, 1996), as a 'dead zone' (Kenway, Kraack, and Hickey-Moody, 2006), devoid of culture, resources, or value. The rural becomes a place to escape from (to the progressive embrace of the city) or a place to escape to (from the incessant speed, inauthenticity, and performative demands of claustrophobic urban centers). Neither narrative recognizes the way in which both often operate simultaneously and in conflict within the same place, and both fail to consider the implications of place as fundamentally *in flux*. Place as verb rather than noun.

The Department of Public Transformation tour

Part documentary art project, part mobile artist residency, the *Department of Public Transformation* began with a six-week tour of rural America, visiting towns with populations under 10,000 where Hanson spoke with artists and community residents about the realities of rural placemaking. With minimal seed funding from a Kickstarter campaign, as well as support from Art of the Rural, the McKnight Foundation, and Minnesota non-profit, Springboard for the Arts, Hanson purchased a converted school bus, lovingly named 'Gus the Bus,' and set off from central California, heading south through the pueblos of the southwest, cattle ranches and oilfields of Texas, the bayous of Louisiana, along abandoned Appalachian coal mines, across the rust belt and over the great plains, ending in the town of Byers, Colorado, population 1,160. What Hanson discovered along the way was a picture of the American rural that is, to quote Michael Woods (2010), 'hybrid, co-constituted, multi-faceted, relational, [and] elusive' (p. 265).

As an iterative, multidisciplinary experiment in creative placemaking, *Public Transformation* was both a survey of the field and a discrete artwork in and of itself. During the six-week journey, Hanson was joined by artists-in-residence, Hannah Holeman, Ellie Moore, and Randi Carlson, each for different legs of the tour. Along the way, they had conversations with 127 artists in 24 towns. Some of the stops were to connect with artists Hanson knew or met at the Summit, while others were through mutual connections or from organizations she found online. Her ability to 'authentically' perform her claim to the rural by rooting her identity in 'five generations of bad ass rural women,' as well as her capacity to 'code-switch' between the urban and rural as a dually placed, dually invested artist, lent her what cultural geographer Tim Edensor calls rural 'performative competency' (2006, p. 485). Geographer Mirek Dymitrow and social scientist Rene Brauer (2017) contend that taking into account the position and performance of the researcher/artist adds a significant dimension to our understanding of 'rural performativity.'

It importantly 'acknowledges that concepts and categories take shape through processes influenced by history, discourses, ecologies, and power relations (Dahlberg, 2015, p. 207).

The rural is a powerful conceptual framework that has historically struggled to address its multiplicity. Rural studies as a discipline has struggled to reconcile a definition of 'the rural' that simultaneously takes into account the varied dimensions of place. Drawing on Lefebvre's spatial triad, geographer Keith Halfacree (2006) offers a three-fold model of rural space that accounts for: the spatial practices of rural locality (in relation to their production/consumption activities); the formal representations of the rural as expressed by capitalist, governmental – and I would argue – artistic interests or forms; and the lived experiences of the rural, which include both individual and sociocultural experiences, as well as their 'interpretation and negotiation.' Performance, in turn, offers a bridge by which to connect the materiality of the rural to its discursive and representational forms in a way that recognizes its embodied and affective dimensions (Edensor 2006; Woods, 2010), including the performance of the scholar or artist who documents, and therefore intercedes into rural space (Dymitrow and Brauer, 2017).

The challenge for *Public Transformation* and other placemaking projects hoping to capture and stage the stories of place thus becomes how to highlight rural heterogeneity without dismissing the political potential of 'the rural' as a unifying theoretical concept (Cloke, 2006). Hanson's project, therefore, attempts to disrupt the binary narratives of place by *performatively* enacting the flows of mobility that mark the rural as 'hybrid and networked' (Woods, 2009, p. 851), shaping the project as a kind of experimental listening session, or what curator Mary Welcome calls the art of 'the deep hang.' Hanson's aim was to bring inquiry into the place of the familiar, to embrace the wide and varied cultural traditions of rural communities, to strip instrumentality from both the research and artistic process, to meet local artists on their own terms, and see community through their eyes – 'to come to your kitchen table, and hear the sounds of your place, to smell the smells of your place, and watch you work.'

Recognizing that, as Edensor (2006, p. 488) argues, rural space is 'highly stage-managed' in a way that contributes to the sense of 'authenticity' characterizing many depictions of the rural (as well as the 'inauthentic' packaging and commodification of rural culture), Hanson took part in practice ranging from the improvisational and unconscious performances of everyday life (Goffman, 1959) to scripted practices of art-making and cultural tradition. Hanson went to people's homes and places of business; traveled with them as they took her on multiple-hour tours of their historic landmarks and local haunts; held story circles and participated in local media broadcasts; attended barbecues, hootenannies, bluegrass jams, and potluck suppers, traditional performances of native ritual, and exhibitions of contemporary native American art. Embedded in deep cultural tradition, the conversations were acts of creative exchange, pushing against cultural assumptions and comfort zones.

The performative aspect of these encounters offered a deep and resonant engagement otherwise reserved for longer ethnographic projects. As such, the experimental practice was concerned more with the dynamics of the interview than the questions being asked (Riley and Harvey, 2007), proposing the question of whether deep and lasting connection can be created through ephemeral encounter. While it is easy to dismiss the kind of breath-over-depth approach taken by Hanson, it is worth recognizing the affective resonance that can result from the ethno-poetic practice of 'visiting.' 'To visit,' is a cultural performance specific to rurality (Stewart, 1996), a reimagining the situated interview as a kind of 'walking and talking' that takes place not just in the field but literally *in the fields*. This performative and mobile engagement can evoke memories that might not otherwise be conjured or reveal the 'neglected rural others' – stories and people who exist outside our generalized narrative of the American rural (Riley and Harvey, 2007, p. 392).

Staging the stories of place

Following the tour, Hanson and the *Public Transformation* artists-in-residence, Holeman, Moore, and Carlson, spent two weeks at M12's Feedstore in Byers, Colorado developing original works based on the stories and ephemera collected on their journey. During the tour, they compiled nearly 100 hours of video and audio footage. The interviews shifted with the presence of each artist, recognizing the encounters were as much about what was being brought into place as what was collected. To that end, Hanson collected an artifact at each stop and asked a representative from each town to write a letter to an artist at the next stop, which she delivered on her way. These letters, though for the most part short and sweet, were a way to connect the communities directly – a tangible and analogue response to the loose bonds typically created by digital or formal networking.

Later they would reconvene in Winona, Minnesota, with filmmaker Nik Nerburn, who developed two short films about the project, and citizen artist Mary (Welcome) Rothlesberger, to plan the exhibition to take place at Art of the Rural's Outpost Gallery. The result was *Department of Public Transformation: Art in Rural America*, an exhibition of art and ephemera, which ran from October 19, 2017, to December 16, 2017. It was kicked off by a weekend of dialogues, screenings, performance, and deep hangs, in which the artists, residents from the community of Winona, as well as rural and urban communities near and far, came together to discuss the role of the artist in rural communities and to debate the meaning of community all together.

If the tour was an open-ended experimental listening project, the exhibition was a highly curated reflection, not just of what was heard, but of how the artists' positionally as multiply-placed female traveler-artists shaped their reception within the communities they visited and how the stories they captured might be recirculated and mobilized within the space of the gallery and beyond. The journey of the *Public Transformation* tour also offered a useful metaphor for reading the structural and thematic organization of the exhibition. The spatial and temporal trajectory of the journey was translated in the space of the exhibition by a series of interconnected gestures. The flow of the display physically launched the visitor along a temporally organized circuit, moving from the site of the artist visits, to the resident artists' works created during their post-tour residency in Byers, to the present space of the gallery, where visitors lingered, socializing and taking Polaroids in front of Gus the Bus before meandering to a local brewery for a bluegrass jam. Morning strolls by the river and even a restaurant crawl moved participants back and forth between the temporally and spatially situated narrative of the tour and the present reflection about the future of rural communities.

There is a risk in curating the story of place of peddling 'intensity' or contrived meaning. It is a challenge that is amplified by the curation of decontextualized objects alongside interview snippets or soundbites. In many cases the challenges and opportunities of place, the deeply engrained trauma of racial, class, and gendered oppression were alluded to but left vague. An original artwork from Whitesburg, Kentucky, for example, that bore the phrase 'I can build my dreams here,' hinted at the idea of building a life amid rural challenges but did not directly address the context of its production, the Appalachian opioid epidemic. These stories did make their way out, however, in the conversations that took place in the space as part of the weekend's dialogues and during the 'deep hangs.' In this context, objects in the gallery ceased to be self-contained. They served to prompt conversation and activate the gallery within the context of sited exchange.

Performing place

For the closing event of the opening weekend Hanson gave an interactive performance lecture, narrating the story of the *Public Transformation* tour. Moving performance into the space of the gallery is not a new practice. Even the curator's talk has become institutionalized in the context of the monographic or thematic show. In fact, as curator Clementine Deliss explains, the power of an ethnographic collection is in its mobility, the possibility of multiple reconfiguration and the new and varied meanings they produce. For Hanson, however, the practice went a step further. In a gesture that sought to shrink the spatial, political, and ideological distance between the audience and communities on view, Hanson individually removed the artifacts from the visual frame that rendered them immobile, and circulated them through the audience, into an imagined future where they might be shared once again as a way to reactivate the idea of community and the collective power of the 'rural.'

While it could be argued that the artifacts functioned like props for the performance, and therefore in service of the performer rather than their site of origin, their circulation carried a specific affective resonance. As the items were passed from person to person, Hanson echoed Tom Kennedy of the Zuni Art and Visitor Center when he asked 'mainstream' visitors to 'look deeply' into works of native art. The audience for Hanson's performance was asked not just to look into the objects but to feel their significance. There was a kind of deference paid to the movement, delicate but curious, like the most precious object at a child's class of show and tell.

This was not the first time Hanson gave the performance. Hanson had previewed it following the trip and at the Rural Arts and Culture Summit but, unlike previous performances, Hanson shifted from a chronologically organized narrative to one that affectively linked the disparate rural communities to 12 archetypal roles she found 'bubbling to the surface' in the interviews. As Louis Stewart (1987), scholar of Jungian psychoanalysis explains, the archetype is a way of knowing the world and while not all affects are archetypes, all archetypes convey affect. This is not to say that all archetypes will be read in the same way but rather they are intended to draw on personal narratives or understandings of the self that connect individuals to collective human experience. The translator, storyteller, listener, mentor, rabble rouser, curator, host, connector, architect, dreamer, archivist, and energizer spatially situated the artist and their story in relation to their community and what Holly Barcus and Stanley Brunn (2010) call, the 'elasticity of place' (p. 284) or the bonds to place, permanence, and portability, allowing connections to place to remain even as they are altered by the relations of mobility.

In a move that brought her narrative full circle, Hanson began the story with the role of the translator, referencing the many 'unofficial' stops made on the tour to speak with other 'translators' Gabriela Munoz from the Arizona Commission for the Arts and Savannah Barrett of Art of the Rural who are working in urban spaces while trying to bridge the rural–urban divide. Hanson progressed through her tale linking the labor performed by 'mentor' Warren Montoya of Resilience and Resonate Art in Santa Ana Pueblo, New Mexico (pop. 479) through that of artist Calvin Phelps, 'architect' of community dialogue with the Pike School of Art in Summit, Mississippi (pop. 1682).

At the most (sadly) serendipitous moments, the artifacts being circulated among the crowd reflected not just their community of origin but the narratives-as-*felt*, the deep stories of place. A small sculpture of two antlers locking horns from Show Low, Arizona symbolized the impasse faced by writer and theatre artist Lisa Jayne, who had recently given up her art practice all together after her attempts to slowly introduce progressive content into her rural community, 90 per cent of which were comprised of members of the Latter Day Saints, by staging the musical *Grease* was met with scandal and an empty house. 'I think as an artist, the struggle is just being able to talk about new ideas in our community and the traditional doesn't always appreciate that,' she explained. 'The traditional wants to invest in the status quo, so if it is not traditional, it is considered a threat. It is a new idea, it is anti-"what the community is about" and it is therefore suspect.'

In some cases, the connection between object and affective narrative was even more literal. As Hanson recounted the story of Rachael Ensor of Murphysboro, Illinois (pop. 7,970) audience members felt the weight of a brick from the small home where Ensor runs the Murphysboro School of Art. In her community, Ensor explained that there is very little access to the arts, and even less access to social services. During the interview, she joked that art gave her relief from the stress of rural life, 'It means I don't have to take my Xanex.' Though said in jest, the weight of the brick collided with the subtext of Ensor's reflection, and lingered as Hanson went on to describe the complicated feelings different artists on the tour had about their work serving as stop gaps for desperately needed services in rural areas.

The combination of these performative strategies was most effective when used in combination. During a particularly poignant moment, Hanson began to tell the story of visiting Whitesburg, Kentucky, home of the long-standing arts organization, Appalshop, and meeting with Institutional Development Director, Ada Smith. Hanson began telling the story of driving around Whitesburg in the rain for hours, because as Ada had explained, 'to understand Appalshop you first have to understand this place,' when Ada's image and distinct Kentucky accent appeared on the screen. Walking up a gravel path, Hanson's camera bouncing in time, Ada led Hanson, and in turn the viewer, through a hole in a barbwire fence onto the site of an abandoned coal mine. As Ada explained the history of the site and the way the shutting of the mine transformed the town, she picked up a piece of coal and presented it to the camera in her outstretched hand. At the same time, the smooth, black rock was passed through the audience, collapsing the distance of space and time. Here Hanson picked the narrative back up. She didn't reflect on the number of lost jobs or the community exodus but Ada's journey of leaving and coming home, of the mobility and affective narratives that bind us to place.

Conclusion

To evaluate Hanson's project requires us to consider if and how ephemeral encounter can produce an affective resonance capable of connecting people through the 'deep story' of place. It can be argued that to situate this project within any predetermined disciplinary framework would inevitably deem it a 'failure.' It does not easily slide into parameters we understand. It is not extensive enough to produce the substantiated narratives of place we associate with traditional ethnography. It is not aesthetically specific or developed enough to be clearly read within a singular artistic discipline. The end product does not translate into valued metrics – network nodes created, project visibility gained, and impressions calculated, opinions changed.

Following the tour and exhibition at the Outpost Gallery in Winona, Minnesota, Hanson initially planned to do a second iteration of the project that would include fewer but more sustained visits to rural communities. Upon reflection, however, Hanson noted that time and time again, what community members on the *Public Transformation* tour said they really needed was more lasting investment by artists in rural communities. At the same time, the opportunity to create a long-term community partnership arose in the form of a previously vacant Main Street storefront in Granite Falls, Minnesota (pop. 2,734). Here, the Department of Public Transformation now exists in a permanent site, developing the Granite Falls City Artist in Residence,

> embedding arts and cultural workers in the City of Granite Falls to design and implement arts and cultural strategies to increase civic participation and community engagement in policy-making, planning and public processes. placing temporary artists-in-residence within city agencies to develop creative solutions to local problems.
> *(City Artist-in-Residence, n.d.)*

It could be argued that this 'tangible outcome' of the *Public Transformation* project – a program leveraging artistic labor for 'creative solutions' to community problems – clearly demonstrates that placemaking is inevitably the handmaiden of neoliberalization, asking art practice to stand in for badly needed community services, cut by state and federal austerity measures. This argument is not necessarily wrong. Still, as we become more globally and digitally linked, while simultaneously further ideologically and affectively divided, it feels misguided to call for *more* boundaries – socially, culturally, disciplinarily, or otherwise. It also feels overly simplistic to accuse the artists, community agencies, or any of the myriad other creative placemaking stakeholders of being either unwitting dupes or complicit collaborators of the project of neoliberal capitalism (Jackson, 2020).

Much like the monolithic narratives that undergird political polarization, the reality is far from a simple binary of moral positions. For many, a relationship to art, or lack thereof, is the result of what land artist and farmer Nikiko Masumoto, of Masumoto Family Farm in Del Rey, CA (pop. 1,639), calls a 'creative wound.' Whether the result of an artistic 'failure' at a young age, alienation from art discourse as an adult, or even as the cumulative toll created by ceaseless neoliberal demands for creativity and innovation in all aspects of our personal and professional lives, these 'creative wounds' personalize and internalize narratives about the sociopolitical, spatial, *and* structural conditions of art.

In the same way projects such as Hanson's question both the *assumed* politic of rural artists, their artworks, and communities in which the art is situated, they also force us to ask what it really means to frame such creative placemaking efforts as 'art for the common good' (Deutsche 1996; Massey 1994). This subsequently requires us to recognize critiques from both progressive and conservative critics who ask, 'what public?' and 'whose good?' Creative placemaking, therefore, in all its interdisciplinary forms, cannot simply be concerned with the ways in which art may open the hearts and minds of those resistant to a progressive truth. Rather it must consider how the contexts and circumstances of those thought to be 'the opposition,' open space to question the essentialist narratives that exacerbate political divides.

References

Barcus, H.B. and Brunn, S.D. (2010). 'Place elasticity: Exploring a new conceptualization of mobility and place attachment in rural America', *Human Geography*, 92(4), pp. 281–295.

Bishop, B. and Cushing, R.G. (2008). *The Big Sort : Why the Clustering of Like-Minded America Is Tearing Us Apart*. Boston, MA: Houghton Mifflin.

Cloke, P. (2006). 'Conceptualizing rurality', in Cloke, P., Marsden, T. and Mooney, P. (eds.) *Handbook of Rural Studies*. London: SAGE, pp. 18–28.

Dahlberg, A. (2015). 'Categories are all around us: Towards more porous, flexible, and negotiable boundaries in conservation-production landscapes', *Norwegian Journal of Geography*, 69(4), pp. 207–218.

Darby, W.J. (2000). *Landscape and Identity: Geographies of Nation and Class in England*. Oxford: Berg.

Davis, B. (2013). 'A critique of social practice art: What does it mean to be a political artist?', *International Social Review*, 90.

Department of Public Transformation (n.d.) *City Artist-in-Residence* [online]. Available at: https://www.publictransformation.org/ (Accessed: 9 June 2020).

Deutsche, R. (1996). *Evictions. Art and Spatial Politics*, Cambridge, MA: MIT Press.

Dymitrow, M. and Brauer, R. (2017). 'Performing rurality. But who?', *Bulletin of Geography. Socio-Economic Series*, 38(38), pp. 27–46.

Edensor, T. (2006). 'Performing rurality', in Cloke, P., Marsden, T. and Mooney, P., (eds.) *Handbook of Rural Studies*. London: SAGE, pp. 484–495.

Galston, W.A. (2017). 'The populist moment', *Journal of Democracy*, 23(2), pp. 21–33.

Goffman, E. (1959). *The Presentation of Self in Everyday Life*. New York: Anchor Books.

Halfacree, K. (2006). 'Rural space: Constructing a three-fold architecture', in Cloke, P., Marsden, T. and Mooney, P., (eds.) *Handbook of Rural Studies*. London: SAGE, pp. 44–62.

Herring, S. (2010). *Another Country: Queer Anti-Urbanism*. New York: New York University Press.
Hochschild, A.R. (2016). *Strangers in Their Own Land: Anger and Mourning on the American Right*. New York: New Press.
Johnson, V. E. (2008). *Heartland TV: Prime time television and the struggle for U.S. identity*. New York: New York University Press.
Kenway, J., Kraack, A. and Hickey-Moody, A. (2006). *Masculinity Beyond the Metropolis*. New York: Pelgrave Macmillan.
Kintz, L. (1997). *Between Jesus and the Market: The Emotions that Matter in Right-Wing America*. Durham, NC: Duke University Press.
Leiss, W., Kline, S. and Jhally, S. (1986). *Social Communication in Advertising: Persons, Products and Images of Well-being*. Toronto: Metheun.
Little, J. and Austin, P. (1996). 'Women and the rural idyll', *Journal of Rural Studies*, 12(2), pp. 101–111.
Markusen, A. (2003). 'Fuzzy concepts, scanty evidence, policy distance: The case for rigour and policy relevance in critical regional studies', *Regional Studies*, 6(7), p. 701.
Massey, D.B. (1994). *Space, Place and Gender*. Cambridge: Polity.
Massey, D.B. (2007). *World city*. UK: Polity Press.
Ratcliffe, M., Burd, C., Holder, K. and Fields, A. (2016). *Defining Rural at the U.S. Census Bureau*, ACSGEO-1. Washington, DC: U.S. Census Bureau.
Riley, M. and Harvey, D. (2007). 'Oral histories, farm practice and uncovering meaning in the countryside', *Social and Cultural Geography*, 8, pp. 391–415.
Robinson, J. (2016). *Theatre and the Rural*. New York: Palgrave.
Stewart, K. (1996). *A Space on the Side of the Road : Cultural Poetics in An 'Other' America*. Princeton University Press.
Stewart, L. H. (1987). 'Affect and archetype in analysis', in N. Schwartz-Salant & M. Stein (eds.) *The Chiron clinical series. Archetypal processes in psychotherapy*. Chiron Publications, 131–162.
Szewczyk, M. (2009). 'Art of conversation, Part I', in *E-flux Journal*, 3.
Woods, M. (2009). Rural geography: blurring boundaries and making connections. *Progress in Human Geography*, 33(6), pp. 849–858.
Woods, M. (2010). *Rural*. [electronic resource]. Routledge.
Woods, M. (2010). 'Performing rurality and practicing rural geography', *Progress in Human Geography*, 34(6), pp. 835–846.
Woodward, R. (1996). '"Deprivation" and "the rural": An investigation into contradictory discourses', *Journal of Rural Studies*, 12(1), pp. 55–67.

Further reading in this volume

Chapter 3: An annotated history of creative placemaking at the federal level
Jen Hughes
Chapter 14: Experts in their own tomorrows: placemaking for participatory climate futures
Paul Graham Raven
Chapter 17: 'I am not a satnav': Affective placemaking and conflict in 'the ginnel that roared'
Morag Rose
Chapter 19: Placemaking in the ecology of the human habitat
Graham Marshall
Preface: The radical potential of placemaking
Cara Courage
Chapter 20: Displacemaking 2015 and 2020
Catherine Fennell and Daniel Tucker
Chapter 26: Sculpturing sound in space: on *The Circle and the Square* (2016) by Suzanne Lacy
Trude Schjelderup Iversen
Chapter 29: The solution is in the problem; the art of turning a threat into an opportunity by developing resilience using a creative placemaking critical praxis
Anita McKeown
Chapter 30: Ecological selves as citizens and governance as ethical placemaking
Lisa Eckenwiler
Chapter 31: Seven generations: a role for artists in Zuni PlaceKnowing

Theodore S. Jojola and Michaela P Shirley
Chapter 32: *The Hollywood Forest Story*: Placemaking for the Symbiocene
Cathy Fitzgerald
Chapter 34: Reconnecting *cité* and *ville*
Philip Graus
Chapter 40: Transforming community development through arts and culture: a developmental approach to documentation and research
Victor Rubin
Chapter 41: Rituals of regard: on festivals, folks, and findings of social impact
Maribel Alvarez
Chapter 43: A theory of change for creative placemaking: the experience of the National Endowment for the Arts' *Our Town* program: an interview with Patricia Moore Shaffer, PhD
Maria Rosario Jackson

13

SENSING OUR STREETS

Involving children in making people-centred smart cities

Sean Peacock, Aare Puussaar, and Clara Crivellaro

Introduction

Digital technologies pose significant opportunities and challenges for the future of urban life. Many regeneration schemes now come packaged as 'smart city' programmes, which integrate Internet of Things (IoT), sensor technologies, and data collection devices into the urban fabric to guide city officials in making their decisions (Karvonen et al., 2018). However, these programmes have been criticised due to their limited potential for transforming socioeconomic conditions, threats to privacy and public space, and their top-down imposition on communities (Cardullo and Kitchin, 2018; van Zoonen, 2016). Such criticism has pushed scholars and practitioners developing processes and systems to critically engage citizens in the visions and technologies driving the smart city agenda and open up civic participation in shaping the way smart technologies could improve cities (Foth, 2017; Fredericks et al., 2018; Hunter et al., 2018). Here, we are concerned with children's participation and the significance of engaging children with smart technologies, to prepare them for inheriting smarter cities and equip them with the tools to hold decision-makers to account.

In this chapter, we ask: how can we critically engage children with smart technologies and through this support their participation in placemaking? Building on our recent work (Heath et al., 2019; Peacock et al., 2018) we collectively define placemaking as a practice of reimagining public spaces with the input of citizens. Here, we report on a pilot engagement that took place in a primary school in Newcastle upon Tyne, UK, that we designed for children to engage critically with smart city tools. Drawing on literature from placemaking, urban planning, and Human–Computer Interaction (HCI), we provide rich qualitative insights and methodological reflections from our engagement and offer an incremental but significant contribution to the literature on this burgeoning topic. In doing this, we seek to attend to children's absence in placemaking and smart cities by proposing implications and strategies for addressing their exclusion.

Background work

Technology's greater role in the management of public spaces poses significant implications for placemaking (Karvonen et al., 2018; Tenney and Sieber, 2016). The companies responsible for these technologies claim that providing public servants with data leads to 'better' decision-making, but scholars have argued smart cities pose privacy concerns and offer little value for ordinary citizens (DiSalvo and Jenkins, 2017; Krivý, 2016; van Zoonen, 2016). The overarching neoliberal logics that accompany smart cities pose urgent questions of who smart cities are for and who has a right to shape them (Heath et al., 2019). In response, calls for people-centred smart cities seek to involve those without a voice in city-making processes and create spaces for critical engagement with the visions and technologies driving smart cities (Hunter et al., 2018). This would help move us beyond 'cookie-cutter' approaches (Townsend, 2013) towards more inclusive and sustainable smart cities (Heitlinger et al., 2019).

The vision of smart cities involves integrating technical sensors into the urban fabric to gather precise data about the environment, infrastructure, and resources (Ratti and Claudel, 2016). These amplify the abilities of human senses, e.g. to 'see' air pollution invisible to the eye (DiSalvo and Jenkins, 2017). Scholars suggest community-led investigation, using portable and low-cost versions of these tools (Balestrini et al., 2017), could open new avenues of constructing knowledge in smart cities and involve a broader range of citizens in their realisation. However, to date, limited engagement has taken place with politically marginalised groups (Balestrini et al., 2017; Gabrys, 2016).

Children and the smart city

Despite increasing recognition of the important role of citizens in crafting smarter cities (McKinsey Global Institute, 2018; Wilson and Chakraborty, 2019), children are rarely involved in built environment projects supported by smart technologies. Understanding children's relationships with public space, digitally augmented or otherwise, is fundamental to our holistic understanding of cities (Horschelmann and van Blerk, 2013). Yet only a handful of scholars have explored the role of children in the smart city at all (e.g. Nijholt, 2019; Scholten, 2017; Wolff et al., 2019), and no known work has provided opportunities for children to engage with smart technologies in public spaces.

In the last 15 years, digital methods such as computer simulation games have gained traction for youth engagement in city-making (de Andrade et al., 2020; Mallan et al., 2010). Previous work from two of the authors has involved children using bespoke digital tools to specifically involve children in placemaking (Peacock et al., 2018). This involved using open-source mapping software in combination with school activities to translate formal placemaking processes into meaningful and creative engagements with children. Here, adults and children working together in a community placemaking project provided new insights and opportunities for improving their neighbourhood; but it also opened new spaces of conflict as children's views were 'disruptive' to the process (ibid, p. 8). This work reiterated the need to configure placemaking processes that legitimise children's contributions and support their participation alongside adults (Freeman and Tranter, 2012). Building on this work, we set to involve children in a critical interrogation of smart city tools and the value of such tools for participatory placemaking.

Research context

Newcastle upon Tyne has recently launched its own smart city initiative, centred on deploying sensors to monitor waste disposal, traffic, parking, and air pollution, and pushing this data to a publicly accessible online dashboard (Urban Observatory, 2020). Meanwhile, air pollution is found to breach safe legal limits on some major roads in the city. In response, residents in a neighbourhood two miles from the city centre founded an action group to campaign for measures to reduce private car use (e.g. new segregated cycle routes.) Seeking ways to place children at the centre of their efforts, this group asked for our help to gather evidence of air pollution with the involvement of local children. This led us to co-design a pilot engagement in collaboration with a local primary school. This school fronts a busy road close to the neighbourhood centre, and over half of its students live in walking distance. The headteacher also corroborated air quality, traffic, and noise pollution as issues affecting them. We worked with one class of Year 4s (third graders) across the three workshops, at the headteacher's recommendation, and designed the workshops in collaboration with their class teacher to ensure it would be appropriate for students' needs and abilities.

Designing our engagement

We structured activities around using handheld versions of environmental sensing tools to gather evidence and come up with placemaking ideas for their neighbourhood. For our pilot, 27 children took part in three consecutive workshops: a sensory exploration of their neighbourhood close to the school, achieved through a guided walk using handheld sensing tools, a voice recorder, and an iPad as tools for observation recording; thinking through the findings from their walk through creative activities (e.g. story-writing); and generating placemaking ideas to improve their neighbourhood and communicating these through posters, drawings, 3D models, and demonstrations. The children we worked with were 8–9 years old, and there was a close-to-even gender split. The class was demographically diverse and of mixed ability. At the teacher's recommendation, children worked in pre-set teams of five to six throughout. We delivered our three workshops in normal school time – the first two lasting two hours each, and the third three hours with a break. Prior to running these, we sought full ethical approval for our data collection through our university's ethics committee.

The three different sensing tools we obtained could monitor and display live readings of air quality, environmental sound levels, and the speed of oncoming traffic. We structured our engagement around questioning what they could 'sense' in their environment using the five senses – sight, hearing, smell, touch, and taste (Goodchild, 2011). Curious about the creative possibilities of what our participants might be able to sense with, and without, the tools, we structured the engagement as a comparison between the two. Using their human (embodied) senses, and using the (technical) sensing tools, how would our participants 'make sense of' the environment and how would this contribute to their evidence-gathering? What could they see, hear, smell, and feel through these two mediums?

Once the children had brought their ideas to life, we invited four representatives from the city council to join with the children in an 'Ideas Carousel.' This was an opportunity for the children to exhibit their evidence and ideas to give peer-to-peer feedback (i.e. what did you like most about this group's idea?), discuss the implications for the wider community (i.e. what would need to happen locally to make it a reality?), and influence city decision-making with the city officials present. Throughout this activity, the class teacher emphasised students (and the visitors) giving positive and constructive feedback, as opposed to critique.

Insights on a pilot engagement

Our insights centre on three observations from our engagement. First, how the contrasting of embodied and technical sensing generated issues of trust in relation to data production. Second, the exploration of playful possibilities for sensing their environment. Third, the ways in which the tools supported the generation of ideas and responses to local issues.

Contrasting embodied and technical sensing

Our first observation was that the children approached their exploration with a degree of caution, comparing both their observations and the readings from the monitors before reaching any conclusions. But our walking activity surfaced interesting tensions between whether to place more trust in the sensor readings, or their own embodied feelings. For example, the air quality monitor reported higher levels of pollution the further they stood from the busy 'front road':

> I think [here] has more pollution in than the front road, but it's pretty hard to tell why. Because [...] the front road, had lots and lots of cars, but there's hardly any cars here.
> *[voice recording, workshop 1]*

However, just because the sensors told them it was not polluted or noisy, they did not reject their initial observations out of hand. Here, in the absence of the sensors providing clarity, it was the contextual readings of their neighbourhood – knowing the issues of traffic outside their school – that instead helped them to reach these conclusions. This encourages us to think through the respective roles of our embodied senses and of digital technologies: just as there are limits in what we can see and hear, there are limits to how sensors can generate useful information about our neighbourhood.

While these tensions surprised us, they also transpired as an opportunity and a resource for the children; not only to think critically about smart city tools, but also more generally about the trustworthiness of environmental data. Children reminding themselves to trust their own judgements (as opposed to those made by the sensor tools) showed the potential for the tools and the results they generate to serve as a resource for critical comparison between human and machine-generated data. Rather than the latter just serving to validate the former, the latter actually strengthened the former and convinced the children that the assessments they made using their own senses were more likely to be accurate, encouraging us to question what it means to engage children in using these kinds of sensing technologies.

Exploring possibilities in their environment with the sensors

Our second observation was that the children's use of the sensors revealed the intrinsic potential that such tools hold for fun, open-ended, and creative interpretation of their possible uses. For example, participants tried recording the speed of other things in their environment and shouting into the sound-level meter to see if it would report a higher decibel level. Their motivation for doing so appeared to relate to a desire to play and experiment (as children do) with the sensors, as opposed to obtaining accurate readings.

Throughout the walk, we were reminded of the obscurity of environmental sensing tools – within and beyond our engagement – by the limited ways that the children could interact with them. Only the traffic speed monitor allowed for two-way interactions through its red button to take a reading, while the other monitors simply allowed for passive observation of numbers and graphs. It was also unclear how the monitors calculated these numbers (i.e. how the algorithms behind them worked.) We were encouraged by one of the participants using this ambiguity as a resource (Gaver et al., 2003) to imagine what the inner-workings of the air quality monitor looked like. Questioning how an air quality monitor decides how good or bad the air is, they wrote a reflective story to personify what might happen to an air pollution particle being scanned by the monitor:

> I am a small micro-meter sized dirt particle [...] I was following [the team] around their air investigation, but as they were seeing how bad a bus was, I was sucked into

> a green machine brown as a frog [...] I was scanned all over and it TICKLED soooo much. The frog gave me a big SHOVE! And I was back in the air.
>
> *[reflective story, workshop 2]*

Imagining air-quality sensing as a playful interaction between the environment and the tools (being 'tickled' and 'sucked in') exemplified the potential to engage in fun activities with children while making sense of serious issues – the opaque, 'black box' configuration of technologies used in smart cities.

Generating place-based ideas and responses to issues

Our third observation was seeing the tools used by the children to gather evidence about serious issues in the environment. One of the most pressing issues among the groups was litter, which they conflated with air pollution as causing a blight on their neighbourhood and a priority for the whole community to tackle. One child took issue with various discarded items on a bench outside their school gates:

> You can see this purple rag, and then there's a blue towel, and a soft toy... this is hopeless, and it is not representative [of the community]! We can do better.
>
> *[voice recording, workshop 1]*

Using this evidence, they responded with creative ideas that would remediate this issue and others to transform their neighbourhood. The ideas they came up with were: The Travelator, a moving walkway to replace all roads in the neighbourhood; Duodecacycle, a solar-powered 'pool bike' carrying 12 people to different destinations; Bin Bus, with a 2-minute frequency for passers-by to dispose litter easily; International Robot Bins, a colourful solar-powered bin with vacuum-cleaner hands, detectors and drones; and Moving Bin, a similar idea that comes with an integrated 'pooper scooper.'

These ideas embody the playful approaches we saw earlier, but they still deal with serious issues grounded in the evidence they had collected about air pollution, traffic, and litter. They constitute meaningful reimagining's of the civic realm and how citizens might go about their lives differently, which if put into practice would contribute new street furniture (mobile bins), alternative modes of shared, sustainable transport (Duodecacycle) or more comprehensive rethinking of how we travel (Travelator). Their placemaking ideas may seem distant from reality; but we could see them as creative amalgamations of several existing technologies. Moreover, all of these spoke to the value of technology solving 'wicked' problems of litter, pollution, and traffic in the absence of care and attention given by the wider community, or the city council.

Thus, using the sensing tools seems to have encouraged them to think about how other technologies could be used to improve their neighbourhood as part of a wider placemaking initiative. Such radical ideas for reforming waste management and polluting travel behaviours would not feel out of place in a smart city vision either. One of the city officials who saw the children's ideas confirmed this:

> I really like the fact that everyone's started to think about robots and how technology can help [...] I think we're still looking at robots as what we think of in cartoons, but I think there is a space for using the technology behind that to think of having some of that in bins.
>
> *[voice recording, workshop 3]*

Their ideas to take action and make a positive change reiterated the value of our engagement, in that the sensing tools had their limitations, and were subject to significant critique, but they still proved useful for the gathering of evidence to remediate important issues in their neighbourhood. It also hinted at what could be possible if children and city officials were to collaborate in more productive ways – a fruitful direction for future research.

Designing approaches to support children's inclusion

Our engagement and activities supported children as young as eight to question the legitimacy of data, and whether it can ever be regarded as an exhaustive or objective measure of truth (Gabrys, 2016; Gitelman, 2013). The activities offered children an opportunity to compare and contrast situated technical sensing against their embodied senses observed at the same time and location, thus supporting them to explore the spatial, temporal, and social dimensions of data bound up with the material specificities in which they were gathered – what Taylor et al. (2015) term 'data-in-place.' Our case points to the promising potential for processes that support children developing critical skills to question algorithms bound up in smart cities too (Wolff et al., 2019) – e.g. making sense of the air quality monitor's 'black box' through story-writing. This is all the more critical when such algorithms play an increasingly significant role in the configuration and management of public space (Tenney and Sieber, 2016).

The ambivalences surrounding the sensor data transpired as valuable for introducing children to the processes and technologies involved in smart cities and opening a space for them to contribute their own ideas. But we also recognise the limitations of our pilot, and the need for institutional processes to change to accommodate children's voices if their ideas are to result in transformative action (Hautea et al., 2017; Jenkins et al., 2016). In our engagement, the dialogue with the city officials showed in some ways that data as evidence is still powerful to advocate for changes in the city. But while they warmed to the children's ideas, our engagement sits alongside the growing body of evidence that emphasises the political pressure and structural changes required for parity in children's contributions to city-making processes – in other words, commitments from decision-makers to act on their evidence and implement their proposals (Jenkins et al., 2016; Nordstrom and Wales, 2019; Peacock et al., 2018).

Drawing on our insights, we propose three concrete ways that placemaking scholars and practitioners might build on our work to develop socio-technical processes to include children in making people-centred smart cities.

Give prominence to context and subjectivities in smart cities

Context played an important role in the children's data gathering. For these children, it was not necessarily about collecting the quantitative data they needed – it was about obtaining a deeper understanding of their neighbourhood to decide how to improve it. Our methods gave them scope to draw on their own, contextual knowledge, e.g. expressing concerns about pollution from cars despite the confusing air-pollution readings. In our engagement, they collected experiential data that they could not have obtained using the city's data dashboard and put this to use in their ideas. But despite its apparent value in our engagement, this kind of subjective, citizen-generated data is rarely favoured over 'objective' numerical data in the smart city (Townsend, 2013). Scholars and practitioners might wish to give prominence and value to these subjectivities. This would help to reframe smart cities on the capabilities of the citizen and their respective contributions to realising smarter cities, especially from politically marginalised groups like children (Balestrini et al., 2014). Placemaking practitioners could draw inspiration from critical pedagogies to increase knowledge and awareness of smart city technologies and create a space to question their value, while eliciting contextual and technical evidence for productive contributions to city-making processes. bell hooks (1994, p. 12) suggests framing learning spaces as 'radical space[s] of possibility' to empower children to co-construct urban futures, while Shor (1992, p. 10) emphasises activities that place students and their lived experiences at the centre of critical inquiry.

Expose the limitations and seams of smart technologies

Another way of facilitating critical reflection on the design of smart cities might be the deliberate exposing of gaps and limitations of smart technologies. We witnessed the children experiment and test the limits of what they could sense, such as the noise-level of shouting – exposing what Human–Computer Interaction scholars call the 'seams' of possible interactions with the tools (Chalmers et al., 2004; Sengers and Gaver, 2006). Deliberately exposing these 'seams,' as opposed to concealing them, can draw attention to the limits of technologies and open spaces for new possibilities for their use. In civic contexts, Korn and Voida (2015) argue this can expose power relations embedded in technologies to create a space for citizens to subvert, manipulate, and imagine alternative possibilities for cities. Our deliberate contrasting of sensors and embodied feelings communicated to the children that technology is fallible and should not be taken at face value. Future work could experiment with embracing the gaps, in order to promote critical thinking around how people and technologies could work together to shape places and what the value might be of each for crafting smarter cities (Balestrini et al., 2017; DiSalvo and Jenkins, 2017).

Open playful spaces for designing cities and technologies

Running through our engagement was a desire for ludic interactions with the tools and their environment. We kept possibilities for using the tools open to create a design space where children could have fun in the process (Dix, 2003). Civic designers highlight the importance of scaffolding collaborative, creative, and open-ended engagements for children that allow for self-expression and use of their imagination (Wood et al., 2014). Our walking, storytelling, and ideation activities respectively 'pushed the envelope' and opened possibilities for creative responses in the context of placemaking. Such creative methods have a significant tradition in the social sciences (Bates and Rhys-Taylor, 2017; Brooks et al., 2020); here we adapted these to encourage critical interaction not only with the politics of placemaking and city planning as explored previously (Crivellaro et al., 2015), but also with the politics of data-collection processes and technical sensing. These methods resonate particularly with young children, whose natural curiosity, creativity, and enthusiasm can serve as an 'untapped' resource for the transformation of the built environment (Peacock et al., 2018).

The contrasting of experiential and technical evidence also had the effect of introducing them to complex debates around the reliability of data and algorithms behind smart technologies (i.e. how did the machine arrive at that number?) Here, we build on the work of other scholars who have attempted to do this through dialogical (Heath et al., 2019) and kinaesthetic (Andersen and Wakkary, 2019) methods, such as exploratory making and discussions, and demonstrate the value of doing this with children. However, we saw children struggle when they could not see how the technical sensors arrived at their readings. We suggest that future work should meet children closer to where they are at in their lives by being playful, interactive, and drawing on their lived experiences as learners (Lui et al., 2014) and as citizens (Cockburn, 2013;

Wood et al., 2019). Such design activities could involve opening up 'black boxes' of smart city technologies (DiSalvo and Jenkins, 2017), making smart city data tangible (Nissen and Bowers, 2015) or appropriating technologies to help them address placemaking issues in their neighbourhood (Jenkins et al., 2016).

Conclusions

If smart technologies are increasingly implicated in the making of future cities, important questions remain as to how we can prepare children for such a future. Through the comparisons of sensory and technical data, our insights suggest our activities effectively facilitated critical evidence-gathering and analysis among children. Aside from some light-touch scaffolding to frame the engagement, children worked to gather evidence and mobilise this to produce ideas for the future of their neighbourhood. This gave us a window into the possibilities for children to have greater agency in smart cities and smart technology design. While our pilot study is a modest initial experiment, it shows that there is significant value in engaging children with smart city technologies as a way to both generate ideas for changes in the places where they live and to develop critical thinking in relation to data and sensor technologies. However, while understanding the workings and limitations of smart technologies is a start, harnessing the potential of technologies to advocate for changes they want to see is an altogether bigger challenge. It is clear that we have a long way to go before children are recognised as competent actors in placemaking processes (Freeman and Tranter, 2012). For city planners and designers to involve children in the making of inclusive, sustainable, and people-centred smart cities, we must create the conditions to enable children to exercise their right as citizens to help shape the cities and technologies of the future. The recent youth climate strikes show the potential for addressing this challenge, as children across the world mobilise compelling data around the climate emergency to influence environmental policy (Singh et al., 2019). We hope that the insights and ideas we offer here will help placemaking scholars and practitioners to address the challenges of children's participation in making people-centred smart cities.

Acknowledgements

We wish to thank all the children, teachers, adult facilitators, and representatives of Newcastle City Council for their participation and support in carrying out this research. This research was funded through the EPSRC Centre for Doctoral Training in Digital Civics (EP/L016176/1). Data supporting this publication is not openly available due to ethical considerations. Access may be possible under appropriate agreement; please contact the corresponding author for further information.

References

Andersen, K. and Wakkary, R. (2019). 'The magic machine workshops: Making personal design knowledge', in *Proceedings of the 2019 CHI Conference on Human Factors in Computing Systems*. Chicago 2019. New York: ACM, pp. 112:1–112:13.

Balestrini, M., Diez, T. and Marshall, P. (2014). 'Beyond boundaries: The home as city infrastructure for smart citizens', in *Proceedings of the 2014 ACM International Joint Conference on Pervasive and Ubiquitous Computing: Adjunct Publication*. UbiComp 2014 Adjunct. New York: ACM, pp. 987–990.

Balestrini, M., Rogers, Y., Hassan, C., Creus, J., King, M. and Marshall, P. (2017) 'A city in common: A framework to orchestrate large-scale citizen engagement around urban issues', in *Proceedings of the*

2017 CHI Conference on Human Factors in Computing Systems. Chicago 2017. New York: ACM, pp. 2282–2294.

Bates, C. and Rhys-Taylor, A. (2017). *Walking Through Social Research*. Abingdon: Taylor and Francis.

Brooks, R., Lainio, A. and Lažetić, P. (2020). 'Using creative methods to research across difference. An introduction to the special issue', *International Journal of Social Research Methodology*, 23, pp. 1–6.

Cardullo, P. and Kitchin, R. (2018). 'Being a "citizen" in the smart city: Up and down the scaffold of smart citizen participation in Dublin, Ireland', *GeoJournal*84, pp. 1–13.

Chalmers, M., Dieberger, A., Höök, K. and Rudström, Å. (2004). 'Social navigation and seamful design', *Cognitive Studies*, 11, pp. 171–181.

Cockburn, T. (2013). *Rethinking Children's Citizenship, Studies in Childhood and Youth*. London: Palgrave Macmillan.

Crivellaro, C., Comber, R., Dade-Robertson, M., Bowen, S.J., Wright, P.C. and Olivier, P. (2015). 'Contesting the city: Enacting the political through digitally supported urban walks', in *Proceedings of the 33rd Annual ACM Conference on Human Factors in Computing Systems*. New York: ACM, pp. 2853–2862.

de Andrade, B., Poplin, A. and Sousa de Sena, Í. (2020). 'Minecraft as a tool for engaging children in urban planning: A case study in Tirol Town, Brazil', *ISPRS International Journal of Geo-Information*, 9, p. 170.

DiSalvo, C. and Jenkins, T. (2017). 'Fruit are heavy: A prototype public IoT system to support urban foraging', in *Proceedings of the 2017* Conference *on* Designing Interactive Systems, DIS '17. New York: ACM, pp. 541–553.

Dix, A. (2003). 'Being playful: Learning from children', in *Proceedings of the 2003* Conference on Interaction Design and Children, IDC '03. New York: ACM, pp. 3–9.

Foth, M. (2017). 'Lessons from urban guerrilla placemaking for smart city commons', in *Proceedings of the 8th International Conference on Communities and Technologies*, C&T '17. New York: AC Med, pp. 32–35.

Fredericks, J., Hespanhol, L., Parker, C., Zhou, D. and Tomitsch, M. (2018). 'Blending pop-up urbanism and participatory technologies: Challenges and opportunities for inclusive city making', *City Culture Society Innovation and Identity in Next Generation Smart Cities*, 12, pp. 44–53.

Freeman, C. and Tranter, P. (2012). *Children and Their Urban Environment: Changing Worlds*. Abingdon: Routledge.

Gabrys, J. (2016). *Program Earth: Environmental Sensing Technology and the Making of a Computational Planet, Electronic Mediations*. Minneapolis, MN: University of Minnesota Press.

Gaver, W.W., Beaver, J. and Benford, S. (2003). 'Ambiguity as a resource for design', in *SIGCHI Conference on Human Factors in Computing System*. New York: ACM, pp. 233–240. https://doi.org/10.1145/642611.642653.

Gitelman, L. (2013). *Raw Data Is an Oxymoron, Infrastructures Series*. Boston, MA: MIT Press.

Goodchild, M.F. (2011). 'Citizens as sensors: The world of volunteered geography', in Dodge, M., Kitchin, R. and Perkins, C. (eds.) *The Map Reader*. Hoboken, NJ: Wiley-Blackwell.

Hautea, S., Dasgupta, S. and Hill, B.M. (2017). 'Youth perspectives on critical data literacies', in *Proceedings of the 2017 CHI Conference on Human Factors in Computing Systems*, CHI '17. New York: ACM, pp. 919–930.

Heath, C.P.R., Crivellaro, C., and Coles-Kemp, L. (2019). 'Relations are more than bytes: Re-thinking the benefits of smart services through people and things', in *Proceedings of the 2019 CHI Conference on Human Factors in Computing Systems*, CHI '19. New York: ACM, pp. 308:1–308:12.

Heitlinger, S., Bryan-Kinns, N. and Comber, R. (2019). 'The right to the sustainable smart city', in *Proceedings of the 2019 CHI Conference on Human Factors in Computing Systems*, CHI '19. New York: ACM, pp. 287:1–287:13.

Hooks, B. (1994). *Teaching to Transgress: Education as the Practice of Freedom*. New York: Routledge.

Horschelmann, K. and van Blerk, L. (2013). *Children, Youth and the City*. Abingdon: Taylor and Francis.

Hunter, T., Worthy, P., Viller, S. and Matthews, B. (2018). 'Designing for citizen-centric smart cities: Supporting people's needs of a community garden', in *Proceedings of the 30th Australian Conference on Computer-Human Interaction*, OzCHI '18. New York: AC Med, pp. 162–166.

Jenkins, H., Shresthova, S., Gamber-Thompson, L., Kligler-Vilenchik, N. and Zimmerman, A. (2016). *By Any Media Necessary: The New Youth Activism, Connected Youth and Digital Futures*. New York: New York University Press.

Karvonen, A., Cugurullo, F. and Caprotti, F. (2018). *Inside Smart Cities: Place, Politics and Urban Innovation*. Abingdon: Taylor and Francis.

Korn, M. and Voida, A. (2015). 'Creating friction: Infrastructuring civic engagement in everyday life', *Aarhus Series on Human Centered Computing*, 1, p. 12.

Krivý, M. (2016). 'Towards a critique of cybernetic urbanism: The smart city and the society of control', *Planning Theory*, 17(1), pp. 8–30.

Lui, M., Kuhn, A.C., Acosta, A., Quintana, C. and Slotta, J.D. (2014). 'Supporting learners in collecting and exploring data from immersive simulations in collective inquiry', in *Proceedings of the SIGCHI Conference on Human Factors in Computing Systems*, CHI '14. New York: AC Med, pp. 2103–2112.

Mallan, K., Foth, M., Greenaway, R. and Young, G.T. (2010). 'Serious playground: Using second life to engage high school students in urban planning', *Learning, Media & Technology*, 35, pp. 203–225.

McKinsey Global Institute. (2018). *Smart Cities: Digital Solutions for a More Liveable Future.* New York: McKinsey and Company.

Nijholt, A. (2019). 'Playable cities for children?', in Fukuda, S. (ed.) *Advances in Affective and Pleasurable Design, Advances in Intelligent Systems and Computing.* Cham: Springer, pp. 14–20.

Nissen, B. and Bowers, J. (2015). 'Data-things: Digital fabrication situated within participatory data translation activities', in *Proceedings of the 33rd Annual ACM Conference on Human Factors in Computing Systems*, CHI '15. New York: AC Med, pp. 2467–2476.

Nordstrom, M. and Wales, M. (2019). 'Enhancing urban transformative capacity through children's participation in planning', in *Ambio*, 48, pp. 507–514.

Peacock, S., Anderson, R. and Crivellaro, C. (2018). 'Streets for people: Engaging children in placemaking through a socio-technical process', in *Proceedings of the 2018 CHI Conference on Human Factors in Computing Systems*, CHI '18. New York: ACM, pp. 327:1–327:14.

Ratti, C. and Claudel, M. (2016). *The City of Tomorrow: Sensors, Networks, Hackers, and the Future of Urban Life, The Future Series.* London: Yale University Press.

Scholten, H. (2017). 'Geocraft as a means to support the development of smart cities, getting the people of the place involved: Youth included', *Quality Innovation Prosperity*, 21, 119–150.

Sengers, P. and Gaver, B. (2006). 'Staying open to interpretation: Engaging multiple meanings in design and evaluation', in *Proceedings of the 6th Conference on Designing Interactive Systems*, New York: ACM, pp. 99–108.

Shor, I. (1992). *Empowering Education: Critical Teaching for Social Change.* Chicago: University of Chicago Press.

Singh, M., Oliver, M., Siddique, H., Zhou, Paul, K., Pulkkinen, L., Cagle, S., Phillips, D., Daniels, J.P., Luscombe, R., Phillips, D., Aratani, L., Cagle, S. and Phillips, T. (2019). 'Global climate strike: Greta Thunberg and school students lead climate crisis protest—as it happened', in *The Guardian*, 21 September 2019 [online]. Available at: https://www.theguardian.com/environment/live/2019/sep/20/climate-strike-global-change-protest-sydney-melbourne-london-new-york-nyc-school-student-protest-greta-thunberg-rally-live-news-latest-updates (Accessed: 9 June 2020).

Taylor, A.S., Lindley, S., Regan, T., Sweeney, D., Vlachokyriakos, V., Grainger, L. and Lingel, J. (2015). 'Data-in-place: Thinking through the relations between data and community', in *Proceedings of the 33rd Annual ACM Conference on Human Factors in Computing Systems*, New York: ACM, pp. 2863–2872.

Tenney, M. and Sieber, R. (2016). 'Data-driven participation: Algorithms, cities, citizens, and corporate control', *Urban Planning*, 1, pp. 101–113.

Townsend, A.M. (2013). *Smart Cities: Big Data, Civic Hackers, and the Quest for a New Utopia.* New York: W. W. Norton.

Urban Observatory. (2020). *Urban Observatory* [online]. Available at: http://www.urbanobservatory.ac.uk/ (Accessed: 8 November 2019).

van Zoonen, L. (2016). 'Privacy concerns in smart cities', *Government Information Quarterly*, 33, pp. 472–480.

Wilson, B. and Chakraborty, A. (2019). 'Planning smart(er) cities: The promise of civic technology', *Journal of Urban Technology*, 26, pp. 29–51.

Wolff, A., Wermelinger, M. and Petre, M. (2019). 'Exploring design principles for data literacy activities to support children's inquiries from complex data', *International Journal of Human Computing*, 129, pp. 41–54.

Wood, G., Dylan, T., Durrant, A., Torres, P.E., Ulrich, P., Carr, A., Cukurova, M., Downey, D., McGrath, P., Balaam, M., Ferguson, A., Vines, J. and Lawson, S. (2019). 'Designing for digital playing out' in *Proceedings of the 2019 CHI Conference on Human Factors in Computing Systems*, CHI '19. New York: ACM, pp. 679:1–679:15.

Wood, G., Vines, J., Balaam, M., Taylor, N., Smith, T., Crivellaro, C., Mensah, J., Limon, H., Challis, J., Anderson, L., Clarke, A. and Wright, P.C. (2014). 'The department of hidden stories: Playful digital storytelling for children in a public library', in *Proceedings of the SIGCHI Conference on Human Factors in Computing Systems*, CHI '14. New York: AC Med, pp. 1885–1894. https://doi.org/10.1145/2556288.2557034.

Further reading in this volume

Chapter 25: 'If you can make it there, you can make it anywhere…': cultural placemaking at the heart of cities
Sherry Dobbin
Chapter 35: Planning governance: lessons for the integration of placemaking
Nigel Smith
Chapter 36: Facilitator skills for effective collaborative placemaking
Husam AlWaer and Ian Cooper
Chapter 37: *The Neighbourhood Project*: a case study on community-led placemaking by CoDesign Studio
Lucinda Hartley, Eliza Charley, Sama Choudhury, and Harriet McKindlay

SECTION 3

Problematizing placemaking

Section Editor: Louise Platt

PREFACE

The problem with placemaking

Louise Platt

The work presented in this section questions whether placemaking as a concept needs to be retheorised or reconceptualised according to different contexts. Is the idea of placemaking as employed in practice or academic debates 'fit for purpose'? Discourses of placemaking are becoming ubiquitous in the contemporary urban landscape. It is often presented as the panacea for urban ills or as the way in which to fulfil the aspiration for metropolitan lifestyles. For example, it is instructive to capture the views of Bruntwood (urban property owner in England's North West region) on the placemaking potential of one of their latest developments in Manchester, Circle Square, which will include apartments, retail, and 'green public' spaces (the inverted commas my own addition as I question the extent to which these spaces will be fully accessible to the public on private land):

> Circle [S]quare has a masterplan that masters the art of using physical space, scale, density and proximity to create a new kind of collaborative community... this beautifully designed, architecturally interesting, green and pleasant place brings forward-thinking people and progressive businesses closer together, on a human scale. So that in its offices, apartments, shops, bars and restaurants, minds will meet. Randomly. Intentionally. Sharing. Co-operating. To enable the extraordinary... Thoroughly researched, painstakingly designed, Circle Square takes an intelligent approach to placemaking.
>
> *(Bruntwood, n.d.)*

However, backlash against the 'top-down' placemaking of urban developers or city councils is evident in both practice and theory. Indeed, so tied up together are narratives of gentrification and neoliberal urban development that when Bristol City Council appointed a Director of Placemaking, British tabloid newspaper, the *Daily Express*, proclaimed it a 'gobbledegook job' (Stevens, 2011). Where placemaking is seen as something that can be 'intelligently' *done to* a place is illustrative of the very problem of placemaking which is critiqued through the chapters presented here. By examining who is included or excluded from placemaking practice either by choice or through structural inequalities, the section brings together a diverse range of work that addresses issues faced by those marginalised in the urban experience or contexts often not accounted for in placemaking theory and practice.

With reference to a historical place-based procession, I have, elsewhere, argued that this binary between top-down and bottom-up (or community-led) placemaking need not be in conflict. There might be a potential for a more collaborative and ethical approach to placemaking which accounts for the ongoing and 'throwntogetherness' (Massey, 2005, p. 140) of place over time:

> This is not to say that organisations and institutions cannot, or should not, constructively engage in place-making activities, or introduce new cultural events and festivities into urban spaces. Rather, these kinds of interventions must, we suggest, embrace a positionality at the middle, adopting a concomitant understanding that place-making is not simply started or brought to neat strategic conclusions.
>
> *(Platt and Medway, 2019)*

In this work we propose that a 'Deleuzoguattarian' emphasis on the 'middle' might offer a way to reconceptualise placemaking as an always-becoming process rather than something that is *started* or *finished*, which is more reflective of the nature of messy nature of place, as Massey proclaimed.

Further, the emphasis on placemaking as an organisation or event community-led endeavour has not meant that everyday placemaking processes have been dismissed in the academic literature. Indeed, these everyday experiences of people in their own spaces, whether private or public, have been on the agenda more recently, which further leads us to the question on the relevance and value of received wisdom on placemaking, especially how it has been enacting in policy and under neoliberal urban agendas (e.g. Dyck, 2005; Edensor and Millington, 2018; Platt, 2019). The work in this section goes some way to problematise placemaking in both theory and practice.

The most pressing concern in the contemporary context that has been given little attention in placemaking literature is that of climate change. In his thought-provoking chapter, Paul Graham Raven, positions placemaking as, 'a living laboratory for the participatory production of new practices, as well as for the reconstitution of the places in which those practices are situated.' His chapter raises big questions around adaptation to the Anthropocene but places this within the context of the everyday activities of communities and placemaking practitioners in order to consider how placemaking can provide potential solutions 'on the ground.' This work is an important addition to the critiques of placemaking as they have been thus presented in academic literature so far, and, in the way that it has been so freely enacted as a 'solution' for community problems on a local scale, without due consideration of the global.

The idea of context in terms of both the spaces themselves and the users of these spaces is raised by Claire Edwards. Edwards' chapter addresses the way in which placemaking needs to account for disabled people's everyday and localised experiences. Specifically, Edwards examines the fear of violence that can be engendered in public spaces especially when placemaking interventions have not been enacted to meet the needs of specific groups. By reframing un/safety as event, as part of the assemblage of the urban experience, Edwards complicates the relationship between placemaking and disabled people's perspectives on place. This problematisation is reflected in the participants responses whereby they:

> contest notions of disabled identities grounded in dependency, are continually utilising and (re)making space in the 'small and ordinary' (Friedmann, 2010, p.162) places of their lives, and are engaging in proactive and sometimes resistive strategies to generative affectual connectivity with place grounded in feelings of comfort and safety.

Here, un/safety is an embodied and affective way in which place is experienced, which is intertwined with processes of placemaking in both an everyday and interventionist perspective.

Further emphasising the importance of context is the work by Samantha Edwards-Vandenhoek. The voices of indigenous communities are often marginalised in accounts of placemaking and this work seeks to redress this. Through an art-led placemaking project, *Art in the Streets of Warmun* was utilised by the Gija community in Australia in order to negotiate identities of place as a result of crisis. Indeed, the very notion of what 'place' is being made is problematised here:

> This research contributes to understandings of the complex, political and locally nuanced discussions around placemaking connected to place-related meanings in First Nations community development, as there was indeed no 'place' to be made here, but rather enacted, reinforced and made visible.

As mentioned above, placemaking needs to reflect the messiness of place as a socially constructed idea. Edwards-Vandenhoek's work challenges us to think differently about what place means for marginalised communities who have faced times of conflict and crisis.

Edwards-Vanderhoek's work reflects on her own role in the *researcher-assemblage*, an idea which is extended in the chapter by Morag Rose. Blurring the academic and activist perspective, Rose presents the case of 'the ginnel that roared' – the battle over Library Walk, in Manchester UK. Rose's chapter gives voice to the campaigners who fought to save a cherished 'ordinary place rendered extraordinary when it was threatened.' In this account, Rose examines her own role as an activist/academic and how we might position ourselves in placemaking debates in order to drive change at a policy level for the good of the communities who use these seemingly mundane and everyday spaces. Indeed, Rose's chapter evidences the need for an understanding of placemaking as part of an ongoing processes in the lives of our cities and the impacts that decisions made without a fair and robust, two-way dialogue with the people that care of the spaces in question. Her chapter symbolises the battle for our city streets and how placemaking, when enacted unequally, can galvanise communities to recognise how places, no matter how small (in this case a ginnel – a Mancunian word for a walkway between two buildings), contribute to a sense of belonging.

This question of belonging and inclusivity is raised by Martin Zebracki in his chapter contribution which takes the form of a dialogue with Thijs Bartels, author of *Dancing on the Homomonument* (2003). The *Homomonument* situated in Westermarkt, Amsterdam provides the focal point for the dialogue and, 'reveals queer placemaking through the questioning, or "que(e)ring," of the monument's "stone matter" in contexts of "lived matter": memories of past and present experience of inclusionary vis-à-vis exclusionary processes and realities.' Whilst Zebracki concludes that people make places, he suggests that physical structures, such as memorials and the lived experiences of these can contribute to a problematisation of placemaking as embedded practices open to 'queer(y)ing.'

Finally, presenting a practitioner perspective spanning an entire career, the chapter presented by Graham Marshall examines how people need to be at the heart of placemaking. Marshall, with a wealth of place-based policy and community experience, examines the way in which the problem of placemaking has led to an approach within our communities that has often disregarded wellbeing of the inhabitants. Despite the policy fixation on the wellbeing agenda, how this has been embedded in neoliberal placemaking strategy and interventions, according to Marshall, falls short. In the context of Liverpool, UK where he has spent much of his career, Marshall suggests that intervention after intervention has failed communities through poor urban design which creates conditions of 'unhealthy' places. Reflecting on his experiences he identifies, 'the failure of process in contemporary urban planning and explore how an understanding and application of evolutionary psychology and human ecology could provide better models.' Placemaking as it has been put into practice needs to be problematised in order for communities to thrive in the future.

References

Bruntwood. (n.d.) *Circle Square Brochure* [online]. Available at: https://bruntwood.co.uk/media/2199/circle-square-brochure.pdf (Accessed: 4 October 2017).

Dyck, I. (2005). 'Feminist geography, the everyday, and local–global relations: Hidden spaces of place-making', *Canadian Geographer / Le Géographe canadien*, 49(3): 233–243.

Edensor, T. and Millington, S. D. (2018). 'Spaces of vernacular creativity reconsidered', in Courage, C and McKeown, A. (eds.) *Creative Placemaking: Research, Theory and Practice*. London: Routledge.

Friedmann, J. (2010). 'Place and placemaking in cities: A global perspective', *Planning Theory and Practice*, 11(2), pp. 149–165.

Massey, D. (2005). *For Space*. Thousand Oaks, CA: Sage.

Platt, L. C. (2019). 'Crafting place: Women's everyday creativity in placemaking processes', *European Journal of Cultural Studies*, 22(3).362–377.

Platt, L. and Medway, D. (2019). 'Sometimes… Sometimes… Sometimes… Witnessing urban place-making from the immanence of "the middle"', in Space and Culture [online]. Available at: https://doi.org/10.1177%2F1206331219896261 (Accessed: 20 April 2020).

Stevens, A. (2011). 'There are right and wrong ways to place mark', in *The Guardian* (11 April 2011) [online]. Available at: https://www.theguardian.com/public-leaders-network/2011/apr/04/place-making-criticism-value (Accessed: 8 January 2020).

Further reading in this volume

Chapter 1: Introduction: what really matters – moving placemaking into a new epoch
Cara Courage
Preface: Placemaking in the age of COVID-19 and protest
Jason Schupbach
Chapter 2: Placemaking as an economic engine for all
James F. Lima and Andrew J. Jones
Chapter 3: An annotated history of creative placemaking at the federal level
Jen Hughes
Chapter 4: A future of creative placemaking
Sarah Calderon and Erik Takeshita
Chapter 5: Making places for survival: looking to a creative placemaking past for a guide to the future
Jeremy Liu
Preface: 'Disastrous forces, accidental actions, and grassroots responses'
Tom Borrup
Chapter 11: *Free State Boulevard* and the story of the East 9th Street Placekeepers
Dave Lowenstein
Preface: The radical potential of placemaking
Cara Courage
Chapter 23: Routing out place identity through the vernacular production practices of a community light festival
Gail Skelly and Tim Edensor
Preface: Towards developing equitable economies; the concept of *Oikos* in placemaking
Anita McKeown
Chapter 35: Planning governance – lessons for the integration of placemaking
Nigel Smith
Chapter 37: *The Neighbourhood Project*: a case study on community-led placemaking by CoDesign Studio
Lucinda Hartley, Eliza Charley, Sama Choudhury, and Harriet McKindlay

Chapter 39: *Translating Outcomes*: laying the groundwork for interdisciplinary evaluation of creative placemaking
Jamie Hand
Chapter 40: Transforming community development through arts and culture: a developmental approach to documentation and research
Victor Rubin
Chapter 42: Creative Placemaking and Placekeeping evaluation challenges from the practitioner perspective: an interview with Roy Chan
Maria Rosario Jackson

14

EXPERTS IN THEIR OWN TOMORROWS

Placemaking for participatory climate futures

Paul Graham Raven

Introduction: the inevitability of the Anthropocene as a mandate for change

This chapter is framed within three points. The first is that climate change is a *fait accompli*, and it will transform the ecosystems in which we live. The second is that reconfigurations of our immediate environments are thus also inevitable, and indeed already underway, as states and institutions scramble (or idle) their way toward the making of systemic changes. These systemic shifts may express themselves as new technologies and systems but are perhaps more likely to manifest as more mundane and subtle changes in the less glamorous aspects of the sociotechnical landscape – or what we might think of as the 'built environment.' The third point, upon which this chapter's argument will be founded, is that amidst all of this change, the basic goals (and pleasures) of life are unlikely to change significantly; however, the ways in which those goals might be fulfilled will have to change, as the incumbent constitution of the sociotechnical landscape changes around them.

I must make a statement of positionality: this chapter is as much a polemic as a research paper, if not more so, and as such my epistemic biases should be acknowledged. I am a theorist of sociotechnical change, with a particular interest in the role played by infrastructural systems in the historical reconfiguration of consumptive practices. I am also a writer and scholar of science fiction, and as a result I analyse sociotechnical change from the perspective of characters trying to fulfil goals by overcoming obstacles to their aims. My positionality is thus a hybrid of the qualitative social sciences, the humanities and the arts – and it is a positionality of considerable relative privilege, too. As such, I stand in critical opposition to the conceptualisations of sociotechnical change which dominate the domains of policy and governance. But my intention here is not to eradicate these technocratic supply-side perspectives, so much as to counterbalance them with an understanding and appreciation of the demand-side dynamics which their epistemic positionalities render effectively invisible. This is a project in which I see placemaking as a potentially powerful ally.

People, places, practices: re-narrating transformative adaptations for the Anthropocene

The sociotechnicality of adaptive transformations

The most fundamental quality of adaptive transformations is that they are *sociotechnical*. In the context of adapting to the effects of climate change (an ontological fact that merits the declaration of emergency), transformations in both the social and the technical/technological dimensions of human patterns of living are necessary – but neither is sufficient in isolation. The term sociotechnical emphasises that the ontological distinction between social and technical phenomena is entirely illusory, and that the two dimensions cannot be disentangled – though their interrelationship may be analysed and understood. Without wishing to be waylaid by long-running theoretical debates about scale, it can be observed that there are two levels at which sociotechnical transformation must occur. These levels are neither hierarchical or separate, but rather form a sort of dialectic of mutual influence and transformation: each is the context of the other.

The first of these is the systemic level. Whether the goal is the mitigation of climate change or adaptation to its effects, changes must be made – indeed, *are* being made, albeit piecemeal and slowly – to the systems of technology and ideology which govern our relationship with the ecosystems in which we are embedded. The necessity is two-fold: mitigation is rooted in the recognition that we must prevent further disruption to ecosystems, while adaptation is rooted in the recognition that the effects of the damage already done will put an end to 'business as usual,' whether we like it or not. The second level of sociotechnical transformation is that of more intimate and locally specific changes in the everyday practices of individuals and communities. Here too, the necessity is two-fold: it will be necessary to adapt to changes in the environment itself, as the impacts of climate change make themselves felt in people's lives, but it will also be necessary to adapt to the changes being wrought at the systemic level already discussed. It bears noting that, due to the differentials of power within the system of global capitalism, these local transformations are perforce primarily adaptive, if not exclusively so. But they are not necessarily reactive – and the possibility of proactive adaptations is of particular interest.

The socialisation of responsibility

At this point it bears repeating that, while the countless individual acts of consumption performed every minute of the day combine to cause the carbon emissions and other forms of pollution which are perturbing the global ecosystem in which we live, those individual acts are performed within cultural, economic, and sociotechnical contexts which make them difficult, if not sometimes impossible, to avoid. Those who worship Marx and those who worship markets are in implicit agreement on this point; no one sat down and planned this system, but we all play our part in its emergent effects, because there is no outside to which we might flee to escape the choices or, as we are finally beginning to understand, the consequences thereof. 'Climate change,' then, is an inadequate label for the cluster of those inescapable consequences of our inescapable choices – but it's the label we have. And when it comes to climate change, while none of us are to blame, we are all of us complicit. All of which is to say: mitigation of, and adaptation to, climate change is not the exclusive responsibility of individuals. Many of the changes that must be made to slow the rate of global temperature increase, environmental degradation, and resource depletion are at the systemic level: reconfigurations of the networks of people and things and ideas that comprise our world; us included. We may find ourselves with roles to play in those reconfigurations, too – but that is a matter of politics rather than ecology. For our purposes here, the important thing to note is that those systemic sociotechnical reconfigurations – alongside other changes, both global and local, due to carbon already long since emitted – will inevitably impact the options available to individuals regarding the things they can do and the things they might make use of in doing them, and that those options (and those *changes* of options) will vary considerably across human timespace. We will find our worlds transforming around us, but we will not all experience the same changes at the same speeds.

Thus, we will all be presented with different choices and possibilities, dependent on upon who we are and where we live.

Management and markets: the behaviourist perspective on sociotechnical change

The study of (and attempts to manage) sociotechnical reconfigurations is currently dominated by a collection of models and heuristics with an implicit basis in behaviourist conceptualisations of human agency. As pointed out by Shove and Walker (2007), among many others, these models biased toward a top-down 'managerial' conception of sociotechnicality, are often linked to free-market dogmas and metrics of economic growth, and paradoxically rely upon entrepreneurship and technological novelty as the 'solution' to a crisis which is largely the consequence of untrammelled entrepreneurship and technology. Seen from this perspective, climate change can only be slowed or mitigated through encouraging 'behavioural change' (Shove, 2010) in the population of consumers, which can be achieved by a combination of ensuring that the 'right' products and services are available to compete in the market against the incumbent options, and informing consumers as to which choices are 'best.' Evidence has long been mounting that the information-deficit model of behavioural change (which holds that people don't change because they don't know why they should and recommends sustained campaigns of informational bombardment to combat the problem) is fundamentally flawed (see e.g. Potter and Oster, 2008). Another part of the problem is a conceptualisation of people as an undifferentiated mass of ignorant but nonetheless impeccably rational *homo economicus*; the behaviourist perspective homogenises not just the things that people do but, crucially, the specific ways in which they do them. However, other, more subtle and flexible models of sociotechnical change are available, which explicitly do away with behaviourist assumptions of homogeneity, and which emphasise the particularity of place in human practices.

Situated systems: the particularities of practices in place

I now invoke Donna Haraway's notion of 'situated knowledges,' which makes explicit an opposition and alternative to behaviourist perspectives. To reiterate, part of the problem with behaviourist perspectives is that they are inherently *placeless*: the 'right' behaviour is assumed to be universally applicable to pretty much anyone, pretty much anywhere. Against such hegemonising assumptions, Haraway argues for the particularity of specific stories as an important gateway to understanding what actually happens when people interact with the devices and systems that surround them:

> Situated knowledges are about communities, not about isolated individuals. [...] The science question in feminism is about objectivity as positioned rationality. Its images are not the products of escape and the transcendence of limits (the view from above) but the joining of partial views and halting voices into a collective subject position that promises a vision of the means of ongoing finite embodiment, of living within limits and contradictions – of views of somewhere.
>
> *(Haraway, 1988, p. 590)*

Haraway sometimes refers to the 'view from above' as 'the god trick': a narratorial positionality analogous to the managerial perspective of behaviourism, in which the analyst (or manager, or policymaker, or developer) is assumed to have an omniscient and objective perspective on the system under study as if seen from the outside. Situated knowledges, by contrast, see things and relationships close up, from a position embedded within (and thus influenced by, as well as upon) the system under study. Furthermore, situated knowledges privilege partial voices, quiet voices, voices that are otherwise lost in the long tail of the behaviourist bell curve. By combining Haraway's situated knowledges with the social practice theory model of sociotechnicality, we have a way of looking at the things people do, and the ways in which those things might (or might not) change, which returns an authentic agency to a heterogeneous population of communities, while also embedding their practices in a sociotechnical landscape which is rich in variations that are influential upon the specific performances made in fulfilment of particular goals.

Performance in place: the plurality of practices

In social practice theory, behaviourist homogeneity is replaced with an understanding that geographical variations in the constitution of sociotechnicality are ineluctably formative of not only the things that people do, but also the very specific and individual ways in which they do them. Practices do not simply happen in place; rather, practices are constitutive of place, and place is constitutive of practices:

> [T]here is no such thing as 'just' doing. Instead, doings are performances, shaped by and constitutive of the complex relations – of materials, knowledges, norms, meanings and so on – which comprise the practice-as-entity.
>
> *(Shove et al., 2007, p. 13)*

The point is that practices which the managerial behaviourist perspective assumes to be homogeneous are in fact highly heterogeneous; this is one reason that technopolitical interventions intended to foster large-scale transitions to 'sustainability' are rarely successful. The obduracy of practices is a function of their particular sociotechnical constitution for each individual or community – and that constitution is, at least in part, a function of place.

Materials, methods, and meanings: elements of the practice

As its name implies, social practice theory focusses on *the practice* as its unit of analysis. A popular categorisation of the components comprising a practice describes them as a combination of meanings, competencies, and materials (Shove et al., 2012); in other words, this model conceptualises the things that people do as being shaped by their reasons for doing them, by the techniques for doing them which they have inherited or acquired from their peers, and the material things with which they work to get them done.

Part of social practice theory's project is to refuse the prioritisation of any one element of a practice over the others: neither the social nor the technical elements are 'more important.' But I think it fair to focus briefly here upon the materials element, because within that term are subsumed all the technologies and infrastructural systems which underpin and enable the performance of consumptive practices – and in so doing, enable us to consume the finite resources of the world while expelling pollutants and 'waste.' This extension of the *materiality* of our most intimate and mundane practices out to the global scale of infrastructures of extraction, production, distribution, and consumption serves to emphasise the inevitability of transformative changes in our immediate environment *in addition to* those direct impacts of climate change itself; if efforts toward mitigation and adaptation are to be made at the systemic scale, then it is upon the infrastructural metasystem that they must be enacted, and through that same metasystem that those changes will be distributed to and experienced by citizens.

A living lab, a lab for living: extrapolative and speculative practices

Why do situated perspectives and particular practices matter in the context of climate change, and what has that to do with placemaking? I would argue that situatedness is a pioneering concept in the development of the notion of place. Many disciplines have come to their own definitions of place by different theoretical and empirical routes, but in terms of sociotechnicality – and transformative changes thereof – Haraway in particular is a fine guide for those of us working not only to make change, but to make change that works for those whom it is made for, in the places to which they are connected. While the 'wicked problems' identified by Rittel and Webber (1973) will likely remain insoluble, we might in this manner reduce the distinctions (and barriers) between those who plan and those who are planned for, and achieve something closer to the 'model of planning as an argumentative process in the course of which an image of the problem and of the solution emerges gradually among the participants, as a product of incessant judgement, subjected to critical argument' (p. 162) which they sought. If Haraway can be seen as providing the ethical impetus of this approach, then the 'three elements' model, with the material element extended out into the global infrastructural metasystem, provides a simple heuristic which affords us analytical purchase upon the things people do, and which assumes that variance in and transformation of practices is not only possible, but rather inevitable in the context of a local environment being transformed by both climate change and systemic-scale mitigation and/or adaption. In other words, it permits us not only to conceive of practices being different in the future, but to engage in grounded and extrapolative speculation as to how they might (be) change(d). Adherents of the behaviourist paradigm for understanding sociotechnical change have criticised social practice theory for its overattentiveness to the small, domestic, and mundane (see e.g. Geels, 2011), but I argue that this molecular approach to thinking about sociotechnical change is exactly what makes it capable of talking productively to placemaking concepts.

Having shown that social practice theory can allow us to explore the situated specificity of the things people do and the way they do them, and argued that we might think of this as being a place-based conception of the practice, I will next show how placemaking might operate as a method for extending or extrapolating that situated model of practice into futurity. This is to see placemaking as a mode of enquiry that allows for the surfacing of the intangible cultural meanings bound up in practices, and – perhaps most importantly – a creative praxis through which communities might intervene in and reconfigure the material infrastructures and affordances upon which their practices are necessarily based.

A lab for living: placemaking for sociotechnical transformation in the context of climate change

The exemplification function: extending the practice model into futurity

My vision of a role for placemaking is rooted in my understanding of the exemplification function. According to McCormack (2013, pp. 12–13), exemplification is 'a mode of presenting a sense of how participation within relation-specific affective spacetimes might be considered to make a difference to the sensibility through which thinking takes place'; glossing this definition, Courage (2017, pp. 46–47) notes that in social practice placemaking interventions, and particularly in urban contexts, 'exemplification aids an articulation around change [and] complicates the term of encounter by presenting specifics in tandem with a transformative potential; what is exemplified is both what is within and without the spacetime conditions.' More simply, exemplification is the process of making apparent the possibility that things could be different in the future of a specific place. As such, there are two roles for exemplification in the context of climate change. The first is to express the likelihood that the environment of a given place will be transformed by climate change in concrete and relatable terms, as opposed to the abstractions (e.g. 'two degrees of warming') common to climate change science communications efforts. The second is to express the potentiality of *chosen* change in and of the place itself – of transformations wrought through the efforts and imaginings of the 'micropublics' (Amin, 2008) comprised of people who identify with that place. Through these roles, I see social practice placemaking as an extension of the social practice model of sociotechnical change into futurity: a speculative, creative, and coproductive articulation of that model which, rather than imposing a predetermined 'solution' in the manner of a behaviourist intervention, invites a community to begin the process of imagining their own future practices in place. This recalls Rittel and Webber's desire for 'an argumentative process in the course of which an image of the problem and of the solution emerges gradually among the participants' (1973, p. 162), with the intervening artist or practitioner deliberately avoiding an expert positionality: the placemaker enables the micropublic to see that things could be different, but they do not set out how things could be different. The how – and indeed the why – is a matter for the community itself.

But what might such a placemaking-powered intervention into climate futurity actually look like? The first step, by necessity, would be to collect and communicate the local specificities of climate change for the place in question. This would likely require some degree of engagement with the science, and with the scientists who produce it, in order to tease out a translation of the models and statistics into concrete impacts that citizens can grasp as functions of their local environment: what temperature might it be at midday in May? What familiar flora and fauna might disappear, and what new forms of life might replace them? The key to this translation will be a parallel engagement with the community aimed at uncovering what it is about the place that is valued and noticed by them and seeking as much as is practically possible to express the predicted changes in those terms. To draw upon a theatrical metaphor, if we consider the placemaking process to be a form of dramatic enactment of futurity, then there needs to be a stage and a set against which the future drama might be performed. The second stage would be to start thinking about the infrastructural element of the social practice model – with 'infrastructure' here signifying the built environment more broadly, as well as the more fundamental systems-of-distribution sense of the term (Raven, 2017). The environmental impacts expressed in the first stage will result in changes to the infrastructural constitution of the place, whether by accident (e.g. through damage, decline, or neglect) or through deliberate attempts at mitigation or adaptation to the environmental shifts. It is at this point, and particularly with regard to the latter category of deliberate changes, that one would hope to see the placemaking participants beginning to contribute and express their own imaginings of the place with which they identify.

The line between the second and third stages might well be impossible to discern, as the latter would involve a shift to thinking about the meanings and competencies elements of the social practice model – and as the model makes clear, meanings and competencies are not just entangled with one another, but also with the material infrastructures underpinning the practices which they comprise. Through the process of imagining one element of their practices having changed, it would be expected that at least some of the participants would inevitably begin to reimagine those practices *in toto*, tweaking the entangled meanings and competencies in such a way as to maintain the fulfilment of the original ends in the context of changed circumstances.

Relative expertism: placemaking as (ethno)methodology

I'm leaning hard on the social practice model here because it offers a fairly concise explanation of sociotechnical change as seen from the 'demand side' rather than the 'supply side,' but also because it seems to me that social practice placemaking practitioners are enviably well-placed to make use of it. My work, and that of other scholars in the same field, might be described as a sort of infrastructural ethnography: an exploration of how people live with the devices and systems that surround them. But due to the way our roles as researchers are currently constituted, while we might fairly easily access the *materials* and *competencies* involved in a given practice (e.g. through participant observation and interviews), it is rare that we can engage in research that gives us more than the most fleeting glimpses of the *meanings*. By contrast, through their long periods of embedded engagement with a place and its micropublics, placemaking practitioners may encounter or uncover those meanings in surprising and affecting ways, thus gaining authentic insights into situated knowledges à la Haraway. Furthermore, placemaking practitioners are in a position to encourage those meanings to change – not in the instrumentalised 'sustainability education' sense of the behaviourists, but rather the sense of making the possibility of self-directed change apparent to people for whom such transformations of self and place may have always been assumed as the exclusive domain of experts, be they scientific, gubernatorial, or otherwise.

Finally, the placemaking process appears to offer a framework through which coproductive research might be made somewhat less intimidating for all concerned, by positioning the placemaking practitioner and their praxis as a nexus through which expert knowledges might flow as and when explicitly requested by the citizen participants. Courage describes social practice placemaking as 'a material, place-led social practice that is co-produced from a level position of *relative expertism* between artist and non-artist' (2017, p. 182; my emphasis). My contention is that placemaking which engages with the challenges of climate change might extend the dynamic of relative expertism to incorporate the sciences and social sciences within the process. In other words, the placemaking process might form an arena within which expertise might be made available rather than imposed; the successes of some coproductive research projects on community energy generation (e.g. Krzywoszynska et al., 2016) suggest that, once a community realises *for itself* a utility of academic expertise as applicable to their own context and concerns, they will ask for it, and receive it gladly.

Catalyst of practice, not author of work: placemaking for resilience

Another way to describe my vision of a role for placemaking might be for it to foster place-based resilience in the face of climate change. Much like 'sustainability,' the concept of resilience has been co-opted and commercialised so thoroughly as to have become all but meaningless. Nonetheless, the core definition of the term as applied to people – namely 'the ability to be happy, successful, etc. again after something difficult or bad has happened' (Dictionary.cambridge.org, 2020) – gets us grasping toward something which, in the context of climate change, we might describe as a grass-roots competence in and engagement with thinking about neighbourhood futurity. This stands in stark contrast not only to the behaviourist approach to climate transitions, but also to more commercial forms of developer-led placemaking, wherein a profit-oriented corporation aims to 'parachute in' a sense of place (however notionally defined) in order to observe a compensatory uptick in economic metrics (and, not at all coincidentally, property values.) What I would hope to see are 'resiliences' which may in some cases be quantifiable – fewer grams-per-capita of carbon emitted, for instance, which can be extrapolated from other measures along the way – but in other cases may be all but impossible to measure, such as a personal and communal sense of feeling (and acting) capable of coping with the difficulties arising from climate change.

The true measure of success of any such intervention would be that the process of communal engagement with futurity and its possibilities for transformation should continue long after the artists and academics have packed up and gone away. It is said of placemaking that the artist should be a catalyst of the process rather than the author of 'a work' (Courage, 2017). This modality of intervention is a long way from the deficit model of science communications, and perhaps further still from the patronising 'help them cope' benevolence inherited from the religious roots of anthropological research; it's not about fixing things for people, nor is it about fixing the people themselves. Rather, it's about awakening in people the understanding that they're capable of making these changes for themselves, because they're experts in their place and their lives in a way that no intervening expert could ever hope to be.

In summary, I see social practice placemaking as having the potential to provide a living laboratory within which the social practice model of sociotechnical change might be extended into futurity in the context of climate change; through the central function of exemplification, placemaking brings the potential of transformation alive among the people who identify with a given place. Furthermore, the process opens a window on the meanings and values that underpin practices as currently constituted in a way that is rarely (if ever) accessible to ethnographers of sociotechnicality, as well as providing the impetus and encouragement for those meanings to change in an organic and authentic fashion that emerges from the participants themselves, rather than from some theory or paradigm imposed from above by experts. And finally, I see placemaking as generative of a form of cultural infrastructure, provoking a reflexivity and autonomy within communities with which they might begin and sustain their own engagement with the futurity of the places with which they identify, as well as of themselves.

Having suggested what placemaking might do for the various disciplines working to foster adaptive sociotechnical transformations in the context of climate change, I will now turn around to ask what we in those other disciplines (and elsewhere) might do for placemaking in return.

Strategies of support: ways to work with and for placemaking

Science and sci-comms: the work of worldbuilding

My first exhortation goes out to scientists, but also to science communicators. This chapter has focussed on climate change, and so climate science is my main target – but given the interlocking nature of the sciences (and of the adjacent practical disciplines such as engineering), which in turn mirrors the interlocking nature of the multifaceted ecological challenges which face us in the years ahead, it might be thought of as a more general suggestion to anyone working on the supposedly 'hard' side of the scholarly spectrum. What would be of most value to placemakers, and to the participating communities with which they work – and, indeed, to people in general – would be some effort expended on expressing climate change (and the more plausible adaptations and mitigations thereof) in terms of concrete, situated impacts and consequences. To borrow a term from science fiction, this is the work of worldbuilding: working downwards from the abstractions of General Circulation Models (GCMs) and Integrated Assessment Models through scientific induction in order to build out a detailed view of what the predicted climatic shifts will look like on the ground in particular places. This is speculative work, to some extent – but it will be grounded by your having extrapolated from the best science available and, crucially, having halted before you get to that messy and hard-to-model social stuff.

This will benefit placemakers and their participants because it will make the placemaker's first task of setting the stage for a place's future that much easier. It's of course understood that you cannot be expected to provide a detailed future almanac for every location on the planet. Instead, be ready for the arrival of artists (and others) with questions, and do your best to answer them, or to direct them to others who can answer the questions you can't. (If that sounds like a distraction from the task at hand, well, perhaps you can find some way to write up your input to a placemaking project as an innovative form of research impact.) This communicative effort will have more general benefits, too – because if you can begin to describe a climate-changed future in terms that an artist can use, you'll likely find a lot of other people can get a grip on it, too. Again, the bias in climate science toward quantitative and abstract outputs is understandable, given that your main audience is the policy machine, which is largely incapable of parsing anything which can't be expressed in cash-money terms – and that work keeps you busy. And so, I say to the science-communicators: dare to drop the long-discredited information deficit model and try thinking like an artist-placemaker. Treat people as experts in their own lives and locations: find out what they actually want to know, rather than what you think they need to know, and then tell them it. Nothing less, and nothing more.

Social sciences and humanities: articulating consumption from local to global

For my colleagues in the social sciences and the humanities who would like to support placemaking (and perhaps other forms of co-productive research, as yet methodologically unformulated), perhaps the most vital task at hand is to analyse and describe the interactions and connections between, on the one hand, the complex and entangled global metasystem of resource extraction, consumption, and distribution, and on the other hand, the hyperlocal specificity of daily sociotechnical practices. This project has of course been underway for many years, from myriad different positionalities and through many different theoretical lenses – and I don't need to tell you it's urgent, because you already know that. My epistemological biases have already been thoroughly exposed, and so I feel no shame in advocating models of these interactions which are rooted in so-called 'flat ontology' theories, and which generate fundamentally relational (and situated!) accounts of agency and operation. But there are many rooms in the mansion of theory, and I would argue that the topic of focus is more important than the way you choose to approach it. Albeit armed with the best of intentions, placemaking has generally struggled to articulate the role and influence of ecological and infrastructural complexity in the situations in which it is trying to intervene. If we can provide models and frameworks that make plain the linkage between practices and the systems of provision upon which they depend, then we can help in the development of new methodological strengths in this regard. Also, where possible, join forces with placemaker-artists: seek them out, write them into your project bids, and start thinking of placemaking as a new paradigm for action research and co-production. As I have shown, placemaking can not only give you a window on the facets of sociotechnical practices that are all but impossible to reach through regular research, but also presents the possibility of nurturing the grass-roots change we long to see, but are ethically constrained from pushing for. It will be a struggle, the hardest of hard sells to the neoliberal gatekeepers of the funding system – but if enough of us try, then some bids will make it over the wire, and pioneer new ways of working for futurity in the context of climate change.

Citizens of futurity: the embrace of exemplification

Finally, I want to recommend that all of us, academics and practitioners and activists and everyone else, find ways to argue for placemaking (and activities like it) in our own domains of specialisation and action. I recommend that we join in chorus with the arts, with artists, and with communities in demanding more funding (with fewer economic evaluation measures), greater autonomy and devolution, and greater voice and agency at the grassroots. At the behest of an arguably well-intended but ultimately unguided (and increasingly amok) system of global socioeconomic organisation, we have tried to open up communities to policy, with results which I think we might fairly describe as mixed. By supporting placemaking as a practice, but also an expression of an ethical principle of working together with people to build a liveable world in an increasingly uncertain and unstable futurity, we might instead open up policy to communities. I invite you to see placemaking as a kind of meta-exemplification: as a counter to the hegemony of capitalist realism (Fisher, 2009), as well as a demonstration of the possibility of transformative change not just in specific places, but in each and every place on this beautiful, damaged planet.

References

Amin, A. (2008). 'Collective culture and urban public life', *City*, 12(1), pp. 5–24.

Courage, C. (2017). *Arts in Place: The Arts, the Urban and Social Practice*. Abingdon: Routledge.

Dictionary.cambridge.org. (2020). 'RESILIENCE | meaning in the Cambridge English Dictionary [online]. Available at: https://dictionary.cambridge.org/dictionary/english/resilience (Accessed: 2 May 2020).

Fisher, M. (2009). *Capitalist Realism*. London: Zero Books.

Geels, F.W. (2011). 'The multi-level perspective on sustainability transitions: Responses to seven criticisms' *Environmental Innovation and Societal Transitions*, 1(1), pp. 24–40.

Haraway, D.J. (1988). 'Situated knowledges: The science question in feminism and the privilege of partial perspective', *Feminist Studies*, 14(3), pp. 575–599.

Krzywoszynska, A., Buckley, A., Birch, H., Watson, M., Chiles, P., Mawyin, J., Holmes, H. and Gregson, N. (2016). 'Co-producing energy futures: Impacts of participatory modelling', *Building Research & Information*, 44(7), pp. 804–815.

McCormack, D.P. (2013). *Refrains for Moving Bodies: Experience and Experiment in Affective Spaces*. Durham: Duke University Press.

Potter, E. and Oster, C. (2008). 'Communicating climate change: Public responsiveness and matters of concern', *Media International Australia*, 127(1), pp. 116–126.

Raven, P.G. (2017). '(Re)narrating the societal cyborg: A definition of infrastructure, an interrogation of integration', *People, Place & Policy Online*, 11(1), pp. 51–64.

Rittel, H.W. and Webber, M.M. (1973). 'Dilemmas in a general theory of planning', *Policy Sciences*, 4(2), pp. 155–169.

Shove, E. (2010). 'Beyond the ABC: Climate change policy and theories of social change', *Environment and Planning. part A*, 42(6): pp. 1273–1285.

Shove, E. and Walker, G. (2007). 'CAUTION! Transitions ahead: Politics, practice, and sustainable transition management', *Environment and Planning. part A*, 39(4), pp. 763–770.

Shove, E., Pantzar, M. and Watson, M. (2012). *The Dynamics of Social Practice: Everyday Life and How It Changes*. London: SAGE.

Shove, E., Watson, M., and Spurling, N. (2015). 'Conceptualizing connections: Energy demand, infrastructures and social practices', *European Journal of Social Theory*, 18(3), pp. 274–287.

Further reading in this volume

Chapter 1: Introduction: What really matters – moving placemaking into a new epoch
Cara Courage
Chapter 4: A future of creative placemaking
Sarah Calderon and Erik Takeshita
Chapter 5: Making places for survival: looking to a creative placemaking past for a guide to the future
Jeremy Liu
Chapter 11: *Free State Boulevard* and the story of the East 9th Street Placekeepers
Dave Lowenstein
Preface: The radical potential of placemaking
Cara Courage
Preface: Towards developing equitable economies; the concept of *Oikos* in placemaking

Anita McKeown
Chapter 22: Embedded Artist Project: Epistemic Disobedience + Place
Frances Whitehead
Chapter 28: Integral placemaking: A poiesis of sophrosynes?
Ian Wight
Chapter 29: The solution is in the problem; the art of turning a threat into an opportunity by developing resilience using a creative placemaking critical praxis
Anita McKeown
Chapter 30: Ecological selves as citizens and governance as ethical placemaking
Lisa Eckenwiler
Chapter 32: *The Hollywood Forest Story*: Placemaking for the Symbiocene
Cathy Fitzgerald
Chapter 33: Conceptualizing and recognizing placemaking by non-human beings and lessons we might learn from Marx while walking with Beaver
Jeff Baldwin
Chapter 40: Transforming community development through arts and culture: a developmental approach to documentation and research
Victor Rubin
Chapter 43: A theory of change for creative placemaking: the experience of the National Endowment for the Arts' *Our Town* program: an interview with Patricia Moore Shaffer, PhD
Maria Rosario Jackson
Chapter 45: How the city speaks to us and how we speak back: rewriting the relationship between people and place
Rosanna Vitiello

15

UN/SAFETY AS PLACEMAKING

Disabled people's socio-spatial negotiation of fear of violent crime

Claire Edwards

> I have focused on the small and ordinary because small and ordinary are mostly invisible to those who wield power, unless, when stepped upon, they cry out.
>
> *(Friedmann, 2010, p. 162)*

Introduction

Disability and disabled people's lives have been largely neglected in studies or understandings of placemaking. Albeit a contested and power-laden concept, placemaking – whether in the context of official planning practices, processes of community participation, or localised meaning-making – often suggests an inherent agency, an ability to shape, mediate, or indeed, resist articulations of place. For disabled people, whose lives have been shaped by multiple socio-spatial exclusions, this agency has never been assumed. Whether through lack of access to the built environment or public transport, inappropriate housing, or discriminatory attitudes which result in mundane and everyday acts of oppression in public space, spatial configurations have served to 'keep disabled people "in their place"' while at the same time making them feel 'out of place' (Kitchin, 1998, p. 343; see also Soldatic et al., 2014; Chouinard et al., 2010). If, as Friedmann (2010, p. 159) suggests, 'making places is everyone's job,' how are we to make sense of disabled people's absence in discourses of placemaking? How do we understand the complexities of placemaking in the 'small and ordinary' (p. 162) places of their everyday lives?

There is little doubt that in the past 20 years, campaigns for disability rights have prompted policy changes in many Western nations which have had significant spatial consequences for disabled people's lives. The development of international human rights tools, such as the United Nations Convention on the Rights of Persons with Disabilities (UNCRPD, 2006) has (re) asserted the rights of disabled people to lead autonomous, independent lives in the community, which has seen the closing down of segregated, institutionalised settings (Power and Bartlett, 2018a, 2018b), as well as modifications designed to promote access to the built environment. In the official placemaking discourses of planners and professionals, claims to disabled people's equity and inclusion in placemaking can be seen in city plans and public realm strategies which seek to promote liveable and accessible urban design; through principles such as Universal Design, for example, planners seek to create spaces that are safe and accessible for inhabitants 'irrespective of their age, gender or ability' (Dublin City Council, 2012, p. 30; see also Imrie,

1996). While these developments are much needed, there is a danger that placemaking becomes focused on high-profile public-realm projects, without recognising how other places, including the home and local neighbourhood are significant, albeit more hidden, sites of placemaking (see for example, Dyck, 2005; Anderson, 2012; Power and Bartlett, 2018b). They also have potential to obscure from view disabled people's 'affective connections' (Jones and Evans, 2012, p. 2321) with place, which may incorporate, but also exceed, physical modifications to the built environment.

In this chapter, I seek to uncover some of the everyday understandings of disabled people's placemaking which often remain hidden in discourses about access rights, particularly in urban spaces. I do this in the context of qualitative research conducted in Ireland which sought to explore disabled people's understandings of safety or unsafety in the context of fear of hostility and harassment as they navigate their everyday lives and geographies. Recent international studies have shown that people with disabilities are more likely than their able-bodied counterparts to be subject to violence (Mikton et al., 2014). Simultaneously, there has been a growing awareness of everyday harassment and hate crime as it has been experienced by people with disabilities (Hall, 2019; Hall and Bates, 2019; Power and Bartlett, 2018b; Roulstone and Mason-Bish, 2013; Roulstone et al., 2011; Thomas, 2011). As Beebeejaun (2017) notes, fear of violence is a common experience of women and minority groups in the city and has the potential to challenge and disrupt the daily practices and sense of belonging of diverse groups. As an affectual response shaped by gendered (or disablist) socialisation, spatial habits, memories, human and non-human relations, fear of violent crime (FOVC) demands that we explore the co-construction of body and place, and the ways in which feelings – about fear, comfort, safety or unsafety – are implicated in everyday material practices of placemaking.

In this chapter, I follow Friedmann's (2010, p. 162) call for a focus on the 'small and ordinary' places of disabled people's lives, by exploring how un/safety mediates, shapes, and is shaped by, their everyday processes of placemaking. In particular, I focus on the narratives of three participants with mobility and visual impairments – Martin, Aoife, and Carol – to examine how people with disabilities give meaning to places in the context of un/safety. Drawing on their testimonies, I explore the ways in which people with disabilities negotiate feelings of fear and safety in their everyday geographies, and in challenging dominant discourses which associate disability with vulnerability, highlight the proactive strategies they employ to 'take possession' (Koskela, 1997, p. 308) of space in the context of unsafety and fear of hostility. In so doing, I suggest that notions of places as safe or unsafe – and disabled people's feelings of fear or safety – cannot be fixed as neat binaries; rather, we must attend to the spatio-temporalities of un/safety in different contexts, and recognise the complex interactions of identity and environment which produce affective encounters with place.

Placemaking and un/safety: geographies of FOVC

Since the 1990s, a significant body of work has sought to explore the geographies of fear of violent crime (FOVC). Geographical work emerging out of a behavioural geography tradition was concerned with mapping the geography of crime rates 'on to' space or understanding FOVC as an individual behavioural issue enacted in space. Feminist geographers, who have focused on women's experiences of violence and fear of violence in public space, have utilised space as a social category and construct to highlight the complexity of the experience and fear of violence as an interaction between power relations, space, and social identities (Koskela, 1997; Pain, 1997, 2000, 2014; Valentine, 1989). This research has been significant in exploring not only how women find themselves excluded or restricted from particular spaces as a consequence of fear (for example, the fear to go out at night time), but how fear itself reflects a set of gendered and other power relations and attitudes about certain bodies being seen as 'out of place.' It has also drawn attention to the ways in which certain places become socially constructed or symbolically represented as 'safe' or 'dangerous,' and the implications of this for the everyday mobilities and geographies of women and other groups.

The work of Koskela (1997) and, more recently, Brands et al. (2015) has drawn attention to FOVC as an affectual, embodied encounter with place (see also Brands and Schwanen, 2014). As Koskela (1997, p. 304) notes, 'feelings are not a mathematical function of actual risk but rather highly complex products of each individual's experiences, memories and relations to space.' Thus, while objective measures of safety, such as crime rates, may build into feelings and perceptions of un/safety and constructions of particular places as risky, there is no neat correlation between actual incidences of violence and subjective feelings of safety or fearfulness. Indeed, Koskela (ibid.) points to the 'spatial mismatch' which characterises the gendered dynamics of FOVC: while women express most fearfulness about public space, it is in the private space of the home that most violence takes place (see, for example, Pain, 2014). For Brands et al. (2015), affectual encounters with fear and safety are best understood as produced through complex assemblages of human and non-human interactions, which are spatio-temporally specific, and are in a constant process of becoming. Dynamics of socialisation, memories, and spatial habits all have a role to play within these configurations and help to explain the complex ways in which people layer and construct meaning about fearfulness and safety in their everyday lives.

The observations of feminist geographers – and relational understandings of FOVC – have particular pertinence in the context of understanding disabled people's experiences of fear and safety. Pain's (1997, 2000) geographical study of fear of sexual violence among women included a sample of disabled women, who described feeling more vulnerable to attack, and often employed specific spatial strategies in order to avoid particular areas and people. These experiences have to be understood in the context of disabling attitudes and inequalities which serve to exclude disabled people from particular spaces and places, including socialisation from 'concerned others' about the need to protect disabled people from potential risks (Edwards and Imrie, 2003). Historically, disabled people have been perceived as vulnerable and dependent on others, having lived their lives in segregated, semi-private spaces. Shifts in disability policy in Western societies which have led to deinstitutionalisation are changing the spatial practices of people with disabilities, but while community inclusion and belonging may be stated policy goals, questions remain about what this means to disabled people in the context of their material and affectual connections with place. Power and Bartlett's (2018a) work on people with learning disabilities in the UK, for example, has shown how the closure of formal day-care settings has led people with learning disabilities to create and seek out 'safe havens' or everyday spaces in the 'post-service landscape'; this however has taken place against a backdrop of resource scarcity, in which many people with disabilities live in areas of relative disadvantage where harassment, and anxiety about encounters with others, are an on-going part of their everyday lives (see also Power and Bartlett, 2018b; McClimens et al., 2014; Hall, 2019; Hall and Bates, 2019).

It is important to note that studies exploring disabled people's place-based belonging – many of which focus specifically on people with intellectual disabilities – do not point to the inevitability of fear, harassment, or victimisation in disabled people's lives, but rather understand it as a facet of a broader range of place-based encounters and practices, which also include safety, comfort, and welcome. Hall and Bates' (2019) work on relational geographies of disability hate crime, for example, points to the need to look beyond the individualised understandings of the disabled person as victim, and to explore the micro-spaces and places in which hate 'events' take place, through spatially and temporally specific encounters in the city which connect with broader socio-political relations and contexts. Based on such a reading, they argue, we cannot presume a priori who is vulnerable and where (Hall and Bates, 2019), but rather need to explore the intricate social relations which make up the everyday places of disabled people's lives – the home, streets, local shops and other amenities, public transport – in order to make sense of fear and safety.

Situating disabled people's socio-spatial negotiation of fear and safety in place

In the rest of the chapter, I draw on empirical material to explore how FOVC mediates disabled people's everyday geographies, and how they read and negotiate safety and fear in their daily lives. The material is drawn from a two-year qualitative study conducted in Ireland, which sought to explore how the experience and fear of hostility and harassment shaped disabled people's everyday geographies and mobilities. The study engaged some 54 disabled people with different impairments across three case study sites in Ireland (one inner- city urban, one rural, and one small-town). Aiming to investigate localised, embodied, and affective understandings of place, both in-situ and 'go-along' interviews (Carpiano, 2009) were conducted with participants in their local neighbourhoods; the interviews sought to investigate how people with disabilities understood and defined safety and unsafety, their experiences of hostility, fear, safety in different places, and what the notion of 'safer space' might mean to them in both conceptual and practical terms. As part of the study, interviews were also conducted with national and local disability organisations, policymakers, and practitioners, including the *Gardaí* (Irish police force), although these are not reported here. In this paper, I draw on the narratives of three of the participants in the study – Martin, Aoife, and Carol – as a way of exploring disabled people's everyday encounters with fear and safety.

Martin, who is in his late 40s and has a visual impairment, lives alone in rented accommodation in an inner-city area of a large urban centre, which, by the objective metrics of crime statistics, is considered an area of relatively high crime and social disadvantage. Martin does not work but is studying part time. Separated from his partner, he regularly travels across the city to look after his two children, using public transport. Martin uses a white cane to navigate around the city. Aoife, who has a mobility and visual impairment and uses a wheelchair, is a 23-year-old student who lives at home with her family in the suburbs of a large city. She regularly travels by taxi to the university to study and enjoys socialising with her friends at restaurants and bars in the city centre. She generally travels accompanied by a friend, family member, or personal assistant when out in public spaces. Aoife is active in disability advocacy in her local area and a member of a number of disability organisations. Carol, who is in her mid-50s and has a visual impairment, lives alone in a suburban housing estate of a small satellite town near a bigger urban centre. She describes how she and her former husband bought the house some years ago because of its quiet location but also proximity to amenities. Having formerly taught disability awareness at her local university, today Carol does not work, but looks after her grandchildren after school each day. Carol uses a guide dog to navigate around her neighbourhood. She has a personal assistant once a week who comes to help her with shopping, but she also regularly goes out for walks by herself.

My aim in drawing on these narratives is to reflect on fear and safety as it is produced in and through corporeal and affective engagements with place – engagements which have the potential to both (re)shape disabled people's subjectivities, and (re)shape places themselves. In so doing, I seek to explore the difference that physical and sensory diversity makes to disabled people's everyday encounters and understandings of safety-in-place, something that has received relatively little attention in geographies of FOVC, or in wider debates about placemaking. Feelings of fear and safety relate directly to concerns about social equity and diversity in encounters with place, and yet as a range of commentators have highlighted, these issues are frequently effaced in official placemaking discourses and practices (Fincher et al., 2016; Nejad et al., 2019). While recognising the broader socio-political relations and discourses which (re)produce ideas of disability as dependency, I also draw on the narratives to challenge the idea that disabled people are inherently vulnerable, and therefore should feel fearful in the places they live their lives.

Feeling fear, feeling safety

In her study of women's fear of violence and spatial confidence in Helsinki, Koskela (1997, p. 304) suggests that women's feelings of fear can best be understood as a form of 'sensible incongruence'; that is, while feelings do not always correlate with actual risks, and can appear irrational, women often have a clear sense of 'when and where to be careful or confident.' For the participants, there was not a clear fixity to feelings of fear or safety, which were time and place specific, and could emerge as momentary and fleeting based on the fluid dynamics of being in place (Brands and Schwanen, 2014). Acknowledgement of fears could often co-exist with independent, and apparently confident, uses of space; meanwhile, participants described feeling safe in different places despite constant reminders from professionals and indeed strangers that disabled people should be fearful, particularly in public spaces. As someone who lived in an area of relatively high crime and had experienced several acts of hostility, including an assault near his home, Martin did not describe himself as inherently fearful (although, as I will show later, neither did he always feel safe.) However, he noted how the rationality or 'congruence' of some of his place-based behaviours had been questioned by the *Gardaí* when he reported the assault. As he stated of the Garda (*Gardaí*) response: '"You shouldn't be out by yourselves" – that's what I was told... especially after dark, you should not be out of yourself."' Martin recounted how the Guards (*Gardaí*) had suggested that he move from the place he was living, having been told '"No [sic] of all the streets to be living on, Mount Street is not the one to be living on, it's the one to avoid". Anyway, I don't buy it, I said "It's not that easy for us, I can't just move, housing is much more difficult when you have a disability."' Such attitudes reflect the paternalistic identities often ascribed to people with disabilities as users of public space, and the ways in which disabled people have to manage these identities in negotiating their everyday geographies (Edwards and Imrie, 2003; McClimens et al., 2014).

Living in a housing estate of a small satellite town, Carol's concerns about safety focused on situations in the home space, particularly when she was alone. Countering associations of safety with the domestic sphere, she said, 'I really think my home is less safe than being out in the main street, you know? Now, it mightn't be, but that's my perception.' Carol described fearing the emptiness of her suburban housing estate during the daytime, and the challenge for her in identifying who might come to her door now she lived alone with her guide dog. She related this fear in part to getting older, but also a previous incident in which a workman had stolen tools from her garage, reflecting the way in which memories come to shape current practices and meaning-making (Koskela, 1997; Brands et al., 2015). Describing the seeming irrationality of her fear, she spoke about the micro-spaces of her home: 'Like this is crazy, but I have a dog-run out the back for the dog to relieve himself. Sometimes at night I'm saying, "Now what's to stop anybody coming in my back door while I'm going to the dog-run and back?" I mean I know it's foolish but that's the way I feel, you know? And I'd be going to lock all the doors at night, the internal doors, which is a thing I used to never do.' Carol's account bears witness to Imrie's (2004, pp. 745–46) assertion that 'disabled people's domestic experiences are, potentially, at odds with the (ideal) conceptions of the home as a haven, or a place of privacy, security, independence and control.' Like Martin, Carol had a clear sense of self-in-place, and ascribed identity as a potential victim or 'easy target.' Such fears were nevertheless described as co-existing with an active life in which Carol recounted going for a 3-mile walk every day, visiting the local shops with her personal assistant, and having a network of friends around the neighbourhood: 'I have my church which is very important. I'm a very people-oriented person. I'd have a lot of visitors.' As in Koskela's (1997) study, peopled places were central to Carol's feelings of safety; as she said: 'From my point of view, there has to be people around. There has to be activity.'

For all three participants, awareness of self in different places – and of the need to manage the self in relation to others – was part of the complex nexus of factors contributing to feelings of fear or safety. Describing herself as independent and a user of many different (public) spaces in the city, Aoife noted the constant paternalistic remarks from others – '"aren't you great to be out"' and '"you've a cross to bear"' – when she was out and about in the city. While Aoife described laughing off such comments, which she said did not make her feel fearful or threatened, she described contexts in which the gendered dynamics of FOVC more vividly emerged as sinister and threatening. Describing how men in a nightclub 'come up to you and go "You're gorgeous. Can you have sex?" Like literally come out with stuff like that. Or people come up and try and kiss you and you're like "get away from me, like!",' she described how the intimidation of such inappropriate questioning had affected the places she and her friends utilised. As she stated: 'So I would never go into a nightclub on my own or a pub on my own. We'd go in as a group and it's fine, but I would never feel comfortable going into a pub on my own… because some of the questions you get asked are unbelievable.'

Such encounters with strangers are intimately bound up with perceptions that disabled people's embodied difference renders them 'out of place' in particular spatio-temporal contexts (Kitchin, 1998). What is notable in the participants' accounts is how the adaptive supports and objects associated with disability also significantly intertwine with experiences and feelings of fear and safety. For Gibson (2006), these supports cannot be seen as separate to the body, but an extension of it, an intimate form of connectivity that necessarily shapes disabled people's encounters with the world. Martin, for example, described how the white cane – and the sense of feeling your way along – acted like a warning alarm to people, but was also a source of stigma that could make him feel more vulnerable in certain contexts. As he said, 'My daughter tells me that when I'm walking along, people actually jump out of the way. It's as if the stick has some infectious disease like the plague… You're just trying to see it from society's perspective and it's a freak thing.' In contrast, he suggested, 'guide dogs humanise a blind person and they make them more easily approachable.' For Carol, while her guide dog was not necessarily a conduit for encounter or socialisation, it could make her feel safer; as she said: 'It makes a difference what speed you're travelling. You can travel faster and if you were in trouble you could get away quicker.' Describing how she might react if she sensed fear in a particular situation, or if she felt someone was following her, she said, 'But if I think somebody's around me… I might talk to the dog and say "No, dog, you need to be careful now and make room for others," you know. You kind of get crafty.' These intimate practices and relations – with animals and other technologies – illustrate the continual shifting of disabled subjectivities which get played out in different places.

If feelings of fear and safety emerged out of the interrelationship between place, the body, and 'objects' of disability, they were also crucially a function of the material environment – outdoor spaces such as pavements, roads, as well as buildings and public transport systems – which, through poor design, have been shown to symbolically and materially reinforce disabled people's sense of marginalisation (Imrie, 1996). Accessible design and easy navigability could make participants feel safer in different places, by reducing feelings of uncertainty and insecurity. Aoife, for example, described often feeling lost if she was in the city centre by herself, but comfortable in her local out-of-town shopping centre because of its design: 'Probably shopping centres would be a place where I would feel particularly safe because it's all contained. Like there's only a certain amount of space – you can only get so lost in a shopping centre. And for people with mobility issue and visual impairment... they can be very safe because you can find your way around relatively easily.' In other cases, physical barriers in the environment or on transport systems led to downright experiences of exclusion. Aoife, for example, noted the lack of availability of wheelchair accessible taxis after midnight, which meant she could not stay out late. As she stated: 'It's ridiculous. It's impossible to get a taxi home from a place after 12 o'clock. It's like, you're in wheelchairs, you should be in bed.' She also described the inaccessibility of local buses which meant she had to rely on her parents or taxis to travel around, while Martin explained how the lack of audible announcements on trains had led to him feeling disorientated.

Physical environments, then, could heighten feelings of uncertainty and discomfort, and cause anxiety about potential encounters or incidents with others. Martin raised the issue of the difficulty of crossing the road at complex junctions near his home and being uncertain about moving around the city centre given a significant traffic redevelopment programme that was underway, which led him to feel unsafe. Similarly, Carol talked about avoiding wide open spaces, dark alleys, and also lines of trees in her local town when she was out and about, which made her feel fearful. As she said: 'I'd prefer if it was a solid wall than a big noisy load of trees. That just freaks me out. And I think too, it's important to know that blind people do not cope in the wind... Yeah, they actually get lost.' As Bell et al. (2019, p. 274) note, these 'elemental haptics' have an important role to play in shaping how people with sight loss or impairment negotiate the dynamics of space and place in their everyday lives.

In responding to and negotiating affectual encounters with fear and safety, the participants had developed multiple spatial strategies to promote feelings of safety, and (re)claim places as safe(r). Echoing studies which explore processes of community-belonging among people with learning disabilities (Wiesel and Bigby, 2014; Power and Bartlett, 2018a, 2018b; Hall and Bates, 2019), establishing places and local social networks where people are known and welcomed was crucial. Carol, for example, spoke of the reassurance of routine and repeat encounters by returning to the same local shops to buy groceries. As she said:

> I'm in a very fortunate position that I live in a place where everybody knows my name when I go into a shop... And that's very important that we're integrated into the community. And I make a point of asking people their names on the tills and telling them who I am, you know... I tend to use the same places all the time because of that.

Aoife too, spoke about identifying restaurants and bars in the city that were welcoming to herself and her friends, a number of whom also had disabilities. In this case, comfort and safety lay in the fact that these places were:

> physically accommodating and being welcome. Like, for instance, it's the small things that help. Like they put straws in my drinks without me having to ask, or cut up my food, or, you know – I would have a lot of friends in similar boats, so they do their best to accommodate.

Again, the significance of being known, and of staff who had an awareness of disability, were key to the comfort felt in these spaces.

In other instances, the participants had taken more resistive stances to feelings of fearfulness in place. Following incidences of hostility in public spaces, Martin described how he now wore a GoPro, a bodycam, when out and about, which worked as a pre-emptive strategy to change the way strangers responded to him. As he said, 'It's not actually on, it's a deterrent more than anything.' Both he and Carol had also challenged their local authorities about inaccessible places in their localities, and the Guards about significant challenges in the way they responded to people with visual impairments; while Carol stated that there was a local Guard whom she knew and could contact if she was under threat; for example, she said 'the Guards, they're not really equipped to deal with disability either… it's really bad that the Guards don't have a form of ID that we can work with and haven't worked one out.' Describing the difficulties she would face in being able to identify a Guard if they came to her door, such issues point to the broader response and engagement of agencies concerned with community safety, who often fail to take into account diverse sensory encounters with safety in the micro-spaces of people's everyday lives.

Conclusion

If the notion of fear – and indeed safety – is best understood as an event, a coalescing of a number of different elements of an assemblage, then the accounts of Martin, Carol, and Aoife reflect this complex interplay of body, identity, and environment that gives rise to embodied, affectual understandings of FOVC. What is clear from their narratives is that these feelings are not fixed, but ebb and flow, and are spatially and temporally specific. Koskela's (1997, p. 315) assertion that 'The streets of fear and boldness are 'elastic': different by length according to the time of the day, to whom is passing by and to how you feel at that moment' has pertinence here. The participants were only too aware of how the character of places changed at different times of day, and of the places and times where they might be seen as particularly 'out of place'; self-regulation was a key part of their engagement with fear and safety in place (McClimens et al., 2014).

It would be easy to read off a narrative of fear, exclusion, and discriminatory attitudes from the participants' accounts, and indeed, the dominance of socio-political relations which construct disability in terms of dependency and vulnerability cannot be ignored. People with disabilities face constant reminders of these constructed subjectivities at every turn in their daily geographies and mobilities (Hall and Bates, 2019). Yet as I have shown, Carol, Aoife, and Martin frequently contest notions of disabled identities grounded in dependency, are continually utilising and (re) making space in the 'small and ordinary' (Friedmann, 2010, p. 162) places of their lives, and are engaging in proactive and sometimes resistive strategies to generative affectual connectivity with place grounded in feelings of comfort and safety. To this end, we cannot presume disabled people are passive victims, or that they will always feel unsafe in certain places. Indeed, a relational approach enables us to understand how a 'shifting constellation of coordinates' (Stephens et al., 2015, p. 201) can make certain places feel more unsafe – for example the home space for Carol, or the nightclub for Aoife – but which are in a constant process of becoming that denies fixity.

This analysis is significant because, much like the gendered constructions which have divided public and private spheres in terms of FOVC by underplaying the association between familiarity and harm in women's experience and fear of violence (Pain, 2014), it calls for an examination of often hidden places of disabled people's lives. Disrupting assumptions of the home space as a place of safety or refuge, for example, Thomas' (2011) analysis has shown that it is often people known to people with disabilities, including relatives and carers, who are perpetrators of hate incidents (what she terms 'mate crime'), rather than strangers in public spaces.

For those agencies concerned with placemaking, which foregrounds safety, moreover, it also means acknowledging the complex affectual interrelationships which emerge in place. The issue of lack of access to the built environment is not just a technical point of exclusion, for example; it can lead to feelings of disorientation, uncertainty, and fearfulness, particularly where other elements of an assemblage align, such as a failure of public transport to arrive on time, an absence of people, or conversely, threatening interactions with strangers. As campaigns to promote women's safety in cities have highlighted (see, for example, Gulati, 2015), official practices of placemaking constituted through planning processes or urban renewal initiatives, for example, need to take account of safety as a priority concern, and not as something that emerges as an afterthought. This means taking seriously diverse affectual understandings of place, by recognising that safety in place is often about feeling safe, but also recognising safety as an issue of socio-spatial justice; in other words, all groups should have a right to access, and feel safe in, a range of different places, including the public realm (Beebeejaun, 2017; Fincher et al., 2016). As I have sought to suggest in this chapter, foregrounding safety in placemaking means paying attention to the range of human and non-human relations, such as disabling attitudes, physical environments, social networks, that make up localised feelings of fear and safety in the places of disabled people's lives. This, in turn, means accessing, and listening to, diverse sensory voices and narratives – or to return to Friedmann's (2010, p. 162) term, the 'small and ordinary' – which are frequently absent in top-down discourses of placemaking.

Acknowledgements

The research was funded by the Irish Research Council under its 2016 Research for Policy and Society scheme. I would like to thank all the organisations and participants who gave up their time to take part in the research.

References

Anderson, V.R. (2012). '"Homes" and being "at home" in New Zealand: Women's placemaking in internationalised higher education', *Gender, Place and Culture*, 19(3), pp. 327–343.

Beebeejaun, Y. (2017). 'Gender, urban space and the right to everyday life', *Journal of Urban Affairs*, 39(3), pp. 323–334.

Bell, S., Leyshon, C. and Phoenix, C. (2019). 'Negotiating nature's weather worlds in the context of life with sight impairment', *Transactions of the Institute of British Geographers*, 44, pp. 270–283.

Brands, J. and Schwanen, T. (2014). 'Experiencing and governing safety in the night-time economy: Nurturing the state of being carefree', *Emotion, Space and Society*, 11, pp. 67–78.

Brands, J., Schwanen, T. and van Aalst, I. (2015). 'Fear of crime and affective ambiguities in the night-time economy', *Urban Studies*, 52(3), pp. 439–455.

Carpiano, R.M. (2009). 'Come take a walk with me: The "go-along" interview as a novel method for studying the implications of place for health and well-being', *Health & Place*, 15, pp. 263–272.

Chouinard, V., Hall, E. and Wilton, R. (eds.) (2010). *Towards Enabling Geographies: 'Disabled' Bodies and Minds in Society and Space*. Farnham, NY: Ashgate.

Dublin City Council. (2012). *Your City, Your Space: Dublin City Public Realm Strategy*. Dublin: Dublin City Council.

Dyck, I. (2005). 'Feminist geography, the "everyday", and local-global relations: Hidden spaces of placemaking', *The Canadian Geographer*, 49(3), pp. 233–243.

Edwards, C. and Imrie, R. (2003). 'Disability and bodies as bearers of value', *Sociology*, 37(2), pp. 239–256.

Fincher, R., Pardy, M. and Shaw, K. (2016). 'Place-making or placemasking? The everyday political economy of "making place"', *Planning Theory & Practice*, 17(4), pp. 516–536.

Friedmann, J. (2010). 'Place and placemaking in cities: A global perspective', *Planning Theory and Practice*, 11(2), pp. 149–165.

Gibson, B.E. (2006). 'Disability, connectivity and transgressing the autonomous body', *Journal of Medical Humanities*, 27(3), pp. 187–196.

Gulati, N. (2015). 'How can placemaking help create safer cities for women?' in Project for Public Spaces, 28 October 2015 [online]. Available at: https://www.pps.org/article/un-women-forum (Accessed: 20 April 2020).

Hall, E. (2019). 'A critical geography of disability hate crime', *Area*, 51(2), pp. 249–256.
Hall, E. and Bates, E. (2019). 'Hatescape? A relational geography of disability hate crime, exclusion and belonging in the city', *Geoforum*, 101, pp. 100–110.
Imrie, R. (1996). *Disability and the City: International Perspectives*. London: SAGE.
Imrie, R. (2004). 'Disability, embodiment and the meaning of home', *Housing Studies*, 19(5), pp. 745–763.
Jones, P. and Evans, J. (2012). 'Rescue geography: Place making, affect and regeneration', *Urban Studies*, 49(11), pp. 2315–2330.
Kitchin, R. (1998). '"Out of place", "knowing one's place": Space, power and the exclusion of disabled people', *Disability and Society*, 13(3), pp. 343–356.
Koskela, H. (1997). '"Bold walk and breakings": Women's spatial confidence versus fear of violence', *Gender, Place and Culture: A Journal of Feminist Geography*, 4(3), pp. 301–320.
McClimens, A., Partridge, N. and Sexton, E. (2014). 'How do people with learning disability experience the city centre? A Sheffield case study', *Health and Place*, 28, pp. 14–21.
Mikton, C., Maguire, H. and Shakespeare, T. (2014). 'A systematic review of the effectiveness of interventions to prevent and respond to violence against persons with disabilities', *Journal of Interpersonal Violence*, 29(17), pp. 3207–3226.
Nejad, S., Walker, R. and Newhouse, D. (2019). 'Indigenous placemaking and the built environment: Toward transformative urban design' *Journal of Urban Design*, 25(4), pp.433–442.
Pain, R. (1997). 'Social geographies of women's fear of crime', *Transactions of the Institute of British Geographers*, 22(2), pp. 231–244.
Pain, R. (2000). 'Place, social relations and the fear of crime: A review', *Progress in Human Geography*, 24(3), pp. 365–387.
Pain, R. (2014). 'Seismologies of emotion: Fear and activism during domestic violence', *Social and Cultural Geography*, 15(2), pp. 127–150.
Power, A. and Bartlett, R. (2018a). '"I shouldn't be living there because I am a sponger": Negotiating everyday geographies by people with learning disabilities', *Disability and Society*, 33(4), pp. 562–578.
Power, A. and Bartlett, R. (2018b). 'Self-building safe havens in a post-service landscape: How adults with learning disabilities are reclaiming the welcoming communities agenda', *Social & Cultural Geography*, 19(3), pp. 336–356.
Roulstone, A. and Mason-Bish, H. (eds.) (2013). *Disability, Hate Crime and Violence*. London: Routledge.
Roulstone, A., Thomas, P. and Balderston, S. (2011). 'Between hate and vulnerability: Unpacking the British criminal justice system's construction of disablist hate crime', *Disability and Society*, 26(3), pp. 351–364.
Soldatic, K., Morgan, H. and Roulstone, R. (eds.) (2014). *Disability, Spaces and Places of Policy Exclusion*. London: Routledge.
Stephens, L., Ruddick, S. and McKeever, P. (2015). 'Disability and deleuze: An exploration of becoming and embodiment in children's everyday environments', *Body and Society*, 21(2), pp. 194–220.
Thomas, P. (2011). '"Mate crime": Ridicule, hostility and targeted attacks against disabled people', *Disability and Society*, 26(1), pp. 107–111.
UN (United Nations). (2006). Convention on the Rights of People with Disabilities. New York: UN.
Valentine, G. (1989). 'The geography of women's fear', *Area*, 21(4), pp. 385–390.
Wiesel, I. and Bigby, C. (2014). 'Being recognized and becoming known: Encounters between people with and without intellectual disability in the public realm', *Environment and Planning. part A*, 46(7), pp. 1754–1769.

Further reading in this volume

Chapter 2: Placemaking as an economic engine for all
James F. Lima and Andrew J. Jones
Chapter 13: Sensing our streets: involving children in making people-centred smart cities
Sean Peacock, Aare Puussaar, and Clara Crivellaro
Chapter 16: More than a mural: participatory placemaking on Gija Country
Samantha Edwards-Vandenhoek
Preface: The radical potential of placemaking
Cara Courage
Chapter 20: Displacemaking 2015 and 2020
Catherine Fennell and Daniel Tucker

Chapter 24: Artists, creativity, and the heart of city planning
Tom Borrup
Chapter 28: Integral placemaking: A poiesis of sophrosynes?
Ian Wight
Preface: The only thing constant is change
Kylie Legge
Chapter 35: Planning governance: lessons for the integration of placemaking
Nigel Smith
Chapter 36: Facilitator skills for effective collaborative placemaking
Husam AlWaer and Ian Cooper
Chapter 37: *The Neighbourhood Project*: A case study on community-led placemaking by CoDesign Studio
Lucinda Hartley, Eliza Charley, Sama Choudhury, and Harriet McKindlay
Chapter 38: Public seating: a small but important place in the city
Kylie Legge
Chapter 45: How the city speaks to us and how we speak back: rewriting the relationship between people and place
Rosanna Vitiello and Marcus Willcocks

16
MORE THAN A MURAL
Participatory placemaking on Gija Country

Samantha Edwards-Vandenhoek

My journey to Gija Country began in the hot and sticky coastal town of Broome, over 3,000 kilometres from my home in Melbourne. As a middle-class woman of Celtic-European heritage and privilege, the material and psychological scars of the legacies of invasion, colonisation, dispossession set against the rugged and formidable backdrop of the Kimberley region's spectacular geological formations, afforded a challenging and confronting visual landscape that called into question my sense of belonging, cultural identity, and place, well before I reached my final destination.

I am deeply indebted to the Gija Elders and their families for their support and guidance. This project would not have been possible without the endorsement of the Gija community and its project partners: Warmun Arts, Muralist Tom Sevil (aka Civil), Warmun Council Inc. and the Warmun Indigenous Justice Program, Gija Rangers, Purnululu School, Ngalangangpum School, Former Principal Leanne Hodge and teacher Cimony Vanderpol, Warmun Arts Assistant Manager and Curator Alana Hunt, Community Programs Coordinator Anna Crane, and Linguist Frances Kofod. Special thanks to Rusty Peters, Morris Peters, Desma Juli, Imran Daylight, Nancy Daylight, Andrew Mung, and Terry Mosquito for their contributions and guidance with regard to the location, design, and cultural appropriateness of public artworks. This project acknowledges the traditional custodians of the land on which this research has been conducted, the Gija people, Elders past and present, and the unique diversity of the Indigenous community in the Kimberley region. Despite the lack of constitutional recognition, this project recognises the sovereignty of Aboriginal and Torres Strait Islander people. This research is grounded in protecting, recognising, and acknowledging the continuing Indigenous ownership of the traditional knowledge, cultural expressions, and intellectual property rights of its participants (AIATSIS 2012). This project has been funded by Swinburne University of Technology and approved by Swinburne's Human Research Ethics Committee in line with the National Statement on Ethical Conduct in Human Research.

Introduction

Warmun is an Aboriginal community in the East Kimberley region of Western Australia. The town is built on Turkey Creek which flows into the Ord River near Kununurra, a major service town close to the Northern Territory border. The main languages spoken are Gija, English, and

Kimberley Kriol. From the 1880s, Gija people were put to work by European farmers who came to exploit natural resources and run cattle on their traditional lands. This marked the beginning of the traumatic spiritual and social displacement of Gija people, the loss of their homelands and their cultural practices (Pelusey and Pelusey, 2006). In the 1960s changes to Government laws meant that the pastoralists were required to pay Indigenous people the same as non-Indigenous workers. Turkey Creek was established in the 1970s after Gija people were moved off surrounding cattle stations when some pastoralists refused to abide by these laws. Gija people renamed Turkey Creek 'Warramun' after the Ngarranggarni (dreaming story) of Warrarnany (wedge-tailed eagle) (Kofod, 2016). For the Gija people, the Ngarranggarni explains how things came to be and are continually occurring and ever-present. It is a way of being and knowing that explains and defines social structures, complex kinship systems, language, law, mythologies, creative expression, and place-based relations (Kofod, 2016). Today, the community comprises nearly 100 homes, a school, health clinic, police station, roadhouse, recreation centre, community store, sporting ground, and the internationally celebrated Gija-owned Warmun Art Centre, who partnered on this research.

> The red/muddy squiggly lines represent dirty water. The top and bottom lines show houses, cars and other material things which were washed away in the flood. The figures in the middle represent us. The broken lines represent our spirits, which were broken after everything was taken away. Moving into town, living 'on top' of each other in a different land area, affected us emotionally, mentally and physically.
>
> *(Bessie Daylight, describing her painting entitled 'Warmun Flood,' as cited in Government of Western Australia, 2011, p. 75)*

This chapter reveals a placemaking initiative, known locally as *Art in the Streets of Warmun*, centred on how the Gija community could reassert a positive relationship with their environment through the reclamation and visual activation of public spaces in the wake of the 2011 flood that ravaged the township and Government-led rebuild. The broader research examines the role of socially inclusive participatory public art in community building and healing, shaping new spatial encounters that foster belonging, trauma recovery and pride, cultural continuity and renewal. It rests on the premise that First Nations identities can be embedded in public art and architecture, enabling people to realise their stories and power and challenge-imposed structures, systems, and processes through assertions of cultural identity and connections to place (see also Edmonds, 2012).

Written five years on, this discursive chapter reflects on the outcomes, learnings, and challenges associated with the participatory placemaking processes employed. As a settler-colonial researcher, my understanding of what we did together has evolved and deepened as a result of the enduring relationships I have developed with the Gija community through subsequent collaborative creative endeavours. This project created a unique space when Indigenous and non-Indigenous worldviews, cultures, lived experiences, knowledge traditions, and placemaking practices came together, with often unplanned and serendipitous outcomes, impacts, and insights.

Through the 'doing' of this project, *Art in the Streets of Warmun* became less about 'placemaking' and more about understanding the interconnected relationship the Gija people have to place – what is known as 'Country' to Australia's First Nations peoples and translates as 'Daam' in the Gija language. There are over 500 Indigenous clan groups or nations around the Australian continent, with distinctive cultures, beliefs, and languages. As a grammatical convention, the use of a capital 'C' in 'Country' is intentional, as it makes reference to a specific nation, (e.g. Canada), or in this case, Gija Country. It also distinguishes the term and its embedded associations from Eurocentric perceptions of country, place, and land to be unpacked in the following section. This chapter seeks to contribute to understandings of the complex, problematic, and locally nuanced discussions around placemaking connected to place-related meanings in First Nations community development, as there was indeed no 'place' to be 'made' here, but rather enacted, reinforced, and made visible.

Berrema daam ngarag noonamenke ngagenybe daam

Placemaking is 'the set of social, political and material processes by which people iteratively create and recreate the experienced geographies in which they live' (Pierce et al., 2011, p. 54). It can also be understood as a process of reimagining the use and character of physical environments and structures, for example, and the focus of this chapter, public art, as well as, for example, market gardens, pop-up spaces, walking paths, that are more responsive to their inhabitant's cultural, emotional, spiritual, and social needs (Hazen, 2013). The goal of placemaking, understood thus, is to bring about some form of change or transformation in which the focus rests on the skills, cultural traditions, knowledge, and aspirations of the community of interest (Hung et al., 2006). *Art in the Streets of Warmun* was not about imposing urban settler-colonial placemaking practices and processes on the Gija community. Working within a relational knowledge framework which centralises Indigenous voices, narratives, and perspectives, this project sought to understand what place, placemaking, and public art meant to the Gija people of Warmun:

> Gija people have lived in the lands around Warmun since the Ngarranggarni, or creation time, when spiritual beings roamed the land and created everything in it. Their Country contains traditional hunting grounds, ceremonial sites and resting place of their ancestors, who embody the past and define the future. (Government of Western Australia 2011 p. 11)

Archaeologists have confirmed that Aboriginal people have occupied and thrived in the Kimberley region for 60–80,000 years. It was not until 3 June 1992, that the High Court of Australia announced its decision to overturn the legal doctrine of 'terra nullius,' meaning 'legally unoccupied,' the term applied by the British to land ownership of Australia, providing significant insight into the attitude of the denial of Aboriginal occupation and, by extension, the consequences enforced placelessness have had on the livelihoods, health, and wellbeing of the fundamentally place-based, sustainable, and state-free social order of Australia's First Nations peoples (Havenmann, 2005).

Gija Country encompasses impressive geographical formations, including Purnululu National Park, and is the site of the Daiwul (Barramundi) Ngarranggarni, currently Rio Tinto's Argyle Diamond Mine. However, as Deborah Rose (1996, p. 7) explains, 'Country' does not just mean the creeks, hills, rock formations, and waterholes. 'Country is multi-dimensional'. It consists of people, plants and animals. It also embraces the seasons, dreaming stories, and creation spirits. 'People talk about country in the same way that they would talk about a person: they speak to country, sing to country, visit country, worry about country, feel sorry for country and long for country' (Rose, 1996, p.7). In Aboriginal mythopoeia, 'an understanding of the relationship between an ancestor and a place – developed through learning verbal and graphic stories, songs and dances – is necessary before one can read an ancestor in a place' (Fantin, 2003, n.p.). Indigenous placemaking is understood thus as a form of land stewardship and learned and lived psychogeography, also known as 'Caring for Country' by Australia's First Nations peoples. In contrast non-Indigenous people often see land as something they own, a commodity to be bought and sold, an asset to make profit from (such as the Argyle diamond mine), a living off it, or a 'home.' However, 'for Aboriginal people the relationship is much deeper. Connection to Country is central to maintaining health and wellbeing (of both Country and community), cultural life, 'individual autonomy and Indigenous sovereignty' (Ganesharajah, 2009, p. 6). Even today, the pivotal role 'caring for Country' plays in the lives of Indigenous people and communities cannot be underestimated. In Australia's First Nations cultures, identity and Country is inextricably and externally linked to the Ngarranggarni (in the case of the Gija people) of that place. Moreover, within an Indigenous framework, knowledge is also situated, dynamic, shared, and alive and mirrors the symbiotic and equitable relationship between 'individuals, communities, generations, the physical environment, and other living creatures' (Mann, 1997, cited in Kennedy, 2015, p. 28; see also Chow, 1995). This sits in stark contrast to Euro-Western knowledge frameworks grounded in empirical positivism. Accessing, harnessing, and honouring this interconnectedness was requisite to the participatory placemaking processes of *Art in the Streets of Warmun*.

Sadly, the denials and whitewashing of Indigenous land stewardship practices and forced removal of people from their homelands and families is ongoing. There was minimal consultation between the Government of Western Australia and the Gija community in terms of the design, suitability, and location of new homes after the 2011 flood. Despite Indigenous and non-Indigenous protest in April 2014 (around the time of my first visit to Warmun), the Western Australian Government decided that it may no longer accept responsibility for the provision of municipal services to remote Aboriginal communities (Harrison, 2014). *Art in the Streets of Warmun* afforded a timely and politically charged 'on-Country' intervention for people living on their homelands, enabling Gija residents to publicly reassert their connection to Country through the highly visible reclamation of their streets via public art. Along the Kimberley's Great Northern Highway, protest banners with powerful slogans were hung from trees, serving as reminders that, despite the apparent invisibility and perceived remoteness of some Indigenous communities, 'this is [our] home'; 'money is nothing, Country is everything' and 'living on Country is not a lifestyle choice.'

Prior to the commencement of the project, a Gija language workshop was held at Warmun Art Centre with artists, linguists, and artworkers to consider and develop Gija phrases and expressions to understand and describe what art 'outside' means, how it translates, and how it is connected to the Gija experience of 'daam,' and informs the project's intentions, outcomes, and locations. Through participating in this process (rather than reading about it in textbooks), I came to understand that Gija identity, culture, and creative expression is 'not separate from external forces and influences and architecture is one of those influences' (Fantin, 2003, n.p.). The language developed to describe *Art in the Streets of Warmun* mirrored this interconnectedness – Berrema daam ngarag noonamenke ngagenybe daam, which translates as 'This my Country, I'm painting here.' Reminiscent of a deictic expression, 'I'm painting here', is literally said, while pointing to a specific 'place,' surface, or place in time (e.g. creation time). Significantly, the Gija expression for painting on canvas and painting on buildings is interchangeable; in Gija culture, all forms of creativity and the making of artworks (e.g. printmaking, photography, pasted-up paper art with wheat paste glue, known as paste ups, and murals) are considered to be the caring, making (e.g. painting, drawing), celebration, and re-enactment of Gija narratives, traditions, Ngarranggarni, as well as specific sites and geographic formations.

The guiding principle of the participatory framework I worked with was one of decolonisation, 'a process of conducting research in such a way that the world views of those who have suffered a long history of oppression, trauma and marginalisation are given space to communicate from their frames of reference' (Chilisa, 2012, p. 23). This project also drew upon the longstanding legacy of Community Based Participatory Action Research (CBPAR) (Kendall et al., 2011; Ritchie et al., 2013; Lucero 2013; Wallerstein and Duran, 2006), especially Transformative Participatory Action Research (TPAR), as introduced by Bagele Chilisa (2012), as prioritising self-determination, social and personal transformation and emancipation, and encouraging participants to become engaged citizens through developing a sense of individual and shared accountability for their decisions and outcomes. Crucially, TPAR affords a means whereby the focus of the research originates in the community and the problem (such as legacies of the rebuild) is defined, analysed, and resolved by the community.

The challenge I faced was providing the space to build the trust required to initiate these kinds of intimate exchanges and community-led processes given the project's restricted time frame. By chance, I drove into Warmun in a bright-yellow hire car. Clearly, no one was going to miss me. What had been an initial source of embarrassment (for me) turned out to be a means of connecting with people. The car became an identifier, a conversation starter, and the butt of many jokes (e.g. Uber/taxi service.). It was only then that people started to open up to me. Time was spent 'going for a cruise' with old people, young folk, and artworkers handing out and putting up posters, providing opportunities to assess and photograph potential locations that could be documented and discussed in the planning sessions. Through 'cruising,' I learned that official street signs (as locators) were largely redundant, as were roads. The community divided itself into six distinct camps, known as 'Top Camp,' 'Garden Area,' 'Middle Camp,' 'Big Bottom Camp' (circular cul-de-sac), 'Little Bottom Camp' (smaller circular cul-de-sac), 'Overflow,' and 'Other Side' (of the creek). This kind of counter-normative relational spatial awareness defined the playful and political physical dynamic of Warmun and played a central role in the assessment and location of public art outcomes. Passengers would point out sites of cultural significance, Ngarranggarni, 'sorry business,' and massacres. Much to the amusement of my travelling companions, cruising involved numerous 'cut-throughs,' 'side-tracks,' and 'short cuts' that meant going literally off the 'colonial' road, and onto the red dirt of Gija Country. I found you could traverse the entire township (on both sides of Turkey Creek) without touching bitumen. This kind of transgression correlates to the notion that Country is experienced physically and is deeply rooted to belonging and wellbeing. Much like official street names, the bitumen is a barrier to the way that place and home is sensed and experienced by the Gija people.

With funding secured, the project was framed, planned, and enacted over a period of six months in the 2015 dry season (March to August). Climate (temperature and conditions) played a critical role in the timing of the project. A staged approach with more 'formal' meetings at the Art Centre or Rec Shed (as well as 'cruising', which involved picking up and dropping people home) was taken to understand and incorporate the voices, interests, aspirations, priorities, and cultural knowledge of the people who would drive the form and content of the outcomes. Elders, senior artists, artworkers, schoolteachers, and Gija linguists were brought in as guides to facilitate the process. Tom Civil, a Melbourne-based street artist who had developed ongoing relationships with the art centre and with Elders and their families, was commissioned to instruct and oversee the making of large-scale mural-based artworks.

Art in the Streets of Warmun

This section outlines the activities that shaped the substance of *Art in the Streets of Warmun* as it unfolded. This is because the working processes and visual outcomes can be better understood as critical spatio-temporal junctures of conversation, negotiation, exchange, collaboration, and co-creation to reveal new knowledge about Indigenous approaches to placemaking. High-school students were responsible for the first large-scale public art installation in Middle Camp that heralded the start of *Art in the Streets of Warmun* in July 2015. It was hoped that this installation would generate broader community interest in the project and give young people the chance to make their mark on the spaces and surfaces where they spent time together would, over time, generate a sense of respect, ownership and self-empowerment. Affixed to the handball wall in front of the recreation centre, this artwork took the form of a large-scale 'paste-up' (3m × 3m) of digitally produced self-portraits which comprised graphic silhouettes of the student's profiles filled with photographic images of textures taken from the surrounding bushlands. The digital files were emailed to Melbourne, assembled, printed, and brought back in my hand luggage. It resulted in a striking and contemporary visual exposé exploring the interconnectedness of identity and place, while disguising the identity of each person represented. It also covered up a proliferation of tags and faded scribbles on the concrete handball wall. Students commented on how they felt pride seeing their artworks publicly displayed, without feeling exposed and at risk of being tagged over or defaced by their peers.

The Indigenous research method of yarning (also referred to as a 'yarning circle') at the recreation centre (affectionately known as the 'Rec Shed') in Middle Camp provided a space for Elders, artists, artworkers, and interested community members to gather, drink nalaga (tea), contest styles, thematic and potential location of public artworks. As described by Melissa Walker, 'yarning is a conversational process that involves the telling and sharing of stories and information. Yarning is culturally ascribed and cooperative; yarns follow language protocols and result in some acquisition of new meaning,' insights, and knowledge (Walker et al., 2014, p. 1217). Each morning, I would set up a number of chairs in a loose circle in the open-air interior expanse, put the urn on to boil, layout large tin camp-style cups, open a packet of biscuits, and wait. People would visit, stay for a yarn, and leave, then return the following day to talk through ideas or to paint. Specific meeting times were set in keeping with prevailing temperatures when decisions (about what to paint, where, how, and why) were required. In these yarning circles valuable conversations took place around visual expression, communication modes, appropriate acknowledgements of Country, and reinterpretation of images from rock art sites, paintings, and Ngarranggarni. In line with the participatory methodology, the decision-making processes were deliberative, ongoing, and shared. A3 printouts of photos taken of buildings and potential surfaces were used to stimulate discussion and to sketch out and share ideas. The idea of painting a colourful interactive path in the shape of the rainbow serpent (a recurring figure in Aboriginal mythopoeia) on the concrete floor of the Rec Shed came out of a discussion about how to encourage young people to utilise the space. It stemmed from the idea of painting the game of 'hopscotch,' naively put forward by myself, and quickly squashed with much laughter because, as I was told, 'black fellas don't play hopscotch.'

After much deliberation and reconnaissance, the Rec Shed's spaces, structures, and surfaces were chosen for the enactment of this initiative as part of a broader strategic redevelopment plan to reactivate youth areas in Warmun. The building itself and surrounding toilet blocks, gym, canteen, basketball court, handball wall, fences, boulders, road signs, poles, and paths were covered in an proliferation of typographic inscriptions comprising long lists of first names, family names, and crew names (e.g. 'Roo Boy's,' 'Sisters 4 Life'), doodles, rude words, and slang codes, some of which predated the 2011 flood. I learned that, much like Western forms of 'tagging,' this kind of naming graffiti was not well regarded by the community, and that Elders, in particular, frown upon it. However, for young people, rather than disconnection and isolation, these illicit writings evidenced strong family bonds, skin and kinship ties, and place-based connections and belongings, as well as boredom, disenfranchisement, and frustration. Despite this, the Warmun Council hoped to encourage young people to more creatively and caringly engage with these spaces and surfaces. The Rec Shed was also chosen because of its scale and significance as a central gathering place (in Middle camp) and site of communal performance, celebration, and play, as well as its high visibility to residents and the visiting public; however, budget and time constraints also played a role. By focusing on one central location, efforts could be consolidated, and the visual impact of public art outcomes maximised.

Garnkiny

The production of the 'Garnkiny' (Moon Dreaming) mural developed organically from a conversation that unfolded over several days with Rusty Peters, one of Warmun's senior artists. Pointing to the sky with his walking stick and then to the interior open-air corrugated expanse of the Rec Shed, where movies are screened at night, Rusty suggested creating a mural based on recurring cosmic imagery in his own paintings – the Milky Way. Rusty's paintings recreate 'Garnkiny,' a moral tale of forbidden love (wrong skin) from his father and grandmother's Country where he was born on Springvale Station (Peters, n.d.). Working with Rusty, we began by creating a miniature version of the proposed mural on paper. Rusty instructed us (Desma Juli, Nancy Daylight, Tom Civil, and me) in the technique he used to create the stars, guiding, pointing, and directing. Involving Desma was essential; as Rusty's granddaughter she has permission to translate the story and re-enact the visual expression of it. The completed mural blurred ontological, astronomical, and cultural distinctions and insider and outsider relations in its collaborative undertaking to realise a shared vision of our galaxy, extending across a narrow strip of corrugated iron around two internal shed faces, with the crescent moon and blur of the Milky Way forming the centrepiece.

Always was, always will be Aboriginal land

Conscious that it may have appeared a cliché or cursory choice or even an unwelcome(ing) political sign, considerable time and community consultation went into making an informed decision to paint a 4m × 4m reproduction of the Aboriginal flag on the rear wall of the Rec Shed that faced the main entrance to the town. Fran Edmonds (2012) has suggested that 'murals provide alternative approaches for Aboriginal people … to assert their Aboriginality and provide a visual language for "re-membering" history from an Aboriginal perspective' (p. 21). A painting of the Aboriginal flag can then be understood as a way of performing this kind of re-membering. It also makes a clear and recognisable statement about land rights and unseeded sovereignty of Australia's First Nations peoples to outsiders, service providers, and tourists visiting the Warmun Art Centre. During such a politically charged time, in the midst of proposed Government forced closures, the flag was (and is) not only a powerful symbol of cultural pride that enables Indigenous communities to realise their stories as assertions of identity and connections to place, but also supports a 'rhetoric of difference' (Edmonds, 2012). It also serves as a reminder that this country has a rich history that precedes European colonisation. Its large scale and uniform shapes and colours meant that people of all ages could be involved in its painting, allowing a broader spectrum of Warmun youth to be engaged, with school buses bringing in kids of all ages from Frog Hollow community (20 minutes down the Great Northern Highway) to assist in its production.

Warrrarnany Gooningarrim-Noongoo

With the smaller murals completed, the focus shifted to the 15m corrugated expanse of the Rec Shed – the largest, longest, and most visible public space in Warmun. Consideration was given to the reproduction of iconic paintings incorporating the visual styles of senior Gija artists. Considering the centrality and size of this interface, the planning committee wanted to ensure that the mural was not just participatory in its making, but also in its communication. Consequently, the Gija Ngarranggarni, Warrrarnany Gooningarrim-Noongoo (wedge-tailed eagle and crow dreaming), which speaks to how Warmun came to be, was chosen. The wedge-tailed eagle is also the mascot of the Warmun Eagles, the much revered Warmun football team, and remains a cultural force and binding figure in everyday Gija life.

The main challenge was deciding which parts of this complex and multi-layered Ngarranggarni of Eagle Hawk and Crow, who were once human in form and husband and wife, were to be communicated without losing its core messages. In Katie Cox's powerful retelling on canvas,

> Eagle Hawk asked Crow to help him make spearheads to hunt kangaroo, but she was a lazy woman and said 'No.' Eagle Hawk built a fire for cooking, then went hunting and brought back a kangaroo to the camp. He found Crow still sleeping. He was proper angry and put hot rocks in her eyes and burnt her all over with the coals from the fire. They turned into birds and live in the trees at the top of the hills behind Warmun. The white band of quartz in the hillside is their campfire – and this is why the crow is black with white circles in her eyes.
>
> *(Cox, n.d.)*

A narrative-based style that focused on key elements and visual references (e.g. hills, kangaroo, eagle, and crow) was agreed upon. Knowledge-holders led the interpretation of the narrative in mural form with extracts of the written text in Gija language that points to the three main characters. The chosen text reads Moolarriji thoorroob wananyjinde jiyirrem miyalgaleny! (He's the best hunter, he never misses a kangaroo – Number one hunter!); Danya garayi wiyinji miyaleboorroo biyaya wiyinji (He's flying around, looking for meat); Ngeleli Wanggarnal, Ngajigal-Noongoo (This is the crow; she is his sister) and Jiyirriny Nginiyin Goorrngam-Boorroo roord nginji yilag (The kangaroo has come for water and he is sitting down there). Linguists, Elders, Gija Rangers, and local community members were central to its production. Tom Civil utilised a digital projector at night to trace out the location of the text and placement of key elements. The final outcome was impressive, immediate, aesthetically pleasing, and responsive to its scale and location, as it faces the hills behind Warmun with a depiction of the Eagle flying over the town. The constant re-experiencing of this mural on a daily basis by people walking, riding, or cruising by has provided a way to remind and reinforce the importance of its motifs and messages over time, instilling feelings of happiness and pride to those who experience it on a daily basis, and by visiting Gija community (from surrounding outstations, such as Bow River) who come to play or support the Warmun Eagles or Bow River Blues. It has also prompted conversations with tourists about its meanings and ongoing significance. The inclusion of exerts of the narrative in Gija in such a prominent location affirms the centrality of language to Gija life. In these ways, it is much more than a mural.

A place of reconciliation, a reconciliation of place

> Thank you for asking permission. Thank you for listening to us. No-one has come here and done that before.
>
> *(Terry Mosquito)*

A community barbeque was held at the Rec Shed before Tom and I left Warmun, providing an opportunity to celebrate, yarn, and reflect on the outcomes and our collective experiences. From the outset it was made clear to us that this had been the first time the community had been given the opportunity to participate in and direct a project that impacted on their own lives and livelihoods. Terry Mosquito's feedback spoke to the unforeseen negative and positive effects of divergent and oppositional modes of placemaking practice. The Government had come in after the flood and made decisions on the community's behalf concerning the rebuild;

Art in the Streets of Warmun had enabled local residents to reclaim their community and buildings and make more visible their rights and place-based connections via street art, transgressing top-down government decision-making.

Gija residents commented that they felt the murals made their community look beautiful and that this enhanced their experience of their own public spaces. Residents expressed satisfaction and pride in the positive image the murals projected of Warmun to people visiting the community for work or social reasons and to service providers. In this respect, the project enhanced a sense of community, building resilience, pride, and place. Such large-scale collaborative works had not been attempted in Warmun before, opening up the possibility of further projects exploring the transformation of public space (Personal communication with Anna Crane, 25 August 2015). Five years later, the Rec Shed remains a highly activated space. The large-scale murals have become popular locations for selfies and family portraits that are circulated widely via personal social media networks. Crucially, the painted murals remain undefaced, except for one micro penis symbol on the body of the eagle, which was quickly painted over:

> Today places like the Kimberley are rich with creative projects bringing together methods historically linked to anthropology and social research with those from the domains of arts and community development, engaging diverse groupings of collaborators to facilitating the telling of local stories and furthering a range of social and political agendas.
>
> *(Havilland, 2016, p. 44)*

Looking back, *Art in the Streets of Warmun* went onto develop a life of its own in ways that were unforeseen or predicted. In 2016, Ngalangangpum School students transformed the handball court into a rotating gallery of ephemeral paste-up art. I was invited back to the school to instigate a place-based design education program, providing opportunities for high-school students to explore the connection between culture, art, design, media, and sustainable local employment (Edwards-Vandenhoek, 2018). In 2016, Warmun Council secured Western Australia Government funding to instigate a community-led wayfinding project to replace existing and irrelevant public signage. Tom Civil returned to Warmun in 2017 to work with artworkers and families of senior artists to reproduce their artworks on the newly opened aged care facility. In 2018, I was engaged by the Warmun Art Centre Board to co-produce two collaborative community films focused on the revitalisation of language and archiving of stories connected to the practice of ochre extraction, production, and use for the Western Australian Museum.

Conclusion

Art in the Streets of Warmun sought to understand how place, placemaking, 'art outside,' and, by extension, meaning-making, is understood and enacted by the Gija people of Warmun. As I would slowly come to understand, to Australia's First Nations peoples, the landscape is the source of their identity – inseparable to place. Connectedness to Country permeates all facets and layers of Gija life – from the Ngarranggarni, law, familial relations, and creative expression, to cruising, drinking nalaja, and watching football. As a problematisation of placemaking, *Art in the Streets of Warmun* was an act of decolonisation, a conscious attempt to destabilise power relations connected to race, identity, and land, and prioritise Gija culture, values, and place-based meanings. It afforded a space where relations between participants, operating systems, methodological processes, knowledge systems, and places could be reimagined. Contributing to Indigenous perspectives on placemaking, surfaces were transformed into access points and conduits to hidden meanings, new knowledge, and deeper relationships. Questioning settler-colonial practices of placemaking and its effect on claims for land ownership, coupled with the knowledge that no formal treaties are in place with Australia's first peoples, this research also highlights the deeply troubling nature of attempts to 'make' place on lands (urban, regional, or bush) that have, in effect, been stolen from Australia's First Nation's people.

Contributing to more just Indigenous–settler relations, *Art in the Streets of Warmun* demonstrated that as an outsider and an ally, it is possible to develop relationships and provide resources that can be utilised by individuals and communities in ways that allow them to draw from their own spiritual references and cultural practices (i.e. caring for Country) to add value their own lives. Its enactment required attending too, listening, taking time and care; accepting that not everyone wanted to be involved; respecting internal conflicts and tensions (tags vs. murals); and creating unexpected spaces (like a yellow car) for dialogue. As intimated by Lucero (2013), building trust was paramount to all endeavours. Often, it was in the space between what I thought I should be doing and what I was doing that people revealed information (e.g. side-tracks and cut-throughs) and stories that formed the substance of *Art in the Streets of Warmun*. Crucially, this process involved understanding that Indigenous knowledge and placemaking research is an 'engagement within a field of powerful and often hidden cultural, environmental, historical and social relations' (Sheehan and Walker, 2001, p. 14; see also Maddison, 2020). There are registers of relationality that I may never truly comprehend as a non-Indigenous person.

A recent independent report by the Australian Human Rights Commission highlighted the continued failure of the Commonwealth of Australia's 'Closing the Gap Strategy' to address Indigenous disadvantages and noted that the most successful attempts to improve health, education, and socio-economic outcomes for Aboriginal and Torres Strait Islander peoples have taken place at the local and grassroots level (Holland, 2018; Hunt, 2013; see also Behrendt et al., 2012). Key to the success of these endeavours has been the centralisation of culture and language and 'the formation of relationships between community leaders and trusted outsiders, and the shared understanding and new knowledge they derive' (Moran, 2016, n.p.). While imperfect, hard to fully articulate, and small scale, I believe that this project's decolonising mindset and attempts to honour it through our actions – shifting power relations, promoting cultural revitalisation and development plans – provided a space for Gija people to shape and assert their own identities while realising forms of collective agency. *Art in the Streets of Warmun* speaks to the power of Indigenous approaches to placemaking in supporting healing, reciprocity, and knowledge acquisition, leading to more socially resilient and autonomous local First Nations communities.

References

Australian Institute of Aboriginal and Torres Strait Islander Studies. (2012). 'Guidelines for ethical research in Australian Indigenous studies', Australian Institute of Aboriginal and Torres Strait Islander Studies [online]. Available at: https://aiatsis.gov.au/sites/default/files/docs/research-and-guides/ethics/gerais.pdf (Accessed: 1 November 2013).

Behrendt, L., Larkin, S., Griew, R. and Kelly, P. (2012). 'Review of higher education access and outcomes for Aboriginal and Torres Strait Islander people final report', Department of Industry, Innovation Science, Research and Tertiary Education [online]. Available at: https://docs.education.gov.au/system/files/doc/other/heaccess andoutcomesforaboriginalandtorresstraitislanderfinalreport.pdf (Accessed: 23 November 2015).

Chilisa, B. (2012). *Indigenous Research Methodologies*. Berkeley: SAGE.

Chow, W. (1995). *Three-Partner Dancing: Placing Participatory Action Research Theory into Practice Within an Indigenous, Racialized and Academic Space*. Master's thesis. University of Victoria, Canada.

Cox, K. (n.d.). *Eagle Hawk and Crow*. Artlandish Aboriginal Art Gallery [online]. Available at: https://www.aboriginal-art-australia.com/artworks/katie-cox-eagle-hawk-crow-2a// (Accessed: 1 October 2019).

Edmonds, F. (2012). 'Making murals, revealing histories: Murals as an assertion of Aboriginality in Melbourne's inner north', in Kleinert, S. and Koch, G. (eds.) *Urban Representations: Cultural Expressions, Identity and Politics*. Canberra: AIATSIS Research Publications [online]. Available at: https://aiatsis.gov.au/sites/default/files/products/monograph/kleinert-koch-urban-representations-cultural-express-identity-politics_0.pdf (Accessed: 1 May 2019).

Edwards-Vandenhoek, S. (2018). '"Over there, in the future": The transformative agency of place-based design education in remote Aboriginal communities', *International Journal of Art & Design in Education* (Special Issue iJADE Conference, 'Art and Design as Agent of Change,' 17-18 November 2017. Dublin, Ireland: National College of Art and Design), 37(4), pp. 622–637.

Fantin, S. (2003). 'Aboriginal identities in Architecture', in *ArchitectureAU* [online]. Available at: https://architectureau.com/articles/aboriginal-identities-in-architecture/ (Accessed: 1 May 2019).

Ganesharajah, C. (2009). *Indigenous Health and Wellebing: The Importance of Country*. Native Title Research Report No. 1/2009. Native Title Research Unit, Australian Insititute of Aboriginal and Torres Strait Islander Studies.

Government of Western Australia. (2011). *Warrambany of Warrmarn: The Flood of Warmun*, 13 March 2011, Department of Housing [online]. Available at: http://www.housing.wa.gov.au/HousingDocuments/Warrambany_of_Warrmarn_The_Flood_Of_Warmun.pdf (Accessed: 1 March 2014).

Harrison, D. (2014). 'Indigenous communities fear trauma and dislocation as bulldozers roll in', *Sydney Morning Herald*, 9 November 2014. Available at: https://www.smh.com.au/politics/federal/remote-indigenous-towns-fear-trauma-and-dislocation-as-bulldozers-roll-in-20141128-11vvl9.html

Havenmann, P. (2005). 'Denial, modernity and exclusion: Indigenous placelessness in Australia', *Macquarie Law Journal*, no. 57 [online]. Available at: http://classic.austlii.edu.au/au/journals/MqLawJl/2005/4.html (Accessed: 9 January 2020).

Havilland, M. (2016). *Side by side?: Community Art and the Challenge of Co-Creativity*. New York: Routledge.

Hazen, T. (2013). 'The participatory design process', in Marcus, C.C. and Sachs, N.A. (eds.) *Therapeutic Landscapes: An Evidence-Based Approach to Designing Healing Gardens and Restorative Outdoor Spaces*. Hoboken, NJ: Wiley.

Holland, C. (2018). 'A ten-year review: The closing the gap strategy and recommendations for reset', *Close the Gap Campaign Steering Committee* [online]. Available at: https://www.humanrights.gov.au/sites/default/files/document/publication/CTG%202018_FINAL-WEB.pdf (Accessed: 25 February 2018).

Hung, D., Tan, S.C. and Koh, T.S. (2006). 'From traditional to constructivist epistemologies: A proposed theoretical framework based on activity theory for learning communities', *Journal of Interactive Learning Research*, 17(1) [online]. Available at: http://link.galegroup.com.ezproxy.lib.swin.edu.au/apps/doc/A142475978/AONE?u=swinburne1&sid=AONE&xid=b07689d6 (Accessed: 20 April 2020).

Hunt, J. (2013). 'Engaging with Indigenous Australia—Exploring the conditions for effective relationships with Aboriginal and Torres Strait Islander communities', in Issues Paper No. 5. produced for closing the gap clearinghouse, Australian Government, Australian Institute of Health and Welfare [online]. Available at: https://www.aihw.gov.au/getmedia/7d54eac8-4c95-4de1-91bb-0d6b1cf348e2/ctgc-ip05.pdf.aspx?inline=true (Accessed: 1 May 2015).

Kendall, E., Sunderland, N., Barnett, L., Nalder, G. and Matthews, C. (2011). 'Beyond the rhetoric of participatory research in Indigenous communities: Advances in Australia over the last decade', *Qualitative Health Research*, 2(12), pp. 1719–1728.

Kennedy, R.J. (2015). *Designing with Indigenous Knowledge: Policy and Protocols for Respectful and Authentic Cross-Cultural Representation in Communication Design Practice*. Ph.D. thesis. Swinburne University of Technology, Melbourne, Australia.

Kofod, F. (2016), *Gija-Kija – English Dictionary*. Australian Institute of Aboriginal and Torres Strait Islander Studies.

Lucero, J. (2013). *Trust as an Ethical Construct in Community Based Participatory Research Partnerships*. Ph.D. thesis. University of New Mexico, Albuquerque, NM.

Maddison, S. and Nakata, S. (eds.) (2020). *Questioning Indigenous-Settler Relations*. Singapore: Springer.

Mann, H. (1997). *Indigenous Peoples and the Use of Intellectual Property Rights in Canada: Case Studies Relating to Intellectual Property Rights and the Protection of Biodiversity*. Report submitted to Industry Canada, Intellectual Property Policy Directorate, and to the Canadian Working Group 8(j) of the Convention on Biological Diversity [online]. Available at: http://nativemaps.org/files/Mann.pdf (Accessed: 1 May 2015).

Moran, M. (2016). 'How community-based innovation can help Australia close the Indigenous gap', in *The* Conversation [online]. Available at: https://theconversation.com/how-community-based-innovation-can-help-australia-close-the-indigenous-gap-54907 (Accessed: 4 March 2017).

Pelusey, J. and Pelusey, M. (2006). *Life in Indigenous Communities. Warmun, East Kimberley, Western Australia.* Melbourne: Macmillan Education Australia.

Peters, R. (n.d.). 'Rusty Peters, defying empire', in Third National Indigenous Art Triennial [online]. Available at: https://nga.gov.au/defyingempire/artists.cfm?artistirn=23224 (Accessed: 1 December 2019).

Pierce, J., Martin, D.G. and Murphy, J.T. (2011). 'Relational place-making: The networked politics of place', *Transactions of the Institute of British Geographers*, 36(1), pp. 54–70.

Ritchie, S.D., Jo Wabano, M., Beardy, J., Curran, J., Orkin, A., VanderBurgh, D. and Young, N.L. (2013). 'Community-based participatory research with indigenous communities: The proximity paradox', *Health Place*, 24, pp. 183–189.

Rose, D. (1996). *Nourising Terrains: Australian Aboriginal Views of Landscape and Wilderness.* Australian Heritage Commission, Canberra.

Sheehan, N. and Walker, P. (2001). 'The Purga project: Indigenous knowledge research', *The Australian Journal of Indigenous Education*, 29(2), pp. 11–17.

Walker, M., Fredericks, B., Mills, K. and Anderson, D. (2014) '"Yarning" as a method for community-based health research with indigenous women: The indigenous women's wellness research program', *Health Care for Women International*, 35(10), pp. 1216–1226.

Wallerstein, N. and Duran, B. (2006). 'Using community-based participatory research to address health disparities', *Health Promotion Practice*, 7(3), pp. 312–323.

Further reading in this volume

Chapter 1: Introduction: what really matters – moving placemaking into a new epoch
Cara Courage
Chapter 4: A future of creative placemaking
Sarah Calderon and Erik Takeshita
Preface: 'Disastrous forces, accidental actions, and grassroots responses'
Tom Borrup
Chapter 8: Queer placemaking, settler colonial time, and the desert imaginary in Palm Springs
Xander Lenc
Chapter 13: Sensing our streets: involving children in making people-centred smart cities
Sean Peacock, Aare Puussaar, and Clara Crivellaro
Chapter 18: '*Homomonument* sounds like a poem': queer placemaking 30 years on: a conversational dialogue with Thijs Bartels, author of *Dancing on the Homomonument* (2003)
Martin Zebracki
Preface: The radical potential of placemaking
Cara Courage
Chapter 22: Embedded Artist Project: Epistemic Disobedience + Place
Frances Whitehead
Chapter 31: Seven generations: a role for artists in Zuni PlaceKnowing
Theodore S. Jojola and Michaela P Shirley
Chapter 32: *The Hollywood Forest Story*: Placemaking for the Symbiocene
Cathy Fitzgerald
Chapter 35: Planning governance: lessons for the integration of placemaking
Nigel Smith
Chapter 36: Facilitator skills for effective collaborative placemaking
Husam AlWaer and Ian Cooper
Chapter 37: *The Neighbourhood Project*: a case study on community-led placemaking by CoDesign Studio
Lucinda Hartley, Eliza Charley, Sama Choudhury, and Harriet McKindlay
Chapter 40: Transforming community development through arts and culture: a developmental approach to documentation and research
Victor Rubin
Chapter 41: Rituals of regard: on festivals, folks, and findings of social impact
Maribel Alvarez
Chapter 45: How the city speaks to us and how we speak back: rewriting the relationship between people and place
Rosanna Vitiello and Marcus Willcocks

17

'I AM NOT A SATNAV'

Affective placemaking and conflict in 'the ginnel that roared'

Morag Rose

> It's possible, for a moment, to be overtaken by the place and forget where you're going or why but feel part of a shared heritage and cityscape that exists on a grand scale. Library Walk is a place that is unlike any other in Manchester.
>
> *(Bradbury, 2012)*

Introduction

This chapter explores Library Walk in Manchester, an ordinary place rendered extraordinary when it was threatened. Community action transformed an everyday path into a space of imagination and a conduit for wider debates around heritage, consultation, and public space. Most significantly for this Handbook, Library Walk illustrates how municipal placemaking strategies may fail to anticipate or appreciate how emotionally and viscerally attached citizens can be to the built environment. The discrepancy between the corporate agenda and lived experiences of those who used, and treasured the space, resulted in a protracted conflict.

In 2011, Manchester City Council (UK) revealed ambitious plans to redevelop key buildings and public realm in St Peters Square. This is a conservation area which houses several significant civic buildings and the proposed scheme included refurbishing the Central Library and Town Hall. It was broadly welcomed until, in 2012, an additional announcement was made: the intention to erect a new reception area linking Central Library and The Town Hall extension. This necessitated the enclosure of Library Walk, a curved path that bisects the two buildings. A campaign group quickly emerged challenging the official description of the path as dangerous and unloved; they reframed Library Walk as a site of civic pride, beauty, and considerable affective resonance. Friends of Library Walk mobilised in a variety of ways, including an action 'beating the bounds' of the contested area utilising vernacular tradition to reassert community ownership. The Friends failed to stop planning permission being granted. However, in 2014 they mobilised again to object to the 'Stopping Up Order' required for the completion of work. This legislation would remove the public right of way along Library Walk and the volume of complaints meant a Public Inquiry was held over eight days in 2014. Campaigners revealed the multiplicities of a space the council defined as merely an accidental void between buildings.

This chapter gives voice to those campaigners and offers an autoethnographic account of community placemaking. It will discuss lessons learned when folk came together to define, defend, and celebrate Library Walk. It explores why this place became so important to so many people, exemplifying their sense of belonging in the contemporary urban landscape. There was profound anger and disconnection felt when that connection to their local environment was ruptured. This revealed hidden power dynamics, issues of land ownership, and the fissure between planning regulations and citizen passion. Library Walk became, to use local Manchester dialect for an alleyway – ginnel – 'the ginnel that roared,' and this is part of its place narrative. I am writing here not just as an academic but as a citizen and a campaigner, implicated and embedded in this work as one of the instigators of Friends of Library Walk. Although this interpretation and account is my own, and omissions are mine, the narrative itself is co-created. A huge debt of gratitude is owed to everyone who contributed to the campaign to save Library Walk, and in particular to those who have contributed directly to this chapter. I will begin by explaining more about the social, cultural, and political context of Manchester before outlining who Friends of Library Work were and what we did. This chapter is necessarily partial and partisan as it explores the tangled meanings and conflicting demands placed on an everyday place.

Context and case study

This case study is situated in Manchester, a city in North West England. Its population in 2015 was 530,300 and it is one of 10 metropolitan boroughs which comprise Greater Manchester, the second largest UK conurbation after London. It was hailed as 'Cottonopolis' when it became '*the* city of Britain's industrial revolution (1840s–1920s): a mythic time of city prosperity, change and growth' (Hetherington, 2007, p. 632, emphasis in the original). Between 1930 and 1980 it suffered from deindustrialisation and decline, before implementing a range of urban regeneration initiatives facilitated by national, regional, and local policies (Parkinson-Bailey 2000; Bayfield, 2015). In the latter decades of the twentieth century these transformed the perception of Manchester from 'grim' to a place which 'more than London or any other British city, has been represented as "cool"' (Hetherington, ibid.).

Manchester has frequently been cited as 'the perfect example of a city which symbolised the trajectory of progress … from urban decay to urban renaissance' (Minton, 2009. p. 39) and a variety of placemaking strategies were implemented to achieve this. These include – but of course are not limited to – cultural events, festivals, and marketing campaigns (Bayfield, 2015). Manchester City Council (MCC) employed graphic designer Peter Saville to devise a brand that would encourage tourism and he designated us 'the original modern city' (Marketing Manchester, 2009) to emphasise dynamism and to position Manchester on a competitive global stage. Bayfield (2015) provides a pertinent analysis of the 'Original Modern' concept and conflicts around culture, branding, and placemaking in 'a city that is never short of event, spectacle or opinion.' She also discusses the local political background and how elite figures produce dominant narratives of the city.

Manchester's current status has been achieved embracing neoliberalism as defined by Harvey (2007) with Haughton et al. (2016) and Ward et al. (2015) amongst many demonstrating how this has manifested in a Manchester context. Labour have been in control of the council since 1974 when the city was reconstituted as a metropolitan borough, and since 1996, Sir Richard Leese has been the council leader. Concerns about the possible implications of this extended period of uncontested power are apparent in the conflict this chapter discusses. Leary (2008) documents shift in local politics from municipal socialism to entrepreneurism, Minton links this to a new culture of 'authoritarianism and control' (2009, p. 40) where democracy gives way to profit and grassroots voices are seldom heard.

Placemaking in Manchester has come at a cost. Hatherley states Manchester 'has neatly repositioned itself as a cold, rain-soaked Barcelona' (2010, p. 115) but is very clear that this involves 'a transfer of assets from the poor to the affluent.' (2010, p. 141). The myth of Manchester's 'success' has been challenged from its inception when Engels (1845 / 2009) documented a place of poverty and horror. Of course, the landscape has changed dramatically since Engels was here, but Manchester remains a place of stark inequality. More recently the impact of austerity has been documented by Greater Manchester Poverty Commission (2013), Goulding and Silver (2019), Folkman et al. (2016) amongst many others, and arguments about economic priorities were central to many objections to building on Library Walk. Hatherley (2010) and Minton (2009) both demonstrate the direct and tangible impact local, national, and global economic policy has on the urban landscape in Manchester. This encompasses the key civic area of St Peters Square and Library Walk, which was deemed in need of a makeover. It must be asserted that campaigners were not nostalgic or anti-change per se; indeed many were actively involved in supporting other developments in the city, but they felt strongly about losing Library Walk. The aesthetics of the neoliberal city are beyond the scope of this work, but the glass, chrome, and security-conscious design proposed embodies those values and was perceived as 'culturally hollowed out' (Davis, 1990, p. 78) in comparison to what was previously there.

Manchester Town Hall is classic Victorian Neo-Gothic, embodying success, prestige, and civic pride. Designed by Alfred Waterhouse it was completed in 1877. Between 1934 and 1938 an extension was built; its architect E. Vincent Harris was also responsible for the neighbouring Central Library. Manchester Central Library opened in 1934 and at the time it was the largest public library in the country. It is a Classical-style building with a circular plan inspired by the Pantheon in Rome. The entrance is a huge portico, emphasising the importance of St Peter's Square. Both of Vincent Harris's buildings are Grade 2* listed and in adjoining conservation areas. The side of the Town Hall extension adjacent to the library has a concave curve, forming a 200-metre-long path. This is Library Walk. The plans included not just a link building at the St Peters Square end of the walk but gates at the other. These were necessary because of the cul-de-sac created by the new structure.

The Twentieth Century Society North West was one of many organisations alarmed at plans to damage that area and, in their expert view:

> Library Walk is probably unique in that instead of merely separating these two great civic buildings, it creates a sense of tension between them. In architecture, as in art, literature and music the absence of something and the space between can be as powerful as form itself.
>
> Library Walk is a true walkway and can only be experienced on foot. Because of the curve, it reveals itself gradually, introducing an element of mystery and surprise, with something of the character and atmosphere of film noir cinematography. It is that kind of kind of urban space which creates the distinctive character of a city and one that should be cherished. In its original form it is also an excellent example of how the city can be perceived as series of visual sequences responding to different spatial elements, light and even memory, as argued by Gordon Cullen in his influential The Concise Townscape (1966).
>
> *(Email to the author 2019)*

Many residents of, and visitors to, Manchester also cherished Library Walk and people mobilised to object almost as soon as the plans were announced. Initially the conversations were online, but when it became clear there was sufficient motivation a meeting was organised at a local bar whose manager was sympathetic to the cause. I was the convener and recall being amazed at the passion, energy, and diversity of the people in the room. From the beginning the focus was on local political action, lobbying, and building a broad-based community coalition to demonstrate wide public support. The strategy was multi-faceted, and a decision was made to focus on a positive campaign message rather than a negative or defensive approach. This was exemplified in the run-up to the original planning committee meeting. The loose collective opted to call themselves Friends of Library Walk, taking a cue from friends of parks and green spaces. These are organisations MCC have always been keen to facilitate and we felt it demonstrated an ethos of care and communality. The Friends were never formally constituted as they were purely a task-and-finish campaign, feeling bureaucracy would be an unnecessary burden. Pertinently, we never engaged in fundraising or needed to apply for grants etc., so a bank account or incorporated status was deemed unnecessary. The cost of campaign materials – postcards and badges – were met by individuals, particularly Mancunians who were living too far away to attend events but who wanted to contribute. The main tactics used were: social media, crucial to group identity and message dissemination, with Facebook and Twitter being particularly important; petitions and letter-writing to lobby Manchester City Council; mobilising support and representation at the planning committee meeting, later scrutiny committee meetings, and the planning inquiry; researching to ensure our arguments were robust and factually sound; events, mainly informal meetings but also a larger more carnivalesque event, 'Beating the Bounds,' before the planning meeting; and producing promotional literature, including a celebration booklet, badges, and postcards.

There was an alternative suggestion that more radical, direct action tactics could have been utilised but there was little appetite to do so – although that in no way diminishes their power in other campaigns. Those most active in Friends of Library Walk included people with caring responsibilities, disabilities, employment, and other circumstances that meant they were uncomfortable risking arrest. There had been a suggestion to occupy Library Walk, but there was also a major logistical obstacle. Ongoing renovations in Central Library meant Library Walk itself was in the middle of a building site hidden by large hoardings. Active numbers of the group fluctuated and several supporters who were urban design, architecture, and built environment professionals wished to remain anonymous. They were concerned about a negative impact on their careers if they were viewed as troublemakers; whether this was justified is a moot point, but it indicates concern about power structures within the city.

Over 1,300 people signed a petition and 137 letters were received by the Planning Committee (in contrast, and to contradict any notion that people were merely opposed to any change, no other application for changes to St Peters Square received more than two). Some of the comments were published in a celebration pamphlet (Friends of Library Walk, 2012) and comments included:

> This walkway really is a hidden gem in Manchester. I remember over ten years ago walking down it for the first time, I was rushing across town, double busy and it actually stopped me in my tracks. It made me appreciate a whole lot more what Manchester is and was, a great city with lots of history.
>
> These squeezed thoroughfares between some of Manchester's most historic architecture give us a sense of our city's history – and can inspire daydreams and creative ideas as we pass through every day and evening.
>
> It is the most beautiful spot in Manchester.

The tactics utilised by the group were based on consensus view about what they felt were the most likely to be effective. They were influenced by participants' previous experiences as well as by limitations of time and money. One tactic, the 'Beating of The Bounds,' drew on traditional ritual to designate ownership and establish communal boundaries. It was intended to be a convivial spectacle with plenty of noise and colour. Placards were made using a variety of slogans and many featured Aidan O'Rourke's photographs. Aidan became involved because for him it was a 'matter of principle and aesthetics… A respect for the architecture's vision.' He had celebrated Library Walks qualities several times in print over the years, and likened passing through it to 'being in a time machine, virtually unchanged since it was built.' Giant book props were borrowed from The Book Bloc and participants were encouraged to bring musical instruments. We gathered in Piccadilly Gardens and processed along Mosley Street to St Peters Square, a distance of approximately 650m. We then circled the hoardings surrounding Central Library and the Town Hall extension, an area encompassing Library Walk itself. As we walked, instruments were struck and noise was made to assert our right to be there and own our space. The group paused at the west end of Library Walk to pose for photographs and peer through the gates of the building works. We then carried on with the circling, coming to stop at the St Peters Square entrance. There, several people and organisations spoke about why they were Friends of Library Walk. Some of these speakers had been invited, others were impromptu. We finished with a song, specifically written for the campaign by Matt Hill, aka Quiet Loner.

This event served several purposes. It raised the campaign's profile, engaging press and social media. It alerted the planning committee to the energy and passion and it also brought campaigners together and was undoubtedly an enjoyable evening. This mattered because the research and writing takes its toll and respite is needed. Furthermore, the convivial atmosphere strengthened a sense of purpose and cohesion prior to the planning committee meeting. This took place in the formal atmosphere of the Town Hall. When the committee met, protocol dictated only one representative was able to speak in opposition to the plans and that they only had three minutes to speak. Tom volunteered for the task and read a speech collectively written by the group. Supporters filled the room, but the experience was dispiriting for many reasons. Members of the planning committee played with their phones, left the room, and generally appeared disinterested. They passed the plans, and this was upheld by the scrutiny committee.

However, this was not the end. There was a lull but in 2014 Manchester Council applied for a Stopping Up Order to remove the public right of way on Library Walk. The Friends mobilised again. Sufficient objections were received for the Secretary of State to announce a Public Inquiry, overseen by the Planning Inspectorate. Unlike during the planning committee, anyone who wished to make a representation to the Inquiry was able to do so and the Friends spread the invitation widely. We did not have any funds or means to legal representation and so many individuals supported each other to speak. The Friends were joined by Don Lee of The Open Spaces Society and Gloria Gaffney of The Pedestrian Association. Describing themselves as 'veteran campaigners' with an impressive track record of legal action to protect rights of way, their expertise in navigating the Inquiry process was invaluable.

The initial Inquiry was due to last two days but the volume of people wanting to speak meant it was extended to eight days. Many local residents spoke up alongside representatives from Manchester Disable Peoples Access Group (MDPAG), Manchester Women's Design Group, Manchester Modernist Society, The Open Spaces Society, The Greater Manchester Pedestrian Association, The Ramblers Association, The Twentieth Century Society, academics with specialisms in lighting, architecture, and urban design and, unexpectedly, a local councillor. Supporters were always present in the public gallery, many coming in on their lunch hours. The experience of giving evidence in an Inquiry such as this is worthy of its own chapter; it was an incredibly intense and demanding experience. Flick Harris of MDPAG remembers it as 'daunting, because it was so formal. A lot of stamina was needed, but the support was incredible, and we were a team.' Several Friends of Library Walk have commented about how access to judicial processes such as this is contingent on many personal factors, most notably time, energy, and literacy. It was also noted who was absent; privately practicing architects expressed support but felt unable to do so publicly because of their relationship with Manchester Council.

The end came in a brown paper envelope, and despite our best efforts and an extensive dossier of evidence, we lost, because Inspector Yates said he was 'not satisfied that it has been shown that any disadvantages to the public arising out of the proposed stopping up are sufficient to outweigh the benefits of the Order'(Yates, 2014). Many of our arguments were deemed 'not material' to his decision because 'people will still be able to access and look at the remainder of Library Walk during permitted hours.' There was, of course, a profound sense of dismay and disagreement with many of the points made in his report, in particular a lack of appreciation of the key limits of permissive access rather than a genuine public space. Many felt the whole process was more attuned to a rural context than urban civic space. The Friends contemplated a Judicial Review and informal advice suggested some grounds to believe it could succeed. However, the risk would be high and the potential costs in time, energy, and money considerable. Reluctantly consensus felt it would be too expensive proceed and, sadly, I remain convinced this was the correct choice. The scale of the space involved mattered too; although we cared passionately, this was a micro-level incident and lessons learned could benefit other struggles. Alliances were formed which had lasting resonance, although there is also a lingering feeling of being cheated, and of power relationships being laid bare. As Don Lee, of The Open Spaces Society, says, it felt like a 'very shoddy deal.'

Key placemaking issues

The struggle for Library Walk focused on three main issues; process, access, and the quality of the built environment. The first of these is exemplified by questions the case raises about democratic and bureaucratic processes in Manchester specifically and the UK planning system more generally. Space and focus here preclude detailed examination of these issues, but they broadly encompass criticism of the public consultation and whether public engagement genuinely seeks to value diverse voices. Manchester resident Peter Castree had attended the very first public consultations about the development of St Peters Square, because:

> I was concerned for the future of Library Walk in particular and made a point of asking the relevant representatives whether there were any plans to roof it over or otherwise interfere with it. I was reassured to be told that there were no such plans, the proposed underground link between Central Library and the Town Hall extension having been designed to allow people to pass easily from one to the other. When I learned later that Library Walk was to be obstructed by the addition of a superfluous link building, clearly dreamed up well after the close of the 'public consultation' period, I felt betrayed and incensed – enough to want to attend the resulting inquiry and speak as a citizen witness.

There were also questions raised about what constitutes valid evidence, for example around crime statistics for the area. There had been claims from the council that Library Walk was dangerous, but no actual figures were ever produced to support this. In one particularly arcane interlude during the Inquiry there was a debate about whether or not the extension constituted a building and MCC could claim it exempt from regulations. Finally, there were concerns about fairness and power. Building work on the link building commenced prior to the Stopping Up

Order being granted, thus demonstrating an assumption that it would be approved. Melissa Moore was studying for her LLB (Bachelor of Laws) at the time she gave evidence to the inquiry and said she used her legal knowledge

> to expose the deliberate misquoting of legal precedent by the Council in their application. This has still never been explained – copy and paste was used as an excuse by the City Solicitor but… the falsehoods on the part of the Council throughout the campaign made and still make my blood boil.

The second key issue is around public space and its importance to civil society. I extend this to include the maintenance of the rights of way and the need for fully accessible architecture. The privatisation of public assets is central to neoliberal discourse, although Worpole and Knox review research and conclude that 'public spaces play a vital role in the social life of communities' (2007, p. 5) and 'public spaces facilitate the exchange of ideas, friendships, goods and skills' (ibid., p. 7). The scale of Library Walk may not fit the conventional image of a public space, but its civic role is demonstrated by the many accounts that pay tribute to the respite it offers visitors, a sense of being both within and yet apart from the city. It is a place of private epiphany and personal attachment, but it has no monogamous relationships. Library Walk was open to anyone who chose to walk down it. Public space is where many bodies mingle and includes not just agora but everyday sites such as pavements, footpaths, and bus stops. Cities have always been cosmopolitan places where diverse people encounter each other in 'light-touch gatherings' (Thrift, 2007, p. 217). These encounters may not hold individual significance, but their cumulative effect is an awareness of other actors in the environment that come to 'constitute a binding affective force' and sustain the cosmopolitan city (Thrift, 2007, p. 218). Jane Jacobs' (1961) vision of the street as a site of conviviality, cohesion, and natural surveillance also addresses the positive emotional and material impact of sharing space. Meeting with people breaks down barriers, challenges stereotypes, and turns 'the other' into an individual that can be related to on a personal, human level. The street is a physical manifestation of social, economic, and cultural forces which fluctuate over time.

The Library Walk link building did not simply destroy public space for all equally. The particular design had a number of flaws which made it particularly difficult to use for neurodiverse people and those with sensory impairments. MDPAG cited numerous breaches of both policy and law. Flick was particularly concerned about the uneven surfaces in the link building, the disorientating qualities of the architecture, and the poor acoustics in the new structure. During the inquiry MCC redesignated Lloyd Street as an alternative route to Library Walk, but Flick demonstrated that the kerbs posed a serious risk hazard for those using this street. Equality was also ironically evoked when MCC claimed women were threatened by Library Walk, an idea robustly challenged by campaigners who wryly noted cuts to domestic violence services was a far bigger problem. Another recurrent refrain from MCC was an assertion the Walk was a prolific site for public urination. This is a widespread problem in the '24-hour city' but no evidence was ever given this was a specific hot spot. The Friends countered areas with lots of night-time entertainment suffered disproportionately, and that improved street cleaning and provision of public toilets was a more sensible solution.

The third issue was around the affective and aesthetic components of the cityscape. Competing definitions of heritage, space, beauty, and fear led to conflict between the municipal authorities and citizens. The two groups valued different things and perceived Library Walk in contrasting ways. This was apparent from the very beginning when the Link Building was first announced and was exemplified by a tweet sent on 10 February 2014 by Sir Richard Leese, Leader of

Manchester City Council. There was a widespread perception that he had a personal interest in completing the project, and local news website Manchester Confidential came to refer to the building as Leese's Folly (Schofield, 2014). Sir Richard had tweeted about a meeting with the Commission for Architecture and the Built Environment (CABE) to discuss urban design in Greater Manchester and Eddy Rhead, of Manchester Modernist Society, asked him about The Cenotaph (moved during the renovation of St Peters Square) and the closure of Library Walk. At 12.45pm Leese posted, 'And hate 2 tell you but Library Walk was just the space left between buildings' (Twitter).

This tweet, whilst an ephemeral, probably spontaneous remark, gets to the very crux of the matter. Ignoring the disregard it shows for design – buildings do not accidentally curve together – this illustrates the profound conflict between interpretations of, relationship to, and feelings about a place. To Leese, and his 'Official Manchester' team of top-down bureaucracy represented by the city council, planners, regeneration professionals, and others with concern for the macro view, Library Walk was not really worthy of consideration. On a metaphysical level it did not even really exist. (I do not wish to suggest this was a homogenous mass of people or a singular view, and neither do I attribute negative intent to all those holding this perspective; I am simplifying due to word constraints). There is an economic logic to their unsentimental view which is also shaped by institutional practice, professional training, and a focus on a particular, neoliberal vision of what Manchester is for. Library Walk had no commercial value, no tangible purpose, and building on it did not directly impact on anyone's work or home life. However, this does not take into account emotion, affect, and attachment.

The testimonies from Friends of Library Walk and its allies revealed a radically different perspective on this space between buildings. The value inscribed here is not about money or commercial use; indeed, Library Walk was often framed as respite from the urban and a breathing space away from the more normal business of the city. Where Sir Richard saw a place that was threatening, dirty, and to be avoided, others saw sanctuary, wonder, and beauty. No one supported the idea of Library Walk as threatening and spoke instead of how they chose to walk down it, often necessitating a detour. Joan Rutherford is a retired town planner, design expert, and a member of many key local organisations. She said she 'was shocked [MCC] hadn't read work on Lynch and Cullen which provides background information regarding the value of cityscapes. The beautiful curve is not an accident, it's a space for quiet.' She also questioned the need for a link building on the grounds that it 'creates an entrance, because the grand portico to the library already does that.' The affection for, and a collective sense of ownership and belonging in, Library Walk is underlined by its informal description as 'a ginnel.' Its scale and splendour mean that designation is tender rather than strictly accurate; this is not a domestic space or a narrow gap between houses.

There appeared to be genuine surprise within MCC about the uproar precisely because these perspectives on place are so different. This was summed up in a memorable moment at the public inquiry which was recalled as a vivid memory by many that were there. Matthew Schofield spoke as a member of The Society of Friends, The Quakers, whose meeting house is opposite the Mount Street end of Library Walk. He was being questioned about why his preferred route between the Meeting House and his home was via Library Walk when this necessitated a detour which made the trip longer. He spoke quietly and passionately when he explained 'because I am not a sat nav' and expounded on the need for beauty and joy as well as utility.

The incongruency leads to a conflict, because the physical imposition of the link building obliterates the possibility of alternative values. It closed down the imagination and removed potential; it also removed the personal connection to the affective joys of the glimpse of sky. The walk was saturated with emotion and stories of personal connection and collective memories, but these could not be quantified or commodified. It is perhaps significant that Library Walk adjoins a beloved public amenity. Libraries have been hard hit by austerity but remain a manifestation of a civil idyll, a public resource, open to all for learning, self-empowerment, and recreation. They have themselves been sites of community building and fighting. The centralisation agenda in Manchester has led to the closure of local neighbourhood amenities and an assumption that people can and will travel to a more central hub. The implication of this policy is beyond the scope of this chapter, but the survival of Central Library is of huge significance to the psyche of the city. Many Friends of Library Walk spoke of the Library itself, whilst they rarely spoke with such affection about the Town Hall, despite this being the site of many personal and public ceremonies. One particularly memorable letter read at the Inquiry was written by Doris Hardy, a woman who remembered as a child seeing King George open the library; she was indignant at the slander on the landscape.

The background hum of austerity had another role to play in the rhetoric of campaigners. The link building was roundly condemned as a waste of money, and local journalists spoke later of how it had become an informal shorthand for public profligacy with money. Jonathan Schofield, Editor of Manchester Confidential, reported the cost as £3.5million and dubbed the project 'Leese's Folly,' highlighting concerns about an autocratic council leadership style. Jonathan says he stands by those words now, telling me he still believes it

> was ultimate folly, a double folly because of the need for gates at the other end too. It wasted a massive amount of money in a time of austerity for something that was not needed. (It was) a vast vanity project… a waste of time and a flagrant twisting of authority.

He remains very critical of the justifications given for the work 'the engineering of facts to make a case for it – I hated that.' (It is worth noting Manchester Confidential is neither a radical voice nor inherently critical of developments within the city.) The controversy led to a widespread sense of disenfranchisement. Just after the Inquiry the power imbalance implied by this critique was pulled into sharp relief by a collective of homeless people who set up camp in St Peters Square and were later evicted.

There is always complexity to topophilia and there is a particular constellation of relationships at play in Manchester. I've discussed this elsewhere, using hipcholia (Rose, 2018) to refer to an ambivalent nostalgia, where people are attached to places such as redundant mills and warehouses despite knowing they were sites of hardship and exploitation. However, this does not explain the hostility to the changes in St Peters Square. The Friends of Library Walk were not anti-development per se; they just felt this particular development was unnecessary, damaging, and wasteful and they wanted to protect a cityscape they loved.

Conclusion and legacy

At the time of writing I find myself once more campaigning on issues of access, inclusion, and public space alongside several members of Friends of Library Walk. The catalyst this time has been artist Jeremy Deller's Peterloo memorial. This commemorates the two hundredth anniversary of the Peterloo Massacre, a pivotal moment in the fight for democracy and equality, and something dear to many of us in Manchester, and UK-wide. Unfortunately, the interactive monument is a flight of steps, so reifies inequality, segregation, and exclusion of disabled people. The debacle has many similarities: an inadequate consultation process, a failure to listen, and an idiosyncratic approach to what constitutes a building for legislative purposes. There is another connection too – embedded in the floor of the Library Walk link building are the names of those murdered at Peterloo. Initially the gates were intended as a memorial, but the irony was too rich, and the plan scrapped. Placemaking still remains a source of conflict in other parts of Manchester, perhaps most prominently in debates around human rights abuses in United Arab Emirates and whether investment from the country should be so pivotal in current regeneration projects. Writer Kate Feld makes this explicit in her powerful work 'Ethiad,' about persecuted journalist Ahmed Mansoor: 'Placemaking is happening here, right now. Our city's being sold out from under our feet and its streets are filling with the bodies of people who have no place. Do you see them? Pay Attention' (Feld, 2018).

The Link Building itself is empty of any furniture and only scantily used. It looks visually dissonant and was a runner-up in the 2015 Carbuncle Cup. Although the fight to save Library Walk was lost, the campaign itself has had a legacy for many of those involved. Individuals involved learnt new skills, formed new relationships, and deepened existing connections. They also believe it had an impact on the local environment in a number of ways. Emma Curtin was a key part of the team and told me one of the legacies ofThe Friends for her was a demonstration of how 'activism functions as pedagogy, as learning and teaching which is important and often overlooked.' Reflecting on why she gave so much time for the cause she said:

> [Library Walk] has always been special to me, genuinely, from my lived experience not as an architect or academic. I would get off the bus early, it felt like a different city, felt a bit special. Now it's a symbol of bad decision making and activist solidarity.

Peter sensed there had been an impact on a number of local campaigns and had motivated him to become involved in other groups. He mentioned a key lesson was the need to get involved in the planning process early on and 'citizens need to be much more vigilant about what is being done supposedly in their name.' Aidan notes regretfully the physical changes to St Peters Square and fears much has been lost, although people still use it: 'It's destroyed, erased… Possibility was erased too. They've won – obliterated the past, … its depressing really – most people using the space will have no memory [of what it was like before].'

Jonathan Schofield says: 'It's not too offensive now if you can't remember it, but I can,' although he thinks the bigger legacy is more intangible. Despite the failure it was 'a morale boost for campaigners that was of benefit … Library Walk was a good measure of people power and those who want to get their voices heard.' He feels the engagement with planning paved the way for bigger and more successful protests, although that is of course unverifiable here. The solidarity within the Friends of Library Walk was powerful and Melissa told me:

> The campaign itself means so much – that so many disparate groups and people came together… the fact it was a space, not a building per se, is also an incredible thing and still shows that humans need aesthetics to grow and that they are prepared to fight for something so ephemeral.

She felt the publicity around Library Walk engaged people who were not generally aware of planning issues and inspired them to get involved more, citing successful objections to developments on Bootle Street as an example.

The place called Library Walk remains and there are parts of the passageway that look much the same as they did before the link building. However, everyone I spoke to felt the atmosphere has changed, the sense of escape has been lost, and the knowledge we are only there on sufferance has had a profound psychic impact. The passageway is saturated with stories, with contrasting and conflicting perceptions of the value of a space between buildings. Library Walk became a conduit for wider debates around, and beyond, placemaking. It demonstrates the intense attachment people can, and do, feel for well-designed civic space. The emotional and physical movements of people may challenge dominant narratives; vernacular creativity contrasts with planning processes. To borrow from de Certeau (1984), The Friends of Library Walk animated place through their everyday practices, bringing life in unexpected ways and making the space their own. Powerful affective and sensory connections were generated, and when they were threatened, people joined together to fight for their rights to a beautiful city. A chasm between top-down and grassroots trajectories for the same location was revealed and corporate placemaking visions were challenged. Library Walk was, is, cherished, and hope remains that in time the Link Building will be demolished, and Library Walk returned to its publics. Until then it remains 'the ginnel that roared.'

Acknowledgements

Heartfelt thanks to everyone who supported The Friends of Library Walk, and particularly those who spoke to me for this work: Peter Castree, Emma Curtin, Gloria Gaffney, Flick Harris, Don Lee, Melissa Moore, Aidan O'Rourke, Joan Rutherford, Jonathan Schofield, Howard Smith, and Aidan Turner-Bishop. Thanks also to John Hawes, Marc Hudson, Steve Millington, Louise Platt, and Maureen Ward for critical and inspiring conversations which fed into this chapter.

References

Bayfield, H. (2015). *Mobilising Manchester Through the Manchester International Festival: Whose City, Whose Culture? An Exploration of the Representation of Cities Through Cultural Events*. University of Sheffield Thesis [online]. Available at: http://etheses.whiterose.ac.uk/ (Accessed: 20 April 2020).

Bradbury, N. (2012). 'Help save library walk! in the shrieking violet [online]. Available at: http://theshriekingviolets.blogspot.com/2012/06/help-save-library-walk.html (Accessed: 20 April 2020).

Certeau, M. de. (1984). *The Practice of Everyday Life*. Berkeley, CA: University of California Press.

Davis, M. (1990). *City of Quartz: Excavating the Future in Los Angeles*. London,:Verso.

Engels, F. (2009). *The Condition of the Working Class in England* [1845]. London: Penguin.

Feld, K. (2018). *Etihad* [online]. Available at: http://www.manchizzle.com/2018/06/etihad-by-kate-feld.html (Accessed: 20 April 2020).

Folkman, P., Froud, J., Johal, S., Tomaney, J. and Williams, K. (2016). *Manchester Transformed: Why We Need a Reset of City Region Policy*. Manchester: CRESC.

Friends of Library Walk. (2012). [online]. Available at: https://friendsoflibrarywalk.wordpress.com/ (Accessed: 1 September 2019).

Goulding, R. and Silver, J. (2019). *From Homes to Assets Housing Financialisation in Manchester Update for 2018/2019*. Manchester: Greater Manchester Housing Action.

Greater Manchester Poverty Commission. (2013). *Recommendations Report*. Manchester: Greater Manchester Poverty Commission.

Harvey, D. (2007). *A brief history of neoliberalism*. Oxford, UK: Oxford University Press.

Hatherley, O. (2010). A *Guide to the New Ruins of Great Britain*. London:Verso Books.

Haughton, G., Deas, I., Hincks, S., & Ward, K. (2016). Mythic Manchester: devo Manc, the northern powerhouse and rebalancing the English economy. *Cambridge Journal of Regions, Economy and Society*, *9*(2), 355–370. doi: 10.1093/cjres/rsw004

Hetherington, K. (2007). 'Manchester's urbis: Urban regeneration, museums and symbolic economies', *Cultural Studies*, 21(4–5), pp. 630–649.

Jacobs, J. (1961). *The Death and Life of Great American Cities*. New York: Random House.

Leary, M.E. (2008). 'Gin and tonic or oil and water: The entrepreneurial city and sustainable managerial regeneration in Manchester', *Local Economy*, 23(3), pp. 222–233.

Marketing Manchester. (2009). Original Modern. Manchester: Marketing Manchester.

Minton, A. (2009). *Ground Control: Fear and Happiness in the Twenty-First-Century City.* London: Penguin Books.

Parkinson-Bailey, J.J. (2000). *Manchester: An Architectural History.* Manchester: Manchester University Press.

Rose, M. (2018). *Women Walking Manchester: Desire Lines Through The 'Original Modern' City.* PhD thesis. University of Sheffield.

Schofield, J. (2014). *Sleuth* [online]. Available: https://confidentials.com/manchester/new-mcr-restaurant-library-walk-leeses-folly-the-britannia-is-hel (Accessed: 4 August 2019).

Thrift, N.J. (2007). *Non-Representational Theory: Space, Politics, Affect.* Abingdon: Routledge.

Ward, K., Deas, I., Haughton, G., and Hincks, S. (2015). 'Placing greater Manchester', *Representation*, 51(4), pp. 417–424.

Worpole, K., & Knox, K. (2007). *The Social Value of Public Spaces.* York, UK: Joseph Rowntree Foundation.

Yates, M. (2014). *Order Decision for Inquiry Opened on 21 October 2014* (Planning Inspectorate) [online]. Available at: https://friendsoflibrarywalk.files.wordpress.com/2012/07/library-walk-order-decision.pdf (Accessed: 20 April 2020).

Further reading in this volume

Chapter 2: Placemaking as an economic engine for all
James F. Lima and Andrew J. Jones
Chapter 5: Making places for survival: looking to a creative placemaking past for a guide to the future
Jeremy Liu
Preface: 'Disastrous forces, accidental actions, and grassroots responses'
Tom Borrup
Chapter 7: Conflict and memory: human rights and placemaking in the city of Gwangju
Shin Gyonggu
Chapter 9: From the dust of bad stars: disaster, resilience, and placemaking in Little Tokyo
Jonathan Jae-an Crisman
Chapter 10: From moon village to mural village: the consequences of creative placemaking in Ihwa-dong, Seoul
Jason F. Kovacs and Hayun Park
Chapter 11: *Free State Boulevard* and the story of the East 9th Street Placekeepers
Dave Lowenstein
Preface: The radical potential of placemaking
Cara Courage
Chapter 23: Routing out place identity through the vernacular production practices of a community light festival
Gail Skelly and Tim Edensor
Chapter 25: 'If you can make it there, you can make it anywhere…': cultural placemaking at the heart of cities
Sherry Dobbin
Preface: Towards developing equitable economies; the concept of *Oikos* in placemaking
Anita McKeown
Preface: The only thing constant is change
Kylie Legge
Chapter 35: Planning governance: lessons for the integration of placemaking
Nigel Smith
Chapter 42: Creative Placemaking and Placekeeping evaluation challenges from the practitioner perspective: an interview with Roy Chan
Maria Rosario Jackson
Chapter 45: How the city speaks to us and how we speak back: rewriting the relationship between people and place
Rosanna Vitiello and Marcus Willcocks

18

'HOMOMONUMENT SOUNDS LIKE A POEM'

Queer placemaking 30 years on: a conversational dialogue with Thijs Bartels, author of *Dancing on the Homomonument* (2003)

Martin Zebracki

Prelude

The *Homomonument* (Figure 18.1), inaugurated at Westermarkt square in central Amsterdam on 5 September 1987, is widely regarded as the world's first commissioned monument devoted to, and empowering, the lives and rights movement of gay men and lesbian women. Many LGBT monuments around the world followed in the footsteps of the *Homomonument*, such as the *Gay Liberation Monument* in New York (see Zebracki, 2019) and monuments to homosexuals persecuted under Nazism in Berlin and Sydney (see inventory in Orangias et al., 2018). Over recent decades, the monument's underpinning Homomonument Foundation, co-founded by local gay activist Bob van Schijndel, along with annual on-site events – primarily comprising commemorations and festivities around key Dutch public holidays and special dates including Pride Amsterdam – have aimed to also embrace bisexual and transgender people (i.e. LGBT) as part of the monument's mental map and its public uses. This has been resonating with international advocacy for the inclusion of a much wider diversity of sexual and gender characteristics (LGBT+) and allied identity forces (e.g. 'Gay–Straight Alliances) (see, e.g. Ferentinos, 2014). Sometimes 'queer' is added to the LGBT acronym, which some construe as an identity category. As adopted in this account, queer can rather be rendered as a stance to question, or 'que(e)ry,' sexual and gender identity categories altogether and the social norms that underpin them (notably white heteropatriarchy), whilst unravelling the critical intersections between social identities and identity expressions (see, e.g. Zebracki, 2020a, 2020b).

The story of engaging a monument for and through sexual and gender minorities should be seen as a grounded practice (Zebracki, 2019). Such engagement is in constant flux and involves an amalgamation of collaborative or opposed actors, ranging from policymakers to members of public communities. This implicates a politics of inclusion around identity, which 'makes' places inclusive for some but at the same time perhaps not for others (see, e.g. Ghaziani, 2011; Gieseking, 2016). As such, queer placemaking wants to problematise ambiguous processes revolving around sexual and gender inclusivity and the (trans)formation of accompanying place

Figure 18.1 *Homomonument* (1987), Westermarkt, Amsterdam. Aerial photo taken and colour-highlighted by Geert-Jan Edelenbosch (CC BY-SA 4.0). Images: author's own.

identities. In Thijs Bartels, author of *Dancing on the Homomonument* (2003), I found an obvious interlocutor for debating this matter for the monument in question.

Thijs Bartels (born in Maastricht, 1960) studied Dutch language and literature and philosophy at the University of Amsterdam. In 1988, he pursued his career as editor at Bert Bakker Publishers. Subsequently, from 1993 to 2006, he acted as freelancer for various literary publishing houses and other institutions, including the Ministry of Foreign Affairs, the Royal Concertgebouw, and the Rijksmuseum. Since 2006, he has worked as a non-fiction editor at Meulenhoff-Boekerij Publishers. In 2011–12, Bartels was a guest lecturer in the Media, Information, and Communication department at the Amsterdam University of Applied Sciences. Bartels edited with Jos Versteegen the *Homo-encyclopedie van Nederland* (Gay Encyclopaedia of the Netherlands), issued by Ambo/Anthos Publishers in 2005. Central to the conversational dialogue in hand is Bartels' monograph *Dansen op het Homomonument*, issued by Schorer Boeken Publishers along with the translated edition of *Dancing on the Homomonument*, an accessibly written and evocative key resource about the genesis and social values of this monument.

On 5 July 2019, I entered into an in-depth conversation with Bartels to explore the idea of inclusive placemaking through the lens of the *Homomonument*. This monument consists of three pink granite equilateral triangles (Figure 18.1). Together, these triangles form a larger triangle, which in the conception of the designer, Karin Daan, reflect the past, present, and future –

which I interpreted in an earlier piece as 'a spatial commemorative constellation' (Zebracki, 2017, p. 346). The 'sunken' triangle with seating steps along the canal, almost situated in the water's daily stream, symbolises the present. The elevated triangle, used as seating furniture and event podium, expresses uplifting values associated with the future. And, finally, the triangle at street level embodies the past, resembling a kind of tombstone with the inscription of Jacob Israël de Haan's verse 'such an endless desire for friendship' [translated from the Dutch] (see Bartels, 2003; Zebracki, 2017). During the interview, Bartels and myself sat on the steps of the monument's 'sunken' triangle – or in the conceived present, whilst we reflected on the past and ruminated about the future.

The *Homomonument*'s tripartite triangle-shaped design is carried through in Bartels' monograph *Dancing on the Homomonument*, subdivided into three parts. The first part engages the past, that is, the monument's provenance. It documents the rationale for establishing the *Homomonument* to primarily remember the discrimination, persecution, and eradication of homosexuals by the Nazis. The monument's triangular form makes visual reference to the pink concentration camp badge that homosexual inmates were identified with, which after WWII turned into a symbol of queer pride. The book's second part is about the importance of LGBT visibility and recognition (e.g. Gorman-Murray and Nash, 2016) and the social value of the *Homomonument* in its present-day context (at the time when the book was published). In the present, as Bartels (2003) conveyed, the *Homomonument* holds the equivocal identity as a place to celebrate and a place to remember – or even a 'site of pilgrimage', Binnie (1995, p. 175) observed. Drawing from the LGBT rights and emancipation movement, the book's third part is a projection onto the future, where we have arrived now, including the challenges of a transcultural society, as Bartels highlighted.

As networked through the past and the path ahead of us, what place does the *Homomonument* take and 'make' today? As Bartels concluded his monograph, LGBT discrimination is not over, still warranting the fight for an LGBT-inclusive society. A 'sleeping or petrified monument' does not help here (Bartels, 2003, p. 100); the book's back cover text reminds the reader, therefore, that 'the *Homomonument* is a call for constant vigilance'.

The three-part setup of the *Homomonument* is also extrapolated in the conversational Acts on the pages of the account in hand. The subsequent Acts question how the *Homomonument*, about 30 years on, has opened up the space for putting LGBT minorities in place, pursuing inclusivity beyond symbology, and imagining inclusive alternatives. The dialogical narrative hereinafter reveals queer placemaking through the questioning, or 'que(e)ring,' of the monument's 'stone matter' in contexts of 'lived matter': memories of past and present experience of inclusionary vis-à-vis exclusionary processes and realities.

Act I: Past – Placing encounters

Martin Zebracki (MZ): While we are sitting here at the *Homomonument*, could you briefly tell me something about your background and motivation for writing your book *Dancing on the Homomonument*?

[NB: This conversational dialogue has been translated from the Dutch. Typeface in bold signposts the start of a conversational thread.]

Thijs Bartels (TB): Well, the reason for the book is quite simple. I knew, or I still know, Richard Keldoulis, through a gay swimming club. He established the info kiosk 'Pink Point,' and he was very involved in the *Homomonument* and the whole [LGBT] scene. My profession is editor, I studied Dutch, and he asked me to make a leaflet and then … he suddenly said, 'there ought to be a book. Would you be willing to do that?', he asked. The leaflet actually never came out how we wanted it, but then it became a book.

MZ: Back then, was it clear what the book should be about?

TB: It had to be about the *Homomonument* itself. It has been translated [from Dutch] into English. It needed to be very informative, aimed at a very wide public, so it's in simple language, simple layout, and it was published by Schorer Boeken, which at that time was, let's say, a lesbian and gay publisher. The book includes photographs and illustrations. I divided it up in three parts: the past, present, and future [see Prelude].

MZ: How would you describe the *Homomonument* in three key words?

TB: I think it's very beautiful. It's a living monument, that's really the case for me, or what I discovered throughout writing the book. I think this is the most beautiful part about the monument, and the most successful thing. Just look around you, people are sitting on it, they have a party, all kinds of things happen here, demonstrations … while it's actually a kind of memorial monument. It doesn't often happen with monuments that people dance on it and throw parties, and everyone – tourists and foreigners – pass by it … [It stands for] gays, lesbians, transgenders and you can list the whole caboodle, and that's the beautiful thing about the monument.

MZ: To summarise your answer: the *Homomonument* is beautiful, a living monument, and what's the third key word?

[NB: One of the founding members of the Homomonument Foundation, Pieter Koenders, also conceived of the Homomonument as 'living monument,' in lieu of a kind of 'misery on a pedestal' (Koenders, 1987, p. 29).]

TB: I think that it's well thought-out … It has become a piece of Amsterdam, it's part of the city. That's also one of the good things about it.

MZ: How often do you come here?

TB: Not very often … I used to come here more often, but … you no longer feel like it, you get a bit older. My work was based here along one of the canals and then I would cycle past it twice a day … And then I'd see that there were always flowers put on it … but other than that, I live in Amsterdam-West now, so you just don't come as often. And my work is in Amsterdam-Oost at present, so I don't get to be in the city [centre] that much … In the past I used to go to demonstrations on 4 May [i.e. WWII Remembrance Day] and 5 May [i.e. Liberation Day]. Once I made a speech. At a certain moment I had enough. So, I don't come as often now … It has nothing to do with the monument itself. Not at all, because I still like to keep up to date about it. If something happens, I think, great, it's going well. No, I think it's a good thing that there is a new generation that is taking over the helm. Also, I know nothing about the music of today, so why should I come [to the music party events] here?

MZ: Who is that new generation then?

TB: I don't know, but they do have a very beautiful place [i.e. *Homomonument*], and they're a group of people who take that on … The threat remains the same, the suffering, the discrimination.

MZ: What feeling do you get at the *Homomonument*?

TB: Um, I think it's great it exists. That's just it for me. That it [i.e. being LGBT] is possible and that it's allowed … For that, I think it's an important monument

MZ: Can you remember the first time you saw the *Homomonument*, if you dig for your first memories of it in 1987?

TB: Well … I did think it meant some kind of acceptance, which was there already of course, at least for me, you do get that feeling. People see you, you become visible. I had nothing like, oh what an ugly monument, I thought it was great from the start. I do have memories like that. What I do remember is that there was a lot of moaning about it [i.e. the granite] not being pink enough … And I did think that myself in the beginning, I thought it didn't stand out very much. I thought it could stand out a lot more … At the opening [the media asked] what is that poo-brown coloured thing? … The granite had to come from Italy and there were lots of problems with it. Later at the Rokin [i.e. canal and major street in Amsterdam's city centre], there came a building of the stock exchange [which is no longer in existence] and that was really pink marble … They had been saying that it [the *Homomonument*] couldn't be pinker, and then you got this really pink building. But perhaps the monument was a different type, more porous, and it is of course granite and not marble.

MZ: Was there any further commotion when the monument came into existence?

TB: The papers were full of that it shouldn't be allowed, not possible, and lots of letters were sent in. Why should gays have their own monument? Well, you will be familiar with the arguments that come then. Financing was very difficult, it was really pushing and pulling, and there were fundraising concerts.

MZ: What role has the *Homomonument* been playing for the local LGBT community?

TB: Look, I think it has changed a lot. Of course, when it came it was a big celebration. The acceptance of the monument by the municipality, the authorities, that was all very important at that time. At a certain moment, I wrote, I compiled, a gay encyclopaedia [see Prelude]. About the time that was published, people had something like, we can relax now, it'll all be alright, we are almost there. That feeling. At least that was one of the voices that also emerged. And now with [religious] conservatism and shifts to the right … it's all necessary again to ask for attention … You [i.e. LGBT people] always remain a minority and that means you always have to fight, in any case. All of us.

MZ: Has the *Homomonument* provided some educational encounters? I mean, what role could it play in sexual education, school visits, etcetera?

TB: I do find that a quite difficult question, you know. Let's say, if anything occurs with gays, put a stop to it, or talk about it. You could create special lessons about it [i.e. the *Homomonument* and its values], but you cannot put all the social issues on the shoulders of teachers. Visibility on TV, in film, art, and so on are also very important.

MZ: In what way do you think the *Homomonument* has played an important role in the city of Amsterdam?

TB: It's about visibility on its own. I was once chairman of a swimming club … There were people that didn't come for the swimming alone, but just because they wanted to feel accepted. And that's the same for such a monument. People come here and say what a wonderful monument we have here. And everyone can see it and be part of it. I think that's very important, also for people from abroad, for people from oppressed countries … let's say, yes, I think it's very important … This is a monument that makes you think.

Act II: Present – Placing beyond inclusive symbolism

MZ: In terms of events, design, and location, do you think the *Homomonument* reaches its inclusive potential?

TB: Yes, yes. Do you know why? Because you can sit on it, and because you can dance on it. If there had been a kind of pink triangle on sticks, then all of this wouldn't have happened. Or at least, um, it would have probably been less living than it would be now. You have three things [i.e. granite triangles] to sit on, to be able to walk around, to organise parties around, including commemorations, and you can put a wreath down ... and you don't have all that with a statue. What's also interesting is that you have to explain the triangle [i.e. Nazi reference; see Prelude] ... In the past you didn't have a rainbow flag, you just didn't have that [as LGBT symbol]. So, when it [the *Homomonument*] was designed, that rainbow flag didn't exist – I mean it wasn't quite visible here as yet.

MZ: Speaking of the rainbow flag, often visible at LGBT celebrations at the *Homomonument*, how do you read its message of inclusivity?

TB: Let everybody just join in, that's great. Look, everyone wants different things, and you'll have to compromise. You have to let people do what they want to do. I do think that you are stronger together. With the swimming club you always had that lesbians wanted different things than the [male] gays, but, in the meantime, together you form a swimming club. And you ensure that people are accepted. I think that's the most important thing.

MZ: Do you think that the *Homomonument* is open to everyone?

TB: I do hope so, I mean, well, of course. It's the same as with swimming. The swimming club is meant for gays, and we go to gay tournaments but also to regular ones ... Heterosexuals could come and swim, they could become members, that's no problem ... And, yes certainly, there were a number doing that ... As a gay person I can join a heterosexual club, so the heterosexuals can also join a gay club. And what about all the mothers and fathers, they have to be accepting as well ... they're there, and they're all heterosexuals, or well, nowadays, a bit less.

MZ: These days you hear the word 'queer' quite often as a kind of all-embracing term. What does 'queer' mean to you?

TB: You know, for me it means 'gay.' I am a bit old-fashioned. Regarding gays, lesbians, and the rest, I always say ... LGBQT [sic] ... I always stumble over that term.

MZ: 'Q' in this acronym often refers to 'queer.' I find that an interesting one. In the way I see it, queer challenges the whole idea of using identity labels, so then it's perhaps a bit problematic when it appears in such an acronym. Some would see the 'Q' referring to 'questionables,' those exploring or questioning their sexual orientation and sexual or gender identity. For me, 'queer' means 'to query,' or 'to question,' to ask questions about norms and to challenge forms of exclusion and systematic oppression in the grander scheme of things [see also Prelude].

TB: Yes, a wider view, embracing all differences.

MZ: Indeed, and 'queer' is also relevant to memorialising the *Homomonument*. Also, I appreciate that *memorial* is not a Dutch word, but how do you see the difference between a *monument* and *memorial*? – as I think this is relevant here.

TB: If you were to make a difference, I think that 'memorial' means that you have to keep it alive – think of a statue reminding us of the Second World War. What does that add? As we should, wreaths are laid on the Dam on the 4th of May, and then we, as ordinary citizens, just carry on with our lives again.

MZ: If I understand this correctly, you see a monument as a kind of stone creation, a statue for example, and a *memorial* more as a process, as *memorialising* [i.e. gerundial use of the word] or a happening, which brings the monument to life.

TB: Yes, okay, if you look at it that way … The *Homomonument*, for me, is a living monument. The *memorialising* part is more something that is done on the 4th of May, when we reflect on the past [i.e. WWII], and on 5 May [i.e. Liberation Day] we are thinking about totally different things again. That's approximately what I mean … But that difference [i.e. between *monument* and *memorial*] doesn't exist for me in Dutch at all … You see, remembering without a future is no use at all. That's kind of what I mean. It doesn't have to involve something fanatical like campaign, campaign, campaign, but something has to stay in our minds.

MZ: How do you feel about the rainbow-coloured zebra crossings that appear in so many cities these days, such as on the footbridge at the train station of Sloterdijk? [i.e. city district of Amsterdam]

TB: Very good, isn't it? It's about what I said: being visible, again being visible, everything helps. It's playful and funny, as long as everyone understands that it's a zebra crossing.

MZ: Relatedly, what do you feel about the Pride? Then, everything is about the rainbow, you know, do you think that's just playful and fun also?

TB: No, no … Again, I'd like to bring up the swimming club.

MZ: Yes, the swimming club keeps popping up.

TB: Yeah, but I was way more active in the past … We wanted to join the [Amsterdam Pride] parade, and the problem for us was that participation was so expensive … And then all those companies came with large boats which had really nothing to do with LGBT, that was embarrassing … All of a sudden, the ABN Bank came by on a boat, well that's just too much! That's a marketing tool, it has nothing to do at all with discrimination, liberation, and with acceptance. It's marketing.

MZ: Whom do you think the *Homomonument* is intended for?

TB: For all the gays, lesbians, the whole community, the LG ... well you know what I mean: the rest … And Amsterdam, it's an Amsterdam monument.

MZ: Is it inclusive?

TB: What do you mean by that? I think that's such an awful word.

MZ: Let me 'unpack' it: I mean, does this monument speak to people other than 'homo'? So not only gays and lesbians but also bisexuals, transgenders and intersections with people of colour, older and younger generations, people living with disabilities, the whole spectrum.

TB: Yes, well, if you put it like that. I do think it goes a long way. You have to roll up your sleeves, do something, and it will happen. If people join in who want to commit, then I can't imagine that anyone would say that something wouldn't be okay. I can't imagine that. What I can imagine is that they [i.e. organisers of events at the monument's site] are all volunteers, who all have their own ideas. If there's no volunteer who is black, for example, then they don't feature high on the agenda. If there's no elderly person, with any clear idea, then nothing is done for the elderly. I do think you have to roll up your sleeves … We have all been there. We weren't allowed to marry, because we were gay … Now we're all allowed to marry, but we had to do something for that, and put ourselves in a certain kind of fear, or embarrassment. If you don't take the outstretched hand, then you shouldn't complain … I am not that much of an activist. My contribution was my book about the *Homomonument* – and the gay encyclopaedia. That's where my strength lies. And, yes, it's the same with swimming, people have to take turns and now it's my turn. I wanted to join in, and they wanted me to do it and I said fine … My gay club isn't that impressive, but you're there, you are visible, you defend your club and you take decisions in the interest of a gay club. Well, this is how you ought to become inclusive.

MZ: The *Homomonument* apparently has a lot of personal associations and memories for you. As of today, how has the importance of this monument as an inclusive place changed for you?

TB: Well, the [social and political] scene has changed ... It has all shifted to the right. That's really not good, you know, for gays. And the whole idea now with the internet is openness, but coming out there has also become a lot harder for a lot of people. That's not pleasant.

MZ: When you're here, how do you relate to the *Homomonument*? Do you feel it's an inclusive atmosphere?

TB: When I come here, I do feel I know a lot about this place. I researched it fully, so that's kind of funny ... I'm always pleased that there're flowers on the monument, so it's being used, it has helped somebody, and it offers recognition ... At the same time, you know what's also happening, my husband, or my partner, and myself, we actually no longer hold hands in Amsterdam that readily these days. We used to do that, which was quite normal back in the days, but now we no longer do that ... Well, you don't see it at all anymore.

MZ: Why is that?

TB: Everyone gets back in the closet ... There's something going on here, but look, I didn't do research into that. There is conservativism and the shift to the political right ... Now, when there are boobs, they are covered up. In the past you didn't have that.

MZ: You've written a lot about the *Homomonument* – have you also captured this place by any photographs yourself perhaps, which may give a visual impression of the atmosphere?

TB: I did take photographs but not that I was very pleased with them ... Nowadays everything can be found on the internet, so you can easily find beautiful photographs there ... Well, with parties I did take photos, also when Job Cohen [former mayor of Amsterdam] was interviewed here on the monument. I took photos of those kind of things here, also of friends sitting on the monument.

MZ: As an experiment I brought this 'old school' disposable camera. Imagine that you'd like to show the *Homomonument* to someone who hasn't been here before, and that you're allowed to only one photo with this camera? What would you take a picture of as a postcard? There you go! (Figure 18.2).

MZ: Do you get to talk with people around you about what the *Homomonument* means to them? They might know that you have authored a book about it – how does the monument come up in conversations with them?

TB: I've not been really asked about it, but it does come up occasionally indeed.

MZ: In what sense?

TB: Well, recently, someone said to me, which was not about the *Homomonument* though, do you know that homosexuality appears in a lot of animal species? And then I had something like, why are you telling me this? How come? Have I got the word 'gay' written on my forehead? I find this interesting, very bizarre actually ... Regarding the *Homomonument*; no, people haven't directly asked about it. But if they talk about Amsterdam, they may well mention that they've been there, and then it's talked about.

MZ: Relatedly, is the *Homomonument* a kind of tourist attraction, as it were? Do people specifically go there like 'I'm in Amsterdam and I must go and see'?

Figure 18.2 Participatory photography: the picture that Bartels took of the *Homomonument* (with Zebracki's disposable camera), imagined as postcard for someone who would not have encountered this monument in real life. Photo courtesy of Bartels.

TB: Yes, I do believe there're people like that. Sometimes you do see people on the monument who are very emotional.

MZ: What does that mean to you then?

TB: I'm very pleased to observe that. But for myself, like what I said already when you asked me for this, I'm no longer part of it in the same way as in the past.

Act III: Future – Placing inclusive changes?

MZ: Suppose you'd have all the power in the world, how would you change the *Homomonument*?

TB: To change it?

MZ: Yes, it can be as subtle or as rigorous as you want it to be.

TB: What an awkward question!

MZ: Simple answer please!

TB: [laughing loudly] What else could be done here? You know what I would really like … There used to be a gay library in Amsterdam-West, which was later added [as IHLIA LGBT Heritage archive] to the public library [OBA, the Netherlands' largest library nearby Amsterdam Central Station]. It would be great to bring together the gay library and the gay museum, well everything, and then preferably here around the corner near the *Homomonument*.

MZ: What is your view on the name of the *Homomonument*? – perhaps 'homo' might sound 'gay-exclusive' or it may have a negative connotation (see also Zebracki, 2017). In other words, does the name '*Homomonument*' need to be changed?

TB: No … why would you suddenly now have to change a monument that was called that? Perhaps you can give it an English subtitle, think of 'Gay Memorial.' But it's locally so well-known as *Homomonument* and it has become such an Amsterdam word, that you shouldn't change it … I think it's too much part of Amsterdam.

MZ: Should it include the full acronym LGBT?

TB: I am too much of an editor and work with language … I feel LGBT monument doesn't sound good, does it? *Homomonument* sounds like a poem. *Homomonument*, that just easily slips off the tongue.

Coda

People make places, monuments in themselves probably don't. But monuments might evoke memories of people and places, as they are intertwined with our lived experiences across different walks of life. Monuments may have the power to lend visibility to those living on the margins of society, who have been chronically subjected to the exclusive forces of oppression, prejudice, bigotry, and systematic discrimination in the here and there. The dialogical narrative presented in this account should be, by no means, taken as representative of the possible complexity of community engagement with the *Homomonument* as well as of the generations of writings and embodied knowledges thereof. I also realise that much social justice work remains to be done to address, and redress, structural exclusions and inequalities. Queer, in theory and practice, I think holds strong potential in inclusive placemaking. It thus manifests a stronghold that demands activism and solidarity beyond sexual and gender identity alone to shed light on, and deconstruct, multi-layered privileges (see, e.g. Zebracki, 2020b). LGBT struggle should therefore not be at the expense of any other minorities across ethnicities, classes, ages, abilities, creed, and so on. Queer placemaking, as such, entails an extensive intersectional commitment and investment of hope in a perpetual social justice project. More than a name or 'poem' alone, my journey has prompted me to envisage the *Homomonument* as an activist project: a potential action space for pursuing such intersectional solidarity in placemaking practice.

Acknowledgements

I would like to extend my gratitude to Thijs Bartels for his time and insights and for perusing the draft version of this dialogue, whilst any errors remain my own. Bartels has given consent for being named here, as it would be impossible to anonymise this type of account. The interview has been conducted as part of a case study on the *Homomonument* in my role as Principal Investigator of the research project 'Queer Memorials: International Comparative Perspectives on Sexual Diversity and Social Inclusivity' (QMem), supported by a grant awarded by the UK Arts and Humanities Research Council (AHRC), http://www.queermemorials.org.

References

Bartels, T. (2003). *Dansen op het Homomonument*. Amsterdam: Schorer Boeken.

Binnie, J. (1995). 'Trading places: Consumption, sexuality and the production of queer space', in Valentine, G. and Bell, D. (eds.) *Mapping Desire: Geographies of Sexuality*. London: Routledge.

Ferentinos, S. (2014). *Interpreting LGBT History at Museums and Historic Sites*. Lanham, MD: Rowman and Littlefield.

Ghaziani, A. (2011). 'Post-gay collective identity construction', *Social Problems* 58(1), pp. 99–125.

Gieseking, J.J. (2016). 'LGBTQ spaces and places', in Springate, M. (ed.) *LGBTQ America: A Theme Study of Lesbian, Gay, Bisexual, Transgender, and Queer History*. Washington, DC: National Park Foundation [online]. Available at: https://www.nps.gov/articles/lgbtqtheme-places.htm (Accessed: 23 June 2020).

Gorman-Murray, A. and Nash, C.J. (2016). 'LGBT communities, identities, and the politics of mobility: Moving from visibility to recognition in contemporary urban landscapes', in Brown, G. and Browne, K. (eds.) *The Routledge Research Companion to Geographies of Sex and Sexualities*. London: Routledge.
Koenders, M. (1987). *Het Homomonument*. Amsterdam: Stichting Homomonument.
Orangias, J., Simms, J. and French, S. (2018). 'The cultural functions and social potential of queer monuments: A preliminary inventory and analysis', *Journal of Homosexuality*, 65(6), pp. 705–726.
Zebracki, M. (2017). '*Homomonument* as queer micropublic: An emotional geography of sexual citizenship', Tijdschrift voor economische en sociale geografie (Journal of Economic and Social Geography), 108(3), pp. 345–355.
Zebracki, M. (2019). 'Queerly feeling art in public: The gay liberation mo(nu)ment', in Boyd, C. and Edwardes, C. (eds.) *Non-Representational Theory and the Creative Arts*. Singapore: Palgrave Macmillan.
Zebracki, M. (2020a). 'Public art and sex(uality): A 'wonky' nexus', *Public Art Dialogue,* 10(1), pp. 1–10.
Zebracki, M. (2020b). 'Public artivism: Queering geographies of migration and social inclusivity', *Citizenship Studies*, 24(2), pp. 131–153.

Further reading in this volume

Chapter 5: Making places for survival: looking to a creative placemaking past for a guide to the future
Jeremy Liu
Chapter 8: Queer placemaking, settler colonial time, and the desert imaginary in Palm Springs
Xander Lenc
Chapter 20: Displacemaking 2015 and 2020
Catherine Fennell and Daniel Tucker
Chapter 23: Routing out place identity through the vernacular production practices of a community light festival
Gail Skelly and Tim Edensor
Chapter 24: Artists, creativity, and the heart of city planning
Tom Borrup
Chapter 25: 'If you can make it there, you can make it anywhere…': cultural placemaking at the heart of cities
Sherry Dobbin
Chapter 33: Conceptualizing and recognizing placemaking by non-human beings and lessons we might learn from Marx while walking with Beaver
Jeff Baldwin
Chapter 42: Creative placemaking and placekeeping evaluation challenges from the practitioner perspective: an interview with Roy Chan
Maria Rosario Jackson
Chapter 45: How the city speaks to us and how we speak back: rewriting the relationship between people and place
Rosanna Vitiello and Marcus Willcocks

19

PLACEMAKING IN THE ECOLOGY OF THE HUMAN HABITAT

Graham Marshall

Author preface

The author's perspective on the three themes discussed in the chapter reflects his first-hand experience as a seasoned urban designer. He has contributed to the national, regional, and local design policy referenced; was a director of the pilot Urban Regeneration Company Liverpool Vision; a Built Environment Expert with Design Council CABE contributing to the Building for Life programme; and, an active member of several regional design review panels and the Commission for Architecture and the Built Environment (CABE) national panel. The most important thing learnt on that professional journey is that an optimistic perspective is an essential attribute for both experts and communities alike in successful placemaking; more so than any concepts or theories about placemaking. We'll come back to this.

Introduction

While the following discussion is UK focussed, the economic shapers of contemporary urban settlements are generic across the globe along with the urban design theory underpinning urban policy-making. The understanding of 'place' as a complex socio-emotional experience is captured by Cresswell's (2004) definition of place as 'space endowed with meaning.' However, 'place' is a fluid state where 'meaning' can ebb and flow, influenced and directed by complex changes generated by individuals, communities, and wider external forces. Sarr and Palang (2009) present a noteworthy explanation of this complexity in defining place and placemaking through an exploration of theoretical approaches and methodologies on the relationship between humans, environment, and landscape. Within this chapter, I will work with Cresswell's simple definition, and focus on why 'meaning' is important from a placemaking perspective. The physical environments that surround us are no longer the wild ones that served our basic foraging needs. Instead, they are human habitats crafted by us to provide surpluses. This habitat is still ruled by the evolutionary and ecological processes of our social species, but the scale at which we cooperate with each other has enabled us to reshape and colonise most environments in the world. While this cooperation makes us supremely adaptable, our ability to survive extremely harsh environments also, paradoxically, makes us susceptible to exploitation. Examples include our endurance of prolonged warfare, lifetime impoverishment, and the many negative impacts of disadvantaged urban living. Though people can survive in these environments, their lives are often shortened by sickness and typified by non-cooperative community relationships. As a model aiming for thrival and not just survival, our urban settlements should feel like eutopia ('well-place'), but in practice this is far from the case. Many forms of exploitation make places fragile and therefore unsustainable for individuals, groups, and sometimes whole societies (Diamond, 2005). The aim of this chapter is to identify the failure of process in contemporary urban planning and explore how an understanding and application of evolutionary psychology and human ecology could provide better models. In short, the ethos of any urban model should start and end with people in mind.

Theoretical underpinnings of urban planning

I began formulating this chapter on an aeroplane over Northern England on a clear November evening. As we approached from the sea, settlement patterns became clearly defined against the black landscape. Cities blazed and villages twinkled, with traffic pulsating in capillary ribbons of red and white around and between them in endless streams. There is a purity in this systemised image of our human habitat from the night sky. There is also a terrible beauty in its scale, sprawl, and traffic volume that can only be appreciated from above. It's a living plan from this perspective, unlike our everyday experiences on the ground, in place. This is the strategic view of our contemporary environment where we can reflect on the complexity of our habitat while not being overwhelmed by the complications of living in it. We will start our exploration at this strategic level, noting that there are many players, from administrators to users, that determine how well the system functions.

The education of every urban planner and designer starts with the primary theories and concepts of architecture, town planning, landscape, and urban design. While these ideas are not usually applied as tools in practice, they form the bedrock of professional opinion and belief. Perhaps the most enduring of these from a physical design perspective is the work of Vitruvius, famous for his *Ten Books on Architecture* written in the first century BC (transl. Morris, 1960). Known as the 'Vitruvian Virtues,' he proclaims that structures must exhibit the three qualities of *firmitatis, utilitatis, venustatis* – stability, utility, beauty. While we continue to perceive cities as collections of buildings rather than spaces for people and togetherness, architecture remains the dominant narrative among urban designers.

However, beyond the Vitruvian Virtues, several urban theorists published works in the mid-twentieth century that questioned the modern practice of large-scale city development and set in train the contemplation of how we design cities well for people. Reflecting on the theories and concepts underpinning this movement, Marshall published a paper titled 'Science, pseudo-science and urban design' (2012), introducing his paper with a quote from Jane Jacobs: 'As in the pseudoscience of bloodletting, just so in the pseudoscience of city rebuilding and planning, years of learning and a plethora of subtle and complicated dogma have arisen on a foundation of nonsense.' While asking if this is important, Marshall concludes that there is a need for 'more systematic verification and critical assimilation of scientific knowledge within urban design theory.' In his paper, Marshall critiques four classic urban design narratives in detail: *The Image of the City* (Lynch, 1964); *The Concise Townscape* (Cullen, 1961); *The Death and Life of Great American Cities* (Jacobs, 1961); and *A City Is Not a Tree* (Alexander, 1965). All published in the 1960s, Marshall notes that they remain uncritically accepted after over half a century by many educational institutions and practitioners alike.

Building on this emerging urban design platform, architects Colin Rowe and Fred Koetter (1978) published *Collage City*, a critical analysis of Modernist city planning. The title of the book conjures an attractive visualisation of the above theories, legitimising a *bricolage* of different utopias in contrast to single-handed city models. However, they argue that the ad hoc assembly of historic fragments (what some consider best practice today) provides temporary solutions to problems, with the risk that they become dated as quickly as any totalitarian, fresh-slate approach. They argue for a middle ground where the fragments are applied strategically but not prescriptively by urban designers, with emphasis on the quality of the public realm. Their final paragraph reads: 'Utopia as metaphor and Collage City as prescription: these opposites, involving the guarantees of both law and freedom, should surely constitute the dialectic of the future…; and, possibly, even common sense concurs.'

A common thread in these theoretical approaches is acknowledgement of the human experience of place beyond the educated architectural narrative. They innately look at the science of placemaking, although they do not employ a scientific method in place analysis. While the consideration of cognitive maps, affective navigation, art of relationships, social understanding, and managing complexity are all recognised psychological matters in the context of human habitats, the disciplines of psychology have yet to make inroads into the contemporary training, practice, or policy guidance of built environment professionals. This invites the question whether urban design is merely big architecture, where placemaking is an emerging discipline more concerned with people in place. Instead of conceiving places as the 'built environment,' should we instead conceptualise them as the 'living environment' (Corcoran and Marhall, 2017), recognising that science is no less creative than art, and that both disciplines are complementary requirements for successful placemaking?

This ecological lens underpinned by psychology and sociology does frame the practice and teachings of some of landscape architecture, with Lawrence Halprin an early exponent. Unlike his contemporary urban theorists, Halprin's leading concept, the *Ecology of Form* (1982), is an exploration of natural processes and ecological relationships in the context of the methodologies of Gestalt psychology. Encouraging people to participate in the creation of their own environments, Halprin brought abstracted natural form and process into the urban fabric, translating them into everyday lives, and developed a methodological approach to participation and co-design in *Taking Part* (1975) and *The RSVP Cycles: Creative Processes in the Human Environment* (1970). Halprin was a Modernist and thus connected to an international critical conversation, unlike the largely uncritical bubble containing urban design and placemaking. Halprin states:

> To be properly understood, Modernism is not just a matter of cubist space but of a whole appreciation of environmental design as a holistic approach to the matter of making spaces for people to live… Modernism, as I define it and practice it, 'includes' and is based on the vital archetypal needs of human beings as individuals as well as social groups.
>
> *(Walker and Simo, 1994)*

But perhaps Halprin was an outlier, and the resource system of place is too wide to be contained by a single movement. Is this why the optimal model of place is so elusive? Or is it because the complexity of place gets in the way of a simplistic notion of cities as engines of economic progress?

Emergence of urban design and regeneration in the UK

In practice, academic theories are usurped by a consensual professional narrative revolving around client briefs, the management of development projects, and navigation through the planning system. This is the business world where products are sold, where fashion, style, and conservative clients and planners' rule. In response to the negative impact of this practice, urban design emerged as a discipline, building on the UK Conservative Governments regeneration agenda for the 1980s post-industrial urban wastelands. Within a decade of Government putting regeneration on the agenda, and providing significant funding and opportunity for designers, everyone from architects to planners was describing themself as an urban designer. However, they promoted different codes according to their professional backgrounds. Planners generally espoused old Garden City principles, others the architecturally led New Urbanism concepts from the US, while many architects simply rebranded their masterplans as urban design. Despite the creation of the cross-disciplinary Urban Design Group and its call for joined-up thinking and holistic methods, emerging multidisciplinary practice simply created uneasy silo thinking under a single roof. Increasingly it became dominated by architectural concepts and styles.

The era of the 'built environment' was born, encompassing planning, development, and regeneration. Initially referred to as an 'industry,' it was soon elevated to 'sector' status. This terminology and referencing reflect the increasing scale of operation and economic importance of this activity consistent with the way government viewed its function. Since the 1980s, governments have encouraged public authorities to move departments concerned with the built environment into the private sector. The first discipline outsourced was engineering, leading to the growth of international super-sized 'one-stop-shop' consultancies and the stripping out of professional built environment expertise from local authorities. The immediate consequence of this is that there was no intrinsic and embedded knowledge of place, pride, or sense of ownership in the people that administer the stewardship of places. All that remains is a statutory skeleton staff of planning and highway officers. In contrast, public health has not been outsourced in the same way, retaining a credible critical mass of excellence and knowledge for population health within the public sector. In this way, any residual idea of urban theory underpinning the work of the public sector is squeezed out alongside any sense of duty to provide environments conducive to thriving for communities, and with little space left for theory in the business plans of the private sector. Traditionally, local authorities prepared bespoke guidance on elements of place-making with the purpose of benchmarking quality aspirations for their local areas. This included conservation policy, guidance to building owners, planning briefs to guide areas of change, and development briefs for specific sites etc. These proactive tools now need to be bought in from boutique urban design consultants with no locally vested interest, accountability or opportunity to monitor and nuance the guidance over the long term. Quality outcomes rely on the experience, expertise, and capacity of the client manager within authorities.

This creates a contradiction. While government seeks improvement to place design on the one hand, it has set design activity remote from place and community. Focussed on economic outcomes rather than a plural model including wellbeing and sustainability, it is destined to fail. The idea that 'authority' and strangers are doing things to 'communities' instead of with them becomes clear. Artist Amy Casey (2020) captures a sense of this: 'In my paintings, I have been trying to find stability in a landscape with no land. How can we find our footing in a world that seems to be constantly shifting?' The recognition of the ecological flow of cities in this statement, contrasted with the almost universal lack of agency people have over this process, makes the city just like any wild forest we previously foraged; not such a bespoke human habitat after all.

An urban renaissance – improving design

When the Blair Labour Government took office in 1997, they commissioned a Task Force of built environment professionals to establish a vision based on design excellence, social wellbeing, and environmental responsibility for declining urban places. Their report 'Towards an Urban Renaissance' (1999), provided the basis of the 'Urban White Paper: Our Towns and Cities – the Future' (Department of the Environment, 2000). This was the beginning of a new millennium and new aspirations for urban places. The Town and Country Planning Act 1990, which controls the development of all land, was reformed by the Planning and Compulsory Purchase Act 2004 and again reformed by the Planning Act 2008. The Coalition Government subsequently introduced further reforms to the primary Act with the Localism Act 2011. These are significant changes over a short period of time within a planning landscape that needs stable and long-term processes for places and their communities to thrive. In practice, much of this change centres on deregulation aimed at stimulating local economic growth, which is not always complementary to the claimed objectives of increasing design excellence, social wellbeing, and environmental responsibility.

The first action from the Task Force recommendations was the establishment of the Commission for Architecture and the Built Environment (CABE) and for CABE to publish urban design guidance in support of the Planning Act targeted at local authority planners, private developers, and their design teams. The result of this, *By Design: Urban Design in the Planning System – Towards Better Practice* was published in 2000. The national regeneration agency, English Partnerships (EP), concurrently published the *Urban Design Compendium* with similar guidance targeted at the agency's project managers who were responsible for all public regeneration investment. Both documents were enthusiastically welcomed by planners in the public sector, because like the project management surveyors in EP, they had little or no design training or experience and yet both groups had become the front line of this Urban Renaissance aiming to drive up design quality in all new developments. In the private sector, feelings were mixed about these documents. The emerging urban design practices embraced them, partly because they were authors or contributors, but also because they provided a supportive touchstone for their specialised practice, helping to drive up the aspiration of their public and private sector clients. Architects on the other hand were less enthusiastic, perceiving the guidance as rules that would suppress their creativity. They also levelled similar criticism at the new national design review panel directed by CABE and the regional panels established by the new Regional Development Agencies.

CABE went on to create the *Building for Life* (2008) protocol for residential development proposals which posed a checklist of questions for design teams and planning officers to use as the basis for discussion and negotiation through the planning application process. The first edition contained 20 questions, each worth zero or one point. The CABE rule of thumb was that a scheme needed to score a minimum of 14 points to have satisfied the requirements of the Planning Act and thus be eligible for planning consent. To strengthen the protocol, CABE set up an independent panel of accredited assessors and ensured that every planning authority in England had at least one accredited assessor within their department. However, while it is a good idea to set out clear principles as to what constitutes excellence in design, with peer review panels articulating those principles and tools like *Building for Life* filtering out poor design, in practice it is very difficult to define terms like 'well designed,' 'well planned,' 'good design,' or 'design excellence.' These terms are used throughout the urban policy documents and design review discussions, but when applied to real places with many players involved, the terms are wide open to interpretation and challenge. The problem of defining 'good design' and of reaching consensus among professionals is most evident during the design review process and in the negotiation of planning applications. Design and Access Statements are key documents in planning applications that set out how the proposal will deliver 'good design' in line with national and local design guidance and other statutory requirements. Residential schemes often include self-reported Building for Life Assessments too. Not infrequently, the two sides to the negotiation hold opposing interpretations of the generic guidance. Less frequently, design review panel members can also disagree among themselves.

From a placemaking perspective, this strategic context of urban space planning and design determines whether 'places of meaning' can be delivered or sustained. As Marshall notes, without 'any systematic verification and critical assimilation of scientific knowledge' the theoretical basis of good place design is based largely on an uncritical professional consensus of architectural practice. Essentially, it becomes a subjective value judgement on how people respond to places. Thus, when design guidance is put to the test, consensus fades. Our intrinsic human need to connect and interact with others effectively relies upon a public realm that affords this safely. In statutory planning instruments, this domain is the primary responsibility of highway engineers who define urban success in terms of free-flowing transport, a convention that often conflicts with urban design best practice and acts in direct opposition to the need to interact face-to-face in our public realms. Just like planners and development surveyors, highway engineers have no formal training or experience in urban design, concentrating instead on a technical agenda. While the design guidance documents described earlier explicitly deal with movement, they are largely ignored by this powerful group. Recognising these shortcomings, the Department for Transport commissioned their own design guide, *Manual for Streets* (2007). Unfortunately, its scope is restricted to lightly trafficked new residential roads, which undermines the effectiveness of design guidance on the quality of existing places.

An urban renaissance – regeneration in the UK

The second theme of this chapter is urban development and the disconnect from strategic thinking. The Urban Renaissance brought a positive focus to existing urban areas and communities, with emphasis on tackling difficult issues that were perceived to be beyond the resources of local authorities. These included failing and complex city centres and extensive areas of perceived housing failure in post-industrial cities such as Liverpool, Manchester, and Sheffield. Urban Regeneration Companies (URCs) were partnership quangos between local authorities, English Partnerships, Regional Development Agencies, and Government Offices. They had independent chairs and a board containing senior executives from the public, private, and third sectors, and a staff of executives with specialist skills, supplemented with secondees from the partnership organisations. The largely economic role of these URCs was to prepare visionary regeneration frameworks for their areas to raise aspirations and direct public spending to catalyst projects that would attract and support private inward investment. The authors involvement in the pilot URC in Liverpool was edifying from a placemaking perspective.

The brief prepared by EP called for the development of a city centre masterplan but citing their draft *Urban Design Compendium* enabled us to take an alternative framework approach better suited to addressing the strategic issues. However, it was a revelation to discover that most people working on the regeneration had no design training and did not see design as their responsibility. Effectively, it was business as usual, with the URC seen as another funding opportunity and delivery agent. All authorities are economically dominated, and although they see the 'value' in architecture, they struggle with placemaking because its measures are not simple. Unfortunately, this short-term approach is blind to the considerable costs associated with not delivering or nurturing placemaking appropriately. Taking the framework approach, the pilot URC in Liverpool city centre successfully stitched the fabric of the city streets back together, and in the retail expansion at Liverpool One, the fabric was appropriately extended to re-join city quarters, producing more 'social places.' We achieved the delivery of 'good design' because it was a condition of funding or development land access, but we could not advance strategic urban thinking or embed a positive placemaking ethos for the future. Political leadership is a fundamental issue in this. In short, places 'choose to fail or succeed' (Diamond, 2005).

The Housing Market Renewal Initiative (HMRI) was a second significant programme in the Urban Renaissance. Developed by the Centre for Urban and Regional Studies (CURS) University of Birmingham, the National Housing Federation used their research to successfully lobby the government's 2002 Comprehensive Spending Review. The programme was a fresh-slate, one-size-fits-all approach, on an immense scale applied in many established but unique urban areas, conceived without any urban design underpinning. Authorities partnered with volume house builders largely without appraisal, analysis, or design consideration of the selected renewal areas, assuming these private companies were the only or the best agents of urban renewal. The decision-makers in the new public–private partnerships then concluded it would be uneconomic to regenerate the existing fabric, making mass demolition a fait accompli, despite government insistence that clearance should be a last resort. This mirrored the 'clean slate' approach of the 1960s where terraced housing was replaced with towers in open green spaces in the inner-city, and similar clearances in the 1970s where inner-city communities were moved to suburban New Towns. This was a third wave of the same mistakes where communities were destroyed, while new 'places' were rarely created. Partnering with developers on a purely economic basis bound the authorities to delivering a suburban built form within the inner city. This top-down economic process imposed the antithesis of best practice urban design, which cannot be overcome by any amount of community engagement or participation. In the context of Collage City, this is more like jigsaw planning without common sense or wisdom. Despite community opposition from London to Liverpool to Glasgow, the HMRI programmes implemented what they believed people wanted and needed (National Audit Office, 2007).

The main issues in the HMRI areas were social ones, exacerbated when communities were 'cleansed' from their areas, and the impact of that extended into the wider urban fabric, predictably damaging adjacent communities in turn. However, within the Urban Renaissance, New East Manchester URC illustrated a more sustainable approach to community-led regeneration. When the failure of places becomes too toxic to ignore, programmes like the recent national 'Sink' Estates Strategy or NHS Healthy New Towns are initiated, but seldom address underlying and complex social issues of place. However, addressing these issues is vital to the social sustainability of existing places and the shortcomings that make places toxic. It therefore raises the question: can 'placemaking' be the bridge between strategic urban planning and development led urban design?

Placemaking – a social science approach

Shifting our perspective from the strategic overview to the lived experience on the ground, we immediately find ourselves foraging in our environment for resources, drawing on all our evolutionary instincts, responses, and behaviours that have helped us to survive to this point of human history. This foraging includes resources such as trade, companionship, and cooperation in an environment of trust and safety. To understand our implicit needs, how we respond to our environments, and how these things effect our behaviours we must turn to sound science for models, not to architects or economists.

Psychological research over the past century has provided insight into our being, the significance of our individuality, and the evolutionary importance of living cooperatively with others. It provides an understanding of life-span human development through the cycle of our responses to environment, how this affects our behaviour, and in turn how our behaviours impact on the environment. When we find places harsh or suboptimal, it is often the result of coercion from 'outside' rather than enduring cooperation from 'within.' However, even 'harsh' cities will endure because humans have evolved to adapt and cope with threat, change, and disaster, although neither cities nor their citizens will be sustainable if they remain in a chronic survival mode. Public health research identifies the negative impacts of cities on health and wellbeing as the 'urban penalty,' or the 'urbanicity effect.' Social disintegration and social isolation are identified as key factors, and where there are urban–rural differences, they survive controls for broad socio-economic factors. For example, Huxley and Rogers (2001) showed that communities characterised by high quality of life have greater sense of belonging, access to leisure opportunities, neighbourliness, sense of security, and less isolation. Conversely, communities whose residents report lower quality of life perceive that their neighbourhoods are failing to thrive, less neighbourly, and facilitate fewer leisure opportunities. Furthermore, Ellaway et al. (2001) reported that people living in under-resourced communities have lower levels of self-esteem, tend to feel lonelier and have less sense of control over their lives compared to those living in better-resourced neighbourhoods.

In the aftermath of recent global disasters (Hurricane Sandy; Chicago heatwave; Kobe earthquake; Tamil Nadu tsunami), studies illustrate that 'social ties can be a matter of life or death – neighbourhoods with strong community connections regularly had the highest survival rates' (Sampson, 2013). This extends to the prosocial behaviour they support which is central to the wellbeing of social groups across a range of scales where empathy is a strong motive in eliciting prosocial behaviour but is harder to sustain in disadvantaged harsh environments. Thus, poorly performing places are 'toxic assets' that generate their own negative costs, and placemaking to support resilient communities becomes of interest to everyone. While these findings put 'placemaking' at the top of the wellbeing agenda, there is disconnection between what people really need from their places and the practice of urban design.

The work of biologist David Sloan Wilson illustrates a way forward that embraces the human need to support ourselves and others. He has been applying modern evolutionary theory to understand and improve his home city of Binghamton, NY. In Darwin's City (2011, p. 146) he says: 'I really wanted to see a map of altruism… I saw it in my mind.' Wilson contends that his pioneering approach can be applied to any city because 'evolution takes place not only by small mutational change – individuals from individuals – but by groups becoming so well integrated that they become higher-level organisms in their own right – individuals created from groups' (Wilson and O'Brien 2009, p. 156). Among the findings of Wilson's group is that perceived neighbourhood quality is a statistically significant predictor of individual and group prosociality. A teenager's tendency to behave in prosocial ways correlates with their own perceptions of neighbourhood quality and with the quality ratings of other residents, demonstrating that our prosocial cooperation is dynamic and contingent upon our perceptions of environment. Sustainable placemaking must understand the physical attributes of place that determine this conditional prosocial behaviour in order to produce environments that maximize citizens wellbeing and quality of life. Until we develop a model to affect positive change in both community and place, the creation of healthy and sustainable places will remain a slogan. The first step seems clear: we must design and maintain psychologically informed environments that reduce feelings of 'threat' and so optimise opportunities to interact and cooperate. The application of Psychologically Informed Environments (PIE) is already mainstream in hospitals and some workplaces. It needs to expand into urban design and placemaking.

Taking the PIE or human ecology approach in a world of rapid urbanisation and climate change will deliver and maintain optimum environments to suit the human condition and maintain the dynamic and symbiotic relationship between place, man, and culture. Several countries, including Wales and New Zealand, have put wellbeing at the top of their national agendas, demanding action to deliver this in all policy areas. Perhaps it is time to reinvigorate the Urban Renaissance – there was nothing wrong with a vision based on design excellence, social wellbeing, and environmental responsibility. However, perhaps it should be redrafted in a way that emphasises the primacy of social wellbeing as an outcome delivered through good placemaking and resulting in environmental responsibility. The intuitive theories and concepts underpinning urban planning should be assimilated and verified by scientific knowledge, and new ones developed for prosocial places. Perhaps Collage City and its emphasis on the quality of the public realm is a useful metaphor, along with Halprin's Ecology of Form. To reconnect the three themes of strategy, development, and experience requires more movement back and forth between the world of theory, the world of practice, and the world of experience, in cooperative action over our commons. This movement needs to be based on building optimism, because that is the only route out of languishing into a future of thriving. A socio-ecological approach is about more than a horizontal growth strategy for declining places. It is a vertical strategy to move places and communities from languishing to flourishing.

References

Alexander, C. (1965). 'A city is not a tree', Architectural Forum, 122(1), pp. 58–62.

CABE. (2000). *By Design: Urban Design in the Planning System: Towards Better Practice.* London: The Stationery Office.

CABE. (2008). *Building for Life: The 20 Criteria.* London: CABE.

Casey, A. (2020). Artist statement in online catalogue [online]. Available: https://20x200.com/products/amy-casey-extended-city (Accessed: 20 April 2020).

Corcoran, R. and Marshall, G. (2017). 'From lonely cities to prosocial places: How evidence-informed urban design can reduce the experience of loneliness', in Sagan, O. and Miller, E. (eds.) *Narratives of Loneliness: Multidisciplinary Perspectives from the 21st Century.* London: Routledge.

Cresswell, T. (2004). *Place: A Short Introduction.* Hoboken, NJ: Wiley-Blackwell.

Cullen, G. (1961). *The Concise Townscape.* Abingdon: Routledge.

Department for Transport. (2007). *Manual for Streets.* London: Thomas Telford Ltd.

Department of the Environment, Transport and the Regions. (2000). *Our Towns and Cities: The Future: Delivering an Urban Renaissance,* Vol. 4911 of Cm Series. London: The Stationery Office.

Diamond, J. (2005). *Collapse: How Societies Choose to Fail or Survive.* London: Penguin Books Ltd.

Ellaway, A., Macintyre, S. and Kearns, A. (2001). 'Perceptions of place and health in socially contrasting neighbourhoods', *Urban Studies,* 38, pp. 2299–2316.

English Partnerships. (2000). *Urban Design Compendium.* London: The Stationery Office.

Halprin, L. (1970). *The RSVP Cycles: Creative Processes in the Human Environment.* New York: George Braziller Inc.

Halprin, L. (1982). *The Ecology of Form* (audio book). Pidgeon Digital [online]. Available at: https://www.pidgeondigital.com/talks/the-ecology-of-form/ (Accessed: 20 April 2020).

Halprin, L. and Burns, J. (1975). *Taking Part: A Workshop Approach to Collective Creativity.* Boston: The MIT Press.

Harris, E. (2011). 'Darwin's City', *Nature,* 474, pp. 146–149.

Huxley, P. and Rogers, A. (2001). 'Urban regeneration and mental health', *Health Variations Programme Newsletter,* (7), pp. 8–9.

Jacobs, J. (1961). *The Death and Life of Great American Cities.* New York: Random House.

Lynch, K. (1964). *The Image of the City.* Boston: The MIT Press.

Marshall, S. (2012). 'Science, pseudo-science and urban design', *Urban Design International,* 17(4) pp. 257-271.

National Audit Office. (2007). *Housing Market Renewal.* London: The Stationery Office.

Rowe, C. and Koetter, F. (1978). *Collage City.* Boston: The MIT Press.

Saar, M. and Palang, H. (2009). 'The dimensions of place meanings', *Living Reviews in Landscape Research, 3.* pp.1-24

Sampson, R.J. (2013). 'When disaster strikes, it's survival of the sociable', *New Scientist,* 218(2916), pp. 28–29.

Urban Task Force. (1999). *Towards an Urban Renaissance.* London: Department of the Environment, Transport and the Regions - London: The Stationery Office.

Vitruvius, P. (transl. Morris, HM, 1960). *The Ten Books on Architecture*. New York: Courier Dover Publications.
Walker, P. and Simo, M. (1994). *Invisible Gardens: The Search for Modernism in the American Landscape*. Boston: The MIT Press.
Wilson, D.S. and O'Brien, D.T. (2009). 'Evolutionary theory and cooperation in everyday life', in Levin, S.A. (ed.) *Games, Groups and the Global Good*. Berlin: Springer.

Further reading in this volume

Chapter 1: Introduction: what really matters – moving placemaking into a new epoch
Cara Courage
Chapter 5: Making places for survival: looking to a creative placemaking past for a guide to the future
Jeremy Liu
Chapter 14: Experts in their own tomorrows: placemaking for participatory climate futures
Paul Graham Raven
Chapter 15: Un/safety as placemaking: disabled people's socio-spatial negotiation of fear of violent crime
Claire Edwards
Chapter 17: 'I am not a satnav': Affective placemaking and conflict in 'the ginnel that roared'
Morag Rose
Chapter 18: '*Homomonument* sounds like a poem': queer placemaking 30 years on: a conversational dialogue with Thijs Bartels, author of *Dancing on the Homomonument* (2003)
Martin Zebracki
Preface: The radical potential of placemaking
Cara Courage
Chapter 20: Displacemaking 2015 and 2020
Catherine Fennell and Daniel Tucker
Chapter 21: Placemaking through parkour and *Art du Déplacement* (ADD) as a Singaporean applied performance practitioner in London
Adelina Ong
Chapter 22: Embedded Artist Project: Epistemic Disobedience + Place
Frances Whitehead
Chapter 24: Artists, creativity, and the heart of city planning
Tom Borrup
Chapter 25: 'If you can make it there, you can make it anywhere…': cultural placemaking at the heart of cities
Sherry Dobbin
Chapter 28: Integral placemaking: A poiesis of sophrosynes?
Ian Wight
Chapter 33: Conceptualizing and recognizing placemaking by non-human beings and lessons we might learn from Marx while walking with Beaver
Jeff Baldwin
Chapter 30: Ecological selves as citizens and governance as ethical placemaking
Lisa Eckenwiler
Preface: The only thing constant is change
Kylie Legge
Chapter 34: Reconnecting *cité* and *ville*
Philip Graus
Chapter 35: Planning governance: lessons for the integration of placemaking
Nigel Smith
Chapter 36: Facilitator skills for effective collaborative placemaking
Husam AlWaer and Ian Cooper
Chapter 37: *The Neighbourhood Project*: a case study on community-led placemaking by CoDesign Studio
Lucinda Hartley, Eliza Charley, Sama Choudhury, and Harriet McKindlay
Chapter 38: Public seating: a small but important place in the city
Kylie Legge
Chapter 39: Translating Outcomes: Laying the groundwork for interdisciplinary evaluation of creative placemaking

Jamie Hand
Chapter 40: Transforming community development through arts and culture: a developmental approach to documentation and research
Victor Rubin
Chapter 41: Rituals of regard: on festivals, folks, and findings of social impact
Maribel Alvarez
Chapter 42: Creative Placemaking and Placekeeping evaluation challenges from the practitioner perspective: an interview with Roy Chan
Maria Rosario Jackson
Chapter 44: Creative Placemaking and comprehensive community development: rethinking neighborhood change and evaluation
Maria Rosario Jackson
Chapter 45: How the city speaks to us and how we speak back: rewriting the relationship between people and place
Rosanna Vitiello and Marcus Willcocks

SECTION 4

Art, artists, and placemaking

Section Editor: Cara Courage, Handbook Editor

PREFACE

The radical potential of placemaking

Cara Courage

... the art is in what we're doing. This is art.

(Homebaked participant, 2013)

From Jane Jacobs' (1961) street ballet as the art form of the city, to the quote above, from a participant in the Liverpool, UK arts-led regeneration project, *Homebaked*, arts and placemaking have been intertwined as performative metaphor and practice for some time. It is the explicit inclusion of the arts in placemaking that is one of its differentiating factors, when stood next to its built environment sector siblings. Art interventions transform cities every day, at an intimate and micro scale, to a public and macro scale. Art in the city cannot be othered from any other built environment process. A product of historical processes, any creative intervention or practice in a place context exists in an interrelational matrix ecology of interdependent actors and organisations – its culture. Culture is not a performative gesture in cities. People live, create, and recreate culture in cities on a daily basis. The city does not 'create culture,' nor is culture in its service. The city is more than 'of' culture. The city *is* culture.

A list of the outcomes of an arts-based approach to place is seemingly endless, as fundamental to its being. Art in place intersects with the economy, innovation, health and wellbeing, education and learning, and can be attributed to (variously and not exhaustively): engendering creativity, understanding, self-reflection, and empathy; respite and pleasure; attracting and retaining residents; employment and income generation; the promotion and actualisation of positive and equitable neighbourhood change; social cohesion and inclusion and active citizen participation; (re)imagining new narratives for individuals and communities; acting as a catalyst for cross-sectoral collaboration and networking; and having cognitive, psychological, and material impact on a place's form and function.

Art and placemaking will forever be associated with 'creative placemaking,' a term created by Ann Markusen and Anne Gadwa in their 2010 Creative Placemaking White Paper for the USA's National Endowments for the Arts. Creative placemaking, as 'defined' on its third page, is understood as something that marries public and private, that has fiscal benefits as well as social and cultural ones; there is a public–private investment mix and a concern with communities thriving; and an overt mention of arts and culture as a tool of and mechanism for placemaking. Since 2010, creative placemaking as presented by Markusen and Gadwa has been critiqued, not least by themselves (Markusen and Gadwa Nicodemus, in Courage and McKeown, 2019, amongst others prior to this) and keenly along lines of race and power, and vis-à-vis regeneration and gentrification.

In my own research, I saw artists engaged in a social practice art work in the placemaking context as social practice placemaking (Courage, 2017). Place-based artists, while not necessarily calling themselves placemakers, hold a relational concept of space that encompasses social structure and material and embodied dimensions. They work in the space that is at the intersection of object, structure, and action, and that interjects with the material, social, and symbolic dimensions of space in the context of the politics of the public realm and urban development. Examples abound of artists taking into their own hands the means of production and assets of development – the land and buildings – and keeping them for creative and community use, keeping the community in place and using the gentrifying multipliers of the arts to community benefit. What is key here is the intentionality and the very singular skill that a socially engaged artist can bring to a place-based context. These artists don't need to be 'of' the place in which they are working, but they do need to understand places and communities based on their own lived experience. There is a specific skillset of the artist that is able to broker conversations between different stakeholders in a process, the art process becoming a relational object by which to talk through contested issues. This is what Amin (2008) would call the micropublic – no one group will ever be in total agreement, but through a facilitated process, here, using the art process, a group can come to a settled consensus, where, even if their own view isn't taken forward into action, they will they have been listened to and been active in the decision-making.

Through this lens in particular, placemaking, as a mode of arts-based activity in the public realm, is a co-produced practice, one that is also concerned with process as much as any material outcome. The artist might be an instigator and a catalyst of activity, but they will work in equanimity with the community to create shared outcomes and outputs, and their aim will be to 'hand over' as it were, the life and legacy of the project to the community. It demands a working in a relative expertism (Courage, 2017, and this volume) that is not of sole, singular authorship, and it demands you get out into the public realm and are hands-on with it, conversing with the people of it, questioning the politics of it. This is a practice aligned with social justice, and with artist Tania Bruguera's *Arte Útil*, 'useful art,' that suggests art as a tool or device and draws on artistic thinking to imagine, create, and implement tactics that change how we act in society. Here, projects should: propose new uses for art within society; use artistic thinking to challenge the field within which it operates; respond to current urgencies; operate on a 1:1 scale; replace authors with initiators and spectators with users; have practical, beneficial outcomes for its users; pursue sustainability; and re-establish aesthetics as a system of transformation. The similarity between *Arte Útil* and the placemaking of this, and indeed other, sections of this Handbook is self-evident.

None of this is to say that arts-led or creative placemaking should not be met without critique. Arts in place has been used as a salver to smooth over social cleansing and to give a veneer of community-located authenticity or acceptability, a placewash (Pritchard, 2019) as much as an artswash, where the arts are used as a tool to gloss gentrification and used as an outward badge of good placemaking practice. In the context of placemaking or arts in the public realm more generally, this begs a number of questions: are – are artists complicit in a gentrification agenda? Is their role to protest in such places, for and with the people of those places? Or, does the artist have an agency to take place development into their own hands and make it work for a social and cultural agenda?

Placemaking, as understood in this Handbook, does not work with or towards the ideal, but with the real – it's a messy process and it's conflicted and it's open to a healthy critique. Herein lies its radical potential. We know how to do the stuff of urban planning and design differently now and we have a whole sector of artists and creative practitioners that specialise in just that. Arts-led and socially concerned placemaking celebrates the capacity of the arts to address the city as a complex ecosystem of vibrant material and symbolic creativity that is ever in flux. Such projects can integrate culture and community and can result in boosted local economies and increased levels of social connection and civic engagement. Such projects can also intersect and advance missions in transportation, housing, employment, health care, environmental sustainability, and education.

The barriers to inclusion that many communities of place face do not exist in isolation from each other and we can't collectively address complex place-based issues – such as environmental collapse, discrimination, social inequity, or inequitable access to quality public spaces – with a singular agenda in mind. This means being cognisant of, and actively supporting, the arts in offering quotidian spaces and practices of vernacular creativity. Not only will working with all that artists know (Whitehead, 2006, and this volume), and with making art useful, and seeing the people of place as creative placemakers, result in a more progressive and integrated approach to cities, but it also makes possible a rethinking of the synergies between the arts, community, and our urban place. The intentionality of a project matters. This is the start, middle, and end and the golden thread that joins it all, and it is in the sensibility of the arts practice and process and in working with artists in equanimity that the radical power of placemaking can begin to be realised.

This section begins with the artists voice, moves through scales of artist and arts practice, and closes with what all placemakers should be primarily concerned with, the experience of place in the first person. It is curated to make it useful for any reader, from any placemaking purview, to both recognise their practice in others and to extend their practice through learning from the similar and the different.

The section opens with a reflective and philosophical questioning and probing of arts and placemaking practice, from Catherine Fennell and Daniel Tucker ('Displacemaking 2015 and 2020'). In this two-part interview, the two consider their own experiences and encounters with placemaking projects. They pay close attention to how these encounters tease out the contradictions and complexities facing the field at large and take up several notable developments, including placemaking's professionalisation and the standardisation of familiar tools and techniques, such as storytelling. What, they ask, can such developments tell us about how placemaking projects address, or might be better positioned to address, ongoing equity issues at a contemporary moment?

Adelina Ong continues the consideration of an emplaced practice, in 'Placemaking through parkour and *Art du Déplacement* (ADD) as a Singaporean applied performance practitioner in London.' This chapter reflects on the parkour and *Art du Déplacement*-inspired place practices used as part of a session on place and placemaking in London, UK. Ong posits that the 'lively nomad' might use placemaking as a way of resisting assimilation and being transformed into a 'zombie migrant'. Ong reflects on a placemaking that prompts reconsideration of the failure encouraged in creative learning environments and an understanding of death as an invitation to contemplate an understanding of living that recognises our interconnectedness to the more-than-human in place.

Artist Frances Whitehead continues this consideration through the lens of her practice and projects in her '*Embedded Artist Project*: epistemic disobedience + place.' The chapter includes global case studies of Whitehead's work, epistemologically driven practice experiments under the concept of the Embedded Artist. These experiments aim to explore the role of culture in sustainability and demonstrate how multiple values (social, cultural, environmental, and economic) can inform a net benefits model of development.

Staying with a case study, Gail Skelly and Tim Edensor, in 'Routing out place identity through the vernacular production practices of a community light festival,' focus on a community festival in the north of the UK in the wake of its loss of funding. The chapter considers three aspects of this adaptation to unpromising circumstances: the ongoing reproduction of the mythic story on which the festival is founded; the ways in which the parade route is devised, organised, and managed; and the inventive selection of an annual theme that both marks historical identity and undergirds a shared sense of place. Under conditions of austerity, longstanding local inhabitants have managed to keep the festival going while fostering an inclusive sense of participation amongst recently arrived, more affluent residents and have developed their own social, organisational, and creative skills while extending a strong, shared sense of place amongst participants.

Moving into a still-wider terrain of practice, Tom Borrup ('Artists, creativity, and the heart of city planning') considers the role of arts, placemaking and planning. This chapter focuses on ways artists have enhanced the meaning and outcomes of the public participation process in planning in the United States. It traces precedents from the 1960s, and describes some approaches used. While some literature on public participation in city planning cites the value of storytelling, sketching, and other creative techniques, calls from planners for artists to join them are absent. Artists who come from a tradition of community activism, on the other hand, have been quicker to engage in planning. The author calls on planners to go beyond the consultative function of public engagement: to see it as an opportunity to build on cross-cultural relationships and working capacities, co-creation and collective problem-solving, and to promote participatory democracy.

The section then turns to its largest scale – the placemaking-led regeneration of Times Square, and an in-depth account of this transformation process, from its then-Founding Director of Times Square Arts, Sherry Dobbin ('"If you can make it there, you can make it anywhere...": cultural placemaking at the heart of cities'). Dobbin reveals how she structured the programme and presents lessons learned in practice and experimentation. Project examples are included to demonstrate how to raise artistic ambition, while simultaneously contributing to the stakeholders' and communities' strategic objectives for place. Dobbin's subsequent methodology and framework were shared internationally amongst city- and town-centre leadership and cultural civic agencies, as well as implemented into guidance documents. The chapter also includes how the cultural placemaking approach was adapted for a London BID and concluding principles that can be applied across city centres of all scales.

Lastly, philosopher, curator, and critic, Trude Schjelderup Iversen, offers, in 'Sculpturing sound in space: on *The Circle and the Square* (2016) by Suzanne Lacy,' a first-person account of the experience of arts in place and placemaking. Reflecting around three key concepts – place, space, and (emphatical) experience – it is found that Lacy's work is often interpreted as early examples of community-based activist and participatory art practice. From the purview of an embodied experience of the artwork and conversation with the artist, the author posits that one must create one's own experience and this creation is related to an aesthetical pursuit in Lacy's work and is fundamental in how Lacy approaches and prioritises within community-driven collaborations.

References

Amin, A. (2008). 'Collective culture and urban public life', *City*, 12(1), pp. 5–24.

Bruguera, T. (n.d.) *Arte Útil* [online]. Available at: https://www.arte-util.org/about/colophon/ (Accessed: 12 April 2020).

Courage, C. (2017). *Arts in Place: The Arts, the Urban, and Social Practice*. Abingdon: Routledge.

Homebaked [online]. Available at: http://homebaked.org.uk/ (Accessed: 21 April 2020).

Jacobs, J. (1961). *The Death and Life of Great American Cities*. New York: Random House.

Markusen A. and Gadwa, A. (2010). *Creative Placemaking* [online]: National Endowment for the Arts. Available at: https://www.arts.gov/publications/creative-placemaking (Accessed: 20 April 2020).

Markusen, A. and Gadwa Nicodemus, A. (2019). 'Creative placemaking: Reflections on a 21st-century American arts policy initiative', in Courage, C. and McKeown, A. (eds.) *Creative Placemaking: Research, Theory and Practice.* Abingdon: Routledge.

Pritchard, S. (2019). 'Place guarding: Activist art against gentrification', in Courage, C. and McKeown, A. (eds.) *Creative Placemaking: Research, Theory and Practice.* Abingdon: Routledge.

Whitehead, F. (2006). 'What do artists know?', in Embedded Artist Project [online]. Available at: http://embeddedartistproject.com/whatdoartistsknow.html (Accessed: 8 June 2020).

Further reading in this volume

Chapter 1: Introduction: what really matters – moving placemaking into a new epoch
Cara Courage
Chapter 2: Placemaking as an economic engine for all
James F. Lima and Andrew J. Jones
Chapter 4: A future of creative placemaking
Sarah Calderon and Erik Takeshita
Chapter 5: Making places for survival: looking to a creative placemaking past for a guide to the future
Jeremy Liu
Chapter 9: From the dust of bad stars: disaster, resilience, and placemaking in Little Tokyo
Jonathan Jae-an Crisman
Chapter 10: From moon village to mural village: the consequences of creative placemaking in Ihwa-dong, Seoul
Jason F. Kovacs and Hayun Park
Chapter 11: *Free State Boulevard* and the story of the East 9th Street Placekeepers
Dave Lowenstein
Chapter 14: Experts in their own tomorrows: placemaking for participatory climate futures
Paul Graham Raven
Chapter 15: Un/safety as placemaking: disabled people's socio-spatial negotiation of fear of violent crime
Claire Edwards
Chapter 16: More than a mural: participatory placemaking on Gija Country
Samantha Edwards-Vandenhoek
Chapter 17: 'I am not a satnav': Affective placemaking and conflict in 'the ginnel that roared'
Morag Rose
Chapter 27: Is 'tactical urbanism' an alternative to neoliberal urbanism?: reflections on an exhibition at the MoMA
Neil Brenner
Chapter 29: The solution is in the problem; the art of turning a threat into an opportunity by developing resilience using a creative placemaking critical praxis
Anita McKeown
Chapter 39: Translating Outcomes: Laying the groundwork for interdisciplinary evaluation of creative placemaking
Jamie Hand
Chapter 40: Transforming community development through arts and culture: a developmental approach to documentation and research
Victor Rubin
Chapter 42: Creative placemaking and placekeeping evaluation challenges from the practitioner perspective: an interview with Roy Chan
Maria Rosario Jackson
Chapter 43: A theory of change for creative placemaking: the experience of the National Endowment for the Arts' *Our Town* program: an interview with Patricia Moore Shaffer, PhD
Maria Rosario Jackson
Chapter 44: Creative Placemaking and comprehensive community development: rethinking neighborhood change and evaluation
Maria Rosario Jackson

20
DISPLACEMAKING 2015 AND 2020

Catherine Fennell and Daniel Tucker

Introduction (2020)

Five years after our original dialogue on placemaking was published in *Avery Review* (Fennell and Tucker, 2015), we reconvened in 2020 to revisit our initial interest in the topic and reflect on how the discourse and practices have evolved. Drawing on our diverse reference points as an artist and as an anthropologist, we consider the ways in which our field experiences have introduced contradictions and complexities facing the field at large. Starting out with a dialogue focused on recent experiences in Chicago and broadening in 2020 to include reference points in Philadelphia and New York City, this interview considers developments in the critical literature of placemaking as well as a number of specific case studies including Rebuild Foundation and the Village of Arts & Humanities. In conclusion, trends regarding storytelling and concerns about how these practices confront issues of equity during the public health crisis are considered.

Displacemaking (2015)

As two former Chicagoans who think about the politics of urban development in and beyond Chicago, we decided in the early summer of 2015 to stage the kind of dialogue we, and our publishers at the Avery Review, would be eager to hear at the Chicago Architecture Biennial. The starting point was our shared concern with the now firmly established set of practices known as 'placemaking' and we branched out into art, philanthropy, and planning. In order to ground our conversation, we read two recent White Papers on the subject, 'Places in the Making' (Silberberg, 2013) and 'Creative Placemaking' (Markusen and Gadwa, 2010). Placemaking, as one of these papers defined it, involves 'the deliberate shaping of an environment to facilitate social interaction and improve a community's quality of life' (Silberberg, 2013).

Catherine Fennell (CF): One can't follow the politics of urban redevelopment in Chicago without coming across conversations about the importance of 'place' or 'placemaking.' These conversations seem to have picked up especially in the last five or six years, in tandem with the 'Great Recession.' Yet these terms don't seem that prevalent in New

York, the city where I currently reside, at least not in everyday talk. What should we make of the tractability of these terms in some cities and not others?

Daniel Tucker (DT): Chicago is a city that uniquely combines (some of) the wealth of Manhattan with the abandonment of Detroit. This opens up the city as a policy lab where there are resources to test ideas, and disinvested contexts and populations to test them on. That combination accounts for top-down policy experiments as well as some of the more quasi-grassroots interventions. I say 'quasi,' because the grassroots is a bit more embedded within power structures in Chicago than in a city with a more thorough abandonment. So placemaking might emerge in Chicago as a rhetorical bridge between those different sectors. It seems like NYC has plenty of placemaking projects, but they are much more integrated into the logic of planning there as opposed to being layered in as a special project like in the context of Chicago. Bloomberg could be considered a placemaker, while Giuliani was a 'broken windows' guy (that might pave the way for placemaking?). Bryant Park, Project for Public Spaces, and Times Square are all proto-placemaking efforts, no? Maybe the term is less commonly used because of the ubiquity of the concepts?

CF: Chicago's long been a space for producing knowledge about urbanism and experimental interventions that would retool urban life. You see it already in the writings of the Chicago School sociologists. More recently, you see it in the city's ambitious public housing reforms of the past 20 years, in which public officials, private developers, and philanthropic foundations partnered to refine redevelopment approaches that might prove to be useful in other cities. So, for instance, consider how staff and trustees of Detroit's Skillman Foundation recently looked to Chicago's MacArthur Foundation for 'placemaking strategies' that would mitigate the effects of the foreclosure crisis at the neighborhood level. Skillman was looking at MacArthur's 'New Communities Program.' This program emerged when that foundation sought to stimulate community-based quality-of-life planning initiatives as public housing came down all around that city. Terms like 'community' and 'place' resonate emotionally, but they also need to be situated within an inter-urban economy of knowledge and development practices.

What's certainly ubiquitous in New York is just how much private interests are able to capitalize on novel development opportunities. The case studies we read were all clear that placemaking projects and the 'vibrant places' they create have quite quantifiable returns. For instance, we learned that 'in just the two years following [Bryant Park's] restoration, rental activity in the area increased by 60 percent' (Silberberg, 2013, p. 31). Seems like a winning prospect.

But it's important to note here that a private corporation drove this restoration, not a city government's parks and recreation department. And here's where imaginations around development, placemaking, and urban or regional competition seem to overlap: if cities that attract tourists and a young, talented workforce have enticing public places like an elevated rail line reimagined as a park (New York's High Line), or a derelict city park reinvented as a premier events and gathering space (New York's Bryant Park), wouldn't it be good for other cities to have that too, especially if they will become competitive in an era of inter-urban competition? In the US, we need to recognize that placemaking is first and foremost a development activity that emerged during the very decades we saw federal and state investments in urban infrastructure, housing, and social programming diminish.

I want to pick up on your observation about placemaking and broken-windows policing. All the cases we read emphasized the need for 'vibrant' spaces and talked about a diversity of users and uses. But who and what counts as 'vibrant'? We read that Bryant Park 'provides much-needed amenities to anyone who wants to use them,' including 'award-winning restrooms.' We read that it 'welcomes the homeless – assuming the vast number of other visitors will diminish any negative effect the homeless may have.' After a day of working in the public library in Bryant Park, I can go outside into the park that surrounds it, sit on one of those folding green chairs, and unwind. But attempt to stretch out on the lawn? I promise you, someone will be by to tell you to sit up. Bryant Park Corporation might officially welcome the homeless, yet it doesn't welcome homeless-like activities and their potential 'negative effect.' 'Vibrant for whom, and toward what end?' need to be questions we ask of all placemaking projects.

But I also think we should avoid simply reducing this very complex story to big interests, real estate developers, or philanthropic donors. Clearly, placemaking projects strike a chord for many people, in many different cities. Otherwise they wouldn't take hold or be so exportable. Those chairs all over Bryant Park? They're light. I can pick them up. I can reposition them to get the best view of what's going on. I feel like I have some say, some autonomy, some ownership over how I use, or in the language of the cases we read, 'activate' that space. But it's important to ask about the relationship between the demand for places that (for some) feel authentic and flexible, and the demand for urbanites – citizens, workers, neighbors, etc. – who bring flexibility and authenticity to the table. It's especially important to ask these questions as the kind of large-scale public works projects or social investments we associate with a different era of city building recede further into the past.

The pieces we read together all argued that artistic practice can be harnessed toward placemaking, and several made a strong case that placemaking is inextricable from the activity of redevelopment and small-scale entrepreneurialism. You're an artist who has thought long and hard about development, but you're also an artist who has been pulled into institutions with concerted commitments to 'creative placemaking.' I'm thinking of your residency at The University of Chicago's Gray Center for Arts and Inquiry, just one of the new initiatives happening on campus along with its 'Arts Incubator.' After 50-plus years of cutting itself off from its surroundings in ways that advanced the devaluation of areas to the west and south, my alma mater has recently been capitalizing on real estate investment and development opportunities in its own backyard. One way it is doing this is by promoting 'creative placemaking' projects on the peripheries (or frontiers) of its expansion.

For example, one of the first instructions I got as a newly arriving graduate student in the fall of 2001 at an orientation session for new students was to avoid waiting at the 'desolate' Green Line train and bus stop on Garfield Boulevard and Washington Park. Today, if I need to pass some time while waiting for the bus to campus, I can stop in at a quirky storefront café right next door to the Arts Incubator. It is named 'Currency Exchange,' after a business that used to occupy this site – a one-stop financial services station for people too poor and too disenfranchised to have a regular bank account. Such businesses typically charge hefty fees for services like cashing a paycheck or paying a utility bill. While waiting for the bus, I can get a nice cup of coffee or a really good biscuit, a nod to culinary traditions of Black migrants who settled the South Side of Chicago just as ethnic whites abandoned it. While there, I might take in the work of local artists, or an event featuring discussions about community-driven arts initiatives. Theaster Gates, a well-known artist who is also on faculty, is the brainchild of this space and its programming, but the university actually owns the building. How can or should we draw firm lines between development, entrepreneurialism, and artistic practice?

DT: There are many conceptions of what it means to be an artist, and certainly the identity of an artist is different than a coherent definition of art, despite many definitions leaning toward art being anything that is made by a self-identified artist. I recently read a piece by Willa Cather from 1920 on 'The Art of Fiction,' where she explains:

> Writing ought either to be the manufacture of stories for which there is a market demand – a business as safe and commendable as making soap or breakfast foods – or it should be an art, which is always a search for something for which there is no market demand, something new and untried, where the values are intrinsic and have nothing to do with standardized values.
>
> *(Cather, 1920)*

I think Cather's range of what writing ought to be corresponds with the broad definition of art that is utilized in placemaking literature. For those advocating placemaking, great importance is not placed on the content or form of art – the point is to get artists' bodies into particular contexts. Literally described in Creative Placemaking (Markusen and Gadwa, 2010) as 'an entrepreneurial asset ripe for development,' those bodies represent latent capital, with the caveat later acknowledged that 'artists are twice as likely as workers overall to have completed college degrees… *yet artists' median annual income lags behind that of other professional workers by 19.4 percent.'*

If you are an artist working today, you cannot ignore developers and policy-makers' instrumentalization of your entrepreneurial yet precarious position. You must make an effort to negotiate your relationship to concepts like 'creative industries,' 'placemaking,' and 'civic engagement' in the same way that artists have also (and continue to need to) grapple with categories like 'beauty,' 'politics,' 'identity,' 'community,' and 'autonomy.' For instance, there are those artists that continue to advocate for autonomy, yet their bodies are as functional as any other body as an 'asset ripe for development,' and so it is irresponsible to act as if that isn't taking place. Considering that, I am not sure that a firm line between the noncommercial and the commercial is any more possible than one between the autonomous and the instrumentalized. But I do think that tension requires artists to be strategic about their engagement with creative placemaking and the like in order to both advocate for art and deepen democratic participation and redistribution of resources.

CF: You're someone who has built your artistic practice around the exploration of popular or grassroots social movements. Much of the placemaking discourse positions placemaking as a bottom-up, radically democratic, or civic activity, in marked contrast to more top-down interventions. Is creative placemaking akin to a social movement?

DT: To address this, let me dig into the texts a little bit. Rhetorically, placemaking is described as 'iterative, process-oriented, combining tactics' and 'decentralized' – which all sound a lot like the descriptions of the open source software and global justice movements of recent decades – and is directly compared with the environmental sustainability movement that emerged in the 1970s.

The texts we read seem to waiver between claiming that the placemaking professional of the past is gone and making numerous references to 'the placemakers' as if they are a class of people. In my reading of the texts, most projects strongly rely on some kind of 'creative initiator' taking the lead role. They seem to correspond with social theorist Michael Albert's 'coordinator class,' who are often the people performing most of the creative and empowering parts of a job. A veteran of the New Left, Albert has described the social movements of the 1960s and 1970s in the US and Europe as being shortsighted about the role of the coordinators – saying that these movements put a lot of energy into direct democracy but never addressed that fundamental division of labor, which became reproduced in the leadership of many of their organizations.

While my gut tells me that placemaking and social movements are different on the basis of professionalization, I have to admit that there is as much professionalization in social justice activism as there appears to be in placemaking. I think that movements and creative placemaking both need to struggle against this professionalization in order to retain what is really most important about both – people taking control of their own lives and environments.

CF: One of the assumptions driving the discourse on placemaking is that far too many Americans move through spaces that lack a coherent or fulfilling 'sense of place.' These soulless spaces are ready to be 'activated' through artistic or design-based activities into vibrant and inclusive places where one might stroll, sit, and talk, or maybe even eat their lunch in the company of others. It's fair to ask 'who' is included and left out when we talk about this 'lack,' but harder to entertain the aspirations driving it. How/why should we do both?

DT: There are a few moments in the two texts we read where the authors articulate how cities arrived at the state they are presently in (which, presumably, they're trying to get out of via placemaking.) The general narrative is that big infrastructure projects messed up the intimacy of cities and then deindustrialization changed their physical, social, and economic landscape. The lineage of thought traced as a canon of placemaking includes Jane Jacobs (1961) and Kevin Lynch (1996) among others. Henri Lefebvre (1968) and David Harvey (2013) were also invoked, to point to a more critical take on the redistributive goals that placemaking projects might take on in relationship to the concept of the 'right to the city.' If placemaking as a discourse can hold some of these critiques alongside lighter and more positive practices consistent with the definition offered in *Places in the Making* – 'The practice aims to improve the quality of a public place and the lives of its community in tandem' (Silberberg, 2013) – then it could be a really powerful and useful framework for creating more participatory and equitable cities.

Back to your reference about the University of Chicago. I see these major institutions like University of Chicago or University of Pennsylvania doing a dance where they direct resources toward repairing damaged relationships with surrounding communities with art and social justice programming at the forefront. Cynically I'd say these instances of community building, outreach, and resource sharing are serving as multicultural Trojan horses for the ongoing expansion of the university's development agenda. But I also recognize that institutions are people as much as they are structures for people to pass through. And incorporations of the concepts of the 'right to the city' could facilitate some kind of small-scale but meaningful reparations to neighboring communities.

I am curious about the role conflict can play in placemaking. For instance, when we started this dialogue in early June, an article came out on a local website about neighborhood residents opposing a new pop-up beer garden in the Point Breeze neighborhood of Philadelphia. The Center City – adjacent neighborhood is historically African American and is experiencing a high rate of displacement. When the developer who is backing this business was defending his plan, he explained that 'the beer garden is really just the backdrop… We have food trucks, we have a farmstand, we have a CSA (Community Supported Agriculture) drop-off, flea markets, yoga classes' (Henninger, 2015). Sounds like a laundry list of vibrancy. What do you think is the endgame for this kind of sharing economy, a locavore placemaking-driven development?

CF: Asking about endgames is important. I think we need to not only be clear about what the endgames are, but also about for whom they're conceived. Point Breeze's story is particular, but there's something very familiar here for those who think about the transformation of American cities over the past 30 or so years.

Point Breeze is in South Philadelphia. Before it was an African American neighborhood it was an area where Irish, Italian, and Jewish immigrants and their children lived. Many of them, like my grandmother, left South Philadelphia for the suburbs as soon as they got the chance. Now some of their grandchildren have become interested in such spaces. So, the endgame for one group of people – financially stable, young white people – might be getting the chance to live in dense urban spaces in which they feel rooted or grounded. What is this group so hungry for that makes the idea of locality, of local food boxes, or whatever else is on that laundry list, so appealing? Answering that question would force us into a serious examination of suburban America and the types of social collectivities it promoted. This examination would have to go beyond knee-jerk critiques of 'placeless' suburbs or naive gentrifiers. What is the endgame for current African American residents of Point Breeze or similar neighborhoods in Chicago, Detroit, or Cleveland? They have been waiting for decades for substantial reinvestment. Now that it's finally arrived, it's come as opportunities to consume craft beer and similar products. Asking about the endgame of placemaking for these residents forces a serious reckoning with the unevenness of how goods and services are distributed in urban space, and how that distribution extends long-standing racial and economic inequalities.

What you have in this case are two very different visions of a healthy place and what it takes to sustain it. The first vision seems to suggest that what's necessary is energetic young folks who gravitate toward similar self-improvement projects. So, let's gather together with our yoga mats, or make sure that we, as neighbors, buy from local farmers. Such projects do have a collective ethos, but it seems to me that this ethos is underwritten by the experience of individual or shared consumption. The other vision does not rely on energetic young consumers, but on general resources distributed in ways that would sustain a much broader group of people. What do long-term Point Breeze residents want? The article tells us that they hoped for investment in a recreation center or a library. These institutions are especially important for the care of very young and old people. These groups may not have any substantial spending power, but long-term Point Breeze residents see their care as important. What's more, making such institutions viable usually requires resources that exceed those available in a sharing economy, no matter how chock-full of goodwill it may be.

And that's what is most interesting to me about these calls for 'placemaking' – they do articulate a vision of collective life and obligation. Yet this vision seems somewhat narrow, safe, and conflict averse. Energetic young people or creative, entrepreneurial types passing around local vegetables or selling each other craft beer might activate one particular vision of a healthy place. Yet it is still unclear to me how such gestures will guarantee that all of Philly's, or for that matter Chicago's, rec centers, libraries, schools, and parks are open, well-maintained, well-staffed, and safe places in which a range of people can play, learn, and spend time.

Displacemaking (2020)

In this second interview, conducted in late April 2020, five years after the original specifically for this volume we revisit questions about placemaking practices' relationship to equity and inclusion while also considering the standardization of these practices and the status of 'place' at a moment of pandemic.

Catherine Fennell (CF): In recent years I've noticed some standardized features of 'placemaking' practices, at least those that unfold within smaller-scale urban community economic development efforts. So, in addition to moves to bring a recognizable 'place brand' or 'place identity' to a particular area, I've noticed similarity across social programming undertaken under the banner of 'placemaking': small stages or covered shelters for social meets ups and theatricals, community game or play nights, or the incorporation of 'storytelling' events and tours. The latter are especially interesting to me because they evoke standard methods that a professional ethnographer might employ to investigate 'the lived experience' of our interlocutors – interviews, oral histories, guided walks, etc. Has there been a similar move to draw artists or at least recognizable aesthetic practices into placemaking projects? What do you make of them?

Daniel Tucker (DT): I've seen the same patterns. Professionalization is certainly one culprit. Placemaking has become a cottage industry: consultants fulfill every aspect of placemaking projects, from project management and community engagement to the more open-ended role of artists, who sometimes teeter between those employed for their critical self-expression and those who are effectively designers fulfilling the demands of a client. Often there are filmmakers and even social scientists brought on to document the project. Within a low-budget lab atmosphere, these roles can become hybridized. Certain tried-and-true tropes emerge, especially if you have people either working on multiple projects in multiple cities using a similar or identical set of techniques, or because the field has gradually been codified through funders, or educators like myself teaching these techniques. Most art schools would not have offered courses of this nature a decade ago.

As placemaking has become professionalized, the cost of making an 'engaging' new public space has intensified. It can be cost-prohibitive to design a space that will last. That task might actually be better suited to landscape architects who are prepared for that challenge but also charge appropriately. An artist hired instead might fall within the budget but not build something lasting for the community to take pride in. How many small grant-funded parklets in disrepair does it take before new lines get drawn? And if the line says that pop-ups are all that can be funded, then programming has to be front-and-center. So yeah, sometimes standard tools develop because they look good in documentation, sometimes because they're cheap, and sometimes because they work.

A more critical discourse about placemaking has developed that argues for placekeeping. Thanks largely to the work of writer and arts administrator Roberto Bedoya (Schwartzman, 2017), practitioners have also begun to interrogate what constitutes 'the work' of place. In the ethical minefield of these practices, engaging people around 'stories' or 'lived experience' allows outside practitioners to effectively de-center themselves. People leading this work in communities where they have deeper histories are then able to mobilize their sensitivity to the relationships they've cultivated by doing work that shares hidden histories that would not be immediately apparent from less attuned perspectives.

I'd love to hear more about the ways you see storytelling functioning in relationship to placemaking and 'doing ethnography.' How do you understand the appeal of storytelling within placemaking efforts? Do you see a connection to the equity issues alluded to in the tension between placekeeping vs. placemaking?

CF: I understand why 'placemaking' programming might foreground storytelling or oral history collection. It's a way to incorporate local or resident perspectives into broader development processes that have a history of leaving such perspectives behind. These practices are something of a corrective, as part of a larger turn within planning or development circles to become more sensitive to the thoughts, needs, and wants of long-standing community members. And they can imbue a place with the feel of particularity, of texture, of authenticity.

The question I have at a moment of more formal or even professionalized approaches to placemaking would be the terms on which facilitators would invite and then include these perspectives. When I've observed these sessions, it quickly becomes apparent that some kinds of stories, histories, or perspectives are valued over others. So, for example, take a point often valorized in placemaking discourses, the conviction that robust public social life characterized by a density of diverse persons and the possibility of chance meetings among friends, acquaintances, and strangers is a good. There's a world of difference in asking someone to reminisce or tell a story about general practices of social togetherness, i.e. the things that you and your family or neighbors do or used to do together here, and asking them to reflect upon why some forms of social togetherness are valorized over others, or how their neighborhood became diagnosed with needing more 'place' or 'togetherness' in the first place.

As much as chance meetings or storytelling sessions unfolding in places like a well-designed parklet might be exciting for some, they are, depending on histories of segregation and criminalization, unpleasant or downright dangerous for others. I have seen placemaking approaches acknowledge or welcome histories or stories that are unexpected or even difficult, for instance, narratives about racial discrimination, civil unrest, and the like. But my question is still toward what end? How are such histories and stories framed, edited, archived? What 'takeaways' emerge in the process of crafting and consuming them, and for whom?

Here the distinction that Roberto Bedoya and Jenny Lee draw between creative 'placemaking' and creative 'placekeeping' might be helpful. Let's define 'place' as feelings of social cohesiveness anchored by the material and aesthetic dimensions of a particular geographic location, dynamic feelings that take shape over long stretches of time. If you start with the assumption that 'place' is lacking and needs to be made, then the intervention begins already from the assumption of some kind of deficiency. And because 'place' is so often taken to be a reflection of its inhabitants, that deficiency gets mapped onto the very people associated with a particular location. Yet if you assume that 'place' already exists, then the task is somewhat different. It's less about moving to fill a hole or even 'activate' some nascent force that members of local communities have been unable to marshal. You are not asking community members to choose between or authorize several designs or plans. They are not consultants or even 'stakeholders' among a varied group of equally legitimate interests or perspectives. They do not embody texture, particularity, or history, or perform place for others, including the professional ethnographer or oral historian who might collect, assemble, or relay their perspectives and stories for some audience. You are rather starting from the assumption that they have long built and kept the conditions of local social cohesion and that this invaluable work not only needs to be recognized and honored but centered and continued. If the fruits of this work as well as the conditions that fed it are respected, there will be a better chance that the people at the center of it will likewise be respected. They stand a stronger chance of being able to remain in place, and reap whatever benefits remaining there holds. The chance to remain opens up the possibility of equity. In the United States at least, the possibility of enlarged equity is especially important for communities that have faced systematic disinvestment and neglect from municipal and state governments. How have artists working explicitly in the vein of placemaking or placekeeping thought about equity?

DT: In the five years since we first talked on placemaking, there has been a shift precisely around the term 'equity' across the myriad of groups and voices involved in public art and community development. It can appear like a sea change, but that chance actually emerges from slow-burning relationships and conversations in groups like Art x Culture x Social Justice Network – so, the very groups steered by Lee and Bedoya and likeminded people. This change has unfolded alongside continued struggles and high-profile critiques about the politics of leadership within arts organizations and of course leadership's implications for who gets funded, hired, and exhibited (Sargent, 2018). Professionalized arts administrators, critics, and artists are all confronting the politics of representation and its material consequences. At the same time there is so much community activism taking on the material manifestations of symbolic capital in the city's landscape. A great example of this is the #blackbrunch campaign that disrupted over-hyped Sunday brunch spots with actions about police brutality and the racist policing that often accompanies gentrification (Romney, 2015).

You mentioned the valorization of social togetherness, and I cannot help but think about earlier precedents for this like the 'Third Place' discourse. Can you speak to how that earlier conversation might grapple with the question of equity? How is that discourse getting resuscitated during this pandemic, as some people seem to be mourning the loss of public space and social life, while also mourning the loss of life?

CF: My sense from following conversations within planning about 'place' over the past 20 years is that placemaking discourse's traction within planning but also civically minded art practices in the US at least must be situated within longer standing frustrations with development in late-twentieth-century America. The critiques of geographic isolation and anonymity that you see intensifying in the 1990s – so the very critiques out of which paeans to place and 'third place' emerge – pick up on earlier frustrations concerning modernist planning schemes. In the United States, these frustrations reach back at least to the 1950s.

Jane Jacobs's *The Life and Death of Great American Cities* (1992) is an obvious touchstone for a moment of skepticism about modernist development pitched to single-use zoning and automobility. Skeptics worried that such development flattened everyday social experience, that encroaching suburbanization or hyper-dense urbanism would standardize social exchanges or rob them of variety or spontaneity. For Jacobs, suburbs and hyper-dense housing projects represented the worst development outcomes because they demanded 'togetherness or nothing.' Either one embraced the stifling intimacy of a suburban life in which socializing occurred mostly in private, or one moved through anonymous housing blocs unable to connect with others. She argued that these flattened geographies sapped everyday life of character, spontaneity, and diversity even as they made inhabitants feel less connected, safe, and committed to common projects. Such concerns only grew more trenchant. Writing nearly 20 years after Jacobs, Joan Didion remarks that 'the freeway experience' is 'the only secular communion that Los Angeles has' (1979). In other words, its residents' surrender to a short-sighted, self-interested individualism is the only collective end imaginable in a city experienced as so many disconnected points that drivers shuttle to, mostly alone.

The 1980s and 1990s saw a robust critique of the post-war American built environment emerge, a critique that treated this environment as both a source and a symptom of American anomie and social corrosion. A vocal contingent of planners and urban designers begins calling for a 'return' to a 'golden age' of American cities and towns, an age that apparently prioritized broad interactions through authentic streetscapes lined with parks, storefronts, and civic buildings, mixed-income housing, and multiple-use zoning. It's at that very moment that interest in 'third places' explodes in urban planning and policy circles. These would be spaces beyond the two central nodes of contemporary life – 'work' and 'home.' So, places where a diversity of people gather with some regularity and informality, where one can expect to encounter and engage friends and acquaintances but also strangers in barbershops, neighborhood parks or libraries, cafes and the like. If some of this reminds you of the 'placemaking' conversation, it should. As much as placemakers might like to draw a line between what they do and the world of 'third places' there's an argument to look at them together. Regardless of whether their commercial coordinates are explicit, these are places thought to merge spontaneity and familiarity, to promote contingency alongside cohesion. And it's precisely these places that the closures of the COVID crisis have taken aim at. Have a look online and you'll see laments for lost 'third places' stacking up. But these laments miss several key points that bear on equity, especially at this moment.

First off, 'home,' 'work,' and 'third spaces' only seem to be discrete because of the ways our contemporary socioeconomic order organizes labor. Home has never been the opposite of work. It takes an enormous expenditure of time, money, and effort to create the appearance that home is a space of pure intimacy and private affections. And while the cafe, the pub, the library, or even the parklet constructed by placemakers might seem to be a respite from both 'work' and 'home,' that respite is conditioned by labor relations as well as normative understandings concerning appropriate practices of being present. Even in 'third spaces' that place a premium on accessibility, there are conditions on what counts as appropriate speech and behavior. Those who promote third spaces or placemaking rarely foreground those conditions and their exclusions. This is nothing new. As others have pointed out, it's really hard for Jacobs to entertain the idea that the 'sidewalk ballet' she watches outside her Greenwich Village apartment is an effect of race-based residential segregation. To her it just seems like a delightful and spontaneous mix of schoolchildren, Italian-American grocers, and bohemians. What's so interesting about the pandemic is its capacity to foreground the conditions that made 'the home' seem like a place of pure sentiment or 'the park' a place of public access or spontaneity. Just ask the New Yorkers who find themselves reckoning with the fact that socializing in parks without prescribed social distancing measures or gear could either, depending on who you are, get you a rebuke and a face mask from a patrolling police officer, or a summons to appear in court. And then there's the very real problem that the folks who have no choice but to show up and facilitate the spontaneity of the 'third place' but also prop up the fantasy of a sequestered, private home – baristas, delivery people, and the warehouse workers who package goods, even police officers – are the most likely to contract the COVID virus. That discrepancy puts the conditions of place, its making, and its spontaneity in very stark relief.

But the tacit and implicit conditions of place, and placemaking enterprises, are not a new problem. We talked about it five years ago, and I know it's been a persistent theme within placemaking circles. Do you think placemaking projects are capable of addressing it? Or rather, what would it take for them to address it seriously?

DT: I keep thinking about a project in my neighborhood in Philadelphia called 'Lola38.' Its tagline is: 'A Creative Placemaking Project.' That tagline decorates a sign on the odd triangular-shaped corner lot in between three major streets that is the site of a former bank, and now owned by a People's Emergency Center (PEC) a Community Development Corporation (CDC) founded in 1972. I walk past 'the bank,' as 'Lola38' is commonly known, and in the last two months it has turned into a distribution hub for free food in the neighborhood. Every Monday a physically distanced line forms down the street and food is distributed from the building surrounded by brightly painted colors on the asphalt parking lot, where light strands hang. Until recently, any number of 'place' activations happened to engage neighborhood residents in arts and culture programming. But the site was initially a direct-service organization that then ventured into placemaking practice. As its placemaking programming has been canceled due to lockdown, the site has returned to its direct service origins. For the same reason it is an iconic site in the neighborhood appropriate for arts programming, marking a gateway between university development and neighborhoods, it is also an accessible location well-suited to organizing and distributing aid. But I love that the signage of the building suggests it is still 'a creative placemaking project' despite those activities now on hold. I love how it shows that placemaking projects could, depending on their precursors and prehistories, also be reconceived as rapid response or mutual aid.

Across town in North Philadelphia, the Village of Arts and Humanities, a recipient of numerous 'placemaking' grants has also pivoted toward direct service. The Village has also been involved in #FreeOurYouth and #freeoutmamas campaigns to release incarcerated youth and women during the pandemic. While the history of this organization could be read almost as an inverse to PEC – they are an arts organization that has gradually adapted to filling in the gaps often addressed by CDCs – they can be interpreted together as parallel case studies.

In both instances creative placemaking as a framework was used in a number of ways, opportunistically it was an opportunity to fund the kinds of activities the group was already engaged in under different names. Additionally, it was a way to assert a counter-narrative of what placemaking practice looked like when initiated by organizations with long histories of serving the communities of color that frequently experience the kind of public space policing and impending displacement that you mention above. This work and these sites have a totally different quality when paired with rapid response service and bail-out campaigns. Perhaps that quality offers a direction forward for this field where activism can be centered as an expression of values and potential for long-term trust building. Without it, the practices are at best an outlet for experimental park design and at worst, a kind of parasitic extension of the nonprofit industrial complex's tendency to trade project-based grant funding while also turning a blind eye to the inequities that unfold around the parklet.

References

Cather, W. (1920). *The Borzoi 1920. Being a Sort of Record of Five Years of Publishing*. New York: Alfred A. Knopf.

DIdion, J. (1979). *The White Album*. New York: Simon and Schuster.

Fennell, C. and Tucker, D. (2015). 'Displacemaking', in *The Avery Review*, no. 10 (October 2015) [online]. Available at: http://averyreview.com/issues/10/displacemaking (Accessed: 16 May 2020).

Harvey, D. (2013). *Rebel Cities from the Right to the City to the Urban Revolution*. London:Verso.

Henninger, D. (2015). 'Why point breeze residents are fighting a pop-up beer garden', in *Billypenn.com*, 5 June 2015 [online]. Available at: http://billypenn.com/2015/06/05/why-point-breeze-residents-are-fighting-a-pop-up-beer-garden/ (Accessed: 1 September 2015).

Jacobs, J. (19611992). *The Death and Life of Great American Cities* (orig. publ. 1961). New York: Vintage Books.

Lefebvre, H. (1968). *Le Droit à la ville*. Paris: Anthropos.

Lynch, K. (1996). *The Image of the City*. Cambridge: MIT Press.

Markusen, A. and Gadwa, A. (2010). *Creative Placemaking*. National Endowment for the Arts [online]. Available at: https://www.arts.gov/sites/default/files/CreativePlacemaking-Paper.pdf (Accessed: 1 September 2015).

Romney, L. (2015). '#BlackBrunch brings peaceful protest to Oakland restaurants', *in Los Angeles Times*, January 4, 2015 [online]. Available at: https://www.latimes.com/local/california/la-me-black-brunch-20150105-story.html (Accessed: 16 May 2020).

Sargent, A. (2018). 'To fight racism within museums, they need to stop acting like they're neutral', in *Vice.com*, 21 May 2018 [online]. Available at: https://www.vice.com/en_us/article/pavpkn/to-fight-racism-within-museums-they-need-to-stop-acting-like-theyre-neutral. [Accessed: 16 May 2020].

Schwartzman, A. (2017). *The Cultural Placekeeping Guide: How to Create a Network for Local Emergency Action*. CERF+ in collaboration with South Arts for the National Coalition for Arts' Preparedness and Emergency Response [online]. Available at: https://www.americansforthearts.org/sites/default/files/pdf/2017/by_program/networks_councils/local_arts_network/Cultural-Placekeeping-Guide.pdf (Accessed: 16 May 2020).

Silberberg, S. (2013). *Places in the Making*. MIT [online]. Available at: https://issuu.com/mit-dusp/docs/mit-dusp-places-in-the-making (Accessed: 1 September 2015).

Further reading in this volume

Preface: the problem with placemaking
Louise Platt
Chapter 12: Public transformation: affect and mobility in rural America
Lyndsey Ogle
Chapter 14: Experts in their own tomorrows: placemaking for participatory climate futures
Paul Graham Raven
Chapter 15: Un/safety as placemaking: disabled people's socio-spatial negotiation of fear of violent crime
Claire Edwards
Chapter 16: More than a mural: participatory placemaking on Gija Country
Samantha Edwards-Vandenhoek
Chapter 17: 'I am not a satnav': Affective placemaking and conflict in 'the ginnel that roared'
Morag Rose
Chapter 18: '*Homomonument* sounds like a poem': queer placemaking 30 years on: a conversational dialogue with thijs bartels, author of *Dancing on the Homomonument* (2003)
Martin Zebracki
Chapter 19: Placemaking in the ecology of the human habitat
Graham Marshall

Chapter 21: Placemaking through parkour and *Art du Déplacement* (ADD) as a Singaporean applied performance practitioner in London
Adelina Ong
Preface: The only thing constant is change
Kylie Legge
Chapter 34: Reconnecting *cité* and *ville*
Philip Graus
Chapter 36: Facilitator skills for effective collaborative placemaking
Husam AlWaer and Ian Cooper
Chapter 37: *The Neighbourhood Project*: a case study on community-led placemaking by CoDesign Studio
Lucinda Hartley, Eliza Charley, Sama Choudhury, and Harriet McKindlay
Chapter 41: Rituals of regard: on festivals, folks, and findings of social impact
Maribel Alvarez
Chapter 42: Creative Placemaking and Placekeeping evaluation challenges from the practitioner perspective: an interview with Roy Chan
Maria Rosario Jackson
Chapter 45: How the city speaks to us and how we speak back: rewriting the relationship between people and place
Rosanna Vitiello and Marcus Willcocks

21

PLACEMAKING THROUGH PARKOUR AND *ART DU DÉPLACEMENT* (ADD) AS A SINGAPOREAN APPLIED PERFORMANCE PRACTITIONER IN LONDON

Adelina Ong

At this time of writing in early 2020, COVID-19 has prompted violent expressions of racism towards people who look Chinese in the UK. Amanda Rogers' articulate critique of Royal Shakespeare Company's *The Orphan of Zhao* (2014) and Bryony Lavery's *More Light* (2009) notes that the use of yellowface in theatrical representations of Asians 'support[s] the idea that all Asians look alike' (2014, p. 457). As a Singaporean Peranakan, I am reminded that the celebration of diversity will not vaccinate against the tendency to use cultural differences against East Asians in the UK. Perhaps, in telling themselves that COVID-19 is a virus associated with the unhygienic or exotic food practices of the Chinese, people convince themselves that they can reduce the risk of being infected by alienating those who appear ethnically Chinese. The exoticisation of cultural difference will inevitably fall short of intercultural understanding. On 9 February 2020, a Thai tax consultant, Pawat Silawattakun, was called a 'coronavirus' whilst being filmed by two teenagers (Silawattakun, 2020). He was then robbed and punched in the nose (ibid.). It was late afternoon in West London, and there were bystanders, but no one tried to intervene (ibid.). On 24 February, Jonathan Mok, a Singaporean student studying in the UK was assaulted by five teenagers on Oxford Street (Lau, 2020). One said 'I don't want your coronavirus in my country' during the attack which left Mok with fractures on his face that now require surgery (ibid.). These recent incidents have prompted me to reflect on placemaking through parkour and *Art du Déplacement*-inspired (*ADD*-inspired) applied performances, without feeling settled, in London, despite having lived here for seven years. In this chapter, I will extend Sara Ahmed's ideas of the 'melancholic migrant' (2010, p. 142), suggesting that the lively nomad might use placemaking as a way of resisting assimilation and being transformed into a zombie migrant. Drawing from Roisin O' Gorman and Margaret Werry's (2012) observations on failure, I will reflect on placemaking that prompts reconsideration of the failure encouraged in creative learning environments. Finally, I reflect on Nishitani Keiji's (1982) understanding of death as an invitation to contemplate an understanding of living that recognises our interconnectedness to the more-than-human in place.

Troubling the narratives of place

The parkour- and *ADD*-inspired place practices that I have developed extend Sally Mackey's place practices, which 'can trouble the meanings of place, destabilising suppositions of locality, dwelling, inhabitation, territory, indigeneity, community, residence, belonging, connection and ownership' (2016, p. 107). It is this troubling of existing meanings of, and relationships with, place that opens up the possibility of initiating alternative relationships with people in place and, therefore, a counternarrative of place. In thinking about place, I have been inspired by geographer Doreen Massey's definition of space and place where 'space is rather a simultaneity of stories-so-far' and 'places are collections of those stories' (2005, p. 130). These stories describe a relationship with a physical site, as well as relationships between the human and the more-than-human in this site. My methodological approach draws inspiration from theatre educator Belarie Zatzman's use of narrative inquiry as a form of 'countermemory' that creates 'anti-redemptory, self-conscious memorial spaces constructed specifically to challenge and resist the certainty of monumental forms' (2006, p. 117). In resisting an approach to history that prioritises one narrative at the expense of others, Zatzman uses narrative as a way of adding complexity to history through personal memories that enrich one's understanding of a time that is past.

As a visiting applied performance lecturer based in London, I often use parkour- and *Art du Déplacement*-inspired place practices as part of a class I teach about place and placemaking. *Art du Déplacement* (the art of displacement, hereafter referred to as ADD) closely resembles parkour. Both parkour and ADD are ways of moving over, on, through, or around obstacles (walls, fences, railings, buildings) and surfaces in the city based on movements developed from an obstacle training course (Chow, 2010, p. 148). Parkour and ADD have slightly different movement philosophies although they have shared origins. Parkour might place more emphasis on moving across the city in a straight line, using the most efficient route possible (Angel, 2011, p. 222; Lisetz, 2014). ADD might explore how playful exploration of a familiar place can transform an overlooked urban feature, like a park bench, into a place of beauty (Piemontesi and Najjar, 2012).

My approach to applied performance practice is inspired by Mackey's 'place practices' (2014) where the creation of 'subversions' and 're-experiences' offer opportunities for the formation of new 'narratives' of, and relationships within, a familiar place (Mackey, 2017, p. 11). Extending Mackey's place practices, the parkour- and ADD-inspired applied performance practices I use create counternarratives of place that open up more nuanced understandings of the world, encouraging the development of more meaningful relationships that challenge existing narratives of place. As an applied performance practitioner who works mostly with young people from low-income families in Singapore, my participants often articulate their concerns in ways that teach me about the inequalities of place (Singapore). The conversations we have about place (Singapore) often extend beyond the performance and workshops. For 15 youths, these conversations have continued for almost 10–12 years now. Two of the youths have become my mentors, helping me understand the future that young people are fighting for in Singapore. Their personal hopes are inextricably linked to the future of place (Singapore.)

This practice of learning from my participants is something that I have carried into my pedagogic practice in London. There are resonances here with Courage's 'social practice placemaking' in terms of how 'urban co-creators' work together in 'a horizontal collaborative process with a deeper level of engagement' (2015, p. 2). We learn from each other, together. In facilitating parkour- and ADD-inspired place practices as part of the session on place and placemaking,

I have taken inspiration from Majid Rahnema's approach to participation (2010). In critiquing the manipulative way in which participation is used to legitimate foreign control over populations in developing societies, Rahnema suggests that instead of 'empowerment', this encounter between activists and participants of development programmes might be approached as an opportunity 'to live and to relate differently' (2010, pp. 135, 140). The notion of empowerment positions the recipient as powerless and misleadingly suggests that the activist's power is superior (p. 135). Instead of empowerment, Rahnema persuasively argues for the co-creation of new knowledge through a process of listening without ideological bias (p. 141). This means attending to needs and negotiating between the values of the people and the values that activists hold, even when these values conflict with what activists believe they need. The process of mutual learning must also be a process of respectful ideological negotiation. A rigorous debate does not need to be resolved with consensus.

The placemaking session that I facilitate usually begins with an invitation to explore moving over and under a chair using basic parkour or *ADD* moves. I demonstrated a vault over the chair and a quadrupedal crawl under the chair. I demonstrated an adapted lazy vault where I sat on the edge of the chair and swung my legs over. The quadrupedal crawl involved moving on all fours, where hands and feet keep to an imaginary line. I invited participants to be creative with what 'over' and 'under' mean. Holding the chair up with my legs, I suggested that this too, can be considered as movement 'under' the chair. After the participants explored moving over and under a chair, I invited them to use the chairs in the room to create a sculpture that represents their fear. In a recent session, I demonstrated this by putting two chairs next to one another, one upright and the other lying on its back with legs facing the participants, as if it had fallen backwards. I sat on the edge of the seat of the fallen chair and looked towards the upright chair with my head bowed. In these sessions, I have chosen to explain what my fear sculpture means in order to disrupt preconceived notions of the teacher as one who knows all and has no fear. I told the participants that they do not need to articulate what their fear sculpture means. I am only articulating my fears so that they might undertake a similar thought process in the creation of their own fear sculptures. The abstract chair sculpture is intended as a means of creating an ambiguity that enables participants to express their fears without feeling too exposed. The chair sculptures are abstract representations of fears that are specific to place. Then, participants were invited to position themselves in relation to the chair sculpture: first avoiding the fear, and then becoming acquainted with the fear.

The lively nomad

On 5 February 2020, before the COVID-19 lockdown was enforced in the UK, I facilitated a session with first-year applied theatre students in a university. One student (P1) kept adding chairs to her fear sculpture. P1 called this a 'new-age' sculpture that would never be finished. This constant movement seemed to suggest a state of perpetual unsettledness, but it is not unsettledness that P1 fears. In a follow-up interview, P1 explained that her fear is 'this constant flow of taking, taking, taking and producing a pile of the same experiences or rhythms' (P1, 2020). For P1, this rhythm is associated with an unthinking, repetitive mode of living in the city. She explains:

> I am not in control... I feel the city is quite mechanic [sic] and I can't really experience being connected to life itself. It becomes habitual and I can see that I have no power upon [sic] my habits or the rhythms I follow. But I start to identify with the rhythms. And I start to experience myself as the rhythms of the city.

P1 is aware of how she has started to take on some of these habitual rhythms of London, and how these habits have shaped her experience of London. P1 distinguishes 'being connected to life' from these 'rhythms of the city'. For P1, these habits hinder a more dynamic responsiveness to living. With each chair (or habit) that P1 added to the stack, her physical mobility became more regulated, more 'mechanical' in its predictability. P1's fear is related to how all the habits she is forming (and will form) in London solidifies a way of being that is not living. There are echoes of Lefebvre's 'arrythmia' here, where a disruption of associated rhythms in the body manifests as illness (2004 [1992], p. 68). P1 suggests that within the city, the body is disciplined to work and move at a heightened pace that she finds inimical to being alive. This mechanical busy-ness resonates with ideas of zombie capitalism where working, despite burnout, is socially validated as self-actualisation through work.

In present work conditions where 'psychological (and psychosocial) identification with the workplace and with the work' is required, Kelina Gotman notes that workers are compelled to achieve self-actualisation through the process of becoming 'resilient to degrading work conditions' (2019, pp. 123–24). Workers are expected to be grateful to employers for this opportunity to realise one's purpose in life through work, even as work conditions become evermore precarious, exploitative, and demanding of time beyond designated work hours. In reflecting on 1930s dance marathons where exhausted dancers kept moving (barely shuffling) to stay in the competition, Gotman notes that 'dance marathons theatricalise an economy that has become anaesthetic… [workers] hover in the zombie state that capital would begin more and more systematically to deploy: grinding us down, so that we believe that we like it – almost (2019, p. 127, p. 145). Workers are not coerced to continue working beyond exhaustion, yet, like the dancers in the dance marathon, we do. To stop would constitute some sort of moral failure where one chooses not to realise some inherent potential identified by one's employer. In recognising that the city compels not only a certain rhythm of movement but also, a certain way of thinking and creating art, P1 fears becoming a zombie.

Extending Sara Ahmed's ideas of the 'melancholic migrant' (2010, p. 142), I suggest that P1 recognises how migrants are expected to assimilate and resists her own transformation into a zombie migrant. For Ahmed, the melancholic migrant is one who refuses to forget the suffering associated with experiences of racism in the UK (ibid.). This migrant resists integration (ibid.). When I told P1 that her constant movement made me think about unsettledness, P1's face lit up as she said, 'Yes, I decided to become unsettled for this period of my life, so I came to the UK.' P1 moved to London, from Hungary, about two months ago. P1 recognises that being in London unsettles her in a way that provides creative stimulation, yet P1 has no intention to apply for settled status in the UK even though she is an EU citizen who could qualify for pre-settlement status. P1 has decided to pursue undergraduate qualifications in London, with an expressed intention of leaving at the end of her studies. She has chosen geographical mobility, believing that this mobility will invite challenging experiences that will be more conducive to learning. Unlike many EU citizens living in the UK, P1's fear is not unsettledness. In defiance of narratives that assume all migrants want settled status in the UK and narratives that associate happiness with settledness, P1 recognises how this narrative perpetuates a form of cruel optimism. She resists becoming a zombie migrant in favour of becoming, I suggest, a lively nomad. P1's decision to unsettle herself resonates with Maurya Wickstrom's description of nomadism as 'a multivalent strategic reference to a collective of practices… that defy the absorption of Travellers as neoliberal subjects' (2012, p. 136). These practices include a resistance to land and property ownership that restricts mobility, an appraisal of formal education as optional while emphasising alternative modes of knowing and the prioritisation of commitments to the extended family over work commitments (p. 139). Wickstrom argues that 'the capacity for indifference to private property and all that it means' implies that nomadism can offer 'a new politics that is based on rethinking everything' (p. 140). Extending Wickstrom's nomadism, P1 demonstrates an indifference to obtaining settled status in the UK and this, I suggest, opens up new ways of thinking about the politics of mobility.

Tim Cresswell notes that 'mobility is a resource that is differentially accessed' and argues that it is not so much the degree of mobility that is differentiated, but the experience of geographical mobility. The degree to which people are able to control the speed, route, comfort, and representation of their own mobility, and the mobility of others, is differentiated by gender, race, and financial resources (2010, pp. 162–66). Cresswell's *Towards a Politics of Mobility* (2010) observed that 'the globetrotter sits in plush velvet seats and chooses from extensive wine lists, while the hobo travels close to death on a wooden plank precariously balanced on the same carriage's axels' (ibid.). While the experience of mobility remains life-threatening for refugees and victims of human trafficking, the experience of comfortable mobility has become more accessible with budget travel options. Cresswell's nuanced appreciation of mobility challenges Terry Eagleton's assertion that 'the rich have mobility while the poor have locality' (2004, p. 22). It may be true that the rich will find it easier to settle in post-Brexit UK, but this may not matter to the lively nomad as settling in the UK is, arguably, not mobility.

In the creation of her fear sculpture, P1 was constantly mobile. I suggest that it is this mobility which materialises a counternarrative of place (London). In the UK, high population turnovers in neighbourhoods have been linked to the erosion of a sense of place attachment for people living in these neighbourhoods (Livingston et al., 2008, p. 56). Implicitly, Mark Livingston, Nick Bailey, and Ade Kearns suggest that people who choose geographical mobility do so at the expense of others' sense of place attachment as they disrupt familiar social networks and limit opportunities for people to establish trust within a place (p. 57). However, Harry Ferguson's account of how going for a drive can facilitate deeper communication between social worker and child suggests that in building a relationship through (auto)mobility, the car can become a significant place (2009, pp. 279, 282). I suggest that these are placemaking conversations that encourage a different relationship to the car. The car is no longer just a vehicle that facilitates commuting between places. It has become a mobile place and attachment to this place grows with each (auto)mobile conversation. Ole Jensen has positioned place as 'a mobility-defined spatio-temporal event that relates to the way we configure narratives of self and other' (2009, p. 147). Place, for Jensen, is not necessarily determined by the length of one's residence in a particular location. The mobility of the lively nomad need not be conceived as an obstacle to placemaking, therefore.

P1's constant mobility materialises a counternarrative of place where the mechanical rhythms of the city (London) temporarily unsettle P1's nomadic approach to living. Each chair that P1 acquires manifests a representation of sedentarism that anchors P1 to a fixed point in the room. Wickstrom defines sedentarism as 'a deeply formative attachment to place, to staying in one place, and to a... belief that being itself can only begin, cohere and persist through being in place' (2012, p. 134). Extending Wickstrom's sedentarism, P1's fear might therefore be interpreted as a fear of gradually conforming to the mechanical rhythms of living, learning, and working that perpetuate settledness in the UK. P1's decision to unsettle herself has prompted me to think more critically about narratives of settledness in the UK. Even though settledness temptingly promises a sense of belonging in place, the racism revealed by COVID-19 reminds me that I will not find belonging through assimilation here. Instead, I have chosen a placemaking mobility. Jenson has asserted that 'none of the poles within the sedentary-nomad polarisation can claim to understand the contemporary mobility phenomenon' (2009, p. 142). Extending Jensen's call to think beyond dualities in order to think critically about contemporary mobility, nomadism should not be understood as the opposite of sedentarism. In mobility, there is both nomadism and sedentarism. In mobility, one can still form deep attachments to places that have prompted reconsideration of the way one lives. Though not physically in place, these places become a part of who we are, manifest in how we relate to others. These places are carried within us, and the practices formed in these places are shared with others in the places that we travel to. This conversation about fear and the contemplation of unsettledness that emerged from a parkour-inspired applied performance practice will stay with me as I travel to other places. It is a placemaking conversation that is located in London, but I wonder, is London the place that remains with me? Or is there a trace of P1's Hungary?

Facing the consequences of failure

During this session, another participant (P2) decided to hang a chair from the metal frame of a stack of bleachers which could be rolled out for lectures. He then stood directly beneath the suspended chair. When I asked him about his fear sculpture, he explained that in addition to creating a sculpture of his own fear, he 'also wanted people to experience some bit of that fear… that's why [he] decided to hang a chair on a height… it's very reasonable that it might fall'. I told P2 that I was indeed worried that the chair might fall on him. I was relieved when he moved to the left of the chair and stared at it when prompted to acquaint himself with his fear.

Then, the participants were invited to create a movement path through the room, using the parkour- and ADD-inspired movements they had explored to respond to the fear sculptures created by others before returning to their own sculpture. They were invited to make one change to their fear sculpture before positioning themselves to create an image that represents a first step towards working through their fears. The movement paths were intended to encourage an exploration of how we might negotiate the fears of others in this shared place, even as we are working through our own. I joined the participants in navigating the sculptures of fear they had created. I placed my cheek on a cool, metal leg of one of P1's chairs. I vaulted over an arrangement that looked like an empty casket. Using quadrupedal crawling movements, I stood next to P2's chair and placed my ear onto the metal scaffold that was holding it up, listening. After moving through the fear sculptures of others using some quadrupedal crawling movements, P2 returned to his fear sculpture, took the chair down, and held on to it as if he was embracing it. Later, when I asked P2 to reflect on his response to the prompts, he said:

> No matter what you do, you're always negotiating with your fear. Sometimes you're winning over it, sometimes it's winning over you. It's always there, always on top of you. Sometimes you face it, and sometimes you ignore it. And it's always hanging there.

When I asked what his fear was, P2 said his fear, on that particular day, was 'not being aware of his fear'. When I invited elaboration, he said:

> It's so boring not to have any fear. I'm not talking about that one traumatic fear… but just day to day, not having a fear to prove your fear wrong. That gives you a different kind of confidence and boost, that I can prove my fear wrong. Fear is a tricky thing. You're not afraid of fear, you're afraid of the presumption [sic] of fear. And in that sense you are not afraid of falling. Actually, you like falling, but you're afraid of what comes after that.

This fear of not having a fear seems to suggest that P2 uses his fears as a prompt for learning. P2 believes that learning can only happen when one attempts something daunting which could have painful consequences. P2 recognises that even though one may say that one fears the task, the fear is really of the consequences of failing, not the task itself. This appreciation of the consequences of failing resonates with current observations within schools where a fear of failure impedes learning (Lahey, 2015).

Róisín O'Gorman and Margaret Werry note that 'we live in the depressive ruins of the university, an entity dedicated to the rabid pursuit of illusory success when any substantive mission that might give that success substance has long since been mortgaged to market values' (2012, p. 3). Students know that failure is a necessary part of learning. They want to learn by failure, but they have also learned that academic failure has a significant impact on their employability. While the student might be persuaded against building a narrative of self-worth based on their present academic performance, the consequences of academic failure on graduate employability have become prohibitive. In this environment, O'Gorman and Werry observe that even the 'strategic, emancipatory or experimental use of failure – however much it is still necessary – is freighted with risk, danger and difficulty' (ibid.). In this context, P2's placemaking practice offers a counternarrative of place that demands the reconsideration of the romanticisation of failure in creative learning environments. Is the encouragement of failure in creative learning arguably made from a place of privilege which is insensitive to the precarity of graduates?

Then, I invited the participants to write down one thing they had learned, about themselves or their fears on a small slip of paper without signing off. These reflections were then collected and shuffled before being redistributed to all the participants. This process seems to encourage openness as the writer's identity remains anonymous. The responses read out by the participants ranged from 'seeing others interact with your fear and not being scared makes me feel less scared of it' to 'when I face my fears, I don't defeat them, I join them.' One participant noted that 'there was a feeling of coexisting and letting it fall away like a leaf (decoupling).' Another wrote, 'My fear is inevitable so I don't know why it still scares me.' I asked P2 what he had written when invited to reflect on what he had learned about his fears. P2 said he had drawn a face and wrote, underneath it, 'FACE.'

Liveliness is a conversation about death

There were quite a few chair sculptures that suggested a fear of death. One participant had arranged the chairs around her such that they evoked a casket. She was lying down in the middle of the chairs with her hands folded over her chest. I had initially wondered if the sculpture P2 created also represented a fear of death. When I shared this with P2, he said: 'there was a period in my life where I was very aware of death, but you cannot fear it. That's what I learned in a very shocking way which very disturbed me, but that's part of life, to die.' I invited P2 to elaborate on his present perspective on death. He spoke of the loss of a close friend in India to suicide. He said:

> When I was looking at my friend, lifeless… I just remembered that yesterday night at 3 am we were laughing and talking about all the best movies. I just [clicks his fingers] "This could happen to me." Which then, for a very long time, made me very suicidal and depressed… I was so much afraid of dying that I stopped living… nothing in life is certain but dying. And that changed the way I look at death. I think, "Okay, I'm going to die. Doesn't matter when, but we are not going to live 500 years. So until I am done, I have to promise myself that I live. And that will make my time worthful" [sic]. Otherwise I am already dead.

P2's acceptance of death as a certainty of life, and subsequent acceptance of life as impermanent, resonates with Graham Parkes' ideas on living with conscious awareness of the impermanence of life (1999, p. 97.) In accepting that life is brief and that it is impossible to avoid death, one begins to live differently (ibid.). For Parkes, this involves 'renouncing the immortality of the soul and also the substantiality of the ego by seeing through the illusion of duration' (ibid.). When one accepts that there is no soul that endures, then one begins to re-evaluate the desire to leave legacies to remind future generations of our achievements. For P2, this appreciation of life as impermanent has renewed his commitment to living. He notes that:

> The beauty of death is that when it is going to happen, you will be not there. I won't be there... And being sad, being depressed, being fearful is a certificate that you are living, which is a good thing.

This acceptance of one's mortality seems to prompt a different relationship to fear. P2 seems to understand fear without being controlled by fear. In a sense that echoes Parkes when he noted that 'it is all over... but somehow – so far, at least – it is all back, too' (1999, p. 97). Fear has become part of P2's renewed commitment to living. I resonate with P2's reflections on death. Accepting my own mortality has enabled me to let go of the desire to establish legacies. In contemplating my own death, I have become more attuned to the impermanence of all living things. Nishitani Keiji observed that death should not be understood as an anticipated event that removes one from the world of the living (1982, p. 3). Rather, death is a part of life, and both death and life are coexistent in all living things (p. 93). Death is 'something that we bring into to the world with us at the moment we are born' (p. 3). Extending Nishitani's understanding of death and life as coexistent in all living things, I suggest that this understanding of living as death-in-life can be understood as another dimension of mobility. This brief passing through life is enriched by these conversations about death. Although these connections are momentary encounters that can never be repeated, even when we meet again, they are placemaking conversations that deepen our connection in the present. This contributes to the liveliness of living.

At the end of the workshop, I invited the participants to create chalk graffiti haikus reflecting on their hopes for the year ahead. A few participants wrote poems that expressed their concern about the ongoing climate crisis. One participant wrote:

> A red tree dying
> The ridges on your back rise
> Rain rustles the leaves

Another wrote:

> Black chairs stacked up high
> Heart hopes for a saved planet
> Rain in wild torrent

Considered in the light of these haikus, P2's reflections on death have led me to a slightly different contemplation of impermanence. While the realisation of our own mortality has offered humans the privilege of considering how we might realise the full potential of our limited life on Earth, the realisation of these ambitions often deprives the more-than-human of living out their limited life on Earth. Perhaps, contemplating the impermanence of all living things is the beginning of compassion. Jason Luger notes that 'when gentrification or failed urban redevelopment projects tear neighbourhoods and human souls apart, it is often art that remains to tell the story – representations of memories, of dreams, of hope' (2017, p. 230). These chalk graffiti haikus have been washed away by the rain, but the counternarratives of place that have emerged from our placemaking will remain. These chalk graffiti haikus capture the impermanence of all life. This is a contemplation of impermanence that has prompted further understanding of how humans hasten the death of the more-than-human as we live. To what extent does this human renewed commitment to living hasten the demise of the more-than-human? This understanding of death emphasises a more-than-human understanding of interconnectedness.

On the evening of 23 March 2020, a COVID-19 lockdown across the UK was announced. Many international students have returned to their home countries, but I have chosen to stay in London. I know this will not be recognised as solidarity. This fear of people who look Chinese will linger, long after the lockdown is lifted. This parkour- and ADD-inspired placemaking practice has troubled narratives of settledness and assimilation. This conversation about death has prompted me to explore how the contemplation of impermanence in all living beings might encourage a more compassionate placemaking. A placemaking that enables liveliness of all living beings.

References

Ahmed, S. (2010). *The Promise of Happiness*. Durham, NC: Duke University Press.

Angel, J. (2011). *Ciné Parkour: A Cinematic and Theoretical Contribution to the Understanding of the Practice of Parkour*. London: Julie Angel.

Chow, B.D.V. (2010). 'Parkour and the critique of ideology: Turn-vaulting the fortresses of the city', *Journal of Dance and Somatic Practices*, 2(2), pp. 143–154.

Courage, C. (2015). 'What are the arts in social practice placemaking?', in *Arts in Society*, University of Brighton [online]. Available at: https://www.academia.edu/14284466/What_are_the_arts_in_social_practice_placemaking (Accessed: 28 February 2020).

Cresswell, T. (2010). 'Towards a politics of mobility', *Environment and Planning D: Society and Space*, 28(1), pp. 17–31 [online]. Available at: https://doi.org/10.1068/d11407 (Accessed: 7 August 2014).

Eagleton, T. (2004). *After Theory*. London: Penguin Books.

Ferguson, H. (2009). 'Driven to care: The car, automobility and social work', *Mobilities*, 4(2), pp. 275–293.

Gotman, K. (2019). 'Anaesthesis: Dance marathons and the limits of sense', in Fisher, T. and Katsouraki, E. (eds.) *Beyond Failure: New Essays on the Cultural History of Failure in Theatre and Performance*. London; New York: Routledge; Taylor and Francis Group.

Jensen, O.B. (2009). 'Flows of meaning, cultures of movements: Urban mobility as meaningful everyday life practice', *Mobilities*, 4(1), pp. 139–158.

Lahey, J. (2015). 'When success leads to failure', in *The Atlantic*, 11 August 2015 [online]. Available at: http://www.theatlantic.com/education/archive/2015/08/when-success-leads-to-failure/400925/ (Accessed: 13 October 2016).

Lau, J. (2020). 'More shock than anger: S'porean student opens up about covid-19 racist attack in London', in *The Straits Times* [online]. Available at: https://www.straitstimes.com/singapore/singaporean-student-in-london-seeks-eyewitnesses-after-coronavirus-related-taunt-and (Accessed: 11 March 2020).

Lefebvre, H. (1992). *Rhythmanalysis: Space, Time, and Everyday life*. Translated by S. Elden and G. Moore. New York: Continuum.

Lisetz, A. (2014). 'David Belle: This is Parkour', in The Red Bulletin [online]. Available at: https://www.redbulletin.com/za/en/lifestyle/david-belle-this-is-parkour (Accessed: 23 December 2015).

Livingston, M., Bailey, N. and Kearns, A. (2008). *People's Attachment to Place: The Influence of Neighbourhood Deprivation*. Coventry, UK: Chartered Institute of Housing.

Luger, J. (2017) 'Conclusion: towards the worlding of art and the city?', in Luger, J. and Ren, J. (eds) *Art and the city: worlding the discussion through a critical artscape*. New York; Oxfordshire: Routledge, pp. 229–238.

Mackey, S. (2014). 'Outline of place practices', in Performing Places [online]. Available at: http://www.performingplaces.org/docs/outline.pdf (Accessed: 29 December 2014).

Mackey, S. (2016) 'Performing Location: Place and Applied Theatre', in Hughes, J. and Nicholson, H. (eds) *Critical Perspectives on Applied Theatre*. Cambridge, United Kingdom: Cambridge University Press, pp. 107–126. Available at: http://dx.doi.org/10.1017/CBO9781107587977 (Accessed: 6 May 2016).

Mackey, S. (2017). 'Performing local places: Working with councils to reach new communities and facilitate wellbeing in living environments', in *Performing Places*. London: Royal Central School of Speech and Drama, p. 44 [online]. Available at: http://performingplaces.org/docs/ppbrochure.pdf (Accessed: 5 May 2017).

Massey, D.B. (2005). *For Space*. London: SAGE.

Nishitani, K. (1982). *Religion and Nothingness*. Berkeley: University of California Press (Nanzan studies in religion and culture).

O'Gorman, R. and Werry, M. (2012). 'On failure (on pedagogy): Editorial introduction', in *Performance Research*, 17(1), pp. 1–8.

Parkes, G. (1999). 'Death and detachment: Montaigne, Zen, Heidegger and the Rest' in Malpas, J. and Solomon, R.C. (eds.) *Death and Philosophy*. London: Routledge.

Piemontesi, L. and Najjar, R. (2012). *Nothing is Something #1 The Bench* [online]. Available at: https://www.youtube.com/watch?v=uGT0Cpcz7kE (Accessed: 5 August 2015).

Rahnema, M. (2010). 'Participation', in Sachs, W. (ed.) *The Development Dictionary: A Guide to Knowledge as Power*. 2nd edn. London: Zed Books.

Rogers, A. (2014). 'Asian mutations: Yellowface from more light to the Royal Shakespeare Company's the Orphan of Zhao', in *Contemporary Theatre Review*, 24(4), pp. 452–466.

Royal National Theatre (2009) 'More Light'. London, UK, May.

Royal Shakespeare Company (2012) 'The Orphan of Zhao'. Swan Theatre, Stratford-upon Avon, UK, November. Available at: https://www.rsc.org.uk/the-orphan-of-zhao (Accessed: 25 September 2020).

Silawattakun, P. (2020). 'The first time in my life I've been described by the police as a "victim"', in *Facebook*, 9 February 2010 [online]. Available at: https://www.facebook.com/pawatsilawattakun/posts/10157740512042435 (Accessed: 17 February 2020).

Wickstrom, M. (2012). *Performance in the Blockades of Neoliberalism: Thinking the Political Anew*. New York: Palgrave Macmillan.

Zatzman, B. (2006). 'Narrative inquiry: Postcards from Northampton', in Ackroyd, J. (ed.) *Research Methodologies for Drama Education*. Stoke-on-Trent: Trentham Books.

Further reading in this volume

Chapter 5: Making places for survival: looking to a creative placemaking past for a guide to the future
Jeremy Liu

Chapter 9: From the dust of bad stars: disaster, resilience, and placemaking in Little Tokyo
Jonathan Jae-an Crisman

Chapter 10: From moon village to mural village: the consequences of creative placemaking in Ihwa-dong, Seoul
Jason F. Kovacs and Hayun Park

Chapter 18: '*Homomonument* sounds like a poem': queer placemaking 30 years on: a conversational dialogue with Thijs Bartels, author of *Dancing on the Homomonument* (2003)
Martin Zebracki

Chapter 42: Creative Placemaking and Placekeeping evaluation challenges from the practitioner perspective: an interview with Roy Chan
Maria Rosario Jackson

22

EMBEDDED ARTIST PROJECT

Epistemic Disobedience + Place

Frances Whitehead

Introduction

From 2006 to 2016, I undertook a series of epistemologically driven practice experiments under the aegis of the *Embedded Artist Project*, a concept I have engaged in urban design projects of varying scales in cities in the USA and South America. Embedded Artist projects aim to explore the role of culture in sustainability, and demonstrate how social, cultural, environmental, and economic values can inform a 'net benefits' model of development, generating a series of linked civic initiatives that model a whole-systems, place-based approach. As a geo-specific approach, the Embedded Artist strategy is anchored in bioregional thinking and localism, driven by a place-based ethos that aligns with and broadens the social, cultural, and quality-of-life aims of placemaking. Its epistemology questions what Mignolo (2010) terms 'imperial knowledge' and its separation of nature and culture moves this practice to 'epistemic disobedience' (ibid. p. 15), engaging with indigenous perspectives on the land such as the concept of 'place-thought' (Watts, 2013) and the rejoining (from a Western perspective) of epistemology and ontology, thinking and being. The first section of this chapter takes up the origin and implementation of the *Embedded Artist Project* platform chronologically, followed by a section on global project examples, and lessons learned. The third section reflects on its implications on current cultural discourse, artistic practice, and considerations of place and placemaking.

The Embedded Artist: Double Agent

The *Embedded Artist Project* began as a program with the City of Chicago, Department of Innovation, running 2008 to 2012. Artists were 'embedded' in municipal workgroups to bring new perspectives to the daily work of the city. Key to its formation was the insight that the intellectual and creative 'free agency' of artists is key to their ability to contribute to 'possibility'; and that artists' research and working methods can and must be allowed to operate within and alongside the highly structured multidisciplinary and consultative processes typical in public planning. The program was catalyzed by a 'knowledge claim' document entitled *What do Artists Know?* (Whitehead, 2006) which has proven useful as both method and message to elucidate the kinds of (tacit) skills artists deploy with place-based practices.

Sustainability + Agency

The *Embedded Artist Project* operates from the speculative proposition that un-sustainability is at core a cultural problem, emanating from systemic disconnects of Western epistemological specialization and habitus (Fry, 2007, p. 5, from Bourdieu). The aim of the *Embedded Artist Project* is to test the 'cultural hypothesis' that artists can contribute to a more sustainable world by joining the work of multidisciplinary teams and (re)integrating cultural perspectives into the formulation of civic projects. It asks: can art/artists contribute to a culturally informed, and reciprocal, trans-disciplinary method as other disciplines are challenging themselves to do?

OPTing IN: The Diplomacy of Art

In the Embedded Artist transdisciplinary framework there is no focus on artistic autonomy, nor do we operate solely within the symbolic economy of art practice. Long-held conventions around authorship are called into question, alongside ideas about art's usefulness/uselessness and purpose/purposelessness, renegotiating the symbolic and the practical. As Janeil Englestad (n.d.) frames it, to 'make art with purpose'; as Tania Bruguera (n.d.) frames it, '*arte útil*' (useful art). Focused on reciprocity and structured around shared interests, ethics, and goals, we believe that one's voice is amplified not diminished. The artist is embedded to be of service, and thus is content to defer, at least temporarily, the question of 'art' and any limits on possibility. The framework privileges integration, multi-valency, and the creation of new working models, not the maintaining of borders or old modalities. Conventional activist art strategies are therefore extended by this 'opting *in*.' Through this engagement we have learned to speak the languages of other disciplines, both nomenclature and attitude, reflecting multiple intents and values. Cultural geographer Ingram (2012) has called this the 'diplomacy of art,' a symbolic handshake, reaching outside art practice towards the work of others, to become value-added. This diplomacy sometimes disrupts these practices by operating within their sphere differently. Some would claim this as an act of 'generosity' (Purves, 2004), a joining in, dot connecting.

Embedding the Embedded Artist

The structure of how the artist enters the government setting will vary, and will reflect assumptions about 'art,' 'art-making,' and contested ideas about the role of the artist in society and 'socially engaged art.' Ideas differ concerning what can be achieved by having an artist in the government. Simply put: is the artist there to make 'art? Are they there to make 'change'?

We might begin by recognizing that an 'embedded' or 'placed' artist differs from other kinds of city engagement strategies such as the Artist-in-Residence model, and the City Artist or Town Artist. In the Artist-in-Residence model, an artist might primarily reflect on the milieu around them but remain outside the principal tasks of the city workgroup, and instead maintain artistic autonomy to create artworks from, with, and about the city systems. Some cities are framing their artist engagements as City or Town Artist, which in some cases blends the two strategies, or perhaps leaves it to the artist to navigate and experiment. Each type of engagement has historical precedents, for example, Artist Placement Group (Tate, n.d.); David Harding (n.d) as Glenrothes Town Artist; Merle Ukeles as NYC Sanitation Department Artist-in-Residence (Kennedy, 2016) and each negotiates and models ideas about artistic integration and/or autonomy and reflects different theories of change.

In contrast, the Embedded Artist is a new kind of problem-solver, or sometimes problem-finder. Problem articulation and problem definition are key challenges in sustainability planning and are areas where artists can be adept due to their criticality and lateral thinking. At their most basic, Embedded Artists seek to take a seat at the collective table; to work upstream from the domain where most art projects operate and aim to impact the everyday work of the government and policy, even while 'art' may (or may not) be made. Importantly, for Embedded Artists working on complex city projects, we have found that without the support and buy-in at the Commissioner level, it is hard for city staff to prioritize these collaborations, and less can happen.

It is crucial to recognize which city departments and individuals are receptive to engagements with artists, and why. Reaching into established networks can help identify imaginative partners willing to embrace these experiments, or whose departments face such challenges, and with such urgency, that they are open to new ideas and unorthodox methods. Oftentimes individuals may be locked into older modalities and highly prescriptive art genre definitions. Those charged with civic innovation or those who face intractable social problems and are hungry for new ideas are often the best prospects. In addition to receptivity, matching the expertise and interests of the artist to the 'placement' proves important to advance the mutual steep learning curves and allow for meaningful work. With its reciprocal framework, city workers and artists learn from each other, and duration is important to this process: a rule of thumb indicates that all placements should be for a minimum of two years, longer for big projects.

BOTH/AND Art: Double Agency

A corollary to the Embedded Artist is the concept of the 'Embedded Artwork,' manifesting the possibility that something can be understood *both* as art *and* as something else (remediation, community development, education, etc.) In one part of the artist's multivalent role, they seek free agency; in another, it is understood that some of the transgressive and subversive strategies of artists have to remain unspoken or in some cases, be suspended in order to address urgency and cooperation. We recognize that our role is also intentionally disruptive, that we are present as change agents, for 'redirective practice' (Fry, 2007), or as Sacha Kagan (2007) would say, to 'play on the rules rather than in the rules' or 'entrepreneurship in conventions.' The melding of cultural logics and figurative tropes into the multidisciplinary team model of civic projects produces what we have called, elsewhere, the 'tropological transdisciplinary' (Whitehead, 2015).

But what of the rules of art? In what ways does Embedded Artist also redirect conventional art practice? These *both/and* projects, which form the core of this civic art practice, are not always legible to art worlds as 'art,' and the status of the projects are often contested. Here the strategic knowledge, or metis, of the artist turns on art itself. Using the double agency of this practice to redirect the 'cultural quo,' the Embedded Artist shuttles between worlds like a cross pollinator, border-hopping, changing both sides in equal measure. Beyond the 'free agency' of arts thinking, beyond redirective practice, disruption, and change agency, the Embedded Artist is at core a double agent, working inside and outside conventions, inside and outside worlds, a *double change agent.*

Civic Experiments: Projects Undertaken as Embedded Artist

> When culture is understood to denote the social production and transmission of values, meanings and purpose... and when it is recognized that the expression of social goals and aspirations is at the heart of the public planning process... the connection between culture and planning becomes clear.
>
> *(Hawkes, 2001, p. 1)*

The following consecutive and generative *Embedded Artist Projects* explore the cultural dimension of sustainability and model new cultural strategies for creating change. While there are explicit aims and outcomes in each, additional unforeseen outcomes also arrived, which carry on to the next project.

Slow Cleanup, Chicago

Working as Embedded Artist with the Chicago Department of Environment (2009–2012), we developed 'Slow Cleanup,' a net benefits model for Chicago's 400-plus abandoned gasoline stations. We deployed Hawkes' Four Pillar (2001) social, cultural, economic, environmental model, and that of the Slow Food movement, as a schema for a new approach to brownfields redevelopment. Paired with Chicago's top brownfields expert, by a visionary Commissioner, we focused on phyto- or plant-based remediation to reimagine and repurpose the landscape, revaluing these degraded properties. The program was constructed as a series of *interim* approaches that model time in relation to investments, benefits, and complexity. We also evolved an *in situ* soil prep method for keeping all soils on site, repurposing a road-building tool. Students from four communities of practice: art, soil science, horticulture, STEM learners – have been involved in the project. The purely disciplinary ambition to critically extend the sculptural genre of the 'earthwork' or to engage wholesale in 'social practice' or 'placemaking' remained unspoken. 'Slow Cleanup' was not artist-led, but multivalent and deeply collaborative. I believe that many city workers learned things from working with artists, but it is unclear how they understand these insights. Unfortunately, there was no formal assessment of the program. On our end, we learned that there are many constraints that dampen the energies of the even most creative staffers, and that there is a world of difference between career civil servants and elected officials and 'politicians.'

Lessons Learned

- correct placement of artist's expertise is crucial
- high-level support is required
- plan on a two-year placement minimum, longer for some projects
- open-ended questions work better than prescribed outcome execution
- examine differences between 'practices' and 'policies'
- time is an underutilized asset in low development areas
- community involvement is key for a sustainable program
- volatility of political cycles hinders long-term thinking

Diaspore, Lima

/dī ə-spôr /: n botany, a seed or spore, plus any additional elements that assist dispersal.

We were introduced to the City of Lima, Peru in 2009 through Centro Internacional de la Papa (CIP), which holds over 4,000 varieties of Andean potatoes, the cultural heritage of 10,000 years of hybridization by indigenous growers, and which underpins this multifaceted project that operates as a political and poetic gesture, and also as a knowledge experiment. Out of conversation with the CIP Director, concerning the cultural nature of food, the Four Pillar model (ibid.), and how these ideas might enrich the new urban agriculture program underway in Lima, we assembled a team of Chicagoan artists, designers, and preservationists to provide creative support to the City of Lima in their efforts to integrate architectural conservation in the historic center and food security planning. A crumbling but magnificent UNESCO World Heritage Site built around 1500, the historic center of Lima now houses the urban poor, mirrored at the city edge by informal settlements of a growing population. These populations have inadequate nutrition and food security and are also threatened by diminishing glacier-fed water supplies, and a lack of public open space (most open space is private, in courtyards.) We became aware that all food security programs in Lima must be evaluated against the underlying pragmatic dilemma of sustaining a city that is in the wrong place – a perpetual colonial legacy – an unsustainable settlement pattern. Still we sought integrative solutions to enhance democratic participation, food security, and heritage conservation.

During our work in Lima, a hex pattern emerged as a motif for many of these investigations, moving from a metaphor for participation at City Hall, the Civic Hive, to a space-saving spatial configuration for roof gardens, to a motif for a mobile orchard, reversing the private courtyard and the spatial interiority (Plöger, 2007) of the city. While formal tropes such as the hex shape allowed us to navigate between the symbolic and the practical, everyone understood that urban agriculture in Lima was a short-term proposition, raising as many questions as it answered. Our work in Lima ended in 2012 when the Mayor supporting the initiative was recalled by more conservative political forces. However, this umbrella concept has gone on to inspire other food-based conversations, including with rural collaborators in the Irish countryside. In this way, the 'Diaspore' project is extended through reciprocity to all sites where food production and cultural production are intertwined.

Lessons learned

- artists/designers do not have to operate in a framework of scarcity
- artists and institutions can redistribute their strategic thinking capacity to unlikely partners
- solve more than one problem at a time
- involving students enlarges capacity but limits follow-through
- volatility of political cycle hinders implementation of ideas (again)

The 606, Chicago

From 2012 to 2016 I was the Lead Artist on the Design Team for a 3-mile-long rail adaptation project, 'The 606,' which opened to the public in June 2015. 'The 606' is a civic experiment in every way, a public–private partnership with great ambitions, and for me, an opportunity to actualize the ideas that we had been developing at a more speculative scale. The private partner, The Trust for Public Land, established public engagement as the ethos of the project. Working rhetorically with the values of participation and engagement, the arts became the organizational framework for the project, shifting the multidisciplinary team structure towards the more collaborative (but in this case, more contested) transdisciplinary model. Tacitly, we transformed the Four Pillars into a set of cultural values – expression, participation, innovation, and sustainability – shifting the focus from cultural heritage to cultural futures.

'The 606' has been called 'arguably the most ambitious experiment in placemaking ever undertaken' (Hart, 2014), and prominently features a climate monitoring artwork revealing how large bodies of water like Lake Michigan affect local temperature patterns. Based on a climatological study of the site, a planted line of 453 native, flowering trees, *Amelanchier × grandiflora* (Apple serviceberry) runs its full length, propositionally forming an embedded artwork, a landscape intervention, achievable only by proclaiming it 'art.' Modeled after the Japanese cherry blossom festival, the five-day bloom spread of this flowering line makes legible Chicago's famous

Lake Effect in Spring and Fall, and a living data visualization in time and space. As a form of speculative artistic activism, 'Environmental Sentinel' explores the potential of the cherry blossom festival to be replicated elsewhere. Is it a 'transferable model'? Will this work in Chicago with native plants? Can beauty be catalytic and educational? While a participatory observation program links academic and citizen scientists, most encounters will be informal, by regular trail users who engage this slow spectacle in other ways. This synthetic approach blends new participatory art practices, climatology, and the expressive potential of public infrastructure to create what we are calling 'pink infrastructure.'

Lessons learned

- public infrastructure projects require artists who can think at the scale of the city
- artists must be supported by the client to have any effect
- artists' scope of work and compensation must be adequate to keep up with the team
- art world aspirations linked to economic development may be at odds with community-driven arts programs and at odds with anti-gentrification sentiments
- there are multiple artworlds and multiple rulebooks
- artists can help navigate social issues with new social practice skills and this requires a level of trust and political courage on behalf of the commissioner who must accept risk
- due to the short-term political cycle, it is difficult to bring long term ideas to public projects

Disobedience Beyond Disruption: Linking Embedded Artist Project To Social Justice + Place

Since 2015, I have been involved in two contexts that both extend and challenge *Embedded Artist Project* strategies and reveal the complexities of thinking about place and placemaking in the Western and non-Western context. It is to this emerging discourse that this chapter now turns, focusing on two projects: Fruit Futures Initiative Gary (FFIG) – a civic fruit-growing initiative and community orchard undertaken with a largely African American community, in Gary, Indiana – and engagement with the Kei Uta Collective – artists, designers, and community members operating within the semi-rural agricultural 'hinterlands' (Kei Uta) of Kuku, Horowhenua, Aotearoa (New Zealand.) The latter is seeking to explore how the mātauranga Māori (Māori knowledge or episteme) might link to other knowledge systems, in order to envision climate adaptation strategies for this longstanding, Māori coastal community (Smith, 2019). These new situations have allowed and required me to see Embedded Artist in a broader way; a more geo-political proposition beyond working with municipalities and multidisciplinary team-based civic projects.

Embedded Artist UnEmbedded: Co-creativity

As success with the Embedded Artist model grew, so did the reality that this is a 'top-down' practice best suited to large-scale endeavors; in response to this, and to practice at a more intimate scale, we moved our studio to Gary, Indiana. Here, while in contact with city government (what remains of it), we are working most directly with a handful of long-term Gary residents, 'bottom-up,' replacing expertise with 'co-creativity.' Lisa Grocott (2005, p. 10) has written about the importance of transferring knowledge in a divergent and speculative manner rather than a convergent and directive manner. While this model may create uncertainty, it may also create 'possibility' and agency. Thus, in Gary, we asked a new set of questions regarding knowledge: could we transfer the knowledge already captured by previous Embedded Artist placements? And could we transfer the 'agency' held within that knowledge in order to build capacity in the community? This is the Embedded Artist *un*embedded, a new kind of transfer engagement, deployed through divergence and uncertainty; through 'deep hanging out' (Geertz, 1998).

FFIG also poses questions about time and scale, which impede our understanding of other 'natures,' ongoing but invisible and uncounted by the Western mind. Recognizing the larger bioregional ecology, FFIG cultures a pan-animistic worldview, offers non-anthroponormative regional futures, and alternative economic possibilities. FFIG and the liminal spaces of post-urban Gary are a monument to the failure of specialization, and embody critiques of Western 'rationality,' opening space for Afro-futurism, deep localism, poetry, participation, 'tactical magic,' and a 'pluriverse' of wonder.

Artist Embedded in Kaupapa Māori (or not)

This critique of Western rationality and the split between nature and culture, as part of the Western episteme, links directly to the dynamic, bi-cultural context emerging in Aotearoa/New Zealand which is actively indigenizing (de-colonizing) art, research, and the discussion about knowledge and place (Ndlovu-Gatsheni, 2017).

In early 2019, I was 'embedded' at the Tukorehe Marae, a traditional spiritual and community center, in a cross-cultural *wānanga*, an intensive forum and collaborative process based in *Kaupapa Māori*, a holistic Māori methodological approach to research. As part of the 'Deep South Science Challenge-Vision Mātauranga Programme,' the *wānanga* aim is to 'ground science in culture, and to communicate complex knowledge and data through art and design strategies' (Deep South Challenge, n.d.). Key features of this approach include the use of *hīkoi*, walking together on the land, as an embodied, kinesthetic form of learning that is simultaneously an act of political demonstration and solidarity. Daily *hui* meetings and active *korero* discussions complement the introduction to core, integrative, Māori concepts such as *whakapapa* – the genealogical linkage of people and their connections to all things. The *tiro ā-Māori ki tōna ake ao* or *Te Ao Māori*, the Maori worldview, sees knowledge as shared, passed down, ancestral, accumulative; not 'produced', not industrial or instrumental – it is a value proposition. Mignolo (2010, p. 15) refers to 'knowledge making' rather than 'knowledge production' to acknowledge the shared authorship and ancestral processes that inform indigenous 'knowing.'

Mātauranga Maori, the Māori knowledge model, is integrative, and like other indigenous perspectives, contrasts with the Western view that dis-integrates the world into disconnected disciplines. Here nature and culture are not separated, and nature is also not conceptualized as the place conveniently lacking humans, ready for Western colonization. This worldview does not embrace the Anthropocene concept as this concept reflects the underlying Western assumption which universalizes the 'human' as responsible agent in the climate crisis, when it is a product of Western thought and action (Todd, 2015). It is also crucial to recognize that the Maori indigenous worldview suffered under colonialism and is undergoing a process of reclamation and revitalization – a process that is simultaneously cultural, political, and epistemological. To be 'embedded' within this bi-cultural knowledge experiment is to reflect on these underlying epistemological differences regarding knowledge-making and its meanings, to wade into contested territorial and historical narratives, requiring the Embedded Artist to be an epistemic 'diplomat' as Stengers (2005) would say, to 'turn contradiction (either/or) … into a contrast (and, and),' similar to the 'both/and art' mentioned earlier. Can you be 'embedded' within an integrative worldview where all elements are already linked; where there is no inside/outside? The inter-epistemic and inter-cultural initiative underway in Kuku, Horowhenua is a site for exploring these questions.

Place-Thought + Epistemic Disobedience

> Place-Thought is the non-distinctive space where place and thought were never separated because they never could or can be separated.
>
> (*Watts, 2013, p. 21*)

The view that nature and culture are not separate connects the Māori perspective to land-use practices and policies in Gary, Indiana. Gary is dominated by steel production and, like all company towns, this unsustainable economy has faltered under globalization and automation, leaving behind population loss, environmental devastation, rampant suburbanization, and institutionalized racism.Vacant land, including fragments of native landscape, does not legally qualify as 'natural' in land use and tax code. Here 'nature' exists only in the largely white suburban counties that surround Gary.Through this inconsistent public policy, 'nature' and 'ecology' are not available to Gary's largely African American residents, an example of the widespread racial inequality in the area. Gary's history as an industrial center also obscures its prime geologic location; Gary sits within the moderate microclimate of Lake Michigan and directly on a seabed of ancient sand deposits, providing the perfect drainage required for excellent fruit-growing.The available (vacant) land in Gary clearly messages new possible futures for the region – the land speaks for itself, *if we can only bring ourselves to listen*.

North American Anishnaabe and Haudenosaunee scholar Vanessa Watts, describes the indigenous concept of'Place-Thought' and its challenge to Western attitudes. Place-Thought is based upon 'the premise that land is alive and thinking and that humans and non-humans derive agency through the extensions of these thoughts' (ibid.).The transposing of Indigenous cosmologies through a Eurocentric, Western purview, make a distinction between place and thought, where 'The result of this distinction is a colonized interpretation of both place and thought, where land is simply dirt and thought is only possessed by humans.'

> The division between 'knowing' (epistemology) and 'being' (ontology) is inherent in much of Euro-Western world views. These theoretical frameworks guide many Western beliefs of creation and agency, viewing indigenous histories and worldviews as mythical stories, or as an 'alternative' way to perceive and understand the world, rather than as real events that actually occurred.
>
> (*Ibid, p. 22*)

The projects in Gary and Aotearoa, furthermore, also challenge simplistic thinking about 'place-making,' as places already exist historically and cosmologically. More broadly, the challenge to the separation of nature and culture, ontology and epistemology lands us squarely in the domain of de-colonial theorist Mignolo and the concept of 'Epistemic Disobedience.' Mignolo proposes a purposeful 'de-linking' with the Western knowledge model and mindset and likens this delinking to the intentional resistance of'civil disobedience':

> Epistemic disobedience is necessary to take civil disobedience (Gandhi, Martin Luther King) to its point of non-return. Civil disobedience, within modern Western epistemology … could only lead to reforms, not to transformations. For this simple reason,

> the task of de-colonial thinking and the enactment of the de-colonial option in the 21st century starts from epistemic de-linking: from acts of epistemic disobedience.
>
> *(2010, p. 15)*

Mignolo proposes that this is accomplished through de-linking with the institutions of Western modernity, which are based on a monocultural and exploitative mindset, and with their silences, specializations, and disciplines. In this way Mignolo parallels much of the thinking that informs the *Embedded Artist Project* platform around the limitations of the Western knowledge model and the necessity of integrative, whole-systems approaches that 'reconnect to the realities of place' (Wilhelm, 2012).

Conclusion

In *We Have Never Been Modern*, Bruno Latour (1993, p. 56) explores the multiple fallacies of the dis-integrative binary thinking that Watts and Mignolo address, calling it out as 'The Great Divide.' In these recent *Embedded Artist Projects* these deeper epistemological 'divides' also cleave around strategies of 'place' but these issues have not as yet received much attention. What questions lay buried here for those of us working with place-based responses to climate change, and in what ways are these questions relevant to the approaches to community development we know as 'placemaking'? It is too simplistic (and too binary) to say that these approaches are antithetical? Currently they are operating within different discourses and, perhaps, at different scales and time frames. Further, the connections between 'place-making' and 'place-thought,' between the discourses of indigeneity and the Anthropocene, and other emerging imperatives, must be brought together (critically) with ideas about creative 'placemaking.' The *Embedded Artist Project* started as a change strategy, motivated by seeking to operate somewhere else: upstream or downstream, or inside, outside or *be-side*, some perceived boundary or limitation so that we can know (or understand) something else. Clearly the making of knowledge, and perhaps the 'making of place,' are not neutral propositions. What began as a disruption of disciplinary boundaries within the Western professional system of expertise has grown into a reassessment of the philosophy that underpins this entire system. More than mere institutional critique or disruptive innovation, this view of knowledge, of agency, of being, and of place, challenges not only the Western system of thought but also its metaphysics. Perhaps the Embedded Artist is a rule-breaking epistemic disobedient, as well as a lateral thinking double-agent, useful for reconnecting ecologies of practice (Stengers, 2005), expanding imaginaries and possibilities. For the twenty-first century, the *Embedded Artist Project* may also become a platform for broader geo-political aims, the decolonization of knowledge, justice, land, and place, through epistemic disobedience and place-based strategies!

References

Bruguera, T. (n.d.) 'Arte útil', in *Arte Util* [online]. Available at: https://www.arte-util.org/about/colophon/ (Accessed: 12 April 2020).

Deep South Challenge. (n.d.) *Vision Matauranga* [online]. Available at: https://www.deepsouthchallenge.co.nz/programmes/vision-matauranga (Accessed: 12 April 2020).

Engelstad, J. (n.d.) 'About', in *Make Art with Purpose* [online]. Available at: https://www.makeartwithpurpose.net/about.php?id=2 (Accessed: 12 April 2020).

Fry, T. (2007). 'Redirective practice: An elaboration', *Design Philosophy Papers*, 5(1), pp. 5–20 [online]. Available at: https://pdfs.semanticscholar.org/3a3f/2a4a1c096d697a27bdcc33e1ee08c7c67877.pdf (Accessed: 12 April 2020).

Geertz, C. (1998). 'Deep hanging out', in *The New York Review of Books*, 22 October 1998 Issue [online]. Available at: https://www.nybooks.com/articles/1998/10/22/deep-hanging-out/ (Accessed: 12 April 2020).
Grocott, L. (2005). 'Promoting potential: The dissemination and reception of practitioner-led design research', in *Design Perspectives Conference*, 29 September 2005. Mexico: Universidad Iberoamericana.
Harding, D. (n.d.) *Glenrothes Town Artist 1968–78: Part of an unpublished memoir* [online]. Available at: https://www.davidharding.net/?page_id=37 (Accessed: 12 April 2020).
Hart, J. (2014). 'Artist as leader: Whitehead on the 606', in *Public Art Review*, Issue 50, Spring/Summer, 2014 [online]. Available at: https://issuu.com/forecastpublicart/docs/par50_final (Accessed: 12 April 2020).
Hawkes J. (2001). *The Fourth Pillar of Sustainability: Culture's Essential Role in Public Planning*. Melbourne, Australia: Common Ground Publishing, in association with the Cultural Development Network.
Ingram, M. (2012). 'The diplomacy of art: What ecological artists offer environmental politics', in *Annual International Conference of the Royal Geographical Society*, 31 August 31–2 September 2012. London.
Kagan, S. (2007). Art effectuating social change: Double Entrepreneurship', in *Conventions 1* [online]. Available at: https://www.semanticscholar.org/paper/Art-effectuating-social-change%3A-Double-in-1-Kagan/1fcfd36001674474d35a31d5bb972da7d3e55267 (Accessed: 12 April 2020).
Kennedy, R. (2016). 'An artist who calls the sanitation department home', in *New York Times*, 21 September 2016 [online]. Available at: https://www.nytimes.com/2016/09/22/arts/design/mierle-laderman-ukeles-new-york-city-sanitation-department.html (Accessed: 12 April 2020).
Latour, B. (1993). *We Have Never Been Modern*. New York: Harvard University Press.
Mignolo, W.D. (2010). 'Epistemic disobedience, independent thought and decolonial freedom', *Theory, Culture and Society*, 26(7–8), pp. 159–181, p. 18 [online]. Available at: https://doi.org/10.1177/0263276409349275 (Accessed: 12 April 2020).
Ndlovu-Gatsheni, S. (2017). 'Decolonizing research methodology must include undoing its dirty history', in *The Conversation*, 27 September 2017 [online]. Available at: https://theconversation.com/decolonising-research-methodology-must-include-undoing-its-dirty-history-83912 (Accessed: 12 April 2020).
Plöger, J. (2007). 'The emergence of a "city of cages"' in Lima: Neighbourhood appropriation in the context of rising insecurities', *European Journal of Geography*, June 5, 2007 [online]. Available at: http://eprints.lse.ac.uk/3610/1/The_emergence_of_a_city_of_cages_(LSERO).pdf (Accessed: 12 April 2020).
Purves, T. (2004). *What We Want Is Free: Generosity and Exchange in Recent Art*. New York: State University of New York Press.
Smith, H. (2019). *Drawing Ecologies 2019–2020—Project Plan: Planning for Climate Change Impacts on Māori Coastal Ecosystems and Economies in the Horowhenua-Kāpiti Rohe, Aotearoa/New Zealand: Phase 2, 2019* [online]. Available at: https://www.mfe.govt.nz/more/iwi-m%C4%81ori/te-manat%C5%AB-m%C5%8D-te-taiao-hui-%C4%81-rohe (Accessed: 12 April 2020).
Stengers, I. (2005). 'Introductory notes on an ecology of practices,' in *Cultural Studies Review*, 11(1), pp. 183–196 [online]. Available at: https://epress.lib.uts.edu.au/journals/index.php/csrj/article/view/3459 (Accessed: 12 April 2020).
Tate. (n.d.). *Artists Placement Group* [online]. Available at: https://www.tate.org.uk/art/art-terms/a/artist-placement-group. (Accessed: 12 April 2020).
Todd, Z. (2015). 'Indigenizing the anthropocene', in Davis, H. and Turpin, E. (eds.) *Art in the Anthropocene: Encounters Among Aesthetics, Politics, Environment and Epistemology*. London: Open Humanities Press.
Watts, V. (2013). 'Indigenous place-thought and agency amongst humans and non-humans', in *Decolonization: Indigeneity, Education & Society*, 2(1), pp. 20–34 [online]. Available at: file:///C:/Users/cc-06/Downloads/19145-Article%20Text-45987-1-10-20130504.pdf (Accessed: 12 April 2020).
Whitehead, F. (2006). 'What do artists know', in *Embedded Artist Project* [online]. Available at: http://embeddedartistproject.com/whatdoartistsknow.html (Accessed: 12 April 2020).
Whitehead, F. (2015). 'Civic experiments: Tactics for praxis', in *Plenary Keynote, EU Cost Action: Investigating Cultural Sustainability Conference*, May 2015. Helsinki, Finland. http://www.culturalsustainability.eu/
Wilhelm, G. (2012). In phone conversation with author.

Further reading in this volume

Chapter 1: Introduction: what really matters – moving placemaking into a new epoch
Cara Courage
Chapter 4: A future of creative placemaking
Sarah Calderon and Erik Takeshita

Chapter 5: Making places for survival: looking to a creative placemaking past for a guide to the future
Jeremy Liu
Chapter 14: Experts in their own tomorrows: placemaking for participatory climate futures
Paul Graham Raven
Chapter 16: More than a mural: participatory placemaking on Gija Country
Samantha Edwards-Vandenhoek
Chapter 18: '*Homomonument* sounds like a poem': queer placemaking 30 years on: a conversational dialogue with Thijs Bartels, author of *Dancing on the Homomonument* (2003)
Martin Zebracki
Chapter 19: Placemaking in the ecology of the human habitat
Graham Marshall
Preface: Towards developing equitable economies; the concept of *Oikos* in placemaking
Anita McKeown
Chapter 27: Is 'tactical urbanism' an alternative to neoliberal urbanism?: reflections on an exhibition at the MoMA
Neil Brenner
Chapter 28: Integral placemaking: A poiesis of sophrosynes?
Ian Wight
Chapter 29: The solution is in the problem; the art of turning a threat into an opportunity by developing resilience using a creative placemaking critical praxis
Anita McKeown
Chapter 30: Ecological selves as citizens and governance as ethical placemaking
Lisa Eckenwiler
Chapter 31: Seven generations: a role for artists in Zuni PlaceKnowing
Theodore S. Jojola and Michaela P. Shirley
Chapter 33: Conceptualizing and recognizing placemaking by non-human beings and lessons we might learn from Marx while walking with Beaver
Jeff Baldwin
Chapter 39: Translating Outcomes: Laying the groundwork for interdisciplinary evaluation of creative placemaking
Jamie Hand
Chapter 40: Transforming community development through arts and culture: a developmental approach to documentation and research
Victor Rubin
Chapter 41: Rituals of regard: on festivals, folks and findings of social impact
Maribel Alvarez
Chapter 42: Creative placemaking and placekeeping evaluation challenges from the practitioner perspective: an interview with Roy Chan
Maria Rosario Jackson
Chapter 45: How the city speaks to us and how we speak back: rewriting the relationship between people and place
Rosanna Vitiello and Marcus Willcocks

23

ROUTING OUT PLACE IDENTITY THROUGH THE VERNACULAR PRODUCTION PRACTICES OF A COMMUNITY LIGHT FESTIVAL

Gail Skelly and Tim Edensor

Introduction

Lighting the Legend, a community light festival in Ordsall, a neighbourhood in inner-city Salford, UK, has taken place each November for the last 26 years. This parade of hand-held, locally made paper lanterns generates an exuberant burst of energy in the November night, as residents walk through the housing estate towards a firework display in Ordsall Park. A small band of young children led by a professional drummer spiritedly play and sing throughout, attracting other residents, some long-standing, some new, who wave or join in the parade. The new residents are welcomed just as warmly as existing friends and neighbours.

Situated between Manchester city centre and Salford Quays, Ordsall's residents have experienced momentous changes over the past five decades. In this chapter, we explore how their participation and the creative vernacular practices they mobilise in *Lighting the Legend* resist threats to their sense of place amidst fears that current levels of development will eventually displace them. Such concerns emerge from a historical context in which many recall the obliteration of around 18,000 dwellings in the 'slum clearances' of the 1960s and 1970s. These demolitions blighted the tight-knit community 'long noted for a strong sense of community based on close association through family, friends, workplace as well as social and cultural activities' (Norris Nicholson, 2001, p. 43). Though some families returned to replacement social housing in the 1980s, only a 'minority of the original population were re-housed in the immediate area' and there was considerable population loss (Cassidy, 2012, p. 169). A subsequent period of regeneration between 2001 and 2011 has since increased the population of the Ordsall ward by 111 per cent (Salford City Council, 2019).

Lighting the Legend emerged from a grassroots partnership of residents and artists in the early 1990s, harnessing a strong local sense of community pride despite prevailing social and economic conditions. For while certain areas of Ordsall's rapid upmarket development are prospering, its social housing estates are some of the most deprived parts of the UK; Salford is the twentieth most deprived local authority area and deprivation on the Ordsall estate is within the lowest decile nationally, with health deprivation among the most disadvantaged in the country (Department for Communities and Local Government, 2019). The complex effects of such living conditions, along with gang violence, drug use, and gun crimes, can lead to stigmatization (Wallace, 2015). In responding to these conditions, Salford City Council has provided an annual subsidy to support *Lighting the Legend*. However, since 2014 austerity-driven cuts to local authority spending have been passed on to services deemed 'non-essential,' including the festival. This loss has provoked the development of organisational and creative skills amongst participants in attempting to sidestep their former reliance on the skills of paid professional artists. In exploring these creative adaptations, we first highlight academic accounts of contemporary light festivals, including smaller, community events. We then explore how making, displaying, and carrying the lanterns creates 'spatial, temporal and human interstices' (Petcou and Petrescu, 2015, p. 259) within the frequently harsh everyday experiences of living in Ordsall. We subsequently discuss how participating in the festival and its production generates valuable individual outcomes and enables the forging of neighbourliness between disparate adjoining communities, reproducing a shared place-identity.

Light festivals

To date, research on light festivals has tended to focus on large, spectacular, metropolitan showcase events, often led by professional artists or cultural organisations. Contributions have explored the role of festivals in the formation of cultural policy and in boosting night-time economic strategies, their lack of place-specificity, the particular sensory and affective affordances of illumination, and the ways in which they may contribute to fostering a sense of place (Edensor and Millington, 2013; Alves, 2007; Giordano and Ong, 2017; Evans, 2011). More critically, while some accounts focus on how the process of 'festivalisation' (Häussermann and Siebel, 1993) can contribute to delivering economic outcomes, this economic instrumentality exposes them to criticisms that all too often, formulaic, homogeneous spectacles empty of meaning are produced that are passively consumed by audiences. However, other accounts suggest that a more nuanced analysis can reveal alternative responses and potentialities in placemaking (Edensor and Sumartojo, 2017; Edensor, 2012; 2017; Jeong and Almeida Santos, 2004). Moreover, festivals of all kinds can be understood as social and cultural experiences that diverge from the mundanity of everyday life and have the capacity to temporarily transform ordinary places into spaces that offer new meanings and sensations (Edensor, 2018; Rofe and Woosnam, 2015). In particular, community-centred festivals have the potential to offer bottom-up, locally oriented approaches to placemaking that veer away from the top-down economic and cultural strategies of city marketers (Friedmann, 2010). However, such potentialities have been somewhat overlooked, perhaps because many occur in marginal places and are small in scale. Derrett (2003) contends that community-based festivals can evolve over time to reflect shared values and foster a sense of community and place, partly by employing networks of local residents (Jarman, 2018), while Alves (2007) claims that a festive opening up of community spaces for local audiences can lead to a renewed sense of place and re-territorialization. Other accounts discuss how community festivals can permeate neighbourhoods, encourage more intensive use of places, generate memories for local residents, and offer scope for the evolution of new meanings and practices while sedimenting shared community values (Edensor, 2018; Rofe and Woosnam, 2015). In this context, we focus on how one small light festival, *Lighting the Legend*, acts as a crucible for initiating the formation of shared community identities.

In considering lantern-making and parading as the symbolic crux of this festival, we elucidate Alves' (2007, p. 1252) contention that light can play 'a central role in the creation of cultural events that may energise cities' nocturnal lives. Throughout history, communities have used light to celebrate and commemorate, mark agricultural growing seasons, protest and unify

(Fox, 2002). Bonfires, torchlight processions, Christmas lights, and lantern parades have a long history (Stevens and Shin, 2014) and may provide significant occasions in people's lives. Such festive events often have roots in carnivalesque rituals, away from the burden of working life and social control (Croose, 2014), facilitating ludic behaviours and expressions that disrupt mundane experience, outcomes often neglected by criticisms of instrumentality (Edensor, 2014). In some cases, these events have developed from a mythical story or a celebrated local legend (Fox, 2002; Edensor, 2018; Quinn, 2013). Here, we consider how residents involved in the *Lighting the Legend* festival, itself inspired by a local legend, develop skills in developing and staging the festival.

Crucially, we draw on a longitudinal study which documents the production practices, creative programmes, festive event itself, and the annual cycle of organisation of *Lighting the Legend*, an event that annually engages approximately 200 participants, predominantly children under 12, who create their own lanterns. The festival takes place around the UK's Bonfire Night, 5 November annually, and is the most significant event of the year for many local residents, some of whom are also part of the festival production team that primarily comprises female residents with deep roots in the community. They are currently grappling with the challenges of replacing resources once provided by the state and other public agencies to deliver the festival, and we explore how they develop the new skills required to sustain this place-based festival.

From cultural policy to vernacular creativity?

In order to examine the development of the placemaking capabilities of *Lighting the Legend*, we review its evolution as an outcome of cultural policy initiatives from the 1970s onwards. The deployment of artists as intervention agents to support struggling communities followed the radical period of the alternative arts and theatre movement that emerged from the global student insurrections of 1968. Due to its rates of high crime and deprivation, the Ordsall estate was the recipient of local political intervention in this era. The City Reporter (10 November 1978) details that revenue funding for Ordsall Community Arts was financed over a five-year period under Phase 17 of the Urban Programme. This government aid package gave 'financial help for social and educational projects in urban areas with particular social needs and problems, for which Salford ... qualified' (Salford CVS Report, 2016), and Ordsall Community Arts employed artists and practitioners to develop culturally engaging programmes. Influenced by Welfare State International, who pioneered lantern-making practices in the UK by importing paper and willow technology from China (Fox, 2002), the artistic team utilized lantern- and prop-making as part of the processional 'folk-art' that 'came to form the bedrock of participatory alternative theatre' (Croose, 2019).

In 1992, many UK cities experienced summer rioting; Ordsall's ferocious reputation was sealed in the summer of 1992. Riots broke out across Salford, but it was Ordsall, just across the road from the gleaming Salford Quays, which hogged local and national headlines. A fireman was shot as he tried to douse a blazing Job Centre and a police dog handler survived a bullet ricocheting inside his vehicle. The Carpet World warehouse was destroyed in an arson attack (Keeling, 2004).

By this time, community artists had already formed strong partnerships with local residents and in order to counter further unrest, they continued to develop arts practices that particularly sought to engage with young people. Two years later, the idea of a community light festival as a strategic intervention to avert further anti-social behaviour was proposed, led by Ordsall Community Arts. According to Bennett (2018, p. 161), the community arts movement took as its 'starting point the material circumstances of those perceived to be disadvantaged or oppressed.'

Utilising a broader historical perspective, Matarasso (2013) attests that since their beginnings in the late 1960s, community art practices have constantly changed; at Ordsall, continued funding from national and local authority revenue allowed the organisation and the festival to flourish but this came at a cost, with the radical roots of community arts, forged during the early years, left behind. Reviewing the community arts movement's efficacy following policy interventions of the 1970s, Jeffers and Moriarty (2017) claim that its lifespan ended in the late 1980s, following a trend from 'radicalism to remedialism' (cited in Matarasso 2013, p. 216). As Courage (2017) contends, changes have also been provoked by the dilemma for cultural organisers in employing artists as key agents, particularly pertinent in this case with the withdrawal of state subsidy to Ordsall Community Arts under austerity conditions, soliciting a change from a professional to a voluntary arts organisation.

Formerly managed by a professional festival manager, *Lighting the Legend* is currently organised by a group of local women, initially recruited as volunteers to support art activities, who schedule regular meetings to plan the festival. By participating as volunteers and in workshops and events between 2016 and 2019, we gained privileged access into these women's experiences as they have taken up the reins of festival organisation and followed a steep learning curve to manage multiple administrative, financial, logistical, technical, and creative tasks. They constantly juggle the organisation of a creative programme of workshops with their busy lives as family carers in an area where extended families are common and residents experience levels of 'health and wellbeing that is worse than the national average' (Greater Manchester Health and Social Care Devolution, 2017).

In 2017, approximately 200 children and young people were invited to make lanterns in school or community environments. The women's group recruit other adult volunteers to help prepare materials which they often take home to spend hours doing 'in front of the telly'; 'plabour' (Hawkins and Price, 2018, p. 4), rather than the hard graft of more formal work. They also initiate and manage fund-raising activities, attract local business sponsors, run raffles, and organise a fundraising 'Ordsall's Got Talent' contest. They begin work on the forthcoming festival early in the year, commencing with the organisation of a steering group of local residents to discuss the parade route and selection of the annual theme. In following the ongoing emergence of *Lighting the Legend* since 2016, we consider how residents are reconfiguring a bottom-up approach to building community resilience and explore how emergent forms of vernacular creativity are remaking everyday space (Edensor, 2018). In so doing, we focus on the creative retelling of the mythical tale upon which the festival is founded, the inventive route-making practices that are organised, and the imaginative processes through which *Lighting the Legend*'s annual theme is produced.

The salience of the mythic narrative of Lighting the Legend

The original inspiration for *Lighting the Legend* is the myth that in the lead-up to the Gunpowder Plot in 1605, Guy Fawkes stayed at Ordsall Hall, a fifteenth-century landmark manor house. The Hall was home to the Radclyffe family, recusant Catholics who were potentially close to the conspiracy. A nineteenth-century romantic serial by William Harrison Ainsworth, *Guy Fawkes* (1842), imagines Fawkes escaping from the authorities through an underground tunnel directly from the Hall, assisted by the Radclyffes. This work of fiction continues to ignite a proud association within the community to one of England's most infamous revolutionary plots, and this has been accentuated by the annual 5 November celebrations. The nationwide commemorative celebrations following the failed plot eventually lost their anti-Catholic fervour, evolving into the vernacular, secular tradition of Bonfire Night. Before the 1960s demolitions, this was a popular night for residents of Ordsall's closely packed terraced houses, which afforded plenty of local places for the night's activities, as local resident Madge Bown recalls:

> Bonfire night was a big thing here, nearly every street had a bonfire – the one outside our house; we had it there that long that … [laughs] you could walk down the street, any street and you could tell where the bonfire had been, and it had damaged the cobbles in that area. And then of course you had fireworks, and the bonfire [laughs] … everybody brought chairs out, and if you had anything you wanted to get rid of, anything, it was round the bonfire, and gradually, somebody would say 'I want that, that'll go on the bonfire,' so you ended up with nothing [to sit on] [laughs]. You know, people would make things [to share] and chestnuts would go in, and somebody would bring toffee apples; it was all just for the one street, everybody was round that bonfire. But then you could go round and look at different bonfires, some were bigger than others.

Former residents directly link 'the decline of family bonfire nights, [the cessation of] of children begging for pennies for the Guy, the diminution of bonfires on street corner bomb sites' (Fehler, 2006, p. 4) with the loss of the old terraced streets. Yet *Lighting the Legend* echoes the community spirit described above and has provided at least a temporary space for the loss of small, vacant plots and cobbled streets after 1980s demolition and rebuilding. Reiterating the tradition founded on Bonfire Night has been effective in reinforcing place-identity in an area once renowned for its close-knit community. Battling with anti-social problems well into the 1990s, the pride released by making and carrying lanterns in this annual tradition in honour of their mythical history provided a defence against negative stereotypes for local children. Perhaps the alluring association with the romantic imagery of a revolutionary chased through the tunnels of Ordsall (Sharp, 2005, as cited in Fehler, ibid.) has been significant enough to fire local imaginations in spite of the lack of any historical evidence. Indeed, one parent vehemently complained that there was no reference to this story at her son's 5 November primary school assembly, a perceived threat to the continued reiteration of this local myth that provoked strong feelings.

Guy Fawkes Street, which runs alongside Ordsall Hall, is a constant element of the parade route, suggesting a shared social dimension to remembering, not only in the case of the myth but the memory of the old estate. For Ordsall Hall was among a small number of buildings left standing following the slum clearances of the 1960s and 1970s. Resonating with Deacon's claim that 'the shared dimension of remembering, and the equally social nature of how space is produced' are intertwined, the formation of the festival has provided alternative symbolic spaces to replace what had once existed (Deacon, 1998, cited in Hoelscher and Alderman, 2004, p. 348).

Creative route-making and managing the parade

The predominant ingredient of *Lighting the Legend* is the carrying of hand-made lanterns through the streets by children. Aside from the release of fun and excitement, the route through the estate is also a vehicle by which the wider community can participate, support, and witness the parade, as local resident Sylvia Sharples articulates:

> It is one of the largest events in our community and people of all ages participate during the event and in the preparations leading up to it. It helps keep our community together and is one of the few events during the dark nights at winter.

The route itself is an important part of this illuminated, night-time walking tradition, as the parade claims space, makes noise, and draws attention. Parading through the streets was already part of local cultural practice with Roman Catholic and Church of England Whit Walks. According to Entwistle (2012, p. 209), these 'processions not only represented the visible symbols of religious, political, and community identities, but were also an integral part of the common culture of working-class streets, the sites of most social interaction,' though today there is little evidence of this once strong tradition in Ordsall. Residents reiterate the reasons for the decline of Bonfire Night in asserting that the loss of the streets is physically and socially dismantling the community, including the decline of a walking tradition, yet *Lighting the Legend* is ameliorating the loss of the 'glorious spectacle' of community expression in the public domain (ibid., p. 214). Practically, organisers must consider temporary access limitations, logistics of the support vehicle's ability to access the whole route, and closeness to the main roads. Accordingly, they must negotiate between 'carnivalesque' rights for local residents and the upholding of Salford City Council's obligations for ensuring residents' safety. Jarman (2018) contends that irrespective of such impediments, it is the participants that collectively contribute to the place-making possibilities of a procession.

These potent affects generated by walking the streets have been made more effective by the changing routes that are devised from year to year. Recent changes to the route have been devised by the emerging leadership of the festival group, who organised a walking activity during school half term. The walking party consisted of two local parents and a group of seven children between the ages of 8 and 11. The adults had previously reviewed previous year's routes in their meetings, with concerns focusing on the practical realities of the length of the walk and the weather conditions, one opining that 'if it rains, it's always too long, and people start to go home.' *Lighting the Legend* is fuelled by children's energy at the front and, frequently, a battle with the elements to get to the end. Along the walk, the group bumped into neighbours and friends who were greeted in a loud sing-song style, and the high-spirits of the children led to them joining in as a chorus. From then on, every person was greeted with a sung chorus of 'Hiya,' with all joining in and laughing. The experience of this impromptu choral performance inspired an idea for the community theatre at the end of the *Lighting the Legend* parade. This theatre piece expands the parade narrative, and in this case featured the recordings of the street names. The young people involved in the walking exercise were then invited in a separate workshop to create poems for each of the special lanterns that represent buildings, as detailed later, and during the creation of the lanterns this poetry was incorporated into a soundscape played for the audience present during this event. The familiar street names and daily paths through the estate were therefore elevated to something more than the everyday during the festival. The recordings allowed them to share their unique place experience, as well as entertain neighbours and audiences.

The 2017 parade walked past a developer's hoarding, erected to enclose a large piece of land earmarked for housing development, the final phase of the Ordsall Development Framework (Manchester Evening News, 2017). The site was formerly a shopping centre which once provided local amenities, as well as a much-used path from the north of the estate to the schools and the doctor's surgery. As they walked past the hoardings on their night-time parade carrying their symbols of place identity, the loss of access rights for local residents was accentuated. Following a decade of new housing developments, such hoardings are familiar, with new construction occupying 73 hectares of land (LPC Living, 2019). The public narrative is of regeneration success:

> a once feared council estate, synonymous with trouble in the early 1990s, has been transformed into hundreds of modern, new homes. The £100m transformation of the

> derelict Ordsall estate has taken more than a decade but the last few homes have now been built.
>
> *(Manchester Evening News, ibid)*

For local residents however, the newly built houses across the estate do not register as belonging to them, and they are still subject to the 'vagaries of regeneration politics' (Wallace, 2014, p. 536) and continue to refer to construction projects on the sites of demolished houses as taking place on 'our land.'

Despite these local sentiments, a study in 2014 recognised that a lively community spirit persists in Ordsall, with many residents 'engaged and active in their community, working to stimulate a sense of pride in the area' (Symons, 2018, p. 208). This positive, forward-thinking disposition was echoed during festival steering group discussions where the organising group actively encourage new families to take part in the festival, providing an example of 'hidden' care-giving which reveals the nature of 'women's work in the urban ... landscapes that daily act to transform the understanding of the spatiality of everyday life' (Dyck, 2005, p. 243). A generosity of inclusion is expressed even though the small group of women acknowledge the social divisions being created in the estate, where they feel that there are now two separate communities. Until recently, the festival manager leveraged public funding on the community's behalf by foregrounding the festival's promotion of social integration and the potential reduction of social isolation. Now, unsupported by such professional roles, the residents group demonstrates a magnanimous intention to enable the two communities to come together as participants in their community light festival, even if for a brief symbolic moment.

They also take considerable responsibility during the parade. For instance, one of the residents, Jean, volunteered to help out on the parade by handing out lanterns and stewarding. The energetic, older, unsupervised children at the head of the parade can in their ebullience occasionally bump into the drummers who lead the procession. Jean took it upon herself to take up a position between the drummers and the children and, as a strong athlete, using her body to prevent the children from coming forward too rapidly. In filling this responsible leadership void, Jean joined the group of volunteers who annually support the festival. The longevity of the festival allows for intergenerational interactions, with parents and grandparents accompanying young children and fostering family and communal connections. In supporting the children's lantern-making, adults such as Jean show how 'the relations among the generations are not just interpersonal; they are caught up in the material and cultural processes that give rise to places' (Mannion, 2012, p. 395).

Creating the parade theme

2017 was the first year that the residents group decided upon the parade theme without the influence of a festival manager, and they drew on their own experiences for inspiration, beginning with a discussion of Ordsall's lost heritage; between 1960 and 2015, as well as thousands of homes lost, demolition laid waste to schools, nurseries, churches, job centres, pubs, clubs, and factories. However, what began as a lament for these now-vanished sites developed into an acknowledgement of still-existing but uncelebrated sites in Ordsall, usually unreflexively apprehended through daily routines but integral to the habitual lives of these residents. They focused upon the importance of celebrating everyday sites at which they met friends and neighbours, joined in classes or workshops, experienced celebrations like fundraising events and parties, and where they collectively experienced a sense of community spirit. Accordingly, the route was designed to honour these everyday sites and buildings as symbols of Ordsall, and it was decided that lanterns fashioned in the shapes of shops, schools, local institutions, churches, and the Hall would be created.

Once the list of local buildings was decided upon, the group discussed issues around the fabrication of lanterns; one previous lantern of Ordsall Hall, made in the 1990s, 'took ages to make,' a luxury which the group no longer had the resources to pay an artist to spend days intricately making. Consequently, they discussed the making of basic and symbolic shapes to represent, for example, a logo for the school and a stained-glass window for the church. Gail participated in workshops at the Community Cafe and at Salford Girls Club. These sessions were co-curated by an adult maker and young people and discussed both the importance of particular buildings and the suitability of its structure for lantern construction. At the Girls Club workshop, the Lads Club building was deemed not especially recognisable in itself, but what has garnered attention has been the door and the sign above it, made famous by the iconic photograph of four members of band 'The Smiths' standing outside, taken by Steven Wright in 1984, and now exhibited at the National Portrait Gallery. The Club has experienced a renaissance that grew out of the enormous appeal of Wright's photograph, with tourists travelling from all over the world to photograph themselves in front of the door and sign. However, the Girl's group decided only to make the door sign in lantern form, changing the iconic text to 'Salford Girls Club,' to reflect their Tuesday night girls-only use of the club. They were asserting their identity in place, a shared pride that was highly evident on the night of the parade. However, in the session during which the lantern was fashioned they were presented with two identical sides, which they decided to label as 'Boys' and 'Girls.' This suggests nuanced and multiple layers of feelings of local pride, to a specific sense of belonging to the Girls Club as well as membership of the broader community of Ordsall. The lantern-making process is thus revealed to be far from static, but dynamic, replete with improvisation and negotiation.

As well as the small lanterns carried by children, a series of large lanterns are created and carried to boost the visual impact of the parade and act as thematic motifs; they are also used as props for a short community theatrical piece at the parade's finale. As a creative placemaking tool, this short drama is designed to explore the annual theme. For the 2017 festival, as discussed above, these lanterns took the form of 'buildings' regarded as symbols of community vitality and were proudly carried by the youths and adults responsible for making them. At the end of the parade, the larger lanterns were laid out one by one to reproduce an image of a symbolic Ordsall street. As each lantern was ceremonially added, it was accompanied by a poetry reading devised by the young people to express its importance in the community:

> With these Ordsall lights of ours
> The community takes part
> To celebrate what's ours
> From the Lads Club to the Park
> With these lights of ours
> Shining brightly like the stars
> We're the Ordsall community and proud

Here, illuminated forms were utilised to express the importance of places that had endured as key elements of Ordsall's built environment, while also enabling a forward-facing, inclusive approach that acknowledged the continuously expanding community, allowing old and new residents to explore their environs together.

Conclusion

In this chapter, we have explored the processes through which the *Lighting the Legend* lantern festival in Ordsall has been produced since its initiation as a partnership project between the state and local residents in the early 1990s. Since then, the festival has grown in tandem with funding support, and its recent withdrawal has created huge challenges for the festival's continued sustainability, as the residents involved in its production must cope without the support of the professionals who were previously employed. While their conversations may reflect Salford City Council's language of community development priorities, their exclusion from official cultural policy means that their actions are predominantly led by their need to be together on such an important night for them, the 'hard to reach' community (Symons 2017, p. 208). Accordingly, though the residents have experienced many State-supported efforts to 'engage them' in the past, they must now take the lead, and are developing their own vernacular skills, expressing their own sense of place-identity in devising the festival's route, managing the parade and selecting the annual theme. Concurrent social changes in the population of the estate have challenged a sense of community, and yet the residents have welcomed new residents, overcoming their own insecurities. By generating their own festive practices that are based on their own experiences of living in Ordsall, they are rebuilding a sense of community which is inclusive of all those willing to engage with it, including those attracted by Ordsall's regeneration. The lantern-making programme generates excitement and is a symbolic, material expression of local identity, while the parade unites the whole community. Taking into account 'a massive sense of loss of sense of place in Salford over the last forty years, as the vast majority of the architectural fabric has been erased in all urban districts' (Cassidy, 2012, p. 177), these local people have been active in the construction of a renewed sense of place through the creative and innovative production of a local light festival.

References

Ainsworth, W. (1842). *Guy Fawkes*. London: Routledge.

Alves, T. (2007). 'Art, light and landscape: New agendas for urban development', *European Planning Studies*, 15(9), pp. 1247–1260.

Bennett, O. (2017). 'Memories, dreams, reflections: Community arts as cultural policy' in Jeffers, A. and Moriarty, G. (eds.) *Culture, Democracy and the Right to Make Art: The British Community Arts Movement*. Bloomsbury Publishing, pp 161–182.

Cassidy, L. (2012). 'Recalling community: Using material culture and digital archives in Salford Gateways', *International Journal of Community Research and Engagement*, 5, pp. 166–182.

City Reporter. (1978). *Community arts for Ordsall*. Salford Local History Library.

Courage, C. (2017). *Arts in Place: The Arts, the Urban and Social Practice*. London: Routledge.

Croose, J. (2014). *The Practices of Carnival: Community, Culture and Place*. PhD thesis. University of Exeter.

Croose, J. (2019). 'The Official Feast': Cultural tensions in U.K. Carnival', *Social & Cultural Geography*, 20(4), pp. 551–574.

Department for Communities and Local Government. (2019). *Indices of Deprivation* [online]. Available at: https://www.gov.uk/government/statistics/english-indices-of-deprivation-2019 (Accessed: 20th November 2019).

Derret, R. (2003). 'Making sense of how festivals demonstrate a community's sense of place', *Event Management*, 8(1), pp. 49–58.

Dyck, I. (2005). 'Feminist geography, the "everyday", and local-global relations: Hidden spaces of placemaking', *Canadian Geographer*, 49(3), pp. 233–43.

Edensor, T. (2012). 'Illuminated atmospheres: anticipating and reproducing the flow of affective experience in Blackpool', *Environment and Planning D: Society and Space*, 30, pp. 1103–1122.

Edensor, T. (2014). 'The aesthetic, social and cultural effects of festivals of illumination: Contrasting transformations of space and place by lighting', in Hassenhorl, U., Krause, K., Meier, J. and Pottharst, M. (eds.) *Urban Lighting, Light Pollution and Society*. London: Routledge.

Edensor, T. (2017). *From Light to Dark: Daylight, Illumination and Gloom*. Minneapolis: Minnesota University Press.

Edensor, T. (2018). 'Moonraking: Making things, place and event', in Price, L. and Hawkins, H. (eds.) *Geographies of Making/making Geographies: Embodiment, Matter and Practice.* London: Routledge, pp. 60–75.
Edensor, T. and Millington, S. (2013). 'Blackpool Illuminations: Revaluing local cultural production, situated creativity and working-class values', *International Journal of Cultural Policy*, 19(2), pp. 145–161.
Edensor, T. and Sumartojo, S. (2018). 'Reconfiguring familiar worlds with light projection: The Gertrude Street Projection Festival, 2017', *GeoHumanities*, 4(1), pp. 112–131.
Entwistle, D. (2012). 'The whit walks of Hyde: Glorious spectacle, religious witness, and celebration of a custom', *Journal of Religious History*, 36(2), pp. 204–233.
Evans, G. (2011). 'Hold back the night: Nuit Blanche and all-night events in capital cities', *Current Issues in Tourism*, 15(1–2), pp. 35–49.
Fehler, T. (2006). 'Remember, remember: A cultural history of Guy Fawkes Day', *History: Reviews of New Books*, 34(4), pp. 121–122.
Fox, J. (2002). *Eyes on Stalks.* London: Methuen.
Friedmann, J. (2010). 'Place and placemaking in cities: a global perspective', *Planning Theory and Practice*, 11(2), pp. 149–165.
Giordano, E. and Ong, C. (2017). 'Light festivals, policy mobilities and urban tourism'. *Tourism Geographies*, 19(5), pp. 699–716.
Greater Manchester Health and Social Care Devolution. (2017). Locality Plan for *Salford*, Executive Summary [online]. Available: https://extranet.salfordccg.nhs.uk/application/files/8715/5369/5130/Salford_Locality_Plan.pdf] (Accessed 26 November 2019).
Häussermann, H. and Siebel, W. (1993). 'Festivalization of urban policy: Urban development through major projects', *Leviathan*, 13.
Hawkins, H. and Price, L. (eds.) (2018). *Geographies of Making, Craft and Creativity.* London: Routledge.
Hoelscher, S. and Alderman, D. (2004). 'Memory and place: Geographies of a critical relationship', *Social and Cultural Geography*, 5(3), pp. 347–355.
Jarman, D. (2018). 'Festival community networks and transformative placemaking', *Journal of Place Management*, 11(3), pp. 335–349.
Jeffers, A. and Moriarty, G. eds., (2017). *Culture, democracy and the right to make art: The British community arts movement.* Bloomsbury Publishing.
Jeong, S. and Almeida Santos, C. (2004). 'Cultural politics and contested place identity', *Annals of Tourism Research*, 31(3), pp. 640–656.
Keeling, N. (2004). 'Recovery from the Riots', in *Manchester Evening News*, 13 August 2004.
LPC Living. (2019). *Ordsall Regeneration* [online]. Available at: https://www.lpcliving.co.uk/developments/ordsall-regeneration/ (Accessed 26 November 2019).
Mannion, G. (2012). 'Intergenerational education: The significance of reciprocity and place', *Journal of Intergenerational Relationships*, 10(4),pp. 386–399.
Matarasso, F. (2013). 'All in this together': The depoliticisation of community art in Britain, 1970–2011', in *Community, Art, Power.* Rotterdam: International Community Arts Festival, pp. 214–240.
Norris N.H. (2001). 'Two tales of a city: Salford in regional filmmaking, 1957–1973', *Manchester Region History Review*, 15, pp. 41–53.
Petcou, C. and Petrescu, D. (2015). 'R-URBAN or how to co-produce a resilient city', *Ephemera: Theory and Politics in Organization*, 15(1), pp. 249–262.
Quinn, B. (2013). 'Art festivals, urban tourism and cultural policy', *Journal of Policy Research in Tourism, Leisure and Events*, 2(3), pp. 264–279.
Rofe, M. and Woosnam, C. (2015). 'Festivals as a vehicle for place promotion: Cars, contestation and the creative city ethos', *Landscape Research*, 41(3), 344–359.
Salford City Council Partners in Salford. (2017). *Annual Public Health Report 1617.* Salford City Council, (2019) Regeneration:Housing/Ordsall [online]. Available at: https://www.salford.gov.uk/planning-building-and-regeneration/regeneration/regeneration-housing/ordsall/ (Accessed 19 November 2019).
Salford CVS. (2016). *A Brief History in Time* [online]. Available at: https://www.salfordcvs.co.uk/sites/salfordcvs.co.uk/files/Salford%20CVS%201973-2016%20FINAL%20-%20v7%20%28002%29.pdf (Accessed on line 26 November 2019).
Sharp, J. (2005). *Remember, Remember. A Cultural History of Guy Fawkes Day.* Cambridge: Harvard University Press.
Smithers, D. (2017). '£100m transformation of Ordsall estate is complete', in *Manchester Evening News*, 12 February 2017 [online]. Available at: https://www.manchestereveningnews.co.uk/news/greater-ma

nchester-news/100m-transformation-ordsall-estate-complete-12592319 (Accessed 26 November 2019).

Stevens, Q. and Shin, H. (2014). 'Urban festivals and local social space', *Planning Practice and Research*, 29(1), pp. 1–20.

Symons, J. (2018). '"We're not hard-to-reach, they are!" Integrating local priorities in urban research in Northern England: An experimental method', *The Sociological Review*, 66(1), pp. 207–223.

Wallace, A. (2014). 'Gentrification Interrupted in Salford, UK: From new deal to "limbo-land" in a contemporary urban periphery', *Antipode*, 47(2), pp. 517–538.

Further reading in this volume

Chapter 6: Listen, connect, act
Kim Cook
Chapter 9: From the dust of bad stars: disaster, resilience, and placemaking in Little Tokyo
Jonathan Jae-an Crisman
Chapter 12: Public transformation: affect and mobility in rural America
Lyndsey Ogle
Chapter 14: Experts in their own tomorrows: placemaking for participatory climate futures
Paul Graham Raven
Chapter 18: '*Homomonument* sounds like a poem': queer placemaking 30 years on: a conversational dialogue with Thijs Bartels, author of *Dancing on the Homomonument* (2003)
Martin Zebracki
Chapter 21: Placemaking through parkour and *Art du Déplacement* (ADD) as a Singaporean applied performance practitioner in London
Adelina Ong
Chapter 24: Artists, creativity, and the heart of city planning
Tom Borrup
Chapter 40: Transforming community development through arts and culture: a developmental approach to documentation and research
Victor Rubin
Chapter 41: Rituals of regard: on festivals, folks, and findings of social impact
Maribel Alvarez
Chapter 42: Creative Placemaking and Placekeeping evaluation challenges from the practitioner perspective: an interview with Roy Chan
Maria Rosario Jackson
Chapter 45: How the city speaks to us and how we speak back: rewriting the relationship between people and place
Rosanna Vitiello and Marcus Willcocks

24
ARTISTS, CREATIVITY, AND THE HEART OF CITY PLANNING

Tom Borrup

> Strategy making in the planning field requires complex imaginative, intellectual and technical work, involving a wide range of sources of understanding and imaginative power.
>
> *(Patsy Healey, 2010, p. 188)*

Introduction

Artists are celebrated for their imaginative powers – creative inclinations and skills they act on and foster throughout their lives. To provide the power British planning scholar Patsy Healey calls for above, some artists have applied themselves to city planning. In my work as a planning consultant in the United States, I've had opportunity to bring artists into many planning processes. Their skills are applicable in virtually all stages of planning, but my focus, and that of this chapter, is with artists in the public participation process. Projects I've been involved with span 25 different US states and have included artists from theater, design, visual, and fabric arts, writers, choreographers, vocalist/songwriters, amongst others. While some artists take their work into other arenas of public service, the focus in this chapter is based on my experience in neighborhood, district, creative economy development, and cultural planning. In these settings, artists have worked as co-designers, facilitators, and collaborators in data collection and interpretation. Projects are carried out directly with municipal agencies as well as independent actors such as community foundations, chambers of commerce, or other service organizations. Sometimes we work on stand-alone plans, sometimes on integrated plans side-by-side with city planners or other consultants with expertise in areas such as transportation, parks, and economic development. Cities have included Worcester, Massachusetts, South Salt Lake City, Utah, Lawrence, Kansas, Des Moines, Iowa, Marquette, Michigan, Oklahoma City, San José, California, Dublin, Ohio, Minneapolis, St. Paul, and Bloomington, Minnesota, as well as many smaller towns. In this chapter I explore, from a first-person and practitioner perspective, theory, history, and experiences related to the roles of artists in advancing these practices. While this work is not unique – and I'll reflect on a few of its earlier practitioners – it remains nascent and is slowly finding acceptance within the formal city planning profession.

City planning and art making

The field and practice of city or urban planning is increasingly critical to the future of humanity as cities try to find solutions to a growing list of challenges, ranging from basic infrastructure maintenance, to climate emergency adaptations, tensions related to global population movements, to place-based health and housing disparities, and transportation, to name a few. With public sector resources being squeezed while cities confront increasingly complex challenges, it is important to find ways to do more than one thing at a time in civic work. Expanding the imaginative capacities of the public sector and promoting engagement in local democracy present enormous challenges, especially in a neoliberal environment bent on dismantling people-centered institutions and policies. Creative engagement practices are among the contributions artists can make to city planning, community development, and civic institutions generally. In normative public process, planners try to inform participants and engage them in sometimes-heated discussions in response to professionally developed proposals. Artist-led, culturally attuned creative practices when applied to the public participation process can result in multiple outcomes. These include expanding the diversity of people involved, eliciting higher-quality local knowledge, and making participation more meaningful, enjoyable, productive, and less contentious. Well-planned and facilitated public process can build social connections, gather more and different data, approach challenges with fresh eyes, activate group imaginations, find common ground, lead productive co-design activities, reframe complex problems, strengthen participatory democracy, among other things. These are tall orders and not every artist is inclined or prepared to jump into situations calling for such outcomes. Although still far from mainstream, artist-led creative engagement represents a growing practice among cities in the United States.

Over the past decade cities such as Los Angeles, New York City, Boston, Minneapolis, and St. Paul have been among two-dozen cities (at the time of writing) employing artists in various roles and departments including planning. These range from regional and comprehensive city plans to neighborhood and district planning to special topical plans including parks, culture, and transportation, for example. Artist Jackie Brookner, with the city planner in Fargo, North Dakota, launched the *Fargo Project* (2017), a multi-year process in 2010 to address recurring flooding. Brookner engaged residents in creative and social activities that led to the development of an 18-acre site as a storm-water detention system that doubles as a multipurpose neighborhood commons. Multiple artists began working in the City of Minneapolis planning department in 2013. Through a program known as *Creative CityMaking* (2019), launched by the nonprofit Intermedia Arts, artists and planners devised and deployed a variety of techniques to increase public involvement in local planning. A mobile engagement theater pulled behind a bicycle entertained people while soliciting comments; a mobile tracing unit involved passersby at street fairs and other community events to observe and trace the urban landscape as a way to see it differently. Another engaged young people through making zines and comic books that told stories of their neighborhood. Through an artist-in-residence program in New York City launched in 2015, artists engage across many city agencies (Alliance of Artist Communities, 2019). Tania Bruguera was the first selected to help build trust between immigrants and public service agencies. A similar program in Boston began in 2016 engaging seven artists to address racial equity across multiple city platforms (City of Boston, 2019). A year later, the Boston-based Metropolitan Area Planning Council, a regional planning agency, brought on a playwright and performance artist to help staff improve strategies for involving the public across the 101 towns and cities the agency covers (MAPC, 2019).

Artmaking involves repurposing raw materials to create value, beauty, and new meaning as well as to identify unique and different purposes for those materials. Moreover, artists needn't limit themselves to conventional materials. They might work with sound, movement, color, objects, buildings, spaces, neighborhoods, relationships, sense of identity, or social systems, just to name a few. Applying artmaking practices to such tangible and intangible assets of a neighborhood or city can change the game and open new possibilities. Creative engagement can take stakeholders on symbolic or metaphorical journeys engaging in learning, change-making, and community-building together. This enables and empowers community members and public officials to learn more about each other and themselves. It can move them to a new level where they make things together, find new connections, form new narratives and new relationships, and experience collaborative action.

Planning and public process

Public participation in city and community planning has its own complex history. After half a century as an institutionalized part of municipal planning processes in the United States, it is still gestating and still seeking ways to be more engaging and effective. Working alongside city planners it is evident that, to some, required public processes are perfunctory, annoying requirements, and for others an opportunity to 'sell' their ideas to a skeptical or uninformed public. Some planners find public responses challenging, while others discover useful information to help iterate and improve plans. Creative processes led by someone outside the formal municipal structure can take public participation to another level. Veteran Canadian cultural planner Greg Baeker attests: 'The tools of the artist are an essential part of how we imagine cities: through stories, images, metaphors, exploring possibilities as well as critiques' (2002, p. 24). Tools range from techniques to find shared vision, exercise voice, articulate ideas through song and movement, share stories, and celebrate the collective making of something new. Notable urban theorists of the late- nineteenth and mid-twentieth century, including Patrick Geddes (1915/1949) and Lewis Mumford (1961), advocated citizen participation in planning before it was generally accepted in the practice, promoting 'civic exhibitions on urban and regional issues, surveys, and through input to the creation of planning alternatives or scenarios. Mumford saw plans as "instruments of communal education"' (Baeker, 2002, p. 23). Early city planning required engineering and organizational skills to coordinate resources and materials to implement expert-designed schemes. In contemporary planning, it is now recognized that planners need social and political skills more than, or in addition to, technical skills. In the 2006 *Just Planning*, British scholar Heather Campbell wrote: 'The recent history of planning thought has seen the replacement of the planner as instrumental rationalist by the planner as facilitator' (p. 103).

Some urban planning literature touches on creative techniques but largely ignores partnerships with creative practitioners (aka artists). American planning scholar John Forester, a long-time proponent of public participation and of planning as a learning and deliberative process, references theatrical terms so often in his writing, it's surprising he doesn't advocate recruiting theater professionals. In *The Deliberative Practitioner* (1999), Forester described 'the staging of deliberation in planning' (p. 236). Forester called for 'creativity and finesse' as well the importance of improvisation. Like Campbell, Forester calls on planners to reframe part of their work as 'process managers,' to engage people in idea generation and problem-solving. Forester also wrote about, 'the power of the sketch, the power of visual inquiry' (1999, p. 109) as potent tools in public engagement. In more recent writing, Forester (2018) described planning as, 'real drama – drama no less powerful, no less moving, no less instructive, and no less illuminating for being set on the stages of our cities and neighborhoods' (p. 15). He continues to call on planners to create environments conducive to productive idea generation, discussion, and deliberation of solutions to planning challenges – precisely the things artists do well. Creative environments artists facilitate change not only the expectations of participants but how they relate to planners and to one another. In her advocacy of 'unlearning the colonial cultures of planning', British planning scholar Libby Porter (2016) asserted one step is to 'place people in a different relation with each other: one of service, not of winning the argument' (p. 158). Planners, in their important role of 'process managers' (Forester, 1999), 'exert real influence ... by shaping processes of inclusion and exclusion, of participation and negotiation' (p. 90). In Forester's experience, some of the most productive community deliberations took place in meetings that 'brought them all together, not abstractly by appealing to "community building" or to some unitary public interest, but *performatively* – practically, interactively – over cards, drinks, music, camaraderie and dance' (2018, p. 21; emphasis in original text). For Forester, these cultural practices served only to form social connections. Relationships are an important dimension of the work. However, the creative activities Forester describes did not fully integrate creative thinking and cultural difference into the process of gathering information, devising solutions, and deliberation. Forester simply observed that sociability greased the wheels for planners to get on with their work.

Planning as storytelling

Among the most articulate advocates of the importance of story and storytelling in planning is Australian planning scholar Leonie Sandercock, now based in Canada. Sandercock cited growing diversity in what she calls 'mongrel cities' and sees stories, as well as other creative narrative practices, particularly filmmaking, as ways to open planning processes to the increasingly diverse people and cultures inhabiting cities of all sizes. This brings threefold benefit: the expansion of practical tools, the sharpening of critical judgment, and the widening of democratic discourse. Supporting such a need for a multivariate input and criticality in the planning process, Porter further called for planning to be 'self-aware as well as world-aware' and working in the service of intellectual and emotional connection (2016, p. 158). Building connections between people may, in fact, be the root purpose of planning. Borrowed from ethnography, the practice of collecting stories and bringing 'local knowledge' into the planning process has emerged as a critical part of good planning. Local knowledge is best derived through storytelling and focused listening. Australian planning scholar Paul Maginn (2007) described how techniques of applied ethnography offer planners a way forward in achieving more effective community participation. Steven Dang, a Canadian scholar-practitioner, characterized artists as the storytellers of their communities: 'They can provide a planner not only deep insight into a community, but ready-made and powerful means of communicating them' (2005, p. 124). Dang advocates that planning education should include mining the rich meanings in stories and cultural artifacts. Planners need to take advantage of their communities' story-collectors and storytellers.

Studies related to creativity and innovation consistently cite play, experimentation, and risk-taking as critical elements for new discovery. Those elements are rarely described in municipal procedure manuals. Creative city theorist, consultant, and writer Charles Landry focuses his more recent work on opening municipal agencies and processes to creativity (see: *The Creative Bureaucracy and its Radical Common Sense*, with Margie Caust, 2017). Conducting public engagement and planning work using routine settings and formats along with cut-and-paste solutions is sure to fulfill the colloquial definition of insanity. City bureaucracies are notoriously risk-averse, necessitating new ways of thinking in order to just open themselves to creative process. New thinking doesn't come from the center or mainstream of any profession. Computer science scholar Gerhard Fischer reminded that the edges 'are where the unexpected can be expected, where innovative and unorthodox solutions are found, where serendipity is likely, and where old ideas find new life' (2005, p. 5). This is another way artists can be helpful. They tend to be comfortable on the edges, outside the mainstream, outside the box. Innovation, play, experimentation, and risk-taking are core to their approaches.

Art as data

Artists' training and practice emphasize observation, listening, questioning, and intuiting the shape, meanings, stories, and symbols in their surroundings and in human actions, as well as the possibilities in materials. Planners are not trained to approach their work this way, or to understand, or even value such data. This is where extended dialogue between artists and planners is necessary. Rich and multi-layered data revealed through creative process is not always readable by planners. British-born design scholar living in Australia Paul Carter (2015) characterized planners and designers as dramaturgs. Dramaturgy is the practice of contextualizing, interpreting, and adapting story for the stage, giving structure and representation of the key dramatic elements appropriate for the audience and the time. Carter also constructed the term 'choreotopography' to blend movement and spatial thinking. There is much work for artist-generated data to have meaning for planners and these conversations are increasingly taking place. Creative methods employed by artists can offer needed dimensions in planning, listening to and appreciating the nuances in diverse cultural traditions of the residents who make up contemporary cities. Such methods also help engage people of more diverse ethnic and economic backgrounds in public process and draw people more deeply into contributing to the analytical and visioning work of planning. Artists who are culturally attuned to – better yet, part of – intersectional traditions, cultures, and ways of thinking can involve diverse communities sharing geographic places in neighborhoods and cities in the co-creation of solutions to community challenges, engaging people in devising more viable alternatives.

Perhaps because urban planning as a profession grew from technical roots in engineering and architecture as well as data-driven policy development, it has resisted the creative disciplines or working with artists as partners. California planner/artist James Rojas is among those who are making the case that data are not limited to things that can be counted. Among urban planning students at MIT, where he graduated in 1990, Rojas described how his creative inclinations made him the 'odd duck.' Rojas made models and analyzed space not through quantitative methods but as a creative visual practice. To Rojas, the city 'is a spatial, visual, and material culture that ignites memories, emotions, and aspirations' (personal communication, January 9, 2019). This, Rojas said, has to be understood through storytelling. In Los Angeles in the early 2000s, while involved with the arts, Rojas developed his practice of putting model-making to work to help people tell their stories. Rojas wrote in a 2018 *Planetizen* blog, 'Stories increase public involvement in city planning, increasing awareness of how the built environment shapes people's lives and how they can shape it in return.' Rojas estimated he has conducted 500 workshops in cities across the United States and Mexico, yet his practice is far from embraced by mainstream planning: 'I ask participants to collaborate in building their ideal community using familiar objects such as hair rollers, popsicle sticks, pipe cleaners, and other material' (ibid.). Rojas considers this immersive storytelling that 'allows participants the physical activity of reflecting, touching, moving, and playing with objects on imaginary maps' (ibid.). Rojas explained that participants are able to, 'quickly inquire, discover, prototype, and experiment with solutions. They learn their city is occupied by thousands of stories, seen or unseen, and all these stories collectively have power to promote physical change' (Rojas, 2018). People gain a greater sense of agency, common purpose, and shared vocabulary when they make something together than when they simply try to talk about it.

Similarly, in his work on deliberative process and through his research with planners, Forester observed cases in which planners who draw rendered informal sketches or concepts for community members to respond to. This encourages curiosity and play, what he called 'interactions of inquiry before argument' in service to stimulating participants' imaginations of possible futures. Sketches, collages, sculptures, photographs, and other creations of participants in planning processes provide valuable data. Planners, however, are not trained to interpret or analyze them, and as such are often inclined to dismiss them. Stories, sounds, and movements elicited in response to prompts or to community challenges can also serve as valuable data as is standard in sociological research. How to best harvest and analyze such data in planning is an emerging practice – one rich in possibilities. Artists and planners can borrow a page from qualitative research, especially ethnographic and phenomenological studies, to systematically collect and code data – stories, interpretations of drawings, photographic images, movement scores – to find hidden meanings and patterns that can lead to meaningful findings and conclusions. Planners spend considerable time in universities learning to read data expressed through numbers. Artists and art historians learn to work with different symbols. Ethnographers and dramaturgs find meaning in nuances of stories. There are many interpreters who can read more than one language. More are needed.

Diversity and cultural competency

One challenge facing most city planners in the United States is growing ethnic and racial diversity that brings different culturally based ways of working, different relationships to local government, and different uses of public spaces, among others. Diversity both complicates and enriches planning work as Sandercock (2003, 2004) has written about extensively. Planners need more ways to foster dialogue and creative problem-solving while the environments they work in become more complex. Canadian planning scholar Neil Bradford (2004) described the planning field as largely unprepared for many challenges including diversity, creative thinking, economic inequity, and issues around environmental sustainability. As current social and natural environments elevate the importance as well as potential value of greater participation, new skills are called for to manage and maximize the meaning and value of the involvement of more and different people. Such skills are absent from the technical and analytical training most planners are equipped with (Kunzmann, 2004). Planners and policymakers often set up planning processes and local partnerships with insufficient knowledge of local cultures (Maginn, 2007). These increasingly include the presence of 'global' cultures in 'mongrel cities' (Sandercock, 2003, 2004). American urban planner Leo Vazquez (2009, 2012) asserted that most planners lack 'cultural competence.' He argued that most planners lack reflective understanding of even their own cultural practices and biases, let alone appreciation and understanding of the cultures and practices of others.

The goal in public process is for community members, planners, and other public officials to come together in a deliberative environment to, 'learn together and craft strategies to act collaboratively' (Forester, 1999, p. 4). Before this can happen, these parties have to achieve basic trust and confidence in the process. This is complicated, Forester asserted, 'when they distrust or even detest one another, when they inherit painful histories of eviction or terror, racial hatred or sexual violence' (ibid., p. 4). Cultural competency – also known as intercultural competency – in planning education is one building block to work through racism and troubled histories. American planning scholar Julian Agyeman and planner Jennifer Sien Erickson argued that, 'recognizing, understanding, and engaging difference, diversity, and cultural heterogeneity in creative productive ways requires cultural competency' (2012, p. 358). They built on Vazquez's definition describing it as 'the range of awareness, beliefs, knowledge, skills, behaviors, and professional practice that will assist in planning for, in, and with "multiple publics"' (2012, p. 361). This enables people to appreciate difference, to work in cross-cultural situations, and to proactively engage with diversity and promote intercultural dialogue.

For a planner to develop an awareness of their own cultural vantage point and biases is at least as critical as learning about others, asserted Agyeman and Erickson. This impacts 'how their own conscious and unconscious assumptions, beliefs, knowledge, and desires affect their ability to listen well and understand other cultures' (2012, p. 359). However, planners, need to understand more than just the people in the neighborhood or in the room. They need to understand the cultural specificity of existing structures and processes they employ and how their attitudes towards nonprofessionals in communities impact on the experiences of different groups (Maginn, 2007). Meanwhile, planning practitioners find themselves bound by established practices that sometimes include prescribed ways of conducting meetings. Like everyone, they are captive of their own biases and experiences and often pull from a limited vocabulary of aesthetic choices, land use patterns, and ways of involving people in civic dialogue. Motivating engagement of those who feel, and actually have been, excluded requires more than impressing on them the importance of choices over street widths and zoning variances. 'Deliberative practice and participatory processes will fail,' wrote Forester, 'if planning analysts pay so much attention to technique or "substance" that they ignore and dismiss the history and culture, the self-perceptions and deeply defining experiences, or the citizens involved' (1999, p. 245). When co-designing and leading these processes, artists have demonstrated remarkable abilities to surface unique experience, honor histories, bridge differences, and move people to constructing something together, even if symbolically.

Creative engagement: setting the stage

This is not the first published writing to argue the merits of artist-led engagement practices in the planning arena or to reflect on such experiences. While by no means a history of this work, a few notable forerunners and trends in the practice are reviewed here. Among early documented practitioners who integrated creative disciplines with formal spatial planning were American avant-garde dancer Anna Halprin and her landscape architect husband, Lawrence Halprin. Based in the San Francisco Bay Area, their 1960s and 1970s experiments brought together their respective practices. British geographer Peter Merriman described how the Halprins attempted 'to rethink landscape architecture and dance through the understanding of the spaces of choreography and performance, and the performativities and choreographies of spaces' (Merriman, 2010, p. 431). The Halprins not only tried new ways to conceptualize spatial planning but new ways to involve a wider mix of people in articulating spatial thinking. They combined their practices and progressive politics to democratize choreography and dance performance, to include people of color and other marginalized people in collective creativity to democratize landscape and urban design. According to Merriman, they developed and practiced community processes 'grounded in principles of group therapy and an emphasis on participation, direct personal experience and play' (ibid., p. 439).

Australian planner Wendy Sarkissian began her creative practice in the 1970s, publishing in 2010 a reflective book, *Creative Community Planning: Transformative Engagement Methods for Working at the Edge*, with artist-planner Dianna Hurford. In summing up their experiences as creative planning consultants, they described their efforts, 'to meet at a place of creation that calls new, informed and meaningful ideas into existence through rationality, integration, community knowledge and experience' (2010, p. 7). Their book traces a wide-ranging body of work in communities of all types and sizes across Australia. They experimented with performance, music, visual arts, and stories, worked with local artists and activists in planning, 'Listening to stories, identifying common goals and forming partnerships in action: this is creative community engagement – engagement that is as much about learning as doing' (ibid., p. 154). Since the late 1990s Steven Dang, a scholar-practitioner also worked with communities in Canada in creative community engagement. Dang uses the artistic process to 'tap the creative collective potential of participants,' (2005, p. 123), continuing, 'While the planning profession may be reluctant to engage in community cultural development work, community-based artists are hard at work in community planning' (ibid.) Dang's own work focuses on activating local artists and imaginations. Dang described gathering stories as well as images from local artists (artists defined loosely) – images that tell profound stories into community values and categorizes three ways to employ art in planning: 'Art as Window,' 'Art as Dialogue,' and 'Art as Doorway' (pp. 123–25). In this construct, creative processes help participants see, discuss, and move to new places together.

Artists of many disciplines working in community settings, since at least the 1960s, have developed an extensive array of techniques and practices to foster community building and promote engagement in social issues and civic affairs. American cultural activists and writers Bill Cleveland (2000) and Arlene Goldbard (2006) both chronicled community-based arts and 'community cultural development' through which artists across disciplines developed approaches to activate engagement in community through culturally based creative work. Their writing surveyed many of the ways artists impact communities from spatial change and policy-making, to social network building and organizing around social justice issues. This work prepared the ground for much of the artist activism, and artist involvement in planning and civic work of the current era. Dang observed how 'community-based arts practice often demonstrates community planning at its best: strengths-based, capacity-building, participatory, inclusive, communicative, reflective, innovative and adaptive' (2005, p. 124). This work is also part of a wider 'turn to community in the arts' described by Australian sociologists Danielle Wyatt, Lachlan MacDowall, and Martin Mulligan. They described a trend beginning in the late twentieth century, 'a surge of new or renewed interest in the idea of community across all art forms' (2013, p. 82). Multiple forces, they assert, have contributed to this turn to community, one they observe stems from 'artists' attempts to bridge relationships between aesthetic and activist practices; a dissolution in the boundaries between cultural, social, political and economic domains; and the increasing instrumentalization of the arts' (ibid., p. 81).

Process as product in planning: four case studies

Four brief stories follow based on some of my recent work involving artists in community planning contexts. The investment needed to include artists varied widely from one-time facilitation activities to extensive 12–18-month involvements in conceptualizing, implementing public engagement, and harvesting relevant planning outcomes. Most of these artists had extensive backgrounds in community-based work and/or teaching and were eager and adept at welcoming and mentoring less experienced artists. Most of the planners involved readily admit they learned much from the artists. One planner vowed he would never conduct community planning again without involving artists.

Community building as patchwork: a North Dakota case

In the Spring of 2015, the planning team I was part of was commissioned jointly by the City of Grand Forks, North Dakota and a local foundation, and invited local quilting artist Sarah Heitkamp to lead an eclectic group of over 80 community members through the steps of cutting and arranging fabric into small squares. Invited to a community planning meeting, some participants were caught unaware, but all engaged constructively and ultimately appeared to enjoy the experience. This was 18 years after a massive flood and fire devastated this fiercely independent community. Residents had worked hard to put their city back together, yet early in our planning process we sensed a piece still missing. Residents had done the long and sometimes painful work of physical and economic reconstruction, yet they seemed to have lost the joy in building a community together. Along with quilt-making, a local music and dance group performed while residents assembled their squares into a large, collective patchwork on the floor. As the work came together, a shift was evident. Young and old passed around the microphone to explain the meaning of their contribution to the quilt and how their piece contributed to the community. Without judging the artistic merit of any square, they applauded each other celebrating their creative achievement. Making artwork together under the guidance of a locally known, accomplished artist, while they discussed their vision for the community's future, brought a critical element of joy in collective achievement back to community work.

Learning from diversity in an urban center: a Minneapolis case

A year-long downtown Minneapolis district planning project I led in 2011–12 was designed to engage a diverse mix of stakeholders in a variety of ways in accordance with the complex nature of the linear district. The project was funded by grants and supported by the Hennepin Theatre Trust, operator of several large downtown theaters, and the City of Minneapolis. While the process leaned largely on local knowledge, it included outside expertise to push the boundaries of thinking and to inform the process. Visitors included New Orleans-based artist Candy Chang, British creative city visionary Charles Landry, American urban geographer Don Mitchell, and Los Angeles community development leader Chanchanit Martorell. Set a month apart, their presentations were complemented by a parallel set of artist-led public-participation workshops each month. Participation activities also addressed long-term capacity-building by creating and/or strengthening relationships among stakeholders in the practice of problem-solving. The project drew wide attention and discussion to both assets and challenges of the district and built understanding of how resources and creative energies of various institutions and businesses could be collectively employed to take on challenges from homelessness to street crime to unwelcoming urban design to the demonization of teenagers. A challenge facing the district was one of connectivity – connectivity between institutions, businesses, and a wide range of stakeholders. With a wealth of cultural assets and historic meaning, Hennepin is considered the city's 'main drag,' the heart of its nightlife.

Moving beyond physical assets to the collection of stories was an important step. For part of this work the project engaged several youth organizations working with artists. Youth interviewed, videotaped, photographed, wrote poetry and music, and created radio spots highlighting stories of people in the district. Products were presented in numerous venues on the Avenue including the central library, art museum, office building atrium, and public school as well as through social media. Among the challenges with the project was devising participatory activities to involve the diverse mix of people who make the avenue their space. Stakeholders ranged from white suburban families attending the Disney Lion King, LGBTQ club-goers, basketball fans from the adjacent arena, and black teenagers hanging out. The district represents a kind of urban space and experience unfamiliar to many Midwesterners.

To design and assess the activities we conducted multiple meetings with four locally based lead artists, urban designers, and architects, with regular input from three institutional project partners and a representative of the City. Artists included theater director, Harry Waters Jr, visual artist, Ta-coumba Aiken, choreographer, Leah Nelson, and vocal artist/songwriter, Mankwe Ndosi. All the artists were African American. They used facilitation techniques attuned to diverse cultural sensibilities appropriate to this urban place. The core team also included urban designer Peter Musty and a team of architects led by landscape architect and planner Bob Close. The challenge for the team, similar to that described by Sarkissian et al., was to create 'spaces of trust for different kinds of stories to emerge and for people to express themselves in their own vocabularies' (2010, p. 13). Activities exercised every voice singularly and in unison, practiced listening to others, moved in relation to others, and drew visions of the future to activate and bring forth a tapestry of ideas that were incorporated into the plan. As importantly, the Hennepin Theatre Trust continues a robust program of placemaking and public art engaging stakeholders on an ongoing basis.

Testing the limits: a suburban case

In a 2014–15 project in a growing Minneapolis suburb of Bloomington, I was joined by a team of four artists and two planners, including artists from the Hennepin Avenue project, Aiken, Nelson, and Waters, and urban designer Musty, along with graphic artist Witt Siasoco and landscape architect Carrie Christensen. The team conducted multiple projects across 18 months. These included a 7-day charrette under tents in a central park location. It was more of a festival of artist-led discovery and planning that included walking and bike tours, an idea competition, dialogues with historians, planners, and developers, in addition to food, music, dance, ice cream socials, and more. One small project I'll describe responded to the municipality's articulated desire to incorporate creative placemaking and public art in new development and existing neighborhoods. This mini-project tested the local government's permit review procedures through what they call the Design Review Committee (DRC). This group included city planners, fire marshal, police department, economic development officials, risk manager, and others. We selected several projects proposed by community artists for a mock review by the DRC on the stage of a small auditorium with an audience. The most revealing and simplist artist proposal was to place a soapbox – a small raised platform – in a public space to be used at will by members of the public with something to say. Multiple members of the DRC found significant problems with the proposed soapbox. Citing existing regulations and public safety concerns, the DRC declined the proposal. This did not bode well for the potentials of a public art program! This semi-staged theatrical exercise shed light on the gap between desired ends and existing policies. That said, the City established an ongoing creative placemaking commission and program and the community has since become rife with public art.

Finding stories of place: a rural Midwest case

The story circle is a vehicle long employed by community-based theater practitioners and served as a useful tool for a 2012 project. An 8,000-square-mile rural watershed area spanning parts of Wisconsin and Minnesota, known as the St Croix River Valley, was identified by leaders in the region for possible formation of a US Park Service National Heritage Area – a locally managed network of cultural, educational, and touristic resources organized around a meaningful historic or cultural theme. A foundational task was to identify whether people across the region shared a substantive connection – a shared story – that was broadly meaningful, well-supported by existing historical assets and relevant to the region's future. With oversight of a stakeholder-based steering committee representing the vast geography, our team was invited to assemble artists and planners to help identify that critical story. The heritage project was more than a simple listening tour and required core artists and teams of volunteers to conduct each of 11 gatherings in a consistent fashion. Considerable care went into devising a methodology and preparing teams of planners, artists, and volunteers. This enabled the process to engage people more deeply through participation in story circles and careful listening with systematic documentation of every story.

In each of the 11 counties in the region, community members attended workshops that resulted in 438 discrete stories grounded in an accumulated 10,000 years of experiences. One of the warm-up activities employed at each gathering was to ask every one of the round tables populated by six to eight participants to add up their total years living in the region. This allowed some to claim many years of wisdom and others to claim the freshest perspectives – all valued equally. The next stage involved distilling data from 438 stories to identify four most prominent themes for further consideration. We then chose a process that represented some risk and had at least one vocal skeptic. A youth theater troupe based in one of the small towns in the region was engaged to work with an area playwright to turn the four themes back into stories through three-minute scenes. The youth troupe acted out each of the prospective themes at four strategically located subregional gatherings. Rather than a presentation by a historian or economic development specialist, teenagers illustrated the themes through theater. As a clear majority of participants in all the workshops and summits sported gray hair, this brought fresh thinking as well as participation of a new generation. The response was phenomenal. Theatre brought the ideas to life in wonderful ways simple words or PowerPoint images could not. The cross-generational relationship-building added a critical dimension not only to the planning but to the meaning of the prospective Heritage Area.

Finding joy in civic engagement

In his 100-year survey of urban planning, American planning scholar William Rohe examined the evolution of local and neighborhood-based approaches citing important lessons. One of those is 'that local social relations and networks matter greatly to people and should be given great weight in revitalization planning' (2009, p. 216). Building, rebuilding, and maintaining social networks that are open, equitable, functional, and leave community members a sense of fulfillment and even joy are essential for every community. I learned during the 1980s and 1990s, through my involvement in community-based arts that artists can be highly effective organizers, problem-solvers, and literal, as well as metaphorical, builders of alternative social, organizational, and physical community networks. For over 20 years I led an arts organization that served as ground for convening, conducting dialog, forming partnerships, and staging action. I learned more and more about the capacities of creative processes and of individual artists to effectively assess group dynamics, identify solutions, and build bridges.

Through my more recent two decades as a community planner, I've come to see how creatively activating people in planning can help stakeholders better tap their thinking and use multiple expressive forms, feel part of a group, and create a more constructive or positive 'vibe' around public process. I've come to appreciate that how planning is conducted significantly influences the outcomes. And, I learned that every community activity must include food and beverage for participants to share. At the same time, it's worth noting that artists' efforts, participatory processes, and even food for community members can be a smoke screen. For those involved, it can become something that New York planning critic Samuel Stein described as cover, 'designed to make them feel good about losing' (p. 190). He described how major real-estate interests have inordinate control of city governments and how planners find themselves boxed in with few options but to serve the interests of capital over residents. It's imperative, therefore, that planners and their creative partners work towards building community capacity – helping diverse people in their communities find common ground, build their imaginative and collaborative muscles, learn how the wheels of government work, and find joy in the work so they're more likely to do it again!

Conditions conducive to fostering positive interactions, activating imaginations, and facilitating deliberative practices critical in urban planning are precisely the conditions I have experienced through artist-led practices. Orchestrating or staging new and meaningful ways for people to be together and to share fulfilling experiences is exactly what theater directors, choreographers, poets, or musical conductors do brilliantly. Dang asserted that 'as a means of conversation, the arts are often more accessible and inclusive than the standard town hall meeting or open house' (2005, p. 124). Creative environments are less intimidating, less judgmental, and often provoke a good degree of simple social enjoyment and sense of satisfaction. Artists can bring skills to help people find and amplify their voice especially for individuals less skilled at verbal debate or reluctant to stand up in a community meeting. 'Art can be that important initial point of entry, transcending language and providing opportunities for residents to learn to work together on shared projects,' wrote Dang (p. 125). Unlike planners, artists are not expert in the technical details of planning, nor do they often have direct access to the levers of decision-making in municipal government. As such, they provide an ideal complement to planners as they focus on the human interactions, deeper meanings, and problem-solving processes. As Baeker observed, 'the tools of the artist become key to the participation of all' (2002, p. 24). Artists draw on local knowledge and how it can be relevant to the questions at hand. They are inquisitive and deconstruct things to reframe issues and bring participants on a journey together to new ways of understanding the challenges their communities face. Artists can't provide answers to all the challenges planners and communities face. They can be potent collaborators and facilitators to help communities work together to devise solutions.

References

Agyeman, J. and Erickson, J.S. (2012). 'Culture, recognition, and the negotiation of difference: Some thoughts on cultural competency in planning education', *Journal of Planning Education and Research*, 32(3), pp. 358–366.

Alliance of Artist Communities. (2019) [online]. Available at: https://www.artistcommunities.org/public realm-NewYork (Accessed: 21 April 2020).

Baeker, G. (2002). *Beyond Garrets and Silos: Concepts, Trends and Developments in Cultural Planning*. Quebec, Canada: Department of Canadian Heritage. Ontario Ministry of Culture, Quebec Ministry of Culture and Communications.

Bradford, N. (2004). 'Creative cities structured policy dialogue backgrounder', Background Paper F/46 in *Family Network*. Ottawa: Canadian Policy Research Network.

Campbell, H. (2006). 'Just planning: The art of situated ethical judgment', *Journal of Planning Education and Research*, 26, 92–106.

Carter, P. (2015). *Places Made After Their Stories: Design and the Art of Choreotopography*. Perth: University of Western Australia Press.

City of Boston. (2019) [online]. Available at: https://www.boston.gov/departments/arts-and-culture/bos ton-artists-residence-air (Accessed: 21 April 2020).

Cleveland, W. (2000). *Art in Other Places: Artists at Work in America's Community and Social Institutions*. Oakland, CA: New Village Press.

Creative CityMaking. (2019). *City of Minneapolis* [online]. Available at: http://news.minneapolismn.gov/ 2019/04/26/3814/ (Accessed: 21 April 2020).

Dang, S.R. (2005). 'A starter menu for planner/artist collaborations', *Planning Theory & Practice*, 6, pp. 123–126.

Fargo Project. (2017). Available: https://www.thefargoproject.com/tag/jackie-brookner/ (Accessed: 21 April 2020).

Fischer, G. (2005). *Distances and Diversity: Sources for Social Creativity*. Boulder, CO: University of Colorado, Center for LifeLong Learning and Design (L3D), Department of Computer Science.

Forester, J.F. (1999). *The Deliberative Practitioner: Encouraging Participatory Planning Processes*. Cambridge: MIT Press.
Forester, J.F. (2018). 'Deliberative planning practices without smothering invention: A practical aesthetic view', in *The Oxford Handbook of Deliberative Democracy*. London: Oxford Handbooks.
Geddes, P. (1949/1915). *Cities in Evolution*. London: Williams and Norgate.
Goldbard, A. (2006). *New Creative Community: The Art of Cultural Development*. Oakland, CA: New Village Press.
Healey, P. (2010). *Making Better Places: The Planning Project in the Twenty-First Century*. Houndmills, Hampshire: Palgrave Macmillan.
Kunzmann, K. (2004). 'An agenda for creative governance in city regions', *The Planning Review*, 40(158), pp. 5–10.
Landry, C. and Caust, M. (2017). *The Creative Bureaucracy and Its Radical Common Sense*. Gloucestershire: Comedia.
Maginn, P.J. (2007). 'Towards more effective community participation in urban regeneration: The potential of collaborative planning and applied ethnography', *Qualitative Research*, 7, pp. 25–43.
MAPC. (2019) [online]. Available at: https://www.mapc.org/resource-library/artist-in-residence/ (Accessed: 21 April 2020).
Merriman, P. (2010). 'Architecture/dance: Choreographing and inhabiting spaces with Anna and Lawrence Halprin', *Cultural Geographies*, 17(4), pp. 427–449.
Mumford, L. (1961). *The City in History: Its Origins, Its Transformations, and Its Prospects*. New York: Harcourt, Brace and World.
Porter, L. (2016). *Unlearning the colonial cultures of planning*. Abingdon, UK: Routledge. https://doi.org/10.4324/9781315548982
Rohe, W.M. (2009). 'From local to global: One hundred years of neighborhood planning', *Journal of the American Planning Association*, 75, pp. 209–230.
Rojas, J. (2018). 'Let me tell you a story! storytelling to enhance urban planning engagement', *Planetizen*, February 14, 2018 [online]. Available at: https://www.planetizen.com/features/97224-let-me-tell-you-story-storytelling-enhance-urban-planning-engagement (Accessed: 21 April 2020).
Sandercock, L. (2003). *Cosmopolis II: Mongrel Cities in the 21st Century*. London: Continuum.
Sandercock, L. (2004). 'Towards a planning imagination for the 21st Century', *Journal of the American Planning Association*, 70 (2004), pp. 133–141.
Sarkissian, W. and Hurford, D. (2010). *Creative Community Planning: Transformative Engagement Methods for Working at the Edge*. London: Earthscan.
Stein, S. (2019). *Capital City: Gentrification and the Real Estate State*. New York: Verso.
Vazquez, L. (2009). 'Cultural competency: A critical skill set for the 21st century planner', in *Planetizen*, December 2009 [online]. Available at: http://www.planetizen.com/node/42164 (Accessed: 21 April 2020).
Vazquez, L. (2012). *Creative Placemaking: Integrating Community, Cultural and Economic Development*. Unpublished paper. Bloustein School of Planning and Public Policy, Rutgers University
Wyatt, D., MacDowall, L. and Mulligan, M. (2013). 'Critical introduction: The turn to community in the arts', *Journal of Arts and Communities*, 5(2–3), pp. 81–91.

Further reading in this volume

Chapter 1: Introduction: what really matters – moving placemaking into a new epoch
Cara Courage
Preface: Placemaking in the age of COVID-19 and protest
Jason Schupbach
Chapter 3: An annotated history of creative placemaking at the federal level
Jen Hughes
Chapter 4: A future of creative placemaking
Sarah Calderon and Erik Takeshita
Chapter 5: Making places for survival: looking to a creative placemaking past for a guide to the future
Jeremy Liu
Preface: 'Disastrous forces, accidental actions, and grassroots responses'
Tom Borrup

Chapter 11: *Free State Boulevard* and the story of the East 9th Street Placekeepers
Dave Lowenstein
Chapter 15: Un/safety as placemaking: disabled people's socio-spatial negotiation of fear of violent crime
Claire Edwards
Chapter 19: Placemaking in the ecology of the human habitat
Graham Marshall
Chapter 28: Integral placemaking: A poiesis of sophrosynes?
Ian Wight
Preface: The only thing constant is change
Kylie Legge
Chapter 34: Reconnecting *cité* and *ville*
Philip Graus
Chapter 36: Facilitator skills for effective collaborative placemaking
Husam AlWaer and Ian Cooper
Chapter 37: *The Neighbourhood Project*: a case study on community-led placemaking by CoDesign Studio
Lucinda Hartley, Eliza Charley, Sama Choudhury, and Harriet McKindlay
Chapter 38: Public seating: a small but important place in the city
Kylie Legge
Chapter 42: Creative Placemaking and Placekeeping evaluation challenges from the practitioner perspective: an interview with Roy Chan
Maria Rosario Jackson
Chapter 45: How the city speaks to us and how we speak back: rewriting the relationship between people and place
Rosanna Vitiello and Marcus Willcocks

25

'IF YOU CAN MAKE IT THERE, YOU CAN MAKE IT ANYWHERE...'

Cultural placemaking at the heart of cities

Sherry Dobbin

Progressively, there are more Business Improvement Districts (BIDs), town centres, and urban districts that are seeking cultural placemaking strategies and delivery plans. These dense urban places already have strong footfall, businesses, and tourists, yet they cannot build 'community' cohesiveness the same way as a residential or mixed-use development. From 2012–16, I served as the Director of Times Square Arts and Creative Director for Times Square Alliance. The experience has provided a robust framework for cultural placemaking elsewhere. This is my journey.

Times Square context

Every day, 380,000 international visitors, American tourists, and New York City residents pass under the mesmerising electronic billboards of Times Square, cross its asymmetrical collection of five pedestrian plazas, and encounter hawkers, business leaders, and Broadway performers. This area, that is now celebrated for being home to the most vibrant theatre on Broadway and a commercial epicenter of global commercial retail, finance, and creative sectors, was once notorious for its proliferation of porn shops, peep shows, and Kung-Fu movie houses. In the late 1970s and early 1980s, Times Square represented a bankrupt and broken New York City. From the late 1990s into the early 2000s, as antidote to Times Square's rapid change, and in honour of their early intentions for the area, independent arts organisations found ways to infiltrate the commercial mecca. Economic development and city incentives to businesses (such as Disney) soon encouraged investment in the area and the City worked to clear the crime. New business boomed, rents soared, and tourists began to flood the thoroughfares. As a result, the majority of artists and local residents and businesses were eventually relocated in exchange for high-commercial tenants, including internationally branded retailers and restaurants. After its earlier theatre-ticket-booth design competition, the Times Square Alliance latterly came to want a cultural programme as this new Times Square identity was overshadowing its longstanding edgy, creative, and flashy quirkiness. New Yorkers' favourite sport was to 'love to hate it' and there was a constant claim that it had lost the authenticity of place.

Taking back the city centre: experimental cultural capital of public space

My vision of success was to make Times Square New York City's cultural hub and to geographically and soulfully regain its role as City Centre. We needed visitors seeking spectacle as well as authenticity; we needed a diverse exploration of narrative of place for businesses to attract world-class talent to headquarter there. It was necessary to raise the ambition of cultural delivery in the programming to showcase the city's cultural infrastructure prowess through institutional partnerships in order to gain loyalty from the jaded New Yorker. As a Business Improvement District (BID), the Times Square Alliance needed to set firsts, take risks, and generously share its learning as Times Square, by default, is an exemplar for cities of all scales. Our precedent would indirectly encourage other cities to be bolder.

Times Square Arts: getting started

In 2010, Times Square Alliance initiated a public art programme to bring the creative community in to counterbalance the commercial visual cacophony. A democratic, non-curated approach was taken that relied upon widely distributed open calls (Requests for Proposals or RFPs), so that the BID was seen to be supporting the presentation of arts, not curating them. An impressive Arts Advisory Committee (including cultural stalwarts as Thelma Golden, Head of Studio Museum of Harlem, Barry Bergdoll, then Chief Curator of Architecture at Museum of Modern Art, Anne Pasternak, then Executive and Artistic Director of Creative Time and later Executive Director of Brooklyn Museum, and Tom Finkelpearl, then Executive Director of Queens Museum and later the Commissioner of Cultural Affairs for New York City) alongside Times Square Alliance Board Members and representatives from the Department of Transportation, Public Art Commission, New York City Parks) was established to inform direction, and a subset Selection Committee chose proposals from the open call. The programme was vulnerable however, with uncertainty over how this would deliver against key District objectives, how it might make a difference in local residents' perception of place, or how it would differ from commercial marketing events. The perception was that these were nice arts efforts when the presentation could be supported. However, The Times Square Alliance wanted to invoke cultural placemaking principles that were visible more regularly and consistently, and courting more ambitious proposals from artists, designers, and architects.

Listening is learning

When I joined the Senior Management Team of Times Square Alliance in 2012, it was a company with 130 employees and an annual operating budget of $15m. My job was to ensure that cultural delivery was valued as much as operations and economic development. I had come to the role with over 20 years of professional experience working in the arts across all forms and it was (and still is) rare to have a dedicated fulltime and senior-level cultural professional within a BID. It was very uncharacteristic to have an arts person sit within the Senior Management, under direct dialogue with the President of the Alliance, and collaborate with the Vice Presidents of Policy, Research & Economic Development; Operations and Security; Events/Marketing; Finance; Human Resources; Communications and Sponsorship Development.

Working outside an arts institution opened perspective and access to wider industries that make a city. I learned that our tourists were visitors and representing well-educated, high to medium income, which helped to redefine the tourist as an adventurer, who sought surprise. Our pedestrian-counting cameras taught me patterns of behaviour according to times of day at each plaza. I was exposed to exceptional issues in global place that was a gathering spot for important political and social protests, as well as capitalistic activity. I was invited to the Board Meetings (up to 55 board members) representing some of the largest developers and property leasers, financial and legal firms; 25 per cent of the hospitality industry in New York City; and entertainment services and theatres, as well as churches and major transportation authorities. I was present at the Executive Committee of eight of New York City's most influential Chief Executive Officer's (CEOs) in the New York Times headquarters. Listening to the challenges of the District became my creative inspiration and direction; I sought general cultural collaborations that would also positively counter the negative perceptions or challenges about the place. Giving artists and curators the problem was actually providing them with the intrigue and dramatic tension that was attractive. In a sophisticated city where the audience is fickle, one needs to surprise the locals to capture their attention. The problem, thus, became the advantage.

I inherited some past programmes from the open call proposals, and I had to begin renegotiation conversations with restrictions and questions, which meant that every question was perceived as criticism and reductive, rather than collaborative. Staff struggled to understand the strategic rationale for the programme without a strong mission. Communications were focused on individual stories and not making an impact. Funding applications didn't apply to well-financed BIDs. Consultation showed that there was suspicion amongst the serious artists and institutions for fear that working with a business-led entity would only exploit marketing-led activation, but without large budgets to support commissions. It was assumed that the audience needed simple. Quickly, I realised that we needed to frame strategic objectives for non-arts collaborators that also served as curatorial focus for the creatives. At the same time, I needed to start with the impossible. For two years, there had been attempts to create a Video Programme on the electronic billboards across the five blocks of the canyon that makes up the icon of 'Times Square.' While public art organisations Creative Time, Public Art Fund, and Performa had run programmes or projects on one screen at a time, there had never been a consistent commitment across multiple screens. Requesting free time on the individually owned and operated billboards that are expensive to run, maintain, and replace required exceptional content, time, and some form of return on investment. There are property owners, screen owners, subcontractors who sell the advertising. There are competitors who need to be collaborators. Curators had presented the information in formats familiar to their field but almost incomprehensible to the billboard owners. In order to make a big shift, one needs to take a big risk, very publicly. I had to expand my knowledge by working with the largest global monitors with the highest visitor numbers. There is no time to wonder if it can work, only time to figure out how it can work. As the billboards are the most unique aspect of Times Square's authenticity, they are the greatest symbol of the place for over 100 years.

Start with who and what you know

Our official opening May 2012 for the electronic billboards was Robert Wilson's *Video Portraits*. I had run Wilson's The Watermill Centre and knew that he wanted to see these high-definition-shot portraits of celebrities and animals in Times Square. The saturated colours would make the monitors look gorgeous and an avant-garde downtown artist in the commercial space would gain attention. Our Sunday, 11.45pm opening event in Times Square attracted some of the world's most notorious collectors, arts patrons, funders, artists, and celebrities – who self-confessed that they came because it was 'so odd to see that work, there.' Things began to change.

I worked with trusted curators from the cross-disciplinary Crossing the Line Festival presented by French Institute Alliance Français, who brought festival funding and accomplished international artists who wanted to engage with the city. I combined video programme and live performance to the visual arts offer, so we could leverage more press. We attracted a major funder, ArtPlace America, to enable more ambitious works and share our findings through blogs, talks, convenings, and dedication to online documentation. Rockefeller Brothers Fund funded more substantial visual arts installations with Cuban Artists Fund, due to our active curatorial role in developing the artist work, mentoring artist practice outside Cuba, and because of my past professional portfolio of producing complex, experimental works. Press communications became more aligned and the partners' cultural standing gave legitimacy to the programme and intrigue to expanded press outlets. I chose to align with the annual arts season(s) and imitate the cultural protocol of New York City. Our 'first season' beginning in September 2012 had an announcement of programmes staged across the electronic billboards and plazas, first creating visual posters of artist Bel Borba's work across the plazas. Our first tipping point was our partnership with Art Production Fund to present Yoko Ono's *Imagine Peace* (2012) on the billboards for the month of December 2012 and then creating a Happening where we invited the public to stop the world from ending (predicted to end on 12.12.12) by singing *Imagine* on the iconic Red Steps. We gained over 600,000 press impressions from 63 countries from a three-minute art event. We began to garner trust from the Board, billboard stakeholders, the staff and the creative community, and so it was time to...

... Structure it!

This vision required a more focused and proactive governance, and strong curatorial leadership that actively solicited the rich cultural organisations, festivals, and artists of New York City and from across the globe to challenge and illuminate the richness of heritage that lives amongst the mysterious, pulsating, chaos. We worked with a consultancy agency and created with them an articulation that informed the mission, vision, and core values, which have led the programme for seven years now. Times Square has always been a place of risk, innovation, and creativity, and thus Times Square Arts needed these qualities to remain central to the district's unique identity. Our Mission made a commitment to be daring and site-specific: Times Square Arts collaborates with contemporary artists to experiment and engage with one of the world's most iconic urban places. We established clear and ambitious goals to focus creative proposals as well as provide clarity to stakeholders. We established our Vision to create a more complex place: through Times Square Arts, Times Square aspires to be: an internationally recognized leader in integrating art into the urban fabric, a must-visit destination to experience ground-breaking, site-specific contemporary art that can't be found anywhere else, a place where boundaries are pushed and the spirit of creative risk-taking is embraced and celebrated – in business, entertainment, and art, and a catalyst for new and unexpected connections and conversations.

We used five Core Values to focus all creative proposals that would ensure keeping the ambition high and focused on our specific place, which doubled as the selection committee evaluation criteria: collaborative – producers, artists, and audience collaborate, not compete; responsive to place – the work is site-specific and could not happen elsewhere; boundary-pushing – Times Square is a place where risks are taken and boundaries are pushed; conversational – questions, not statements, encouraging dialogue, not assault; and transparent – Times Square is never shut, so incorporate the process into presentation. We then determined that all of our experiments had to be easily grouped according to the distinctive platforms we had to showcase art: *Midnight Moment*, on electronic billboards; *At The Crossroads*, on public plazas; *Hidden Assets*, in district venues off the Broadway plazas. To best understand how the simple structure can deliver extensive variety, I explain each platform to define it and give project examples that demonstrate the strategic curation

Midnight Moment

Once the Mission and Core Values were established, it was evident that the strongest place attribute could foster the programme as the world's largest digital art gallery displaying synchronized, cutting-edge creative content every night at 11.57pm; with each month showcasing a new artist work. We partnered with the non-profit entity Times Square Advertising Coalition (TSAC) to have one body representing independent billboard owners, so that we established a producing partnership whereby TSAC presents and Times Square Arts curates. The presentation form as created, and still adhered to, has been successful; many established artists and artist estates have made exceptions in their own specifications to meet the conditions. The approach needed to consider how the screens were part of the architecture, and therefore a part of the scenography, and that the programme needed to consider the viewing experience very differently from typical screenings, whereby the viewers are also performers at the centre of the scene.

The curation needed to be a combination of aesthetic diversity and partnering curators to link to major moments in the city or exhibitions so that we could draw their networks, audiences, and press contacts. We became the outreach location for museums so that they could reach new audiences and we could draw attention to Times Square as a central place to know about culture in New York City. Rather than commission to reflect holidays, we sourced existing works that indirectly celebrated that time. Alex Prager's *Applause* (2016), which is a video portrait of people in in auditorium audience clapping, corresponded to June, the month of the Tony Awards. For February celebration of Valentine's Day, we commissioned Tracey Emin's *I Promise to Love You* (2013), her first animation of her neon sculptures of love messages, and Alex Da Corte's *Blue Moon* (2017), in conjunction with his upcoming exhibition at Whitney Museum. When we felt that the novelty was wearing off, we developed live actions on the plaza with *Midnight Moment* to draw attention and worked around late-night sound restrictions to deliver innovative sound components. We pinched the concept of 'Silent Discos' to create immersive, fine-art moments for general audience through headphones. We held a special concert of Ryoji Ikeda's *test pattern [times square] (2014)* (2014) with a live mix to accompany the intense binary code strobing of his visuals; a 'silent orchestra' or orchestral machines electronically plugged with Jhereck Bishoff; and Laurie Anderson performed a solo, live *Concert for Dogs* on the Red Steps, alongside her film *Heart of a Dog* (2016.) We let any restriction be a creative inspiration.

At the Crossroads: *'the gallery that can be seen from space'*

We set out to create installations and performances on the five Broadway plazas as stages of urban life, capitalistic trade, theatres, 24/7 culture, and one of the greatest cultural melting pots. Ordinary performance would not work – the sites are loud, distracting, and asymmetrical. We needed actions, participation, oddness and everyday relevance. French artist JR and his *INSIDE OUT PROJECT* made their first government-sanctioned intervention with *INSIDE OUT NEW YORK CITY* (2013). For three weeks, people took their self-portraits from a mobile Photo Booth Truck and pasted the resulting photos on Duffy Square, between Broadway and Seventh Avenue. To gain interest and trust, *INSIDE OUT* began in the outer four boroughs to bring their representation into the centre of the city. We timed this with a *Midnight Moment*, and another documentary film about JR that was playing in the Tribeca Film Festival, making a city-wide cultural moment home in Times Square. When the plazas were threatened to be removed and place authenticity was under fire, we invited artist Kimou 'Grotesk' Meyer, *Victory Journal*, and *Juxtapoz* magazine to install *T.S.Q. Newsstand* (2015). The artists created an imitation of classic New York newsstand, equipped with wooden pigeons, cigarettes, and other artist objects, and then graffitied by street artists. Each day, artists such as Charlie Ahearn, were on hand to sign their work, sell limited edition prints, books, and rare pieces from their personal archives. Each day, the artist appearances were tweeted, and queues generated, and the publication editors decided to open professional crits onsite while people waited. For 10 days, an artist and design colony populated organically and took back Times Square's old reputation as place for 'grit' and street artists. Each year, there is a competition to encourage emerging architects to create a Valentine Heart that counters the starchitect presence in the city. That experimental commitment inspired architect, H. Juergen Mayer, to propose another project: *XXX: From Times Square with Love* (2016) seating that referenced the signage of old porn theatres that would display 'XXX' on marquees to indicate content and the 'xxx' people often sign off on letters and texts now. The bright pink seating shaped as 'X' are recliners for enjoying the light show of the electronic billboards. People were worried about sleeping on the loungers, so we set a limit of time to allow enough people to enjoy, rather than telling people 'don't.' The temporary installation became permanent and is a symbol of place.

Perhaps the greatest tipping point for the Board of Directors of Times Square Alliance was when we found that German Expressionistic opera could overturn base entertainment. At a time when the district was under siege by costume characters harassing people for paid photos, we were searching for ways to clear the performers indirectly. Performa (Performance Arts Biennial) brought Arnold Schönberg's *Erwartung – A Performance by Robin Rhode* (2015) as a full, live opera. The one-act monodrama for a solo soprano was originally set in the moonlit, wilderness of the forest, and Rhode's reproduction interpreted Times Square as the contemporary forest, filled with light emanating from the electronic billboards (moon.) People were fascinated to watch this *gesamtkunstwerk* take place in front of them. They chose to follow a character who was battling her place in a loud and oppressive world through extraordinary artistry, and as a result, the costume characters soliciting money left the area due to lack of business. When the work is true to place and mission, programming can challenge assumptions of audience 'taste.'

Practically, we learned that people will queue for almost anything if the queue is organized; artists will come anywhere for great content; seeing oneself in a famous, valuable place of real estate has immense impact on positive self-confidence; people engage openly in a populated space; signage and setup will reframe 'don't's' into 'do's.' Programmatically, we were pleasantly surprised to find that quality wins; opera can be for everyone; and people will surprise you with their level of serious debate in a public space.

Hidden Assets

Stakeholders wanted people to discover the unique assets of the full 40-block District, outside of the Broadway plazas, so we developed this strand to explore scale and approach. We focused on themes of 'discovery' and 'intrigue' by showcasing experimental performance in expensive and quirky commercial venues and creating content through online or app-based projects. With Clocktower Productions & Radio, we created *After Hours* (2013–16), a series of ephemeral installations and performances hidden amongst the Times Square district's diverse venues, recorded for radio and video dissemination. The ask for the venue was simple: one night (ever) of their weakest sales night plus one cheap cocktail to offer plus their trickiest audience demographic. We would provide the content, as well as video and still documentation to be used for their promotion. Clocktower paired musicians and visual artists with each of these unique environments, specifically designed to draw on the neighborhood's history as a social and artistic melting pot. Exclusive lenticular postcards were created as souvenirs as valuable keepsakes. As the venues had tight capacities, word of mouth turned the events into small legends. We generated imagery and mystery that resurrected a vintage feeling of elicit discovery and people were competitive about getting in. The results were so positively, sold-out, successful, that many venues invested in creating their own programming and attracting new audiences (and artists) and taking over our job. We developed a relationship with the AMC Empire, the USA's largest grossing cinema, requesting free evenings for experimental performance that would not be able to normally afford the facilities. While Jay Schieb's experimental Chekhov theatre performance of *Platanov* was live at The Kitchen, the experimental theatre on the lower west side, Scheib filmed the actors onstage to broadcast to AMC Empire audiences as the First Real-time Feature Film called *The Disinherited* (2014). Soundwalk Collective created, *JUNGLE-IZED: A Conversation with Nature*, a virtual soundscape that superimposed the Amazon Jungle ecosystem upon an eight-square block area that was transmitted via an app. They geo-located the sounds to correlate to the heights within the Amazon to the manmade jungle. As one walked and moved their phone through volumetric space of Times Square, they were entering a phonic 3D jungle. At the end of my tenure, we created a *Residency at the Crossroads* (2015–16): a cross-disciplinary and collaborative residency program, in which artists were invited to experiment and engage with Times Square's unique urban identity and users. Luke DuBois, Joseph Keckler, Okwui Okpokwasili and Joshue Ott, and Kenneth Kirschner all came with different disciplines, needs, and end results. We gained support from the National Endowment for the Arts as policymakers wanted to prove that successful (and expensive) real estate also needs artists creating to keep a creative soul of place alive and that artists genuinely want to make dialogue with their cities.

We learned that despite the concept, sound art installations are challenging to market and deliver; apps require fast and free wi-fi. Artists will always teach venues about creative potential they hadn't tapped; free access to exclusive events attracts audiences to try anything; and lenticulars are highly valued. We surprised authenticity sceptics by discovering that elicit feeling can be delivered without elicit activity; subtle and informed security allow more risk and encourage better behaviour; and mirage-like transformations create indelible memories.

Evaluating in real-time

In the spirit of Times Square 24/7 transparency, we actively, internally analysed and externally exposed our learning lessons about behaviour in public space. Internally, we had several approaches to test our programme development. In addition, The Times Square Alliance Board of Directors indirectly provided guidance of the District's objectives and the Times Square Arts Advisory Committee (meeting one to three times a year) gave cultural guidance. We had different selection committees for *Midnight Moment* (half curators and half advertisers) and then specialist curators and association leaders across design, art, and performance for *Residency at the Crossroads* and *At the Crossroads*. We developed a curatorial committee to round out the cultural, gender, and artform diversity. We always sought to reflect representation of our 'contemporary' world and never get lazy or predictable. For four years, 2013–2016, I was an Assistant Professor teaching The Artist and the Arts in Urban Revitalisation course for graduate policy, planning, and arts administration at the New York University Wagner School and used students' critique to measure aspects of the programme. Additional dissemination platforms included national and international panels, conferences, writings, and exhibition: as an exemplar for BIDs, public art, art and technology, and cultural partnerships. Academics, city leaders, city servants, International Downtown

Association, and related, global BIDS and town centre managers constantly visited, asked advice, and requested participation. Architecture, Arts, and Design communities requested my presence to explain, advise, and inspire their ambition through talks, review panels, and academic critiques. We used our live programme for others to dissect and keep our self-scrutiny sharp.

Establishing cultural sustainability

In 2016, I was invited to create the inaugural exhibition for 'The New Space,' of The Theatre of Nations of Moscow that would expose a local Moscow audience to the 'West's approach to public art.' I titled it *Times Squared: Theatre of the Absurd*, because like all great absurdist dramas, Times Square has subjective time and no clear proscenium; everyone is simultaneously audience, participant, and performer. Strelka Institute also invited me to give a keynote, exposing the process of cultural placemaking in Times Square, as an example of how one may redefine major sections of cities. These opportunities provided an opportunity for me to analyse the specifics of the programme and turn the findings into lessons for major cities. I realised the five-year experiment of developing cultural programming and placemaking could be formalised into a framework that could be scaled for any place. The Framework for Vision, Mission, and Core Values across three key platforms is still used, as it is robust enough to focus its role and pliable enough for artists, curators, and cultural leaders to interpret it freshly. While still at Times Square Arts, I was asked to contribute a case study for an international research project between Arts Council England, Mayor of London, and Kings College, which later resulted in the publication, *Improving Places: Culture & Business Improvement Districts: Thriving Partnerships*. We chose to concentrate on the *Hidden Assets* programme of Times Square Arts, to highlight how we used defied stereotypes to attract new audiences and how this contributed across the cultural placemaking themes of Place Branding and Tourism, Public Realm, and Increasing Footfall. UK BIDS first developed in 2005, and by 2018, there were more than 280; 60 of which are based in London and often have leveraged culture to deliver various objectives of economic development, by using this activity to create distinctive places. This publication was designed to provide case studies and key lessons to encourage BIDs to adopt cultural partnerships within their remits.

From Times Square to London Bridge: transference to other business improvement districts

Soon after I moved to London, I was appointed to produce a Culture Strategy and Delivery plan for Team London Bridge, the BID for the area between London Bridge and Tower Bridge and nestled between Borough Market and Shad Thames. It includes the area of the Greater London Authority and Mayor of London offices, the new transport interchange of London Bridge Station, medical Kings College campus with the oldest London hospitals, and contemporary corporate offices in More London. Their main objective was to shift 'the area as a place that one went through' to 'to a destination that people came to.' They simply needed to reframe their opportunity to see their neighbour's assets within view as their assets, also. My approach for the strategy and delivery framework was based upon the model of Times Square Arts, and they were open about their interest in applying that success of methodology to establish the trust with their stakeholders.

Using the framework

Team London Bridge wanted to raise the ambition of their cultural delivery to further the place as a destination. As its location on the river front also faces iconic heritage sites such as

Tower Bridge, Tower of London, and the new skyline of City of London skyscrapers outside its boundaries but within sightline, we valued the neighbour's assets as beneficial in visitor experience. We crafted a Vision that establishes London Bridge will be the front stage that creates our favourite memories of the city through innovative and contemporary cultural programming. We determined a Mission of engaging the city's cultural sector to creatively and continuously reveal the diverse landscape of London Bridge as the epitome of a global capital. This commitment gave purpose to leveraging commercial, cultural, and civic stakeholder partnerships. As much of the public realm is under private ownership and management, the partnerships' first responsibilities were centred around identifying the cultural programming characters for the Platforms and related permissions and invitations. The authenticity of the place is the diversity of scale and extremes of contemporary and historic. Therefore, we had a more complex space categorisation than Times Square, but maintained the framework, none the less. *The Front Stages of London* grouped open spaces that have space for spectacle and large statements to reflect the important moments for London. Temporary artistic innovation of static and performing arts positioned for great visibility are located *At the River*; framing of iconic landmarks took place *At Gateways*, and large courtyarded spaces were grouped as *For Gatherings*. We recognised the small yards and plazas as places to present unexpected heritage stories as *Intrigue*; the vital connectors as *Corridors*; the commitment to cultural institutions as *Insiders*; and a street-front-dedicated indoor lab space for talk, exhibition, and workshops as *The Hive*.

London Bridge core values

The area is the longest continually inhabited part of London, as well as home to the architectural marvel of The Shard, and thus needs to reflect the extremes. In this case, we formed the values as nouns. *Innovation* is reflected in the ingenuity of government, medicine, transport, and commerce, that can translate to thinking and scale; London as *diversity*; *bridges* literally define the area's boundary and connect people metaphorically; *heritage*, commitment to be preserved as innovation thrives; contemporary issues, perspectives, and design to be showcased on its *Front Stages*. We aligned a commitment to cultural sustainability through all four priorities of the area's London Bridge Plan. Rather than having 'Culture' as a distinctive deliverable, we chose to have a distinctive role of Cultural Director working against all areas of statutory delivery of Team London Bridge, to ensure the missions cross-referenced one another. To further demonstrate that commitment, we produced a dramatic launch event that invited stakeholders and press to newly opened London Bridge Station. The launch doubled as an action against the Culture Strategy Delivery Plan, enacting the new central concourse as a performance platform with synchronized video and choir. On 1 March 2018, despite a snowstorm, Team London Bridge announced and distributed and toasted first success to a packed audience. Their new Cultural Director was announced, and the next day, they began discussing the new programme with clear focus.

Concluding principles

I have brought this framework into my role as the Managing and Cultural Director of Futurecity, a cultural placemaking agency, where it has informed new architectural design in Melbourne to purposefully accommodate public art platforms; an Events Strategy for Exhibition Road Cultural Group (made up of cultural and academic institutions along the site of the 1851 Great Exhibition) in London; the creation of a dedicated cultural programme in Wembley Park, London; as well as arts commissioning in Boston. Each project remains authentic and distinct to place. The importance is copying the robust framework, which establishes questions, rather than copying others' case studies or answers. While many places may not have the same 24/7 intense culture as Times Square or other large city districts, the main rule for curating in any public realm is this: one can never control the entire scene; therefore, the best approach is to collaborate with place. In creating or altering any governance for implementing cultural placemaking programmes, consider the following guiding principles.

Be authentic and distinctive through ambition

Authentic means being true to your site and its heritage or character; being distinctive means that you generate a new or celebrate an existing USP (unique selling point). Consider how you will tell the story and be ambitious in accordance with your context and resources. Always use this as a chance to be more daring, surprising, and inspirational.

Create structure to assist not limit

First, create your Vision for success of the place and then, identify the Mission (reason) for the cultural placemaking approach. Establish Core Values that serve as criteria or guidance for creative content. The structure should facilitate objective conversations and help foster better creative concepts. Audit your place and identify the platforms for presentation and their related characteristics that inform cultural producers. Vision first ensures you have the right structure to inspire.

Set up stages, permissions, and systems for collaboration

Begin partnerships before you need anything. Generate a positive relationship with the owners and managers of spaces to establish key contacts and respect their process or limitations. Develop a way of working that empowers them to take risks, by respecting their fears and finding ways to mitigate unnecessary risks. Set up communications protocol at the beginning of the project, so that everyone is clear on the partnership priorities.

Use cultural professionals: respect practice, networks, and professional trust

Hire cultural professional(s) who know how art is made, as well as presented, and want to work with other creators and producers, rather than just create their own work. Develop relationships with cultural institutions and entities to build trust; show up at their events and know about their work. Seek presenting partners who bring diversity of audience and press in order to expand your own.

Address your blind spots: invite expertise

Never assume anyone can do this on their own. Set up an advisory council or curatorial committees or selection committees to round out your own knowledge. Seek out what you do not know; ensure you explore work outside your previous knowledge. Ask advice of those you respect; it's flattering to them and you learn from their experience. Art in public space is not about single authorship; it is about collective representation and idiosyncratic contemporary taste. Surround yourself with those who can help.

Your stakeholders, residents, and staff very often want to feel protected and respected. Once you establish trust through clarity of Mission, process, and details like budget and insurances, they will take risks. We develop strongest bonds with those who have succeeded through shared risk; and those stakeholders will become allies beyond the life of a project. Most importantly, never underestimate your audience or assume that people only want what they know or feels familiar. We bring artists to surprise us, inspire us and make us think or feel more fully. If the choices are guided by the Vision and the place's Core Values, the projects are grounded in context and therefore allow the artists' concepts to fly safely.

Further reading in this volume

Chapter 3: An annotated history of creative placemaking at the federal level
Jen Hughes
Chapter 9: From the dust of bad stars: disaster, resilience, and placemaking in Little Tokyo
Jonathan Jae-an Crisman
Chapter 10: From moon village to mural village: the consequences of creative placemaking in Ihwa-dong, Seoul
Jason F. Kovacs and Hayun Park
Chapter 11: *Free State Boulevard* and the story of the East 9th Street Placekeepers
Dave Lowenstein
Preface: The problem with placemaking
Louise Platt
Chapter 17: 'I am not a satnav': Affective placemaking and conflict in 'the ginnel that roared'
Morag Rose
Preface: Towards developing equitable economies; the concept of *Oikos* in placemaking
Anita McKeown
Chapter 27: Is 'tactical urbanism' an alternative to neoliberal urbanism?: Reflections on an exhibition at the MoMA
Neil Brenner
Chapter 29: The solution is in the problem; the art of turning a threat into an opportunity by developing resilience using a creative placemaking critical praxis
Anita McKeown
Chapter 43: A theory of change for creative placemaking: the experience of the National Endowment for the Arts' *Our Town* program: an interview with Patricia Moore Shaffer, PhD
Maria Rosario Jackson
Chapter 44: Creative Placemaking and comprehensive community development: rethinking neighborhood change and evaluation
Maria Rosario Jackson
Chapter 45: How the city speaks to us and how we speak back: rewriting the relationship between people and place
Rosanna Vitiello and Marcus Willcocks

26

SCULPTURING SOUND IN SPACE

On *The Circle and the Square* (2016) by Suzanne Lacy

Trude Schjelderup Iversen

> A lived-in landscape becomes a place, which implies intimacy; a once lived-in landscape can be a place if explored, or remain a landscape if simply observed.
>
> *(Lippard, 1997, p. 7)*

Having an experience (introduction)

In September of 2016, place-based and socially concerned artist Suzanne Lacy's collaborative art project *The Circle and the Square* (2015–17) took place in Brierfield, North West England. It was the culmination of 18 months of work, and took place as a three-day event, with a song performance from Sufi chanters and shape-note singers, an exhibition, and a banquet with 500 guests. This chapter, written in the first person, describes an embodied experience of the artwork as an experience of placemaking from the personal purview, and, in conversation with the artist, explores the aesthetical pursuit in Lacy's work, fundamental in how Lacy approaches and prioritises within community-driven collaborations.

The American philosopher John Dewey argued in his 1934 book *Art as Experience* that having an *experience* differs from the ongoing stream of experience as such; our distractions, amusements, and desires. When having *an* experience, in contrast, the material of the experience reaches a completion; a problem reaches its solution, a chess game is completed, a book is written. An experience is therefore not something that just ended or stopped but is a fulfilment and a whole. We are used to thinking of the artist as the creator of this whole, through making choices, simplifying and amplifying experience based on their interest and subjective perspectives. In the context of the artwork, for Dewey, having an experience in a real sense is to go through all these same organisation procedures. The viewer, in order to have a real experience of the artwork, must – just like the artist who created it – arrange the elements in a whole. They are both, in Dewey's terms, 'doing and undergoing' an experience.

For me, not trained in placemaking theories – quite the contrary in fact, as schooled in the philosophical direction of the Frankfurter school in general and in Adorno's concept of autonomy in particular – Dewey's insights were fruitful when trying to conceptualise the writing of a chapter about Lacy's 2016 performance of *The Circle and the Square*, in a handbook about placemaking. I was interested in finding another 'criticality,' to borrow a term from theorist Irit Rogoff (2003), that was neither critique nor criticism: to classically 'critique' an art project that had been going on for 18 months, with a myriad of collaborations and stakeholders involved I knew would be impossible for me and uninteresting for the reader. As Rogoff writes, 'The old boundaries between making and theorising, historicizing and displaying, criticising and affirming have long been eroded' (ibid.). This said, a clearer 'writing through art' (ibid.) approach also suited lessons learnt throughout my Adornoian education; writing about an art work must take the form of the work very seriously, and to write about art is to let your writing be affected by what you write about, literally (Adorno, 1958/1991, pp. 3–4). Writing not 'about,' but 'through' Lacy's performance in Brierfield is therefore an attempt to contemplate *place* as something other than environment, and to look at the factors that allows a non-distinctive approach to understanding one's self and one's place, as Lucy Lippard (1997) formulates it.

Situating place: arriving in Brierfield

As soon as I got off the train at the unstaffed Brierfield train station a building complex revealed itself as a massive cathedral in its rural Lancashire setting. Sat empty and idle, the former textile mill emphasised a historic significance for the small town built around this complex. I couldn't help but think of losses, prompted by the shocking emptiness of abandoned industrial buildings, and here surrounded by a post-industrial small town with traditional 'two-up two-down' houses for its millworkers and grand detached houses of its mill owners of old along one main road. I have travelled from Oslo, Norway, to the Lancashire countryside, to Brierfield, which forms part of the Borough of Pendle in North West England, to see Suzanne Lacy's then-new performance.

The massive brick mill was one out of many built to develop the textile industry after the Leeds and Liverpool Canal was completed in 1816. When the mill was erected in 1832 the location itself was central, on the slope between the still vital Leeds and Liverpool Canal on the one side that was to supply the new industry's need for water, and the city's main street on the upper side. If you were in the main street you could see the distinctive English hills surrounding the industrial complex that was to lay the foundations for a Pendle of growth and progress. In a Wikipedia article, I learn that Brierfield Mill was one of the first steam-powered cotton spinning mills in the area, and in 1890 over 2,000 looms were in use and 91,000 spools (Wikipedia, n.d.). The Mill closed for production in 2007 after steady cuts and three years later the landmark became one of English Heritage's 'Listed Buildings at Risk.' The area had long been known as the 'dark corner' of the land; the home of troublemakers and disobedient and subversive forces (Poole, 2002, p. 88). The most famous story in the region is indisputably that of the sixteenth-century witchcraft cases, where a nine-year-old girl, Jennet Device, known as the Pendle Witch Child, testified against her entire family and, by doing so, lost them all. Today, this dark past is encapsulated in tourism, aided by the panoramic beauty of the hillsides that surround Pendle Hill, which is the highest point at 557m above sea level.

When I arrived on a cloudy September day, I could still glimpse the intact platforms that, until 1986, had served train rails in both directions. I had booked into the recommended Oaks Hotel, a large and venerable hotel a few miles from the main street. I don't remember if I was carrying a bag or a suitcase, but I do remember the worn-out asphalt on the various pavements that constantly shifted from one side of the road to the other on the way to the hotel. There are dozens of cars driving back and forth, drivers looking straight ahead, eyes on the road. I am a foreigner here, but I am familiar with this type of topography from my upbringing in Northern Norway. Its logic, structure, its everyday and natural aesthetics seem familiar. The hotel had a new entrance, so I had to walk around the entire facility to find the original reception with tall dark oak panels along the walls of the main room, which ended in a stately staircase up to a small gallery where some guests were drinking tea. I had the impression that there were not many guests. I was given the key to my room, which was large but worn and outdated, and got ready to go back to the historic mill where Suzanne Lacy was in charge for the next three days. An 18-month collaboration between the California-based artist and local artists, residents, and government officials was about to finally manifest as a work of art; a song performance, an installation, and a banquet with 500 invited guests.

A place turned space

The sparse information I find about the history of the area makes me feel unprepared for what is about to happen at the mill but I know this from the project text following the invitation to the collaborative event: after the mill closed in 2007, the small place lost its last hub, which means that the now-composed ethnic group of residents of the village no longer had their natural meeting place. 1962, the year of the Commonwealth, marked a start for the area as Pakistani workers moved from their home country to work in the textile industry along the Liverpool and the Leeds Canal. Together with the local white population, a working community was formed in the many mills along the canal. The Pakistani community that arrived in the 1960s worked with the local white working class, and the daughters and sons of the factory workers went to school together. All Brierfield residents had experienced great attachment and optimism had seen decline in recent years.

I feel a little nervous and expectant as I walk down the steep hill towards the mill. There is a group of people inside the factory gate. I don't quite know where to go, where to direct my steps. I choose to go to one of several entrances and am thinking that the people I see live in the area, that they are local, here to watch the song performance orchestrated by the artist. They also look a little confused. Where are we going and when are they filming, we wonder, after reading various handwritten posters that give indications on how to stay out of camera shot? As we go up the stairs it becomes easier, arrows and tape on the floor indicate the direction. We understand that we are also in a film set. Then I see Suzanne. She smiles and gives me a hug and introduces me to several people standing around her. I remember no names afterwards, but I feel included. We are in the main room now, in the hall where the song performance is about to take place. The hall is as large and monumental as factory halls can be, with a long wall of windows looking onto the valley and the beautiful, picture-postcard hills. The concrete floors are worn, and the sun that floods through the glass of the old windows creates streaks of dust between the sharp white columns that fill the room. The walls are light turquoise, seemingly painted a long time ago. What a strange choice of colour, I think. Rows of foldable white chairs are placed in four rows against each other to form a square with a centre. When I arrive, one young man stands in the middle of the stage and more come to take a seat around him. Several cameras are placed around the room and microphones hang from the ceiling.

The performance is set to begin in two hours, and Suzanne says I can join her in the closed Town Hall they use as a working base and for meals, for the latest preparations or I can hear investors' presentations of future plans for the mill, in one corner of the main room. I am not prepared for this 'ingrediency' in these – as I see it – aestheticized settings, but I choose the presentation, thinking it gives me a context that is not just historical. I stay through the presentation and understand that there is large investment to be made in Brierfield. The mill will be transformed into a hotel and apartments, recreational homes, *pieds-à-terre* for the large neighbouring cities. The beautiful view of the hills does not make it difficult to assume that the investment will yield a good return.

I learn that PEARL (Pendle Enterprise and Regeneration (Brierfield Mill) Ltd) owned the mill complex after Pendle Council bought it for £ 1.5 million from a private investor who had originally intended to build a new school on the site, following the closure of the local school in 2007. The year in which local government bought the facility to develop it through the semi-public organization PEARL was the same year Brierfield Mill landed on English Heritage's Listed Buildings at Risk in 2007.

The performance

I sit down with the small audience next to the stage placed in the middle of the main room. Now 40 men and women have entered the arranged lines of chairs into a square. They are shape-note singers from all over England. Shape-note is a singing tradition designed to facilitate congregational and community singing, popularised in the north of England from 1742 on by preacher John Nutall's Lancashire *sol-fa* (sight) singing sessions (Willsdon and Sanromán, 2019, p. 240). The group face each other and start singing. Their voices are the only thing to be heard in the hall and the sound creates an instantaneous pressure of sound. One hand goes up and down while they sing, forming an accompaniment of harmonious movement to the naked, but strong, voices of the members of the choir.

I can sense the chair I sit on and lean into it, but I also feel the sound of the voices in my body. There is something happening now in the empty mill, with its peeling paint, the smell of mould, and the cold damp filled with sound. The song cuts through the surroundings and the body in those surroundings and touches some inner strings that suddenly tense up and recall some previous experiences of singing in similar settings, maybe it was a school aula, maybe it was a church. There's something soothing about it, and familiar. I am reminded that we all have a voice – my neighbours, the person who lives down the road, my school mates, and my teachers; not just those who are authorized.

Now local Sufi chanters, all old and young Muslim men, are taking the stage after arranging the chairs in concentric circles, as opposed to the hollow square of the shape-note format. The young men choose to sit in the second row, with the oldest in the first. Before we hear anything, there is a dramatic visual change on the stage. The two groups of singers have their own distinct appearance. We are witnessing a quieter song, a prayer. It sounds like a kind of humming, but the words are: *La ilaha illAllah, La ilaha illAllah* (There is no deity but Allah). The acoustics created by the high ceiling and concrete floor accentuate the voices, and unlike the shape-note singing, they never go high in tone. It feels intense and intimate to be in the same room, as if the walls have crawled closer together to facilitate something private and personal. The song escalates, and the singers' bodies move calmly and rhythmically from side to side, while breathing in in silence, and out with song.

There is something deeply spiritual about the song and how it is sung with closed eyes. Maybe it's the voices together – the young, the old; that they share these intimate moments regularly, share spirituality. I think I have too few spiritual experiences. I am reminded of similar spiritual involvement. The view from the mountain, that allows being able to see really far, when the eye strives to see variations far away, to identify what you see; the eye muscle, unused to it; the cornea's experience of something unexpected that sends other signals to the brain, and changes it.

It is always hard to say how much one needs to know about the context to have an experience of an artwork, but it helps that I know that Suzanne saw singing as a key tool to rebuild relationships in the now-divided community. It feels poignant, and it touches me on an emotional level, primarily through the voices and the two groups of singers coming together in the abandoned mill. I understand immediately how the song – from both traditions performed here – can serve as a common ground. It is hard – I would say almost impossible –to write about this without using a pretentious or idealistic language; then again, art history is full of praise of art works that managed to cut through history, to operate beyond the here and now.

My point of view becomes anthropological, I observe, while I suddenly also seem to participate to a small extent. Participation is unexpected, and my role becomes unstable. This reflection is shared by all art reviewers present; at least from *Hyperallergic* (Sheerin, 2016) and *Frieze* (Robertson 2016). They both describe the feeling of participating abruptly, getting dragged, to a greater or lesser extent, into the process of a project that has been going on for a long time, a close collaboration with local artist groups and commissioners (In Situ and Super Slow Way, arts organisations). I am also reminded of Palestinian artist Sandi Hilal, who through the work *Living Room* (2016–ongoing; visibleproject.org) changes the host–guest relationship by facilitating the migrant's ability to take on the host role and thus gain power. The work is informed by Hilal's long experience of living in exile.

The banquet I do not go to feels like an important component of the art production, as are the various activities that have taken place during the 18 months prior to this performance. Another available component is the 65 interviews conducted by mill employees and their family members, shown on screens in the mill. In the main room I can see about a dozen, and they give me – through the testimonies – a more direct access to the experience of working conditions and culture in the mill: the soundscapes from the time when it was at full activity, including the sounds that filled the streets of the village just before the mill would open its doors in the early morning.

Reflecting on the performace, London, March 2019

Notes to self:

> Every time she [Lacy] gathers and talks about her work, it is as if a kind of excellence is emerging, as if (for Europeans) the American unfiltered directness is stepping aside and a genuinely expertly founded commitment takes place throughout the body. The back is straight, the voice is lower, strangely enough, but clearly, something is at stake. This repeats, I observe, every time we talk about her art. In transitions from more practical tasks like ordering a meal, finding the itineraries (always more options!). Wherever we go, Lacy can appear more impatient, yes, unpalatable. But as soon as we talk about her art, she puts herself in another notch. It impresses me. We sat down in a hotel café to do the interview about the Brierfield project. We spend time on the recorder, we are unsure if it will be strong enough since the cafe is busy and noisy. For a moment I wonder if we should find ourselves another place. We stay. Decide that it will work well enough.
>
> She says I can't write down what she says in the text. She recommends that I write it myself, base my article on the interview as a material of preparation. She is going through this change I have seen several times now, and it is as if she is giving a lecture to me, where each sentence creates a deeper and deeper understanding of the audience. I note and nod, asking my questions: I ask her how much of the 'result' she envisions beforehand, when she started a project like this. I ask her how she orientates herself, how does she read and understand the place to work with and from? Is her approach intuitive or strategic or both? What is she looking for, who are typical informants and who are typical collaborators? How does she inform herself about various agendas, here the client, the developer, the local production group?

It is now almost four years since I was present at the performance, so how do I look back on it? What do I remember and what have I forgotten? I am currently preoccupied with Walter

Benjamin's *Angel of History* (in Arendt, 2012). I find Benjamin's angel figure relevant when remembering and looking back at works of art that are by nature more or less objectless, such as Suzanne's performance. According to Benjamin, access to what has happened, the memory or the 'story,' is formulated through the gaze of the angel Angelus Novus in Paul Klee's painting: 'Where we see a number of incidents, the one and only catastrophe is seen as incessantly tossing ruin upon its feet' (ibid.). Articulating the past historically is not the same as acknowledging 'what it really was,' Benjamin says. 'It is to gain a memory when it strikes like a lightning bolt at a critical moment' (ibid.). *Lightning bolt at a critical moment.* It reminds me that Lacy is constantly returning to 'the image' in her negotiations and discussions with the involved process that takes place before something is realised as art. I have the impression that she knows this image is something that does manifest as a result of a collective negotiation, and exchange of ideas where she sees herself as the student and the local community she is engaged with are her teachers, keeping in mind that being an 'artist' can both turn off local engagements and be productive. The lesson from her friend Lucy Lippard seems to be lived: she knows that many communities don't think artists are as indispensable as plumbers or doctors; then again, most artists don't see themselves as service providers.

When I ask her if her many years of experience with collaborative art projects, not to speak of her role as a university professor for many years, makes it difficult for her to enter the role of a student, she reminds me that in the actual context she is always the one who has least knowledge about the place, about the lives lived there. The logic that arises in the production stages that take place is a result of the involved methodology Lacy has developed and uses. She never has a completed 'image' to be realized as a work of art, despite the fact that her works are related and perhaps recognisable as something that can be referred to as a version of an 'emphatic aesthetic.' This term, however, understood in a normative fashion is inadequate for Lacy's art production or for understanding *The Circle and the Square*. But if one were to find descriptive terminology to help us think about a production like this, a closer look at Lacy's process that resulted in the present image could be useful.

All art that seeks to be 'contextual,' 'site-specific,' or 'placemaking' will have to consider questions of involvement. While Lippard prefers an art that grows from a deep experience with people and the places where they live, Lippard knows how seldom this can be realised in public art projects. Being 'local' is often seen as a disadvantage for commission committees. Lippard (1997) introduces the idea of 'multicenteredness,' a term explaining a person of many places (as the author herself, and a description valid for many artists) who is responsive to and responsible for the place where one is at the time. These insights are embedded in Lacy's methodology. While maybe not concerned with the history of the place per se, but with what Lippard formulates as 'the historical narrative as it is written in the landscape or place by the people who live or lived there' (ibid, p. 7). The historical narratives in the place can only be accessed through time spent together that allows for building intimate relationships. For half a year, when my son was an aspirant in the Norwegian Boys' Choir, I was present during the weekly exercises. They practiced for about 50 minutes in a small chapel with the conductor. The boys were on average five to six years old and all the songs were hymns, some in Latin. On one of these rehearsal evenings, the conductor turned to the parents sitting on the spectator chairs. 'Don't worry,' he said, 'if your child doesn't sing. There is a lot going on in an exercise session.' He continued: 'the child hears, stay mute, before one day it opens its mouth and sings. Don't pressure, it's coming.' Now I understand what he meant. In Brierfield I had a similar experience, I stayed mute. But I can sing now, when I think back on the experience.

Conclusion

In bringing together two song traditions, both local but with very different traditions and cultures, Lacy took an active approach to negotiating space – what we can see from outside – and place – from where we act, from where we work. The mill used to be a place but became an (empty) space. By opening it up again for the former workers and the community at large, a restart of a conversation was made possible, in a limited period of time before developers took over. Lippard empathises, in *The Lure of the Local*, that if reflections on social affiliation are incorporated into participatory art projects, then they have the opportunity to gain collective and shared experiences about the place. It is, for Lippard, not to see the place as an 'environment' or something around ourselves, but that we ourselves (also) create the place. Not accepting a distinction between one's self and one's place is a beginning. Lacy's methodology never goes easy on audience participation nor the co-creation of the 'image' (the two song traditions coming together: yet it is hard to overlook the fact that it is the artist's sense of space and aesthetic intuition in orchestrating that space (co-)creates the deep emotional possibility of identification and human solidarity among both local participants and transient arts professionals.

References

Adorno, T.W. (1958/1991). 'The essay as form', in *Notes to Literature, Volume One* (trans., Sherry Weber Nicholsen). New York: Columbia University Press.

Arendt, H. (2012). 'Theses on the philosophy of history', in *Illuminations* [online]. Available at: https://seansturm.files.wordpress.com/2012/06/benjamin-theses-on-the-philosophy-of-history.pdf (Accessed: 8 June 2020).

Dewey, J. (1934/1980). *Art as Experience*. New York: Routledge.

Lacy, S. (2015–17). The circle and the square, Brierfield, UK [online]. Available at: https://www.suzannelacy.com/the-circle-and-the-square (Accessed: 8 June 2020).

Lippard, L. (1997). *The Lure of the Local: Senses of Place in a Multicentred Society*. New York: The New Press.

Poole, R. (2002). *The Lancashire Witches: Histories and Stories*. Manchester: Manchester University Press.

Robertson, L. (2016). 'Shapes of water, sounds of hope', in *Frieze*, October 2016 [online]. Available at: https://frieze.com/article/shapes-water-sounds-hope (Accessed: 29 June 2020).

Rogoff, I. (2003). *From Criticism to Critique to Criticality* [online]. Available at: https://transversal.at/transversal/0806/rogoff1/en (Accessed: 8 June 2020).

Sheerin, M. (2016). 'In a postindustrial ruin, Suzanne Lacy inspires a captive audience to chant and eat together' in *Hyperallergic*, 23 November 2016 [online]. Available at: https://hyperallergic.com/340349/in-a-postindustrial-ruin-suzanne-lacy-inspires-a-captive-audience-to-chant-and-eat-together/ (Accessed: 8 June 2020).

Visibleproject.org. (n.d.) *Award 2019 – Longlisted. Al Madhafah – The Living Room, Sandi Hilal* [online]. Available at: https://www.visibleproject.org/blog/project/al-madhafah-the-living-room/ (Accessed: 8 June 2020).

Wikipedia. (n.d.) *Brierfield, Lancashire* [online]. Available at: https://en.wikipedia.org/wiki/Brierfield,_Lancashire (Accessed: 8 June 2020).

Willsdon, D. and Lucia Sanromán, L. (2019). *Suzanne Lacy We Are Here*. San Francisco: San Francisco Museum of Modern Art in association with DelMonico Books, Prestel.

Further reading in this volume

Chapter 9: From the dust of bad stars: disaster, resilience, and placemaking in Little Tokyo
Jonathan Jae-an Crisman

Chapter 10: From moon village to mural village: the consequences of creative placemaking in Ihwa-dong, Seoul
Jason F. Kovacs and Hayun Park

Chapter 11: *Free State Boulevard* and the story of the East 9th Street Placekeepers
Dave Lowenstein

Chapter 12: Public transformation: affect and mobility in rural America
Lyndsey Ogle

Chapter 14: Experts in their own tomorrows: placemaking for participatory climate futures
Paul Graham Raven

Chapter 16: More than a mural: participatory placemaking on Gija Country
Samantha Edwards-Vandenhoek
Chapter 17: 'I am not a satnav': Affective placemaking and conflict in 'the ginnel that roared'
Morag Rose
Chapter 27: Is 'tactical urbanism' an alternative to neoliberal urbanism?: reflections on an exhibition at the MoMA
Neil Brenner
Chapter 29: The solution is in the problem: the art of turning a threat into an opportunity by developing resilience using a creative placemaking critical praxis
Anita McKeown
Chapter 32: *The Hollywood Forest Story*: Placemaking for the Symbiocene
Cathy Fitzgerald
Chapter 40: Transforming community development through arts and culture: a developmental approach to documentation and research
Victor Rubin
Chapter 44: Creative Placemaking and comprehensive community development: rethinking neighborhood change and evaluation
Maria Rosario Jackson
Chapter 45: How the city speaks to us and how we speak back: rewriting the relationship between people and place
Rosanna Vitiello and Marcus Willcocks

SECTION 5

Placemaking, environment, and sustaining ecologies

Section Editor: Anita McKeown

PREFACE

Towards developing equitable economies; the concept of *Oikos* in placemaking

Anita McKeown

Competitive economic capitalist principles have been the predominant drivers in shaping our current world, with collective and collaborative, locally scaled alternatives deemed unfeasible or considered as nostalgic utopias. Financial capital is consistently valued over social or environmental justice. The chapters in this section have been selected for their contribution to a more diverse and expanded understanding of economics within placemaking that draws on the concept of *Oikos*; the Greek word for home offers a foundation for a different approach within a generative and more than human placemaking practices. Understanding *Oikos* as the basis for a planetary home economics and the management of 'domestic' resources (McKeown, 2015), this section begins to address aspects of socio-economic and environmentally equitable placemaking concerns. *Oikos*, the etymological root of ecology and economics, provides an opportunity to integrate these often seemingly disparate fields. *Oikos*, with *Logos*, the Greek for Law and Science, forms the basis of Ecology and *Oikos* with *Nomos*, the Greek word for management forms the basis of Economics. Together these concepts can be integrated into an understanding of the systemic management of our home or household resources. Considered within the context of planetary boundaries this new management system becomes a planetary domestic science, that scales from citizen to government, local to global. This offers a concept for an integrated equitable approach to economics in placemaking that is more relevant and appropriate for our time.

Collectively the authors in this section provide a theoretical, historical, and philosophical underpinning that raises issues pertinent to placemaking practices that will need to be addressed for a more sustainable more-than-human process rather than simply humans making geo-locations, driven by economics. Social/environmental justice and financial profits have been in conflict for many years with trade policies often in opposition to environmental needs. Depleted resources and the impact on relationships between aspects of the physical world have been documented historically. Plato's documentation of deforestation and impacts on water supply and the economist's George Perkins March (1864) reference to the 'damage to natural systems by human actions would diminish human welfare show the timespan over which such issues have been discussed' (Johnston, 2018). Concern over the planet's ability to sustain life due to human impact has increased since the sixties and seventies, with global grassroots activism contributing to the establishment of environmental NGOs and political parties with a 'green' agenda.

Social justice as a term emerged in the 1780s (Clark, 2015) with environmental justice emerging as a term during the global rise in anti-environmental policies that favoured fiscal development in the 1980s. The discovery of the depletion of the ozone layer by a group of UK scientists in 1985 paved the way to limiting the free market through establishment of the Montreal Protocol (1989). The protocol banned manmade chemicals (CFCs) responsible for ozone depletion. Despite industry claims that the science wasn't clear, the treaty was agreed in 1985 and adopted and ratified by 1989. The treaty is considered as a successful international response to the control of chemical pollutants, enforcing industry to change their production practices. The global response to the depletion of the ozone layer and its impacts paved the way for the scientific community to raise public awareness of their greater concerns; greenhouse gases and global warming. In 1988, the Intergovernmental Panel on Climate Change (IPCC) was formed from the World Metrological Organisation (WMO) and the United Nations Environment programme (UNEP). During the 1990s, IPCC, in its different guises, produced two assessment reports and a supplementary report (1990, 1992, 1995) increasing awareness of the emerging knowledge of climate change. Collaborative global initiatives began to emerge (Rio Earth Summit 1992; Kyoto Protocol, 1997) in an attempt to share information and develop global action.

Public awareness was slowly growing and Rene Dubos' maxim 'Think Global Act Local' Moberg, 2005) was resurrected to encourage citizens to develop local ecologies that could contribute to a more sustainable economy and contribute to systemic change. As the 1990s drew to a close, voices were growing increasingly louder at the need to reduce carbon emissions while fossil fuel industries (coal, oil, and gas), key contributors to carbon emissions, increased their campaign of denial begun in the 1980s. Institutes and think tanks such as The Marshall Institute or the Heartland Institute, funded by the fossil fuel industry, began campaigning against environmental issues and heavily lobbying governments and influencing policy. As the IPCC (2001, 2007) produced reports with increasingly stark predictions based on the developments in climate science; the think tanks, advocacy groups, and industry associations were using their combined annual income of $900 million (Brulle, 2014; Goldenberg, 2013) to focus on undermining the public perception of climate change science (Goldenberg, 2013; Schultz, 2013).

In November 2018 a special Intergovernmental Panel on Climate Change (IPCC, 2018) report was published reducing the safe limits for global warming from 2 to 1.5°C. The emission pathways and their impacts presented explored 'conditions enabling goal-oriented futures while recognising the significance of ethical considerations, the principle of equity, and the societal transformation needed' (IPCC, 2018: 1.2.3 and 1.5.2.). It is clear climate change, or, more accurately, climate breakdown, will have far-reaching impacts across society; floods (Brown et al., 2011), rising temperature (Horowitz, 2009) fire, droughts, and topically, at the time of writing, disease, pandemics (Diamond, 1999), and more, exacerbating economic crises through their impact on global markets (Moyer et al, 2014). There are other systemic considerations that climate change affects; growth rate of the economy (Lemoine and Kapnick, 2016); Bretschger and Valente, 2011; Eboli et al, 2010), size and productivity of labour stock. Such impacts are increasingly recognised by economists and investors (JP Morgan, 2019; Blackrock, 2019). Yet, approaches that do attempt to value the environment are largely anthropocentric rather than bio-centric and position the physical world in the service of humans. This sees the benefits of a healthy ecosystem as providing humans either indirectly or directly, goods and services. Contemporary reports such as the United Nations Millennium Ecosystem assessment (2005) reflect this perspective by making it easier to 'identify how changes in ecosystems influence human wellbeing and to provide information in a form that decision-makers can weigh alongside other social and economic information' (MA, 2005). Although the report does indicate a 'substantial and largely irreversible loss in the diversity of life on Earth,' it is still written from an anthropocentric perspective. The main thrust of the findings highlights that the ability of the ecosystem of the planet to support social welfare and sustain future generations has been severely compromised.

The complex and interconnected global challenges we face demand alternative strategies. This must include systemic approaches (Gawande, 2014) that focus on an overarching ethos of change, which guides the implementation of any strategies. Further, not only will this require international top-level political commitment but the collective effort of all people, and the engagement of all actors, at all levels. As the awareness of species extinctions, systemic breakdown, and greater understanding of the necessity of interconnectedness grows, calls for approaches that increase biodiversity and acknowledge the obligations to all living things and the environment as a whole become prevalent. Such approaches, known as biocentric or life-centred approaches, see nature as having intrinsic worth beyond its usefulness to human beings. The transformation required to limit warming to 1.5°C demands new models of biocentric practice; socially, environmentally, and economically integrated for the wellbeing of all organisms and ecosystems. This perspective is not prioritised in many of the emerging policies although there are some albeit flawed initiatives. With the launch of the Sustainable Development Goals (SDGs), 2016, a valiant attempt at a coherent strategic global agenda, economic drivers must now share the landscape with social and environmental capitals. The SDGs have been criticised for using existing models to develop transformative practices, as in reality sustainable development is an oxymoron and sustainability alone will not be enough. Criticism of sustainable development lies in the limits-to-growth (Meadows, 1972) model recognising the earth's finite resources, a concept further endorsed by Nicholas Georgescu-Roegen, Romanian-American mathematician, statistician, and economist. In his book, *The Entropy Law and the Economic process* (1971) Georgescu-Roegen argues that economic scarcity is rooted in the physical reality of natural resources and the carrying capacity of the earth and its inability to sustain human population and consumption at some point. Ultimately, any economic model that doesn't acknowledge this and account for it is bound to collapse. Given the increasing population numbers and the resources available, a move away from late-stage capitalism and its life-force, consumption, is necessary. A more ecological approach to placemaking that integrates social, environmental, and economic concerns becomes an imperative.

This section argues for a complex ecological approach to placemaking that considers the relation of living organisms to one another and to their physical surroundings, towards a more sustainable approach. The selected chapters include case studies that share praxis; present applied theory from practitioners, which have evolved through durational projects. The chapters selected, combine to present a range of tangible examples; case studies that share praxis and practitioners' applied theory, which have evolved through durational projects and their professional experience, that woven together present ethical and tangible placemaking alternatives to a capitalist, neoliberal paradigm.

In our current context, placemaking becomes a design opportunity for human planetary health and a different concept of wealth. Developing a multi-stakeholder approach is critical if we are to develop innovative systemic methods with any long-term actions requiring ownership, capacity, and consensus at all levels and across all sectors (Møller, 2016). Our operating system requires a complete conceptual overhaul to encourage a more generative approach. Placemaking, both a process and philosophy (Project for Public Spaces, 1995), brings into contact two distinct requirements: the functionality around the everyday needs of people and more existential requirements that are further shaped by socially constructed processes. Further, within a twenty-first-century context of anthropocentric climate change, as we recognise the importance of being participants in complex dynamic bio-psycho-social systems we need to be attentive to other species' placemaking potential, their interactions and systemic relationships. This will encourage an emergent approach to equitable economic ecologies that fosters diversity and is more in keeping with an ecological system's connectivity and equilibrium.

The section opens with a revised text from Neil Brenner, Professor of Urban Theory at the Harvard Graduate School of Design (GSD) presenting the critical concerns regarding an oft-cited method, 'tactical urbanism' within urban design and placemaking, and offers a foundation from which to explore more integrated approaches. Brenner reflects on MoMA's exhibition on *Uneven Growth* (MoMA, 2014–15) that presented speculative interventions by teams of architects and their design proposals for six of the world's megacities – Hong Kong, Istanabul, Lagos, Mumbai, New York, and Rio. Predicted global changes will not ease many of the similar conditions they share, cited by Brenner as rapid population growth, intensifying industrial re-structuring, inadequate social and physical infrastructures, rising levels of class polarisation, insufficiently resourced public institutions, proliferating environmental disasters, and growing social unrest. Given the growing calls for degrowth, and with 68 per cent of the world population projected to live in cities by 2050 (UN, 2018), what might 'tactical urbanism' have to offer within this context? How might this be considered as a strategic intervention within placemaking and urban design that could contribute to biocentric health?

Following on from Brenner's consideration, Ian Wright reflects on his personal and professional journey as a senior scholar and planner, explored as an ever-evolving praxis as a professional making-place. Using an integral lens, Wright understands placemaking as wellbeing by design, an applied integral ecology. Wright's chapter, in opposition to the scale of megacities, considers a more intimate scale; the I, We, and It/s. As a whole-practice form of integral placemaking that has wellbeing as its 'product,' Wright's combination of place-making and wellbeing 'invokes meshworking' (Wright, 2020). The combination of self-organisation, complexity, and the more-structured hierarchical approach of planning, is an integral placemaking that becomes 'a poiesis of sophrosynes' – professionals balancing praxis, ethos, and poiesis. Thus, Wright seeks to bring into being integral placemaking for the production of physical, functional, convivial, and spiritual places. Such places are beyond 'contemporary conventional notions of growth and sustainability' (Wright, 2020), developed for the no-longer-business-as-usual context, necessary for the twenty-first century.

The third chapter in this section presents the initial findings from a two-year practice-based research project, based in South West Kerry, Ireland, funded by the Irish Environmental Protection Agency. The research utilises an existing methodology, the permacultural resilience praxis (McKeown, 2015) developed for situated artistic practice as a means to explore resilience and co-design. The chapter outlines the transfer and application of the original methodology trialled in three unique geo-political contexts; Deptford, South London, Dublin, Ireland, and Taos County, New Mexico, USA, for its contribution to education and community development within a rural context. The findings form the underpinnings of the development of two toolkits initially aligned to four SDGs; SDG 4 – Quality Education, SDG 11 – Sustainable Cities and Resilient Communities, SDG 14 – Life Below Water, and SDG 15 – Life on Land. The SDGs chosen connect to the specifics of South West Kerry which has one of the longest coastlines in Ireland and are utilised as a shared language between policy and practice. The project's methods operate as a series of tactics developed to enable locally scaled interventions into policy and develop localised resilience through practical action strategically positioned and communicated.

In contrast to the third chapter, Lisa Eckenweiler presents a more philosophical approach, a consideration of ethical placemaking for ecological selves and citizens to inform new modes of governance for more liveable, equitable futures. This chapter makes the case for transitioning from 'the rational, self-sufficient, self-interested, extractive, and possessive creature who has been so well suited to industrial capitalism' (Eckenweiler, 2020) towards political possibilities that place care as central to the development of ecological subjects. Eckenweiler's chapter articulates clearly the need to have placemaking practices that create the conditions that support relations of care, and although focused on human placemaking this has potential for Rosa Bradotti's vision of a 'bio-centered egalitarianism (Braidotti, 2013, p. 19).

Indigenous cultures represent diverse and longstanding approaches to shaping the landscape, both physically and socially, 'enlivened by a sense of group and family history' (Eldrich, 1985). In this context, placemaking is an act of belonging, representing a sense of being expressed in an ever-evolving integral relationship between people and place. Theodore Jojola and Michaela P. Shirley, from the Indigenous Design and Planning Institute (iD+Pi), University of New Mexico and the Pueblo of Zuni's chapter presents how efforts to shape the vision to protect their cultural patronage, as expressed in their traditional arts has contributed re-imagining the role of Zuni artists in diversifying the economy of their Main Street.

Cathy Fitzgerald's eco-social art practice contributes to symbiotic placemaking (McKeown, 2019) that moves beyond the Anthropocene to the Symbiocene (Albrecht, 2015); from a human-centred towards a symbiotic, nurturing approach, more akin to the bio-centric approach this section argues for. 'The Hollywood Forest Story' and its underpinning methodology are offered as an example of placemaking that intersects with complex and dynamic eco-social systems. 'The Hollywood Forest Story' documents the artistic process of transforming a small conifer plantation, to a mixed species forest, using continuous cover forestry, as an alternative to conifer plantations, increasingly prevalent in Ireland as additional 'cash crops.' Fitzgerald offers an approach to placemaking that privileges social and environmental values and practices to 're-think alternatives to extractive, unsustainable industrial land practices' (Fitzgerald, 2019), an example of *Oikos* in action.

The section closes with Jeff Baldwin's consideration of non-human placemaking - in this instance, beavers - as historical actors, indeed political actors, that are self-interested, interactive, and co-producers of place. In Europe, there are numerous initiatives to reintroduce Eurasian beavers as a restorative measure, due to reported benefits to hydrology, vegetation, and animal species. In the United States various groups spend more than $1 billion each year restoring streams, a figure that is encouraging intentional beaver restoration projects. As a flagship species, beavers can also be used to raise awareness of riparian and woodland restoration, with areas that have reintroduced the species seeing increases in tourism. Baldwin, a professor of Geography at Sonoma State University revisits Marx's historical materialism in combination with Massey's work on the production of place as a lens through which to view beavers' contribution to the economics of placemaking within the twenty-first century.

In conclusion, the section presents an initial consideration of a range of practices and approaches that offers contributions to the concept of *Oikos* in placemaking, a more integrated and equitable approach to economics. There are many other authors and chapters that could have contributed to this section; however, the editor hopes that this initiates a broader consideration that seeks to move towards developing more equitable economies that integrate the social, environmental, and the economic through the concept of *Oikos* in placemaking.

References

Albrecht, G (2015) Exiting The Anthropocene and Entering The Symbiocene, Accessed 20.6.2020 Available here: https://glennaalbrecht.com/2015/12/17/exiting-the-anthropocene-and-entering-the-symbiocene/

BlackRock (2019) Getting physical: assessing climate risks Accessed 20.6.2019 Available here: https://www.blackrock.com/us/individual/insights/blackrock-investment-institute/physical-climate risks

Braidotti, R. (2013). 'Posthuman relational subjectivity', in *Relational Architectural Ecologies: Architecture, Nature, and Subjectivity*. New York: Routledge.

Bretschger L. and Valente, S. (2011). 'Climate change and uneven development', *Scandinavian Journal of Economics*, 1134, pp. 825–845.

Brown, C., Meeks, R., Hunu, K. and Yu, W. (2011). 'Hydroclimate risk to economic growth in sub-Saharan Africa', *Climatic* Chang, 1064, pp. 621–647.

Brulle, R. (2014). 'Institutionalizing delay: Foundation funding and the creation of U.S. climate change counter-movement organizations' *Climate Change*, 122(4), pp. 681–694.

Clark, M.T. (2015). 'Augustine on Justice', in *Augustine and Social Justice*. New York: Lexington Books.

Diamond, J (1999) Guns, Germs, and Steel by Jared W. W. Norton & Company

Eboli, F., Parrado, R. and Roson, R. (2010). 'Climate-change feedback on economic growth: Explorations with a dynamic general equilibrium model', *Environment and Development Economics*, 15(5), pp. 515–533.

Erdrich, L. (1985). 'Where I ought to be: A writer's sense of place', in *The New York Times* (28 July 1985) [online]. Available at: http://www.nytimes.com/1985/07/28/books/where-i-ought-to-be-a-writer-s-sense-of-place.html?pagewanted=al (Accessed: 20 January 2020).

Fitzgerald, C. (2019). 'Artists' statement', in *The Hollywood Forest* [online]. Available at: https://hollywoodforest.com/about/artists-statement/ (Accessed 18 December 2019).

Gawande, A (2014) The Century of the System, Lecture 2, The Reith Lectures, The Future of Medicine. Available at: http://www.bbc.co.uk/programmes/b04sv1s5 Accessed: 20/8/2020

Georgescu-Roegen, N. (1971). *The Entropy Law and the Economic Process*. Cambridge: Harvard University Press.

Goldenberg, S. (2013). 'Conservative groups spend up to $1bn a year to fight action on climate change', in *The Guardian* (20 December 2013) [online]. Available at: https://www.theguardian.com/environment/2013/dec/20/conservative-groups-1bn-against-climate-change (Accessed: 24 January 2020).

Horowitz, J. K. (2009). 'The income-temperature relationship in a cross-section of countries and its implications for predicting the effects of global warming', *Environmental and Resource Economics*, 444, pp. 475–493.

IPCC. (1992). *Climate Change: The IPCC 1990 and 1992 Assessments* [online]. Available at: https://www.ipcc.ch/report/climate-change-the-ipcc-1990-and-1992-assessments/ (Accessed: 14 December 2019).

IPCC. (1995). *SAR Climate Change 1995: The Science of Climate Change* [online]. Available at: https://www.ipcc.ch/report/ar2/wg1/ (Accessed: 14 December 2019).

IPCC, 2001: Climate Change 2001: Synthesis Report. A Contribution of Working Groups I, II, and III to the Third Assessment Report of the Integovernmental Panel on Climate Change [Watson, R.T. and the Core Writing Team (eds.)]. Cambridge University Press, Cambridge, United Kingdom, and New York, NY, USA, 398 pp

IPCC, 2007: Climate Change 2007: Synthesis Report. Contribution of Working Groups I, II and III to the Fourth Assessment Report of the Intergovernmental Panel on Climate Change [Core Writing Team, Pachauri, R.K and Reisinger, A. (eds.)]. IPCC, Geneva, Switzerland, 104 pp.

IPCC, 2018: Global Warming of 1.5°C. An IPCC Special Report on the impacts of global warming of 1.5°C above pre-industrial levels and related global greenhouse gas emission pathways, in the context of strengthening the global response to the threat of climate change, sustainable development, and efforts to eradicate poverty [Masson-Delmotte, V., P. Zhai, H.-O. Pörtner, D. Roberts, J. Skea, P.R. Shukla, A. Pirani, W. Moufouma-Okia, C. Péan, R. Pidcock, S. Connors, J.B.R. Matthews, Y. Chen, X. Zhou, M.I. Gomis, E. Lonnoy, T. Maycock, M. Tignor, and T. Waterfield (eds.)]. In Press.

JP Morgan Chase and Co (2019) Understanding Our Climate-Related Risks and Opportunities Accessed 20.6.2019 Available here: https://www.jpmorganchase.com/corporate/Corporate-Responsibility/document/jpmc-cr-climate-report-2019.pdf

Johnston, R. J. (2018). 'Ecosystem services', in *Encyclopedia Britannica* (5 Janaury 2018) [online]. Available at: https://www.britannica.com/science/ecosystem-services (Accessed: 26 March 2020).

Lemoine, D. and Kapnick, S. (2016). 'A top-down approach to projecting market impacts of climate change', *Nature Climate Change*, 61, pp. 51–55.

McKeown, A (2015) Cultivating permaCultural resilience; towards a creative placemaking critical praxis Unpublished thesis National College of Art and Design, Ireland

Meadows, D.H., Goldsmith, E. and Meadow, P. (1972). *The Limits to Growth*. New York: Universe Books.

Moberg, Carol L. (2005). *René Dubos, Friend of the Good Earth*. ASM Press. pp. 160–163. ISBN 1-55581-340-2.

Møller, M (2016) 'Business as usual' is not an option anymore Available on line:https://www.weforum.org/agenda/2016/01/why-do-we-need-a-multi-stakeholder-approach-tosustainable- development/ Accessed: 20/8/2020

Moyer, E. J., Woolley, M. D., Matteson, N. J., Glotter, M. J. and Weisbach, D. A. (2014). 'Climate impacts on economic growth as drivers of uncertainty in the social cost of carbon', *Journal of Legal Studies*, 432, pp. 401–425.

Perkins Marsh, G. (1864) *Man and Nature; or, Physical Geography as Modified by Human Action* (London: S. Low, Son and Marston).

Project for Public Spaces (1995) What is Placemaking, Available at: http://www.pps.org/reference/what_is_placemaking/ (Accessed: 15/4/2019)

Schultz, C. (2013). 'Meet the money behind the climate denial movement', in *The Smithsonian* (23 December 2013) [online]. Available at: https://www.smithsonianmag.com/smart-news/meet-the-money-behind-the-climate-denial-movement-180948204/ (Accessed: 7 October 2019).

United Nations Climate Change. (1997). *Kyoto Protocol* [online]. Available at: https://unfccc.int/process-and-meetings/the-kyoto-protocol/what-is-the-kyoto-protocol/kyoto-protocol-targets-for-the-first-commitment-period (Accessed: 14 December 2019).

United Nations Conference on Environment and Development. (1992). *Rio Earth Summit* [online]. Available at: https://www.un.org/esa/dsd/agenda21/Agenda%2021.pdf (Accessed: 14 December 2019).

Further reading in this volume

Chapter 2: Placemaking as an economic engine for all
James F. Lima and Andrew J. Jones
Chapter 3: An annotated history of creative placemaking at the federal level
Jen Hughes
Chapter 7: Conflict and memory: human rights and placemaking in the City of Gwangju
Shin Gyonggu
Chapter 14: Experts in their own tomorrows: placemaking for participatory climate futures
Paul Graham Raven
Chapter 19: Placemaking in the ecology of the human habitat
Graham Marshall
Chapter 22: Embedded Artist Project: Epistemic Disobedience + Place
Frances Whitehead

27

IS 'TACTICAL URBANISM' AN ALTERNATIVE TO NEOLIBERAL URBANISM?

Reflections on an exhibition at the MoMA

Neil Brenner

1.

What can 'tactical urbanism' offer cities under extreme stress from rapid population growth, intensifying industrial restructuring, inadequate social and physical infrastructures, rising levels of class polarization, insufficiently resourced public institutions, proliferating environmental disasters, and growing social unrest? The recent MoMA exhibition on *Uneven Growth* (MoMA 2014–15) aims to explore this question through speculative interventions by teams of architects whose remit was to make design proposals for six of the world's 'megacities' – Hong Kong, Istanbul, Lagos, Mumbai, New York, and Rio. The exhibition has provoked considerable debate about our contemporary planetary urban condition and, more specifically, about the capacities of architects, urban designers, and planners to influence the latter in progressive ways.

Such a debate is timely, not least because inherited paradigms of urban intervention – from the modernist-statist programs of the postwar epoch to the neoliberalizing, market-fundamentalist agendas of the post-1980s period – are no longer appear viable. Meanwhile, as David Harvey notes in his comment on the MoMA exhibition, 'the crisis of planetary urbanization' is intensifying. Megacities appear to be poorly equipped to resolve the monstrous governance problems and social conflicts that confront them. Under these conditions, Harvey grimly declares: 'We are [...] in the midst of a huge crisis – ecological, social, and political – of planetary urbanization without, it seems, knowing or even marking it' (2019, p. 29).

Against this foreboding background, can 'tactical urbanisms' provide tractable solutions, or at least open up some productive perspectives for actualizing alternative urban futures? Despite the cautiously exploratory tone of its curators' framing texts in the exhibition catalogue, the MoMA project on *Uneven Growth* articulates a strong set of claims regarding the potentials of tactical urbanism (Bergdoll, 2019; Gadanho, 2019, pp. 11–25). In the various documents associated with the exhibition, the notion of tactical urbanism is presented as a robust interpretive frame for understanding a variety of emergent urban design experiments in contemporary megacities. Just as importantly, MoMA curator Pedro Gadanho explains his choice of the concept as a basis for stimulating debate and practical experimentation regarding possible future pathways of urban design intervention, and above all, as a means to promote 'social justice in the conception and appropriation of urban space' (Gadanho, ibid, p. 23). As the search for new approaches to organizing our collective planetary urban future gains increasing urgency, these broadly affirmative discourses around tactical urbanism demand critical scrutiny. This chapter confronts this task, and in so doing, engages with some of the broader questions regarding the capacities of architecture/design as a means of producing and transforming space with which this volume is concerned.

2.

In the exhibition catalogue, Gadanho and several other internationally influential curators and urban thinkers frame their understanding of tactical urbanism. They offer a variety of contextual reflections and interpretative formulations to explicate its essential elements and several points of convergence emerge. Tactical urbanism arises in the context of a broader governance crisis in contemporary cities in which both states and markets have failed to deliver basic public goods (such as housing, transportation, and public space) to expanding urban populations. Tactical urbanism is not a unified movement, but a general rubric through which to capture a broad range of emergent, provisional, experimental and ad hoc urban projects. Tactical urbanism is mobilized 'from below,' through organizationally, culturally and ideologically diverse interventions to confront emergent urban issues. Professional designers, as well as governments, developers, and corporations, may participate in and actively stimulate tactical urbanism. But its generative sources lie outside the control of any clique of experts or any specific institution, social class, or political coalition. Tactical urbanism proposes immediate, 'acupunctural' modes of intervention in relation to local issues that are viewed as urgent by its proponents. Its time-horizon is thus relatively short, even 'impulsive' and 'spontaneous.' Its spatial scale likewise tends to be relatively circumscribed – for instance, to the park, the building, the street, or the neighborhood. Specific projects of tactical urbanism are said to evolve fluidly in relation to broader shifts in political-economic conditions, institutional arrangements, or coalitional dynamics. These qualities of malleability and open-endedness are widely praised in discussions of tactical urbanism, generally in contrast to the comprehensive plans, formal-legal codes, and rigid blueprints that were characteristic of modernist-statist projects of urban intervention. Tactical urbanism generally promotes a grassroots, participatory, 'do-it-yourself' vision of urban restructuring in which those who are most directly affected by an issue actively mobilize to address it and may continually mobilize to influence the evolution of methods and goals. For this reason, tactical urbanism is often presented as an 'open-source' model of action and as a form of 'reappropriation' of urban space by its users.

Most of the commentators involved in *Uneven Growth* present tactical urbanism as an alternative to both modernist-statist and neoliberal paradigms of urban intervention – for instance, because they are grounded upon participatory democracy; because they aim to promote social cohesion; and because they are not formally pre-programmed in advance or 'from above.' However, it is the opposition of tactical urbanism to modernist, comprehensive forms of urban planning that is most cogently demarcated in the wide-ranging narratives associated with the exhibition. Modernist-statist modes of urban intervention, it is argued, have receded due to the ideological ascendancy of neoliberalism and the associated 'disassembling of nation-states' (Saskia Sassen) since the 1980s. To the degree that some elements and offshoots of that tradition are still being mobilized in the megacities of the developing world via holistic, comprehensive planning and 'top-down action,' they are often 'entangled in inefficient politics, corrupt bureaucracy, and economic insufficiency' (Pedro Gadanho). Tactical urbanism is thus presented as a potential palliative for urban problems that state institutions and formal urban planning procedures, in particular, have failed to address adequately.

Despite the affirmations of many of the contributors to *Uneven Growth*, it is less obvious as to how the projects associated with tactical urbanism could counteract neoliberal urbanism. Especially in light of the stridently anti-planning rhetoric that pervades many tactical urban interventions and their tendency to privilege informal, incremental mobilizations over larger-scale, longer-term, publicly financed reform programs, it seems reasonable to ask in what ways they do, in actuality, engender any serious friction against the neoliberal order, much less subvert it. In some cases, tactical urbanisms appear more likely to *bolster* neoliberal urbanisms by temporarily alleviating (or perhaps merely displacing) some of their disruptive social and spatial effects, but without interrupting the basic rule-regimes associated with market-oriented, growth-first urban development, and without challenging the foundational mistrust of governmental institutions that underpins the neoliberal project. The relation between tactical and neoliberal forms of urbanism is thus considerably more complex, contentious, and confusing than is generally acknowledged in the contributions to the debate on *Uneven Growth*. As Figure 27.1 illustrates, it cannot be simply assumed that, because of their operational logics or normative-political orientations, tactical interventions will counteract neoliberal urbanism. No less than five specific types of relation between these projects can be readily imagined, only two of which *might* involve a challenge to market-fundamentalist urban policy. There are at least three highly plausible scenarios in which tactical urbanism will have either negligible or actively beneficial impacts upon a neoliberalized urban rule-regime.

Tactical urbanism may be *narrated* as a self-evident alternative to neoliberal urbanism; but we must ask: is this really the case, and if so, how, where, under what conditions, via what methods, with what consequences, and for whom? Clarification of these issues is essential to any serious consideration of the potentials and limits of tactical urbanism under contemporary conditions.

Tactical urbanism/neoliberal urbanism—5 scenarios

1. *Subversion*. Tactical urbanism interrupts the logics of growth-first, market-oriented urban governance and points towards alternative urban futures based on grassroots democracy and social justice.
2. *Reinforcement*. Tactical urbanism alleviates some of the governance failures and disruptive consequences of neoliberal urbanism, but without threatening its grip on the regulatory framework governing urban development.
3. *Entrenchment*. Tactical urbanism internalizes a neoliberal agenda (for instance, related to a diminished role for public institutions and/or an extension of market forces) and thus contributes to the entrenchment of neoliberal urbanism.
4. *Neutrality*. Tactical urbanism emerges in interstitial spaces that are neither functional to, nor disruptive of, the neoliberal project. It thus co-exists with neoliberal urbanism in a relationship that is neither symbiotic, parasitic nor destructive.
5. *Contingency*. Tactical urbanism opens up a space of regulatory experimentation that, under certain conditions, contributes to the subversion of neoliberal programs. But, in other contexts, with many of the same conditions present, this does not occur. The impacts of tactical urbanism on neoliberal urbanism are thus contingent; they hinge upon factors extrinsic to it.

Figure 27.1 Tactical urbanism/neoliberal urbanism – five scenarios, Brenner, 2020.

Neoliberal urbanism is not a unified, homogenous formation of urban governance, but represents a broad *syndrome* of market-disciplinary institutions, policies, and regulatory strategies (Brenner et al., 2010a, b). While certainly connected to the ideology of free market capitalism, this syndrome has assumed deeply variegated political, organizational, and spatial forms in different places and territories around the world, and its politico-institutional expressions have evolved considerably since the global economic crises and accompanying geopolitical shocks of the 1970s. Across all the contextual diversity and evolutionary mutation, however, the common denominator of neoliberal urbanisms is the market-fundamentalist project of activating local public institutions and empowering private actors and organizations to extend commodification across the urban social fabric, to coordinate a city's collective life through market relations, and to promote the enclosure of non-commodified, self-managed urban spaces.

As Teddy Cruz succinctly notes in his contribution to the *Uneven Growth* catalogue, all this has promoted the 'shift from urbanizations benefitting the many into models of urban profit for the few (Cruz, 2019, p. 51). Whereas the idea of 'urbanizations benefitting the many' broadly corresponds to the now-discredited megaprojects and programming techniques of statist-modernism, the promotion of 'urban profit for the few' has been the predominant tendency since the 1980s, at once in the older capitalist world, the former state socialist world, and across most of the postcolonial and developing world. Despite plenty of variegation, resistance, contestation, and reregulatory pushback, this tendency has persisted, and even intensified, through the many waves of industrial restructuring and financial crisis that have ricocheted across every zone of the world economy since that period, including since the most recent 'Great Recession.' The patterns of 'uneven growth' that are under scrutiny in the MoMA exhibition must be understood as its direct expressions and outgrowths.

It is, then, neither the contemporary urban condition 'as such,' nor the inefficiencies of modernist-statist urban planning, that have most directly triggered the problems to which contemporary forms of tactical urbanism are responding. Rather, contemporary tactical urbanisms are emerging in contexts that have been powerfully ruptured and reshaped by historically and geographically specific forms of neoliberal urbanization, based on the class project of restricting 'the right to the city' (Henri Lefebvre) to the wealthy, the elite, and the powerful, and reorienting major public investments and policy regimes in ways that prioritize that project above all others. Despite its pervasive governance failures, its powerfully destructive socio-environmental consequences and its increasingly evident ideological vulnerabilities, neoliberalism continues to represent the taken-for-granted 'common sense' on which basis urban development practice around the world is still being forged. The question of how designers might contribute to alternative urban futures must thus be framed most directly – and, from my perspective, a lot more combatively – in relation to the apparent resilience and elasticity of neoliberal forms of urban governance.

One important consequence of these observations is the proposition that the architectural and design disciplines could significantly enhance their capacity to make durable, progressive, urban interventions by engaging more systematically with questions of *institutional* (re)design – that is, the systems of collectively binding rules that govern the production, use, occupation, and appropriation of space. The latter are arguably as essential to the broad visions for future megacities proposed in *Uneven Growth* as the tactical, acupunctural projects of infrastructural and physical reorganization with which the bulk of the exhibition is concerned. Indeed, in the absence of an aggressively reasserted role for governmental institutions – publicly funded through an equitable tax regime; democratically legitimated and publicly accountable; legally regulated and transparently monitored; and oriented towards the public interest – it is difficult to imagine how the tactical urbanist proposals put forward in *Uneven Growth* could ever accomplish larger-scale, longer-term impacts.

Herein lies a potentially serious contradiction. The anti-statist, anti-planning rhetoric of many tactical urbanist interventions may, in practice, erode their capacity to confront the challenges of upscaling their impacts. To the degree that advocates of tactical urbanism frame their agenda as an alternative to an activist role for public institutions in the production of urban space, they are at risk of reinforcing the very neoliberal rule-regimes they ostensibly oppose. This is in no way to suggest that tactical urbanist projects should ignore the serious deficits of state action in contemporary megacities. On the contrary, the critique of how market-oriented state policies (including privatization, deregulation, and liberalization) erode public institutions in favor of privatized forms of urban appropriation is essential to any counter-neoliberal, reregulatory project. Just as important, in this context, is the collective demand for more extensive public support for key dimensions of social reproduction – the essential infrastructures associated with housing, transportation, education, public space, health care, and so forth. The point here, then, is simply that there are deep tensions between the project of finding viable alternatives to neoliberal urbanism and a tradition of urban intervention that tends to distance itself from state institutions.

In his contribution to the *Uneven Growth* catalogue, Teddy Cruz offers a hard-hitting formulation of the major challenges associated with that endeavor among architects and designers:

> Without altering the exclusionary policies that have decimated a civic imagination in the first place, architecture will remain a decorative tool to camouflage the neoconservative politics and economics of urban development that have eroded the primacy of public infrastructure worldwide [...] the major problems of urbanization today [...] are grounded in the inability of institutions of urban development to more meaningfully engage urban informality, socioeconomic inequity, environmental degradation, lack of affordable housing, inclusive public infrastructure, and civil participation.
>
> *(Cruz, ibid.)*

This is precisely the dilemma: how can tactical urbanisms do more than serve as 'camouflage' for the vicissitudes, dislocations, and crisis-tendencies of neoliberal urbanism? Cruz's formulation underscores one of the key conditions under which it might begin to do so: through the reimagination of design, not simply as a 'decorative tool' or formal set of techniques-for-hire by the ruling classes, but as a basis for asking critical questions about contemporary urbanism, and as a set of collectively shared, creative capacities through which to 'coproduce the city as well as new models of cohabitation and coexistence to advance agendas of socioeconomic inclusion' (Cruz, ibid.). This goal cannot be realized simply through the redesign of specific physical sites within the city; it also requires the creation of 'a new role for progressive policy, [and] a more efficient, transparent, inclusive, and collaborative form of government' (ibid, p. 55). In other words, the pursuit of alternative urbanisms requires the creation not only of new urban spaces, but of new state spaces as well.

3.

These considerations yield a critical perspective from which to examine some of the design proposals for contemporary megacities that are on display in *Uneven Growth*. MoMA curator Gadanho's remit to the six design teams was not only to propose a tactical intervention for a specific megacity – 'acupunctural outlooks on how change for the better could be induced in diverse urban contexts' – but, in so doing, to offer a new perspective on what a socially engaged architecture might look like. We must thus consider the exhibition materials at once as possible scenarios for a future urbanism, and as visions of how the design disciplines might use tactical approaches to contribute to their realization. Even if they harness the speculative capacities of design, then, the proposals on display in the exhibition are clearly not meant to be pure fictions – they are presented as critical tools 'to reflect upon the problems of today' (Gadanho, 2019, pp. 16–23).

Only some of the design proposals featured in the exhibition respond effectively to this remit. While the exhibition's theorists broadly agree on the contours of a tactical urbanism, there is evidently considerable confusion regarding the meaning and implications of this notion among the designers themselves. Although all of the design scenarios are presented under the shared rubric of tactical urbanism, some bear little resemblance to an acupunctural, participatory, open-sourced intervention. Indeed, several of the design proposals involve large-scale megaprojects and landscape transformations that could only be implemented through a powerful state apparatus; they are difficult to envision as more than partial outgrowths of tactical methods. Meanwhile, other design proposals are consistently framed within tactical parameters, but yield a vision of the urban future that appears entirely compatible with most versions of neoliberalism. Such interventions may be 'tactical,' but they totally bypass the intricacies of exploring real alternatives to the currently dominant system of market rule.

A number of the proposals circumvent questions of implementation entirely, offering decontextualized design 'solutions' to the pressing problems of megacity development – for instance, regarding water scarcity, insufficient land for housing, transportation bottlenecks, or issues of energy supply. Because they bracket the formidable constraints associated with implementation under a neoliberalized rule-regime, these design scenarios remain at a purely hypothetical level – visions of an alternative universe that are utopian in the literal sense of that word; they are located *nowhere*. They put the capacities of design thinking on display, often with striking visual flourishes, but with considerably less traction than if the conditions for their actualization were seriously interrogated.

Among the contributions to *Uneven Growth* that most directly attempt to mobilize tactical interventions as part of a broader assault on neoliberal urbanism, the scenarios elaborated by the Mumbai design team (URBZ/Ensamble-POP lab), the Istanbul design team (Atelier d'Architecture Autogérée/Superpool), and one of the New York City teams (Cohabitation Strategies – CohStra) are particularly generative. Notably, each does so through an engagement with the housing question, which has been a fundamental terrain of design intervention and political struggle throughout the history of capitalist urbanization. In confronting this well-trodden terrain, the teams illustrate how an expanded vision of design – as a set of combined capacities for spatial intervention, social empowerment, and political critique – can contribute to the ongoing struggle for alternative urbanisms.

The Mumbai proposals by URBZ/Ensamble-POP lab mobilize tactical interventions to protect so-called 'slum' neighborhoods such as Dharavi and Shivaji Nagar from the massive land development pressures associated with Mumbai's extensively neoliberalized, financialized economy. This is a multifaceted proposal, perhaps reflecting the different positionalities of the project teams in relation to the slum itself (URBZ is a group of activist designers with strong roots in Mumbai's poor neighborhoods, whereas the POPlab is based at MIT). At core, the project presents a series of incremental design strategies to promote an alternative vision of the 'slum' as a space of productivity, creativity and ingenuity – a '*tabula pronta*,' in the team's formulation, rather than a *tabula rasa* that can be readily razed to make room for new zones of single-function mass housing. Instead of imposing a new prototype from outside, the designers propose to enhance spatial practices that already animate those neighborhoods – specifically, the integration of residential spaces with workspaces or 'tool houses.' By supplying a model of up-building that enables residents to construct new platforms for work and everyday life above their homes, and by creating a network of 'supraextructures' on a plane stretched like a 'magic carpet' above the rooflines, new possibilities for endogenous local economic development and social interaction are envisioned. The developmental potentials thus unleashed would, the designers propose, serve as strong counterpoints to dominant ideologies of the slum as a space of backwardness and pathology, while also stimulating the elaboration of a less polarized growth pattern across the metropolitan fabric.

Thorny questions remain, of course, regarding the degree to which the proposed tactical interventions could protect the most strategically located neighborhoods from land development pressures, especially in the absence of a broader political movement that questions the model of market-driven urban growth to which Mumbai's growth coalition committed itself following the liberalization of the Indian economy in the 1990s. Through what institutional mechanisms and political coalitions could tenure security be attained by slum dwellers living in zones of the city that are considered attractive by growth machine interests? As Neil Smith pointed out some time ago, when local government institutions align with development interests to exploit such a 'rent gap' in the urban land market, organized resistance is likely to be met with considerable vilification, if not outright repression (Smith, 1996). There is no doubt, however, that design has a fundamental role to play in defending vulnerable populations and neighborhoods against further disempowerment, dispossession, and spatial displacement. The proposal for Mumbai by URBZ/Ensamble-POP lab very productively puts this issue on the exhibition's agenda. It will hopefully inspire other designers to take up this project in other megacities, in collaboration with local inhabitants, local social movements, and non-governmental organizations that share their concerns.

While the design proposals presented by the Istanbul and New York teams contain important architectural/morphological elements (pertaining, for instance, to buildings, infrastructures, and neighborhood districts), their creative radicalism is strongly rooted in models for new institutional arrangements that would empower each city's low- or middle-income inhabitants to occupy, appropriate, and regenerate spaces that are currently abandoned, degraded, or being subjected to new forms of vulnerability. In the New York context, the CohStra team focuses on a variety of interstitial or underutilized spaces in the city core – from vacant lots and abandoned buildings to various kinds of lower-density housing provision – in order to propose an alternative framework for land ownership (community land trusts), housing provision (mutual housing associations), building management (cooperative housing trusts), and household financing (community credit unions). In the case of Istanbul, the Atelier d'Architecture Autogérée's design proposal targets the mass housing complexes that were constructed for the burgeoning middle classes during the post-1990s period by Turkey's Housing Development Agency, known as TOKI, which are predominantly located in more peripheral districts within Istanbul's rapidly urbanizing metropolitan territory. Here, the designers propose to retrofit existing TOKI housing ensembles, and their immediate landscapes, in ways that facilitate new forms of communal self-management by the inhabitants – including, as with CohStra's proposal for New York, community land trusts and local credit unions, along with other forms of collectively managed infrastructure such as community farming and gardens, fisheries, workshops, green energy sources, and repair facilities.

As with the Mumbai team's proposal, each of these tactical interventions is framed as a response to a specific, immediate set of threats to urban life that have been imposed by the neoliberal growth model in the city under consideration – the 'crisis of affordability' for working New Yorkers; and the destabilization of the model of middle-class consumerism that had been promoted in Istanbul through TOKI mass housing. Notably, however, CohStra and the Atelier d'Architecture Autogérée move beyond a defensive posture in relation to such issues, offering instead a vision of how the spaces that are being degraded under neoliberalized urbanism could become the anchors for an alternative vision of the city as a space of common life and collective self-management. In both projects, the site of design intervention is viewed as a *commons*, a space of continuous, collective appropriation and transformation by its users. Both teams offer a vision of this commons as a *process* to which designers can contribute in fundamental ways, not only by elaborating spatial proposals for the reorganization of housing functions or other dimensions of social reproduction, but by reimagining how such basic institutions as private property, profit-oriented real estate investment, urban land markets, and municipal bureaucracy might be transformed and even superseded to serve social needs, to empower urban inhabitants, and to contribute to the creation of a genuine urban public sphere.

Although the Istanbul and New York projects are presented in tactical terms, they are clearly intended as more than fleeting acupunctural interventions. Part of their appeal is precisely that they offer a model of tactical urbanism that may be aggressively upscaled and converted into a city-wide counterforce to the neoliberal model. Initially offering a kind of protected enclave for a vulnerable population, each project is then meant to be transformed into a generalizable alternative to the specific forms of housing commodification and accumulation by dispossession that have underpinned 'uneven growth' in their respective megacities. It is this reflexive attempt to connect the methods of tactical urbanism to a double-edged redesign of urban spaces *and* institutions that makes these teams' proposals effective as tools for envisioning alternatives to the neoliberal city. In thus proceeding, however, the proposals by the Atelier d'Architecture Autogérée and CohStra move beyond the realm of tactical urbanism: rather than being a focal point for investigation as such, the latter becomes a kind of launching pad for envisioning and enacting a 'politics of space' (Henri Lefebvre) – that is, a political strategy of large-scale sociospatial transformation.

Here, too, of course, the inevitable questions of implementation loom on the horizon. How can this vision of the commons (and of common*ing* practices) be realized when the dominant class interests in each megacity continue to promote a profit-oriented, speculation-driven growth model? Where are the social forces and political coalitions that could counteract that model, and would they really opt for the level of collective coordination and communal sharing proposed by these design teams? How can local alternative economies be protected from incursions by profit-oriented producers, who may (for instance, through economies of scale, or more rationalized forms of labor exploitation) be able to offer more affordable or desirable products to cash-strapped consumers? Designers cannot answer these questions, at least, not among themselves; they can only be decided through political deliberation, public debate, and ongoing struggle. But, because CohStra and the Atelier d'Architecture Autogérée took the fundamental step of integrating such political-institutional considerations into their spatial proposals, they productively contribute to that process. Just as importantly, their proposals also articulate a more socially engaged, politically combative vision of what the design disciplines have to offer the urban public sphere in an era of deepening inequality and highly polarized visions of our global urban future.

4.

Given the difficulties that some of the design teams appear to have had with the tactical urbanism framework, one cannot help but wonder whether it offered them too narrow a terrain, or too limited a toolkit, for confronting the vast challenges that are currently emerging in the world's megacities. MoMA curator Barry Bergdoll anticipates this conundrum, noting the gap between the 'modest scale of some [tactical] interventions' and the 'dimensions of the worldwide urban and economic crisis that so urgently needs to be addressed (Bergdoll, 2019, p. 12). In the face of these challenges, one can hardly reproach the teams that opted to venture forth with big, ambitious proposals rather than restricting themselves to mere 'tactics.'

But here arises a further contradiction of the *Uneven Growth* project. A pure form of tactical urbanism would have to be systematically *anti*-programmatic; it could only maintain a consistently tactical approach by resisting and rejecting any movement towards institutionalization. Yet, to the degree that the tactical design experiments on display in *Uneven Growth* articulate a broader vision of urban transformation, they necessarily hinge upon the (eventual) articulation of a comprehensive vision of the whole. The generalization of tactical urbanism will thus entail its self-dissolution or, more precisely, its transformation into a project that requires longer-term coordination; stabilized, enforceable, collectively binding rules; and some kind of personnel assigned to the tasks of territorial management – in other words, *planning*. We thus return to the supposedly discredited, outmoded terrain of statist-modernism, the realm of big ambitions, large-scale blueprints, elaborate bureaucratic procedures, and comprehensive plans, in opposition to which the precepts of tactical urbanism are generally framed. Even if one prefers tactical methods over those of top-down bureaucracies (or, for that matter, those of profit-hungry developers and transnational corporations), it would seem that a serious discussion of large-scale territorial plans, institutional (re)organization, and political strategies of implementation is unavoidable, at least if the goal is seriously to envision a future for megacities that is more socially and spatially just, democratic, livable, and environmentally sane than our present global urban condition.

For anyone sympathetic to tactical urbanism *and* the project of large-scale, progressive urban transformation, this contradiction is probably unavoidable. Can it be made productive? Perhaps the radical potential of tactical urbanism lies less in its role as an all-purpose method for designing urban futures, than as a radically democratic counterweight to institutional systems, whether state-driven or market-dominated. Some of the most valuable contributions in *Uneven Growth* serve this very purpose: they point towards the possibility that, rather than being instrumentalized for social engineering, political control, private enjoyment, or corporate profit-making, the capacities of design might be remobilized as tools of empowerment for the users of space, enabling them to occupy and appropriate the urban, continually to transform it, and thus to produce a different city than anyone could have dreamt up in advance.

But even in this maximally optimistic framing of tactical urbanism, the 'big questions' regarding how to (re)design the city of the future – its economy, its property and labor relations, its spaces of circulation, social reproduction and everyday life, its modes of governance, its articulations to worldwide capital flows, its interfaces with environmental/biophysical processes, and so forth – remain completely unresolved. As MoMA's Department of Architecture and Design continues its productive engagement with urbanism, let us hope that such questions will stay on the agenda, and that the creative capacities of designers can be harnessed to confront them with all the critical force, political imagination, and systematic vision they require.

References

Bergdoll, B. (2019). 'Preface', in Gadanho, P. (ed.) *Uneven Growth: Tactical Urbanisms for Expanding Megacities*. New York: The Museum of Modern Art.

Brenner, N., Peck, J. and Theodore, N. (2010a). 'After neoliberalization?', *Globalizations*, 7(3), pp. 327–345.

Brenner, N., Peck, J. and Theodore, N. (2010b). 'Variegated neoliberalization', *Global Networks*, 10(2), pp. 182–222.
Cruz, T. (2019). 'Rethinking uneven growth', in Gadanho, P. (ed.) *Uneven Growth: Tactical Urbanisms for Expanding Megacities*. New York: The Museum of Modern Art.
Gadanho, P. (2019). 'Mirroring uneven growth', in Gadanho, P. (ed.) *Uneven Growth: Tactical Urbanisms for Expanding Megacities*. New York: The Museum of Modern Art.
Harvey, D. (2014). 'The crisis of planetary urbanization', in Gadanho, P. (ed.) *Uneven Growth: Tactical Urbanisms for Expanding Megacities*. New York: The Museum of Modern Art.
Smith, N. (1996). *The New Urban Frontier*. New York: Routledge.

Further reading in this volume

Chapter 9: From the dust of bad stars: disaster, resilience, and placemaking in Little Tokyo
Jonathan Jae-an Crisman
Chapter 10: From moon village to mural village: the consequences of creative placemaking in Ihwa-dong, Seoul
Jason F. Kovacs and Hayun Park
Chapter 11: *Free State Boulevard* and the story of the East 9th Street Placekeepers
Dave Lowenstein
Preface: The problem with placemaking
Louise Platt
Chapter 16: More than a mural: participatory placemaking on Gija Country
Samantha Edwards-Vandenhoek
Preface: The radical potential of placemaking
Cara Courage
Chapter 20: Displacemaking 2015 and 2020
Catherine Fennell and Daniel Tucker
Chapter 22: Embedded Artist Project: Epistemic Disobedience + Place
Frances Whitehead
Chapter 24: Artists, creativity, and the heart of city planning
Tom Borrup
Chapter 25: 'If you can make it there, you can make it anywhere…': cultural placemaking at the heart of cities
Sherry Dobbin
Chapter 26: Sculpturing sound in space: on *The Circle and the Square* (2016) by Suzanne Lacy
Trude Schjelderup Iversen
Chapter 41: Rituals of regard: on festivals, folks, and findings of social impact
Maribel Alvarez
Chapter 44: Creative Placemaking and comprehensive community development: rethinking neighborhood change and evaluation
Maria Rosario Jackson
Chapter 45: How the city speaks to us and how we speak back: rewriting the relationship between people and place
Rosanna Vitiello and Marcus Willcocks

28
INTEGRAL PLACEMAKING
A poiesis of sophrosynes?

Ian Wight

Personal placing – an integral practitioner at work

We come to our practice – more deeply, our *praxis* – from our own experience of place, from our own 'making' of it – literally and figuratively. We may try to cast it objectively, abstractly, impersonally – but in our heart and soul it is anything but. Indeed, it is very personal, *and* inter-personal, *and* trans-personal – if we might care to admit such perspectives. Placemaking has been a formative human practice from time immemorial (Schneekloth and Shibley, 1994). There is no escaping it – and in integral terms it is much more than a mere 'it' and more an exquisite integration of I and We and It/Its; 'it' spans form and consciousness, interior and exterior, individual and collective (Wight, 2005). But the integration is always, ultimately, in our selves (ideally in our larger Selves). It is where we make common sense of Nature and Culture; it is where we experience our evolutionary trajectory... our ever-more-whole-making. The placemakers are us, whole-persons, whole-making, as a matter of our life-course.

I present as an integral practitioner at work in this primal and potent field. My integral 'turn' began around 20 years ago, the latest in a series of 'ever-more-all-encompassings' – geographical, environmental, and ecological in their turn. 'Place' had a place in each, a grounding place, a home-base of sorts, that I keep 'coming home' to. But it has evolved, and I expect it to continue to evolve; 'integral' may best place me at present, but I sense a *meta*-integral at work out around my leading edge, my know(ing)-ledge. It is my emerging 'territory beyond' (Zander, 2017, p. 189).

I have a 'primal place' from my early youth that is still very much 'in' me – local, physical, visceral. It is a gift that keeps on giving, that keeps place alive for me, that grounds and centres me. My academic sense of place emerged in my geography studies, at a time when place was being eclipsed by space, with geography aspiring to be a spatial science. But the place seed was well planted and informed my early work as a planner. It came through again as a city planning and environmental design educator, anchoring my 'professing,' validating space–place transformation, and advocating planning as placemaking – initially as a form of applied ecology (Wight, 2005). And my placemaking has continued to evolve under integral influences, first as 'wellbeing by design' (Wight, 2012) and most recently as 'a poiesis of sophrosynes' (Wight, 2017).

What is your own over-*standing of* your *operating environment?*
How might you personally *'place' your Self?*[1]

Framing place – integrally

What set me off on what became my integral placemaking journey? I recall noticing some provocative Australian practice on 'place not space' (Winnikoff, 1995) and began to wonder about giving place more space in a course I was teaching at the time. Ideally, I sought to align them and mobilize them, together: how might a mere 'space' become transformed into a meaningful 'place'? There was an intriguing integration challenge here – that became an opening into the possibility of a more integral approach. Serendipitously, an architect friend introduced me to a book he had just read, that seemed to address this challenge, effectively setting me on my integral path. The book was *A Brief History of Everything*, by Ken Wilber (1996).

'Integral' has been my default 'operating system' ever since, bequeathing an integral view of place and placemaking – *integral* place and *integral* placemaking – with a particular concern for the placemaker, the person/s in/of the place in question. There is an audacity to integral approaches that can be difficult to countenance; it goes against the grain of much mainstream academic practice, and calls for extraordinary overview capacity – an *over*-standing as much as an under-standing, and an unflinching commitment to integration – as a necessary complement to differentiation, and the often associated reductionism. An integral perspective has been summed up by Wilber (1996) as 'comprehensive, inclusive and balanced – not leaving anything out.' It aspires to be comprehensive in the extreme, always alert to what might be missing, what might be being downplayed. In particular it values 'wholistic' perspectives that honour whole persons at work in whole systems in whole-making endeavours. Integration is the underlying intent. It aspires to be inclusive in the extreme, to be wide open to development, to deeply honour an evolutionary trajectory. One of its key tenets is an impulse to 'transcend and include,' to acknowledge the at-least-partial truths in earlier formulations, and the dignities adhering to past practices. Integration is the underlying intent. It aspires to be balanced in the extreme – between exterior form and interior consciousness; between individual and collective contexts; between objective and subjective, inter-objective and inter-subjective; among different levels and lines of development; across different moral perspectives and worldviews. Integration is the underlying intent.

Another short-hand representation of 'integral' is the integration of Body, Mind, Soul, and Spirit in Self, Culture, and Nature (Figure 28.1). There is much in play where integral is concerned, and much to be made when it is mobilized, when it represents the underlying awareness – especially self/Self-awareness. There is no escaping the dimensions of one's self – the source of the making is in 'You,' in your body, mind, soul, and spirit. And the integral quadrants help 'place' Self, Nature, and Culture.

An integral framing of place seeks to reflect these integrations (Figure 28.2). The quadrants represent four primordial perspectives – captured in the pronouns: I, We, It, and Its. I have come to *integrally* render place as an integration of physicality, functionality, conviviality, and spirituality. As such, we should strive not to reduce place to any one dimension; it is essentially *an integration of all four* – the integration constituting the 'making.'

1 Note to readers: these italicized prompts, at the end of each section, are possible 'pause and reflect' occasions, to sense into any potential arising for yourself. Consider them my attempt to prime, in you, some of the interior conditioning that I have come to believe as essential for authentic placemaking. They may help take you beyond 'hand-book' territory into 'heart-book,' or even 'soul-book,' territory. They are my own constructions, influenced by various experiential workshop efforts and experiences. I hope they might constitute some welcome cause for pause, to go inside, to balance ourselves.

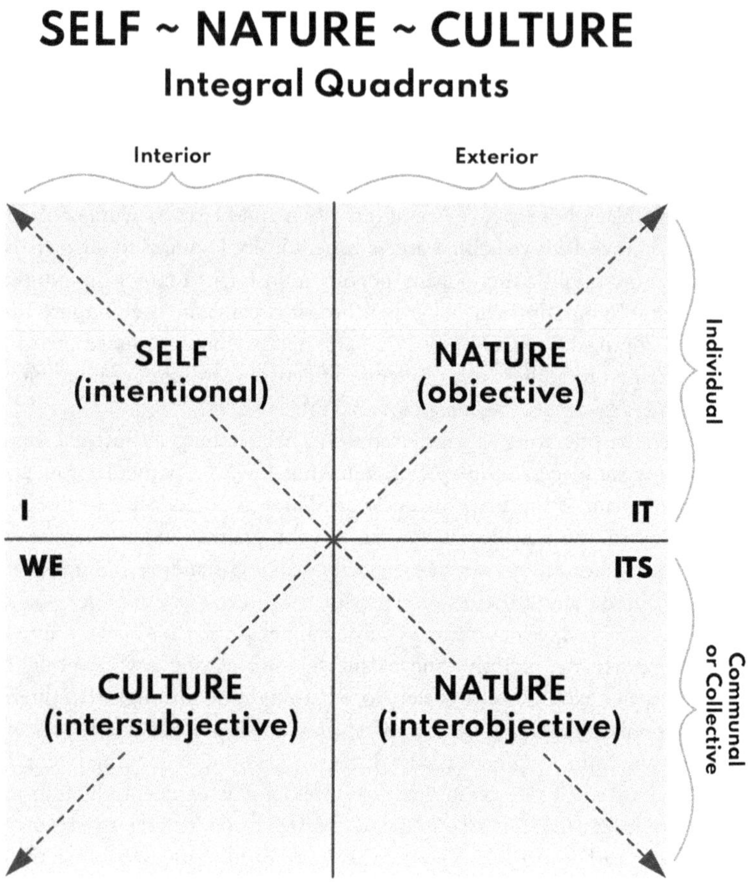

Figure 28.1 Self, nature, culture: the integral quandrants (Wight, 2018).

The qualities of place do not exist independently; rather, they co-exist interdependently – they co-relate. There is obviously some differentiation at work – different place personas depending on the reference quadrant. But the aimed-for disposition needs to be 'honouring all while privileging none.' Biases/preferences are to be expected outwith an integral frame – limiting, qualifying what might be made of place, and especially the associated placemaking. It often seems easy, in academia and professional policy circles especially, to fetishize place, but it is its making, and its makers – the placemaking and the placemakers – that merit prime regard in practice contexts.

> *Reflect on your own integrated-ness – on your inside, and your integration-ability – on the outside.*

Place – through an integral lens: a making, by makers

My exposure to an integral perspective was transformative for my own practice – professionally, educationally, and personally. My outlook became 'through an integral lens' (Wight, 2006), and increasingly my persona became an aspiring 'integral@work.' Inevitably my foundational

Figure 28.2 Integral place: an all-quadrants affair (Wight, 2018; and supplemented by 'place as' distinctions, from Dekay, 2011.

frames of reference were reworked, with integral in mind (and in heart, and in soul). This naturally included place, my enduring anchor, from my formative geography days. I began with some working hypotheses, based on my emerging integral perspective, rooted at the time in a curiosity about possible ties to a post-modern perspective: integral as post-modern; integral placemaking as a post-modernity project – as a post-modernized form of planning (Wight, 2005). Place, it occurred to me, had too often too easily been dehumanized as simply an 'it', as an objective – often spatial – identifier (in what Wilber (ibid.) termed 'Flatland'). I sensed the possibility of a broader Wilberian all-quadrants/all-levels (AQAL) perspective on place, considering the proposition of the actual making of place as a form of the integral practice advocated by Wilber. I hypothesized integrally conceived placemaking as a serviceable bridge between past (basically spiritually bereft) professional and academic agendas and the projected necessary spirit-embracing (constructive) post-modern practice. Place became much more than a simple matter of geography or locality, but a place of makings, by makers – whole persons – sense-making and meaning-making together, hearts and souls and spirit at work.

In terms of the integral quadrants, place may be conceived as *spanning* exterior and interior, communal/collective and individual. It is always at least multiple-quadrant; ideally place is an 'all-quadrants' affair. It is a key venue for the integration of the worlds of *it* or *its* (the physical, functional, geographical, social) and the worlds of *I* and *We* (the makers, making, convivially – their spirit-at-work); i.e. material form *conjoined* with non-material consciousness. Levels, as well as quadrants, are prominent in an integral framing. For example, the levels associated with what Wilber terms the 'Great Nest (or Chain) of Being' are Matter, Body, Mind, Soul, and Spirit. Thus, an integral view of place would embrace all such levels; place is multi-level as well as multi-quadrant, having both primalcy and potency, integrating past, present, and future. Place may be regarded as submergent *and* emergent, a common thread through the 'nestwork.' It is always in flux, in development, being made and re-made, to better ground the development of individuals and collectivities.

Spiral Dynamics (Beck and Cowan, 1996) offers another example of integral levels that can be informative of the evolution of planning – of different 'plannings,' including planning as placemaking. When I encountered this particular framing (see Wight, 2012, for more background) I hypothesized that it might be a potential major bridge between Wilber's thought and the contemporary planning 'mix' that was my world at the time. I began to see an evolution of 'plannings' (as collective culture-specific action) in a developmental progression of integral 'transcending and including': command-and-control, to master-planning and zoning, to strategic planning, to communicative action, collaboration and consensus-building, to ecological/holistic 'second-tier' placemaking. With this framing I began to sense a context for a more integral, constructively post-modern, second-tier planning with placemaking in mind, literally and figuratively, individually and collectively. Planning could be considered to be developing/evolving, occupying new worlds, as integrally informed post-modern placemaking.

While planning – conventionally – might be abstractly conceived as a linking of thinking and acting, and as forward-thinking and intervention-oriented, it might be less easily conceived as always evolving in terms of the level/state of consciousness being manifested. However, I began to sense that we may now be advancing on a period of post-conventional 'vision-logic' planning (as distinct from a planning rooted in more concrete or formal ways of thinking) with a consequential requisite rethinking of place and placemaking. My integral exposure thus emboldened me to suggest that modern(ity) planning has been essentially a linking of first-tier thought with a *differentiating* agenda, dictated by concern for the 'it/s'-world: i.e. *anything but* integral. Instead, planning needs to more consciously become part of a post-modernity project, linking second-tier thought with an *integrating* action agenda, embodying the practice of an AQAL/integral approach (what Wilber would term 'Spirit-in-Action'). Such a planning (integrating the worlds of I, We, and It/its) would clearly be well-served by a renewed consideration of the central place of place - in all-quadrants/all-levels (AQUAL) terms - but an explicitly post-conventional notion of place, rooted in a world-centric vision-logic, transcending while including earlier (lower tier in Spiral Dynamics terms) notions of place.

'Integral' came to represent for myself the new 'comprehensiveness' in planning terms, oriented to transcending while including earlier formulations, with a 'balancing' agenda, balancing the interior and exterior, form and consciousness, the practical and the spiritual (integrally defined). Place became *an integral production*, inclusively interrelating physicality, functionality, conviviality, and spirituality, the latter being a particularly key entrée for planners, and other professionals, to bring spirit into their planning, as a venue for personal and collective self-actualization, honouring interior consciousness as much as material form. I came to profess that 'being professional in a post-modern world is being integral' (Wight, 2005).

What if your practice, and place-conception, were integral?
What would change for you, in you?
What if you could be part of the making of place, of the raising and broadening of collective consciousness, for integrating It and We with the I of the beholder,
tapping soul and spirit?

Placemaking as wellbeing by design: mesh-working and whole-making

When place/placemaking is hitched to wellbeing, within an integral framing of both, a whole new vision opens up (Wight, 2012). Their integral 'meshing' merits a commitment to mesh-working – a notable advance on networking, and the integral common ground becomes whole-making, with an 'ever-more-whole-making' aspiration. What is so special, so uncommon, about wellbeing *and* place/placemaking, not only on their own terms but – perhaps especially – in combination? The two have in common an essential wholeness, goodness and togetherness – resisting easy deconstruction, abhorring insensitive reduction, and confounding simplistic quantification. They transcend individual disciplines and professions; they are – above all – integrations of pretty much all we hold dear – the good, the true, and the beautiful. They range wide and deep; they implicate our whole being – body, mind, soul, and spirit. One, placemaking, is perhaps more means than end, the other, wellbeing, more end than means – but they are indubitably linked. We know this at our core, from everyday – and extraordinary – experience. But it does seem to require a shift in perspective to fully see, and feel, and validate all this. The concern needs to be with the enacting that is placemaking and with the enacting of wellbeing – as ongoing whole-making productions both.

My emerging sense of the place–wellbeing nexus is one that happens to still privilege place (especially its making), but in a way that I feel embodies, ensouls, and inspirits wellbeing; I hypothesize wellbeing as a manifestation of implacement (Schneekloth and Shibley, 2000). But the place I have in mind is much more than simply geographical or locational; it is integral, in the sense of including while transcending the pre-modern, modern, and post-modern notions, or 'senses,' of place. It is also emerging – developing and evolving – with an inherent dynamic; a verb as much as a noun, always in the process of being made and remade, in pursuit of an ever-greater sense of the good (and the true, and the beautiful). Together, in combination, placemaking and wellbeing involve a full engagement of body, mind, soul, and spirit, potentially manifesting a palpable *poiesis*. Wellbeing may be conceptualized as, at the very least, a by-product – if not direct product – of the kind of integrally informed placemaking articulated earlier. If planning can be enacted as integral placemaking, it cannot but play a more significant role in engendering greater wellbeing. But what if wellbeing itself is considered more directly through an integral lens? An integral wellbeing perspective may prove as generative as an integral placemaking perspective, with large dividends (compound interest?) for related planning, policy, and design.

Reaching for a more 'wholistic' view, wellbeing evokes a highly generative interpretation through an integral lens. Conventional etymology needs to be bypassed – 'the state of being comfortable, healthy or happy' (encountered in many dictionaries) falls far short of the meaning being reached for. The more appropriate root word from an integral perspective is 'whole,' which can be seen to encompass 'well' – and much more besides. This reframing more easily invokes whole-system conceptualizations; it also suggests a basic living, life-affirming, orientation in favour of 'wholing,' of seeking ever-greater, ever-more-exquisite, wholeness – in ourselves, in our relationships, and in our environments. Think of 'wholing' as a deeper and wider form of 'healing'; it is the action verb to the wellbeing noun. Its enacting becomes a form of 'wholing,' of making more whole, of 'whole-making' – seeking always to conduce, extend, and embed wholeness as wellness, or thriving as flourishing – the 'wholing' practice yielding wellbeing. This notion of wellbeing as 'wholing' or 'whole-making' engages our need for deep meaning, beyond everyday life. It also encompasses the sense of being part of something bigger, more transcendent, mysterious, but loving – the spiritual dimension of wellbeing (Vernon, 2008). Within an integral perspective there is room for the magical and mythical, as well as the rational and the mystical. The focus is on wellbeing, not ill-being; the framing as 'wholing' recognizes the developmental and evolutionary dynamic, in ever-seeking to become ever-more whole.

Combining placemaking and wellbeing, in integral terms, invokes meshworking – a very new notion, partly a product of the early integral wave, but with its roots in brain science (Hamilton, 2008, Ch. 10). It is built on the more familiar ground of networking, but meshworking takes networking to an almost unrecognizable new level. The attraction here is its response to the challenge at the heart of an integral approach – achieving uncommon 'integrated-ness' in real-time. It provides the context for operationalizing any trans-disciplinary, trans-professional, trans-sectoral endeavour – for those so inclined to seek such communion. Truly an integral-age approach, it is inherently ecological, as well as being highly participatory (Torbert and Reason, 2001), solidly biased in favour of conducing collective action, following the raising of the collective consciousness of all the players in an issue. Think of meshworking as ecological interconnectedness 'inter-personified,' in a complex collective collaborative context. To the extent that all of this can be contained, the associated integral container would have the appearance of a primarily horizontal placemaking (rooted in the quadrants, valuing the balancing of interior and exterior) and a primarily vertical wellbeing (rooted in the levels/lines, valuing the interaction of individual agency and collective communion.) Wellbeing feels like its origins are in the perennial philosophy, the Great Chain of Being (Wilber, 1996); placemaking helps to reframe the 'Chain' as a 'Nest.' The meshworking is the integral interaction, the integrated enaction – a catalyzing agent on a grand scale.

The underlying challenge may therefore be expressed in terms of *meshing* placemaking and wellbeing (as whole-making.) An explicitly integrally informed 'meshing' may be characterized as meshworking, a form of (in Spiral Dynamics terms) second-tier integral collaboration – well beyond mere cooperation or first-tier networking. This will take many – scientists, professionals, citizens – well outside their normal comfort zone, in part because it also necessitates them consciously going well *inside* themselves. An integral engagement of wellbeing and placemaking entails engaging the ineffable, within the realms of consciousness, as much as it entails engaging exterior concrete form. It is an inner work project, engaging not just the mind, but body, soul, and spirit. The 'well' in wellbeing goes back to the original notion of 'whole,' when whole very much referenced body, mind, soul, and spirit. Here is some dignified pre-modernity that deserves to be very much preserved, and sensitively integrated into today's thinking and acting. The coupling with placemaking, and an integral perspective, helps to render wellbeing as a form of 'whole-making,' and it is in this combination that we might all find our post-post-modern calling, our co-mission-ing. Paraphrasing Martin Luther King:

> I have a dream, of professions – and an academy – of servant-leaders, as a community of wellbeings, striving above all for the wellbeing of all, in well-loved places: whole beings, in whole places, tending not just to inanimate matter, but to all that matters –in body, mind, soul and spirit.
>
> *When do you know you are well, and are – in fact – a well being?*
> *When do you know you are in your place – your prime, thick place?*

A poiesis of sophrosynes – the placemaking to come?

Those picking up this Handbook will be at different points on their placemaking journey; they will have to gauge where to engage. I have laid out here some of the main staging points in my own journey – an ever-evolving one, but one rooted in an integral framing, which may not be for everyone – at least not yet perhaps. My operating domain has been professional and academic, often seeking to 'bridge' the two, in the context of planning and design – these being professional fields and academic disciplines. As I began to engage with integral framings, I began to notice an increasing underlying curiosity around 'what I profess' – regardless of professional or academic affiliation. My journey began in geography – still fundamental in the sense of foundational – where my operative sense of place was established. The placemaking I profess took root when my worldview was ecological, which overlapped with my encounter with integral; applied integral ecology became my play-place. As a planner in design-land (in a multi-discipline Faculty of Architecture) I was helped to add design to my planning bow, and my professing became more concerned with my own integration of the design focus on perception, intention, and making. Planning as placemaking became my IOS, my Integral Operating System, and from there it became natural to contemplate more trans-disciplinary, trans-professional, trans-personal dimensions of my professing. Placemaking as wellbeing by design was part of that natural evolution – transcending while including all that had gone before. So where is my journey now taking me? Might you credit 'a poiesis of sophrosynes'? (Wight, 2017).

My placemaking journey has been part and parcel of a larger long-standing concern with evolving what passes for professionalism these days – for evolving professionalism beyond the status quo, and for contemplating the education of the agents of the next enlightenment. The design focus has become *one's professional-self*, involving three particular arenas of 'making' – *praxis* (personal), *ethos* (interpersonal), and *poiesis* (transpersonal.) *Praxis* (Wight, 2015) is particularly associated with sense-making, foundational inner-work; ethos (Wight, 2013) involves meaning-making, in the company of others – a form of inter-work. Poiesis may be regarded as a form of placemaking writ large – *P*lacemaking – intrinsically integral, akin to the 'inter-subjective according in the noosphere' addressed by Wilber in his 1997 book (and elaborated in Wight, 2020). It is the highest form of making that any true, good, and beautiful professional might profess – an *over*-standing@work. Those who might achieve such heights of professional-self design may be considered to manifest a quality or status that the ancient Greeks termed *sophrosyne*. Those ancient Greeks often paired *sophrosyne* with their sense of *poiesis* (Sennett, 1992, p. xiii). I am sensing that the continuing evolution of my placemaking will be in the realms of 'a poiesis of sophrosynes', and this is the placemaking I now seek to advocate.

What this might entail emerged in some recent reflections on the nexus of Art, Place, and Nature (Wight, 2017), when I was moved to consider: might this triad constitute a congenial canvas for a larger integration (a poiesis), of particularly enlightened, evolved, and integrated wellbeings (as sophrosynes) – engaged in a poiesis of sophrosynes? This intriguing trialectic was being set in the context of an explicit interest in 'spiritual and integral ecology perspectives.' It became a rather generative inquiry for myself, and an opportunity for expressing some learning and intuiting, as I explored seemingly relevant 'perspective-taking' – from my own experience. It helps to place where I am currently at – on my ongoing placemaking journey. Art~Place~Nature became perceivable as an integration, aiming for an expression that was: not only ecological, but also integral, in a larger sense, including – especially – the spiritual; an expression full of aspiration – for embodying, enacting, and ensouling the 'integrating'; as a form of 'ever-more-whole-making', integrating truth, goodness, and beauty through love – a living/loving, loving/living expression; an integration grounded in love, for an Earth, beloved and loving. The integration of Art~Place~Nature emerged as 'the poiesis of sophrosynes' – an integration of Self (Art), Culture (Place), and Nature, in Body, Mind, Soul, and Spirit.

One important side-effect of this inquiry has been a rising awareness of the importance of shifting beyond an 'individual' focus towards more explicitly privileging 'the personal' – and the inter-personal, and the trans-personal. The personal is a place of *praxis*, integrating not only my knowing and doing, but also my being and becoming – my essence, my prof-essence (Wight, 2015). The inter-personal scales this up and out, engaging the realm of common-meaning-making – of ethos-making and culturing, while the trans-personal most openly engages the spiritual, calling for poiesis. I am fundamentally curious about 'enacting integral' – in this case – in our artistry, our place-making, and our naturing.

Art, Place, and Nature represent fundamental differentiations within an integral framing – articulations of the elemental 'quadrants', the worlds of I, We, and It/Its that simultaneously demand integration. And the ultimate agent in such integration is the Self, implicating considerations of our integrity as in our integrated-ness on our insides, in our inner work; and as in our integration-ability in relation to our outsides, and our outer work in the world. This is the domain of 'integrals@work,' enacting integral, in one's person, as a whole person – Body, Mind, Soul, and Spirit – well beyond comporting oneself as simply 'an individual,' but fully honouring the nexus of Self, Culture and Nature. For guidance, think of 'Art' as an 'outing' of one's Self, of 'Culture' as our collective 'Place' (manifesting the combination of all our sense-making and meaning-making), and of 'Nature' as the definitive 'co-relate' – the ultimate grounding. At this juncture, in our collective minding, hearting, and souling, it is suggested that we need to privilege: the 'in-siding' and its 'in-sighting' and 'in-tuiting'... (and in their expression through our artistry); the in-personal (rather than impersonal) and the related inter-personal, and trans-personal – (in our artful making, and synthesizing, of Art~Place~Nature); and the 'placing' of our selves 'alongside, inside, and as' nature personified (getting personal about 'our' operative nature, our 'naturing' agency).

The journey from individual to person may help those challenged by the journey from place to poiesis. As Cynthia Bourgeault (2017) has observed, building on the work of Pierre Teilhard de Chardin, these reflect 'distinctly different, progressive evolutionary stages.' Bourgeault defines an *individual* as 'living as an autonomous unit, subject to the old-order laws of "survival of the fittest" and planetary indifference.' By contrast, Bourgeault identifies a *person* as having come to understand themselves as 'belonging to a greater relational field.' Bourgeault expands on the implications:

> They (persons) now sense their identity from a sense of wholeness in an entirely different order of coherence: a whole greater than the sum of its parts...the universe is no longer random, but a system of relationships to which we all belong and are participating in!

What might we therefore attend to, by our intending, in our person so conceived, in any efforts to enact integral? I would nominate for consideration certain aspects of our own artistry, place-making, and naturing: enacting our artistry of possibility (Art); coming home to our Selves (Place); and, naturally realizing our 'third' nature, or 'thirding,' as ever-more-complex nature/Nature/NATURE. And doing so potentially with the necessary integration being pursued by reflecting deeply on the 'eco' in ecology; not simply by problematizing 'ego' but by integrally advancing a transcending, while including, sense of 'eco,' uniting multiple perspectives, transformatively, by design. Such integral enacting begins to capture some of the endeavour that might be distilled in a poiesis of sophrosynes. This framing builds on Richard Sennett's conceptions, discussed in the book foreword (1992, xiii). *Sophrosyne*, for the ancient Greeks, was a quality of being 'balanced and centred, in face of difficulty and diversity, acting with grace and poise, balancing inner and outer life exposure.' *Poiesis*, for the Greeks, associated balancing oneself with acting (perhaps – more accurately – enacting) as well as looking: 'The result of caring about what one sees is the desire to make something – poiesis (poetry – but their word was broader than one art in scope).' The balanced person would want all their makings to engage the same qualities of grace and poise. Sennett concludes: 'As a result of his or her own engagement in making or doing things carefully, sophrosyne and poiesis were intimately related.'

I am suggesting a synchrony between this notion of sophrosynes (in the plural) and what I might allude to as 'integrals' (which may in turn be regarded as an advance on what we currently regard as 'professionals'). I posit 'the makings' of such 'integrals' as a triptych – as a product of some dedicated artistry of possibility – of: *praxis* (personal) – sense-making; *ethos* (inter-personal) – meaning-making; and *poiesis* (trans-personal) – Placemaking. Hence my sensing of integral placemaking as, ultimately, a poiesis of sophrosynes – grounded in a practice as *praxis*, and ethics as ethos. Makings of makers, grounded in place, integrally conceived.

What do you profess, in the academy or in your profession, in relation to placemaking?
What placemaking – elevated, evolved, embodied – might you seek to profess?

References

Beck, D.E. and Cowan, C.C. (1996). *Spiral Dynamics: Mastering Values, Leadership, and Change*. Malden, MA: Blackwell Publishing.

Bourgeault, C. (2017). 'Exploring the mystics: Don't co-exist, coalesce', *Daily Meditation*. Albuquerqe, New Mexico: Centre for Contemplation and Action.

DeKay, M. (2011). *Integral Sustainable Design: Transformative Perspectives*. London: Earthscan.

Hamilton, M. (2008). *Integral City: Evolutionary Intelligences for the Human Hive*. Gabriola Island, BC: New Society Publishers.

Schneekloth, L. and Shibley, R. (1994). *Placemaking: The Art and Practice of Building Community*. Hoboken, NJ: Wiley.

Schneekloth, L. and Shibley, R. (2000) 'Implacing Architecture into the Practice of Placemaking', *Journal of Architectural Education*, 53(3), pp. 130–140.

Sennett, R. (1992). *The Conscience of the Eye: The Design and Social Life of Cities*. New York: Norton.

Torbert, W. and Reason. P. (eds.) (2001). 'Toward a participatory worldview: In physics, biology, economics, ecology, medicine, organizations, spirituality and everyday living', *ReVision*, 23(3–4).

Vernon, M. (2008). *Wellbeing*. Stocksfield: Acumen.

Wight, I. (2005). 'Placemaking as applied integral ecology: Evolving an ecologically-wise planning ethic', *World Futures*, 61, pp. 127–137.

Wight, I. (2006). 'Sensing place through an integral lens: Pointers for a postmodern planning as placemaking', Presentation to the *Senses of Place Conference*, April 2006, Hobart, Tasmania.

Wight, I. (2012). 'Place, place-making and planning: An integral perspective with wellbeing in (body) mind (and spirit)', Chapter 15 in Atkinson, S., Fuller, S. and Painter, J. (eds.) *Wellbeing and Place*. Farnham: Ashgate Publishing.

Wight, I. (2013). 'Exploring inter-being and inter-becoming as ethos-making: The integrally informed pursuit of professional community well-being', in *Journal of Integral Theory and Practice* 8(3), pp. 82–96.

Wight, I. (2015). 'From practice to praxis—as transformative education: Leading at the integral/professional interface', in *Integral Leadership Review – Special 'Canada' Issue* (January/February 2015).

Wight, I. (2017). 'Art~place~nature: The poiesis of sophrosynes?' Paper presented at *Art, Place and Nature: Spiritual and Integral Perspectives*, October 21, 2017. Edinburgh, Scotland: Edinburgh International Centre for Spirituality and Peace.

Wight, I. (2020). 'Making a place we can all call home (inter-subjective according in the noosphere)', in *Integral Mentors/Thriveable Cities*, April 2020: IntegralUrbanHub 20—Accelerating City Change in a VUCA World (curated by Marilyn Hamilton), slides 145–151 [online]. Available at: https://www.slideshare.net/PauljvsSS/urban-hub20-accelerating-citychangein-a-vuca-world-thriveable-cities/1 (Accessed: 7 May 2020).

Wilber, K. (1996). *A Brief History of Everything*. Shambhala: Boston and London.

Winikoff, T. (ed.) (1995). *Places Not Spaces: Placemaking in Australia*. Sydney, NSW: Australian Council for the Arts.

Zander, R. (2017). *Pathways to Possibility: Transforming Our Relationship with Ourselves, Each Other, and The World*. New York: Penguin Books.

Further reading in this volume

Chapter 11: *Free State Boulevard* and the story of the East 9th Street Placekeepers
Dave Lowenstein
Chapter 13: Sensing our streets: involving children in making people-centred smart cities
Sean Peacock, Aare Puussaar, and Clara Crivellaro
Preface: The problem with placemaking
Louise Platt
Chapter 14: Experts in their own tomorrows: placemaking for participatory climate future
Paul Graham Raven
Chapter 15: Un/safety as placemaking: disabled people's socio-spatial negotiation of fear of violent crime
Claire Edwards
Chapter 16: More than a mural: participatory placemaking on Gija Country
Samantha Edwards-Vandenhoek
Chapter 18: '*Homomonument* sounds like a poem': queer placemaking 30 years on: a conversational dialogue with Thijs Bartels, author of *Dancing on the Homomonument* (2003)
Martin Zebracki
Preface: The radical potential of placemaking
Cara Courage
Chapter 22: Embedded Artist Project: Epistemic Disobedience + Place
Frances Whitehead
Chapter 24: Artists, creativity, and the heart of city planning
Tom Borrup
Preface: The only thing constant is change
Kylie Legge
Chapter 34: Reconnecting *cité* and *ville*
Philip Graus
Chapter 35: Planning governance – lessons for the integration of placemaking
Nigel Smith
Chapter 36: Facilitator skills for effective collaborative placemaking
Husam AlWaer and Ian Cooper
Chapter 37: *The Neighbourhood Project*: a case study on community-led placemaking by CoDesign Studio
Lucinda Hartley, Eliza Charley, Sama Choudhury, and Harriet McKindlay
Chapter 38: Public seating – a small but important place in the city
Kylie Legge
Preface: Evaluating creative placemaking: a collection of observations, reflections, findings and recommendations
Maria Rosario Jackson
Chapter 41: Rituals of regard: on festivals, folks, and findings of social impact
Maribel Alvarez
Chapter 42: Creative Placemaking and Placekeeping evaluation challenges from the practitioner perspective: an interview with Roy Chan
Maria Rosario Jackson
Chapter 45: How the city speaks to us and how we speak back: rewriting the relationship between people and place
Rosanna Vitiello and Marcus Willcocks

29

THE SOLUTION IS IN THE PROBLEM

The art of turning a threat into an opportunity by developing resilience using a Creative Placemaking critical praxis

Anita McKeown

The problem landscape

In 2015, the United Nation's (UN) member states adopted the seventeen Sustainable Development Goals (SDGs) formalising a global call to action to 'end poverty, protect the planet and improve the lives and prospects of everyone, everywhere' (UN.org, 2019). Emerging out of the eight Millennium Goals 2000–2015, (Un.org, 2019) the UN's 2030 agenda for sustainable development offered a 15-year roadmap to transform our world. In 2016, the World Economic Forum at Davos discussed the 2030 Agenda for Sustainable Development, where it was internationally recognised that it can no longer be 'business as usual,' given the projected 15 years to achieve the transformative change necessary for the wellbeing of humankind (Møller, 2016).

Critiques of the SDGs are varied (Esquivel, 2016; Gupta and Vegalin, 2016; Hák et al., 2016) yet a key flaw in the goals seems to be offering an approach to transforming the world without using transformative methods. Within the current model, despite being a valiant attempt, establishing commitment within the SDGs to 'harmony with nature,' their underpinning ambition of 7 per cent annual GDP growth in least developed countries and higher levels of economic productivity across the board is in keeping with existing neoliberal models (Hickel, 2015) and as such, sustainable development is an oxymoron. Criticism of sustainable development lies in the limits-to-growth (Meadows, 1972) model, recognising the earth's finite resources (Georgescu-Roegen, 1971). Ultimately, any economic model that doesn't acknowledge and account for this is bound to collapse.

The global challenges we face demand alternative systemic approaches (Gawande, 2014) with real commitment to making equitable change, at the local and global levels. This will require international top-level political commitment and the collective effort of all actors, at all levels. Developing a multi-stakeholder approach is critical if we are to develop innovative systemic methods with any long-term actions requiring cross-sectoral, generative, and consensus-dependent ownership and capacity (Møller, 2016). Placemaking offers a natural arena from which to develop such an approach. Placemaking, both as process and philosophy (Project for

Public Spaces [PPS], 1995), brings into contact two distinct aspects: the functionality around the everyday needs of people and more existential requirements that are further shaped by socially constructed processes.

However, for placemaking to fulfil its potential it must not fall foul of development agendas and commit to inclusive generative practices. Social and environmental decisions are often made based on economic costs rather than social or environmental value. Debate across the varied disciplines engaged in regeneration highlights the need to understand stakeholders in combination with professional and personal interests, yet an eco-systemic outlook is not fully integrated. For development to be considered regeneration and not merely development, there should be long-lasting beneficial impact on the communities and locations that experience the regeneration (Granger, 2010; Roberts, 2000), what Kate Raworth has called 'thriving in balance.' Yet, a majority of development approaches involve large-scale physical infrastructures, formulaic algorithms of urban hardware, such as new-build housing, mixed use and enterprise zones and possibly a work of public art, masking displacement. A more generative creative approach to placemaking should speak to the contemporary inclusion of social and environmental concerns. To be transformative, it should acknowledge multi- and inter-species activities towards positive redesign, reconstruction, or redesignation of existing land use and the set of quantitative planetary boundaries identified in 2009 by Stockholm Resilience Centre director Johan Rockström and 28 internationally renowned scientists. These boundaries are linked to the nine processes that regulate the earth systems and propose limits to enable humanity to thrive for future generations (Rockström et al., 2009). As anthropocentric climate change increasingly becomes a climate emergency these boundaries necessitate considering alternatives such as a 'degrowth' society, also known as a steady-state economy.

This chapter presents research from a pilot project, *CoDesRes*, that explores the evolution of an existing critical praxis for Creative Placemaking; the PermaCultural Resilience (pCr) praxis (McKeown, 2015) that addresses ecological economics, anti-consumerist, and anti-capitalist ideas. Trialled across differing socio-political contexts in London, Dublin, and New Mexico, the pCr praxis has proven to be effective in creating locally relevant and non-formulaic projects that contribute to community development. In the 'century of the system' (Gawande, 2014), and as a creative generative approach to placemaking, the pCr praxis is anchored in the belief that no single individual can change or be expected to change whole systems. As the limits to growth (Meadows et al., 1972; Roth, 1987) are reached, degrowth (Reichel and Seeberg, 2010; O'Neill, 2012; Martinez-Alier, 2012;) is considered the necessary approach to the resultant situation. Such an economy requires locally scaled solutions, rather than the 'ever-increasing levels of extraction, production, and consumption' (Brasuell, in Planetizen, 2015, paragraph 4). *CoDesRes* looks to leverage the SDGs and the 2030 agenda (UN, 2016) as a springboard to reimagine growth, within the context of a de-growth society through developing resilient resistant practices.

From threat to opportunity

Trauma linked to fears and grief in response to climate collapse is increasing. Sudden and abrupt change, particularly from events we have little control over, leave us vulnerable physically and emotionally. Case studies considering the influence of community on mental health show that the communities that fare better under existential and disaster threats are those that not only have resources but demonstrate community cohesion and support (Berry et al., 2010, 2008; Beaudoin, 2007) or a sense of agency. Within the context of resilience, this offers an opportunity for different ways of thinking about situations; by creatively and imaginatively providing opportunities for locally scaled community activities that contribute to survival based on an integrated approach to social, environmental, and economic capitals. Research studies justify tackling resilience and sustainability challenges at the local level (McKeown, 2015) owing to manageable, tangible scales of governance more readily identifiable by citizens.

Maurice Strong, Secretary General of United Nations Conference on Environment and Development (1992 Rio Earth Summit), stated that local government should lead the way, an idea quickly enshrined in local Agenda 21 frameworks and reminiscent of early Rene Dubos' 1972 maxim 'think globally, act locally' (Moberg, 2005) initiatives taking responsibility around tackling sustainability issues is complex and risks the 'governance trap (Hobson, 2010); when citizens expect their elected officials to act, yet simultaneously have no trust in the government's messaging regarding an accurate ecological position or their ability to follow through, with suitable action (see also, Rowson, 2013). Understanding upcoming challenges is further complicated as they are often both spatially and temporally removed and distance individuals from the impact and consequences of their actions.

Yet, if we were to adopt a 'beyond-compliance' culture, then the need for 'culturally situated local approaches that include multiple world views and a systemic design thinking perspective that integrates science' and technology (McKeown in De La Garda and Travis, C., 2018) offers a fertile research opportunity. How can we engage with the needs of multi- and inter-species, necessary for a healthy system critical for our own survival sit within placemaking practices? What could an eco-systemic, situated arts-led approach have to contribute to implementing the SDGs on a local scale in an attempt to create an accessible generative approach, rather than current exploitative approaches? How might we support a sense of agency in our communities that facilitates resilience?

Co-designing for resilience – the role of a Creative Placemaking critical praxis

Visual arts have been identified as an ideal field to support the effort of promoting resilience within young people through developing a sense of belonging, sense of identity, self-awareness and self-esteem and an ability to cope, and learning artistic skills and gaining confidence (Hart et al., 2012). Hawkes envisions the importance of a cultural framework at the heart of planning that enshrined values within policy-making and transcended a simple service delivery model, arguing for 'an active consciousness of the values, which inform our actions' (2001, p. 15). This is necessary to ground productive social action as 'clear, creative and engaging processes for facilitating community expression and debate of those values and their practical application' (ibid.). Hawkes' promotion of the arts acknowledges 'the right to be actively involved in making our own culture' (ibid., p. 24) and its opportunity to contribute to an understanding of being agents in survival, through activities that simultaneously build on collaborative and socially equitable practical actions through creativity and imagination.

CoDesRes operates a bespoke critical praxis for Creative Placemaking, a systemic arts-led situated approach to placemaking while leveraging the momentum around the 2030 agenda to embed resilience and resistance thorough education and community toolkits. It integrates social, environmental, and equitable economic concerns within arts-led actions and place-based STEAM education (Science, Technology, Engineering Arts, Maths) to develop local improvement that acknowledges more than human needs. Through encouraging creativity and imagination *CoDesRes* contributes to a localised exploration of a de-growth economy; addressing social, environmental, and economic inequalities while maximising wellbeing and happiness through non-consumerist means. As an operating system, the 'role' of the pCr praxis presents an Open Source resilient approach to Creative Placemaking that encourages a triad of capitals – social, environmental, and economic, to be integrated into the process as a means to cultivate the creation of conditions towards a healthy resilient eco-socio-cultural ecosystem.

R – Relationships. Using an intensive audit tool to map cultural, economic, socio-political and environmental dynamics, the pCr praxis reveals relationships, resources, and opportunities to help reconfigure and reimagine an understanding of place. Stakeholder consultation is the first stage in the process. The construction of flexible micro-ecologies aids the revelation of multiple knowledge cultures and entities, integrating their place-based knowledge, valued for its potential to contribute to a local resilience.

O: The OBREDIM process log is a three-phase tool that deepens the pCr audit undertaken in advance of developing any project or intervention, adapted from Permaculture and integrated into Design Thinking for situated art practices (where the art emerges out of a specific situation and is context responsive). The first step in the pCr process is to map the skills, activities, and resources of the community using the OBREDIM log; what they already do and how this might contribute to sustainable development and community resilience. This develops the ability to see the localised system, strengths and weaknesses, gaps or bridges that can be built upon. Only then will interventions be created, partnerships brokered, and projects developed, seeding the praxis, adapting and iterating it as necessary.

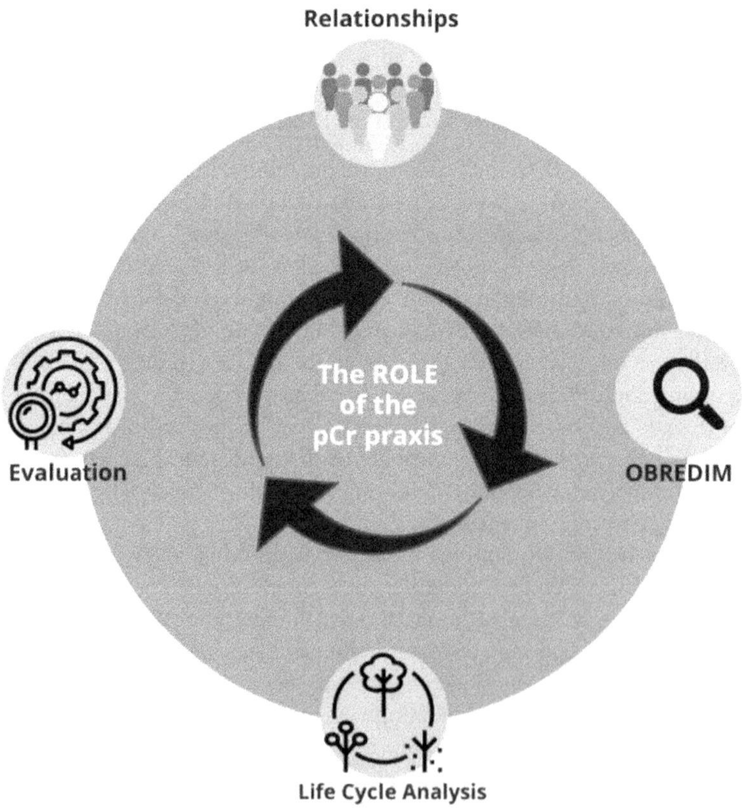

Figure 29.1 The role of the pCr praxis, McKeown 2016.

OBREDIM AUDIT **Activity:** **Date:**

Audit Phase	**Details reference: What to look for and record.**
O – Observation **Phase 1**	Survey all local 'organisms' e.g. organisations, stakeholders, businesses, arts and cultural orgs, community groups, charities, people, animals, vegetation, socio-cultural landscape, history, news / media, politics. Try to create as full a picture of the residency ecosystem.
Ways and things to observe: Patterns of growth, distribution, town layout business layout etc. Traffic flow, people motion, dead spots, flow of information, traffic people, the dynamics; social, cultural physical. Is there an impact? Does it last? Where's it start and stop. Natural system aspects: Weather, Sun, Water sheds, air, flora fauna animals, migration routes or diversions of water, desertification, forest, History – what's changed and why is there a pattern, does this have impact on the future? Communities; connections and relationships, distance/ proximity, inter-species, inter generational. What is successful? What has adapted are there any common traits? Are there any recogniseable patterns, numerical patterns? Are their functions of these patterns? Look at textures / shapes – Draw them, photograph them, record audio, video. Use the senses; What can you see, hear, taste, smell and touch.	
B –Boundaries	The edges / limits of the ecosystem; the location's geo boundaries, organisational boundaries, people's responsibilities, shared values, cross-over of aims, power dynamics. Limits to growth expansion, Laws, regulations and policies. Where do things stop and start? Are there diversity, tensions and encounters? Is there a difference between the edges and centre? Zoning analysis: This can highlight responsibilities, existing partnerships, focus for effort.
R - Resources	Physical and non-physical resources; Time, money, services, skills and knowledge, existing networks and partnerhips, groups, what already exists and how it works (or doesn't). Sample Questions: Venues: what's there, what does it do, how does it function, who sponsors events. Groups: who's doing what, when and with who?
E – Evaluation **Phase 2**	Begin to map a web of relations – using the info from Phase 1 and the Zoning analysis. Evaluate what exists and where the gaps are – how does info flow, notice relationships and communication. Include a SWOT / SMARTER analysis
D - Design	Design on paper, Becomes a map for the implementation stage or if there's an exisiting project in mind re-design in light of information gathered in Phase 1 and evaluation stage.
I – Implementation **Phase 3**	Implementing design: incl logistics eg timelines, production milestones, communication, fundraising, skills needed.
M - Maintenance	Maintaining the project and any maintenance needs or opportunities to evolve the project, handing over passwords, admin details, resource directory – anything needed to move the project forward or maintain its existence and evolve it.

Figure 29.2 OBREDIM Log 1 (adapted McKeown, 2008–15) forming the pCr audit (McKeown, 2015).

L: Lifecycle analysis. The pCr framework offers a simple visual tool that embeds an eco-social commitment within Creative Placemaking by addressing the full life-cycle of a project and beyond. By plotting the position of the inputs, processes, and outputs across Zones 1 to 5 of a project against the proximity to a project's 'Zone 0,' an initial assessment of all production process can be considered.

E: Evaluation. The pCr toolkit includes an evaluative matrix, the pCr Vital Signs Matrix based on the concept of the vital signs of a project and contributing to the vital signs of a place (McKeown, 2015). The vital signs act as indicators of a healthy human and non-human system. The matrix, developed through the research, defined the following foundational characteristics necessary for the pCr praxis of Creative Placemaking: Building Micro-ecologies; Strategic Intervention Tactics; Re-seeding Local Knowledge; and Resituating Arts, Design, and Culture. It further integrates a triple bottom line – earth care, people care, and fair share. An additional indicator, the Inclusive Fitness Theory (Hamilton, 1964, 1963), offers a metric to gauge project impact, used to evidence where the pCr ethos and methods spreads into other organisations or working practices towards long-term behavioural change.

The pCr Vital Signs Matrix tool can also be used as a collaborative project development tool that serves as a foundation to addresses social and environmental equity within a project; short-, medium-, and long-term. The methodology also developed an extended concept of SMART goals, to SMARTER, that sought to integrate goals reflective of the current and future context; S – Socio-culturally specific, Simple, and Sincere; M – Meaningful as well as Manageable and

Measurable; A – Appropriate, Achievable, Aspirational and Ambitious; R – Relevant, Responsive, Reviewed, and Revised; T – Timely and Time-specific; E – Eco-considerate and Ethical; R – Resilient, Resistant, Resourceful and far-Reaching.

A number of evaluative methods were used depending on the context, e.g. Transition Year (15 to 16 years) second-level education or community contexts. Within the school each class was evaluated using a 3-2-1 method; (three things they learnt, two things they would like to know more about, and one comment on their overall opinion of the class or improvements). We also undertook a World Café session with each class (75 students) and a focus group at the end of the year with 20 student participants. The praxis also has a practitioner's reflective log for self-evaluation, providing a systematic approach to gathering data of the lived experience and harnessing the practitioner's knowledge beyond the project context. The community projects were evaluated using the pCr matrix, which clearly showed the project over-delivered within the timeline. Each of the community groups had a different focus, which again offered an opportunity to consider the praxis within different dynamics, foci, and project. These evaluation processes enabled the team to trial ideas and not only consider aspects that were successful but gain feedback on how things could be improved and what didn't work.

Co-designing for resilience on the Iveragh Peninsula, SW Kerry, Ireland

CoDesRes' non-formulaic whole systems approach was transferred to a rural context in South West Kerry, Ireland to develop the creative placemaking praxis as an accessible toolkit and education curriculum that enabled self-organisation for resilience. Funded by the Environmental Protection Agency (EPA, 2018–2020), the project is located on the Iveragh Peninsula, an area defined in the National Spatial Strategy as Rural Area 5, marginal and highly diversified, representing 'an almost "post-agricultural" rural economy with areas of high natural amenity, attracting high levels of tourism, recreational usage' (NSS, 2000) and in some cases non-farming residents, e.g. retirees or second-home owners. However, Kerry is one of the most sparsely populated counties in the state, with low birth rates, high death rates, high aged dependency, and high levels of emigration, creating a population stagnancy that is contributing to demographic shifts, which in turn impact on the region's sustainability.

The research design explored the integration of the existing Creative Placemaking praxis within three fields – Marine Ecology; Environmental Science; the Arts, Media, and Engineering – and through three work packages – Youth, Community, and Media Transition – with each work package working across a number of projects in the education and community settings. This focused the research on SDGs 4, 11, 14, and 15 as a context-responsive approach to key areas pertinent to an Irish rural context; waste and the relationship between the Green and Blue Economies, in County Kerry, which has the largest coastline in Ireland. The interdisciplinary team included: an artist-scholar and educator; two marine biologists, one with a focus on coastal communities; an engineer; a media artist; and a curriculum developer giving the team access to a broad range of prior knowledge and experience. From additional funding from NAISC, Living Iveragh, and Community Social Enterprise the research greatly benefited from the inclusion of an artist and *Seanchai* (storyteller) (O'Laoghaire, 2020) and local resident.

The pCr methodology in earlier trials had shown the importance of encouraging a systemic approach to education and community development that facilitated a creative re-visioning and challenges to dominant narratives through proof of concept. Through formal and informal learning, the project team trialled and iterated the following activities that would share the pCr principles, circulate economic ideas, and develop the foundation for the project, with key outputs including: best practice guidelines, curriculum resources for place-based STEAM educational and community creative placemaking toolkits. The curriculum resources include lesson plans, activities, and teachers' resources for Transition Year students based on the SDGs in particular SDG 4, 11, 14, and 15 with a set of continuing professional development training/resources for second-level teachers and other professionals. The open source toolkit for self-organisation for Creative Placemaking endeavours to create a set of tools for community organisations to integrate the pCr creative placemaking praxis easily and effectively within their organisations and activities. In addition, the project offered media training and rapid prototyping as the foundation of a media and repair café/maker co-op, 'Remade in Kerry,' at the core of a localised innovation and knowledge network. Local residents and volunteers from community organisation, once trained, had access to bookable equipment and ongoing support that enables them to support each other to create media content or develop micro-enterprises.

Although, the project's work packages had distinct aims and objectives, the skills and peer-to-peer knowledge of the whole team are integrated into all packages. Further, the project simultaneously undertakes its research while sharing a proof of concept for place-based rural development and creative placemaking. That is, the rural has a role to play in the development of sustainable communities and cities by developing viable opportunities that can offer realistic alternatives to urban dwelling. By positioning local residents within a satellite research faculty as an intervention into local sustainability and resilience, the project simultaneously explores and develops positive economic social and environmental links between the rural and urban contexts. With recently produced reports (namely Ireland 2040, National Planning Framework (2018b), and the Draft National Marine Planning Framework (2018a)) communities' ability to engage critically with these issues will be an important contributor to personal and ecological resilience and, where necessary, resistance. Ireland, as a small island, must engage with the particular exchange and relationships that occur between land and sea. Both youth and community transition work packages mainly focused on exploring two issues identified as key concerns through a number of local and national government development aims and recommendations (Prescience, 2019; DES, 2015) – waste as resource, and land ecosystem health including freshwater and marine ecologies: the Blue–Green economy.

Outputs and initial findings

During the course of the project and its various events and methods the project delivered 183 events, had 4,927 direct project engagements and 5,467 digital followers, 59 written outputs including papers, blog posts, and articles, 23 working partnerships, and raised an additional €359K funding with €34.7K in kind support over its 2.5 years. At the time of writing the project has another six months until completion, with all resources including the website and links to the evolving toolkits shared on Github from August 2020 and www.codesres.ie until July 2021. The project's research design involved the creation of STEAM place-based learning interventions that would contribute to quality education (SDG 4) formally and informally with thematic content focused on Life below Water (SDG 14) and Life on Land (SDG 15) to contribute to Sustainable cities and resilience communities (SDG 11). This organising principle was applied to the development of the project's delivery, enabling different aspects to be presented and iterated within training and educational contexts, which then informed the foundation of an accessible toolkit for communities and educators.

During the education study 94 per cent of the Transition Year students stated that they were worried about the future. They questioned what the future would be like, if they had one, and stated that lack of information was not helpful to them. They also stated that the adults around them, who they looked to for protection did not seem to be engaged in issues of climate change, adding further anxiety to their already anxious state. They appreciated the knowledge and discussions in the classes and although the classes were challenging at times due to the content, they also understood that without having a realistic understanding of climate change and sustainability it was difficult to make changes. In line with the NCCA review (2016–2020) of the Senior cycle (university entrance exams) students' evaluation clearly showed the interest and needs for citizenship and political education; sustainability and climate change education and opportunities for interdisciplinary learning and applying their knowledge. Students enjoyed the group work, practical and interactive activities, and recognised the importance of the skills embedded in the lessons. Due to the Irish curriculum there are some foundational skills that need to be developed and the work with the students identified this was mostly in technology and how to apply their knowledge to contemporary issues.

CoDesRes' place-based STEAM approach embeds critical thinking, problem-solving, collaborative learning; creativity, innovation, and research skills within a localised tangible context that is connected to global concerns. This enables them to gain skills and competencies beyond the curriculum and opportunities to apply these within real world contexts. Within the community context a number of issues, many not unusual within community contexts, were apparent: high levels of volunteer burnout, highly talented and skilled people, yet full skillsets unknown or under-utilised, and a lack of collaborative partnerships, despite shared aims. Further, with regards to localising the SDGs it was clear that extensive work already existed but was not being captured within SDG reporting. Work Packages 2 and 3's desk-based research, regarding education and community groups, identified clear actions contributing to SDGs 4, 11, 14, and 15 targets but yet non mentioned these goals and, in instances, the groups were unaware of the SDGs. In the community context, using the pCr Vital Signs Matrix, the following initial findings have emerged.

Building micro-ecologies

Through the OBREDIM audit the project developed a web of complex relationships and systems and, by embedding a circular economic approach, facilitated a creative, innovative, participatory development of place that included local under-valued knowledge through 23 working partnerships, including the expansion of the local Tidy Towns activities to include embedding campaigns and activities locally; for example, wild flower planting, beach cleans, and Grow Your Own. *CoDesRes* facilitated place attachment as a motivating factor to encourage agency and action from project constituents. As a situated, networked co-production *CoDesRes* encouraged citizen-led agency within localised tangible activities, evidenced by increased engagement in events, including the St Patrick's Day and Christmas parades and their organising and planning. These are important contributing factors for developing resilience, with varying place-attachment and engagement depending on place, people, and their interactions.

Strategic intervention tactics

CoDesRes was designed to encourage social, environmental, and economic justice into resultant projects, through tools that facilitate the capacity to bounce back through localised tangible activities. The methodology supported the diversity of projects produced, many of which challenged local narratives, for example, '*that* won't work,' offering proof of concept that evolved over time. *CoDesRes* used local festivals to engage with SDG 11, embedding themes and aspects of SDG 14 and 15. An increasing attendance meant events became self-sustaining with many of the performers choosing to donate materials for future community projects and committing to participate. As a deeper, slower, richer (McKeown, 2015), intervention that uses existing and emerging peer-to-peer networks and self-organising methods and commons-based approaches to societal change, participants are encouraged to participate in a more sustained and direct involvement with civil society and governance, becoming more competent and confident in their ability to set agendas and influence policy outcomes.

Re-seeding local knowledge

The 'StoryBank Educational Programme' for primary schools, based on the 'make-do and mend' ethos that is still within living memory, created channels of information and conversation, through the students, into the wider community, further creating a renewed pride in keeping things for longer and learning a new skill set and attitude. The project also developed the concept of the circular economy to include knowledge ensuring useful knowledge is utilised as 'active currency' circulating within the system.

Resituating arts, design, and culture

There is increased understanding of the importance and potential of the arts, design, and culture within community resilience, design and development evidenced by the increased financial and in-kind support that has been agreed by funders and community participation and attendance. Events reimagined the local narrative by increased confidence to try new things in supported ways as well as encouraging organisations to reconsider ways of working. Organisations are now engaging more with an arts-led/creative approach and starting to develop their own initiatives. St Patricks Day Parade's participation increased over two years to 26 floats, many with SDG-related themes, with the local Tidy Towns group taking on the co-ordination in Year Three.

As a proof of concept, *CoDesRes* showed in tangible ways what happens when a systematic Creative Placemaking approach that integrates social, environmental, and economic justice is employed to insure generative outcomes. The research team and local residents, operating as an intervention, showed clearly the potential to generate positive economic social and environmental potential. The resilience, physical and emotional, required to manage future scenarios is not currently embedded in education, community development, or placemaking. The *CoDesRes* research explored the potential of making an existing methodology accessible by repurposing it within these contexts. Ireland is a small island, ranked second worst country in the European Union fighting climate change (CAN, 2018); this means a country with junk status could face annual non-compliance costs of around €500 million if it does not take immediate and efficient action (2018). Physical impacts predicted are sea-level rises, river coastal flooding, water shortages, and impacts on water quality, to name a few. If it is to have any resilience it will need a multi-stakeholder approach, which requires appropriate relevant and agile policy, good governance structures, and 'networks of experience and capacity building' (EPA, 2019). The project conditions are not unique to Ireland, however, and methods proving successful or contributing to the SDGs in these contexts could be transferable to other areas that experience a 'post' economy or migration. The initial findings could also contribute to the European agenda; the uptake and implementation of circular economic principles in their approaches to tackle environmental, societal, and economic problems.

Despite their adoption globally, there are no legal mandates to deliver on the the SDGs and their reach is limited. The EPA funding opportunity enabled CoDesRes to take the nascent pCr praxis, an arts-led bio-psycho-social co-production of place, and explore the potential of localising the Sustainable Development Goals to actualise existing but latent potential. The *CoDesRes* project facilitated increased self-organisation within Iveragh's community development by applying a bespoke critical praxis for Creative Placemaking; an eco-design thinking approach to local concerns; and a process of developing innovative solutions collaboratively and systemically.

Conclusion

CoDesRes' Creative Placemaking praxis is locally scaled and integrates the concept of *Oikos*, the crafting of home, the 'house territory to town cosmos' (Bateson, 2000, p. 187), into placemaking. Through this process, the connection to actions and participants' feedback into the system affecting change, both positively and negatively, becomes perceptible. Creative Placemaking's potential to make and re-make *Oikos*, the ethical management of resources and equity of capitals, transforms Creative Placemaking towards a process of becoming; actualising potential embedded within the knowledge cultures, creativity, and innovation inherent in all communities. The potential value of the concept and practice of *CoDesRes*' Creative Placemaking praxis is in the development of socio-cultural ecologies of participation that are imaginative, creative, and equitable. Such ecologies sustain localised agency, recognise and actualise latent potential, offering possibilities embedded within *CoDesRes*' methods to a return to *Oikos*. By encouraging citizens to 'think globally and act locally' (Dubos, 1972) tangible contributions to systemic change occur though the local ecosystem.

To date the *CoDesRes* project is the only arts-led place-based approach to the Sustainable Development Goals in Ireland. As a pilot research project, *CoDesRes* looked to leverage the SDG 2030 agenda as a springboard to reimagine growth, within the context of a de-growth society through developing resilient resistant practices. *CoDesRes* explored a possibility of moving from the old model of industrial growth and development towards a systemically generative approach for rural areas that is localised. The project was open-source and a proof of concept *CoDesRes* shows there are additional opportunities to develop many of the project's templates to be self-sustaining offering viable realistic alternatives to urban dwelling. Residents and organisations have already begun this process. As part of *CoDesRes*' legacy, additional funding co-supported by a local philanthropic organisation, Living Iveragh and CPL Ltd, Ireland's largest recruitment company, has been secured to continue to build out the place-based STEAM education tool kit and trial it in other locations. In addition, *CoDesRes* also contributed to Kerry County Councils Rural Regeneration Development Fund Bid to repurpose a derelict building to create a centre of excellence for place-based STEAM education, research, and mixed-reality learning centre. The project contributed to the employment of 10 people, built capacity in the local organisations and research assistants, and helped to establish a local business to develop its research reputation. The core team are now working on a national project, MARplas, utilising the pCr praxis as part of a co-design process to develop rural micro-enterprises through a circular economic approach to marine plastic waste (Sea Synergy, 2020).

By asking 'what's at stake?' on a local scale the *CoDesRes* project embeds sustained stakeholder engagement by broadening the concept of stakeholding. This also brings into play underutilised community potential, which aids the facilitation of citizen inclusion in governance and decision-making through active engagement with problem-finding and solving. This in turn leads to a localised resilience, through increasing agency and self-organisation. By encouraging social and environmental equity, the production and sharing of knowledge, the development of opportunities to cope with diverse conditions, and capacity building, the project has initiated the creation of an expanded identity for a number of villages in the region. While *CoDesRes*' approach to implementing the Sustainable Development Goals on a local scale is an accessible generative approach it will not cure all ills, but it does help seed a resistance to neo-liberal development practices, by creating viable alternatives.

References

Bateson, G. (2000) *Steps to an Ecology of the Mind. 2nd edition.* Chicago: University of Chicago Press.

Beaudoin, C.E. (2007). 'News, social capital and health in the context of Katrina', *Journal of Health Care*, 18, p. 418.

Berry, H.L., Bowen, K. and Kjellstrom, T. (2010). 'Climate change and mental health: A causal pathways framework', *International Journal of Public Health*, 55, pp. 123–132.

Berry, H.L., Kelly, B.J., Hanigan, I.C., Coates, J.H., McMichael, A.J., Welsh, J.A. and Kjellstrom, T. (2008). *Rural Mental Health Impacts of Climate Change.* Canberra: The Australian National University.

Brasuell, J. (2015). *Critiquing the UN's Sustainable Development Goals* [online]. Available at: https://www.planetizen.com/node/80183/critiquing-uns-sustainable-development-goals (Accessed: 29 October 2019).

CAN (Climate Action Network). (2018). 'Off target: Ranking of EU countries' ambition and progress in fighting climate change', in Climate Action Network Europe [online] Available at: http://www.caneurope.org/publications/reports-and-briefings/1621-off-target-ranking-of-eu-countries-ambition-and-progress-in-fighting-climate-change (Accessed: 29 October 2019).

CoDesRes. (2018). www.codesres.ie/blog. (Accessed: 12 January 2020).

eds De la Garda, A and Travis, C (2018), The Beautiful Midden, towards an Integrated STEAM Curriculum in The STEAM revolution: Transdisciplinary Approaches to Science, Technology, Engineering, Arts, Humanities and Mathematics, Springer Publishing, NYC USA

Department of Education and Skills. (2015). *Ireland's National Skills Strategy 2025.* Dublin: Government of Ireland.

Department of Housing, Planning and Local Government. (2018a). *National Marine Planning Framework Draft Consultation.* Dublin: Government of Ireland.

Department of Housing, Planning and Local Government. (2018b). *National Planning Framework.* Dublin: Government of Ireland.

Esquivel, V. (2016). 'Power and the sustainable development goals: A feminist analysis', *Gender and Development*, 24(1), pp. 9–23.

Gawande, A (2014) The Century of the System, Lecture 2, The Reith Lectures, The Future of Medicine. Available at: http://www.bbc.co.uk/programmes/b04sv1s5 Accessed: 20/8/2020

Georgescu-Roegen, N. (1971). *The Entropy Law and the Economic Process.* Cambridge: Harvard University Press.

Granger, RC (2010) 'What now for urban regeneration?' Proceedings of the ICE Urban Design and Planning, 163 (1). pp. 9–16.

Gupta, J. and Vegelin, C. (2016). 'Sustainable development goals and inclusive development', *International Environment Agreements*, 16, pp. 433–448.

Hák, T., Janoušková, S. and Moldan, B. (2016). 'Sustainable development goals: A need for relevant indicators', *Ecological Indicators*, 60, pp. 565–573.

Hamilton, W.D. (1963). 'The evolution of altruistic behavior', *American Nature*, 97, pp. 354–356.

Hamilton, W.D. (1964). 'The genetical evolution of social behaviour. I and II', *Journal of Theoretical Biology*, 7, pp. 1–52.

Hart, A., Heaver, B., Macpherson, H. (2012) Building resilience through community arts practice: A scoping study with disabled young people and young people facing mental health challenges, Arts and Humaniteis Research Council

Hickel, J. (2015). 'The problem with saving the world', in *Jacobin* [online]. Available at: https://www.jacobinmag.com/2015/08/global-poverty-climate-change-sdgs/?utm_content=buffer9c605andutm_medium=socialandutm_source=twitter.comandutm_campaign=buffer (Accessed: 29 October 2019).

Hobson, K. (2010) 'Beyond The Control Society: Realist Governmentality And The Analysis Of 'Good Governance' in Administrative Theory and Praxis, 32 (2):252–261.

Martinez-Alier, J. (2012). 'Environmental justice and economic degrowth: An alliance between two movements', *Capitalism Nature Socialism*, 23(1), pp. 51–73.

McKeown, A. (2015). 'Deeper, slower, richer: A slow Intervention towards resilient places', in Courage, C. and Barton, G. (eds.) *Edge Condition*, Vol. 5 January 2015 [online]. Available at: http://www.edgecondition.net/vol-5-placemaking.html (Accessed: 12 December 2019).

McKeown, A (2015) Cultivating permaCultural resilience; towards a creative placemaking critical praxis Unpublished thesis National College of Art and Design, Ireland
Meadows, D.H., Goldsmith, E. and Meadow, P. (1972). *The Limits to Growth*. New York: Universe Books.
Moberg, Carol L. (2005). *René Dubos, Friend of the Good Earth*.ASM Press. pp. 160–163. ISBN 1-55581-340-2.
Møller, M (2016) 'Business as usual' is not an option anymore Available on line:https://www.weforum.org/agenda/2016/01/why-do-we-need-a-multi-stakeholder-approach-tosustainable- development/ Accessed: 20/8/2020
O'Laoghaire, S. (2020). *Seán an Seanchaí* [online]. Available at: https://seananseanchai.com/ (Accessed: 12 January 2020).
O'Neill, D.W. (2012). 'Measuring progress in the degrowth transition to a steady state economy;, *Ecological Economics*, 84, pp. 221–231.
Reichel, A. and Seeberg, B. (2010). 'Rightsizing production: The calculus of" ecological allowance and the need for industrial degrowth' [trans. Romano, O. (2012)) in 'How to rebuild democracy, re-thinking degrowth'], *Futures*, 44(6), pp. 582–589.
Roberts, P. (2000) Evolution, Definition and Purpose of Urban Regeneration. In: Roberts, P.W. and Sykes, H., Eds., Urban Regeneration: A Handbook, SAGE Publications, London, 9–36.
Rockström, J., Steffen, W., Noone, K., Persson, A., Chapin, F.S., III, Lambin, E., Lenton, T.M., Scheffer, M., Folke, C., Schellnhuber, H., Nykvist, B., De Wit, C.A., Hughes, T., van der Leeuw, S., Rodhe, H., Sörlin, S., Snyder, P.K., Costanza, R., Svedin, U., Falkenmark, M., Karlberg, L., Corell, R.W., Fabry, V.J., Hansen, J., Walker, B., Liverman, D., Richardson, K., Crutzen, P. and Foley, J. (2009). 'Planetary boundaries: Exploring the safe operating space for humanity', *Ecology and Society,* 14(2), p. 32.
Roth, A.W. (1987). 'The choice of a technology for the future of the human race', *International Journal of Technology Management*, 2(3–4), pp. 329–335.
Rowson, J. (2013). *A New Agenda on Climate Change Facing up to Stealth Denial and Winding Down on Fossil Fuels*. London: Royal Society of the Arts.
Seasynergy. (2020) [online]. Available at: https://www.seasynergyresearch.org/marplas (Accessed: 12 January 2020).
United Nations. (2016). *The* Sustainable Development Agenda [online]. Available at: https://www.un.org/sustainabledevelopment/development-agenda/ (Accessed: 29 October 2019).

Further reading in this volume

Preface: Placemaking in the age of COVID-19 and protest
Jason Schupbach
Chapter 2: Placemaking as an economic engine for all
James F. Lima and Andrew J. Jones
Chapter 3: An annotated history of creative placemaking at the federal level
Jen Hughes
Chapter 4: A future of creative placemaking
Sarah Calderon and Erik Takeshita
Chapter 7: Conflict and memory: human rights and placemaking in the city of Gwangju
Shin Gyonggu
Chapter 12: Public transformation: affect and mobility in rural America
Lyndsey Ogle
Chapter 13: Sensing our streets: involving children in making people-centred smart cities
Sean Peacock, Aare Puussaar, and Clara Crivellaro
Chapter 14: Experts in their own tomorrows: placemaking for participatory climate futures
Paul Graham Raven
Preface: The radical potential of placemaking
Cara Courage
Chapter 20: Displacemaking 2015 and 2020
Catherine Fennell and Daniel Tucker
Chapter 22: Embedded Artist Project: Epistemic Disobedience + Place
Frances Whitehead
Chapter 24: Artists, creativity, and the heart of city planning
Tom Borrup

Chapter 23: Routing out place identity through the vernacular production practices of a community light festival
Gail Skelly and Tim Edensor
Chapter 36: Facilitator skills for effective collaborative placemaking
Husam AlWaer and Ian Cooper
Chapter 39: Translating Outcomes: Laying the groundwork for interdisciplinary evaluation of creative placemaking
Jamie Hand
Chapter 40: Transforming community development through arts and culture: a developmental approach to documentation and research
Victor Rubin
Chapter 43: A theory of change for creative placemaking: the experience of the National Endowment for the Arts' *Our Town* program: an interview with Patricia Moore Shaffer, PhD
Maria Rosario Jackson
Chapter 44: Creative placemaking and comprehensive community development: rethinking neighborhood change and evaluation
Maria Rosario Jackson

30

ETHICAL PLACEMAKING FOR ECOLOGICAL SUBJECTS

Lisa Eckenwiler

In this chapter I propose that we embrace an ecological conception of persons and, from there, go on to argue that we should understand the fundamental goal of governance in terms of 'ethical placemaking.' Next, I suggest considerations that should guide thinking about governance aimed at generating more livable, equitable futures for ecological subjects through placemaking and call for ecological communicative democracy. In the final section, I identify the ethical and political capacities that ought to characterize ecological subjects and conclude by reiterating the central place of care and places of care in nurturing them.

Ecological subjects-citizens

The independent human being of reason, a conceived being in terms of self-interested calculation and material acquisition who is traceable through Plato, Augustine, Descartes, and Locke (Taylor, 1989), has been subjected to considerable critique in contemporary philosophy. As feminist and other philosophers, such as Communitarians have argued, this idealized account of persons, with its undue emphasis on independence and rationality, ignores the significance of social structures and the body in our lived experience and shaping of our identities, and obscures our interdependence with others in filial and communal bonds, and essential need for relations of nurturing and care (Sandel, 1992; Tronto, 1994; Meyers, 1997; Kittay, 1999).

Yet, bodies, caring relations, and social structures are still not the whole story for a comprehensive account of our radically relational nature. We are 'place-lings,' as Ed Casey puts it, beings who are 'never without emplaced experiences' (2009, p. 321). Lorraine Code, highlighting our social *and* geographical embeddedness, describes us as 'ecological subjects' (2006). Rosi Braidotti goes farther still and argues in favor of a 'zoe-centered subject… [a creature] shot through with relational linkages of the symbiotic, contaminating/viral kind which interconnect it to a variety of others, starting from the environmental or eco-others' (2013, p. 30). Humans are displaced from center on Braidotti's 'bio-egalitarian' account. Each of these conceptions helps capture our profoundly relational experience and possibilities. In addition to understanding our social situatedness, they reckon with our locatedness: our material and atmospheric emplacement and interdependence with non-human forms of life.

We might think about our embeddedness, or 'emplacement,' in at least two ways then: geographical and social. As noted above, feminists and others have written eloquently on our experience as situated within relations of care and dependency and significantly shaped by our membership in communities and by social norms and economic structures (Kittay, 1999; Young, 2011). At the same time, we are emplaced in particular geographic and atmospheric conditions – locatable places – that create the possibility of imagination and thought, survival, flourishing, or failing. Contemporary epistemology, for instance, argues that the environs we inhabit and traverse are integral to the kinds of thinking, knowing creatures we are. Christopher Preston points out that 'organisms that know things about the world are situated beings, beings cognitively grounded in the worlds from which they speak. An important part of this grounding is a physical location among material realities' (Preston, 2003, p. xi).

Contemporary research in social epidemiology using realist methods to better understand societal determinants of health inequalities offers grave but abundant evidence that aligns with these philosophical accounts. Studies show that heightened exposure to hazards (industrial and waste, and weak infrastructure), and diminished access to resources of all kinds (social connection, food, health care, parks and other public space, transportation) harm health (Marmot, 2007). Data suggest that people face increasing social isolation and depression, and restricted mobility due in part to the configuration of the built environment (Kelly et al., 2012). The effect of trade agreements and structural adjustment programs on health systems and, more specifically, facilities and their material and human resources also makes clear the essential relationship between place – our embeddedness in it – and the capability to be healthy (Eckenwiler, 2012).

This underscores the fact that where and how we are situated amid social structures has important implications for our level of exposure to health threats and for the distribution of health and illness within a population. Economic status, gender, race and ethnicity, social connectivity, and citizenship/immigration status all figure in the development and persistence of health inequalities for they operate to locate people in precarious environs such as low-income coastal or dense urban settings. Even 'natural' events like heat waves, famines, and tsunamis, which occur in and affect specific places and people are, in part, attributable to social practices and policy choices. Climate emergency is an especially compelling case of the significance of geographical *and* social embeddedness for health, for as Marmot notes, 'the poor, the geographically vulnerable, the politically weak, and other disadvantaged groups will be most affected' (2007, p. 1156). This belies one further fiction found within the inherited view of subjects: that we are roughly equal or similarly situated.

A final underappreciated dimension of our relational nature that requires a reconceived subject is what geographers call the 'inter-subjectivity' of place (Massey, 2004), the idea that we are constitutive of one another's identities and environs. Code underscores this in arguing that we are 'made by and making [our] relations in reciprocity with other subjects *and* with... (multiple, diverse) locations' (2006, p. 128). Seeing not only our personhood and identities but also *place* in relational terms, as intersubjectively constructed 'highlights the multiplicity of locations [and] the variety of interactions between people who are located differently that go into making places' (Raghuram, Madge, and Noxolo et al., 2009, p. 8). As Iris Young puts it, we 'dwell together' in 'complex and causal' relations of interdependence and in specific atmospheric and material conditions (2000, p. 224). The ethical import of these ideas is that we contribute to the construction of place, ours and others' – often unintentionally – through actions and interactions within a larger context of transnational social structures and processes that serve to enable some people in the realization of their capacities, yet constrain others. These considerations point to the necessity of reimagining our conception of subjects as ecological: embodied, embedded beings who dwell in dense relations with others, including non-human forms of life, all of whom are in need of care and place for the sake of sustenance.

Toward ethical placemaking and equitable cohabitation

Thinking about persons ecologically suggests we should resist a reductive, decontextualized view of individuals as the units of moral and political concern, for we cannot properly be conceived apart from our embeddedness. Social and political responsibility, on this view, might be conceived in terms of 'ethical placemaking,' a phrase first coined by geographers Raghuram, Made, and Noxolo (2009). Place is not a fixed or merely an external thing (Casey, 1997). It can be understood as being *around* us, but also *in* and *with* us (Casey, 1999). The examples above point to tracts of land, waterways, worksites, dwellings, neighborhoods, hubs and vehicles of transportation, hospitals and clinics (or their remnants), bedrooms and bathrooms, kitchens, food markets, makeshift housing, bodies and psyches. They gesture toward what can be called 'transnational space,' which includes, for instance, places of transition for nomads, refugees, and migrant workers – such as borders and immigration offices – or for people getting to and from work and other responsibilities, and places where economic transactions occur amid dense, global financial relations and processes.

To return to the geographers' point on intersubjectivity, places are, in Massey's words, 'constructed out of a particular constellation of social relations, meeting and weaving together at a particular locus. [E]ach 'place' can be seen as a particular, unique point of... intersection' (1991, p. 28). Its specificity is its set and configuration of relations and its position within a wider web of social relations. Moreover, because they are shaped by social interactions, places, like subjects, have plural, shifting 'identities.' They also become, endure and evolve; they are not static.

But what do I mean by 'ethical placemaking'? Placemaking is a set of practices that brings together architects, urban planners, and designers intent on transforming neighborhoods, parks and paths, features of landscape, housing developments, streetscapes, long-term care facilities, and hospitals (Project for Public Spaces, 2016; Silberberg, 2013). Public health leaders point to place-based interventions as 'the new frontier' (Amaro, 2014) and it is on the agendas of the World Health Organization (WHO, 2012), the US Centers for Disease Control (CDC, 2014), even the World Bank (Zhan, 2016). It also figures prominently in a growing number of international documents and declarations including the Sustainable Development Goals (2015) and the UN Habitat's New Urban Agenda (2016).

Elsewhere I have interpreted ethical placemaking – grounded in an ecological conception of persons – as a core component of an enabling, capabilities-oriented, conception of justice (Eckenwiler 2012, 2016, 2018). These conceptions of justice aim at attending to the social and political conditions that support people's capacities for self-development and self-determination. Iris Marion Young's theory of justice as enablement calls for reform of the social and institutional structures that systematically constrain people's capacities for self-development and self-determination (2011). Another example is Carol Gould's notion of justice as 'equal positive freedom,' which requires not only 'the absence of constraining conditions such as coercion and oppression' but also access to the means or conditions for 'self-transformation' and the 'development of capacities and the realization of projects over time.' Justice, here too, is about 'the availability of *enabling* [my emphasis] conditions' for individuals (Gould, 2009). Finally, the widely known capabilities approach, developed by Sen and Nussbaum (2006) emphasizes that support for the realization of key functionings and capabilities is a matter of justice for people everywhere.

Ethical placemaking, I have argued, first involves a negative condition: it takes care to avoid creating, through our actions, interactions, practices, and policies, conditions of deprivation. Conditions of deprivation are those that make it impossible to support and sustain the capacities of social, corporeal creatures whose vulnerability comes from a need for care across the lifespan and also fit within specific atmospheric and material environments. Furthermore, it requires: nurturing relations of care and interdependence; protecting bodily integrity; supporting autonomy, not interpreted in terms of individual self-reliance, but in the relational sense that sees us as originating, persisting, and flourishing within relations of care and interdependence, given ongoing opportunities for self-directed thought and action; promoting stability and a sense of rootedness and, at the same time, supporting generative movement; and finally, where necessary responding to inequities (Eckenwiler, 2016, 2018). The specific targets of intervention are the social processes and practices that create and sustain deprivation and inequality – for ecological subjects and their habitats. The precise meanings of these elements should be worked out in overlapping political processes. While we begin from a theory of needs, human and other, we should shift our attention toward the habitats – and this is key – all of which are distinct, dynamic in the sense of unfixed, and intersecting – where people dwell and move about, for their own particular and shared purposes. If the scope of ethics and politics includes creating enabling conditions for ecological subjects and thus promoting ethical placemaking, what are ideal forms of governance? I now turn to this question.

Models of governance for ethical placemaking

The perils of interest group democracy have been well-highlighted by political philosophers (Young, 1996), so I will not dedicate time to those criticisms here. An ecological version of deliberative democracy might serve as a more ethical alternative to the primacy of a materialistic and consumptive notion of interests. Yet, critics are wary of its potential for perpetuating exclusions and inequalities given its potential to: (1) privilege certain forms and styles of expression (the impartial expert's for instance) and thus discount some contributions and forms of expression sometimes discounted as less credible (storytelling, for example); (2) assume that people are sufficiently symmetrical and that they can reverse perspectives, thus obscuring morally relevant differentiation and particularity in aiming for 'unity,' the 'common good,' and 'shared meaning'; and (3) obscure other possible objectives in political engagement such as recognition or understanding (Young, 1996).

Communicative democracy may better advance efforts to address exclusion and inequality (Young, 2000). On this model of democratic governance, moral and political engagement should be understood as largely dialogical; decisions emerge from the interactions of a plurality of subjects. Communicative democracy does not privilege impartiality but instead, drawing from standpoint epistemology, identifies and incorporates partialities in generating social knowledge to inform decisions. Instead of discounting or being suspicious about the value of personal narratives, for example, it regards them as resources. For ecological subjects, indeed, meanings and expressions of wellbeing, of harm, and so forth are situated, socially and ecologically. Rather than aiming for unity, this model of democratic governance aims for recognition and understanding. Participants express their experiences and perspectives so that others situated differently can learn how it is for them, what the meaning of events, experiences, any given policy is, and perhaps transform accepted knowledge to reshape preferences, even identities. Communicative democracy can also better acknowledge that there may be different understandings of what it is to cooperate or to justify decisions, of what emancipation itself means. Such democracy can embrace different kinds of sites for generating ideas: chosen, even oppositional communities (not those invoked by communitarians), street protests, and marginalized social networks.

Ecological communicative democracy, as I describe it here, goes beyond feminist standpoint theory in situating knowers and knowledge-production efforts in particular places – socially *and* geographically shaped habitats – and conceives of them as

> intersecting… with other locations and their occupants.… It maps locations of knowledge production and demographics of knowledge producers… [and] considers the specificities of the 'habitat' conditions and the inhabitants within and surrounding each location, to discern where analogies can be drawn and where exposed dis-analogies demand acknowledgment and/or rethinking.
>
> *(Code, 2006, p. 52)*

Code invokes James Cheney's notion of the *bioregional narrative* as an essential instrument in facilitating epistemic responsibility, which lies at the core of democratic governance. A bioregional narrative is one 'grounded in geography rather than in a linear, essentialized, narrative self' (Cheney, 1989). Its aim is to 'map ecological relations to discern conditions for mutually sustaining lives within a specific locality… or *the interrelations among them* [my emphasis]' (Code, 2006, pp. 59–60).

Given the nature of ecological subjects, then, ecological communicative democracy, in part through tools like bioregional narratives, aims to produce responsible, epistemically just (Fricker, 2007) assessments of the habitability of particular conditions. Crucially, responding to the intersubjectivity of placemaking and in turn, inequities generated, it also aims to assess these conditions in relation to the conditions in which others dwell (Eckenwiler, 2018). The essential questions under ecological communicative democracy become: 'What would ethical placemaking mean here, or here, and where they intersect?' and 'What is necessary for promoting mutually sustaining, equitable lives?'

Essential capacities for ethical placemaking

If we are to advance from being a people of competitive consumption into ecological subjects working toward ethical placemaking, what sorts of capacities should characterize us? What should be cultivated among us in order that we become and endure in mutually sustaining, equitable relations of cohabitation? Given the concerns raised in the previous section, what capacities might move us toward ideal – that is, maximally inclusive, participatory, and fair – forms of governance?

For humans as ecological subjects, the givens are our embodiment, interdependence, and locatedness. Essential to this understanding of persons, then, is respect for our *animality* (Nussbaum, 2006, p. 54). This has the benefit not only of 'acknowledging that there are many types of dignity in the world'; it also helps to account for our relationality and embeddedness in habitats in a way that the undue emphasis on reason as the principal, if not sole, source of dignity cannot. Additionally, it captures the idea that we consume and excrete, relying on the resources around us in order to be and persist. Rather than starting from an assumption of disinterest in others, ecological subjects, as interdependent, should nurture their capacities for *recognition* of: an individual's unique identity as an autonomous individual; persons as belonging to particular communities or groups; others' needs for relationships, both interpersonal and associative; and finally, of the places and the conditions in which people dwell and their need for fit (Eckenwiler, 2018).

Ideally, ecological subjects should cultivate a *disposition toward empathy*. In Gould's formulation, empathy 'signifies a feeling or imaginative identification with another and that other's perspective and situation' (2007, p. 251). While the notions of 'identification,' 'mutuality,' or 'fellow feeling' are ethically perilous given social and ecological differentiation and asymmetries, the practice of seeking knowledge of another's situation, listening carefully, and trying to develop feeling for people's expressions about their particular plight – which may well intersect with ours and that of people we know – has value for promoting understanding and realizing justice (hooks, 2009). Instead of taking for granted that we are acquisitive creatures, we should call for privileging *critical acquisition*. Starting from an expectation of acquisition, then, we should scrutinize what and how much we are trying to get our hands on, not to mention how we are going about it, whether the resource will continue to have generative capacity, and so forth.

Being able to *take the long view* (temporal and spatial, across terrains and timeframes) is an essential capacity for ecological subjects. For it allows for identifying effects and their sources that may not be readily apparent, and for envisioning interventions that can be sustained over time. In taking the long view, ecological thinking can be understood as emphasizing the future over the past and present. As Elizabeth Grosz argues, unless we develop concepts of time and duration that 'welcome and privilege the future, we will remain closed to understanding the complex processes of becoming that engender and constitute both life and matter' (1999, p. 16). This capacity holds ethical import. Informing our prevailing conception of responsibility is a specific phenomenology of agency that emphasizes, or 'gives experiential primacy,' to short-term effects rather than to remote ones. Thus, when consequences or outcomes are generated by an (often wide) array of agents and unfold over time, our sense of agency diminishes; that is, we see ourselves as implicated very little, if at all (Young, 2004). Ecological subjects should have a sense, then, of individuality but also of solidaristic and long-reaching agency.

Finally, ecological subjects should cultivate *critical social reflexivity*. This is crucial for purposes of assessing our situatedness, the nature and extent of our connections, and, in turn, our responsibilities. While a fuller discussion is beyond the scope of my project here, elsewhere I argue in favor of the view that differences in the nature and degree of these responsibilities align with how one is situated within the structural processes that organize society (and that can generate injustice) and the geography of intersecting, interdependent ecosystems (Eckenwiler, 2012). Moreover, picking up on the earlier reference to solidarity, I have argued that solidaristic recognition requires appreciation of one's contribution to the unsustainable conditions in which others dwell (Eckenwiler, 2018). As a final consideration, it warrants repeating that the work of care is crucial to the foundations of social organization and cooperation in part because it creates the next generation of citizens and shapes their capacities: corporeal, social, intellectual, ethical, and political. If ecological civic engagement is to thrive, societies must thoroughly and effectively support those who work to generate it and attend to the conditions in which it occurs (Christopherson, 2006; Herd and Meyer, 2006).

Conclusion

I have argued that we should embrace a conception of persons as ecological subjects and transform politics into an endeavor aimed at ethical placemaking for the sake of more livable and, above all, just futures. Stated differently, justice requires that we target for reform the social processes and practices that create and sustain deprivation and inequality – for ecological subjects and their habitats – and work to create the conditions and kinds of dwellings and habitats that support relations of care; protect bodily integrity; nurture autonomy; promote stability and generative movement; and finally, respond to inequities (Eckenwiler, 2016, 2018). My focus in this chapter has been on humans, yet, these reflections may well press us further, toward a *post-human* conception of subjects and perhaps even a 'bio-centered egalitarianism' (Braidotti, 2013, p. 19) that shifts humans from the center and assigns value more generously to non-human forms of life. This goes beyond my scope here, however. For present purposes, my modest claim is that we must reckon more thoroughly with our radically relational subjectivity in formulating accounts of what justice requires.

References

Amaro, H. (2014). 'The action is upstream: place-based approaches for achieving population health and health equity', *American Journal of Public Health*, 104(6), p. 964.
Braidotti, R. (2013). 'Posthuman relational subjectivity', in *Relational Architectural Ecologies: Architecture, Nature, and Subjectivity*. New York: Routledge.
Casey, E.S. (1997). *The Fate of Place: A Philosophical History*. Berkeley, CA: University of California Press.
Casey, E.S. (1999). 'The time of the glance: toward becoming otherwise' in Grosz, E. (ed.) *Becomings: Explorations in Time, Memory, and Futures*. Ithaca: Cornell University Press, pp. 79–97.
Casey, E.S. (2009). *Getting Back into Place: Toward a Renewed Understanding of the Place-World*. Bloomington, IN: Indiana University Press.
Cheney, J. (1989). 'Postmodern environmental ethics: Ethics as bioregional narrative', *Environmental Ethics*, 11(2), pp. 117–134.
Christopherson, S. (2006). 'Women and the restructuring of care work: Cross national variations and trends in ten OECD countries', in Zimmerman, M.K., Litt, J.S. and Bose, C.E. (eds.) *Global Dimensions of Gender and Carework*. Stanford, CA: Stanford Social Sciences, pp. 318–323.
Code, L. (2006). *Ecological Thinking: The Politics of Epistemic Location*. Oxford: Oxford University Press.
Eckenwiler, L. (2012). *Long-term Care, Globalization, and Justice*. Baltimore, MD: Johns Hopkins University Press.
Eckenwiler, L. (2013). 'Ecological selves as citizens and governance as ethical placemaking', *Humans and Nature*, 6(2) [online]. Available at: https://www.humansandnature.org/ecological-selves-as-citizens-and-governance-as-ethical-placemaking (Accessed: 6 June 2020).
Eckenwiler, L. (2016). 'Defining ethical placemaking for place-based interventions', *American Journal of Public Health*, 106(November), pp. 1944–1946.
Eckenwiler, L. (2018). 'Displacement and solidarity: An ethic of placemaking', *Bioethics* 32 (November), pp. 562–568.
Fricker, M. (2007). *Epistemic Justice: Ethics and the Power of Knowing*. New York: Oxford University Press.
Gould, C.C. (2007). 'Recognition, empathy, and solidarity', in Bertram, G.W., Celikates, R., Laudou, C. and Lauer, D. (eds.) *Socialite et Reconnaissance: Grammaires de l'Humain*. Paris: Editions l'Harmattan, pp. 247–265.
Gould, C.C. (2009). 'Reconceiving autonomy and universality as norms for transnational democracy', in Langlois, A. and Soltan, K. (eds.) *Global Democracy and Its Difficulties*. London: Routledge.
Grosz, E. (1999). 'Thinking the new: Of futures yet unthought', in Grosz, E. (ed.) *Becomings: Explorations in Time, Memory, and Futures*. Ithaca: Cornell University Press, pp. 15–28.
Herd, P. and Meyer, M.H. (2006). 'Care work: Invisible civic engagement' in Zimmerman, M.K., Litt, J.S. and Bose, C.E. (eds.) *Global Dimensions of Gender and Carework*. Stanford, CA: Stanford Social Sciences, pp. 324–340.
hooks, B. (2009). *Belonging: A Culture of Place*. New York: Routledge.
Kelly, J.-F., Breadon, P., Davis, C., Hunter, A., Mares, P., Mullerworth, D. and Weidmann, B. (2012). *Social Cities*. Melbourne: Grattan Institute.
Kittay, E.F. (1999). *Love's Labor: Essays on Equality and Dependency*. New York: Routledge.
Marmot, M. (2007). 'Achieving health equity: From root causes to fair outcomes', *Lancet*, 370, pp. 1153–1163.
Massey, D. (1991). 'A global sense of place', in *Marxism Today*, 38, pp. 24–29.
Massey, D. (2004). 'Geographies of responsibility', *Geografiska Annaler*, 86, pp. 5–18.
Massey, D. (2006). 'Space, time, and political responsibility in the midst of global inequality', *Erdkunde*, 60(2), pp. 89–95.
Meyer, D.T. (1997). *Feminists Rethink the Self*. Boulder, CO: Westview Press.
Nussbaum, M.C. (2006). *Frontiers of Justice: Disability, Nationality, Species Membership*. Cambridge, MA: Belknap Press of Harvard University Press.
Preston, C.J. (2003). *Grounding Knowledge: Environmental Philosophy, Epistemology, and Place*. Athens: University of Georgia Press.
Project for Public Spaces. (2016). *What Is Placemaking?* [online]. New York: Project for Public Spaces.
Raghuram, P., Madge, C. and Noxolo, P. (2009). 'Rethinking responsibility and care for a postcolonial world', *Geoforum*, 40, pp. 5–13.
Sandel, M. (1982). *Liberalism and the Limits of Justice*. Cambridge: Cambridge University Press.
Silerberg. S. (2013). *Places in the Making: How Place Making Builds Places and Communities*. Boston: MIT.
Taylor, C. (1989). *Sources of the Self: The Making of Modern Identity*. Cambridge: Harvard University Press.
Tronto, J. (1994). *Moral Boundaries: A Political Argument for an Ethic of Care*. New York: Routledge.

U.S. Centers for Disease Control and Prevention (CDC). (2014). *About Healthy Places.* Atlanta: CDC.

UN Habitat. The New Urban Agenda. (2016). United Nations Task Team on Habitat III. Habitat III Issue Papers: *Migration and Refugees in Urban Areas.* New York, June 2015 [online]. Available at: http://unhabitat.org/wp-content/uploads/2015/04/Habitat-III-Issue-Papers-and-Policy-Units_11-April.pdf (Accessed: 7 June 2020).

United Nations. (2015). *Sustainable Development Goals.* New York: United Nations.

WHO (World Health Organization) Regional Office for Europe. (2012). *Addressing the Social Determinants of Health: The Urban Dimension and the Role of Local Government.* Copenhagen: WHO.

Young, I.M. (1996). 'Communication and the other: Beyond deliberative democracy', in Benhabib, S. (ed.) *Democracy and Difference: Contesting the Boundaries of the Political.* Princeton: Princeton University Press.

Young, I.M. (2000). *Inclusion and Democracy.* Oxford: Oxford University Press.

Young, I.M. (2004). 'Responsibility and global justice: a social connection model' *Social Philosophy and Policy*, 23, pp. 102–130.

Young, I.M. (2011). *Responsibility for Justice.* Oxford: Oxford University Press.

Zhan, M. for the World Bank. (2016). 'Investing in better public spaces', Presented at *Future of* Places Leadership Forum, September 15, 2016. Vancouver, BC, Canada.

Further reading in this volume

Chapter 1: Introduction: what really matters – moving placemaking into a new epoch
Cara Courage
Preface: Placemaking in the age of COVID-19 and protest
Jason Schupbach
Chapter 3: An annotated history of creative placemaking at the federal level
Jen Hughes
Chapter 4: Future of creative placemaking
Sarah Calderon and Erik Takeshita
Chapter 5: Making places for survival: looking to a creative placemaking past for a guide to the future
Jeremy Liu
Chapter 14: Experts in their own tomorrows: placemaking for participatory climate futures
Paul Graham Raven
Chapter 15: Un/safety as placemaking: disabled people's socio-spatial negotiation of fear of violent crime
Claire Edwards
Chapter 17: 'I am not a satnav': Affective placemaking and conflict in 'the ginnel that roared'
Morag Rose
Chapter 16: More than a mural: participatory placemaking on Gija Country
Samantha Edwards-Vandenhoek
Preface: The radical potential of placemaking
Cara Courage
Chapter 18: '*Homomonument* sounds like a poem': queer placemaking 30 years on: a conversational dialogue with Thijs Bartels, author of *Dancing on the Homomonument* (2003)
Martin Zebracki
Chapter 19: Placemaking in the ecology of the human habitat
Graham Marshall
Chapter 22: Embedded Artist Project: Epistemic Disobedience + Place
Frances Whitehead
Chapter 31: Seven generations: a role for artists in Zuni PlaceKnowing
Theodore S. Jojola and Michaela P. Shirley
Chapter 41: Rituals of regard: on festivals, folks, and findings of social impact
Maribel Alvarez
Chapter 43: A theory of change for creative placemaking: the experience of the National Endowment for the Arts' *Our Town* program: an interview with Patricia Moore Shaffer, PhD
Maria Rosario Jackson
Chapter 45: How the city speaks to us and how we speak back: rewriting the relationship between people and place
Rosanna Vitiello and Marcus Willcocks

31
SEVEN GENERATIONS
A role for artists in Zuni PlaceKnowing

Theodore S. Jojola and Michaela P. Shirley

Introduction

In 2013, the Indigenous Design and Planning Institute (iD+Pi) at the University of New Mexico was approached by the newly established Zuni Pueblo Main Street organization to assist them in developing a development plan along State Highway 53. This nonprofit held the singular distinction of being the first and only tribal Main Street organization in the US Chartered in 2012, Zuni Pueblo Main Street evolved as an arm of the Tribal Government and was entrusted with developing a master land-use plan along three miles of State Highway 53 adjacent to the old village and built atop a traditional road. The highway serves as a connection to an adjacent community, Black Rock, which was established at the turn of the twentieth century to house administrative offices, a boarding school, and a hospital provided by the Bureau of Indian Affairs (Dodge, 2003). As far as could be determined, the highway was paved sometime around WWII and turned into a public thoroughfare. It connects the village of Zuni to surrounding Navajo and Mormon communities, as well as the nearby city of Gallup. Today, Zuni is a community with a population of approximately 6,300. It is a community that is driven by an informal economy (Bureau of Business and Economic Research, 2014) and although much of its wage economy is supported by tribal government and federal services, it is estimated that 80 per cent of all households produce arts and crafts to augment their incomes. With a median household income of $30,250 USD, these homemade efforts, as embodied by multiple artisans in a given household, are significant (ibid).

As we learned in the course of our engagement, few if any of these transactions go towards supporting local businesses or generating taxes. It has been estimated that the community loses circa $30 USD million in local revenue annually to surrounding businesses outside of the reservation (ibid). Because of this, the Zuni tribal government functions the same way that many other federally recognized tribes do. It relies on trust funding to support local government as well as public services for education, health, and other necessary functions. Due to this dependency on federal funds, little if any input for community development comes from the business or private sector. This is especially the case with the area designated as the Zuni Pueblo Main Street. This commercial strip – if it could be called that – originates at the Halona Market, intersects State Highway 53 at a four-way stop, and continues northward for a few more miles towards the Zuni Visitor Center. Along the three-mile route, there are two grocery stores, four trading posts, an art co-op, three gas stations, a bank, a commercial telephone business, a bistro, the tribal administrative center, a US post office, and numerous homes. The only establishments that house tourist services are the Zuni Visitor Center and the A:shiwi A:wan Museum and Heritage Center, which on occasion, is closed due to a lack of revenues. Despite this situation, tourists are drawn to Zuni Pueblo because of its historic village, its cultural ceremonies, and its world-class traditional artwork. The community of artists, in particular, have evolved distinctive styles of jewelry, pottery, painting, stone animal fetishes (small carvings depicting animals and their holy peoples that are made with various precious nature stones by the Zuni people), and textiles. Zuni art has attained agency among specialized collectors. The demand is so great that it has created a secondary market of so-called 'knock-offs' or fakes. Such opportunists have allegedly fabricated millions of dollars' worth of fake Native American jewelry (Cornell, 2018).

The challenge

The iD+Pi was established in 2011 in the belief that tribal communities should benefit from the best practices that design and planning have to offer, and in a manner that is culturally informed. This community development design and planning process requires that leadership balance the immediacy of action (short-term) with a comprehensive vision (long-term.) Community engagement and meaningful public participation is the key to its success (Jojola and Shirley, 2017). Indigenous planning is an emerging paradigm that uses a culturally responsive and value-based approach to community development. It is a participatory process predicated on establishing a set of principles that are informed by traditional knowledge. At its foundation are a seven-generations planning model that is intended to assist the community in connecting its past, present, and future. The structure is simple and is driven by a demographic model that is situational: older generations are identified as great grandparents, grandparents, and parents; the middle generation are adults who are in their middle years; and the younger generations are indicated as comprising children, grandchildren, and great-grandchildren. This generative model was used to convey to Zuni Pueblo Main Street participants how communities evolve and develop. Activities included mapping the land-uses along Zuni Pueblo Main Street and extended conversations as to why the streetscape provided little opportunity for artists to directly merchandise their wares nor provided places to shop and eat.

At the crux of the challenge were the conventions of ownership as poised by land tenure. Reservation lands are held in trust by the US federal governments for the Zuni people, but land tenure predates this. Because of its trust status, there is no equivalent of private property. Instead, the land is inherited by the right of a matriarchal system, with women inheriting the right to use and benefit from its productivity (Goldman, 2003). Approximately 16 land assignments are either undocumented or documented empty/vacant lands or commercial buildings (UNM Indigenous Design and Planning Institute and UNM School of Law, 2016). Once the land has been inherited and occupied through custom law, there is little or no opportunity to reinvest it for businesses or other non-domestic uses. The notion that the tribe or individuals can designate land uses as real estate is a non-sequitur situation. During the project, it became evident that there was little that could be done to impose a rational Western model for land-use. Although technically, the tribal government has the authority to reclaim lands that had been vacant or deemed unproductive over time, this is rarely imposed. This situation causes a quandary for vacant lands and abandoned buildings where the heirs of deceased owners have left the community for extended periods. Moreover, the secular tribal council cannot reclaim vacant land without working with the traditional religious leaders to determine matrilineal ascendancy.

Even in the event that heirs are disposed to opening businesses, the inability to use land held in federal trust as collateral curtails access to capital through bank loans.

The opportunity

Creative placemaking 'animates public and private spaces, rejuvenates structures and streetscapes, improves local business viability and public safety, and brings diverse people together to celebrate, inspire, and be inspired' (Markusen and Gadda, 2010). This is a concept that has been adopted by the American US mainstream as a way to utilize creative industries in establishing places of economic growth and production. In an Indigenous community such as Zuni, one cannot merely deploy a placemaking paradigm to its community development toolkit. At best, it will take decades, if not generations, to appropriate vacant lands or replace abandoned buildings along its Zuni Pueblo Main Street. The shorter-term solution, as such, is to enlist its artisans in other ways to bring their production to public light by building on locally based cultural practices.

In 2015, iD+Pi, with the help of partners from Zuni Pueblo Main Street and Creative StartUps, leveraged a $225,000 USD ArtPlace America grant to support the 'Solving Real Places for Real People: Zuni Main Street' project. The project was implemented over three years beginning with conversations with local artists, especially those who had marketed their products successfully and who had experience in managing their enterprises. The first set of conversations revealed that these artists lacked a way to collectively merge their efforts in order to bring local development to their village. Lacking this, they relied on outside marketers who handsomely profited by representing their wares elsewhere. The second set of conversations focused on local concerns as to who determined 'what was Zuni art?' Zuni artists regularly innovate by incorporating new techniques and materials. Their designs, however, rarely stray from depicting cultural motifs that are grounded in their beliefs and symbols. As newer generations of artists evolve, they began to introduce new mediums that are not considered to be traditional. It includes photography, spray-can media, digital media, performative arts, 3D printing, 2D screen printing, painting, and kiln-fired ceramics made from greenware (molded ceramic pots that just need to be glazed and fired). It was felt by some traditional artists that these processes devalued Zuni art because they result in wares that can be mass produced and easily replicated. At another level, an issue also came up as to whether performance activities such as music and storytelling, as well as, local village cuisine, could be represented as art.

A final series of conversations emerged as to what aspects of Zuni culture the community wanted to share with the outside and what they did not. This undercurrent has persisted for generations – especially among the religious leadership – who assert that some depictions are shared only among those who have been initiated in the proper way. One example of this controversy are murals that were painted inside the walls of the old Zuni historic mission church that depict the seasonal spirits that are an integral part of the ceremonial cycles of the Zuni moieties. These are the handiwork of renown Zuni artist, Alex Seowtewa (1933–2014), who was enlisted by church, federal, and village officials in 1966 to reintroduce this motif as part of the reconstructed church (Seowtewa, 1992). An uneasy measure exists to limit the public exposition of these murals as the family of Seowtewa, traditional leaders, and the secular tribal government continue to attempt to resolve its cultural patrimony and depiction. As artists continued to reflect on these concerns, the main issue that everyone did agreed upon was how their organizing could control how marketers were misrepresenting their art. These acts were seen as not only degrading the culture, but its impact weighed heavily on maintaining standards of Zuni authenticity. In particular, artists felt that stereotyping was affecting the quality of their craft, cheapening the value of their work and foreshortening demand for higher quality pieces.

As Zuni artists pondered the next action steps, the non-profit Creative StartUps partner was enlisted to provide organizational and cultural entrepreneurship training. As these workshops ensued, the participants turned their attention away from bringing in outside venture capitalism and onto identifying existing assets among them. With no opportunity to move to storefronts along the Zuni Pueblo Main Street, they worked at other solutions. One major point that was made is that many artists worked in studio spaces in or adjacent to their homes. The more accomplished of them had display rooms where they conducted business. A simple solution was to create a map showing where artists lived. This posed its own challenge, however, as the traditional village lacked street signage, house numbers, and reference points. Privacy, especially during ceremonial events, further hampered visitors from moving freely through the village. Moreover, there was uncertainty as to where public egress was allowed. Eventually, artists discussed the possibility of linking their studios along a specified route. This idea would entail enlisting artists who were willing to allow visitation in their homes. This was deemed as agreeable as long as it occurred during specific times or by appointment.

The *Zuni Pueblo Artwalk*

As the discussions ensued, a concept of an art walk emerged. It entailed organizing a daylong event where participating artists would open their working studios to outside visitors, but in a regulated manner. Logistically, two key factors had to be solved. One was creating a loop that could be plied by public transportation or cars. The other was designating a schedule for completing the loop during which visitors could choose which artist to visit and for how long. This solution was a breakthrough. The venture would not only allow the artists to circumvent the need to occupy storefronts on Zuni Pueblo Main Street, but it would also educate the potential buyer through cultural sharing and crafts demonstration. The approach would become the *Zuni Pueblo ArtWalk*.

Leadership from within this group enlisted other local partners to provide assistance. One key partner was a local Zuni video team, ShiwiSun Productions, who created, 'I am a Zuni Artist' artist profiles demonstrating their personal stories, meanings behind their artwork, and the process they used to manufacture these unique pieces. These were then posted on a newly created the *Zuni Pueblo ArtWalk* website. iD+Pi facilitated the wayfinding design-build component of the 'ArtWalk.' Through a Summer studio, university faculty and students from UNM's School of Architecture and Design fabricated metal signposts. Students designed and created symbols to represent each art form. Community volunteers then helped place the signs at the entryway of each artist's home. The destinations were further identified by placing 'breadcrumb' stones that were gathered from a nearby quarry. These were painted blue and yellow to form and serve as a visible reference for those visitors who opted to do their own self-guided tour. Seven Zuni artists founded the event, which was inaugurated on Saturday, October 14, 2017. The date was chosen to coincide with the Zuni Visitor Center's annual Main Street Festival. Visitors registered at the Zuni Visitor Center and took a free hop-on/hop-off shuttle provided by the A:shiwi Transit.

The event proved to be a resounding success. It fostered personal engagement, cultural learning, and resulted in a substantial value-added return on items sold. It built on existing assets and circumvented the lack of storefronts on Zuni Pueblo Main Street. The *Zuni Pueblo ArtWalk* was originally envisaged to be held four times during the year but is now done on the second weekend of each month. In just one year, the number of participating artists increased from the original 7 to 23.

Reflections on PlaceKnowing

> How do we go forward, [and] how do we use those gifts of our intellect of our creator that we have been bestowed with… I've always been of the thought that as native people, Pueblo people, we're bestowed with that gift, but for some reason, somewhere along the way, we've buried it. And we need to extract that… to bring that out, you know, and not be afraid of our gifts; of our intellect and genius.
>
> *(Pueblo PhD cohort, focus group, March 27, 2018)*

One does not choose to be Indigenous. You are born into it (Jojola and Shirley, 2017). A worldview represents the community's understanding of itself and its relationship to the natural world that sustains it. From the first heartbeat to the last, a person is the steward of the culture by ascribing to the Indigenous worldview and attending to their collective responsibility. Although the Zuni Pueblo Main Street project was premised on the assumption that a place could be revitalized through economic opportunity, there were other cultural determinants that factored into the making of place. As learned, a sense of place is embodied by the whole. It was not engineered in a single moment of time but evolved over time and into the present. As seen in the *Zuni Pueblo ArtWalk* project, art supersedes the material basics of color, materials, medium, and form. Instead, its representation carries the spiritual and philosophical meaning of its culture. This understanding is not transferred through a classroom. As people learned, they used their own cultural assets to create a way to both market and educate the visitors to Zuni. It was less about placemaking and more about sharing the meaning around place.

Because the lessons learned from the Zuni project were so important, we wanted to see if other Indigenous practitioners shared the same perspectives about placemaking. In 2017, iD+Pi applied for and received funding from the McCune Foundation to hold a series of five conversations about placemaking held in different parts of the country over the course of a year. The participants ranged from PhD scholars of the Santa Fe Indian School Leadership Pueblo Indian PhD Cohort program, their degrees granted by Arizona State University; Indigenous university students and faculty; ArtPlace America grantees from Indigenous communities; as well as tribal planners, Indigenous architects, and designer-maker practitioners. The iD+Pi team began each meeting with a presentation of Indigenous Planning concepts which was followed by findings from the Zuni Pueblo Main Street and *Zuni Pueblo ArtWalk* projects at Zuni Pueblo. The open discussions that ensued were audio recorded and later transcribed for purposes of doing content analysis. When the final draft document was produced, it was summed in a manner that consolidated the key findings with the voices left intact.

What ensued was a rich discussion that was infused with deep theoretical understanding, deference to traditional knowledge, and qualified by practical application. The major key finding that was shared among all groups is that placemaking, in itself, was a misnomer. The idea that places could be 'made' in time and place was antithetical to indigeneity. Rather, the consensus is that tradition is inherited and that places are rooted, first and foremost, in the culture. Places make people. People do not make places (ArtPlace America Pre-Summit, focus group participant, unidentified, May 21, 2018).

The concept that best exemplified that notion is *PlaceKnowing*. PlaceKnowing is defined as a place whose meaning is derived from a cultural construction. This results in a holistic understanding of how places are evolved through time and space. Although it shares the same conceptual framework as placemaking, the major difference is how traditional knowledge serves to inform how communities give meaning to the cultural landscape they use and inhabit. As the discussions progressed there were four distinct determinants that were identified as being integral to PlaceKnowing. They are kinship, fluidity, culture, and land. The following subsections explain these in more detail:

Kinship

> I think it goes back to when were [are] born as babies; how our placenta is buried and where [it is] rooted. I am always grateful that with my daughter when she was a baby, [had] her placenta rooted there [at the Pueblo]. I think about my father when we were growing up. We had opportunities to go move to different places but I think he wanted to keep us home so that we would know our place... It makes sense now. That's sort of why I'm trying to grapple with the term placemaking – because it is very complicated and it is esoteric.
>
> *(Pueblo PhD Cohort, focus group participant, unidentified, March 27, 2018)*

Despite being away, individuals are always connected to home and the land. A Diné (Navajo) planning student recalled this same custom and suddenly realized that the landscape where she was from has borne countless umbilical cords from numerous generations. She was surrounded by a cultural landscape of placentas:

> I'm a planner, and I know... it always gets to the notion of what is planning for, us as people, and what does that mean? It goes back to [the question of] how do we relate to each other, and how do we relate to the land? In Navajo we say k' e, which is kinship. People lived according to the kinship in a system. It's the way they relate to each other and how they relate to the community.
>
> *(Tribal Planners' Roundtable, focus group participant, unidentified, September 14, 2018)*

The history of the family is another important facet of culture. Knowing one's family's history is an essential reference for the concept of home. Home builds connection as implied by one participant:

> How do we create that space where we recognize the sacredness in one another and that they're connected to the ancestors to where they come from?
>
> *(Pueblo PhD Cohort, focus group participant, unidentified, March 27, 2018)*

This task is invested in protocols. When one introduces themselves to a group or another community, they first relate their clans and the places their families are from. That information links other clan relatives who may or may not have grown up in the same place. Stated in another way:

> [PlaceKnowing] is owning the place, it is somewhere you can always reference to as home... [I]t's a place that you are proud of, even though there may not be that much there. There should be a movement to make people want to go back home... [C]lans are place-based.
>
> *(ASU Faculty and Student, focus group participant, unidentified, April 10, 2018)*

In essence, the discussions got to the root of identity:

> Who are you when you say 'I am Pueblo, Diné, Lakota,' or whatever. It's very interesting to hear the answers to those questions... [T]he next question is, well, that's great [but]

> what do you care about? And then, lastly, … how does [that matter in] how that makes a building or makes a plan? That's our pretty simple [approach for] any conversation.
> *(Pueblo PhD Cohort, focus group participant, unidentified, March 27, 2018)*

As further pointed out by other conversations, when cultural bonds break, it introduces a decline in the quality of life:

> My own mother, whose mother was a healer, went to nursing school. She rejected her [practices], because [she felt] the new way was a better way. I look at art and culture and re-imagine how we [can use it to go back to] a good place… The assumption [is that] we're all going to be able to find those old ways… Remember, the bad part came from the exterior culture, reacting [negatively] to those old ways. So, this [is] how we heal internally.
> *(ArtPlace America Pre-Summit, focus group participant, unidentified, May 21, 2018)*

Kinship is about belonging. It fosters a relationship between people and shapes the places they inhabit. It helps inform designers and artists about the role that cultur plays in giving meaning to what they design.

Fluidity

> Then there's the fluidity. My friend wrote this beautiful piece on time. [It] talked about fluidity. Gosh, Pueblo people probably don't consider themselves urban… because there's fluidity within you. I live in Santa Fe, but there's fluidity to Jemez. And you're probably feeling the same attachment to Cochiti. [It's] deep and I [imagine] that's the same for Indigenous people in [other places of] the United States.
> *(Pueblo PhD Cohort, focus group participant, unidentified, March 27, 2018)*

This is a concept that even though places are fixed, culture is fluid. An Indigenous person, as such, is nurtured in place and as a result of that grounding, they are a reflection of its culture. They are part of a cultural continuum in time and place. Fluidity also brings into perspective the role of intergenerational knowledge and exchange. Youth, adults, and elders interact seamlessly as they participate ceremonially in the societies that they have inherited. Those connections are communicated by traditional languages. As related by one participant:

> Yes, it almost seems like the ones that do succeed are the ones given over to the grandparents of the parents. That does two things. One, it rejuvenates them. It gets them out of their isolation. Second, they're the ones that, in many cases, are the keepers of the cultural knowledge and the language. Children [raised] from those types of situations are the ones that come out speaking and participating.
> *(Pueblo Education Convocation Planning Meeting, focus group participant, unidentified, May 5, 2018)*

Tradition is not static. As Indigenous people migrated throughout the generations, they built interconnections among other places. These movements and the significant things they saw and learned were remembered through storytelling. Knowledge was power and it was manifested in language. The power of stories becomes integral in building bonds of fluidity through the generations:

> The concept of memory – place[s] will evoke certain memories for elders… For me, it is always about the story-work, the memory-work, [and] the kind of communication that [gets] structured into the environment. [This] allows people to leave a record in perpetuity.
> *(ASU Faculty and Student, focus group participant, unidentified, April 10, 2018)*

Culture

> Our hills, not only are they able to heal us as artists, but it also heals any visitors… we want everyone to learn from it. Obviously, [you won't know what it] means what it actually means until you're taught, or it's explained to you.
> *(ArtPlace America Pre-Summit, focus group participant, unidentified, May 21, 2018)*

Culture is the foundation of identity. As is the case among Zuni Pueblo artists, the pieces they fashion depict PlaceKnowing stories. Their pieces become mnemonic objects. Artists who are knowledgeable about them use them to evoke meanings. As one muralist explained:

> Yes, I was just going to say that our Zuni community is very rich in culture. With the [mural] project that's going on with the Zuni Youth Empowerment Program (ZYEP), we're trying to create art that's going to tell the story of our culture. Each mural… tells the story of our migration. [One depicts] the story of how our clan systems came about. [Another] the story about agriculture and just life in general. [This is] how art heals in Zuni community. That's what we think about.
> *(ArtPlace America Pre-Summit, focus group participant, unidentified, May 21, 2018)*

Culture becomes a determinant, especially when it pertains to the youth. Children have a special place in advancing their communities. They are considered the investments in the future and it was generally agreed that their engagement in design and planning sets the stage for community visioning:

> It's the youth that is also pushing for change. [They] are currently in the position of actually changing communities. They have their ideas; they want to be heard. Let's include them as part of that discussion so that at the end, we're better than we were today.
> *(Tribal Planners' Roundtable, focus group participant, unidentified, September 14, 2018)*

Land

> Since I've met [my professor], I've been developing a better appreciation for design... they take into [account] a lot of thought process into their design… As contractors, where do we fit into this? We are not designing the buildings; we're not planning the communities. But we do need to understand where they are coming from. Understanding what their inspirations are, understanding what the values are, and being able to maintain that throughout the construction process; understanding that once we touch the land there are connections to the land we need to have respect for.
> *(ASU Faculty and Student, focus group participant, unidentified, April 10, 2018)*

Land not only provided the sustenance and the resources that their communities needed to survive but it gave them identity. The diverse ecosystems – deserts, coastal areas, mountain forests, wetlands, and plains – are the connections that bond people as one with the land. Land is sacred. If this is not respected, it leads to imbalance and the lack of sustainability. A central conversation among participants dealt with the imbalances created by the exploitation of the land for short-sighted economic development:

> When you insert placemaking, Indigenous placemaking, it really begins introducing a whole new philosophy of how we should govern ourselves. Not in the county's boundaries, but the regional level. And so, when you placemake with an Indigenous philosophy, it always boils down to balance. So to me, placemaking is really critical, and we need to sit down with the traditional ones. They will say at a very fundamental level, [that] the environment, the land, the water, [and] the air cannot sustain the influx of life that does not necessarily belong here.
>
> *(ASU Faculty and Student, focus group participant, unidentified, April 10, 2018)*

Connections with the land also applied to the suitability of materials used building construction. It was indicated that efficiency and energy standards could be exceeded when using traditional methods:

> The idea [is] that the materials are the soul of the building. There are structures that have existed for thousands of years, and there is a reason why they are still standing. There are different elements. Water is the soul, [and it gives] life to every tribe in this state… That is another aspect of placemaking. The materials, how you use them, where you get them, [and] how can you be more efficient with them.
>
> *(ASU Faculty and Student, focus group participant, unidentified, April 10, 2018)*

Indigenous designers moreover indicated that they base their practices on traditional knowledge:

> You sit down with the client to talk about programming and the big picture – up high, 30K feet up. You start to talk about a sense of place and the spirit of place to anchor their identity and history. These are some big, magic words… And [as] I shared about this quote on water from an Edward Abby book, it goes something like this, 'there is no shortage of water in the desert. That there is the perfect ratio of the rocks to rocks and trees to trees.' It goes on to say that there was no shortage of water in the desert because everything is perfect unless you try to establish a city where no city should be. I am still hooked on that. We, the first people, understand this. Our elders understood this notion. This basic principle about the soul of the space [is the] Indigenous way of thinking.
>
> *(ASU Faculty and Student, focus group participant, unidentified, April 10, 2018)*

In summary, PlaceKnowing is comprised of four basic determinates. The first is kinship. Kinship sets the foundation for belonging. It builds an understanding of how individuals grow to assume roles complement the needs of the community. The goal is not merely to fulfill a task, but its role is to build cultural resiliency within the community. Second is fluidity. Fluidity is the concept that culture is fluid and that a person is part of a continuum in time and place. It represents the connectedness to home and the land through the transfer of intergenerational knowledge and exchange. Third is the culture. Culture is the foundation of identity. Elements include history, language, ceremony, and other intergenerational ways of passing understanding from the elders, to the adults, and onto the youth through time and space. The fourth is the land. Land is sacred. Respect for the land builds sustainability. It provides for a community's sustenance and becomes the foundation from which a worldview is borne.

Conclusion

> 'Indigenous ancestor' spirits pray for its peoples to respect all forms of life, family, homes, relationships. Ancestors' spirits maintain faith and assurance for its peoples to thrive because they love unconditionally. Ancestors' spirits offer its peoples strength to carry forward and onward toward brighter and better times to come. Ancestors' spirits secure its peoples' futures for all time.'
>
> *(Shirley, 2020)*

For indigenous communities such as Zuni, a sense of place has evolved over countless generations. Over the millennia, ancestors have given meanings to the places they inhabit. The meaning of place is what they have passed on and each succeeding generation has continued to be influenced by it. For Zuni people, art is regarded as a 'material record of the past' (Pueblo of Zuni, no date). This revelation was employed as a way to reconsider a new approach for how Zuni artists could empower themselves as an artist community and to formulate a vision of how this effort would affect visitors and buyers alike. This became the seed for designing a PlaceKnowing intervention strategy to create a program to stage a 'show-and-tell' experience as a way of educating the public. The *Zuni Pueblo ArtWalk* had the goal of getting visitors to appreciate and respect the work through learning and engagement. In other words, knowing about Zuni art is a requisite to appreciate their culture. However, for all the reasons related to land tenure, prospects of building a more diverse economic presence by occupying or building new businesses on Zuni Pueblo Main Street was not viable. Instead, the Zuni community had to look locally at their assets: home-based art studios which were typical for many artists. Through the exchange, visitors learned how pieces were unique because of the materials and traditional methods employed, but most importantly how each piece carried meanings from the Zuni culture.

As urbanization and modernization continue to create challenges for Indigenous peoples and their communities, PlaceKnowing interventions will become more important. The elders from the Pueblo of Santo Domingo elders said it the best: 'When we no longer walk our land with our children, we lose our culture' (Santo Domingo Comprehensive Plan, 2015).

References

Cornell, M. (2018). 'Biggest fake native American art conspiracy revealed', in *National Geographic*, 15 March 2018 [online]. Available at: https://www.nationalgeographic.com/news/2018/03/native-american-indian-art-fake-forgery-hopi-zuni0/ (Accessed: 23 July 2019).

Dodge, W.A. (2003). *The Meaning of Place at Blackrock: Change and Identity on the Zuni Indian Reservation*. PhD dissertation. University of New Mexico [online]. Available at: http://libproxy.unm.edu/login?url=https://search.proquest.com/docview/305312970?accountid=14613 (Accessed 12 Oct 2019).

Goldman, I. (2003). 'The Zuni of New Mexico', in Mead, M. (ed.) *Cooperation and Competition Among Primitive People*. New Brunswick, NJ: Transaction.

Jojola, T. and Shirley, M. (2017). 'Indigenous planning: Replanting the roots of resistance', in Kenny, S., McGrath, B. and Phillips, R. (eds.) *The Routledge Handbook of Community Development: Perspectives from Around the Globe*. New York: Taylor and Francis.

Markusen, A. and Gadwa Nicodemus, A. (2010). *Creative Placemaking*. Washington, DC: National Endowment for the Arts, p.69 [online]. Available at: https://www.arts.gov/sites/default/files/CreativePlacemaking-Paper.pdf (Accessed 12 Oct 2019).

Pueblo of Zuni. (n.d.) [online]. Available at: http://www.ashiwi.org/index.html (Accessed 14 May 2019).
Seowtewa, K (1992). 'Adding a breath to Zuni life', in *Native Peoples Magazine*, Winter.
Shirley, M. (2020). Personal verse.
UNM Indigenous Design and Planning Institute and UNM School of Law. (2016). *Zuni Main Street Law 584: Economic Development in Indian Country and CRP 598/470: Zuni ArtPlace America*. Albuquerque, NM: UNM Indigenous Design and Planning Institute.

Further reading in this volume

Chapter 4: A future of creative placemaking
Sarah Calderon and Erik Takeshita
Chapter 6: Listen, connect, act
Kim Cook
Preface: 'Disastrous forces, accidental actions, and grassroots responses'
Tom Borrup
Chapter 8: Queer placemaking, settler colonial time, and the desert imaginary in Palm Springs
Xander Lenc
Chapter 10: From moon village to mural village: the consequences of creative placemaking in Ihwa-dong, Seoul
Jason F. Kovacs and Hayun Park
Chapter 12: Public transformation: affect and mobility in rural America
Lyndsey Ogle
Chapter 16: More than a mural: participatory placemaking on Gija Country
Samantha Edwards-Vandenhoek
Chapter 18: '*Homomonument* sounds like a poem': queer placemaking 30 years on: a conversational dialogue with Thijs Bartels, author of *Dancing on the Homomonument* (2003)
Martin Zebracki
Chapter 22: Embedded Artist Project: Epistemic Disobedience + Place
Frances Whitehead
Chapter 25: 'If you can make it there, you can make it anywhere…': cultural placemaking at the heart of cities
Sherry Dobbin
Chapter 40: Transforming community development through arts and culture: a developmental approach to documentation and research
Victor Rubin
Chapter 41: Rituals of regard: on festivals, folks, and findings of social impact
Maribel Alvarez
Chapter 43: A theory of change for creative placemaking: the experience of the National Endowment for the Arts' *Our Town* program: An interview with Patricia Moore Shaffer, PhD
Maria Rosario Jackson
Chapter 45: How the city speaks to us and how we speak back: rewriting the relationship between people and place
Rosanna Vitiello and Marcus Willcocks

32

THE HOLLYWOOD FOREST STORY

Placemaking for the Symbiocene

Cathy Fitzgerald

Introduction

Placemaking can offer engaging, inclusive invitations for diverse communities to live well within their places. In its broadest scope placemaking contributes to 'an ecological turn' toward a life-sustaining era (Fitzgerald, 2018a) recently articulated as the Symbiocene (Albrecht, 2019; Fitzgerald, 2019a, b). Albrecht argues that with a shift to the Symbiocene era, society prioritises functioning symbiotic ecosystems over erroneous economic growth-at-all costs indices that permit the ecocidal atrocities of Anthropocene. In this chapter, I present *The Hollywood Forest Story* and new-to-Ireland continuous cover forestry practices, which I have explored and adopted to transform the small monoculture conifer tree plantation in which I live in rural South East Ireland, to a thriving, mixed species, permanent forest (Woodworth, 2020). Exploring Close-to-Nature continuous cover forestry within an ecosocial practice becomes an act of symbiotic placemaking in creating 2.5 acres of biodiverse landscape that benefits more-than-human flourishing. *The Hollywood Forest Story*, begun in 2008, is a live and ongoing ecosocial art placemaking practice. I introduce the Guattarian ecosophy and action research theory method framework that explains durational ecosocial art practice (Fitzgerald, 2018a, b). Blogging my multi-constituent ecosocial art practice significantly helps my collaborators, my community, and myself develop ecoliteracy to assess this more ecological forestry as a critical alternative to dominant extractive monoculture industrial forestry that is inherently ecocidal in the long term (Fitzgerald, 2018a, pp. 110–14). Therefore, I use extracts from my blog '*The Hollywood Forest Story*' (www.hollywoodforest.com) and the action research part of the ecosocial art practice framework to detail the key stages of symbiotic placemaking. The chapter aims to explain the Earth-aligned values of an ecosocial art practice – for its contribution to symbiotic placemaking – to community development professionals, art educators, creative practitioners, and cultural policymakers.

The marginalisation of ecosocial art practices for placemaking

Spaid (2018) and Kester (2016) note the struggle practitioners (and curators) have in exhibiting and writing about their multi-constituent, long-term endeavours – indicating frameworks for understanding ecosocial art practices are few. Collins and Goto-Collins (2016, p. 88) state that ecological art practice remains little understood and this has significantly hindered development of the emergent art and ecology field, despite the growing awareness of the ecological emergency. My research tries to more simply articulate ecosocial art practices so many more will understand and engage in these critical cultural responses for the Symbiocene (Fitzgerald, 2018a, b, 2019a, b).

Difficulties for ecosocial art practices within placemaking are many. Although ecological principles have been known as the foundation for all life in science since the nineteenth century, they have failed to shift Western culture's historically dominant, yet tragically mistaken, philosophy of human supremacism and perpetual economic growth on a symbiotically sensitive and finite planet. As creative activity in the public sphere is seen as a marginal pursuit, and humanities and art education is historically divorced from the sciences, many art educators and creative practitioners are hindered in appreciating that the ecological emergency is, as systems thinker Gregory Bateson fundamentally understood, 'the result of the difference between how nature works and the way people think' (Borden, 2017, p. 89). Embodying ecological, or perhaps more precisely, symbiotic priorities in cultural activity fundamentally challenge the dominant modern worldview of globalised market economics as the organising basis of modern life. But the challenge is as practical as it is philosophical. Creative practitioners, as much as business leaders and politicians, will not embrace ecological philosophy when they intuit that it curtails current economic freedom and privilege in their work, lifestyle, and cultural pursuits. Many who advance ecosocial art practices at present are self-taught and persevere through conviction and personal resources.

Adopting an 'ecological turn' in the arts sector is difficult. Notwithstanding poor ecoliteracy in the arts, specific challenges for ecosocial art practices exist as they disrupt artistic convention and confront the modern art economy and its related cultural policy. In the 1990s, art critic Suzi Gablik signalled ecological insights as presenting a paradigm shift for humanity (Marriott, 2017) and similarly for the art world as they rebuke modern art's autonomy, its possessive individualism, and its mistaken dualistic depiction of the nonhuman world as separate from human affairs (Gablik, 1991, 1995, 2003). Gablik argued modern artists' works are commoditised for markets that externalise vast environmental degradation and social injustice (Gablik, 2009). In contrast, Gablik characterised unconventional situated ecosocial art practices exhibit an expanded Earth-aligned 'connective aesthetic' (1992a), which help communities address ecological concerns. Gablik's work added to Lucy Lippard's analyses on the steady dematerialisation of the art object that occurred in the rise of creative practices in the late 1960s, pivoting from gallery contexts toward collaborative community engagement on social issues (Chandler and Lippard, 1968; Lippard, 1973). Importantly, Gablik (1991, 1992b) discerned much opportunity to 're-enchant' contemporary art education and practice for a better, more just world because of 'the ecological imperative.'

With sociopolitical detail, Félix Guattari's ecosophy confirms capitalism as a calcified, globalised psychosis. His ecosophy, developed in his last writings in *The Three Ecologies* (1989), *Chaosmosis: An Ethico-Aesthetic Paradigm* (1992a), and his important last article 'Remaking social practice' (1992b), was informed by his life's work on innovative psychoanalytical-art treatment and observations (Philibert, 1997), his extensive socialist and Green political experience, and his keen interest in more Earth-aligned non-Western world views (Melitopoulos and Lazzarato, 2012). In Guattari's view, capitilism's collective psychosis cannot be overthrown by Marxist ideas of mass societal revolution (Elliott, 2012, p. 104). Elliott (2012) argues that Guattari understood that instead of mass revolution, that a 'million minor (molecular) revolutions' – from new social ethico-aesthetic formations – as more likely to resist capitalism's pervasive dominance

(p. 105). Elliott concludes that Guattari believed that 'molecular revolutions' are available to anyone at any time, to undermine the patriarchal, globalised capitalist system (ibid.; Fitzgerald, 2018 a, p. 255). Thus, Guattari's ecosophy presents comprehensive arguments that a radical 'remaking of social practices' is required if we want to advance a just, life-sustaining ecological turn (Fitzgerald, 2018a) in society.

Introducing *The Hollywood Forest Story* (begun April 28, 2008 – ongoing)

> *Hollywood forest blogpost November 26, 2012*: My interest in forests in Ireland only goes back 17 years when I first started to work for crann.ie not long after my move from Aotearoa New Zealand to Ireland in 1995 [the home of my ancestors ...] Initially I was particularly interested and learned a great deal from people involved with the Crann 'Local Project' (a pioneering [community] broadleaved forest establishment project that occurred in Co. Leitrim almost 20 years ago).
>
> *(Fitzgerald, 2006)*

Back then, my friend, Irish-based Australian Jan Alexander (Crann founder and past chair of Pro Silva Ireland) led a workshop for myself, neighbours, and others interested in permanent, continuous cover forestry in the small monoculture conifer tree plantation I and my husband live with. Jan and forester Chris Hayes shared new-to-Ireland Close-to-Nature continuous cover forest management principles and the important observational forestry skill of tree-marking in the plantation-becoming a-woodland, that we call Hollywood Forest. Tree-marking skill enables careful selection of 25–30 per cent of trees to harvest every three to five years to encourage natural regeneration of native tree seedlings in perpetuity, so the forest becomes a permanent ecosystem.

Developing Hollywood Forest through ecosocial art practice within a context of symbiotic placemaking

Wendell Berry's ideas in his essay 'Conserving Forest Communities' (Berry, 1995) of a 'good forest economy', and 'the forest is the basis of a culture', served as an early inspiration for my Crann forest work in Co. Leitrim in 1996 (Fitzgerald, 2020). Years later, from 2008 onwards, echoes of this philosophy assist *The Hollywood Forest Story*, which fosters the symbiotic wellbeing of a small 'forest community.' Additionally, practical advice from Irish and European Close-to-Nature Pro Silva foresters helps my husband and I learn, observe, and practise new-to-Ireland continuous cover forestry, to transform the small 2-acre monoculture conifer plantation destined for clear-felling into a mixed species permanent forest.

> *Hollywood forest blogpost January 8, 2009*: Being in woodland and beginning to understand how to manage it, is all about observation. However, even though I have been over-trained at Art College to develop a visual eye, learning how to see and read a forest is a skill that I have learnt, only by being with foresters. I'm always amazed what foresters see when they enter a wood, they are constantly checking what trees are coming up, looking at the undergrowth to assess the soil type; the other thing that foresters who are into sustainable mixed species forestry, is they are always looking upwards; checking to see if there are light gaps to allow seedlings in the forest floor to come up. This is what we are doing with our thinning – leaving trees to give shelter but creating light gaps for native species to come up, or if we are impatient -more likely, plant in ourselves.

The level of visual observation skills and understanding about forests, so lacking in Ireland and elsewhere, has often struck me over the years. In my previous work I tried to show how local people in Leitrim took to planting native woodlands in my 'local project' film in 2006 – those interviewed revealed how they had become skilled in observing the changes in establishing a woodland; what species grew, what were attacked by pests, how some species thrived in open areas – all very valuable stories, which when brought together showed the beginnings of valuable local knowledge. But there is still a lot to learn and the biggest gap in knowledge that Jan [Alexander] and I observed was how people had little or no knowledge on how to manage forests sustainably into the future, both mixed species and more interestingly, how to manage spruce plantations sustainably, both in terms of ecology and economics.

In 2020, the Hollywood 'forest community' – which I affectionately call 'the little wood that could' – includes all the trees and other inhabitants of Hollywood Forest; but, also the interested friends, neighbours, and others from all walks of life who have stumbled across *The Hollywood Forest Story* blog over the last decade who have been similarly inspired to adopt or advocate ecological forestry. Importantly, my doctoral research on Guattari's ecosophy articulates why my and others' ecosocial art practices routinely evoke ecoliteracy, then agency for safeguarding our environments (Fitzgerald, 2018a). This helps me explain my unexpected political advocacy to advance Close-to-Nature forestry as the key policy objective of the Irish Green Party (Fitzgerald, 2012) that arose from my practice. My practice and the Hollywood Forest community contribute to a national conversation for a radical new direction for Irish forestry. In May 2020, the Irish semi-state forestry organisation Coillte announced, two decades after Pro Silva Ireland formed to promote continuous cover forestry, that nine forests in the Dublin Mountain region are to be managed using continuous cover forestry practices (Coillte, 2020).

A Guattarian ecosophy-action research framework applied within a context of symbiotic placemaking

Guattari's ecosophy can highlight that placemaking is primarily a micropolitical community process that has potential to resist the hegemony of unsustainable capitalism. In Ireland and across much of the world, industrial forestry is a mostly unquestioned, neo-liberal capitalistic operation organised for monetary profit well above material need. For *The Hollywood Forest Story*, Guattari's ecosophy frames the broad socio-political context and overarching aim of ecosocial art social placemaking practices (Fitzgerald, 2018a, b, 2019a, b) toward symbiotic placemaking, toward life-giving symbiotic thriving. Guattari's ecosophy clarifies the potential social power inherent in symbiotic placemaking to evolve valuable localised political agency relevant to specific communities and their bioregions thriving. Specifically, Guattari's ecosophy helps explain why ecosocial art practices are important 'sites of independence, experimental, non-capitalist cultures, and forms of artistic life beyond the dominant structures of economic exploitation, the naturalisation of finance and the hypocrisy of green capitalism' (Demos, 2016b, p. 13). Within the context of a Guattarian ecosophy-action research framework, *The Hollywood Forest Story*, which explores ecological forestry as an alternative to unsustainable clearfell monoculture forestry, bears witness to and directly contests the societal sociopathy of industrial culture (Derber, 2013, cited in Fitzgerald, 2018a, p. 167).

Beyond placemaking: transversal practices for 'worlds yet-to-come'

As complex ecosocial lifeworlds unravel on a rapidly degraded planet caught in the ecocide of growth-at-all costs economics on a finite planet, a deepening understanding of Guattari's transversality – rather than transdisciplinarity – must be a priority in our aims to remake social practices for the Symbiocene. Guattari's transversality explains how individuals–groups collectively observe, learn, and advance ongoing responsible relationship to place, chiefly through a dialogical ethico-aesthetic process of information collation and exchange. Thus, transversality explains how symbiotic placemaking has considerable social power to enact ecological thinking and living.

For *The Hollywood Forest Story*, Guattari's transversality frames my diverse disciplinary activities and lived ecological forestry experience as essential lifeworld knowledge needed to maintain a healthy forest–human community. Transversality confirms the value of my employing various disciplinary knowledge (ecological philosophy and ethics, ecofeminism), social skills of mutuality and social media technologies, my political and artistic activities, as well as my developing ecoforestry practices to foster effective symbiotic placemaking (Fitzgerald, 2018). Moreover, transversality clarifies knowledge creation for symbiotic placemaking is more than an interdisciplinary or even transdisciplinary endeavour. Transversal ecosocial art practices are progressive for symbiotic placemaking when Simon O'Sullivan argues (2006, p. 319), using Deleuze and Guattari's words, that such practices are inevitably involved in 'drawing out the contours for worlds yet-to-come.' However, Guattari's ecosophy, for all its prescient ethico-aesthetic sociopolitical detail does not set out the practical detail of how one develops and maintains transversal practices in contexts other than therapeutic situations. In my doctoral research I successfully argue action research, an overarching, non-prescriptive methodology, significantly articulates the critical method points in Guattarian transversal practices (Fitzgerald, 2018). Action research for symbiotic placemaking is confirmed when leading action researchers successfully use it as a site of extreme ecosocial tension to explain how effective ecological placemaking is initiated, developed, and maintained (Reason and Canney, 2015). Adopting action research's methodological pathway, its critically established terms and concepts, may encourage more practitioners to undertake ecosophical transversal endeavours (Fitzgerald, 2018). This has import for symbiotic placemaking that encompass 'the ecological turn' and the creation of more life-sustaining worlds yet-to-come. Action research works well for organising effective symbiotic placemaking. It clarifies the 'worthwhile purpose,' 'practical challenges,' 'many ways of knowing,' and 'participation and democracy' activities that are commonly involved. Overall, action research gives detailed insight into how one can develop and maintain symbiotic placemaking as a clear cycle-of-action and-reflection to progress ecoliteracy, which in turn, fosters valuable 'communicative' outcomes and agency for community (human and nonhuman) wellbeing (Figure 32.1).

The Hollywood Forest Story – a slow ecosocial art practice as symbiotic placemaking

Critics of ecological art note that some artists rush too quickly to practical measures (Boetzkes, 2013), without first developing a broad eco-philosophical socio-political context, to which Guattari's ecosophy excels, or, underappreciate the value of diverse non-scientific and lifeworld experiential knowings to deepen their responses, which action research methodology encourages.

Action research's 'worthwhile purposes' stage clarifies how symbiotic placemaking is initiated

Characterising *The Hollywood Forest Story*'s 'worthwhile purpose' of exploring alternatives to monoculture, clearfell forestry, identifies how placemaking practices are often initiated by social enquiry into the unsustainability of the 'status quo.' Identifying the history and current state of

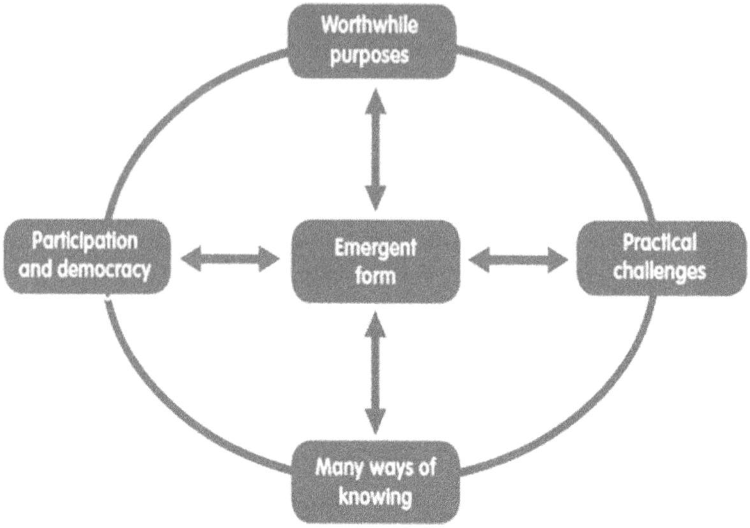

Figure 32.1 The cycle of the five critical dimensions of action research (Reason et al., 2009) can identify the key method stages of symbiotic placemaking.

unsustainable forestry in Ireland, considering why such forestry promotes the global environmental emergency, and why moral reasoning impels me to act for an alternative forestry, clarifies these concerns within symbiotic placemaking practices. Art education researcher Ronald Neperud (1997) characterised that Earth-aligned art practices will be subjective, social, interdisciplinary, and slow – such practices need considerable time for critical reflection and aesthetic translations of place – so as to evolve relevant ecoliteracy, dialogue, and agency with and across a community to adopt new life-sustaining values and practices (Fitzgerald, 2018, p. 56). Many ecosocial art practitioners agree: conducting art-led community practices to understand the ecological emergency, 'translating' relevant scientific analyses through inclusive artistic activity, and encouraging community participation for co-creative restorative responses to place, often requires years.

The 'practical challenges' of symbiotic placemaking

Action research next characterises how the transversal activities in ecosocial art practice readily engage with 'practical challenges,' after establishing their 'worthwhile purpose.' An emphasis on realising 'practical challenges' – the central forest management practices – particularly periodic tree-marking and selective thinning that I employ with forestry professionals to transform our conifer plantation into Hollywood Forest, develops valuable 'real world' knowing of forest restoration and new-to-Ireland continuous cover forestry.

> *Hollywood forest blogpost December 5, 2012*: Our overall and initial aims for our forest are to increase its biodiversity (through natural regeneration) and the resilience of the forest in general. Another side benefit for us, to manage for those objectives, which we achieve by thinning, is that our forest supplies us with a lot of great firewood.
> *Hollywood forest blogpost November 30, 2013*: A big part of transforming and in general terms improving any forest, be it for economic, ecological or aesthetic values, is

marking trees effectively for selective thinning. While some may object to trees being felled, not felling can often have greater detrimental effects to a forest overall.

'Many ways of knowing' as a crucial stage in symbiotic placemaking

As Guattari articulates, transversal practices comprise an ethico-aesthetic sensibility overall. They utilise artistic activity alongside many ways of knowing to promote new values and ways of living so we can live well with our places and others. Here, action research can clarify the importance of inviting 'many ways of knowing' to placemaking. Experiential, artistic, propositional (theoretical), and practical knowing are routinely mobilised in ecosocial art practice to activate practitioners, collaborators, and audiences in a community to reflect on how things are, and how things could be different, for any given place.

> *Hollywood forest blogpost May 1, 2019*: I came to understand that ecosocial art practices, those that are embedded in a particular community over years, foster relevant ecoliteracy for practitioners and their audiences. Ecosocial art practitioners and their audiences become ecoliterate – they quickly understand the connections of what sustains their environments and their lives.

'Participation and democracy' reveals the social skills required for symbiotic placemaking

Guattari's ecosophy identifies the ethical drivers operating within and radiating out of symbiotic placemaking practices. As a psychiatrist, Guattari understood the necessity of ethics to inform all practices involving individual–collective–environmental intervention. Action research usefully clarifies the 'participation and democracy' of all involved in symbiotic placemaking. It usefully alerts practitioners to the value of giving agency (or 'voice') to contributors involved in their practices or situations. Action research underlines this plurality as a methodological priority: notably, in the way we welcome dialogue with non-art collaborators and audience participants, and perhaps with the nonhuman world too. Action research clarifies that I transform Hollywood Forest primarily through maintaining a dialogue with leading continuous cover forestry professionals in Pro Silva Ireland and Europe. I reach out to forest ecologists and forest researchers and sometimes invite them to visit Hollywood Forest. I also attend forest open days and study tours to permanent forest sites across Europe with my Pro Silva forestry colleagues that improve my understanding of how to manage Hollywood Forest in perpetuity. Action research methodology clarifies the value of mutuality (Fitzgerald, 2018a, pp. 295–99) in how we 'learn, and develop genuinely innovative communities of practice' respecting others' lifeworld experience and disciplinary knowledge (Reason, et al., 2009).

> *Hollywood forest blogpost April 1, 2013*: Over the years, I have learned most about Close-to-Nature, continuous cover, permanent forestry management from lessons in the forest itself [... and from Pro Silva Ireland foresters]. Pro Silva Ireland has since 2000 been inviting leading Pro Silva Europe professional foresters to Ireland. They share their knowledge of permanent (non clearfell), continuous cover forestry from their own regions where such forest practices are well understood, and which are often part of their countries established and long-term sustainable forest policy. People coming from many varied perspectives have attended Pro Silva Ireland Open Forest days – Irish foresters, forest owners/managers, harvesters and timber – millers, forest students

and farmers interested in forestry. [...] If you are interested in attending the Kildare Pro Silva Open forest day details are here. The day is pretty informal, coats and boots needed as well as a packed lunch to share in the woods.

Hollywood forest blogpost October 18, 2013: We were very fortunate in the summer to have a visit to Hollywood by a professional ecologist. I have got to know Faith Wilson as she is on the Pro Silva Ireland committee with me [...] Faith also talked about creating 'ecological chimneys.' Areas in the forest to create biological diversity and she stressed favouring 'trees for the future' in this respect. [...] It was a beautiful summers day. I remarked to Faith about the small birds that move in flocks around our forest. Faith said these were coal tits and wrens that in summer fly together for 'family outings' – to teach young birds how to fly. Ecology is Gorgeous!

Understanding the emergent, dialogical form of symbiotic placemaking

Action research highlights how the outcomes of ecosocial art practices will not necessararily be realised in traditional artistic forms. When considered within a context of symbiotic placemaking, scientific recognition and policy can replace paintings and sculptures, as more appropriate restorative and transformative outcomes for the Great Transition. For the Great Transition, *The Hollywood Forest Story*, relates in part with the Indigenous Andean ecocentric philosophy of 'living well,' or *buen vivir*, with the wider community of life. The term characterises the 'fullness of communal life' and implies 'health, education, shelter, food and healthy environment' as rights (Eduardo Gudynas cited in Demos, 2016a, p. 141). *Buen vivir* underpins the paradigm changing and developing international 'Rights of Nature' discourse and developing Earth jurisprudence in the region. Action research clarifies that symbiotic placemaking, as in *The Hollywood Forest Story*, depends on fostering ecoliteracy and inclusive dialogue. Irish forest sector recognition of Hollywood Forest, small as it is, as an accredited example of new-to-Ireland continuous cover forestry, was an important signal of legitimate forest ecoliteracy accumulating around the Hollywood Forest community, which encouraged further dialogue, and then agency to act for forest communities' wellbeing.

Hollywood forest blogpost January 22, 2013: Hollywood forest recognised. Then I got an email last night to say our small forest is to be inspected this Friday! Crikey, I have our forest, all 2.5 acres of it, listed on the new COFORD (Ireland's Forest Research organisation) database of plantations that are being transformed into permanent (non clear-fell), mixed age, mixed species, forests [...] Our wee forest, the smallest listed has been randomly selected and will be visited by some of Ireland's top foresters this Friday – I will report back.

Hollywood forest blogpost January 28, 2013: The inspection was on Friday (25 Jan) [...] when the two foresters, Padriag O'Tuama, Paddy Purser and forest doctoral researcher, Lucy, came to visit [...] I knew the two foresters from the Pro Silva Ireland open forest study days I've attended over the years but it is a different experience when your own forest is been examined by professionals. It was great though, and what a treat to have the areas that I look at so often, looked at by such experienced eyes.

My ongoing cultural activity – giving talks, attending forest open days – helps my own and others' confidence in our developing ecoliteracy for people and forest placemaking.

> *Hollywood forest blogpost March 9, 2015*: I'll be very briefly talking about the 'culture of un-sustainability' that modern industrial society is immersed in. And, how my modest project of transforming a small monoculture plantation into a forest is creating a 'story' that envisions deeper sustainability understandings and agency, to help us move away from the status quo of monoculture, industrial forestry. [... and] about how my own project connects with people from non-art areas: foresters, educators, writers, politicians, and some of the local people in the area.
>
> *Hollywood forest blogpost November 16, 2015*: Great to see ProSilva Ireland's message about Close-to-Nature, non clearfell, continuous cover forestry is getting great press and TV coverage. This is what should have been included in this year's new national forestry policy but we'll just have to keep working at it. Anyway, we all had a great day in Co. Wicklow.
>
> *Hollywood forest blogpost March 27, 2016*: I was surprised and delighted to be asked by Dr. Deirdre O'Mahony to contribute a poster [...] which seeks to commemorate the 100th year of Ireland's Rising and to reflect on developments in farming and country life across Ireland over the last century.
>
> *Hollywood forest blogpost October 13, 2016*: With foresters, ecologists, forest owners, tree-harvesters, forestry students and newcomers, we visited the beautiful broadleaf and conifer-mixed Raheen forest near Scariff in Co. Clare (West Ireland). There are remnants of wonderful ancient Sessile Oak woods there, and they, and much of the conifer areas, are being tended for close-to-nature continuous cover forest (CCF) management (the conifers are selectively thinned to let the native tree species naturally regenerate).
>
> *Hollywood forest blogpost April 7, 2017*: Also, the event in Carlow will be special with the launch of the first full-colour comprehensive and practice guide to Close-to-Nature forest management for Irish forests [...] There will also be discussion of the importance of the freshwater pearl shell mussel in the Blackstairs valley area and how permanent forest management can be a positive management strategy for this endangered Irish species. Please feel welcome to attend if you have an interest in alternative forest management and how forestry is changing in Carlow.

As I was to find and later analyse, as my ecoliteracy about industrial forestry and an ecological alternative grew, likewise my agency to act to safeguard Hollywood forest's future developed: I unexpectedly felt compelled to act for the future wellbeing of my small forest and then for forests everywhere! Guattarian ecosophical understanding explains why such ecosocial practices inevitably become political (Fitzgerald, 2018a).

To shift urgently to the Symbiocene era, Albrecht describes this powerful emotion to protect our places as 'soliphilia' – 'the extent to which one's love of home and kin promotes political action, at all scales, from the local to the global' (Albrecht, 2019, Chapter 4). With others, and the deep forest ecoliteracy afforded by my ecosocial art practice, I contribute to dialogue and progressive national policies for living well with forests and others. This includes prioritising continuous cover forestry as the key aim of Irish Green Party forest policy (Fitzgerald, 2012) and advocacy that the Green Party recognise the crucial yet still developing international law against the crime of manmade ecocide–the destruction of ecosystems (Fitzgerald 2013, 2020b). Due to its inherent unsustainability in the long term I view monoculture, clearfell forestry as ecocide.

> *Hollywood forest blogpost January 20, 2012*: My own personal involvement in policy development was propelled [...] because I have a strong belief that we need to create

> radical new ways of relating to our natural environments, if we and those environments that support us all are to thrive in the long-term [...]

I also wanted to help introduce policy that would finally address the appalling irresponsibility of current Irish policy that ignores the devastation that we inflict on other human and natural communities when we continue to allow the importation of timbers and wood products from countries where unsustainable logging, often still from old growth biodiverse forests, are occurring. For example, short-term returns obscure the fact that clearfell (clearcut) forestry management that relies on serial plantings of monocrops, will lead, in four or five rotations, to severe soil degradation and ever-reducing timber volumes. Such practices also limit and disrupt other species and deaden the social and cultural values of our forests in the meantime.

> *Hollywood forest blogpost February 9, 2013*: I've often noticed the areas in my forest that are regenerating well; 'long live the weeds and the wildness yet' [quote from poet Gerard Manly Hopkins] that is so much a part of a healthy forest.
> *Hollywood forest blogpost April 8, 2013*: In my own writings I point out that ecocide isn't just happening in the Arctic or the Amazon, that the slow violence of ecocide, in our culture and local environments, threads its way through our everyday lives. To me, short rotation monoculture tree plantations are a form of ecocide, leading to eventual soil fertility collapse and limiting severely resilient ecosytems from developing; the very opposite of an ecosystem thriving sustainably in the long term. My work will continue to show alternatives to industrial forestry. Perhaps one day I might even fight for legal standing for the small forest in which I live, a living community that supports me and which I am interdependently connected to.

Conclusion: emphasising the critical outcomes of symbiotic placemaking

Hollywood forest blogpost May 1, 2019: looking back, it has taken much hard work and many conversations over decades to develop the beginnings of sustainable vision for Ireland's forests. Continuous cover forestry is to be welcomed but given how fast the planetary ecological emergency is unfolding, and the short decade deadline climate scientists have announced to change our ways, such integrative forestry practices must be mainstreamed with the utmost urgency. Forestry education will have to recognise the endgame plantation forestry promotes and reinvent itself rapidly. And dialogical creative practices must be recognized for their significant role to reinvent education more broadly, to help communities imagine how living well with forests, lands, rivers, oceans are essential for a sustainable, just, and beautiful world.

At the dawn of a critical decade, a clear theory-method framework as proposed here is vital to make ecosocial art practices for symbiotic placemaking valued and far more accessible. The Guattari ecosophy and action research framework offers comprehensive guidance for effective symbiotic placemaking, that will undoubtedly help us collectively envision and enact the better world we know is yet-to-come.

References

Albrecht, G. (2019). *Earth Emotions: New Words for a New World*. New York: Cornell University Press. Kindle edition.

Berry, W. (1995). 'Conserving forest communities', in *Turning the Crank: Essays by Wendell Berry*. Berkeley, CA: Counterpoint.

Boetzkes, A. (2013). *Should Aesthetics 'Do' Ecology?' Response to 'Ecological Art: What do we do now?'* [online]. Available at: http://nonsite.org/feature/ecological-art-what-do-we-do-now (Accessed: 5 December 2014).

Borden, R.J. (2017). 'Gregory Bateson's search for "patterns which connect" ecology and mind', *Human Ecology Review*, Vol. 23, pp. 87–96.

Chandler, J. and Lippard, L. (1968). The dematerialization of art', *Art International*, Vol. 12, No. 2, Feb. pp. 31–36.

Coillte. (2020). 'Dublin mountains makeover', Coillte [online]. Available at: https://www.coillte.ie/coillte-nature/ourprojects/dublinmountainsmakeover/ (Accessed: 2 June 2020).

Collins, T. and Goto-Collins, R. (2016). 'Defining a practice: With reflection on the sylva Caledonia exhibition', in *Elemental: An Arts and Ecology Reader*. Manchester: Gaia Project Press, pp. 87–105.

Demos, T.J. (2016a). 'Between Rebel creativity and reification: For and against visual activism', *Journal of Visual Culture*, pp. 1–18.

Demos, T.J. (2016b). 'Rights of nature: The art and politics of earth jurisprudence', in *Elemental: An Arts and Ecology Reader*. Manchester: Gaia Project Press, pp. 133–151.

Elliott, P. (2012). *Guattari Reframed*. London: I.B. Tauris.

Fitzgerald, C. (2006). *The Local Project: Revisited 2006* [video]. Available at: https://youtu.be/L3Thj_qeh7M (Accessed: 3 June 2020).

Fitzgerald, C. (2012). *'New Green Party Forest Policy: A Long-Term Sustainable Framework for Ireland's Forests'* [online: press statement: Green Party of Ireland and Northern Ireland website]. Available at: https://www.greenparty.ie/new-green-party-forest-policy-a-long-term-sustainable-framework-for-irelands-forests/ (Accessed: 3 June 2020).

Fitzgerald, C. (2013). 'Greens unanimously adopt motion to end ecocide: A new legal framework to prevent fracking and other pollution' [online: press statement: Green Party of Ireland and Northern Ireland website]. Available at: https://www.greenparty.ie/greens-unanimously-adopt-motion-to-end-ecocide-a-new-legal-framework-to-prevent-fracking-and-other-pollution/ (Accessed: 3 June 2020).

Fitzgerald, C. (2018a). *The Ecological Turn: Living Well with Forests to Articulate Ecosocial Art Practices Using a Guattarian Ecosophy and Action Research Framework*. PhD thesis. National College of Art and Design, Dublin, Ireland [online]. Available at: https://ncad.academia.edu/CathyFitzgerald (Accessed: 27 January 2020).

Fitzgerald, C. (2018b). *The Hollywood Forest Story: Living Well with a Forest to Explain an Ecosocial Art Practice Using a Guattari Ecosophy-Action Research Framework* [online]. Available at: https://hollywoodforest.com/about/the-hollywood-forest-story-ebook-itunes/ (Accessed: 27 January 2020).

Fitzgerald, C. (2019a). 'Goodbye anthropocene—Hello symbiocene: Ecosocial art practice for a new world', in Magdalena Ziolkwska (ed.) *Plasticity of the Planet: On Environmental Challenge for Art and its Institutions*. Ujazdowki Castle Centre for Contemporary Art. Milan: Mousse Publishing.

Fitzgerald, C. (2019b). 'The Hollywood forest story—Ecosocial art practice for the symbiocene', *Minding Nature*, 12(3), pp. 53–58 [online]. Available at: https://www.humansandnature.org/the-hollywood-forest-story (Accessed: 27 January 2020).

Fitzgerald, C. (2020a). *'Reflecting on Wendell Berry's 'the Forest is the Basis of a Culture'* [blog post]. Available at: www.hollywoodforest.com/reflecting-on-wendell-berrys-the-forest-is-the-basis-of-a-culture (Accessed: 3 June 2020).

Fitzgerald, C. (2020b). *'Further steps in Ireland to help end ecocide'* [blog post]. Available at https://hollywoodforest.com/2020/10/09/further-steps-in-ireland-to-help-end-ecocide/ (Accessed: 15 October 2020).

Gablik, S. (1991). *The Re-enchantment of Art*. London: Thames and Hudson.

Gablik, S. (1992a). 'Connective Aesthetics', *American Art*, 6(2), pp. 2–7.

Gablik, S. (1992b). 'The Ecological Imperative', *Art Journal*, 51(2), pp. 49–51.

Gablik, S. (1995). *Conversations Before the End of Time*. London: Thames and Hudson.

Gablik, S. (2003). 'Beyond the disciplines: Art without borders', Position paper from The *MONONGAHELA Conference on Post Industrial Community Development: Art, ecology, and planning with people influencing public places we care about* [online]. Available at: http://moncon.greenmuseum.org/papers/gablik1.html (Accessed: 23 November 2013).

Gablik, S. (2009). 'The art of art: Suzi Gablik discusses art's role and purpose', *Resurgence*, 256, p. 61.

Guattari, F. (1989). *The Three Ecologies*. English translation by Ian Pinder and Paul Sutton (2000). London: Continuum edition 2008; reprint 2010.

Guattari, F. (1992a). *Chaosmosis: An Ethico-Aesthetic Paradigm*. 1995 edn. English translation by Paul Bains and Julian Pefanis. Sydney: Power Publications.

Guattari, F. (1992b). 'Re-making social practices', in Le Monde. Translated by Sophie Thomas (Translation revised by Brian Homes on the basis of the French original) [online]. Available at: http://palimpsestes.fr/ecologie/textes_ecolo/Pour_une_refondation_des_p ratiques_sociales.pdf.) (It first appeared as "Pour une refondation des pratiques sociales" in Le Monde Diplomatique Oct. 1992, 26-27.) (Accessed: 6 November 2013).

Kester, G. (2016). 'Editorial 3 | Winter', in *Field, Journal of Socially-Engaged Art Criticism* [online] Issue 3: (Spring). Available at: http://field- journal.com/editorial/kester-3 (Accessed: 1 April 2016).

Lippard, L. (1973). *Six Years: The Dematerialization of the Art Object from 1966 to 1972*. New York: Praeger.

Marriott, J. (2017). 'Conversations with Suzi Gablik: Living in wartime', in *Platform: Arts, Activism*, Education, Research [online]. Available at: https://platformlondon.org/2017/04/29/conversations-with-suzi-gablik-living-in-wartime/ (Accessed: 27 January 2020).

Melitopoulos, A. and Lazzarato, M. (2012). 'Machinic Animism' (Exhibition essay), *Deleuze Studies*, 6 (2), pp. 240–249 [Online]. Available at: http://www.euppublishing.com/doi/abs/10.3366/dls.2012.0060 (Accessed: 2 June 2020).

Neperud, R.W. (1997). 'Art, ecology, and art education: Practices and linkages', *Art Education*, 50(6), pp. 14–20.

O'Sullivan, S. (2006). 'Pragmatics for the production of subjectivity: Time for probe-heads', *Journal of Cultural Research*, 10(4), pp. 309–322.

Philibert, N. (1997). *Every Little Thing* [documentary film]. Original French title: La Moindre des choses.

Reason, P., and Canney, S. (2015). 'Action research and ecological practice', in Bradbury, H. (ed.) *Sage Handbook of Action Research*. 3rd revised edn. New York: Sage Publications. Kindle edition.

Reason, P., Coleman, G., Ballard, D., Williams, M., Gearty, M., Bon, C., Seeley, C., and Maughan McLaclan, E. (2009). *Insider Voices: Human dimensions of low carbon technology*. Bath, UK: Lowcarbonworks, Centre for Action Research in Professional Practice, University of Bath.

Spaid, S. (2018). *Ecovention Europe, Art to Transform Ecologies, 1957–2017*. Sittard, The Netherlands: Museum De Domijnen Heden daagse Kkunst.

Woodworth, P. (2020). 'Art at the frontline of the environmental crisis: The Hollywood forest in Co Carlow is a source of quality timber and an inspiration for an artistic couple and their network of collaborators' in *The Irish Times*, 14 March 2020 [online]. Available at: https://www.irishtimes.com/life-and-style/travel/ireland/art-at-the-frontline-of-the-environmental-crisis-1.4186881 (Accessed: 1 June 2020).

Further reading in this volume

Chapter 1: Introduction: what really matters – moving placemaking into a new epoch
Cara Courage
Preface: Placemaking in the age of COVID-19 and protest
Jason Schupbach
Chapter 2: Placemaking as an economic engine for all
James F. Lima and Andrew J. Jones
Chapter 3: An annotated history of creative placemaking at the federal level
Jen Hughes
Chapter 4: A future of creative placemaking
Sarah Calderon and Erik Takeshita
Chapter 5: Making places for survival: looking to a creative placemaking past for a guide to the future
Jeremy Liu
Preface: 'Disastrous forces, accidental actions, and grassroots responses'
Tom Borrup
Chapter 12: Public transformation: affect and mobility in rural America
Lyndsey Ogle
Chapter 14: Experts in their own tomorrows: placemaking for participatory climate futures
Paul Graham Raven
Chapter 19: Placemaking in the ecology of the human habitat
Graham Marshall
Chapter 20: Displacemaking 2015 and 2020
Catherine Fennell and Daniel Tucker

Chapter 22: Embedded Artist Project: Epistemic Disobedience + Place
Frances Whitehead
Preface: The only thing constant is change
Kylie Legge
Chapter 43: A theory of change for creative placemaking: the experience of the National Endowment for the Arts' *Our Town* program: an interview with Patricia Moore Shaffer, PhD
Maria Rosario Jackson

33

CONCEPTUALIZING AND RECOGNIZING PLACEMAKING BY NON-HUMAN BEINGS AND LESSONS WE MIGHT LEARN FROM MARX WHILE WALKING WITH BEAVER

Jeff Baldwin

Introduction

This chapter is about how life produces places. While it applies to how humans produce place, the primary point is to clarify that places are also produced by non-human beings; that living beings, all living beings, in their daily negotiation with their wider worlds, produce places. Below, I argue that they can be understood to do so with the intent to increase access to matters they value, i.e. that are useful to them, and to protect themselves from the precarities posed by certain projects of other lives and our wider world (Massey, 2005; Plumwood, 1993). The following discussion elaborates a theory of place, specifically an ontology of place, through a specifically relational lens. The point here is to understand how the more-than-human world produces places that may affect anthropocentric place-producing projects; but more importantly to begin to grasp lessons we can learn from 3.5 billion years of experimentation with producing place more and less successfully by life.

Though place has been variously imagined as a central concept in geographic inquiry (Cresswell, 2014), this study specifically takes up Doreen Massey's formulation of place as the relational product of a dialectic between living beings and the rest of the world. In order to elaborate upon these trajectories, we then turn to Marx's development of the role played by environments in the production of value, here understood as usefulness to various specific living beings. The chapter focuses then upon an example that demonstrates place production by non-human actors working to mutualistically build value which humans also often find useful. Before conclusions asking of potential application to human placemaking, the chapter addresses why this non-human world-building has often remained invisible to contemporary humans, and reviews various attempts to recognize agency among non-human beings.

The production of relational place

Massey argues (2005, p. 74) that geographical studies often use places only as windows into the researchers' main concern, as sites for case studies of widely scaled processes such as neoliberalism, capitalism, and globalization. In these discourses, places are often constructed as disempowered stages upon which wider processes play out with local actors (Massey, 2004, p. 10). Massey offers a rather different way of thinking about place. Working deeply with the spatialization of relationality for two decades, in her book-length address, *For Space* (2005), Massey sets out to conceptualize first space and then place as relational constructs. The latter is our focus here. Massey argues that we may understand space to be constituted by countless trajectories and stories. Trajectories are phenomena that 'may be a living thing, a scientific attitude, a collectivity, a social convention, a geological formation,' while stories are the 'history, change, movement of things [such as trajectories] themselves' (p. 12). Trajectories are ideas, practices, and material processes that can affect people in the conduct of their daily lives, in their quest to live well; they are relationships and processes that affect others and may be authored by individuals, groups, and by non-humans and combinations thereof. It is vital to note that trajectories may operate at any spatial scale. Furthermore, they need not enter place from some global outside; subjects may also author trajectories which may manifest at any scale, as the case example below illustrates.

A focus upon relationships, always through space, offers insights into how we understand place. Massey (2005, Ch. 9) explains that from a relational perspective, places are produced as emplaced actors negotiate relationships with trajectories. This negotiation of relationships with trajectories is the foundation of her relational sense of place. Massey observes that actors have some latitude in their decisions regarding which of the constellation of co-present trajectories they negotiate with, which they may ignore, and which they may try to exclude. Furthermore, emplaced negotiations are not deterministic; actors may be alternatively empowered or disempowered by either prohibiting or allowing articulations with certain trajectories. For each actor, specific trajectories represent value, neutrality, or threat to wellbeing. I want to stress that the term *negotiate* is precisely appropriate here. As Massey (2005, p. 154) explains, and the following case example illustrates, trajectories are neither static nor non-reflexive. Rather, subjects interact with trajectories, and in so doing may modify the trajectories themselves. Those articulations between actors and trajectories are dialectical, they are *inter*relationships, they are negotiations – not navigations.

The production of place by non-humans

Keystone species provide very clear examples of relational place production by non-human beings. In North America, prior to Euro-American exploitation, beaver (*Castor canidensis*) produced certain riparian landscapes from the Sonoran Desert to the arctic. With preference for smaller, low-energy streams flowing through wide valleys or plains, beaver build dams across streams. The resulting places are the result of complex negotiations. The primary driver is to create safety from predation by land mammals. Ponds formed above dams allow beaver to escape underwater to dens built either in mid-pond lodges or bank burrows whose entrances are below the pond surface. Most immediately, beaver create ponds. While the water stored therein is obvious, the effects on adjacent lands are less apparent. In most cases, water from the ponds also charges local aquifers, saturating ground water fields tens of meters to either side of the pond (Westbrook et al., 2006). These saturated aquifers provide water to grasses, shrubs, and trees (Johnston and Naiman, 1990a, b; Pollock et al., 2003; Hood and Bayley, 2007), increasing riparian forest thickness and allowing grasses to grow far into dry seasons. To create dams, beaver cut and assemble woody structures and apply mud and stones from pond bottoms. When optimally situated in low-gradient, broad valleys, beaver dams typically store one- to six-acre feet of water in the pond and a similar amount in groundwater (Westbrook et al., 2006; Müller-Schwarze and Sun, 2003). As a result, beaver dams can diminish high stream flow (floods) through storage (Beedle, 1991) and increase dry-season stream flows for several weeks following spring rains and snowmelt. Further, the water returning to the stream from ground water approximates its temperature at the time of absorption and so can work to lower dry-season stream temperatures, a service vital to salmonids and other cool-water adapted species.

Produced as a negotiation with the trajectories of their predators, beaver ponds are also places which author their own trajectories. Because ponds slow water, fine sediments tend to deposit there. The trajectories produced by that process are multiple. Ponds act as a sort of filter, removing suspended sediment and creating clearer water and less silty stream beds below the dam. The sediments cached behind dams author two notable trajectories as they accumulate and form wetlands and meadows. First, these soils concentrate nutrients which promote plant growth. Through adsorption by fine clay particles, nitrogen, phosphorous, and total carbon may all be significantly (up to four times) higher than in surrounding soils (Naiman and Melillo 1984; Naiman et al., 1986, 1994; Klotz, 1998). Second, wet meadows have been found to sequester up to 12 per cent of all carbon held in some forested areas (Norton et al., 2014). The sequestration of nutrients is not a goal in beaver production of trajectories, yet that sequestration has the same effective agency as an intended trajectory. For those plants able to find use in the places the beaver produce, those places represent valuable trajectories.

As beaver produce place, some of the trajectories they author are intended, and some are unintended. To other plants and animals, intentionality is irrelevant. To non-beavers, the trajectories and places that beavers produce present threats and they present opportunities. To many contemporary humans, beaver works present threats. Amid lands used for logging, beaver will often block culverts (pipes that pass under roads allowing for drainage), using human-made road grades to produce a ready-made dam. The problem for road users occurs during rapid drainage periods which can cause the stream to flow over the road and damage that infrastructure. As a result, in many areas in the American West, beaver are hunted to the point of extirpation (Baldwin, 2017). Beaver trajectories and the places they produce also offer opportunity, value if you will, to many other plants and animals, including humans.

The role of value in the production of place among humans and non-humans

The concept of value provides a frame that allows us to understand place production by non-humans and humans as a unified mode of existence, as the result of exchanges of value. Value, however, is a somewhat fraught concept. Here, extending Marx's work, we take value to refer to that which life finds useful. Presaging Massey's relational ontology, in his development of a *materialist* ontology, Marx argued that, above all, animate and inanimate things exist in material dialectic, partially constituted through their effect on others and the effect of others upon them. Massey (2005) would refer to this as 'negotiating relationships with trajectories authored beyond the self, and authoring trajectories that project beyond the self into the wider world.' Thus, humans are first and foremost understood as sensually acting subjects, that is to say, social agents who are in material interaction (or dialectic) with the objects and subjects of the world, with nature. In the *Manuscripts of 1844*, Marx argues the basis of material dialectic succinctly: no being or object exists of itself, all matters are enmeshed in interrelations, with each affecting (changing) and being affected by others. In *The German Ideology*, Marx explains that our essence lies in material relations with others, in what he calls 'ensembles of social relations' and that social relations are foundationally 'cooperative.'

And here is where we find value. Cooperation may be characterized as interrelationships which produce value for all parties involved. While we may understand exploitation to refer simply to use, in agreement with the Marxist tradition, it may also refer to an interrelationship in which one party takes more value from another than is given (Young, 1990). Among humans, exploitation generally relies upon differentials of power. As exploitative relationships/trajectories/negotiations take more value than they provide, the sites, the places in which exploitation occurs, become impoverished, and living beings unable to prohibit those trajectories from its spaces lose vitality, lose liveliness (Marx, 1973, see p. 489 for discussion). In a sense, this construct of value is central to Marx's theory of our existence, material history.

We can translate Marx's theorized observations of humanity's material historicity ideas into Massey's frame very easily. Marx's most basic formulation of life and value – that we take things from the world and mix our labor with that in order to create more value – is a form of negotiating trajectories. His insistence that we do not live in material autonomy, leads humans to produce value in concert with each other. The use of mutually intelligible words is a fine example of this. We often manage to do this, to create value for ourselves and others at the same time. Marx's historical materiality translates into Massey's conception of making place. By some of our actions/negotiations, we create value, and in so doing create place. We can also translate Marx's observations (1973, p. 489; 1967, p. 290; 1972a, p. 76) about human production of value and place to non-human beings. To be historically correct, the denial of intent among non-human beings was dominant in Marx's time, and it remains so today. That does not mean it is correct, nor is it helpful in understanding the role of value in the production of place by non-human beings. If we look for it, we can find examples of non-human animals and plants producing value through mutually beneficial relationships; indeed, those are the dominant modes of negotiation among most living beings.

Many of the plants and animals that thrive among beaver ponds, wetlands, and meadows are co-adapted to these places that beaver produce. By definition, co-adaptation is mutualistic; each actor finds non-exploitative value in the trajectories and places that others produce. This privileging of mutualism is not native to ecology studies, which for decades focused upon trophic chains – the idea that ecologies can be best understood through entropy, that every time something eats something else, through its metabolic process, it wastes energy. One pound of hawk requires 10 pounds of mice, which require 100 pounds of seed. This ecology was understood through exploitative relations in which one being takes value from the other beyond the value provided. Competition so ruled ecological thinking that the term co-evolution was not coined until 1964 (Ehrlich and Raven). Yet co-evolution, though not as sexy as predation, dominates living relationships.

Through co-evolution, animals have negotiated relationships with carbon dioxide, the unintended trajectories produced by plants; those animal pioneers were able to create value from the by-products of plant life, from their waste. This is one thing life does that non-living processes cannot. Only life values, and only life negotiates, relationships with trajectories. On a co-evolutionary scale dependent on genetic shifts, negotiation occurs over generations and across geographically distinct populations of a specific species. On a co-adaptive scale, changes may occur more quickly, through changes in behaviors.

The danger of not seeing this, and why we did not see this

By dismissing the ability of non-human beings to produce place, we risk a great deal. Over the past 50 years, various American groups have spent hundreds of millions of dollars in efforts to restore streams. However, because we have not recognized the role of non-humans in placemaking, we have generally misunderstood what many American streams looked like and how they functioned. As Fouty (2003) observes, the US conducted its survey of the American West decades after Euro-Americans trapped beaver to near extinction in most watersheds. By the time streams were surveyed, they had already incised into the rich, formerly moist alluvial valleys produced by beaver. Because engineers did not consider the possibility that non-human beings could produce entire landscapes and accompanying ecosystems, they took the state of those beaver-free streams as normal. The models for stream restoration have been built upon those profoundly disturbed, beaver-extirpated landscapes. Thus, the placemaking of most stream restoration efforts is seriously flawed because placemaking by non-human beings has been dismissed (Butler, 1995).

There are reasons we do not see the agency of non-human beings. In the Western tradition, and in many other long-lived cultures, we have conceived of agency as an exclusively human quality. This is linked to a series of qualifiers that we expect agents to meet. First, agency has been located specifically in particularly qualified, until recently adult male, and very specific, human bodies (Hinchliffe, 2007; Braun, 2008a). Often agency is constructed as contingent upon an assumed exclusively human intentionality and/or language use (Callon and Law, 1995). So-called non-representational conceptualizations of agency argue that human perception of another is a prerequisite to agency, thus excluding agency among non-humans and granting human agency over supposedly object-like non-human beings (see Braun, 2008, p. 673). These very anthropocentric constructions of agency are reinforced by Western constructions of humans as essentially different from nature.

Several historians of Western thought argue that Modernity is marked by exclusionary categorizations which reserve agency for humans and humans alone. The Enlightenment (Early Modernity), is in part differentiated from medieval ways of understanding the world by a shift away from searching for likeness (or similitude) to looking for and finding difference (Gurevich, 1985). Indeed, Foucault (1970) and Reiss (1982) both argue that this perception of difference between self and Other is a product of the Modern *episteme*, 'a certain structure of thought [a way of ordering] that the men of a certain period cannot escape' (Foucault, 1972, p. 191). With Early Modernity, abstraction became endemic, formalized by Descartes (1641), who imagined the mind as inhabiting a world separate from the body and worldly objects. In the early eighteenth century, Leibnitz (1989), as a proto Liberalist, argued that we are monads, self-contained beings, essentially separate from our worlds and each other. In his critical history of Western thought, Marx argued that abstraction led not to a true understanding of the world, but to an estrangement of the self from the world. Marx wrote that:

> It is precisely abstract thought from which these objects [i.e. concepts such as wealth, religion, state power] are estranged and which they confront with their arrogation [i.e. false claim to] of reality... The whole history of the alienation-process ... is nothing but the *history of the production* of abstract [ie., absolute] thought—of logical, speculative thought.
>
> *(1972a, p. 110)*

But Marx lost that argument. Instead, binaristic categorization prevailed. In that frame, one categorizes the matters that are not you, are not members of your subset, as not only different from you, but as lacking the aspects that define one's self. If 'we' are rational, intentional, conscious, those who are not us are incapable of reason, intent, and consciousness. This framing is deeply woven into how we understand ourselves and our worlds. Even when contemporary theorists attempt to extend agency and intention to non-human beings, anthropocentrism persists.

Among post-humanists, for example, human subjectivity is taken to arise from a self-world dialectic (Badmington, 2000); however, conceptions of intent tend to still place human awareness at the center of networks constituted by humans and non-humans. This persistent anthropocentrism is evident in Latour's (2004) representative example of non-human agency. There Latour recounts how snail darters, a small fish native to the Little Tennessee River, stopped a major dam project in 1973. However, as Latour explains, it was the new consciousness of the threatened species among human anti-dam activists who employed the Endangered Species Act to stop the dam project (see also, Lorimer, 2006, and Braun, 2008, p. 673). What Latour misses is that the snail darters in fact *did nothing* to stop the dam, other than exist. Cognitive scientists Skarda (1999) and Järvilhto (1998a, b) argue that perceptions of self as separate from the world are an artifact of neural function. From an anthropological perspective Geertz argues that self–world separation is a Western construct and is a 'rather peculiar idea within the context of the world's cultures' (1975, p. 480). More recent research suggests that a uniquely human brain structure, our default mode network, enables perceptions of ourselves as essentially differentiated from our worlds (Pollan, 2019).

We still struggle to see beyond these barriers. Yet, as always already emplaced beings, beaver author their own trajectories and negotiate original relations with the trajectories co-present in their place. Beaver, and potatoes, and assassin bugs do not require human acknowledgment of their agency; they will produce places with or without the awareness of humans or non-humans (Mitchell and Cunjak, 2007). As Callon and Law conclude, 'these Others will ignore us for most of the time. Instead, they will continue, as they always have, to perform their specific forms of agency to one another' (1995, p. 504).

Trying to see placemaking among non-human beings

Scholars studying human–environment relations have sought to transcend this limited conception of non-human beings. However, as environmental historian Radkau (2008) observes, thus far historians have worked under the premise that humans are unique as historical actors. As a result, historians only attend to nature when it plays a part in human histories. Beaver are important in North American histories for the role their pelts played in the exploration and settlement of the continent and for the circuits of wealth those hides supported in Western Europe and Canton. Radkau asserts, however, that 'nature has a life of its own and is by no means only a component of human action, or the topic of human discourse' (ibid, p. 4). Radkau seeks to understand non-human beings as in a mutually constitutive dialectic with humans which is historic in character. However, as a discipline, environmental history has not embraced that assertion.

Geographers also wrestle with non-human participation in placemaking. Inspired by Haraway's work on hybridity theory, Whatmore constructs agency as occurring through potentially far-flung hybrid culture natures in which non-human beings are also agents in the networks they cooperate (2002, pp. 14–15). Hinchliffe is sympathetic with the decentering of agency central to hybridity theory but warns that it 'risks a new division between lived experience and non-living matter' (2007, p. 58), a division he very much opposes. Along these lines both Hinchliffe and Braun refer to work by Deleuze and Guattari (1987) on agency. Braun (2008, p. 671) highlights the authors' notion of *agencement*, which synthesizes agency – the ability to produce an effect – with the configuration (drawing together) of disparate living and non-living elements – configuration being prerequisite for agency. Referring to Law's discussion in *After Method* (2004, p. 41), Hinchliffe explains that through *agencement* (commonly mistranslated as *assemblage*), agency is conceived as 'a suite of stories, practices, technologies, animals and people ... an active combination of technologies, ways of proceeding, their arrangements and their ongoing, unfolding nature' (2007, p. 38). Agency in this sense is still/always in the act of unfolding, becoming. Out of his concern to avoid reimposing the binary categories of agents and non-agents, Hinchliffe argues that:

> we need to unsettle the central role given to humans, and the few carefully chosen species who can also experience, in bestowing meaning on mute objects ... We need other ways of understanding and working with animal spaces, technical spaces and many others besides, that do not return us to Kantian subjects and to such stark differences in kind.
>
> *(2007, pp. 58–59)*

Hinchliffe asserts that differences in agency among humans, non-human beings, and non-living things are of *degree*, rather than of *kind* (ibid, p. 61).

Here I disagree. In terms of agency, living beings are essentially different from non-living matters, not in the sense of agency being manifest in relationship, but in the role they play in producing relationships. After Plumwood (1993), I have argued that unlike non-living matters, living beings work materially to extend themselves and their populations across time and space (Baldwin, 2013). Through those projects, all beings produce relationships imbued with exchanges of value which may be exploitative (as with invasive populations), facilitative, commensal, or mutualistic (Baldwin, 2016.) Unlike a leaf of paper or a computer virus, living beings seek to arrange self-benefitting relationships in configuration with living and non-living others; they organize themselves in space, and thus produce place. In this relational sense, if bereft of living beings, volcanoes, glaciers, and leaves of paper do not produce place.

This assertion is supported within the Academy beyond relational feminists. Foucault himself wrote that 'the living being wraps itself in its own existence ... and constitutes itself as a new space' (1970a, p. 278). The space life creates of itself is an 'interior one of anatomical coherences and physiological compatibilities, and the exterior one of the elements in which it resides and of which it forms its own body' (ibid, p. 274). As Serres (1982) suggests, all things are open to environmental flows; however, living things uniquely reach into surrounding spaces to organize the internalization of the useful and the prohibition of the dangerous. Even Hinchliffe notes that organisms are uniquely 'self-referential, they are autopoietic, self-organizing' (2007, p. 63). Non-living things do not do this. They may affect placemaking, but places are produced by living beings as they negotiate relationships with the trajectories and projects of other living beings and with environmental processes and conditions.

The value of seeing this

This discussion makes four central points. First, place may be understood to be the product of living beings negotiating relationships with living and non-living processes. All life negotiates relationships with projects authored by others, while authoring their own trajectories which work as extensions of their selves into space. Second, all life participates in this production of places. Third, as living beings adapt to/negotiate with the trajectories of others, co-adaptation is among the grand gestures of life on Earth. Fourth, the result of place production by individuals, species populations, and diverse communities is to increase the usefulness, the value, resident in these co-produced spaces. Finally, the intentionality of trajectories has little meaning beyond the realm of the authoring being.

Some of our imaginings of how the world works have not served us well. The Darwinian concept of scarcity as an essential driver of life has shaped every discipline from economics to politics to biology and philosophy. In addition, our denial of agency to non-human beings has long kept us from asking the sort of questions that would have earlier led to a more complete understanding of life on earth. As I show above, behavioral and genetic co-adaptation among intentional and even mindful living beings has produced much of the value life has made of and invested into our biosphere. As we reimagine place, we may also consider ourselves as spatial and relational beings who may work either more exploitatively or more mutualistically to either diminish or increase the value available to our communities. Examples of more exploitative modes are rife, illustrated by every relationship and institution that is biospherically unsustainable; the culture, economy, policy, and nature of automobiles provide a clear case study. We can also find examples of mutualistic place production. Alternative agriculturalists work to find species they can partner with to increase the value of both the food and their lands/places by enlisting an ever-greater variety of lives. Inviting owls and hawks to live nearby who control, but do not eradicate, burrowing mammals and potentially problematic birds which play their own roles when their numbers are not too great. Investing in soil communities increases moisture and nutrient retention and sequesters carbon. Animal waste digesters turn problematic poop into methane gas for cook stoves and excellent, odor and disease-free fertilizer. Urban and industrial spaces present a greater challenge. Lewallen et al. (2017) detail how several city governments in the United States are now experimenting with allowing beaver to live within their places. So-called urban forestry has so far encouraged trees for the ecosystem services they provide. But is there a way to actually cultivate forests, and not only trees in urban spaces. In several large cities, residents are experimenting with rooftop gardens, bringing agriculture into cities while also cooling roofs and surrounding spaces. On a larger scale, cities such as Copenhagen have long worked to find value in waste, using water super-heated by power plants to then run throughout the city, providing heat through residential hot-water heaters. The goal is to find value in waste.

The lesson here will be difficult to translate into our profit-oriented systems. As Marx observed so long ago, exploitation – taking more value than is given – works against the production of mutualistic, plus-sum relationships, against the sustainable making of place. Yet, exploitation is endemic to most for-profit models. Over the past three billion years, living communities worked out the rules for sustainable placemaking. The question now is whether we can learn from our biospheric place-mates … or do we continue to negotiate relationships that benefit ourselves while impoverishing our wider living world?

Bibliography

Agnew, J. (1987). *Place and Politics*. Abingdon: Routledge.

Apple, L.L., Smith, B.H., Dunder, J.D. and Baker, B.W. (1985). 'The use of beavers for riparian/aquatic habitat restoration of cold desert, gully-cut stream systems in southwestern Wyoming', *Investigations on beavers*, 4, pp. 123–130.

Badmington, N. (ed.) (2000). *Posthumanism*. London: Macmillan International Higher Education.

Baker, B.W. and Hill E.P. (2003). 'Beaver (*Castor canadensis*)', in Feldhamer, G.A., Thompson, B.A. and Chapman, J.A. (eds.) *Wild mammals of North America: Biology, management, and conservation*. 2nd edn. Baltimore: The Johns Hopkins University Press, pp. 288–310.

Baldwin, J. (2013). 'What ought I eat? Towards a biospheric political economy', *Environmental Ethics*, 35(3), pp. 333–347.

Baldwin, J. (2015). 'Potential mitigation of and adaptation to climate driven changes in California's highlands through increased beaver (*Castor canadensis*) populations', *California Fish and Game*, 101(4), pp. 218–240.

Baldwin, J. (2016). 'Life, labor, and value: Recreating affective food ecologies through interspecies cooperation', *Visions for Sustainability*, 2016(6), pp. 6–22.

Baldwin, J. (2017). 'Institutional obstacles to beaver recolonization and potential climate change adaptation in Oregon, USA', *Yearbook of the Association of Pacific Coast Geographers*, 79, pp. 93–114.

Beedle, D. (1991). *Physical Dimensions and Hydrologic Effects of Beaver Ponds on Kuiu Island in Southeast Alaska*. M.S. thesis. Oregon State University, Corvallis, USA.
Benton, T. (2001). 'Marx, Malthus and the greens: A reply to Paul Burkett', *Historical Materialism*, 8(1), pp. 309–331.
Braun B. (2008). 'Environmental issues: Inventive life', *Progress in Human geography*, 32, pp. 667–676.
Burkett, P. and Foster, J.B. (2006). 'Metabolism, energy, and entropy in Marx's critique of political economy: Beyond the Podlinsky myth', *Theory and Society*, 35(1), pp. 109–156.
Burnett, K.M., Reeves, G.H., Miller, D.J., Clarke, S, Vance-Borland, K. and Christiansen, K. (2007). 'Distribution of salmon-habitat potential relative to landscape characteristics and implications for conservation', *Ecological Applications*, 17(1), pp. 66–80.
Butler, D. (1995). *Zoogeomorphology: Animals as Geomorphic Agents*. Cambridge, UK: Cambridge University Press.
Buttimer, A. (1980). 'Home, reach, and the sense of place', *The Human Experience of Space and Place*. New York: St. Martin's Press, pp. 166–187.
Callon, M. and Law, J. (1995). 'Agency and the hybrid collectif', *South Atlantic Quarterly*, 94, pp. 481–507.
Castells, M. (1997). *The Power of Identity*. Oxford: Blackwell.
Cresswell, T. (2004). 'Defining place', in *Place: A Short Introduction*. Malden, MA: Blackwell Ltd.
Cresswell, T. (2014). *Place: An Introduction*. Chichester: Wiley.
Deleuze, G. and Guattari, F. (1987). A Thousand Plateaus (trans. Brian Massumi). Minneapolis: University of Minnesota Press.
Descartes, R. (1993/1641). 'Meditations on first philosophy', *The Philosophical Writings of Descartes, 2*, pp. 3–62.
Donkor, N.T. and Fryxell, J.M. (2000). 'Lowland boreal forests characterization in Algonquin Provincial Park relative to beaver (*Castor canadensis*) foraging and edaphic factors', *Plant Ecology*, 148, pp. 1–12.
Ehrlich, P.R. and Raven, P.H. (1964). 'Butterflies and plants: A study in coevolution', *Evolution*, 18(4), pp. 586–608.
Foucault, M. (1970). *The Order of Things: An Archaeology of the Human Sciences*. New York: Random House.
Foucault, M. (1972). *The Archaeology of Knowledge*. London: Tavistock.
Fouty, S.C. (2003). *Current and Historic Stream Channel Response to Changes in Cattle and Elk Grazing Pressure and Beaver Activity*. Ph.D. dissertation. University of Oregon.
Fracchia, J. (1999). 'Dialectical itineraries', *History and Theory*, 38(2), pp. 169–197.
Fukuyama, F. (1989). 'The end of history', *The National Interest*, pp. 3–18.
Geertz, C. (1975). 'Common sense as a cultural system', *The Antioch Review*, 33(1), pp. 5–26.
Gurevich, A.I. (1985/1972). *Categories of Medieval Culture* (trans., G. L. Campbell). Abingdon: Routledge and Kegan Paul.
Hinchliffe, S. (2007). *Geographies of Nature: Societies, Environments, Ecologies*. London: Sage.
Hood, G. and Bayley, S. (2007). 'Beaver (*Castor canadensis*) mitigate the effects of climate on the area of open water in boreal wetlands in Western Canada', *Biological Conservation*, 141(2), pp. 556–567.
Ives, R.I. (1942). 'The beaver-meadow complex', *Journal of Geomorphology*, 5, pp. 191–203.
Järvilhto, T. (1998a). 'The theory of the organism-environment system I: Description of the theory', *Integrative Psychological and Behavioral Science*, 33(4), pp. 321–334.
Järvilhto, T. (1998b). 'The theory of the organism-environment system II: Significance of nervous activity in the organism-environment system', *Integrative Psychological and Behavioral Science*, 33 (4), pp, 335–342.
Johnston, C.A. and Naiman, R.J. (1990b). 'Aquatic patch creation in relation to beaver population trends', *Ecology*, 71(4), pp. 1617–1621.
Johnston, C.A. and Naiman, R.J. (1990a). 'Browse selection by beaver: Effects on riparian forest composition', *Canadian Journal of Forest Research*, 20, pp. 1036–1043.
Klotz, R.L. (1998). 'Influence of beaver ponds on the phosphorus concentration of stream water', *Canadian Journal of Fisheries and Aquatic Sciences*, 55, pp. 1228–1235.
Latour, B. (2004). 'How to talk about the body? The normative dimension of science studies', *Body and Society*, 10, pp. 205–229.
Law, J. (2004). *After Method: Mess in Social Science Research*. Abingdon: Routledge.
von Leibniz, G.W.F. and Gottfried W. and Leibniz, F. (1989). *Philosophical Essays*. Indianapolis: Hackett Publishing.
Lewallen, G., Perryman, H., Scully-Engelmeyer, K. and Wohl, E. (2017). 'Urban beaver population management', in Pollock, M.M., Lewallen, G.M., Woodruff, K., Jordan, C.E. and Castro, J.M. (eds.) *The Beaver*

Restoration Guidebook: Working with Beaver to Restore Streams, Wetlands, and Floodplains, Version 2.0. Bailey's Crossroads, VA: United States Fish and Wildlife Service, pp. 100–114.

Lombard, M. (2014). 'Constructing ordinary places: Placemaking in urban informal settlements in Mexico', *Progress in Planning*, 94, pp.1–53.

Lorimer, J. (2006). 'What about the nematodes? Taxonomic partialities in the scope of UK biodiversity conservation', *Social and Cultural Geography*, 7, pp. 539–558.

Marx, K. (1967). *Capital: A Critique of Political Economy* (trans. Moore, S. and Aveling, E). New York: International Publishers, Vol. 1.

Marx, K. (1972a). 'Economic and philosophic manuscripts of 1844', in Tucker, R. (ed.) *Marx-Engels Reader*. London: W. W. Norton & Company.

Marx, K. (1972b). 'The German ideology', in Tucker, R. (ed.) *Marx-Engels Reader*. London: W. W. Norton & Company.

Marx, K. (1973). *Grundrisse: Foundations of the Critique of Political Economy* (trans. Nicolaus, M). London: Penguin.

Massey, D. (1993). 'Power geometry and a progressive sense of place', in Bird, J., Curtis, B. Putnam, T., Robertson, G. and Tickner, L. (eds.) *Mapping the Futures: Local Cultures, Global Change*. Abingdon: Routledge, pp. 59–69.

Massey, D. (2004). 'Geographies of responsibility', *Geografiska Annaler B*, 86, pp. 5–18.

Massey, D. (2005). *For Space*. Thousand Oaks: Sage.

Mitchell, S. and Cunjak, R. (2007). 'Stream flow, salmon and beaver dams: Roles in the structuring of stream fish communities within an anadromous salmon dominated stream', *Journal of Animal Ecology*, 77, pp. 1062–1074.

Müller-Schwarze, D. and Sun, L. (2003). *The Beaver: Natural History of a Wetlands Engineer*. Sacramento: Comstock.

Naiman, R.J., and Melillo, J.M. (1984). 'Nitrogen budget a subarctic stream altered by beaver', *Oecologia*, 62, pp. 150–155.

Naiman, R.J., Melillo, J.M., and Hobbie, E.J. (1986). 'Ecosystem alteration of boreal forest streams by beaver (*Castor canadensis*)', *Ecology*, 67, pp.1254–1269.

Naiman, R.J., Pinay, G., Johnson, C.A., and Pastor, J. (1994). 'Beaver influences on the long-term biogeochemical characteristics of boreal forest drainage networks', *Ecology*, 75, pp. 905–921.

Norton, J.B., Olsen, H.R., Jungst L.J., Legg, D.E., and Horwath, W.R. (2014). 'Soil carbon and nitrogen storage in alluvial wet meadows of the southern Sierra Nevada Mountains, USA', *Journal of Soils and Sediments*, 14, pp. 34–43.

ODFW (Oregon Department of Fish and Wildlife). (2006). *Oregon Conservation Strategy*. Oregon: ODFW.

Plumwood V. (1993). *Feminism and the Mastery of Nature*. Abingdon: Routledge.

Pollan, M. (2019). *How to Change Your Mind: What the New Science of Psychedelics Teaches Us about Consciousness, Dying, Addiction, Depression, and Transcendence*. London: Penguin Books.

Pollock, M., Heim, M. and Werner, D. (2003). 'Hydrologic and geomorphic effects of beaver dams and their influence on fishes', *American Fisheries Society Symposium*, 37, pp. 1–21.

Pollock, M., Pess, G., Beechie, T., and Montgomery, D. (2004). 'The importance of beaver ponds to Coho salmon production in the Stillaguamish River basin, Washington, USA', *North American Journal of Fisheries Management*, 24, pp. 749–760.

Radkau, J. (2008/2002). *Nature and power: A Global History of the Environment* (trans. Dunlap, T.) Cambridge, UK: Cambridge University Press.

Reiss, T. (1982). *The Discourse of Modernity*. New York: Cornell University Press.

Relph, E. (1976). *Place and Placelessness*. London: Pion.

Ryan, A. (1992). 'After the end of history', *History Today*, 42(10), pp. 8–10.

Sablon, L., Dickens, J., Haubruge, É. and Verheggen, F. (2013). 'Chemical ecology of the Colorado potato beetle, Leptinotarsa decemlineata (Say) (Coleoptera: Chrysomelidae), and potential for alternative control methods', *Insects*, 4(1), pp. 31–54.

Scott, J. (1985). *Weapons of the Weak: Everyday Forms of Peasant Resistance*. New Haven: Yale University Press.

Serres, M. (1982). *Hermes: Literature, Science, Philosophy*. Baltimore: Johns Hopkins University Press.

Skarda, C. (1999). 'The perceptual form of life', in Núñez, R. and Freeman, W.J. (eds.) *Reclaiming Cognition: The Primacy of Action, Intention and Emotion*. Exeter: Imprint Academic, pp. 79–93.

Smith, J.D. (2007). 'Beaver, willow shrubs, and floods', in Johnson, E.A. and Miyanishi, K. (eds.) *Plant Disturbance Ecology*. Cambridge, MA.: Academic Press, pp. 603–671.

Trimble, T. and Albert, S. (2000). 'Beavers are partners in riparian restoration on the Zuni Indian Reservation', *Ecological Restoration*, 18, pp. 87–92.
Tuan, Y.F. (1974). *Topophilia: A Study of Environmental Perception, Attitudes and Values*. Saddle River: Prentice Hall.
Westbrook, C.J., Cooper, D.J., and Baker, B.W. (2006). 'Beaver dams and overbank floods influence ground-water–surface water interactions of a Rocky Mountain riparian area', *Water Resource Research*, 42, pp. 1029–1042.
Whatmore, S. (2002). *Hybrid Geographies: Natures, Cultures, Spaces*. Thousand Oaks: Sage.
Withers, C.W.J. (2009). '"Place" and the "spatial turn" in geography and in history', *Journal of the History of Ideas*, 70(4), pp. 637–658.
Young, I.M. (1990). 'The ideal of community and the politics of difference', in L. Nicholson (ed.) *Feminism/Postmodernism*. New York: Routledge, pp. 300–323.

Further reading in this volume

Chapter 3: An annotated history of creative placemaking at the federal level
Jen Hughes
Chapter 5: Making places for survival: looking to a creative placemaking past for a guide to the future
Jeremy Liu
Preface: The problem with placemaking
Louise Platt
Chapter 14: Experts in their own tomorrows: placemaking for participatory climate futures
Paul Graham Raven
Chapter 18: '*Homomonument* sounds like a poem': queer placemaking 30 years on: a conversational dialogue with Thijs Bartels, author of *Dancing on the Homomonument* (2003)
Martin Zebracki
Chapter 19: Placemaking in the ecology of the human habitat
Graham Marshall
Chapter 20: Displacemaking 2015 and 2020
Catherine Fennell and Daniel Tucker
Chapter 22: Embedded Artist Project: Epistemic Disobedience + Place
Frances Whitehead
Chapter 34: Reconnecting *cité* and *ville*
Philip Graus
Chapter 39: Translating Outcomes: Laying the groundwork for interdisciplinary evaluation of creative placemaking
Jamie Hand
Chapter 41: Rituals of regard: on festivals, folks, and findings of social impact
Maribel Alvarez
Chapter 45: How the city speaks to us and how we speak back: rewriting the relationship between people and place
Rosanna Vitiello and Marcus Willcocks

SECTION 6

Placemaking, urban design, and planning

Section Editor: Kylie Legge

PREFACE

The only thing constant is change

Kylie Legge

I have never met anyone involved in city-making who does not want to make good cities that support happy and healthy citizens. Of course, we all do this through a different lens. The planner focusses on systems of land use and connectivity, the transport engineer believes that making it fast and easy to get around makes a good city; the architect via beautiful buildings, the social planner through equitable access to services and amenity, the landscape architect will use nature as their tool. Our professional expertise is our strength, but it can also be our limitation – if all we have is a hammer, then all we see are nails (Maslow, 1966). In practice, the relationship between placemaking, urban design, and planning is fraught with embedded conflict and tensions that unwittingly create division between our shared purpose and the delivered outcomes in place. It is often the differences in professional priorities and in the range of scales involved in place planning that create the most significant barriers to successful placemaking.

I have often described placemakers as 'expert generalists' or 'discipline agnostics' because place is not transport, nor architecture, nor landscape, art, or events. It is a complex system of relationships between these different aspects of a location and the people it serves and who hold meaningful connections with it. Successful placemakers see past their own discipline, the current rules and incumbent systems. They are collaborators, open and free thinkers who are not bound by the confines of 'this is the way it has always been done.' Fortunately, over the last 50 years, and particularly the last 10, we have witnessed the exponential rise of placemaking as a core tenet of both urban design and planning. This has resulted in the increased value of place across multiple disciplines and a simultaneous increase in openness to innovation in policy, process, and projects. This change is welcome, but more change is still needed if we are to achieve our purpose of cities full of great places for people.

This section shares a spectrum illustrating the challenges across placemaking, urban design, and planning, and the change still required. From the city scale to the intimacy of the public seat, each chapter shares a common theme – the need for the continued evolution of the planning and urban design professions. Each author has identified challenges, biases, or barriers to the successful delivery of placemaking values and objectives in the incumbent planning and urban design systems. And each provides insights, inspiration, and opportunities for the further development of the professions and professionals involved in place-based city making.

In the first chapter, 'Reconnecting *cité* and *ville*,' Phillip Graus explores the disconnection between strategic planning and the delivery of poor physical experiences in place. His premise, that the fine grain urban fabric of local places is being overwhelmed by the large-scale strategy and infrastructure projects of higher levels of government, is illustrated using case studies from Tokyo, Japan and Sydney, Australia. The chapter goes on to identify an interdependent, but three-part proposition, to break down this dichotomy through iterative process, cross-government collaboration, and a re-evaluation of urban economic models.

Nigel Smith's chapter 'Planning governance – lessons for the integration of placemaking,' interrogates Australian state government planning policy case studies to determine their impact on placemaking outcomes. He determines that while placemaking as an objective, and the use of place-associated language, has been incorporated into various planning documents, there is still need for further improvement if effective practice is to be achieved.

In 'Facilitator skills for effective collaborative placemaking,' Husam AlWaer and Ian Cooper step down a level from policy to practice, revealing the implications of the Scottish government's policy document 'Shaping better places together: Research into the facilitation of participatory placemaking' on facilitators and facilitation processes. The findings of their multi-year research project provide valuable insights into the evolution of the required skills of facilitators in place-based and design-led activity, as well as direction on the ideal process development and delivery.

The authors of '*The Neighbourhood Project*: a case study on community-led placemaking by CoDesign Studio' provide a detailed review of their approach to on-the-ground delivery of placemaking projects. It illustrates clearly the inherent challenges and embedded systematic biases against delivering community-led activity. However, the examination and review of these small-scale case studies also indicate opportunities for the evolution of government policy and process that is simplified, more open, and flexible.

Lastly, this Section Editor shares her passion for the smallest place – the public seat. 'Public seating – a small but important place in the city' examines the relationship between cultural bias, design theory, and hierarchical tensions between city-making professionals, and the lack of quality, comfortable, and inviting public seating in urban environments. It too provides direction for opportunities for professional and systematic improvement and invites designers and planners to consider the psychological comfort, physical comfort, and pleasure of the people our places serve, ahead of other considerations.

If change is the only constant, these chapters reveal that many of our current urban design and planning systems are still relatively stuck. In the past, professional evolution has been led by the 'master' planner; the Ebenezer Howards, Le Corbusiers, and Frank Lloyd Wrights (Badger, 2012) – top-down, design-led, righteous, and often lacking local contextual sensitivity or responsiveness. Perhaps placemaking, with its localised, bottom-up, cross-disciplinary, and collaborative approach will provide both the rationale and the tools necessary to underpin the next evolution of city-making, one that is more iterative and more equitable.

References

Badger, E. (2012). 'The evolution of urban planning in 10 diagrams', in *City Lab.com* [online]. Available at: https://www.citylab.com/design/2012/11/evolution-urban-planning-10-diagrams/3851/ (Accessed: 13 June 2020).

Maslow, A. H. (1966). *The Psychology of Science*. London: HarperCollins.

Further reading in this volume

Jason Schupbach
Chapter 4: A future of creative placemaking
Sarah Calderon and Erik Takeshita
Chapter 5: Making places for survival: looking to a creative placemaking past for a guide to the future
Jeremy Liu
Chapter 15: Un/safety as placemaking: disabled people's socio-spatial negotiation of fear of violent crime
Claire Edwards
Chapter 19: Placemaking in the ecology of the human habitat
Graham Marshall
Preface: The radical potential of placemaking
Cara Courage
Chapter 20: Displacemaking 2015 and 2020
Catherine Fennell and Daniel Tucker
Chapter 24: Artists, creativity, and the heart of city planning
Tom Borrup
Chapter 43: A theory of change for creative placemaking: the experience of the National Endowment for the Arts' *Our Town* program: an interview with Patricia Moore Shaffer, PhD
Maria Rosario Jackson
Chapter 45: How the city speaks to us and how we speak back: rewriting the relationship between people and place
Rosanna Vitiello and Marcus Willcocks

34
RECONNECTING *CITÉ* AND *VILLE*

Philip Graus

Introduction

This chapter asks why 'good' strategic plans and urban design often translate into poor physical places. Richard Sennett offers an insight, referring to a '*cité* and *ville*' divide: *cité*, the accumulation of fine-grain human places that people value as an experience; *ville* the comprehensively planned city of top-down plans and policies (Sennett, 2018). In contemporary global cities, placemaking is overwhelmed by metropolitan-scale urban infrastructure and the economic and governance models and processes driving them, narrowing and overwhelming the placemaking process itself. Strategic planning and urban design in isolation cannot deliver successful placemaking. A three-part proposition to deliver improved places is set out here to narrow the divide, in essence, elevating the role of placemaking by adopting its processes into the current economic and political decision-making process. This comprises adopting an iterative, rather than a linear process, more closely aligning and integrating urban governance, and broadening urban economic models including cost–benefit analysis. The chapter provides an overview of contemporary strategic planning and urban design, their urban governance and economic drivers, examining their impact on placemaking with respect to two case studies, one in the author's city, Sydney, the other in Tokyo. The conclusion details the three-part proposition to improve placemaking.

The *cité/ville* divide

This chapter has been prompted by a number of observations and questions arising over experience of several decades of professional practice in planning and urban design. Why do 'good' comprehensive strategic plans and even urban design often translate into poor physical places? Are the process or forces driving comprehensive planning the issue? Or is it poor urban design, or a disconnect between the two, with respect to both process and sequence? Poor outcomes are manifested in two ways: the destruction or erosion of existing places, and the creation of 'placeless' new places.

Sennett provides a series of helpful insights, referring to a '*cité* and *ville*' (Sennett, 2018) divide. *Cité* is the accumulation of fine-grain human places that people value as an experience, that evolve slowly over time, tinkered with and adapted from the bottom up at the local level where we 'dwell'; the messy vibrant places described famously by Jane Jacobs (Jacobs, 1961). *Ville* is the comprehensively planned city of top-down plans and policies. For Sennett, *ville* and *cité* have become divided, as national economic and governance priorities grow in importance in competing global cities in particular. Sennett associates Jane Jacobs, the journalist and later urban activist, with the *cité*, and Lewis Mumford, the regional planner, with the *ville*. The disconnect between the two was famously illustrated by the conflict between Jane Jacobs and Robert Moses, the New York parks commissioner and highway builder, seen as protagonists in a battle between the *cité* and *ville* respectively (Brandes Gratz, 2010). While one might think of strategic planning as *ville*, and urban design as *cité*, the distinction is more subtle.

In contemporary cities, especially global cities, fine-grain human-scale placemaking is overwhelmed by city-scale urban infrastructure and the economic and governance models that drive them. This may not have been the case prior to the early twentieth century where there was little distinction between *cité* and *ville*, as cities were planned and designed at the human scale. Returning to such a preindustrial past is not possible in contemporary cities, especially global ones where 'city shaping' infrastructures are necessary. We cannot therefore choose between Moses' city-shaping infrastructure and Jacob's fine-grain neighbourhoods; rather the two need to be reconnected by elevating the role and process of placemaking to more strongly influence the structure and impact of economic and governance drivers, in particular adopting the more iterative, aligned, and broader processes inherent in the placemaking process into these models. The influence of economic models is manifested in particular by city-scale project infrastructure business cases, reinforced by top-down non-iterative governance models funded at the national level driving infrastructure projects through subsequent state and local levels. Furthermore, national economic priorities, including economic growth favour infrastructure projects such as transport over open space or civic improvement. These forces narrow and overwhelm the placemaking process itself. It is clear that planning and urban design in isolation cannot deliver placemaking or the experience of place.

The three-part proposition

The three-part proposition in this chapter proposes changes to the processes and scope of economic and governance models, more aligned to the finer-grain iterative process of placemaking. The three parts are highly interdependent. First, adopt an iterative, rather than linear, process between comprehensive planning and fine-grain urban design. Second, more closely align and integrate urban governance through genuine partnerships. Third, broaden urban economic models when evaluating cost benefits. This will require a closer alignment of the various levels of government and an expanded context within which urban economic models, such as business cases, are framed to bring placemaking closer to the centre of these processes. The lack of an iterative process or alignment between various agencies and context have separated and narrowed the placemaking process, effectively pushing it to the end of the process where major design decisions have already been set. In this way the placemaking process can integrate *cité* and *ville* more closely together to maintain or create valued places. The first part of this chapter examines the evolution of contemporary strategic planning and urban design, their relationship to placemaking as well as the changing relationship between them, driven to a large extent by the role of urban governance and economic models. The second part examines these impacts with respect to two case studies, one in the author's city, Sydney, and the other in Tokyo (Graus, 2019). The third part describes a more engaged role for urban design in the economic and political decision-making process. The conclusion sets out the three-part proposition.

Contemporary planning and urban design

Ville – *planning, governance, and economic growth*

How cities are governed has significant implications for how cities are planned and the places that result. This has had a particularly significant impact on global or rapidly growing cities. As Clark and others have observed, 'if the 20th century was defined by the nation state, the current century will be driven by global cities' (Clark, 2017). Clark also notes that the relationship between city and central government is complex and will require a much stronger partnership between the two, if both are to benefit. Planners in advanced metropolitan cities since the Second World War in particular have described strategic planning as 'managing growth,' largely to provide rapidly growing urban populations with access to jobs and services as the metropolis expanded. This contrasts with earlier metropolitan plans where civic improvement was an integral part of such documents (Chicago Plan, 1909). In Australia, Spearritt described Sydney as, 'haphazard suburban development of the twenties, a decade in which enthusiasm for expanding the metropolis exceeded interest in improving it' (Spearritt 1978, p. 26). An early case of *ville* overwhelming *cité*.

Over the twentieth century a number of strategic-growth-plan typologies were developed, principally all variant forms of urban containment, from cores surrounded by green belts with satellite centres, to corridors, to the compact city where density is increased, and more recently to economic growth corridors where the focus has shifted to economic productivity. Fainstein (2010) and Sorensen (2002) identified economic growth as the primary driver of contemporary metropolitan planning in the US, Europe, and Japan, respectively. These were seen as the most important elements in ensuring a continually improving standard of living and therefore became important politically. Providing access to jobs and services continues to be a significant driver of strategic planning and requires significant metropolitan-scale transport infrastructure, as in Tokyo where transport infrastructure has been more critical to productivity rather than open space or green infrastructure. Moses, whom many have criticised for his destructive freeway program in New York City, built them to provide recreation access to Long Island's beaches for the average New Yorker who previously had been denied access (Caro, 1974). More recently the focus is shifting to public transport. In either case this results in significant 'city shaping' engineering infrastructures with mixed results with respect to placemaking, especially fine-grain and walkable precincts. Such projects are increasingly evaluated via economic business cases measuring benefits against cost. How the various costs and benefits are allocated obviously impacts on which projects provide the highest cost benefit as well as the nature of the project. Political considerations as well as planning also come into play as different communities either benefit or are disrupted.

Cité – *urban design and placemaking*

The relationship between *cité*, urban design, and placemaking is complex and it is important to define each as well as the relationships between them. Sennett describes *cité* as a person's lived experience of a place, or the act of dwelling rather than building. *Cité* is experienced at the human scale – either static places like squares or streets that one moves through. Hence the interest of many placemakers in spaces where human exchange takes place; the street and marketplace. *Cité* is the end result of successful placemaking. Urban design can be thought of as the process of shaping place, described by Jonathan Barnett as 'the generally accepted name for the process of giving physical design direction to urban growth, conservation and change' (Barnett, 1982, p. 12). It is not necessarily bound to operate at the *cité* or *ville* scales. For the purposes of this chapter, the relevant aspect of urban design is the iterative design process capable of mediating between scales in developing physical three-dimensional urban forms.

Where does urban design sit with respect to *cité* and *ville*? It is generally accepted that urban design occurs at a range of scales from the fine-grain to the metropolitan. Given the divide between *cité* and *ville* it is perhaps not surprising that the term 'placemaking' has come to refer to the finer-grain local scale. Metropolitan-scale processes that have significant spatial impacts on the form of the city, are generally not place based; rather they are abstract and diagrammatic, lacking grain and human scale. For the purposes of this chapter placemaking occurs when metropolitan planning, urban governance, and the drivers of economic growth do not overwhelm the urban design process at the fine-grain local placemaking scale. This is contrasted with *ville*, which is generally conceived by those outside the community – professional planners, bureaucrats, and financiers for example. While *cité* tends to be local, *ville* is increasingly metropolitan in scale.

Prior to the advent of regional- and metropolitan-scale land use and transport planning in the early twentieth century, there was little or no distinction between 'city planning' and 'civic design.' The term 'urban design' did not exist. Barnett (1982, p. 13) traces urban design as a separate technical speciality to the University of Pennsylvania's Civic Design Program begun in 1957, followed by Harvard's Urban Design Program in 1960. Metropolitan-scale plans such as the 1909 Chicago Plan included both scales as 'city planning' and 'civic design' or 'beautification' respectively. The design of civic places was included in metropolitan-scale plans. In these plans, infrastructure was 'civic infrastructure.' The 1909 Royal Commission for the Improvement of Sydney and its suburbs brought together a new city transport plan with plans for the civic improvement of significant Sydney transport places including the Central Railway Station Area improvement scheme and the Circular Quay 'enlargement and beautification of the waterfront' (Brill, 1909). The divide identified by Barnett is echoed by Edward Relph (1976) describing the rupture between strategic planning and placemaking, as the separation of planning space from architectural space, with the latter a deliberate attempt to create spaces, and the former not concerned with the experience of space but rather with function in two-dimensional map space. Relph was one of the first to write about place itself as an experience in *Place and Placelessness* (1976), a reaction against Modernist abstract planning models and theories, preferring the 'lived-world' of day-to-day experience. Barnett offered a similar critique, describing Modernist 'Urbanisme' as a romantic vision of modern technology, freeing the individual from tradition 'admirably suited to mindless bureaucratic repetition, and the cost cutting.' The issues associated with the diminution of placemaking are complex. The case studies below attempt to elucidate rather than simplify the issues at play.

Two case studies – Sydney and Tokyo

Commonalities

Many of the issues discussed above were brought into sharp focus during a visit to a number of Japanese cities in May 2019 where the author was struck by similarities with Australian cities, Sydney in particular. What might a city like Tokyo, with a metro population of 14 million, have in common with Australian cities like Sydney or Melbourne, with populations of less than 5 million, and what insights might such a comparison provide? Reading Andre Sorensen's (2002) *The Making of Urban Japan: Cities and Planning from the Edo to the Twenty First Century*, while catching trains and walking Tokyo, Osaka, Kyoto, Hiroshima, and Nagasaki sharpened and more deeply informed the author's observations, especially the role of both governance and economics in shaping these cities and places. While Japan has been perceived as one of the few genuinely 'foreign' countries by some, Sorensen sees significant similarities as well as differences with Western countries that share similar levels of wealth and urbanisation.

Central governments in both countries have significant control over funding of state city-shaping infrastructure projects. Marcus Spiller refers to this as a 'vertical fiscal imbalance' resulting in significant federal influence over large state projects (Tomlinson and Spiller, 2018). While such large 'city-shaper' infrastructure projects deliver much needed improvements with respect to accessibility to jobs and housing, they do not themselves create the fine, urban-grain fabric and open spaces essential to creating the places we value so highly. Sydney is similar to Tokyo, in that it is the fine, walkable scale grid that pre-dated post-war 'comprehensive planning' that create the valued places. While Japanese cities are no longer growing like their Australian counterparts, one can clearly see the impact on place resulting from decades of rapid and significant population growth as well as strategic plans underpinned by economic growth and infrastructure 'city shapers.' There are also a number of relevant common characteristics, especially the conversion of extensive rural farming areas to greenfield development. The value of the comparison for Sydney is that rapid growth occurred in Japan earlier, and therefore one can see the impacts on place, and potentially apply the learnings to the Sydney context, and perhaps to other places, where extensive growth, especially greenfield, is still occurring, as well impacts on existing areas.

Sydney

Sydney ville

Sustained and strong population growth has been identified as the overarching driver of metropolitan planning for Sydney. From a point of 4.3 million in 2011, Sydney is expected to grow to 6.4 million by 2036 and 8 million by 2056, an average of 85,000 new residents a year (Hamnett and Freestone, 2018, p. 77). Sydney's first metropolitan plan, the 1945 *Cumberland County Plan*, was underpinned by the three pillars of 'co-ordination, consolidation and conservation' (Winston, 1957). While not an expansion plan, it was structured around the orderly planning of a growing population by co-ordinating the ad hoc nature of post-war development, consolidating growth and filling empty pockets, as well as conserving Sydney's natural character. Inspired by Abercrombie's *County of London Plan* of 1943 and *Greater London Plan* of 1944 (Hamnett and Freestone, 2018, p. 35), a green belt was proposed to consolidate growth. Similar to other major cities in the post–World War Two period, the quantum and pace of growth was much greater than predicted. Within a few years after the publication of the *Cumberland Plan*, the green belt was under threat. In 1968 the *Sydney Region Outline Plan* proposed 'growth corridors' along rail lines piercing the green belt. The theme of 'managing growth' has underpinned all subsequent plans. A more recent shift in emphasis has been the focus on economic as well as population growth. The 2014 metropolitan plan, *A Plan for Growing Sydney*, identified nine 'city shapers' to address the city's projected growth. These included global economic corridors, road corridors, and rail corridors. The traditional emphasis on a centre's hierarchy shifted to a combination of corridors, clusters as well as centres reflecting changing economic circumstances, especially economic globalisation manifest in the clustering of financial services and so-called innovation districts. The emphasis on transport, both public and private, reflected an acknowledgement that a catch-up was long overdue.

Urban governance and economic drivers

In Sydney's case there has been an acknowledgement that traditional governance models of state and local planning not aligned to a metropolitan scale cannot effectively manage or plan a growing metropolis that now includes 55 per cent of the entire state's population as well as an increasing contribution to national Gross Domestic Product (GDP). Planners have long asserted that a metropolitan level of governance and planning had been needed, and that Sydney had lacked this since the abolition of the Cumberland County Council. The relationship is complex in Australia where three levels of government each have defined roles and sources of funding. The Australian system of taxation creates a 'vertical fiscal imbalance' where the federal government collects over 80 per cent of taxes but the state is responsible for delivering major urban infrastructure. As a result, the federal government has significant influence in which urban infrastructure is funded. National economic productivity priorities may outweigh state priorities such as health and education, even though there are clearly links between the two. Accordingly, there may be mismatches between the level of government responsible for funding urban infrastructure and the level responsible for planning and delivering. While the states are responsible for delivering major infrastructure, they are dependent on federal funding. While there is a process of independent evaluation of projects, it is inevitable that they are seen through the potentially competing lenses of different levels of government. In many cases federal and state priorities do not align with city or placemaking priorities. This is exacerbated where a city is a major contributor to national GDP and the city in a sense becomes a national and state economic engine. Reflecting the need for a metropolitan level of urban governance, in 2015 the Greater Sydney Commission was established 'to lead metropolitan planning for the Greater Sydney Region.' The Commission released *Region Plan: A Plan for Growing Sydney*, proposing a metropolis of three cities, East, Central, and West, to 'collectively create Global Sydney.' This is currently being supported by a Western Sydney City Deal and a new curfew-free Western City Airport, a major 'city shaper.' The *Region Plan* also includes a series of place-based Collaboration Areas identifying those centres with the greatest potential for jobs growth.

Urban economics has a significant spatial impact on cities, especially the paradigm of economic growth as the means to sustain high levels of employment lifting the standard of living. Cities are becoming stronger economically as capital investment and global businesses coalesce into major centres (Sassen, 2001). A global city may have a stronger relationship with a global city in a foreign country rather than to its own region. Australia is no exception, with capital investment in a relatively small number of metropolitan cities, resulting in their increased economic importance to both state and national economies. Cities are of increasing interest to both levels of government. As a result, federal governance and funding models in particular are oriented towards macroeconomic projects, including regional-scale infrastructure projects subject to business cases that tend to prioritise economic 'value' over social or liveability criteria. Governance and economic structures combine to prioritise metropolitan scale projects over local, in many cases to the detriment of placemaking. In many cases urban transport infrastructure, such as busways, unintentionally divides the urban fabric of centres. The above highlights the complex relationship between urban governance and economics and how they increasingly shape our cities. There is, however, a growing recognition that the most successful global cities are desirable places to live in, and that 'business as usual' governance and economic models may in fact undermine both federal and state economic objectives. The recent Australian City Deals between the three levels of government reflect an acknowledgement of the need to more closely align objectives through a partnership structure.

Tokyo

Tokyo ville

Sorensen argues that Japan's political system, economics, and culture has remained a top-down national structure from the Edo to the present, with a focus on economic growth over liveability following its devastation after the Second World War through to the 1980s. These priorities especially have deeply influenced the form of its contemporary cities. Sorensen highlights a number of impacts on Japanese cities: urbanised land increased from 15 per cent to 80 per cent over the twentieth century (not dissimilar to Australian cities, where both have rezoned very significant areas of rural lands to relatively low destiny residential at the edges of metropolitan cores); the world's best and most heavily trafficked rail system; increasing road congestion and a vast network of elevated freeways; and a lack of public domain. The lack of controls in major metropolitan cities like Tokyo has led to significant sprawl. While roads were delivered as essential to deliver development, open space was not. Like Australian cities, the high cost of urbanised land mitigated against the purchase/allocation of sufficient open space for rapidly growing areas that transition from rural to urban. While Japanese planners introduced measures where a percentage of upzoned land was to be contributed for roads and other public domain, central government was not supportive. Central government's strategy has focused on delivering national economic development projects first, postponing social and civic infrastructure. This is a high-risk strategy for Sorensen as outer areas rarely receive adequate social and public domain infrastructure as value uplifts on rezoned land make postponed acquisition unaffordable.

There are similarities as well as differences with respect to the impact of *ville* in Tokyo and Sydney. In both cases, very significant economic growth and infrastructure have transformed these cities. With respect to infrastructure, major motorways cut though both cities, especially Tokyo. In many cases motorways pay little or no regard to the existing fabric of the city. In the case of Tokyo many motorways were simply placed over existing streets, creating dark lifeless spaces below. Transport infrastructure and filling in of the harbour in cities such as Hiroshima, to create large areas of industrial uses, have separated the city from its harbour and former place of exchange and mercantile activity in a similar way to Sydney's Darling Harbour in the early twentieth century. As Sorensen (ibid.) notes, the push for productivity resulted in industrial and transport infrastructure at the expense of civic infrastructure. This is evident in the lack of public domain, apart from temple, palace, and castle lands converted to parks. Notwithstanding this, in contrast to Sydney, Tokyo has developed an amazing and diverse jumble of uses, places, and built form, due in part to the lack of private domain regulation, but principally (in the view of the author) a much greater retention of the traditional fine-grain street pattern, due in large part to the inability of corporations to amalgamate the many small individually owned parcels of land in built-up areas. While central Tokyo experienced almost total physical destruction during the war, much of the traditional fine-grain street patterns remained, even while the buildings did not. The valued places are the small streets and lanes. This has resulted in high levels of street activity in the brownfield areas. The subsequently planned greenfield areas however lacked such a grain, shaped in the main by metropolitan scale infrastructure. As Sorensen notes: 'the best places escaped "comprehensive planning", the worst places were not so fortunate' (ibid.).

Urban governance and economic drivers

Sorensen (2002) identifies a number of factors that continue to influence the urban form in Japanese cities. First is the dominance of central government that hasn't changed despite postwar attempts to break down the centralisation of power. This is reinforced by the tradition of urban neighbourhood self-reliance for day-to-day needs, without meaningful local government authority or representation. Furthermore, central government funding has favoured economic development over social infrastructure. This has translated into a preference for transport infrastructure over public domain and placemaking investment.

Money, politics, and design

The implications of these models' impact on the design of valued places have not been sufficiently examined to date, nor what changes in policy, or thinking, may be needed to ensure that such forces do not overwhelm or narrow the making of place, especially in greenfield planning. This contrasts with the significant body of research into urban governance and economic models including their negative impacts on the economic performance of cities arising from issues like the federal–state fiscal imbalance and the allocation of urban infrastructure. How these models impact the design of place at the precinct scale and whether some models facilitate more valued places than others deserves greater attention. Such a gap in policy and practice impacts liveability and quality of life, including the creation and maintenance of places that people value. This is reflected to some degree by increased community opposition to development in Australia's major cities, especially Sydney. While a number of policies may appear to address urban design, they do not generally address the more profound structural impacts of both governance and economic models, how they impact and narrow the design process and place outcomes.

While metropolitan-scale infrastructure may deliver much-needed improvements in accessibility to jobs and housing it does not by itself create the sort of urban-grain, fabric, and open space that create the places we value so highly. As reflected in the Tokyo case study, it is the fine walkable-scale grid that pre-dated 'comprehensive planning' that creates the valued places. While this combination of new infrastructure and fine-grain places may occur even in an ad hoc and highly compromised form in existing urban fabric structured prior to car dominance, the real challenge is how to achieve such places in greenfield areas where there is no fine grain, while satisfying the need for significant infrastructure delivered relatively quickly. In the case of the outer greenfield/formerly rural areas of Tokyo and Sydney this does not seem to have been successful, especially with respect to new open space or placemaking. This raises the question of whether it is possible to comprehensively plan the *ville* and 'make' *cité* places at the same time. A growing global city like Sydney does not have the choice between *cité* and *ville* if it is to sustain connected liveable places that people value. Creating a metropolitan *ville* where land use and transport are integrated at the metropolitan scale does not ensure equitable access to the places people value. This creates a significant challenge as current strategic planning processes are essentially linear, structured by state agencies around a hierarchy of spatial plans from metropolitan to local land-use zoning plans. Urban design is then applied largely by local government later in the process.

City planners such as Alex Washburn in New York have recognised the diminution of placemaking associated with the split between *cité* and *ville* advocating effective urban design as the combination of economics, governance, and design or money, politics, and design (Washburn, 2013). Without urban designers being actively engaged in the economic and governance decision-making process, people of importance in government and real estate see design as the icing on the cake. Washburn found that not only were the rich and powerful influencing change, there were also the communities living in the city – energy without consensus. Washburn needed 'a political, a financial, and a design framework to relate to the full range of individual actors with a common good,' providing the opportunity to communicate a common design interest.

Bringing *cité* and *ville* together in the three-part proposition

Sennett once remarked to Jacobs that she was better at *cité*, and Mumford at *ville*. To which Jacobs replied, 'What would you do?' Indeed, what should we do? Can we do both? To move beyond 'business as usual' planning, a three-part approach of aligning, integrating, and broaden-

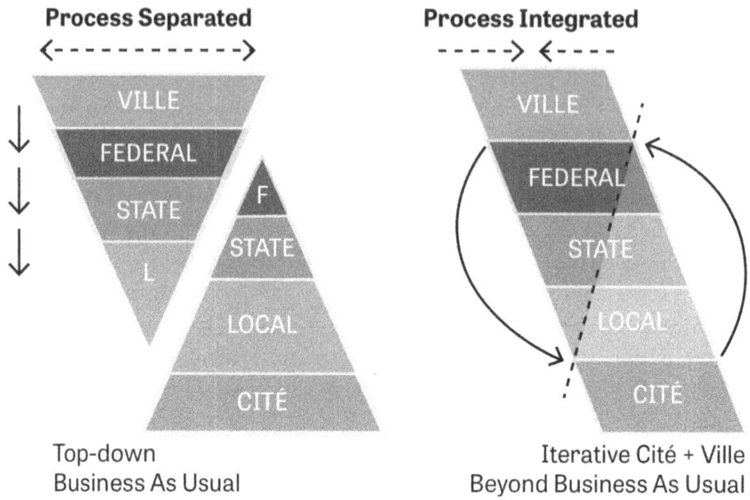

Figure 34.1 An alternative to the current 'business-as-usual' planning process (Graus, 2020).

ing is proposed as follows. The three elements described below are highly interdependent as the diagram attempts to illustrate (Figure 34.1).

An iterative process

The 'business as usual' approach to strategic planning and urban design is hierarchical, both linear and top-down, separating the two with respect to process and sequencing of key decisions. This effectively rules out placemaking, which requires a design process that is iterative and fine-grained as well as three-dimensional. In a linear process, metropolitan-scale decisions are 'fixed' prior to fine-grain placemaking. A more sustainable *cité* and *ville* planning and urban design methodology is to plan and design metro- and finer-grain scales iteratively rather than in a linear hierarchy. This is not to abandon planning hierarchies, but to work at the various scales concurrently recognising that city is not an end-state plan but evolves iteratively. An iterative process brings the more abstract or diagrammatic methodology of planning together with the more concrete place-based concerns of urban design, avoiding the *tabula rasa* view of places that has been common to much unsuccessful urban renewal. Designed iteratively with a finer-grain urban design process, a more strongly place-based outcome is far more likely to be delivered, while also achieving broader metropolitan scale outcomes.

An aligned and integrated process

As noted in this chapter the various levels of government objectives do not necessarily align with city or placemaking objectives. Aligned governance at the geographic scale of metropolitan city is essential in achieving place-based outcomes. Such an alignment provides an opportunity to develop both *cité* and *ville* concurrently as a partnership between the various levels of government, facilitating an iterative process of co-design between *cité* and *ville* as described above. Alignment through partnerships is essential in governance structures such as in Australia where none of the three levels of government – federal, state, and local – covers the geography of the metropolis. This problem is exacerbated by the fact that the economic geography of global cities is metropolitan and is not aligned with any one level of governance. It is therefore essential that partnership structures that create a level of metropolitan governance are established. Such partnerships need to integrate their operations to enable partnerships to be effective.

A broadened process

As noted above, economic models narrow the focus of city-making by limiting the variables of complex infrastructure planning in particular. In this regard the consideration of 'city-shaping' infrastructure needs to be reimagined; broadened as well as aligned. A sustainable city requires the integration of green, transport, and social infrastructures including evaluating the three as part of the same business case with equal weighting. Without broadening the cost–benefit analysis, a transport focussed business case in isolation will seek to maximise rail patronage via higher densitity to improve the project's cost benefit. Such a density analysed more broadly may find that the built form exacerbates urban heat island, reduces the diversity of dwelling types, and thus reduces the opportunity to offer a range of potential price points. A broader analysis may well result in a significantly better outcome.

Conclusion

The Japanese city case studies demonstrate that infrastructure itself is not sufficient to deliver a liveable city. There is an opportunity to learn from this and consider how both the three infrastructures and 'fine-grain' fabric can be delivered to create productive and liveable new places for communities over time to develop both a *cité* and a *ville*. The context and evaluation criteria within which infrastructure and other business cases are developed, evaluated, and prioritised must therefore be broadened, recognising green infrastructure as of equal importance with transport and social infrastructure. Working iteratively at the fine grain concurrently, a walkable public-domain framework can be considered as much a 'shaper of city' as the larger-scaled metropolitan infrastructures.

References

Barnett, J. (1982). *An Introduction to Urban Design*. New York: Harper & Row.

Brandes Gratz, R. (2010). *The Battle for Gotham: New York in the Shadow of Robert Moses and Jane Jacobs*. New York: Nation Books.

Brill, A. (1909). *Report of the Royal Commission into the Improvement of Sydney and Suburbs 1909*. Sydney: Government Printer.

Caro, R. (1974). *The Power Broker: Robert Moses and the Fall of New York*. New York: Random House.

Clark, G. (2017). *World Cities and Nation States*. Oxford: Wiley Blackwell.

Fainstein, S. (2010). *The Just City*. New York: Cornell University Press.

Fensham, P. (n.d.) 'The Sydney metropolitan strategy: Implementation challenges', in Bucek, J. and Ryder, A. (eds.) *Governance in Transition*. Dordrecht: Springer.

Graus, P. (2019). *Observations on Japanese City Making, Sydney: The Fifth Estate*. Our Planet Our Real Estate.

Hamnett, S. and Freestone, R. (2018). *Planning Metropolitan Australia*. Abibgdon: Routledge.

Jacobs, J. (1961). *The Death and Life of Great American Cities: The Failure of Town Planning*. New York: Random House.

Relph, E. (1976). *Place and Placelessness*. London: Pion Press.

Sassen, S. (2001). *The Global City*. Princeton, NJ: Princeton University Press.

Sennett, R. (2018). *Building and Dwelling: Ethics for the City*. London: Allen Lane.

Sorensen, A. (2002). *The Making of Urban Japan: Cities and Planning from Edo to the Twenty-First Century*. Abingdon: Routledge.

Spearritt, P. (1978). *Sydney Since the Twenties*. Sydney: Hale and Ironmonger.

Tomlinson, R. and Spiller, M. (2018). *Australia's Metropolitan Imperative: An Agenda for Governance Reform.* Clayton, Victoria: CSIRO Publishing.
Washburn, A. (2013). *The Nature of Urban Design.* Washington, DC: Island Press.
Winston, D. (1957). *Sydney's Great Experiment: Progress of the Cumberland County Plan.* Sydney: Angus and Robertson.

Further reading in this volume

Chapter 2: Placemaking as an economic engine for all
James F. Lima and Andrew J. Jones
Preface: 'Disastrous forces, accidental actions, and grassroots responses'
Tom Borrup
Chapter 9: From the dust of bad stars: disaster, resilience, and placemaking in Little Tokyo
Jonathan Jae-an Crisman
Chapter 13: Sensing our streets: involving children in making people-centred smart cities
Sean Peacock, Aare Puussaar, and Clara Crivellaro
Chapter 15: Un/safety as placemaking: disabled people's socio-spatial negotiation of fear of violent crime
Claire Edwards
Chapter 19: Placemaking in the ecology of the human habitat
Graham Marshall
Chapter 20: Displacemaking 2015 and 2020
Catherine Fennell and Daniel Tucker
Chapter 24: Artists, creativity, and the heart of city planning
Tom Borrup
Chapter 28: Integral placemaking: A poiesis of sophrosynes?
Ian Wight
Preface: Evaluating creative placemaking: a collection of observations, reflections, findings, and recommendations
Maria Rosario Jackson
Chapter 42: Creative placemaking and placekeeping evaluation challenges from the practitioner perspective: an interview with Roy Chan
Maria Rosario Jackson
Chapter 45: How the city speaks to us and how we speak back: rewriting the relationship between people and place
Rosanna Vitiello and Marcus Willcocks

35

PLANNING GOVERNANCE – LESSONS FOR THE INTEGRATION OF PLACEMAKING

Nigel Smith

Introduction

Placemaking is a new profession compared to the more established sphere of planning. In Australia, planning has a long history of government-driven regulation. Therefore, if placemaking was being successfully integrated into planning governance, its expression in government planning efforts should become more apparent and sophisticated over time. Pre-existing governance systems and documents should be seen to flex to integrate placemaking theory and practice. The effort to integrate placemaking into formal planning governance systems in Australia can be verified by the large number of documents with placemaking content produced by all levels of government over the last 20 years. This chapter seeks to understand to what extent placemaking has been successfully integrated into planning governance and will consider in detail three Australian state government documents that have incorporated placemaking language, theory, and objectives. The research reveals lessons and implications for the future of placemaking practice and its growth as a profession. Recommendations for planning policy writers and placemaking practitioners are drawn from the findings and conclude the chapter.

Authors note: The author has not considered Aboriginal history in this planning research and acknowledges the complex system of relationships to land and country that have long existed between Australian First Nations people and place. Planning records did not exist in a way recognisable to British colonising forces in 1788, but this is not to say that planning was not being undertaken. See for example, Aboriginal fire management methods that are spatially and seasonally planned (Gott, 2005).

A brief history of post-colonisation planning in Australia

Planning governance in Australia is based on the British system, with New South Wales (NSW), the first Australian colony, founded under British military rule in 1788. Post-colonisation executive powers lay with the governor of the day, who would (ideally) make planning and infrastructure decisions in the best interests of the colony – the common good, the common ground, and the common place (Freestone, 1993). Through subsequent years, a three-tier system emerged with states given primary responsibility for planning and related issues such as transport and public works. Federal intervention in city planning was not seen as necessary. The third level of local government was left out of the Australian constitution in 1901, leaving each state to determine how it would enact local government within each state parliament (Kelly, 2011).

From the early twentieth century the former colonial centres that were now state capitals experienced rapid growth. Australia adopted planning structures with broad-based government powers, again borrowed from the British. At this time, the public interests of health, efficiency, and beauty railed against toxic industrial practices and a real estate sector struggling to accommodate burgeoning populations (Toon and Falk, 2003). The planning system continued to professionalise. After World War Two, urban design (then known as civic design) emerged in recognition that good planning and land-use zones would not on their own deliver an amenable city. The increasing prevalence of the private car (and engineering responses to them) was also not helping the urban cause. The top-down command-and-control approach to planning governance naturally demanded increasingly complex state regulations and instruments to ensure equity and transparency (Western Australian Planning Commission, 2019). Adding to complexity, at various times the federal government waded in to planning governance, traditionally managed by the states. This had varying success such as the *Better Cities Program* of the late 1980s (Bryant, 2016). Conversely, local government in the Australian context, has continued to have limited executive planning power, other than to administrate state regulations which in some jurisdictions allows neighbourhood level planning responses.

In 2020, the planning governance systems are fundamentally the same. The Victorian state jurisdiction for example has a planning regulations document over 1,200 pages, and the guide to using the system is nearly 300 pages (Richardson, 2013). Australia has not had easily comprehendible planning governance, nor has it always been equitable, transparent, and efficient. The system is professionally overseen by planning institutes, tertiary education, and planning law, and is, by its very nature, opaque to people who are not from a planning background. Stresses are becoming apparent in planning systems and processes in response to another growth wave in Australian cities. Save Our Suburbs movements (Save Our Suburbs, n.d.), politically motivated planning interference (Bleby, 2018), and wholesale planning reform (Department of Planning, Transport and Infrastructure, 2020) are examples of reactions to a top-down governance system that has little room for the alternative bottom-up placemaking approaches.

The case studies

In considering the integration of placemaking into planning governance, three questions were asked of the chosen state government planning documents: what is the purpose of the document; how has place or placemaking been integrated; and to what extent, if any, does the integration of place and/or placemaking lead to better urban outcomes? Armed with the answers to these questions, the end of the chapter presents some of the dynamics that the integration of placemaking practice brings to planning governance. This is intended to lay out pathways for the next period of placemaking's growth as a movement in Australia, (and may be of help to practitioners and policy writers in other parts of the world). It is important to note that no on-the-ground placemaking projects were studied. In the Australian context, this level of placemaking, while well established, does rely on willing stakeholders to experiment. The research questions reflect this thesis, that government support is needed to integrate placemaking practice as a standard activity in city development processes.

Recognition of the need for placemaking: Western Australian Government, 1998, *Liveable Neighbourhoods Community Design Code*

The *Liveable Neighbourhoods Community Design Code* is currently Western Australian state planning regulation. It was first trialled from 1998 and adopted as planning law in 2007 (Western Australian Planning Commission, 2009). Its geographic focus is Perth, the capital of Western Australia. The evolution of the regulation reveals how planning for place was occurring before it was formally articulated, and then how placemaking and its language became integrated as the theories and practice of placemaking grew in influence.

From the 1960s Perth housing was delivered through car-dependent living in low-density dormitory suburbs (Newman and Kenworthy, 1999). There is limited experience of 'placeness' as a reflection of local landscape, community, and culture; neither is there a mix of housing type, walkability, or vibrant main streets connected to public transport. By 1998, the *Liveable Neighbourhoods Community Design Code* (*The Neighbourhood Code*) had begun a trial period. Its primary purpose was 'the design of neighbourhoods and towns that aims to achieve compact, well-defined and more sustainable urban communities' (Western Australian Planning Commission, 1998). This purpose was in stark contrast to the government's extant development control policies. The new focus was 'to provide for an urban structure of walkable neighbourhoods clustering to form towns of compatible mixed uses in order to reduce car dependence for access to employment, retail and community facilities' (Western Australian Planning Commission, 1998). Initially, the language of place and placemaking was not included, but as the regulation moved from trial to implementation to mandatory requirement, this language became integrated. This is evidence of the emergence of place in the planning governance lexicon: 'This updated edition of Liveable Neighbourhoods also marks several significant milestones in the evolution of the WAPC's policies for the creation of new urban places' (Western Australian Planning Commission, 2009).

The Neighbourhood Code is placemaking as a large-scale strategy. Placemaking is implicit in its planning model of 400m and 800m walkable catchments prioritising the central place of those catchments. It valorises the centre place above its peripheral places, seeing these as vibrant urban main streets supported economically by a walkable suburban hinterland. This is a long lever: that a more sociable place will arise when a car-dependent monoculture is reduced, and walkable diversity is increased. The planning model references the new urbanist transect (Steuteville, 2018) with *The Neighbourhood Code* being awarded the United States-based Congress for New Urbanism Charter Award in 2001 (Congress for New Urbanism, 2001).

Significant to this study and practitioners is that placemaking can occur without the term being used. *The Neighbourhood Code* does not define place, sense of place, or placemaking. It leaves this up to the land developers to self-define and deliver 'a thoughtful sense of place' at the design or tactical level (Western Australian Planning Commission, 2009). This represents a loosening of executive power which has arguably allowed the planning and development industry to innovate and deliver a new kind of Australian urbanism, ironically in the county's most isolated city. The success of *The Neighbourhood Code* can be seen in the increased quality of Perth's suburban development during the mining boom of the 2000s, with new suburbs such as Clarkson in the north (alongside a freeway reserve train station), Ellenbrook's various multi-award winning and evolving villages in the east, and Wellard, centred on a new train station, in the south. The maturation of these planned estates into spaces that encourage walking and a more urban sense of place is occurring (Hooper et al., 2015). This is evidence that the placemaking strategy of *The Neighbourhood Code*, though not explicitly named, has been successful.

When place is used for other purposes: New South Wales Government Architect, 2017, *Better Placed* – an integrated design policy for the built environment of New South Wales

Better Placed is current New South Wales Government policy produced by its Office of the Government Architect. While *Better Placed* is for the whole state, Sydney, like other Australian state capitals, contains the majority of the state's population and economic power and is arguably Australia's only global city.

However, the population is now centred on Parramatta, 20km to the west of central Sydney, and suburban areas have continued to grow a further 20km westward. These areas have a different sense of place to the well-known harbour topography such that the urban centres in and around eastern Sydney continue to attract people for work and housing. Grappling with population growth in established areas, the New South Wales planning system has boldly named dwelling targets for local government to meet (Visentin, 2017). This requires an increase in residential densities in highly valued places, known for their beauty and urban amenity. Planning, architecture, urban design, and placemaking are being called on to apply their expertise to this problem. Implicit here is the idea that good design can help communities to accept new and denser development. The document is reinforcing that growth is inevitable and that local places will change.

Despite its title, the document reveals it is essentially a tool for the promotion and procurement of good design. *Placed* in the context of the title is a play on words, where place becomes a verb, not a noun. At one level it usefully activates the idea of place – *placedness* –but at another level it implies that designers will be *better placed* as a result of the policy. Seventeen of the document's 39 pages explain design, what designers do, and why they should be engaged. It has much in common with the earlier National Urban Design Protocol (Infrastructure Australia, 2011), which it cites as having had 'limited impact in the NSW planning system' (New South Wales Government Architect, 2017, p. 15). However, unlike the national protocol, *Better Placed* does define placemaking:

> Place-making: Proposes a multi-faceted approach to the planning, design and management of public spaces. 'Place-making' looks at understanding the local community with the intention of creating public spaces that promote health and well-being.
>
> *(New South Wales Government Architect, 2017)*

While Objective 1 of the policy is Better Fit, which is defined as 'contextual, local and of its place', of more interest to this analysis of placemaking governance is Objective 6:

> Better Value – creating and adding value. Good design generates ongoing value for people and communities and minimises costs over time. Creating shared value of place in the built environment raises standards and quality of life for users, as well as adding return on investment for industry.
>
> *(New South Wales Government Architect, 2017)*

This is a better definition of placemaking than the document's own definition and more useful to the endeavour of integrating placemaking into planning governance. (Planning, after all, is about negotiating different perceptions of value between the development proponent and the stakeholders for whom the development impacts.) Less effective are the statements that go with this value-added objective as they do not shed light on how to go about creating shared value in a place, but instead defend high up-front construction costs in the light of lower lifetime operating costs (New South Wales Government Architect, 2017, p. 43). As valid as this may be, it echoes common sentiment that good design (vis à vis architecture) costs more. This contributes to elitist perceptions of the architecture profession. This document, of all the documents reviewed, represents the least shift in planning governance and a loosening in executive and professional power. There is mention of the importance of engaging local communities in decisions about the built environment, but it is minor (New South Wales Government Architect, 2017, p. 28). The opportunity to outline how to create shared value through placemaking practice was not taken, which is ironic given the document's action-oriented title. It may be that bottom-up placemaking approaches represent a threat to the top-down power of those who procure architecture, and those who design and deliver buildings and public realm. This is also a recent document, so analysing its impact on integrating placemaking practice in NSW planning is not yet possible. However, given it is light on this detail, the author suspects its impact will be limited.

The place is the reason for planning: Victorian Government, 2019, *Movement and Place in Victoria*

The 2019 Victorian Government policy *Movement and Place in Victoria* (Department of Transport, 2019) is an emerging framework to integrate land use and transport. It is inspired by the United Kingdom's *Link and Place* approach (Jones et al, 2008) and is being deliberately adapted to a Victorian context. The purpose of the new approach is sound – that having an efficient transport network moving large volumes of people at high speed is pointless if this network, at the outset, destroys the very same places and reasons for the movement. This is a concession by government and engineering professionals to the power of people's connection to place.

Melbourne's population growth is driving these shifts in thinking as the city's pre-car movement infrastructure (tram corridors, main streets, laneways) are also some of its most valued places. The mainstreaming of place into transport planning and engineering is evident in the title and throughout the document. Evidence of the role of placemaking is harder to come by as the term itself does not appear in the document. This is an example of how the concept of place can capture imaginations, like the *National Urban Design Protocol* (Infrastructure Australia, 2011) and *Better Placed*. Place is an implicitly understood term, but the notion that making places could be gone about in a deliberate way is not understood, at least in professions that have recently adopted it, like transport planning. Even more concerning is that placemaking here is 'reduced' to placekeeping. Community perspectives are evoked in terms of what is existing, rather than what might be. This raises the conundrums of future or preferred neighbourhood character already extant in the Victorian planning system (Dovey et al., 2010). There are risks that place and placemaking become a proxy for protecting *my* place at the expense of needed urban change. For example:

> The Movement and Place Framework's common language... supports meaningful engagement between the state and local governments and stakeholders and the community. The Framework helps to improve consistency in how transport projects and plans are communicated and discussed with the community. At a time when communities are expecting a greater say in transport and infrastructure decision-making, movement and place provides opportunities to have discussions about how we can address and prioritise our future transport challenges.
>
> *(Department of Transport, 2019)*

Placemaking would seem to have come full circle. Road planning was responsible for the first showdown between government engineers and placemaking, and now placemaking, or at least place, has begun trickling back across to the same profession. The increasing collaboration between, and breaking down of, professional silos is, however, a positive sign for the future of cities to be less about professional power and functional prowess and more about people's wellbeing. The success of this document in establishing better placemaking practice cannot yet be gauged owing to the short period since its publication. As a prescriptive framework, it will need to be translated into legislation to enable development controls that protect and envision current and preferred future places. Melbourne is a city working to reduce car-dependence and has recognised the poor place outcomes that planning for the car can deliver – hence *Movement and Place*. The restructuring and combining of Victorian planning and transport governance and the devolution of traffic engineering voices in transport planning (Carey, 2019) could be the opportunity to empower placemaking in the way this document prescribes.

The integration of placemaking in to planning governance

The following evaluates and organises the extent to which placemaking has been successfully integrated into planning governance through two key findings drawn from the research questions. The first relates to the perceived value of placemaking and the second to whether governance delivers practice.

There is a direct relationship between a high value applied to placemaking and its successful integration into planning governance. When discussing how placemaking has been integrated into planning governance, it is useful to learn from its value, which can be implicit or explicit. Placemaking has been occurring for as long as there have been people, but it is only recently that it has become a legible and valued movement. From this perspective, the earlier *Neighbourhood Code* is an important demonstration of the allocation of a high value to place and placemaking, although it was not named at first. Placemaking was successfully integrated through planning legislation that implicitly, yet powerfully, focused places for people at its spatial core.

The latter case studies explicitly name the theory of placemaking, but there are varying degrees of value allocated. *Better Placed*, for example, poorly defines placemaking and devalues the term in that the concept appears to have been co-opted for political purposes. Placemaking has been used to defend the planning and design elite and bolster professional silos in a growing city rather than deliver the new collaborative governance which is essential to placemaking. In the case of *Movement and Place*, the power of place has given voice to aspirations to change planning in a genuine effort to make places in new suburbs and keep places in existing main streets. This is seen in the radical overhaul to planning governance it represents. *Movement and Place* implies that place is at least as valuable as movement – place is integrated as 50 per cent of the decision-making matrix in planning the future of Melbourne's transport corridors. The case studies reflect that planning is a profession tasked with governing the tensions of city growth. They present place and placemaking as a means to manage these tensions, ranging from the defensive to the aspirational. The defensive posture aligns with a low placemaking value, and the aspirational posture aligns with a high placemaking value. By extension, a high value leads to greater potential for placemaking theory to be successfully integrated into planning governance. This is not to say that this will lead to on-the-ground placemaking practice, as evidenced by the second key finding.

Integrating the theory of place is not the same as delivering the professional practice of placemaking

Placemaking is an ongoing endeavour, so its impact on planning governance, broader professions, and the built environment will cumulatively build, or not, over time. The documents reviewed began to articulate place as a theoretical objective, but less effectively integrated actual placemaking practice to deliver on-the-ground urban outcomes. This is an important distinction. The creation of places through these documents has come about inadvertently – the permission to innovate is not the same as describing the innovation. Governments have a leadership role in the Australian development context, and certainly prescribe planning practice. It follows then, that over time governments may begin to prescribe placemaking practice as it becomes established as a profession.

The oldest document, *The Neighbourhood Code*, has evolved Perth's suburban development with on-the-ground results of sustainable main streets, mixed and denser form, less car dependence, and more opportunity for grassroots-activated public places. *The Neighbourhood Code* was also, paradoxically, the only document that did not use the terms 'place' and 'placemaking' (in the first instance). However, by the time it was a mandatory control, the language of place and placemaking had been integrated. It would be interesting to understand more of how *The Neighbourhood Code* captured a kind of placemaking zeitgeist in the history of Western Australian planning governance. *Better Placed* and *Movement and Place* are new design and planning policies, so evaluating impact on placemaking practice and urban outcomes is difficult. In the former case, success will depend on how much New South Wales' communities, as it states, 'can use Better Placed to understand good design practices and how they deliver benefit to their neighbourhoods, streets, cities and towns (and) help the community to participate in the conversation about design and review processes that affect their local places' (New South Wales Government Architect, 2017). *Movement and Place* is a policy document written as a practice framework. It brings challenges to the traffic engineering and transport planning professions to recognise placemaking as an important tool in responding to the real spatial pressures of narrow streets in the Melbourne place context. How this challenge unfolds on the ground depends on placemaking practice growing in strength and influence in planning governance, which leads to the final discussion.

Three dynamics that the integration of placemaking practice brings to planning governance

Increased cross-sectoral collaboration

Placemaking practitioners will by nature collaborate and avail themselves of all relevant evidence when negotiating development outcomes with planners. In the case of *The Neighbourhood Code*, a planning control, this expertise came from the economic and transport planning sectors in order to integrate sophisticated urban development practice into suburban contexts. It did this by delivering an evidence base of the power of aggregate economies in capital (retail spending) and movement (passing trade) to deliver viable centres (places). This was to challenge conservative planning views of activation potential in well-located places, as much as safeguard against optimistic views of vibrancy in poorly located places. *Movement and Place* shows evidence of the trickle of place theory across into professions parallel to planning. Therefore, as placemaking grows in influence, planning professionals will be challenged to be the collator of a more comprehensive and alternative set of data delivered by the placemaking practitioners. The delivery of this data should therefore be organised in such a way as to be of most help to the planner, such as being sanctioned by planning statute.

The transition from planning trend to standard practice

Understanding that place and placemaking are evocative concepts, the placemaking movement would benefit from practitioners continuing to raise popular awareness and sustain the trend. However, continuing to scope, procure, and deliver placemaking projects is more important. This will challenge and hopefully change conventional planning practice as the positive evidence mounts. Given Australia's modernist planning agenda, there is a real risk, if projects aren't delivered, that the post-modern placemaking movement will be diluted and eventually forgotten. Placemaking will continue to be vulnerable if it does not deliver its projects through planning frameworks, strategies, and policies. Practitioners will need to make the case with the planning profession as to how placemaking is a direct influencer of quality urban outcomes. This case will need to be made differently in every planning jurisdiction as the mechanisms will vary, as per the successful *Neighbourhood Code* in Western Australia. Fortunately, in Australia, the opportunities to move through a challenging professional transition are being created by the ongoing reform of the various state planning systems (Department of Transport, Planning and Infrastructure, 2020).

Ensuring authentic placemaking

The token use of place and placemaking in the *Better Placed* document is a cautionary tale for the placemaking movement and its integration into planning practice. If the movement is to prosper and deliver better places, it should consider becoming organised in its advocacy across professions for authentic placemaking. Making cities is complex work and there are multiple professions challenged (and aroused) by the inclusiveness and innovation of placemaking, including architects and planners. There is also the problem that the term placemaker suggests practitioners 'just make places,' thereby becoming professionally siloed. In the author's experience, placemaking practitioners are imperturbable when it comes to other professions being the Trojan horse in which placemaking enters through the walls of city planning. For example, urban design, architecture, economic development, transport planning, and community engagement all have a role to play. *Better Placed* shows that the built environment professions are capable of advocating for place and placemaking. However, this advocacy will not be consistent or authentic if the placemaking movement does not better define the full breadth of placemaking practice for the planning profession.

Conclusion

It can be seen in this survey of three slices of planning history in Australia that planning governance has, by varying degrees, begun to mainstream place and placemaking. In the highly contested space of city development, the *idea of place* has penetrated planning policies, strategies, and laws, and has been popularised by government codes, principles, and activities. However, integrating the *practice of placemaking* itself has not been as successful. The evidence points to a token understanding that making places could be gone about in a deliberate and strategic manner and that this can be enabled by good planning governance.

In conclusion, the case studies are summarised in three recommendations for placemaking practice if it is to continue to grow and be influential. The recommendations have direct implications for planning governance and how placemaking is described in its movement from policy to effective practice. The recommendations are relevant for jurisdictions outside Australia, especially where placemaking is more nascent. First, monitor and guard the use of the terms 'place' and 'placemaking' in planning documents; be alert to place and placemaking being used inauthentically by governments, communities, and the planning and design professions, especially where they relate to protectionist agendas, in order to avoid the placemaking movement becoming ineffectual. Second, collaborate with a wide range of professionals to deliver placemaking projects; seek collaborative projects with all built environment professions and continue to deliver on-the-ground placemaking projects in order to grow the planning evidence of the benefits of placemaking. Third, integrate placemaking practice in planning frameworks; advocate for the establishment of robust bottom-up placemaking strategies within metropolitan planning systems as they are progressively overhauled in response to paradigm shifts in city-making, especially the devolution of traditional top-down power structures.

As the planet continues to urbanise, the opportunities to make better places and cities will continue to arise and at a faster rate. In Australia, the placemaking journey has only just begun – there is an eager constituency, and the planning profession is taking notice. There are currently, however, limited documented examples of how governance structures are adapting to the power of place and integrating placemaking. This chapter sets some directions for placemaking and planning practitioners everywhere who can rise to the challenge of seeing this occur in order to make places that are meaningful to people.

Bibliography

Beza, B. and Moloney, S. (2011). *Placemaking primer: Outcomes of the Melbourne Placemaking Series*. Melbourne: RMIT University.

Bleby, M. (2018). 'Victoria sets fishermans bend population at 80,000, rejects calls for more', in *Australian Financial Review*, 4 October 2018 [online]. Available at: https://www.afr.com/property/victoria-sets-fishermans-bend-population-at-80000-rejects-calls-for-more-20181003-h16744 (Accessed: 20 January 2020).

Boroondara. (2019). *It's Your Place: Vibrant and Local Places Through Placemaking: A Framework*. Victoria: City of Boroondara.

Brisbane City Council. (2013). *Ideas Fiesta Report*. Brisbane: Brisbane City Council.

Brisbane City Council. (2014). *Brisbane City Centre Masterplan*. Brisbane: Brisbane City Council.

Bryant, L. (2016). *Investing in Australian Cities: The Legacy of the Better Cities Program*. Brisbane: Queensland University of Technology.

Calvino, I. (1974). *Invisible Cities*. Orlando, FL: Harcourt Brace & Company.

Carey, A. (2019). 'VicRoads and PTV to become one mega-agency', in *The Age*, 4 April [online]. Available at: https://www.theage.com.au/national/victoria/big-merger-vicroads-and-ptv-to-become-one-mega-agency-20190404-p51arj.html (Accessed: 18 February 2020).

Centre for Design. (2006). *Ecological Footprint Analysis of Aurora Residential Development*. RMIT University & Global Footprint Network [online]. Available at: http://www.aurora.asn.au/wp-content/plugins/downloads-manager/upload/aurora%20ecofootprint.pdf (Accessed: 13 March 2020).

City of Melbourne. (2016). *Places for People*. City of Melbourne [online]. Available at: https://www.melbourne.vic.gov.au/building-and-development/urban-planning/city-wide-strategies-research/Pages/places-for-people.aspx (Accessed: 15 March 2020).

City of Port Phillip. (2018). *Placemaking Action Plan*. City of Port Phillip, Victoria [online]. Available at: http://www.portphillip.vic.gov.au/precincts.htm (Accessed: 15 March 2020).

Commonwealth of Australia. (2011). *Our Cities, Our Future: A National Urban Policy for a Productive, Sustainable and Liveable Future*. Canberra: Department of Infrastructure & Transport.

Congress for the New Urbanism. (2001). *2001 Charter awards* [online]. Available at: https://www.cnu.org/2001-charter-award-winners (Accessed: 12 February 2020).

Davison, G. (1988). 'Slicing history: Nationhood and the bicentennial', *History Today*, 38(1), pp. 25-32.

Department of Planning, Transport and Infrastructure. (2020). *Planning Reforms: South Australian Planning Portal* [online]. Available at: https://www.saplanningportal.sa.gov.au/planning_reforms (Accessed: 20 January 2020).

Department of Transport. (2019). *Movement and Place in Victoria*. Melbourne: Victorian Government.

Dovey, K. Woodcock, I. and Wood, S. (2010). 'Defending and creating place identities', in *Becoming Places: Urbanism, Architecture, Identity, Power*. Oxford: Routledge.

Dowling, J. (2010). 'VicUrban looks for a modular solution to housing issues', in *The Age*, 4 October [online]. Available at: https://www.theage.com.au/national/victoria/vicurban-looks-for-a-modular-solution-to-housing-issues-20101003-162r9.html (Accessed: 13 March 2020).

Freestone, R. (1993). 'Planning history in Australia: The state of the art', *Planning Perspectives*, 8, pp. 72-91.

Gott, B. (2005). 'Aboriginal fire management in south-eastern Australia: Aims and frequency', *Journal of Biogeography*, 32, pp. 1203–1208.

Hooper, P. et al. (2015). 'Are we developing walkable suburbs through urban planning policy? Identifying the mix of design requirements to optimise walking outcomes from the 'Liveable Neighbourhoods' planning policy in Perth, Western Australia', in International Journal of Behavioural Nutrition and Physical Activity [online]. Available at: https://ijbnpa.biomedcentral.com/articles/10.1186/s12966-015-0225-1 (Accessed: 29 January 2020).

Infrastructure Australia. (2011). *Creating Places for People: An Urban Design Protocol for Australian Cities*. Canberra: Commonwealth of Australia.

James, M. (2010). *Of the Plan: Commonwealth City Planning Systems*. Canberra: Parliament of Australia [online]. Available at: https://www.aph.gov.au/About_Parliament/Parliamentary_Departments/Parliamentary_Library/pubs/BN/1011/CityPlanning (Accessed: 24 January 2020).

Jones, P. Marshall, S. and Boujenko, N. (2008). 'Creating more people-friendly urban streets through link and place street planning and design', International Association of Traffic and Safety Research *(IATSS) Research*, 32(1), pp. 14-25.

Kelly, A.H. (2011). 'The development of local government in Australia, focusing on New South Wales: From road builder to planning agency to servant of the state and developmentalism', in World Planning Schools Congress [online]. Available at: University of Wollongong: Archived (Accessed: 6 March 2020).

Kolson, K. (2001). *Big Plans: The Allure and Folly of Urban Design*. Baltimore: The Johns Hopkins University Press.

Melbourne Placemaking Series. (2010). About Melbourne Placemaking Series, LinkedIn post, April [online]. Available at: https://www.linkedin.com/groups/2973246/ (Accessed: 13 March 2020).

New South Wales Government Architect. (2017). *Better Placed: An Integrated Design Policy for the Built Environment of New South Wales*. Sydney: New South Wales Government.

Newman, P. and Kenworthy, J. (1999). *Sustainability and Cities: Overcoming Automobile Dependence*. Washington, DC Island Press.

Penrith City Council. (2013). *Penrith City Strategy*. Penrith: Penrith City Council.

Richardson, K. (2013). *An Introduction to Victoria's Planning System*. Melbourne: Parliament of Victoria.

Save Our Suburbs. (n.d.). Save Our Suburbs for Sustainable Living [online]. Available at: www.sos.org.au (Accessed: 13 March 2020).

Seddon, G. (1972). *Sense of Place: A Response to an Environment: The Swan Coastal Plain*. Perth: University of Western Australia Publishing.

Steuteville, R. (2018). *The 25 Great Ideas of New Urbanism*. Washington, DC: Congress for the New Urbanism.

Street Plans. (2016). *Temporary Park for a Placemaking Masterplan, Penrith, Australia* [online]. Available at: http://www.street-plans.com/tactical-urbanism-projects/temporary-park-for-a-placemaking-master-plan-penrith-australia/ (Accessed: 17 January 2020).

Toon, J. and Falk, J. (2003). *Sydney: Planning or Politics: Town Planning for the Sydney Region Since 1945*. Sydney University: Planning Research Centre.

Van Eymeren, A. (2018). 'At many tables of discernment: Faith and shalom in the polis', in *Just Faith: Glocal Responses to Planetary Urbanisation, HTS Religion and Society Series*. vol. 3, pp. 253–285. Capetown: AOSIS.

Victorian Planning Authority. (2018). *Practice Note 43: Understanding Neighbourhood Character*. Melbourne: Department of Environment, Land, Water & Planning.

Victorian Planning Authority. (2019). *Victorian Planning Provisions 43.02: Design Development Overlays*. Melbourne: Department of Environment, Land, Water & Planning.

Visentin, L. (2017). 'Commission holds firm on housing targets for Sydney councils', in *Sydney Morning Herald*, 26 October [online]. Available at: https://www.smh.com.au/national/nsw/commission-holds-firm-on-housing-targets-for-sydney-councils-20171026-gz8w4b.html (Accessed: 15 January 2020).

Western Australian Planning Commission. (1998). *Liveable Neighbourhoods Community Design Code: Trial Version*. Perth: Ministry of Planning.

Western Australian Planning Commission. (2009). *Liveable Neighourhoods: A Western Australian Government Sustainable Cities Initiative*. Perth: Department for Planning and Infrastructure.

Western Australian Planning Commission. (2019). *Better Planning: Better Places, Action Plan for Planning Reform* [online]. Available at: https://www.dplh.wa.gov.au/action-plan (Accessed: 14 January 2020).

Yarra Ranges Council. (2019). *Lilydale Place Plan* [online]. Available at: https://www.yarraranges.vic.gov.au/Development/Town-and-urban-centre-planning/Place-planning/Lilydale-Place-Plan (Accessed: 15 March 2020).

Further reading in this volume

Preface: Placemaking in the age of COVID-19 and protest
Jason Schupbach
Chapter 2: Placemaking as an economic engine for all
James F. Lima and Andrew J. Jones
Chapter 3: An annotated history of creative placemaking at the federal level
Jen Hughes
Chapter 5: Making places for survival, looking to a creative placemaking past for a guide to the future
Jeremy Liu
Preface: Evaluating creative placemaking: a collection of observations, reflections, findings, and recommendations
Maria Rosario Jackson
Preface: 'Disastrous forces, accidental actions, and grassroots responses'
Tom Borrup
Chapter 9: From the dust of bad stars: disaster, resilience, and placemaking in Little Tokyo
Jonathan Jae-an Crisman
Chapter 13: Sensing our streets: involving children in making people-centred smart cities
Sean Peacock, Aare Puussaar, and Clara Crivellaro
Chapter 15: Un/safety as placemaking: disabled people's socio-spatial negotiation of fear of violent crime
Claire Edwards
Chapter 19: Placemaking in the ecology of the human habitat
Graham Marshall
Chapter 20: Displacemaking 2015 and 2020
Catherine Fennell and Daniel Tucker
Chapter 24: Artists, creativity, and the heart of city planning
Tom Borrup
Chapter 28: Integral placemaking: A poiesis of sophrosynes?
Ian Wight
Preface: Evaluating creative placemaking: a collection of observations, reflections, findings, and recommendations
Maria Rosario Jackson
Chapter 42: Creative placemaking and placekeeping evaluation challenges from the practitioner perspective: an interview with Roy Chan
Maria Rosario Jackson
Chapter 43: A theory of change for creative placemaking: the experience of the National Endowment for the Arts' *Our Town* program: an interview with Patricia Moore Shaffer, PhD
Maria Rosario Jackson
Chapter 45: How the city speaks to us and how we speak back: rewriting the relationship between people and place
Rosanna Vitiello and Marcus Willcocks

36

FACILITATOR SKILLS FOR EFFECTIVE COLLABORATIVE PLACEMAKING

Husam AlWaer and Ian Cooper

Introduction

Support for more communicative and collaborative approaches to planning through a participatory democracy has been growing over the last 20 years (Healey, 2003; Condon, 2008; Roggema, 2014; Ermacora and Bullivant, 2016; Campion, 2018; AlWaer and Cooper, 2020). Indeed, Allmendinger (2002) suggested that this should be considered as more than just another planning theory, but rather as a new and improved 'world view' that planners, architects, urban designers, and other stakeholders need to adopt. Design-led planning events often involve members of a community working alongside local authorities and developers and other stakeholders to co-create visually planned, agreed action plans and strategies (Campion, 2018). Such events are used to stimulate discussion of place-based issues, promote thinking about community values, and allow consideration of the ways in which assets can be best utilised. However, concerns have also been expressed about the quality of, and skills required to support, effective facilitation in community participation, highlighting an ethical responsibility to include and serve wide-ranging stakeholder groups (Wates, 2014; Campion, 2018; Malone, 2018; AlWaer and Cooper, 2019, 2020). Three issues in particular have been signposted: first, that facilitators may be biased, over-powering, manipulative, or not concerned with meeting community needs; second, that there is often an overdependence on facilitators with 'subject-specific' knowledge; and third, that there may be inadequate attention to process-based facilitation skills, particularly the 'social competencies' needed for process management and stakeholder engagement (Kaner et al., 2007; Wates, 2014; AlWaer et al., 2017; Cooper and AlWaer, 2019).

Ensuring meaningful engagement in collaborative planning has been signalled as a key priority to allow individuals and groups to feel included and valued (Wates, 2014). Facilitation is required to support stakeholder engagement at various points during design interventions in the built environment and stakeholder engagement has become a prominent part of the practice expected from built environment professionals (AlWaer and Illsley, 2017). However, community participation in design is an example of planning complexity (Innes and Booher, 2018) where fragmentation, uncertainty, and 'social problems' are compounded by a multiplicity of stakeholder views, and where the resultant complexity needs to be understood as socially constructed rather than merely a product of complicated processes.

This chapter brings together the diverse expertise and experience of both professional and lay participants of design-led events held across Scotland over the past decade, collated into a report for the Scottish government (AlWaer et al., 2017). This wide ranging report, *Shaping Better Places Together: Research into the Facilitation of Participatory Placemaking*, included advice that equal weighting needs to be given to two skills sets for those involved in mounting community-orientated design-led events: non-context-specific skills for managing the processes involved, and skills required to provide the professional and context-specific information required for such events. The study explored new ideas and shared insights captured from recent experiences across Scotland which have wider international significance. The report's analysis was based on a literature review, a survey to capture relevant experience from both those 'who participated in design-led events' and 'those who facilitated them', and – using the survey outputs to inform the format and focus – a one-day interactive workshop to use the 'aspirations and concerns' they had expressed in the survey (see below) to frame a future agenda for both improving professional practice in community engagement and helping to make the outputs of design-led events more robust and deliverable. The study sought to add to the wider discourse in collaborative community planning about where, when, and how to use engagement opportunities – a topic actively discussed within both practice and academia. It identified that there is no 'one-size-fits-all' approach to the roles and effectiveness of facilitators and other professionals in individual design-led events – let alone in terms of their wider contribution to the ongoing stages of community planning as a whole. Each planning intervention has its own particular context and circumstances, its own journey, rhythm, and hoped-for destination. Understanding these and manipulating them successfully calls for insightful inputs not just from professionals but, aided by well-directed facilitation, from local stakeholders as well. The process employed for collecting and analysing the data on which this chapter is based is detailed in AlWaer et al. (2017). Other non-skills-related findings are reported in AlWaer and Cooper (2019 and 2020).

Recognising professionals' roles and responsibilities (management task)

Facilitation of participatory design-led events is often seen as a team effort, involving a lead facilitator working with co-facilitators called upon to help run active participatory group work sessions. The primary concern of this 'facilitation team' is the smooth operations of participation. Different engagement tools might be used to 'promote meaningful participation, such as generating mutual understanding, inclusive solutions and cultivating shared responsibility' (Kaner et al., 2007). The experience and expertise that members of the facilitation team bring to their roles can vary widely. They may be built environment professionals with some expertise in facilitation, professional facilitators with no built environment expertise, or built environment professionals with little or no expertise or experience of facilitation (AlWaer et al., 2017). Influence from the facilitators and the facilitation team can run upwards (towards regulators and policy-makers) or downwards (towards the community and the end user). Doing this effectively requires combining non-expert knowledge with expert contributions to boost the value of bottom-up experiences and ensure wider impact (Woods et al., 2018, p. 211). The multiple disciplines present in facilitation teams involved in community design processes could make them well-placed to help synthesise such local, context-aware (bottom-up) thinking with national and regional (top-down) 'planning' guidance, legislation, and regulation (Rogers and Leach, 2014). There are suggestions in the literature (Lennertz and Lutzenhiser, 2006; Conrad, 2010; Wates, 2014; Campion, 2018; Malone, 2018) that those who facilitate community-based, design-led events need to address five key dimensions as explained below (Conrad, 2010, p. 47). These dimensions are not presented as being linear. Rather they are regarded as being cyclical in nature, occurring through a collection of processes and strategies made up of a multiplicity of activities and entry points (Ermacora and Bullivant, 2016).

1. Scope: the rationale for involving the public

The purpose of engagement proposed at an event needs to be explicit from the very beginning and the activities and tools chosen for use at it should both support collaborative decision making and be relevant to pursuing outcomes, in order to avoid raising false expectations (RTPI, 2005). It is important that the tools are supportive and do not become a distraction from the objectives of the engagement.

2. Representation and addressing inequalities: the extent to which public involvement in the process is inclusive and represents all those affected.

Those involved in collaborative planning should comprise a broadly representative sample of affected population (Rowe and Frewer, 2000; Mascarenhas and Scarce, 2004). The processes adopted must show how they will help to tackle inequalities and combat disadvantage. Seldom heard and marginalised groups are at particular risk of exclusion; they include children and young people, people with disabilities, ethnic minorities, minority faith groups, people with physical disabilities or mental health problems, and gypsy/travellers (Yellow Book Ltd., 2017). Additionally, it is important to demonstrate how the tools and techniques to be used will support the ongoing engagement of groups or individuals who may face barriers to participation, particularly across areas of deprivation, and areas affected by poverty or people living in remote locations (Policy Link, 2012; Malone, 2018). People living in particularly challenging circumstances, such as the homeless or young carers, also need to be included (Yellow Book Ltd, 2017).

3. Understanding stages in collaborative planning: the extent to which the public is involved both early on and throughout the collaborative planning process.

Participants should be involved from early on in the process, as soon as value judgments become salient, and when there is potential to influence the brief, scope, and methods used for any design intervention in the built environment (AlWaer and Cooper, 2019, 2020). Lay members of the public are often involved late in final stages, only to provide feedback on chosen options, which are already largely complete, to provide buy-in to a preconceived project or vision. Where this happens, collaborative planning is likely to be seen as lacking genuineness and legitimacy, leading to disillusion and even growing distrust AlWaer and Cooper, 2019, 2020).

4. Creating a comfortable and convenient environment: the extent to which the process of participating is rendered easy for the public.

The logistics employed for community-based, collaborative planning are seen as determining whether the process is rendered comfortable and convenient for the wider public. However, there are no hard and fast rules about the amount of engagement activity required, the level of stakeholder engagement that is appropriate, or about the methods to be used. Such issues should be determined jointly by community organisations, facilitation team, planners, and policy-makers (Yellow Book Ltd, 2017). The process has to be focused on how local stakeholders will be involved in the delivery of the outcomes of any design events mounted. Organisers have to set out how they intend to take forward the actions arising from their event(s), and to consider what the challenges and risks might be and how these might be addressed (AlWaer and Cooper, 2020).

5. Influencing outcomes and design interventions: the extent to which collaborative planning delivers concrete results and outcomes.

Whilst the success criteria listed above relate mainly to the process of participation, influence is a measure of the degree of effect that participants have on resultant outcomes such as the quality of life in an area. Whilst influence is manifested in several way, important considerations are held to include: the extent to which the public's contribution will influence the decision-making process; the extent of transparency concerning the incorporation of public views in decision-making as represented in final outputs; and the effectiveness of the process in terms of its effects of outcomes, subsequent performance, and longer-term sustainability, as achieved over the short, medium, and long term (Condon, 2008; Roggema, 2014; AlWaer and Illsley, 2017; Campion, 2018; Malone, 2018; Campbell, 2018).

The personal attributes and skills required for successful facilitation

Those who have been involved in design-led events in Scotland point to two types of skills that facilitators required for their effective delivery. These can be divided into personal attributes and learned skills (AlWaer and Cooper, 2019, p. 209). The personal attributes require that facilitators should be open-minded and supportive of different agendas and views: participants will want to be involved in the community design event for a variety of reasons. This need not be a problem. Diversity of perspectives can assist in developing novel or innovative design interventions. Facilitators must accommodate different, often competing agendas and views in an open and inclusive manner. They should also be approachable and welcoming and respectful of the perceptions, choices, and abilities of all participants. When a facilitator is open and friendly, participants respond. People feel they can talk in an open manner and are more likely give valuable feedback. Facilitators should also be honest, open and trustworthy, and straightforward about the nature of any activity during the event. This is part of managing expectations, both avoiding raising unrealistic aspirations and clarifying what might be achieved. People will participate more enthusiastically if they know that something can realistically be achieved. They also need to be courteous and humble, displaying good conduct before, during, or after charrettes, and the capacity for 'silence.' Where the participants are engaged, a facilitator should remain silent and ensure nobody is disrupting or being left out. Facilitators need to adopt an impartial stance, portraying a neutral attitude. They become compromised where there is potential for manipulating an apparent viewpoint. Above all, they need to be empathetic, able to sense and understand the feelings and concerns of others. This helps them to identify effective means of developing contributions.

There are also learned skills that facilitators need to develop and hone. They need to be able to work flexibly, modifying an event's structure and activities as circumstances dictate and avoiding inflexible methods and strategies, their aim is to help in moving towards common objectives, as agreed during the pre-event preparation. They need to pursue continuous improvement, reflecting and working to improve practice by combining knowledge, skills, and behaviours. Competency grows through experience. They have to empower participants, involving all affected parties as early as possible, seeking to identify participants' needs in terms of perspectives, and abilities to participate effectively. They have to be self-aware, practising self-reflection both during the running of an event and afterwards, recognising the feelings and impacts generated by how they discharge their role. They need to be organised, giving sufficient time and effort to preliminary stages to ensure the smooth running of the event. They should focus on consensus building, supporting all participants so as to search for inclusive solutions. This may require them to mediate, presenting participants with the pros and cons of the positions being disputed, leaving participants to make their own decisions about them. It is not a facilitator's role to choose between or promote one side or another.

They have to communicate clearly, carefully expressing themselves and giving advice or instructions that are unambiguous. Such communications can be given to participants in written form. This will involve them in listening attentively, paying careful attention to what participants are saying, and not letting preconceptions cloud their understanding, and giving careful attention to what isn't being said and, where necessary, reading between the lines about what is. This is also known as 'active listening.' They may have to challenge prevailing assumptions, work to recognise and unpack meanings in a manner that respects the integrity of those who hold them to avoid alienation and work through any conflict. They should seek inclusive solutions, balancing impartial inclusivity and experience-based advice as decisions made from a diversity of inputs are more likely to be sustained by stakeholders.

Faciltators should signal where experience suggests that a proposed course of action is unrealistic or likely to result in failure. When applying their judgement, they need to be transparent if offering advice. This will involve them in being fair and inclusive, creating a safe space where 'truth can be spoken to power,' and where professionals' expertise and lay people's lived experience are both treated as valid. It is essential to give all participants an equal voice, regardless of power, status, education, social capital. Behind all of this is a need to act ethically, respecting participants' confidentiality and acting within a recognisable moral code and in accordance with accepted rules. So, for instance, confidentiality restricts the facilitator from further and unauthorised dissemination of information to which they become privy during their engagement with participants.

The overarching skills within the very broad sets outlined above are: preparation, impartiality and seeking inclusive solutions, and consensus building. To be well-structured, an event will exhibit successful deployment of these skills – which significantly require not just effective communicating but effective listening too. In practice, the features that participants in design-led events see as desirable fall into two categories: those that are predominantly personal attributes, and those that can be taught – but some features contain an element of both. It is worth comparing these categories with what Shulman (1998, p. 525) elaborated as the features of a profession. These comprise, he suggested, 'moral vision, theoretical understanding, practical skills, the centrality of judgment, learning from experience and the development of responsible professional communities.' Shulman's list emphasises both the technical and moral dimensions of preparing for and undertaking professional activity. Shulman later characterised (2005) professional education as a synthesis of three types of apprenticeships: a cognitive apprenticeship where one learns to think like a professional; a practical apprenticeship where one learns to perform like a professional; and a moral apprenticeship where one learns to think and act in a responsible and ethical manner. His conceptualisation of professional education integrates across all three domains. Such a holistic apprenticeship model could help to prepare professional facilitators with what Shulman called 'knowledge for action.'

In the study for the Scottish Government, the professional facilitators and participants in design-led events were asked to rank what they saw as the top six priorities of the competencies, skills and qualities for facilitators. Their first six priorities were: 'effective communicator; organised; good listener; impartial; empowering others; and challenge assumptions.' The first three of these priorities were ranked similarly by both types of respondents, highlighting the shared importance attached to them by both facilitators and event participants. However, impartiality was not as highly ranked as a priority by facilitators as it was by participants. Only a fifth of facilitators signalled this as important in comparison to a third of event participants. Conversely inclusivity and challenging assumptions were not rated as highly as priorities by participants as they were by facilitators. A quarter of facilitators signalled challenging assumptions as a key skill whereas less than a fifth of participants did. These differences, whilst not vast, are important. Those who run design-led events and those who participate in them may share some criteria for assessing how well they are run, but their views are not identical. Clear separating emphases are apparent. Facilitators need to recognise these and understand how to take them on board and adjust their facilitation practices accordingly. Knowing where they and those they facilitate (dis) agree, and making necessary adjustments to how they operate, may prove important for meeting participants' expectations.

The responses from event participants indicate that facilitators require skills relating to 'people management' – coaching, mediation, therapy, and community development, not simply architecture, planning, and urban design. Whatever the personal skills of facilitators, members of the wider facilitation team also need understandings of planning law, policy, local government, action planning, and the production and management of built environment. Such skills are necessary in order to build traction, secure funding, and acquire regulatory approval. A facilitator may have to be neutral – and may deliberately play a 'naive role' – but this does not mean they can be ignorant. They should either know where and how proposed activities are likely to 'end up' or be able to orchestrate the skills of their wider facilitation team to achieve these ends. For example, a facilitator who is a skilled process manager can deliver engaging events – that participants experience as fun – but may result in outputs that cannot be taken forward. However, with some knowledge of built environment processes, they can help generate more plausible/ feasible ideas solutions etc. (e.g. more acceptable to the funding or regulatory system in design terms). A facilitator with a deep knowledge and understanding of design and planning processes, funding streams, and regulatory systems can add to the likely deliverability of an output arising from design-led events. The joint deployment of these two skill sets – people management and professional understanding – do not ensure deliverability but their effective integration makes this more likely.

The facilitation skill set employed in design-led events is seen by respondents as being experientially based, though some suggested that it can be improved through training. Importantly, it is also seen as being learned (honed) through action. Both the confidence required, and the reflexivity deemed important, are personal traits, as is an ability to handle difficult social and interpersonal situations. The study's results indicate that facilitators operating at built environment events have to be capable of deploying a very broad range of skills. When built environment professionals act as facilitator, they need to supplement their knowledge-base of technical domains, of urban design and planning, with social competencies required for effective process management and stakeholder engagement. When trained, non-domain, facilitators do so, they need to bolster their process and engagement skills with domain ones. Since both sets of skills are unlikely to reside in one individual, this bolstering will need to be done by the recruitment of others to balance the facilitation team. This balancing is necessary in order to link spatial planning and community planning, including co-ordination between service provision and physical design considerations. One skill which was missing from the list identified in the survey – but which participants in the workshop suggested cut across all aspects of collaborative community planning – is the nature of the leadership provided by facilitators. This, participants agreed, should be non-dictatorial.

The role of facilitation in the five key stages of collaborative planning

Participants' responses from the survey and at the workshop were analysed and collated in order to identify what they, as individuals or through group/plenary discussions, saw as the imperative actions and activities that need to be undertaken by facilitators throughout each of the key stages of collaborative community planning (see Figure 32.1).

The imperatives respondents offered can be separated into those that occur before, during, and after design-led events. These imperatives are reported below against the stage at which respondents suggested they should occur. These suggestions do not represent a consensus view across all participants in the study: rather they reveal an 'ideal' wish list constructed from the very wide breadth of aspirations expressed by individual and group comments.

Brief and purpose

This stage is to be organised by a 'stakeholder management team' – possibly involving representatives from local authority/public agency, independent consultants, community groups, and the third sector. Briefings are necessary to outline the basics of what will happen, the key issues that an event will explore, and to establish what participants should be aware of, and to agree what background information is required to be established prior to the event. The involvement of the lead facilitator at this stage is desirable, but not essential unless they have acted as the main front-person during previous related activities. There needs to be agreement of who should attend the design-led event and what their actual responsibility and authority will be. Facilitators may not be involved unless they have engaged with the stakeholder management team in writing the brief.

Pre-event facilitation

Facilitators should have a clear plan of action and engagement strategy; bringing together a multidisciplinary team with the appropriate skills, knowledge, and social competences to accomplish

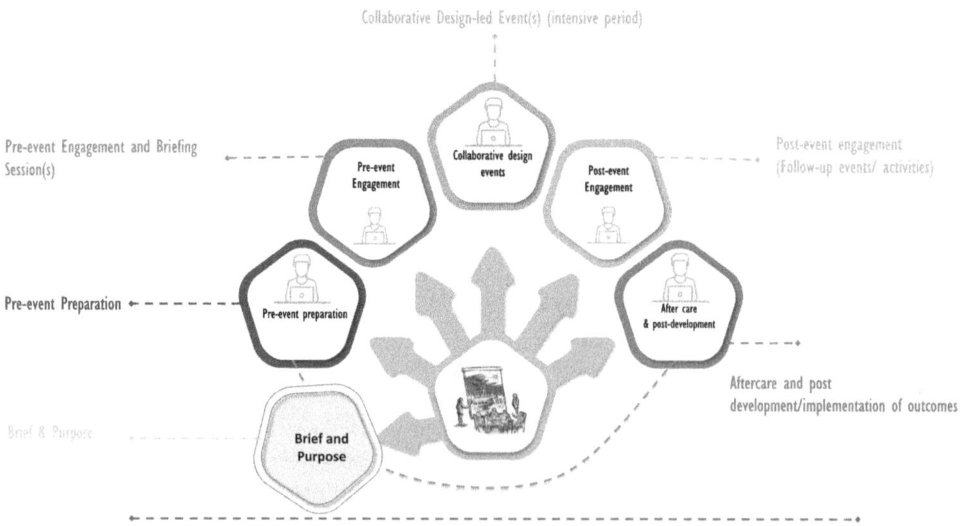

Figure 36.1 Sequence of stages surrounding design-led events – over-simplified linear framework (AlWaer and Cooper, 2020).

this plan; and the logistical organisation of the event itself, e.g. venue, equipment, materials, budget, advertising/publicity, and ensuring key decision-makers are in attendance. According to some engaged through the study, there is a need for facilitator input early on and such early involvement was seen as especially important where: the stakeholder management team does not have a neutral role (i.e. they may be promoting a specific plan/solution or be working to a special agenda); the client or local stakeholder groups are inexperienced – so a facilitator could be helpful in building trust in the process and methods being employed; and there is a pressing need (sometimes because of previous activities) to establish the principles that will underlie the approach to be used, including in how the event will be managed, and how any material arising will be dealt with. The aim here is to try to avoid criticisms levelled at early design-led events that facilitators were 'parachuted in,' then left the community once the event was over.

Pre-event engagement

Facilitators should be involved in agreeing with relevant stakeholders – such as local stakeholder groups – the intended aims, objectives, and outcomes of the design-led event, along with establishing its terms of reference, and detailing the approach used for publicity and engagement. Survey respondents suggested that the facilitation team should get to know where in the locality stakeholder groups congregate, and then take their discussion to these venues; and by facilitators attending community briefings, utilising a wider range of media, and devising innovative engagement approaches specifically targeted to attract under-represented or harder to reach groups, involvement in the event can be promoted to wider stakeholder groups, resulting in a more diverse attendance, and, in turn, the creation of objectives that address the real issues and concerns affecting their community. Facilitated engagement at this stage could be used to: support effective decision-making about the overall structure of the event; identify who with appropriate authority and responsibility should be invited to attend; and clarify which factors could dramatically impact on the success of the engagement process.

Collaborative design events

Facilitators (and, by extension, the facilitation team) create a 'safe space' which can support conflict-free relationships within clear boundaries, so people can freely share their ideas, aspirations, and concerns by jointly working through potentially difficult issues, and eventually translating these into an 'action plan' and 'strategies.' The input of facilitators is essential here. They are expected to direct the whole event. But the rest of the facilitation team also need to be synchronised and aligned in order to manage and deliver a smooth event. Expectations about the facilitator's role, and that of specialists, client, and local stakeholders, should all be made explicit at the beginning of the event in order to effectively manage the following 'live' process. Clear guidelines should be set to empower participants and they should be encouraged to see the issues from the perspectives of other stakeholders. Local stakeholders should be enabled to guide discussion of what are deemed appropriate issues. Facilitators should ensure everyone is given the opportunity to contribute, by encouraging the less confident to speak up whilst managing more vocal individuals and groups. It is important that facilitators bring out the ideas of all the stakeholders assembled and that they draw on the knowledge, expertise, and creativity of the design team and any planners involved. To encourage meaningful participation, the engagement process should be as simple, open, and transparent as possible, with plain English (in an English as first or common language context, naturally) used at all times. Serious consideration also needs to be given to the practicalities of managing community expectations. Building trust by treating local stakeholders as equals through listening and demonstrating that you have done so – by explaining clearly why some of their ideas have emerged as preferred solutions whilst others don't work as well or cost too much – is vital. Lay participants suggested that there were support materials or activities which they thought might have improved experience of their participatory design process. In order of preference they cited: computer-generated 3D visualisations, exploring ideas through art and digital media, physical models, social media, and use of Scotland's Place Standard. They also thought that use of large-scale maps and photographs is helpful. Facilitators broadly agreed here. They added to this list using plans, sketches, diagrams, and sticky notes to develop ideas, using exemplars in the form of visual references or actual site visits, walk-and-talk site visits, storytelling, templates, visual summary boards, and fun activities.

Post-event engagement

Involvement of the lead facilitator at this stage is desirable, but not essential unless they have acted as the main front-person during the previous stages. Continued involvement of the lead facilitator in follow-up events and activities would be beneficial to the community to build momentum, kick-start a stalled action plan, create active community groups, and support communities through the implementation stages of collaborative planning. Facilitators also reported that facilitated follow-up and aftercare were beneficial in supporting local stakeholders through the implementation stages of the collaborative planning and were important in developing a more long-term community-based approach to placemaking. However, they noted that this approach would inevitably increase the work expected of facilitators and that, as a result, it would need to be properly funded.

Aftercare and post-development

Facilitators may usefully be present at this stage. But this may be less crucial if local stakeholders, in the form of trusts, partnerships, or networks, are taking ownership and leadership of delivery phases – often 6, 9, or 12 months after a design-led event. From participants' responses and comments, it is evident that they recognised that the roles of facilitators may change and vary throughout the stages of collaborative community planning. For example, a facilitator may at one point in this process be acting to develop a shared understanding of an issue, at another to explore design possibilities, and at another helping to articulate recommendations or concretise decisions. Clarity needs to be established about whether design input is required from a facilitator or whether this is to be provided by members of the design team participating in events.

Conclusions

The results of this research point to the need to extend the period during which facilitators are asked to make contributions to collaborative planning. This period needs to begin long before the design-led events they are brought in to facilitate and should extend long after these have been held (see Figure 34.1). Specific actions and activities to be undertaken by facilitators were identified by both professionals and local stakeholders that need to be enacted throughout each of the key stages of collaborative community planning.

Facilitators should be involved in pre-event preparation to ensure that a broad range of local stakeholders are approached in their own localities and on their own terms. This is necessary to manage participants' expectations about what design-led events can achieve and to counter concerns that facilitators can have inadequate local knowledge and expertise to guide such

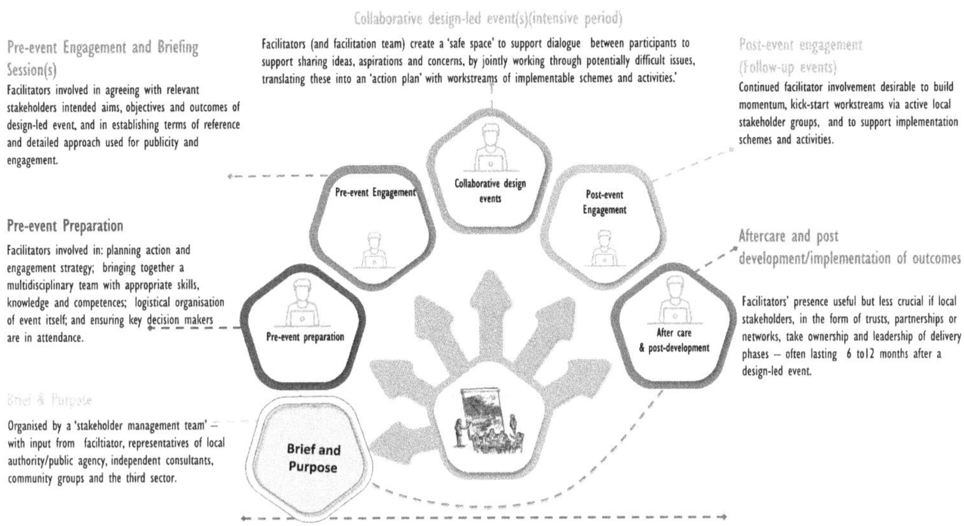

Figure 36.2 Facilitator involvement in stages of collaborative placemaking – over-simplified linear framework (AlWaer and Cooper, 2020).

events effectively, instead pursuing agendas set by those organising them. Developing a more longitudinal approach to the involvement of facilitators would require more extensive funding. It would, however, help build the inclusive solutions participants called for and then carry these forward into the reporting of such events and into subsequent implementation of decisions made at them.

The study undertaken for the Scottish government highlighted strongly divergent opinions on who should facilitate design-led events. Some respondents suggested, indeed insisted, that facilitators need a domain-based understanding of the design of the built environment. Others held that facilitators should be independent and professionally trained in facilitation in order to ensure that the process employed leads to legitimate and socially acceptable outputs capable of delivering not only desired but feasible outcomes. To play their role effectively – whether at a design-led event or through contributions made before and after – facilitators are identified as needing a broad range of skills, running from people management through technical understanding to local knowledge. Since this range is unlikely to be owned by a single individual, this signals the need to build a facilitation team whose members can contribute the range of skills required link spatial planning and community planning, to co-ordinate service provision and physical design considerations. Local stakeholders and professionals have overlapping but different expectations about what are the most important skills that facilitators will exercise in support of collaborative planning. Local stakeholders may stress the need for inclusivity through coaching, mediation, and community development; professionals may emphasise deployment of design and planning expertise. Facilitators need to recognise these differences and adjust their practices accordingly.

To be effective, collaborative planning depends on integration and synergy across professional disciplines, local stakeholders, and process stages. It requires building trust and common purpose between team members and local stakeholders from a wide range of backgrounds and constituencies. Ideally, its aim is to engender a deep, collective understanding of the places where interventions are planned through developing dialogue and deliberative participation.

Enabling this collaborative dialogue, and then empowering implementation of its resultant co-decision-making, are essential. Achieving this may require liberating facilitators from their usual time-limited role in collaborative planning – solely at design-led events – to extend their contributions across the whole time frame of planned design interventions in the built environment.

Acknowledgements

This work was supported by the Scottish Government (Planning and Architecture Division) under the grant: 'The role of the facilitator operating a participatory and community design settings in Scotland' (2017). Our thanks go to our research team colleagues: Frances Wright, Iain MacPherson, and Kevin Murray. We gratefully acknowledge the ideas and inputs received from the wide range of people who took part in the research – in the business and community sectors, academia and practice, and the political and public policy communities –across Scotland and beyond.

References

Allmendinger, P. (2002). 'Towards a post-positivist typology of planning theory', *Planning Theory*, 1(1), pp. 77–99.

AlWaer, H. and Cooper, I. (2019). 'A review of the role of facilitators in community-based, design-led planning and placemaking events', *Built Environment*, 45(2), pp. 190–211.

AlWaer, H. and Cooper, I. (2020). 'Changing the focus: viewing design-led events within collaborative planning', *Sustainability*, 12(8), p. 3365.

AlWaer, H. and Illsley, B. (2017). *Rethinking the Master Planning: Delivering Better Places*. London: ICE Publishing.

AlWaer, H., Cooper, I., Wright, F., Murray, K. and MacPherson, I. (2017). *Shaping Better Places Together: Research into Facilitating Participatory Placemaking*. Dundee: University of Dundee [online]. Available at: https://discovery.dundee.ac.uk/en/publications/shaping-better-places-together-research-into-facilitating-partici (Accessed 20 June 2018).

Campbell, K. (2018). *Making Massive Small Change: Building the Urban Society We Want: Ideas, Tools, Tactics*. Hartford, VT: Chelsea Green Publishing.

Campion, C. (2018). *20/20 Visions: Collaborative Planning and Placemaking*. London: RIBA Publishing.

Condon, P.M. (2008). *Design Charrettes for Sustainable Communities*. Washington, DC: Island Press.

Conrad, E. (2010). *Putting Participation into Practice: The Role of People in Landscape Protection, Planning and Management*. Unpublished PhD thesis. Aberystwyth University.

Ermacora, T. and Bullivant, L. (2016). *Recoded City: Co-Creating Urban Futures*. London: Routledge.

Healey, P. (2003). 'Collaborative planning in perspective', *Planning Theory*, 2(2), pp.101–123.

Innes, J. and Booher, D. (2018). *Planning with Complexity: An Introduction to Collaborative Rationality for Public Policy*. London: Routledge.

Kaner, S., Lind, L., Toldi, C., Fisk, S. and Berger, D. (2007). *Facilitator's Guide to Participatory Decision-Making*. Hoboken, NJ: Wiley.

Lennertz, B. and Lutzenhiser, A. (2006). *The Charrette Handbook: The Essential Guide for Accelerated, Collaborative Community Planning*. Chicago: American Planning Association.

Malone, L. (2018). *Desire Lines: A Guide to Community Participation in Designing Places*. London: RIBA Publishing.

Mascarenhas, M. and Scarce, R. (2004). '"The intention was good": Legitimacy, consensus-based decision making, and the case of forest planning in British Columbia, Canada', *Society and Natural Resources*, 17(1), pp. 17–38.

Policy Link. (2012). *The Community Engagement Guide for Sustainable Communities* [online]. Available at: http://www.policylink.org/sites/default/files/COMMUNITYENGAGEMENTGUIDE_LY_FINAL%20%281%29.pdf (Accessed: 20 June 2020).

Rogers, C. and Leach, J. (2014) Birmingham Policy Commission on Future Urban Living. Birmingham: University of Birmingham. Available at: htt p://www.birmingham.ac.uk/Documents/research/policycommission/future-urban-living/futureurban-living-policy-commission-report.pdf.

Roggema, R. (ed.) (2014). *The Design Charrette: Ways to Envision Sustainable Futures*. Netherlands: Springer.

Rowe, G. and Frewer, L.J. (2000). 'Public participation methods: A framework for evaluation', *Science, Technology, and Human Values*, 25(1), pp. 3–29.

RTPI (Royal Town Planning Institute). (2005). *Guidelines on Effective Community Involvement and Consultation* [online]. Available at: https://www.rtpi.org.uk/media/6313/Guidlelines-on-effectivecommunity-involvement.pdf (Accessed: 20 June 2020).

Shulman, L. (1998). 'Theory, practice, and the education of professionals', *The Elementary School Journal*, 98(5), pp. 511–526.

Shulman, L. (2005). 'The signature pedagogies of the professions of law, medicine, engineering, and the clergy: Potential lessons for the education of teachers', in *Teacher Education for Effective Teaching and Learning Workshop*. Irving: NRC.

Wates, N. (2014). *The Community Planning Handbook: How People Can Shape Their Cities, Towns and Villages in Any Part of the World*. London: Routledge.

Woods, M., Balestrini, M., Bejtullahu, S., Bocconi, S., Boerwinkel, G., Boonstra, M. and Fazey, I. (2018). *Citizen Sensing: A Toolkit*. Dundee: University of Dundee.

Yellow Book Ltd. (2017). *Barriers to Community Engagement in Planning: A Research Study*. Edinburgh: Yellow Book Ltd [online]. Available at: https://www.gov.scot/binaries/content/documents/govscot/publications/factsheet/2017/05/barriers-to-community-engagement-in-planning-research/documents/3210b62e-9f7a-4eb7-8174-4665b21deb5a/3210b62e-9f7a-4eb7-8174-4665b21deb5a/govscot%3Adocument (Accessed: 20 June 2020).

Further reading in this volume

Chapter 2: Placemaking as an economic engine for all
James F. Lima and Andrew J. Jones
Preface: 'Disastrous forces, accidental actions, and grassroots responses'
Tom Borrup
Chapter 9: From the dust of bad stars: disaster, resilience, and placemaking in Little Tokyo
Jonathan Jae-an Crisman
Chapter 13: Sensing our streets: involving children in making people-centred smart cities
Sean Peacock, Aare Puussaar, and Clara Crivellaro
Chapter 15: Un/safety as placemaking: disabled people's socio-spatial negotiation of fear of violent crime
Claire Edwards
Chapter 16: More than a mural: participatory placemaking on Gija Country
Samantha Edwards-Vandenhoek
Chapter 24: Artists, creativity, and the heart of city planning
Tom Borrup
Preface: Evaluating creative placemaking: a collection of observations, reflections, findings, and recommendations
Maria Rosario Jackson
Chapter 42: Creative placemaking and placekeeping evaluation challenges from the practitioner perspective: an interview with Roy Chan
Maria Rosario Jackson
Chapter 43: A theory of change for creative placemaking: the experience of the National Endowment for the Arts' *Our Town* program: an interview with Patricia Moore Shaffer, PhD
Maria Rosario Jackson
Chapter 45: How the city speaks to us and how we speak back: rewriting the relationship between people and place
Rosanna Vitiello and Marcus Willcocks

37

THE NEIGHBOURHOOD PROJECT

A case study on community-led placemaking by CoDesign Studio

Lucinda Hartley, Eliza Charley, Sama Choudhury, and Harriet McKindlay

Tackling process barriers to placemaking

Placemaking, as a collaborative approach to strengthening the connection between people and the places they share (PPS, 2007), has grown in awareness and popularity significantly over the past two decades (Friedmann, 2010). However, as CoDesign Studio has observed, when it comes to community-led placemaking, a number of barriers exist (CoDesign Studio, 2019e). Through practice working on dozens of neighbourhood improvement projects across Australia (Hartley and Lydon, 2014), we have seen that local residents and community volunteers often find it difficult to deliver place improvements in their own neighbourhoods, due to a prevalence of red tape such as permits, fees, and licenses.

In response, CoDesign Studio created *The Neighbourhood Project*, a practice-based program that worked with community groups and local governments in Australia to tackle process barriers while concurrently activating underutilised public land and developing a model for community-led (CoDesign Studio, 2019b). This chapter offers a summative case study overview of the experiences and outcomes of *The Neighbourhood Project*, a four-year community-led placemaking program undertaken by not-for-profit organisation CoDesign Studio, in Melbourne, Australia, from 2015 to 2019. It will discuss *The Neighbourhood Project* as a practitioner case study, and sets out the People, Process, Place (PPP) framework as a tripartite model for placemaking professionals, local governments, and community leaders to adopt to improve their understanding and implementation of community-led placemaking. Our hope, as placemaking practitioners, is that the experience of *The Neighbourhood Project* program in Australia can provide inspiration and insight to community-led approaches in a global setting, and through its practical implementation work towards refining the global understanding of collaborative models of city-making.

The context

Human settlements around the globe are becoming more crowded (UN Habitat, 2016). With 200,000 people moving to cities every day (WEF, 2016) there is increasing pressure on local governments to provide sufficient public space to meet the needs of residents as neighbourhoods become denser. In Australia, social isolation and loneliness are reaching epidemic proportions (Kelly,

2012). This is becoming an increasing concern as rapid urbanisation unfolds (ibid.), coupled with the fact that up to 30 per cent of land in Australian cities is vacant or underutilised (*The Economist*, 2015). Underutilised space offers an untapped resource to meet local needs for public space and social connection. However, local government resources struggle to keep pace with growth.

In this context, placemaking – defined here as a collaborative, human-centred approach to designing, activating, or managing places offers the opportunity to both activate underutilised space and strengthen communities; while tactical urbanism – short-term action for long-term change (Lydon and Garcia, 2015) – creates pathways to accelerate public space creation. Placemaking and tactical urbanism have grown in Australia over the past decade as approaches to transforming underutilised space (Hartley and Lydon, 2014). The challenge is that the majority of these projects are led by practitioners or governments, rather than community leaders and local residents, limiting their scalability and therefore impact.

Participatory approaches to public space improvement offer enormous benefits where they can strengthen social ties as well as improve aesthetics and physical places (Cilliers and Timmermans, 2014). Community-led placemaking (as differentiated from the understanding of placemaking, as above) is defined here as a participatory approach to placemaking where the projects are initiated, led, implemented, or managed by local leaders – local residents, volunteers of community organisations – to impact their own local places. There is potential to activate local communities to take a greater role in leading, delivering, and managing placemaking projects. Furthermore, this further benefits local governments and urban developers in the long-term as in this model local leaders take greater responsibility for sharing the load of public space improvement and management (Healy, 1998).

The problem

CoDesign Studio's work across 132 placemaking projects since 2010 working with local leaders across Australia, often utilising the tactical urbanism approaches depicted in *Tactical Urbanism 4: Australia and New Zealand* (Hartley and Lydon, 2014), revealed that community members were actively trying to establish local neighbourhood improvement projects (such as murals, community gardens, or main street beautification projects) especially in urban areas facing rapid urbanisation. However, they were repeatedly prevented from executing their ideas by certain types of process barriers. These varied from strict permitting procedures and not knowing who to talk to at their council (in the case of community members) to not having appropriate resources or culture to guide people through the process (in the case of councils.)

Some examples include a local business organisation who had been waiting six months for approval for a two-week temporary parklet, and eventually had the permit denied. Another was issued with a fine for drawing with chalk outside their shop in an inner Melbourne suburb. Other community groups trying to build a community garden ran out of steam due to a lack of committed volunteers. Countless other community groups have simply given up due to the complicated or convoluted process of submitting permit applications for even the simplest events and public open-space activities.

The three common categories of barriers identified were: process barriers, knowledge barriers, and network barriers (CoDesign Studio, 2019c). Process barriers include red tape, paying project fees, and land access.

Knowledge barriers includes a lack of practical know-how, from navigating permit applications and liaising with council to practical small-scale construction skills. Lastly, network barriers consider a lack of social capital or not knowing enough neighbours, which may make it difficult for projects to gain traction. Not only did these three types of barriers consistently present a block to local groups, but they are embedded into the systems of council such that they are difficult to overcome on a project-by-project basis. Councils were not willing to change systems to adapt permits as a one-off for a small project, while local leaders (active local residents, local volunteers, and community organisations) didn't have access to tools and resources to help. This led CoDesign to create *The Neighbourhood Project.*

The Neighbourhood Project

In 2015, CoDesign Studio pitched for – and was the successful recipient of – a philanthropic grant from the Myer Foundation for their proposal to improve local neighbourhoods throughout urban areas in Melbourne, using a community-led approach. The proposal was to bring together councils and community members to activate underutilised public space, while concurrently identifying, reviewing, and removing process barriers. This practical program was called *The Neighbourhood Project.*

The program was designed around three key Pillars of People, Process, and Place (PPP). The People Pillar was based on mobilising local citizens to lead community-led placemaking, working with councils to evolve internal systems to enable community-led projects by citizens, building citizen capacity, and mentoring them to mobilise and solve local problems. The Process Pillar facilitated an enabling environment within councils and community governance structures that support communities to take action. The Place Pillar aimed to improve local neighbourhoods thorough community-led projects. Drawing on tactical urbanism (Lydon and Garcia, 2015) as an approach, the program used prototyping and short-term activations as a low-risk testing environment through which councils and communities could work together on practical place-based projects. This was coupled with a six-step workshop program developed by CoDesign Studio to strengthen the capacity of community volunteers and local government to design, construct, and manage small-scale placemaking projects and events. Over the four-year period, CoDesign Studio worked with 8 local governments and over 25 community leaders, their organisations and volunteer networks, on 14 neighbourhood placemaking projects such as community gardens, pop-up parks, street events, community murals, and local art projects. Collectively these projects involved over 60,000 residents and provided early insights into a model for community-led practice by which barriers to community-led placemaking were identified, reviewed, and adapted. As a result of the program, participants noted boosts in neighbourhood pride, social connection, improved land utilisation and local trade, while local councils tested new processes for permit applications and communication (CoDesign Studio, 2019e).

A managed program of training, information, resources, and support

The Neighbourhood Project itself was a managed program that saw council participants and local leaders build their capacity, skills, and knowledge of community-led placemaking through bootcamps, training sessions, expert mentoring, and practical workshops. CoDesign were the program organisers, in partnership with *Resilient Melbourne* of The Rockefeller Foundation's *100 Resilient Cities Program*, and from the Municipal Association of Victoria (MAV). The program furnished participants with fact sheets, templates, and project management collateral to equip them through each step of the program to learn and to deliver projects. Within the program, community-led placemaking projects were delivered by local leaders in partnership with local council staff, alongside the support and guidance of urban practitioners and professional community development staff from CoDesign Studio. Each project within the program took a tactical approach to placemaking (Lydon and Garcia, 2015). For *The Neighbourhood Project* this meant a human-centred approach to designing, activating, or managing places by means of prototyping and testing iterations of community-led projects for engagement and place intervention. This approach mitigates risk through avoidance of large-scale financial or asset investment up-front.

Local leaders were supported through a six-step iterative cycle of project development, with the aim of refining, scaling, building capacity, and ultimately catalysing longer-term systemic change. For each project, local leaders would undertake Step 1, 'ideate together,' creating a project visioning coupled with an ideation process to address how to give underutilised local space a second life, e.g. dog park, playground, or gathering place, create a veggie patch, throw a block party, or commission an art mural. Step 2 is to 'create an enabling environment.' For community leaders this means setting up group governance and project management structures. For councils this means setting up the resources and systems to help Step 1 ideas come to fruition. For example, councils could offer grants or funding specifically aimed at small neighbourhood-based improvement projects; provide a dedicated Placemaking Officer who can help navigate or fast-track permits and processes to allow the idea to happen; create an online application process that is streamlined and user-friendly to submit an idea. Step 3 is to 'mobilise community.' Community leaders reach out and connect with neighbours, volunteers, or a community organisation to support in the project delivery; leading on to Step 4, 'locate resources.' This involved a stocktake of the materials and resources available that might be necessary to implement the project idea. The next stage, Step 5, is to 'test it out' by piloting the project on a small scale and engage the surrounding community response. Lastly Step 6, is 'evaluate and scale' – to reflect on the successes and challenges of the project pilot or trial, assessing how it could be improved and if it is worth replicating. Then return to Step 1 to reiterate and grow the size of the project.

Councils and community leaders were inducted and trained by CoDesign Studio practitioners on how to deliver local projects according to this six-step process. Success of the projects was measured according to the PPP framework. Each group ran one to three iterations of this cycle. Starting with ideation, they co-visioned together as a leadership team to document what they hoped to achieve in their local neighbourhood, highlighting the opportunities and needs. Then they were guided through workshops and training to establish group governance and project management practices as part of creating an enabling environment. Next, project leaders reached out laterally to their community to recruit support, add team members, and generate early interest, as a first step towards mobilising their community. This was followed by an assessment of the resources available to them which included up to 10,000 (AUD) seed money from *The Neighbourhood Project*. Locating resources includes assessing the skills in the team that help shape what is possible and achievable for the project, as well as physical supplies such as reusable goods within the community that can be used for the project. Testing out the project idea is a vital next step in order to ascertain how the team works together, gauging the community uptake of the idea, and receiving early feedback from the broader community through community engagement led by their peers and locals. Finally, the team would evaluate the first iteration of the project and repeat the cycle with the view of scaling the size and impact of the project. Evaluation was conducted according to the Pillars of People, Process, and Place.

People, Process, and Place (PPP) evaluation

The structure of each project and its evaluation was designed to align with the pre-existing CoDesign Studio Pillars of placemaking known as People, Process, and Place (PPP) (CoDesign Studio, 2019b). To create and measure change in a place, PPP was developed to be used as a schema for categorising place conditions, outcomes, and impacts. To validate the PPP model as being fit for purpose for *The Neighbourhood Project* the team sourced local and global best-practice case studies in community-led placemaking and analysed them according to the PPP framework. Additionally, we reviewed the framework through targeted engagement with councils, placemaking practitioners, and community organisations. As indicators of each project, People measured whether the project improved the activation and motivation of local leaders; Process measured whether the project created an enabling and supportive environment at a council level; and Place measured whether the project improved the physical place as well as the softer elements of place attachment such as cohesion, safety, and pride connected to that place.

Surveys and interviews were conducted with council participants, local project leaders, and general community members who attended the projects. These surveys occurred at the conclusion of workshops, on-site at project events and online throughout. In partnership with Melbourne-based social impact consultancy, Think Impact, responses were categorised within the PPP framework according to the following categories, called 'domains': People, including agency (motivation, confidence, influence), connection (to locals, to networks), capability (skills to act, access to resources), and mobility (depth of involvement, willingness to mobilise others); Process, including capability (skills and increased knowledge), systems-(related policy, embedded systems, and inclusion in formal planning), and culture (responsiveness to community, recognition of success, advocacy and leadership for community-led placemaking at council); and Place, including, physical place (amenities, positive feelings, utilisation), cohesion (safety, local connections, belonging), pride (sense of neighbourhood pride), and perceptions (sense of wellbeing). These Pillars were measured at multiple intervals including before, during, and after placemaking projects were delivered. Think Impact was engaged as an independent assessor to assist with designing the specific survey and assessment measures for each Pillar.

The projects

The Neighbourhood Project contained two distinct rounds of community-led projects. While both rounds delivered real-world placemaking action led by local community members, there was a decided difference between whether *The Neighbourhood Project* program was instigated by the council (Round 1) or by the community members themselves (Round 2). Round 1 ran for 12 months across 2015–2016 with three participant council groups who engaged their local citizens to run eight placemaking projects. These were: City of Cardinia – Community Arts Project, Cardinia Lakes Movie Night; City of Whitehorse – Greening the Mall, Town Hall Front Lawn Festival, Community Art Project; and Hobsons Bay Council – Brooklyn Movie Night, Pop-up Dog Park, Laneway Art Project. Round 2 ran for 12 months across 2017 to 2018 with six community groups running multi-iteration projects in partnership with five local councils. These were: Fawkner Food Bowls – Moreland City Council, Edithvale Collective – Kingston City Council, Williams Landing Community Garden – Wyndham City Council, Point Cook Pop-Up Park – Wyndham City Council, Strathmore, Let's Make A Park – Moonee Valley Council, and Thomastown Walk 'N Talk – Whittlesea City Council.

Changing the approach to achieve self-sustaining outcomes

In Round 1, CoDesign Studio ran an Expression of Interest application process and selected six councils. The aim was to work directly with the councils and empower them to enable their own citizens to deliver local placemaking projects. The three participating councils were the instigators to participating in *The Neighbourhood Project* and then sought community participants from their municipalities to deliver placemaking projects on public land as a secondary action. These projects were sought through an open call for ideas where projects could be proposed and voted on by local residents. Three projects were selected in each area, by popular vote. From public planting initiatives, to community-inspired photography exhibitions and a pop-up dog park trial, these three local councils worked closely with their local community members to deliver a series of prototype projects for place improvement. Places were turned green, parks became activated with art and events, and wider community buy-in was effectively gauged by council in order to validate investment in more expensive set ups.

In the evaluation of this first round of projects, *The Neighbourhood Project* organisers noted that community leaders did not continue these events and place interventions after the program ended. It was observed that the ideas and programming had been primarily coordinated at a council level, rather than a local citizen level, and identified this as the possible reason the events and pop-ups did not continue after the initial series was rolled out. The Round 1 projects did however highlight red-tape process barriers for council to address, and saw significant improvements to organisational capability, systems, and culture for community-led placemaking. Participating councils took practical steps to remove process barriers through the program (see below.) While local leaders and councils participated equally in the project, program organisers suggested that because of a perceived 'top-down' influence on the part of council, there was a lower level of license to act perceived on the part of the community.

To address this observation, CoDesign Studio decided that the second round of projects should specifically seek to work with community members directly who already wanted to run local projects in the first place, and then subsequently seek permission from the council. The Round 2 Expression of Interest application process received 91 applications from citizens across the state of Victoria, of which 6 were selected as participants of *The Neighbourhood Project* by a panel of experts from urban design, placemaking, and local government in Australia. As a result of this choice, Round 2 of the program prioritised projects that were both community-led and community-initiated. The Round 2 program maintained the agile and tactical approach of enabling local citizens to run small-scale iterative activations with seed funding, training, and support from CoDesign Studio. It also maintained the aim of highlighting red tape to improve the placemaking process at a council level through a collaborative framework for council–community partnership. In Round 2 community members instigated their participation in *The Neighbourhood Project* and by design were imbued with greater agency to lead the projects for themselves. The councils would still be still actively involved, but the projects themselves were fully ideated, run and 'owned' by the community leaders. This time, by engaging and equipping local community leaders directly through the program, in partnership with their local council's support, five of the six Round 2 community projects continued on after the program concluded in 2018.

The change in approach for Round 2 meant that the program produced a greater number of projects that were self-sustaining and continued after the program ended, while still delivering physical place improvements and increases to social cohesion, community connection, neighbourhood pride, and local economy activation, as measured by the PPP framework (CoDesign Studio 2019e, July). As exemplified in the selected project examples below, this continuance has looked different for each specific group and ranges from the community group becoming an incorporated organisation; securing grants and external funding; becoming a blueprint for larger-scale programs in the community; to being included in council strategies for health and wellbeing plans, pedestrian and street upgrades, community garden planning, and urban design frameworks.

Project Case Study: *Fawkner Food Bowls*

Fawkner is a fast-growing northern suburb of Melbourne transitioning from a predominantly industrial neighbourhood to a thriving and diverse community in need of family-friendly public spaces. Two local community members applied for Round 2 of *The Neighbourhood Project* with the idea that land located at an underutilised local lawn bowls green might be transformed into a community garden. With support from Moreland City Council and from the bowls club, the duo was accepted into the program and began the six-step process from ideating together through to evaluation and scale. The pair was able to grow their project team, locate local resources, collaborate with council, secure $9,460 (AUD) of seed funding, and conduct trial events and construction days as a part of the program. As a result, the bowling green was transformed into a community garden, market, and urban food source where local people grow and share locally grown produce, interact with each other, attend workshops, and learn about sustainable food growing practices. The early activation built a case for greater financial investment, triggering $44,000 (AUD) of external capital works funding after the program concluded. At a council level, the output of the program prompted an internal process review with the aim of putting a more accessible process in place for community project approval in the future, thereby cutting red tape. The project has also been included in the documented Moreland City Council Food Strategy. *Fawkner Food Bowls* is now an incorporated group of eight people, and holds partnerships with Moreland Council Food Network, Merri Health, and local businesses.

Project case study: *Strathmore, Let's Make A Park*

The power of youth-led placemaking is exemplified by the Round 2 project group known as *Strathmore, Let's Make A Park*, located in Melbourne's north-west suburb of Strathmore, which has a rapidly expanding population base under the age of 25. Two university students identified an opportunity to create a new green space near the Strathmore train station by reclaiming a large roundabout, and home to a single tree. They were especially passionate to address growing social isolation in the area and a lack of youth-focussed public space. Through *The Neighbourhood Project* the young leaders recruited a team of peers aged 12–25, quickly establishing a committee of 8 members and a volunteer group of 20. The six-step process helped them to establish group governance, highlight diverse skill sets and existing resources within the team, and run small-scale pilot projects. Through the project iterations, the team planted new greenery, created community artwork to make the space appealing to youth, engaged their community through pop-up engagements, built a street library full of books in partnership with a local Men's Shed, and assessed safety issues around accessibility. The activated space utilised native plants and recycled goods to deliver the lowest costing project of all *The Neighbourhood Project* rounds to date, spending $4,082 (AUD) in program seed funding. They went on to secure an additional $5,000 (AUD) grant from external funding. The resulting new park provides a welcoming green space on the previously underused roundabout land that has endured – when guerrilla gardening techniques may not have – because it was conducted in collaboration with their local Moonee Valley City Council. After the program concluded, the group's analysis of local walkability triggered state government roads agency VicRoads to redesign a master plan of the intersection to increase safety. Later the youth leaders were engaged by their council to contribute to their 20-minute Neighbourhood Strategy.

Project case study: *Williams Landing Community Garden*

The existing Williams Landing Residents Association applied for Round 2 of *The Neighbourhood Project* with the idea to convert an 80 metre by 20 metre patch of vacant land into a green community garden with a sustainability focus. A key desire of this group was to use the agile approach to community-led placemaking to build social connection in addition to creating a greener physical space. Williams Landing has been experiencing rapid urban expansion with a high level of diversity as exemplified by the more than 60 cultures represented at the local primary school. The land itself is located atop an 1890s decommission pipeline stretching 27km in length through Melbourne's western suburbs, which also physically divides the suburb and separates residents from each other. After participating in *The Neighbourhood Project*, the vacant land became a vibrant community garden, complete with a program of local events, picnics, and workshops. By using the PPP framework, social connection and social cohesion was measured to reveal a boost in community connection. Program organisers observed that having a physical task to do at pop-up events was an effective way to help diverse community members to meet and interact. The 27km pipeline forms part of the Greening Pipeline Master Plan run by Melbourne Water to create a linear park in its entirety. As a result of the community-led garden, the process of *The Neighbourhood Project* has been adopted as a 'best-practice standard' to guide the development of the remainder of this piece of public infrastructure to serve communities. The already incorporated Residential Association spent $10,000 (AUD) in seed funding and used the evidence of *The Neighbourhood Project* to secure an additional $7,000 (AUD) from outside sources after the program concluded.

Project findings

The Neighbourhood Project was evaluated in partnership with Think Impact (2019), using online and face-to-face surveys conducted with council workers, community leaders, and project event attendees. The community-led groups also undertook place assessments utilising urban design tools now included in a community-led placemaking manual and the PPP framework. In terms of People outcomes, local leaders and council workers became more active, mobilised, and connected throughout the program. For Process, councils increased their knowledge of placemaking and took practical steps to remove process barriers. All the while, in terms of Place, 14 low-cost low-risk projects were delivered, which resulted in community perceptions of pride around these places becoming more positive.

People outcomes

The People Pillar is used to understand how effectively local leaders and council workers were mobilised to lead community-led placemaking projects by looking at their activation, participation, agency, connection, capability, and mobility. For instance, council workers became more willing to engage with – and developed more connections with – people in their community and within their organisation. At the same time, community leaders became more capable in the delivery of practical place interventions and grew in the number of connections they had with peers, local traders, and other local community organisations. All the while, the delivery of projects through pop-up events and practical engagements meant that the broader community was reached and welcomed into participating in the design and activation of their own local public spaces.

Process outcomes

Overall, council participants across the eight councils noted that they gained improved knowledge and skills (capability) of placemaking and community-led practices across the program; they also had a greater willingness and desire to implement community-led placemaking projects (culture). All of the councils implemented at least one system change. This included one council establishing a central placemaking point of contact, another established a neighbourhood placemaking team to continue to use community-led placemaking projects, and two of the councils created community-led placemaking resources for their constituents.

Place outcomes

The Place Pillar looked at the number and type of projects created, as well as how these low-cost projects influenced participants' social connections, and sense of neighbourhood pride and belonging. Across the projects, 88 per cent of participants felt that the spaces made them feel proud of their neighbourhood, while 86 per cent believed them more friendly, and 87 per cent felt the places were more positively utilised. An important evaluation factor was whether the community groups continued to manage the projects or continue other community-led projects after the program had finished. Participants in Round 1 did not continue, while five out of the six groups in Round 2 did continue. This highlights that having community groups initiate contributes greatly to the sense of ownership and long-term investment in the projects.

Full case studies on individual projects can be found in 'Neighbourhoods made by neighbours: case studies from Round 2 of *The Neighbourhood Project*' (CoDesign Studio, 2019a). For highlighted statistics on project size, participants, and PPP measures see '*The Neighbourhood Project*: outcomes and impact at a glance' (CoDesign Studio, 2019e).

Conclusion

The lessons of *The Neighbourhood Project* offer early evidence for a model of practice for placemaking practitioners and local governments who want to establish more community-led practices. For local leaders, tools and resources have been consolidated into a set of online resources to make it easier to design and deliver placemaking projects. As placemaking grows as an increasing focus for local governments and the property market, there is a risk that activation programs will bring a place to life temporarily, only to leave a space empty and underutilised once the initial investment in programming has come to its natural conclusion. A way to mitigate this is to adopt a community-led approach, using the activations as a means to build community capacity which will be more likely to be self-sustaining in the long term. The People, Process, Place (PPP) model can be adopted as a set of questions by practitioners and local governments in the design of placemaking projects, the objective being to maximise not only the impact of the project on each place but also on the capacity of local people and through the improvement of processes. Put simply, placemaking projects would be much improved if we could simply ask: how will this project create agency (People), what is the placemaking capability of my team or organisation (Process), and how can this project improve amenity as well as perceptions of wellbeing (Place)?

The Neighbourhood Project demonstrated that to increase the likelihood of community-led projects being successful in the long term, it is essential that the projects are initiated by local leaders and funded directly through these local community organisations (Charley, 2019). However, practitioners and authorities can amplify the impact and efficacy of these projects by offering fewer barriers, more resources, and active support. The program highlighted the importance of having a supportive enabling environment within the local government or governing authority. This includes having simple communication – such as through a single point of contact within council, simple forms, and low or waived fees for small-scale projects. *The Neighbourhood Project* is a practitioner case study, yet our hope is that it offers practical insights into evolving approaches to improving community-led placemaking that in turn work towards creating neighbourhoods with more abundant and active public spaces for all.

Acknowledgement

We respectfully acknowledge the Traditional Owners of the land on which our offices reside, the Wurundjeri people of the Kulin nation, pay our respect to Elders past, present, and future, and recognise the oldest living culture.

References

Charley, E. (2019). 'Adapting to rapid urbanization via rapid revitalization that unlocks people, process and place', *Revitalization: Journal of Economic and Environmental Resilience*, Issue No. 104. @eliza@neighbourlytics.com

Cilliers, E.J., and Timmermans, W. (2014). 'The importance of creative participatory planning in the public place-making process', *Environment and Planning B: Planning and Design*, 41(3), pp. 413–429 [online]. Available at: https://doi.org/10.1068/b39098 (Accessed: 5 June 2020).

CoDesign Studio. (2019a). *Better Lives Through Better Places: Empowering Community and Council to Collaborate Through Placemaking, The Neighbourhood Project Research Report 2015–2019*. First published July 31, 2019. CoDesign Studio, Melbourne, Australia [online]. Available at: www.codesignstudio.com.au (Accessed: 5 June 2020).

CoDesign Studio. (2019b). *Community-Led Placemaking Guide*. First published 31 July 2019. CoDesign Studio, Melbourne, Australia [online]. Available at: www.codesignstudio.com.au (Accessed: 5 June 2020).

CoDesign Studio. (2019c). *Locally-Led Neighbourhoods: A Community-Led Placemaking Manual*. First published 31 July 2019. CoDesign Studio, Melbourne, Australia [online]. Available at: www.codesignstudio.com.au (Accessed: 5 June 2020).

CoDesign Studio. (2019d). *The Neighbourhood Project: Outcomes and Impact at a Glance*. First published July 31, 2019. CoDesign Studio, Melbourne, Australia [online]. Available at: www.codesignstudio.com.au (Accessed: 5 June 2020).

CoDesign Studio. (2019e). *The Neighbourhood Project: Round 2 Case Studies*. First published July 31, 2019. CoDesign Studio, Melbourne, Australia [online]. Available at: www.codesignstudio.com.au (Accessed: 5 June 2020).

Friedmann, J. (2010). 'Place and place-making in cities: a global perspective', *Planning Theory and Practice*, 11(2), pp. 149–165 (Accessed: 5 June 2020).

Hartley, L. and Lydon, M. (2014). *Tactical Urbanism 4: Australia and New Zealand* [online]. Available at: http://tacticalurbanismguide.com/guides/tactical-urbanism-volume-4/ (Accessed: 11 June 2020).

Healey, P. (1998). 'Building institutional capacity through collaborative approaches to urban planning', *Environment and Planning A: Economy and Space*, 30(9), pp. 1531–1546.

http://www3.weforum.org/docs/WEF_Shaping_the_Future_of_Construction_full_report__.pdf (Accessed: 4 April 2020).

Kelly, J.F. (2012). *Social Cities*. The Grattan Institute [online]. Available at: https://grattan.edu.au/report/social-cities/ (Accessed: 11 June 2020).

Lydon, M. and Garcia, A. (2015). *Tactical Urbanism: Short-Term Action for Long-Term Change* [online]. Available at: https://islandpress.org/books/tactical-urbanism (Accessed: 5 June 2020).

Project for Public Spaces. (2007). *What is Placemaking?* [online]. Available at: https://www.pps.org/article/what-is-placemaking (Accessed: 5 June 2020).

The Economist. (2015). 'Space and the city', *The Economist* [online]. Available at: https://www.economist.com/leaders/2015/04/04/space-and-the-city (Accessed 5 June 2020).

Think Impact Report. (2019). *The Neighbourhood Project: Round 2 Project Review*. Melbourne, Australia: Think Impact Report.

UN-Habitat. (2016). 'World cities report 2016 from habitat II to habitat III: Twenty years of urban development, in *UN-Habitat* [online]. Available at: https://unhabitat.org/world-cities-report (Accessed: 4 April 2020).

World Economic Forum. (2016). *Shaping the Future of Construction: A Breakthrough in Mindset and Technology, Industry Agenda Report* [online]. Available at https://www.weforum.org/reports/shaping-the-future-of-construction-abreakthrough-in-mindset-and-technology.

Further reading in this volume

Chapter 6: Listen, connect, act
Kim Cook
Chapter 11: *Free State Boulevard* and the story of the East 9th Street Placekeepers
Dave Lowenstein
Chapter 13: Sensing our streets: involving children in making people-centred smart cities
Sean Peacock, Aare Puussaar, and Clara Crivellaro
Chapter 15: Un/safety as placemaking: disabled people's socio-spatial negotiation of fear of violent crime
Claire Edwards
Chapter 17: 'I am not a satnav': Affective placemaking and conflict in 'the ginnel that roared'
Morag Rose
Chapter 23: Routing out place identity through the vernacular production practices of a community light festival
Gail Skelly and Tim Edensor
Chapter 24: Artists, creativity, and the heart of city planning
Tom Borrup
Preface: Evaluating creative placemaking: a collection of observations, reflections, findings, and recommendations
Maria Rosario Jackson
Chapter 39: Translating Outcomes: Laying the groundwork for interdisciplinary evaluation of creative placemaking
Jamie Hand
Chapter 40: Transforming community development through arts and culture: a developmental approach to documentation and research
Victor Rubin
Chapter 42: Creative placemaking and placekeeping evaluation challenges from the practitioner perspective: an interview with Roy Chan
Maria Rosario Jackson
Chapter 43: A theory of change for creative placemaking: the experience of the National Endowment for the Arts' *Our Town* program: an interview with Patricia Moore Shaffer, PhD
Maria Rosario Jackson
Chapter 45: How the city speaks to us and how we speak back: rewriting the relationship between people and place
Rosanna Vitiello and Marcus Willcocks

38
PUBLIC SEATING – SMALL IMPORTANT PLACES

Kylie Legge

Introduction

In late 1972 William Whyte wrote an article 'Please, just a nice place to sit' for the *New York Times* lamenting the lack of places to sit down, even in the context of significant increases in public space:

> But there are still few places to sit – and not entirely by accident, either. Some builders genuinely and sincerely mistrust people – noncustomer people, at least – and are almost obsessive on the threat of hippies. They instruct architects to design features that will discourage 'loitering' – what a Calvinist tract is in that word – and sometimes there is a show of active hostility.
>
> *(Whyte, 1972)*

Nearly 50 years later there is still a paucity of quality, comfortable, and inviting public seating in urban environments, and the motives and subsequent design response remain similar. This chapter aims to uncover the rationale for limiting public seating and to provide an alternate argument for the increased valuation of seating as the smallest increment of place; an opportunity to enhance urban equality, amenity, and to build social connections and community wellbeing.

Of the myriad definitions of placemaking, the objective seems clear: to turn spaces into places that impact human beings positively and that connect people to a location and to each other. This chapter proposes that the smallest increment of place is the public seat. Seating, specifically in public or shared spaces, is an invitation to people; an invitation to enter, a clear sign that staying is welcomed, and that it is OK because you, as an individual, are welcome to not only look at the space but to also be an active participant. As such, public seating should be a significant concern of placemakers and placemaking generally.

Unfortunately, many spaces designed as public places are not sittable. In some cases, sitting is explicitly restricted by a lack of seating, in others the message is implicit: with seating provided but in such a way that it lacks real invitation. Placemakers, planners, urban designers, landscape architects, and other decision-makers involved with the design and delivery of public spaces should have as their primary goal making places for people. However, a range of factors and external considerations continue to inhibit the realisation of this objective. Specifically creating environments that are designed for commercial activity, for marketing, for politics, or for managing behaviours has led to a dearth of sittability and the creation of spaces that are not actually designed for use by people. The ability, and the explicit welcome, to sit improves the sociability, safety, and success of the larger environment, essentially transforming it from an open space to a human-centric place.

As noted above, the value of sitting in public spaces was first highlighted by William Whyte in an opinion piece in the *New York Times* (Whyte, 1972) and then later in his seminal book *The Social Life of Small Urban Spaces* (Whyte, 1980). Since then, there has been minimal interrogation of the role seating plays in successful placemaking (Elet, 2002). The lack of historic or current literature on seating in the public domain can be seen as evidence of the lack of value this key aspect of public space activation holds in current thinking around urban design and planning. The discussions regarding seating that are occurring tend to focus on managing away both undesirable people – whether they be teenagers, non-consumers, or the homeless – and undesirable activity (Schmidt et al., 2011). While public space activation is gaining traction for its economic benefits, there is an equal need for placemakers to consider the role public seating plays in building connections and communities in addition to commerce.

Identified below are four historically established influences that contribute to what may be a professional prejudice against public seating: a cultural bias against sitting; a focus on movement over staying; the primacy of design over usability; and the prevailing value of economic exchange over social connection. The following sections will provide an overview of each of the four influences on the devaluation of public seating before providing a rationale for the prioritisation of places to sit as a key tool of successful placemaking.

A cultural bias against sitting

When placemakers use images to illustrate successful places, they are full of people, sitting around, talking to each other. Yet when a new design is put forward, places to sit are often the last thing on the agenda. Why is sitting deemed so unimportant?

In *A Cultural History of Gesture*, Bremmer considers walking, standing, and sitting in ancient Greek culture, and in doing so defines an inherent hierarchy that may still influence Western city-making today. While public walking in classical literature is seen as essentially male, powerful, and purposeful (Bremmer, 1991, p. 16), sitting is associated with 'postures of inferiority' (Bremmer, 1991, p. 25) or of supplication:

> The presentation of the self in public, then, was often acted out according to the contrast of high (upright carriage) and low (sitting, prostration); the positive side of 'upright' in this contrast is also shown by the fact that the Greek word *orthos* ('upright') and its cognates frequently carry the meaning 'prosperity', 'uprightness', or 'restoration.'
>
> *(Bremmer, 1991, p. 26)*

Sitting down is by definition a lowering of the body into a fixed location. It is a static state, where a person is more vulnerable and less ready to act. Doing 'nothing', whether that be resting, enjoying the sunshine, reading a book, people watching, waiting for someone, thinking, listening to music, or other quiet, passive, and independent activities are seen to be less valuable, even potentially risky behaviours. Inversely vibrancy, movement, and liveliness are all seen as successful outcomes of placemaking and urban design projects. With such an embedded prejudice against passive and sedentary behaviours and inclination towards active and upright activity, it is not surprising to see how a cultural bias against sitting and public seating could be influencing projects today; by purposefully not making places that are seen to encourage purposeless activity. This preference for activity over passivity may also be the foundation for the second influence on placemakers: the inequitable allocation of space and resources for movement over places to stay.

A focus on movement over staying

The prejudice against inactivity has been further supported by the increasing focus on planning for movement in city-making. Over the hundreds of years leading to the eighteenth century, a philosophical shift was occurring in the way that people connected to their environments. Prior to this time the medieval world prioritised place over the spaces between – the 'not' places (Foucault, 2006). These places worked as a hierarchy of dichotomies; 'sacred places and profane places, protected places and, on the contrary, places that were open and defenceless, urban places and country places,' what Foucault calls the 'space of localisation' (Foucault, 2006, p. 176), suggesting that this changed with the rediscovery that the earth's place was not the centre of the universe, but rather only one point in infinite space.

This knowledge diminished place to merely a point defining the connections or movement between two points. The space between these points supplanted the places themselves in importance – 'extension supplanted localisation' (Foucault, 2006, p. 176). Casey (1997) supports this argument in his book *The Fate of Place: A Philosophical History*. The central theme is the increasing neglect of place in favour of space. The philosophic shift described by these authors has dominated spatial planning to the present day with its focus on transportation networks over destinations, cars over people (Tibbalds, 1992). This is apparent at the scale of the city, at the street, and within open space. In much of our city planning, roads, then streets, then paths are used to define and organise space. The space is further classified to prioritise commercial land use, then open space, and lastly places where people might be invited to sit and spend time in their city for free. Movement and the associated allocation of space and resources is clearly dominant, while the humble public seat, a place for pause, is the lowest priority.

The primacy of design over usability

The third influence on public seating is the primacy of design outcomes over the potential for positive social impact. In a world where imagery and creating Instagrammable moments impact design decisions, the look of a place can take precedence over planning for its actual use. The contemporary preferencing of the visual has roots in early-twentieth-century architecture, where there was a move away from the transitory in design due to its perceived irrelevance in the designation of space. The original idea had its proponents in Frank Lloyd Wright and Rudolph Schindler who believed that the furniture of a place was inseparable from the architecture (Leatherbarrow, 2002). As such, non-permanent objects, or those that did not define the architecture or space, were seen as 'transparent,' or essentially irrelevant. Interestingly, furnishing was acknowledged as the 'background for human activity,' but that activity was seen as subservient to the characterisation of space itself (Leatherbarrow, 2002, p. 283). This modernist approach to the furnishing of the public realm is supported by Tibbalds (1992) who propounds uncluttered and simple spaces to generally be more successful, particularly if the 'scale and integrity of the street or space as a whole' (Tibbalds, 1992, p. 47) is maintained.

These architectural theories, added to the romanticisation of the ancient and mostly empty European plaza, has led to a design culture that favours the image of a grand public space over its actual use. Human-centric design features that encourage day-to-day use are often limited and the activation of the space is envisaged to be delivered via commercial activity or often costly programming by placemakers or 'activators' after the design is built. In these types of projects, the objectives of placemaking are subsumed by the desired design outcomes with the result being seating provided in locations and in a format that appear to purposefully limit usability. The location might be determined to make an architectural drawing look more symmetrical rather than providing a seat where a tree will provide shade or where the view towards interesting activity is promised. Materiality and detailing are chosen to fulfil the design ideal rather than being comfortable and climatically appropriate. The outcome is that people do not sit on these benches and we end up with a community asset that must be maintained but is not actually used. Francis, Koo, and Ramirez (2010) call this out as bad design where the result is a kind of 'bench museum' where you can look but are not really invited to sit.

Bad design that focuses on the image versus human use is one thing, but a more insidious rationale for limiting seating is managing users and their behaviours. Removing existing seating, providing no seating, or designing for purposeful discomfort has since the 1960s been an active measure for discouraging staying – particularly for the so-called '"undesirables" such as the homeless, teens and beggars' (White, 1988, as cited in Martin, 2006). A variety of 'hard and soft controls' (Loukaitou-Sideris and Banerjee, 1998) are actively used in planning and urban design, all with reasonable rationales – to reduce crime, to manage homelessness, to improve the perception of safety. The problem for placemakers is that taking away the invitation for one group affects every potential user, changing the culture of the place from one of welcome to one of rejection; from being a place for people, to a non-useable space that is only activated through resource heavy programming rather than through self-sustaining community activity. An extreme example from 1988 was the Sausalito downtown park that was fenced and gated with a sign reading 'This park is for your viewing pleasure. Do not enter' (Francis, 1998). These influences on placemakers and designers, whether theoretical, romantic, or social, have had a direct impact on the amount of inviting, quality, and comfortable public seating in the urban open spaces, and therefore on the number of places generally. Not including spaces to spend time is the antithesis of good placemaking that should put the human experience and the opportunity for connection at the centre of decision-making.

The prevailing value of economic exchange over social connection

The final influence on decisions regarding public seating is fundamentally economic. The management of public space users and their behaviours does not only have social concerns as its foundations. Rather, the economics of place are increasingly the rationale behind public space design and the integration, or not, of public seating. More new public spaces in cities are being delivered as part of private development, either required for public benefit as a developer contribution or utilised as a marketing tool for selling a product – whether that be housing, office space, or retail. In addition, many government-owned open space assets such as urban parklands, national parks, and botanic gardens are seen as economic drains on the public purse and face pressure to become financially self-sufficient. Together, these shifts provide reduced incentive for the free use and extended staying in public spaces that public seating invites. In these cases, placemakers are engaged not to create places but to curate experiences for those audiences deemed acceptable for commercial activity.

In private development particularly, a fear of non-consumer-based staying and the itinerant or the loiterer (someone with no purpose as either a producer or a consumer) results in the design of public spaces, while often high-quality and edged by engaging retailers, being fundamentally exclusionary. Often these urban plazas are designed as 'front yards, signs of status but not for use' (Carr et al., 1992, p. 15). In the worst cases, seating is limited to commercial operations only – if you cannot pay you cannot stay. This also means that the people associated with the place are the same as those that it is being 'sold to.' Thus, with no space for the undesirables, the place become increasingly homogenous and reinforces the segregation of those who cannot afford it. In these economically driven places, public seating may be incorporated but in such a way as to de-incentivise actual use – particularly for those that do not fit into the dominant culture.

Understanding these four key influences on public space design, and specifically the inclusion or not of public seating helps to set a new baseline for placemakers to consider why this prejudice against public seating should be reversed. The rationale for the considered inclusion of inviting, quality, and comfortable public seating can be summarised into three arguments that should help refine the placemakers' arguments for increased sittability in our public spaces: the public seat as the smallest increment of place; equity in access through an invitation to all; and the public seat as social infrastructure.

The public seat as the smallest increment of place

Defining 'place' is a fundamental challenge for most placemakers and therefore for their clients. It is this author's position that space is an environment that holds no meaning for people, and place inversely is one in which meaning is embedded through a connection between those people and the location; this meaning could be historic, cultural, social, personal, or communal.

It could be stated that phenomenologically, a public seat is a place because it is space designated and available specifically for use by a person (Lexico, 2020). Its place-ness is further heightened by the experience it offers the user – an invitation to literally and figuratively connect to a location in the form of a personal investment of time and trust. By sitting we are committing our body and its associated experiences to the specific location. By putting ourselves in a more vulnerable position we are putting our trust in the location. This simple connection satisfies an important human need to make ties to a specific location (Relph, 2016) and is the basic premise of place over space. The invitation to engage physically with the public realm allows for the possible generation of meaningful connections between people and their places. By participating in the life or activity of the place an authentic relationship can be developed and sustained. The lack of these opportunities creates an environment that precludes human engagement – a not-place. Just like the living room of a house, streets and cities must be comfortable and provide a 'home' for the city's inhabitants (Schindler, 1921 in Leatherbarrow, 2002, p. 280) otherwise they simply become storage rooms for cars, trees, play equipment, or just that awkward corner that no one knows what to do with.

For placemaking, perhaps most importantly, public spaces without quality seating opportunities risk never becoming the third places we need for social connection and community wellbeing. The seat can be the generator of place across a wider area, an opportunity to build relationships through repeated sharing of the same location. Inviting, comfortable and well-located public seating encapsulates the characteristics of Oldenburg's third places; they are on neutral ground, are not connected to a person's status, are open and readily accessible, low profile, and support communal ownership and social connections (Oldenburg, 1989). Public spaces that are uninviting and actually unsittable may be attractive to look at, meet quotas for open space allowance or tree planting, but cannot become places if there are no opportunities for people to connect with them. A public seat, considered as a small place for meaningful human experience, can become the seed that generates community, and as such a significant and powerful tool for placemaking.

Equity in access through an invitation to all

The second argument for more high quality, comfortable, and inviting seating is founded on the principle of equity, of ensuring that the public realm really is for everyone. While there are many reasons that a person may choose to sit or not sit in a public space, not actually including inviting, quality, and comfortable public seating will obviously act as a deterrent. However, understanding what constitutes 'invitation' requires an empathetic consideration of the potential users, of all the users – not just the ideal audiences. Public seating plays an important role as a legible encouragement to the community to participate in the life of the place; it reflects the culture of that community, of what behaviours are not allowed and which are encouraged. It can be, and has been, used to implicitly or explicitly limit the invitation to potential users, specifically: rough sleepers, skateboarders, groups of youth, and non-consumers. However, when a place is designed to not be inviting for one group, it is likely to read as unattractive to many more.

The need for unbiased and openly welcoming places to sit reflects a wider need for more inclusive cities where the public realm or shared spaces can be part of the solution 'for addressing and preventing some of the most pressing concerns of contemporary urban life: countering social isolation, negotiating difference, and creating places for all – regardless of age, race, gender, sexuality, or income' (Latham and Layton, 2019). Equity in the access and use of public space is an important and essential asset for all members of the community, but perhaps most specifically for those with the least choices, and therefore should be a key consideration in any placemaking process. For placemakers working on commercial projects it should be noted that while paramount, the benefits for public spaces that both attract and retain people are not just social. Whyte's much quoted adage 'What attracts people most, it would appear, is other people' (Whyte, 1980) talks to the community desire for vibrant places that are interesting and engaging. Local economies benefit from places where people pause, stop, and spend time, because people who don't slow down enough to see what businesses have to offer cannot spend their money there.

So how do you make more inclusive public spaces through inviting public seating? While substantive inroads are being made, the dominant design influencers remain male (Fairs, 2017), formally trained with significant delivery experience but perhaps with lesser insights into the values, needs, and aspirations of others not like themselves. Gender, age, personal preference, background, and many other factors play a role in how people read public space. The less diverse the design team, the less likely they will achieve diversity in users and uses. We need placemaking processes and design teams that act as open and empathetic facilitators of places that attract a range of users through 'deep listening' (Earl E. Bakken Center for Spirituality and Healing, n.d.) and concentrated observation. The people of the place should be valued not only as activators of a finished space or purchasers of the finished product, but also as learned participators in the process of making place. Through a truly collaborative process and an openness to accepting and understanding what makes us all different, we can create places that in turn, accept and welcome the diverse and are an authentic reflection of their local community.

The public seat as social infrastructure

The last argument for placemakers in their quest for more sittable public spaces is in regard to seating as essential social infrastructure:

> Social infrastructure is a set of physical places and organisations that shape our interactions. When social infrastructure is robust, it fosters all kinds of social interactions,

> helps build relationships, and turns community from a vague, fuzzy concept into a lived experience.
>
> *(Florida, 2018)*

Historically urban centres were seeded as spaces for exchange – both economic and social; of goods and information. As discussed earlier in this chapter, a more recent focus on economic exchange by both professional placemakers and designers has dominated to such an extent that other forms, such as social exchange, have become limited or perceived as less valuable. Yet even developers such as Australia's Delfin Lend Lease acknowledge that 'for a place to function as a community, people have to come together' (Delfin Lend Lease, 2004). Public seating provides the basis for the creation of third places that encourage use and connection between people and their places. For placemakers, public space and its design should focus on its role as social infrastructure – a shared investment with the focus of connecting community to their environment and to each other. Seating in public open spaces provides the location for meetings, both planned and unplanned, and the potential for conversation. Placemakers have the power to influence behaviours – to invite connection or support isolation. Sommer (1969) explains that the placement of seating can be designed to enhance interaction between strangers. This is called sociopetal (arrangement allowing people to see and interact with each other) planning as opposed to sociofugal (arrangement allowing privacy of people), which tends to discourage human interaction (Your Dictionary, n.d.).

Human connectedness in our communities is an increasing focus for many concerned with growing trends towards urban loneliness and mental health concerns associated with social isolation. The Grattan Institute's *Social Cities* report summarises the important value of social connections on community wellbeing and why city-makers need to consider a wider spectrum of community needs:

> These needs are both material and psychological. In policy and political terms, material needs tend to be prioritised – in part because they are easier to measure and influence. But psychological needs such as social connection are just as important, and cities play a role in whether they are met.
>
> *(Kelly et al., 2012)*

With quality, comfortable, and inviting seating options, placemakers can deliver on both the physical and psychological requirements for community wellbeing. However public seating design must change its focus to one that aims to maximise use; not bench seating stranded at equal intervals along a road and facing traffic, or ledges designed too high or too low for comfortable sitting, or the thousands of other public seats provided for community amenity that are designed or located in such a way to almost guarantee they will never be sat on. This is a huge waste of public funds both in terms of the cost of delivery and ongoing management of assets that have limited community benefit. Placemakers should consider public seating as a tool for connecting people and as important social infrastructure. A public seat provides the physical space as well as the less tangible but equally important provocation for potential social exchange to occur – the opportunity to connect with other members of the community either directly through a conversation, a nod, a smile of recognition, or indirectly through the simple act of sharing a space.

Creating a more sittable city – a city of small people places

Placemaking is fundamentally about people and building their connection to a location and to each other. This chapter has set out the rationale for and the value of, quality, comfortable, and inviting seating as an important tool for placemakers. This last section provides direction on how improved public seating can be delivered in our cities to help turn public space into community places. Over the last five decades there have been handful of important researchers (Whyte, 1972; Mumcu et al., 2010; Carr et al., 1992; Francis, 1998) who have undertaken studies and provided recommendations for public seating design that both attracts people to use it and provides encouragement for the longer stays that precipitate potential social connection.

This chapter is an attempt to synthesise these works into three key considerations for placemakers to support the delivery of quality, comfortable, and inviting public seating – psychological comfort, physical comfort, and pleasure. Psychological comfort aims to capture the need to consider the diversity of people who may use the space and not only what will attract them to take a seat but also what the barriers to entry may be. Consideration should be given to gender, age, background, and any other factors that may contribute to whether a place is welcoming and perceived to be safe, legible, and inviting for that person. A choice of seating options and location should be provided that offer both refuge (a sense of security) but also prospect (good sight lines) (Mumcu et al., 2010). Physical comfort captures considerations regarding how seating is designed to meet the needs of all types of human bodies in all types of climatic conditions. Here choice is paramount; we need places to sit in the sun or the shade, out of cold winds in winter but with access to cooling breezes in summer. We need ergonomic designs that consider not only the able-bodied adults but people of all ages and abilities; people sitting alone as well as those sitting in groups. Pleasure in public space should focus on creating positive experiences for the users and consider all the senses. The created vista should be visually engaging and diverse. Nature should play an important role in this, providing areas of light and shade and a changing landscape as the seasons change. But more than just the visual experience should be considered. Paley Park in New York is well known for its water wall creating an aural blanket to the busy street noise just outside. The smell of flowering plants, and the ability to touch and interact, can all bring pleasure to the user and should be located with seats where people can sit and enjoy them.

Our objective as placemakers is to make places, to transform space into locations that hold a meaningful connection with the people who use them. Creating places to sit can help deliver that objective. To create a more sittable city, a city of small human places, placemakers, designers, and other decision-makers should prioritise the creation of more places to stay rather than places to spend, or to move through. Designing for sittability is not as simple as providing park benches. Creating places that people will invest both their time and their trust at the most personal level needs the same level of consideration as designing cities at the systems level. A successful public seat is one that a range of people want to and actually do use, that encourages connections between people, and that contributes in its own small way to the making of place at the micro and macro levels.

References

Bremmer, J.N. (1991). 'Walking, standing, and sitting in ancient Greek culture', in Bremmer, J.N. and Roodenburg, H. (eds.) *A Cultural History of Gesture*. University of Groningen.

Carr, S., Francis, M., Rivlin, L. G. and Stone, A.M. (1992). *Public Space*. Cambridge: Cambridge University Press.

Casey, E.S. (1997). *The Fate of Place: A Philosophical History*. Berkeley: University of California Press.

Delfin Lend Lease. (2004). *Yarrabilba Community Strategy*. Brisbane: Delfin Lend Lease.

Earl E. Bakken Center for Spirituality and Healing. (n.d.) *Deep Listening*, University of Minnesota [online]. Available at: https://www.csh.umn.edu/education/focus-areas/whole-systems-healing/leadership/deep-listening (Accessed: 21 September 2019).

Elet, Y. (2002). 'Seats of power: The outdoor benches of early modern Florence', *Journal of the Society of Architectural Historians*, 61(4), pp. 444–469.

Fairs, M. (2017). 'Survey of top architecture firms reveals "quite shocking" lack of gender diversity at senior levels', in Dezeen [online]. Available at: https://www.dezeen.com/2017/11/16/survey-leading-architecture-firms-reveals-shocking-lack-gender-diversity-senior-levels/ (Accessed: 21 September 2019).

Faubion, J.D. (2006) Aesthetics, Method, and Epistemology: Essential Works of Foucault, 1954-1984 (Vol. 2). (Trans. Hurley, R.) New York: The New Press.

Florida, R. (2018). 'How 'social infrastructure' can knit america together', in Citylab [online]. Available at: https://www.citylab.com/life/2018/09/how-social-infrastructure-can-knit-america-together/569854/ (Accessed: 21 September 2019).

Francis, M. (1998). 'Changing values for public space', in *Landscape Architecture*, 78(1), p. 55.

Francis, M., Koo, J. and Ramirez, S. (2010). *Just A Comfortable Place to Sit: Davis Sittable Space Study*. Davis: University of California.

Kelly, J.-F., Breadon, P., Davis, C., Hunter, A., Mares, P., Mullerworth, D. and Weidmann, B. (2012). *Social Cities*. Melbourne: Grattan Institute.

Latham, A. and Layton, J. (2019). 'Social infrastructure and the public life of cities: Studying urban sociality and public spaces', *Geography Compass*, 13(7). Available at: https://onlinelibrary.wiley.com/doi/full/10.1111/gec3.12444

Leatherbarrow, D. (2002). 'Sitting in the city, or the body in the world', in Dodds, G. and Tavernor, R. (eds.) *Body and Building: Essays on the Changing Relation of Body and Architecture*. London: MIT Press, pp. 267–288.

Lexico. (2020) [online]. Available at: https://www.lexico.com/: https://www.lexico.com/definition/place (Accessed: 21 September 2019).

Loukaitou-Sideris, A. and Banerjee, T. (1998). *Urban Design Downtown: Poetics and Politics of Form*. Berkeley: University of California Press.

Martin, M.M. (2006). *Benchmarks: Sensing Their Therapeutic Landscape Qualities Associated with Seating Choice on Terrell Mall on the Washington University Campus*. Washington: Washington State University.

Mumcu, S., Düzenli, T. and Özbilen, A. (2010). 'Prospect and refuge as the predictors of preferences for seating areas', *Scientific Research and Essays*, 5(11), pp. 1223–1233.

Oldenburg, R. (1989). *The Great Good Place*. New York: Marlowe and Co.

Relph, E. (2016). 'Placemaking (and the production of places): Origins', in *Placeness, Place, Placelessness* [online]. Available at: https://www.placeness.com/placemaking-and-the-production-of-places-origins-and-early-development (Accessed: 21 September 2019).

Stephan Schmidt, Jeremy Nemeth and Erik Botsford 2011, The evolution of privately owned public spaces in New York City in URBAN DESIGN International Vol. 16, 4, 270–284.

Shuffield, J.W. (2015). 'Tear down this fence', in *Urban Residue - leftover thoughts from the day* [online]. Available at: https://urbanresidue.blogspot.com/2015/07/tear-down-this-fence.html (Accessed: 21 September 2019).

Sommer, R 1969, *Personal Space: The Behavioural Basis of Design*, Prentice-Hall, New Jersey

Tibbalds, F 1992, *Making People-Friendly Towns; Improving the Public Environment in Towns and Cities*, Longman Group, UK.

William H. Whyte Dec. 3, 1972 https://www.nytimes.com/1972/12/03/archives/please-just-a-nice-place-to-sit.html accessed 21/8/19, Section SM, Page 20

Further reading in this volume

Chapter 15: Un/safety as placemaking: disabled people's socio-spatial negotiation of fear of violent crime
Claire Edwards
Chapter 19: Placemaking in the ecology of the human habitat
Graham Marshall
Preface: The radical potential of placemaking
Cara Courage
Chapter 20: Displacemaking 2015 and 2020
Catherine Fennell and Daniel Tucker
Chapter 24: Artists, creativity, and the heart of city planning
Tom Borrup

Chapter 32: *The Hollywood Forest Story*: Placemaking for the Symbiocene
Cathy Fitzgerald
Chapter 39: Translating Outcomes: Laying the groundwork for interdisciplinary evaluation of creative placemaking
Jamie Hand
Chapter 40: Transforming community development through arts and culture: a developmental approach to documentation and research
Victor Rubin
Chapter 44: Creative placemaking and comprehensive community development: rethinking neighborhood change and evaluation
Maria Rosario Jackson
Chapter 45: How the city speaks to us and how we speak back: rewriting the relationship between people and place
Rosanna Vitiello and Marcus Willcocks

SECTION 7

Researching and evaluating placemaking

Section Editor: Maria Rosario Jackson

PREFACE

Evaluating creative placemaking: A collection of observations, reflections, findings, and recommendations

Maria Rosario Jackson

For more than 25 years, my work has focused on understanding and advancing the roles of arts and culture in comprehensive community development and planning, primarily in the United States. Since 2010, when the term 'creative placemaking' was coined and as funders in government and philanthropy began to pay greater attention to the integration of arts and culture in community development and planning, longstanding relevant work began to be better acknowledged and understood. It is true that creative placemaking can manifest in myriad ways and that, as a result, describing the contours of this 'emerging' field is difficult. That said, if you look deeply at the growing body of work currently available to examine, I argue that at its best, creative placemaking fundamentally seeks to bring greater humanity to our practice of comprehensive community development and planning. It causes us to rethink our practices, raise the bar, and challenge our industry standards. Implicit in all of this is the need to reassess how we think about and gauge systemic and community change. The rise of creative placemaking has underscored that despite the fact that the integration of arts and culture in community development and planning is not altogether new, our prevailing concepts of comprehensive community development and related dominant evaluation practices are woefully inadequate, militate against us doing our best work, and can even do harm.

This collection of chapters

This collection of chapters draws from specific initiatives and organizations as well as general field observations. The collection provides the reader a window into various kinds of creative placemaking practices and documents the evolution of evaluation approaches to the work. It also offers some new ways of conceiving of community development, change processes, and community- and systems-level impacts that we have heretofore not considered sufficiently. The collection lifts up important considerations about the future of creative placemaking evaluation practice – the directions in which the practice needs to go and the identification of barriers to and opportunities for that evolution.

In the first chapter, Jamie Hand, from her vantage point as Director of Research Strategies at ArtPlace America, provides an overview of the different ways that, since 2010, evaluators and others have framed the value of creative placemaking and have attempted to measure expected impacts. Recognizing that initial efforts did not take into consideration the complexity of the work, the possible alchemy of merging fields, and challenges inherent in that confluence, nor the wide-ranging impacts that went beyond primarily economic development concerns, ArtPlace America laid the groundwork for future, more appropriate and effective, research and evaluation across 10 policy sectors that they see as part of community development.

Victor Rubin discusses the documentation and analysis approach and findings from the Community Development Investments initiative, a five-year effort carried out in six places across the United States with the support of ArtPlace America. The chapter provides an account of the developmental approach taken to researching and assessing the initiative and discusses findings from the effort including insights about how the work has had impacts at the community development field and organizational levels. The focus on the field of community development and at the organizational level is significant as these are realms much less frequently considered in creative placemaking program evaluation.

Maribel Alvarez, a folklorist, anthropologist, and founder of the Southwest Folk Alliance provides an account of her experience assessing the impacts of *Tuscon Meet Yourself*, a 47-year-old annual folklife festival in Tucson, Arizona. In her chapter, Alvarez explains the concept of 'festival' as a manifestation of equitable creative placemaking, offers insights into the roles that folklorists and folk life can and do play in relation to creative placemaking, interrogates the ways in which festival success is typically assessed, and offers new constructs for assessing the possible contributions of festivals through an expansive community development and inclusion lens.

From a practitioner's perspective working at the neighborhood scale, Roy Chan, community planning manager at Chinatown Community Development Center, examines the origins and characteristics of creative placemaking practice and his work to assess and convey the power of the work in the Chinatown neighborhood of San Francisco, California. Chan discusses the limitations of dominant community development and housing aspirations and corresponding methods of evaluation. He also offers insights about approaches to community development and corresponding evaluation that account for a community's cultural assets as central to its resilience and progress.

The National Endowment for the Arts has been a seminal player in the advancement of creative placemaking through its *Our Town* program. As a federal agency with the obligation of evaluating its public programs, the NEA has been actively searching for ways to deliver on that obligation and do justice to the work they are funding. Patricia Moore **Shaffer** Deputy Director of the NEA Office of Research and Analysis and her team along with staff from the *Our Town* program recognized that creative placemaking work takes many forms, is difficult to describe and that assessment of its impacts is frequently not straightforward. To address this, the agency embarked on an effort to develop a theory of change for creative placemaking. In conversation with Maria Rosario Jackson, **Shaffer** describes the theory of change development process and the challenges and discoveries along the way.

The last chapter draws from my examination of foundation-supported creative placemaking initiatives and applied research on inequity and the roles of arts and culture in community development that preceded the advent of creative placemaking. In the chapter, I present a framework for understanding the diversity of ways in which equitable creative placemaking practices manifest as elements of comprehensive community development, share some insights about pitfalls of our current evaluation methods, and offer concepts and promising examples of developments in evaluation practices that bode well for raising the bar and shifting industry standards in the community development and evaluation fields.

The current climate in the United States and globally – with COVID-19 and civil unrest as the result of entrenched, systemic racism and inequity – is both heartbreaking and exhilarating, full of energy and opportunity. For many, this is a moment both of reflection and of action; a moment in which what we thought was impossible or out of bounds might actually be possible. As we have the opportunity to reassess and reimagine out systems and institutions, to act on what we have learned and to take some risks, the sustained and bold integration of arts and culture into efforts to expand opportunity and redress historic and current harm may be more viable than ever before. With a new and renewed social consciousness, there is a chance that industry standards, with continued intention and hard work, can be transformed to reflect the kinds of arts and culture-based practices that better recognize our full humanity and what it takes to create healthy places where all people can truly and fully thrive. Our progress with reconceptualizing evaluation practices is crucial. This collection of examined experience, observations, insights, findings, and recommendations is a contribution to that end.

Further reading in this volume

39

TRANSLATING OUTCOMES

Laying the groundwork for interdisciplinary evaluation of creative placemaking

Jamie Hand

Introduction

When I became ArtPlace America's Director of Research Strategies in 2014, the tension around how to measure creative placemaking success was palpable. Practitioners and communities doing arts-based community development work had been defining their own success for decades, but the then-recent formalization of the field – through the National Endowment for the Arts (NEA) *Our Town* grant program, the creation of ArtPlace, significant investments from national philanthropic entities, and numerous other policy and funding shifts – had happened seemingly overnight, without a corresponding or unified theory of change about the actual impacts of creative placemaking. The possibilities – and the risks – were both abstract and endless.

The NEA was in the midst of its multiyear *Validating Arts and Livability Indicators* (VALI) *Study*, which tested a suite of proposed metrics that could demonstrate the positive effects of creative placemaking on various dimensions of community 'livability' – the north star and conceptual through-line across many federal agencies' place-based investments during the Obama Administration. The NEA's candidate indicators relied on national, publicly accessible data and fell into four categories: residents' attachment to communities; quality of life; local economic conditions; and arts and cultural activity (specifically the infrastructure supporting artists and arts organizations). ArtPlace itself had recently released a hotly contested set of indicators that positioned creative placemaking as contributing to the 'vibrancy' of a place (Taylor, 2012). With measures that focused on variables like property values, population density, and creative industry jobs, many critics felt the Vibrancy Indicators privileged traditional economic development strategies over more nuanced approaches to building wealth, wellness, and equity in communities through the arts. Respected researchers and scholars who had spent their careers studying the social or economic impact of the arts were increasingly vocal, calling out the shaky foundation upon which funders and policymakers were attempting to build a field (Moss, 2012). Renowned economist Ann Markusen (2013), for example, noted that 'Evaluation by external generic indicators fails to acknowledge the experimental and ground-breaking nature of these creative placemaking initiatives and misses an opportunity to bolster understanding of how arts and cultural missions create public value.' In a 2014 speech, Mark Stern of the Social Impact of the Arts Project (SIAP) at the University of Pennsylvania stated:

> Creative placemaking needs to work on clarifying its conceptual foundation but at the same time be open to experimentation in developing methods for understanding its outcomes. With a clear set of concepts, one can continue to improve one's methods over time.

To further complicate matters, individual approaches to creative placemaking were rapidly evolving and adapting, necessarily, both to community context and to changing social and political dynamics across the country. Artists in the Midwest were collaborating with residents to address stormwater management in low-income neighborhoods, while longtime affordable-housing developers in Harlem were integrating children's museums into their financial model for supportive housing. An esteemed dance organization was working with local transportation officials to redesign a suburban commuter corridor for pedestrian residents, while community organizers in rural California were turning to local artists to help build community cohesion and public space in an unincorporated migrant farming community. The sheer diversity of practices that fell under the 'creative placemaking' umbrella was both inspiring and dizzying, and the tent was only going to get bigger.

It was against this backdrop that, in 2015, we launched ArtPlace's *Translating Outcomes* initiative – a five-year, multi-disciplinary research strategy designed to establish connections between arts and cultural activities and the countless community development goals that were surfacing in the ArtPlace grant portfolio and across the field. At its simplest, *Translating Outcomes* can be understood as a series of 10 'deep dives' intended to expand our understanding of creative placemaking impacts by adopting and adapting existing research and evaluation methods from other sectors. Taken as a whole, however, the initiative served as an incremental, segmented approach to building creative placemaking knowledge for and with a diverse range of community development practitioners – one that has highlighted the critical importance of interdisciplinarity, and laid the groundwork for the creative placemaking field to embrace a multidimensional array of success measures that are as nuanced and complex as the practice itself. This chapter lays out the methods and values driving ArtPlace's *Translating Outcomes* work, followed by emerging reflections on what this explicitly interdisciplinary approach has revealed about creative placemaking research and evaluation.

Value-driven methods

Segmentation, not siloes

Recognizing that the comprehensive community development field is made up of many professional disciplines, ArtPlace developed a diagram (Figure 36.1) to illustrate 10 segments of the field that are commonly understood as discrete sectors. These sectors are often separated out *vis à vis* distinct municipal agencies, university departments, or funding streams. They are Agriculture and Food, Economic Development, Environment and Energy, Health, Housing, Immigration, Public Safety, Transportation, Workforce Development, and Youth and Education. While ArtPlace used this matrix in multiple ways over the course of its grant-making and other work, the *Translating Outcomes* effort took this matrix as its road map and set out to analyze, make legible, and give language to how arts and cultural practitioners have long been partners in helping to achieve each of these sectors' goals.

Each sector has its own terminology, conceptual frameworks, success measures, and disciplinary cultures to navigate, and as we dove into the first three sectors in 2015 – housing, community safety, and health – the challenge of rigorous segmentation became increasingly clear.

	CIVIC, SOCIAL & FAITH	COMMERCIAL	GOVERNMENT	NONPROFIT	PHILANTHROPY
AGRICULTURE & FOOD					
ECONOMIC DEVELOPMENT					
EDUCATION & YOUTH					
ENVIRONMENT & ENERGY					
HEALTH					
HOUSING					
IMMIGRATION					
PUBLIC SAFETY					
TRANSPORTATION					
WORKFORCE DEVELOPMENT					

Figure 39.1 Community development matrix.

Defining clear boundaries for any given sector was exceedingly difficult given the increasing interconnectedness of housing, health, environment, and more, yet it was crucial to breaking down the complexity of the systems in which creative placemaking operates.

One critique of the initiative is that this segmentation was 're-siloing' the work in a way that undermines the lived experience and reality of both communities and community development work that attempts to be comprehensive. The segmentation, however, simply served as a methodological tool that allowed us – in the tradition of interdisciplinary research – to build understanding, capacity, and nuance in each of the sectors.

> Vocabulary differences are at the heart of translation activities, but research conventions and cultures must also be learned and navigated. Valid interdisciplinary research is necessarily based on a deep understanding of how concepts, methods, and results fit in the body of discourses and practice in which they were developed. Only then can judgments be made about how ideas can legitimately be applied in a new area.
>
> *(Palmer, 2010)*

Taking a page from strategic communications research as well, the segmentation also allowed us to treat each sector as a distinct audience or stakeholder – building discipline- or industry-specific frameworks and resources in language that resonated for each. And ultimately, when viewed as a series, the material can be applied or combined in any number of ways that make sense for a given project, organization, or community context.

Practitioner-led

Our methods in each sector included three distinct phases of work, throughout which we maintained a commitment to highlighting and centering the voices of artists and practitioners – particularly those of color – who have long been pioneers in the field. Our research scope intentionally included people and places who had long been doing arts-integrated community development work, regardless of whether they referred to it as 'creative placemaking' and regardless of whether they had received ArtPlace, National Endowment for the Arts, or other creative placemaking-specific funding.

A traditional philanthropic approach might have been to hire expert researchers to conduct impact evaluations of grant projects within the ArtPlace portfolio, or to revise the set of Vibrancy Indicators as a resource for grantees to use in their own assessments. But how could researchers (or funders) possibly know more about the emerging practice and outcomes of creative placemaking than those who were leading the work in communities? It seemed critical, at this point in the evolution of the field, to begin with practitioners' wisdom and engage them directly in shaping any future frameworks or resources.

In all three phases of work, we also explicitly engaged with the histories of systemic racism, exclusion, and injustice that have shaped each sector – from redlining and displacement in the housing sector, for example, to longstanding discrimination against black farmers in the agriculture and food sector; from the measurable public health impacts of racism, to a growing consensus among environmental advocates that the effects of climate change impact low-income communities of color first and worst. It is often precisely these inequities that community development practitioners are seeking new solutions to, and that artists, designers, and culture-bearers are drawn to solving in their own communities. As ArtPlace sought 'to position arts and culture as a core sector of community planning and development,' it was also critical to acknowledge both the positive and negative legacies within community development practice in order to situate creative placemaking in a values-driven, equitable way (a phrase taken directly from ArtPlace America's organizational mission statement).

Iterative research design

The first phase of work in each sector centered around a field scan – a written report that described and assessed the arts and cultural activity that was already happening in the field. Conducted in partnership with individual researchers or teams uniquely positioned to investigate the intersection of arts and culture and a specific sector, each field scan represented an exploratory first step intended to surface: key goals or needs in that community development sector that arts and culture might address; a typology or framework for understanding the ways that arts and culture might partner with that community development sector; and barriers to integrating arts and culture within that community development sector. Research methods for each field scan included interviews with artists, practitioners, community members, researchers, and thought leaders – some deeply immersed in the intersection, others considering it for the first time; reviews of reports, media, peer-reviewed literature, and other publications on trends and policies in a given sector; a survey and meta-analysis of arts-integrated community development projects; development of a typology about 'what the arts can do' in language resonant to that sector; and creation of qualitative case studies highlighting each approach within the typology.

Prior to being published, each field scan served as a framing document for the second phase of work in each sector: an exploratory working group charged with taking the analysis and findings a step further. Each working group reinforced our commitment to highlighting the perspectives of artists, practitioners, community members, and thought leaders who are closest to the work, positioning them as the experts on whom we were relying to review and critique our findings.

Working groups participants were highly curated to achieve an explicitly interdisciplinary and diverse constellation of voices. Community-based artists shared ideas and inspiration alongside municipal policymakers, community development intermediaries, researchers and evaluators, funders, and organizers. We intentionally balanced local and national perspectives, 'members of the choir' with people who were 'agnostic' or new to the idea of creative placemaking, and those who hold power and privilege with those who seek to shift power. Everyone was asked to serve as peer reviewers for the field scan research, ambassadors for this cross-sector way of working, and advisors on next steps in that sector. While translation and identification of common outcomes were a primary goal of this initiative, an equally important objective was to build communities of practice that could learn from each other, and continue to advance approaches to creative placemaking practice and assessment long after ArtPlace's sunset in 2020. We also engaged each working group in explicit discussion about barriers to collaboration with the arts and culture sector, many of which centered around questions of measurement and evaluation: are there specific policy restrictions or industry metrics that simply cannot accommodate arts and cultural work? What does it take to convince a housing developer, for example, that a slight increase to her financial bottom line will produce an exponential return on investment, albeit one that we don't yet know how to measure?

Co-creation

The third phase of work in each sector, determined primarily by each working group, was the creation of a tailored program, resource, or initiative that would help overcome those barriers and catalyze more collaboration in that sector. The field scans were shared publicly but were never intended as the primary deliverable (ArtPlace America). Instead, ArtPlace funded delegations of artists to attend non-arts conferences, convened cohorts of community-based practitioners for peer learning visits across the country, piloted the first-ever artists-in-residence in state government, developed continuing education courses for affordable housing developers, and more. Seldom were these 'phase three' efforts focused explicitly on evaluation practice; most working groups, regardless of sector, recommended that more intentional learning, collaboration, and experimentation happen as a necessary precursor to defining common measures of success than could be taken up by researchers and evaluators. Practitioners were both the authors and the audience for this collaborative knowledge-building effort – and by focusing on discrete sectors, we were able to create learning resources that were more likely to be both useful and used.

It is important to note that strategic non-arts partners were vital to the success of the *Translating Outcomes* work. ArtPlace built relationships with peer organizations in each sector – some of whom co-convened the working groups and others who co-created learning resources – for a number of reasons that over time became core to ArtPlace's overarching theory of change: to build credibility within the sectors we were seeking to influence, to develop distributed leadership within the evolving field of creative placemaking, and to ensure sustainability after ArtPlace's sunset by embedding the work within other institutions. At the time of this writing, work in all 10 sectors is underway. In 8 of the 10 sectors, we have published clear frameworks for practitioners and engaged strategic partners who will carry the knowledge forward within the specific sphere of community development that they serve. Partners to date include NeighborWorks America and Enterprise Community Partners (affordable housing); Local Initiatives Support Corporation (community safety); Transportation for America (transportation); US Water Alliance and Grist (environment); Welcoming America (immigration); Rural Coalition, Farm Credit Council, and DAISA Enterprises (agriculture and food systems); the University of Florida Center for Arts in Medicine (public health); Jobs for the Future and NORC at the University of Chicago (workforce development); Creative Generation (youth development); and Common Future (economic development).

What interdisciplinarity has revealed: we keep measuring the wrong things

The *Translating Outcomes* series will culminate in a meta-analysis that brings the sector-specific learnings into a comprehensive whole; however, several insights have already emerged that have implications for the next chapter of creative placemaking research and evaluation. Most significantly, it has become clear that we keep measuring the wrong things. After being immersed in the cross-sector work for six years, I offer the following reflections and observations in support of this idea.

The arts can help define more authentic, human-centered outcomes

Established metrics for success in other sectors do not capture the most meaningful impacts of creative placemaking work, or even of community development work. We interrogated each sector to better understand its existing systems of measurement and evaluation, and to draw connections between arts and cultural strategies and those established measures or outcomes. What we encountered, however, was far more complicated. Time and time again we heard in interviews and in working groups that the majority of people doing community change work – even those in fields strongly grounded in evidence-based practice, such as public health and community safety – felt that the measures they relied on were insufficient. Many were not meeting their own goals, or if they were, some deeper, more authentic measure of success was still lacking. In our affordable housing working group, for example, participants acknowledged the housing sector's reliance on quantitative metrics and tendency to measure quantity of housing units, but over two days the group came to a consensus that creative placemaking approaches had the potential to shift the sector's focus to quality of housing – a metric that more holistically incorporated dimensions of resident attachment, tenant turnover, and the long-term wellbeing of a housing community. Nearly everyone we have engaged in this research has requested help from the arts and culture sector: how can we more authentically connect with the people we are trying to serve, and how can we incorporate the social and emotional dimensions of the human condition into our measures of success? Concepts like trust, wellbeing, belonging, and collective efficacy are increasingly valued by those working in communities and are understood intuitively to be the domain of artists and culture-bearers. New (largely social science–based) methods for measuring such concepts will be central to understanding the true impacts of creative placemaking.

Practitioners create new, contextual measures all the time; funders and policymakers need to listen

There is a decades-long history of promoting 'evidence-based practice' in community development, in philanthropy, and in many of the sectors we explored through the *Translating Outcomes* initiative, where 'rigorous tests to assess the efficacy of a given intervention' are required before an investment can be made or an approach is deemed worthy of replication (Brooks, 2016). But as we convened more and more working groups, a pattern became clear: there is a profound disconnect between the work happening on the ground, and the evidence-based decision-making structures that have been embraced by policy and funding institutions for accountability and assessment. Our working groups became opportunities not just to translate across sectors, but within each one.

In our workforce development working group, participants experienced a site visit to the Sweetwater Foundation on Chicago's south side and heard presentations from leadership at Coalfield Development Corporation (based in Wayne, West Virginia), Grace in Action Collectives (based in Detroit, Michigan), and Rolling Rez Arts (based on the Pine Ridge Indian Reservation in South Dakota), among others. These arts-integrated, place-based efforts to advance quality employment opportunities were human-centered, sustainable, and successful in ways that nearly everyone in the room celebrated, yet the more traditional workforce stakeholders were quick to point out that their institutions could never support – or even see – this work outside the context of our convening. Either it wasn't happening at a large-enough scale, or they were using measures that didn't align with commonly accepted standards in workforce development. These local organizations were, however, inventing new approaches and developing new indicators that resonated deeply with where the policymakers and funders knew their own field needed to go. At one point, a participant even said, 'We've got experts in the room and no one thinks that workforce development is working' (quote taken from January 31, 2020, meeting notes (unpublished)). DAISA Enterprises, one of ArtPlace's partners in the agriculture and food systems sector, framed this same disconnect as a unique opportunity for systems change, acknowledging that

> many people within philanthropic and community development fields recognize the constraints of siloed and prescriptive funding strategies, as well as the structures that keep those strategies in place. A key desired outcome of this (research) process is to amplify the realities and needs of practitioners as lessons which inform future availability of resources. Acknowledgment of the authentic, integrated, and multi-faceted ways in which community practitioners work and need to have agency and flexible support is an important step towards holistic systems change.
>
> *(Drew et al., 2019)*

The *Translating Outcomes* process, in hindsight, created liminal spaces not only for artists and community development practitioners to interact and learn across disciplines, but for grassroots practitioners and community members to engage on equal footing alongside policymakers, funders, and others in positions of power or privilege. The more opportunities there are for funders and national entities to listen to and learn from practitioners at the leading edge of creative placemaking work – moving from evidence-based practice to 'practice-based evidence,' as it were – the sooner we will align on definitions of success that truly meet community needs.

Summative evaluation is (still) premature

When we launched *Translating Outcomes* in 2015, the kind of work that ArtPlace was supporting – and the majority of creative placemaking initiatives overall – simply did not lend itself to summative evaluation. Community change is slow; it can take years if not decades to achieve the kinds of outcomes that many creative placemaking efforts are aiming for (and the arbitrary period of time defined by grant funding doesn't magically accelerate those outcomes; starting in 2016, applicants to ArtPlace's National Creative Placemaking Fund were able to identify a grant period of up to four years that best met the needs of their community, organization, or project). Social impact measurement, too, is notoriously challenging, no matter whether you are working on childhood education, poverty, housing security, civic engagement, or racial justice:

> As you widen your scope to deal with a major social problem, the harder it becomes to measure your impact because it is tougher to isolate cause and effect. It's no longer a simple linear relationship, but a complex set of relationships.
>
> *(Hanna, 2010)*

Given the sheer number of geographic and political contexts, demographics, artistic disciplines, community goals, and overall approaches to creative placemaking noted at the beginning of this essay, we acknowledged the limited utility of summative evaluation and instead invested in a broader cross-sector learning agenda for the field.

Now, six years later, one might argue that the creative placemaking field can begin to pursue summative evaluation. Through not only the *Translating Outcomes* initiative but ongoing research efforts at the National Endowment through the Arts and elsewhere, we have identified a much broader array of measures than previously put forward, and are moving closer to consensus on the kinds of outcomes noted above – equity, belonging, wellbeing, and the like. And indeed, organizations who have been working this way for decades, such as Appalshop in Whitesburg, Kentucky or Community Musicworks in Providence Rhode Island, have boldly undertaken the complex work of impact evaluation for some initiatives (see, All Access AKY, Kline and Wolf and Community Musicworks for example). But what I have learned from the artists and community leaders we engaged in our field scan research, working groups, and resource creation is that creative placemaking work often doesn't 'end,' and qualitative, story-based, and participatory methods of assessment remain a better fit for both understanding and communicating the impact of their efforts.

Conclusion

Frans Johansson's 2007 book, *The Medici Effect: What Elephants and Epidemics Can Teach Us About Innovation*, has been an important reference for this notion of intersections, and the powerful opportunity that lies in an unlikely convergence or association across disciplines: 'When you step into an intersection of fields, disciplines, or cultures, you can combine existing concepts into a large number of extraordinary ideas. … Once there, [you] have the opportunity to innovate as never before, creating the Medici Effect.' There is so much more work to do within each of the intersections we have initiated in the *Translating Outcomes* research. As the creative placemaking field continues to evolve and grow, critical and longitudinal evaluation of projects – done in collaboration with community members and residents – will be crucial to understanding the full range of outcomes, as well as the risks and limitations, of arts-based strategies. But creating and holding interstitial space for dialogue and connection has been a vital research method in and of itself, core to the work of bridging multiple disciplinary knowledge bases and honoring the complexity that creative placemaking represents. The artists and practitioners who are leading this work in communities are working at the edges, in non-traditional spaces, and in ways that don't always align with funders' program boundaries or the metrics that have been established in any given field.

A central characteristic of creative placemaking practice is that it is deeply collaborative. And, as anyone who has ever been in a partnership knows, whether it's professional, organizational, civic, or as personal as a marriage, collaboration can be messy. Establishing and aligning values, learning and accepting each other's strengths and weaknesses, sharing responsibilities and accountability, and communicating constantly about all of the above are baseline requirements for any successful partnership. When it comes to research and evaluation of creative placemaking work, the same and more holds true. Within the *Translating Outcomes* work, there has been a multi-directional learning curve for everyone involved – regardless of which sector we were investigating, and regardless of whether we were working with an individual researcher or a larger team with varied skill sets. At its simplest, creative placemaking research requires a strategic and deliberate merging of existing evidence bases and methods, bringing anthropologists, sociologists, geographers, planners, participatory action researchers, artists, and more into dialogue with public health scholars, criminologists, economists, infrastructure engineers, and other such specialists, as well as with community members directly affected by the work. More often than not, however, it also requires a unique combination of rigor and flexibility – with methods that honor both the linear and the nonlinear, the established and the experimental, the known and the unknown, the logic model and the lived experience. Future creative placemaking research and evaluation efforts will require unexpected configurations of expertise; we must proactively structure and support such collaborations with the time and resources it takes to learn from each other and to align different ways of knowing.

Acknowledgments

The reflections and insights outlined in this essay would not be possible without the collaboration, wisdom, and imagination of the field scan researchers and authors to date: Caroline Ross; Sarah Kavage; Danya Sherman (Sherman Cultural Strategies); Ben Stone and Mallory Nezam (Transportation for America); Alexis Frasz and Holly Sidford (Helicon Collaborative); John Arroyo; Christa Drew and Maria Elena Rodriguez (DAISA Enterprises); Jill Sonke and Tasha Golden (University of Florida); and Jennifer Novak-Leonard and Gwendolyn Rugg (NORC at the University of Chicago).

The author is also indebted to Monitor Institute, who was engaged in 2015 as an advisor to ArtPlace's overall approach to convening cross-sector working groups. Together we created facilitation templates, meeting agendas, and strategies for holding space for divergent perspectives and conversation that were largely inspired by lessons captured in a guidebook developed by Monitor Institute and the Rockefeller Foundation (Flower et al, 2013).

This chapter is an adaptation of the essay 'Multiple ways of knowing: translating outcomes between the arts and community development,' which originally appeared in Volume 14, Issue 2 of the San Francisco Federal Reserve's *Community Development Innovation Review* (November 2019) (Hand, 2019).

References

All Access EKY. [online]. Available at: https://www.allaccesseky.org/what-we-do. see All Access EKY (https://www.allaccesseky.org/what-we-do) (Accessed: 25 June 2020).

ArtPlace America. *Translating Outcomes* (research papers) [online]. Available at: https://www.artplaceamerica.org/areas-of-work/research/translating-outcomes (Accessed: 25 June 2020).

Brooks, J. (2016). 'Making the case for evidence-based decision-making', in *Stanford Social Innovation Review*, December 6, 2016 [online]. Available at: https://ssir.org/articles/entry/making_the_case_for_evidence_based_decision_making# (Accessed: 25 June 2020).

Drew, C., Rodriguez, M., Ross, D. and Hand, J. (2019). 'Cultivating creativity: Exploring arts and culture in community food systems transformation', in *DAISA Enterprises*, LLC/*ArtPlace America*. Brooklyn, NY.

Flower, N.R., Muoio, A. and Garris, R. (2013). 'Gather: The art and science of effective convening', in Deloitte Development LLC. New York: The Rockefeller Foundation.

Hand, J. (2019). 'Multiple ways of knowing: Translating outcomes between the arts and community development', San Francisco Federal Reserve's *Community Development Innovation Review*, 14(2) [online]. Available at: https://www.frbsf.org/community-development/publications/community-development-investment-review/2019/november/multiple-ways-of-knowing-translating-outcomes-between-the-arts-and-community-development/ (Accessed: 25 June 2020).

Hanna, J. (2010). 'The hard work of measuring social impact' in *Harvard Business School Working Knowledge* [online]. Available at: https://hbswk.hbs.edu/item/the-hard-work-of-measuring-social-impact (Accessed: 25 June 2020).

Johansson, F. (2007). *The Medici Effect: What Elephants and Epidemics Can Teach Us About Innovation.* Cambridge: Harvard Business School Press.

Kline C. and Wolf, D. (2016). 'Sharing the table with strangers', in Community Musicworks [online]. Available at: http://communitymusicworks.org/wp-content/uploads/2016/07/CMW_SharingTables-2016.pdf (Accessed: 25 June 2020).

Markusen, A. (2013). 'Fuzzy concepts, proxy data: Why indicators won't track creative placemaking success', in *Createquity.com*, November 9, 2012 (later published in *International Journal of Urban Sciences*, 17(3), 2013, pp. 291–303.
Moss, I.D. (2012). 'Creative placemaking has an outcomes problem', in *Fractured Atlas*, May 9, 2012 [online]. Available at: https://blog.fracturedatlas.org/creative-placemaking-has-an-outcomes-problem-97686ba491cb (Accessed: 20 June 2020).
Palmer, C.L. (2010). 'Information research on interdisciplinarity', in Frodeman, R., Klein, J.T. and Mitcham, C. (eds.) *Oxford Handbook of Interdisciplinarity*. Oxford: Oxford University Press.
Stern, M.J. (2014). 'Measuring the outcomes of creative placemaking'. In *The Role of Artists & The Arts in Creative Placemaking, May 30–31, 2014, Baltimore, MD—Symposium Report* (pp. 84–97). Washington DC: Goethe-Institut and EUNIC. (Accessible at https://repository.upenn.edu/siap_placemaking/1/).
Taylor, A. (2012). 'Vibrancy by proxy', in *The Artful Manager*, October 9, 2012 [online]. Available: https://www.artsjournal.com/artfulmanager/main/vibrancy-by-proxy.php (Accessed: 20 June 2020).

Further reading in this volume

Preface: Placemaking in the age of COVID-19 and protest
Jason Schupbach
Chapter 3: An annotated history of creative placemaking at the federal level
Jen Hughes
Chapter 4: A future of creative placemaking
Sarah Calderon and Erik Takeshita
Chapter 6: Listen, connect, act
Kim Cook
Chapter 10: From moon village to mural village: the consequences of creative placemaking in Ihwa-dong, Seoul
Jason F. Kovacs and Hayun Park
Chapter 11: *Free State Boulevard* and the story of the East 9th Street Placekeepers
Dave Lowenstein
Chapter 12: Public transformation: affect and mobility in rural America
Lyndsey Ogle
Chapter 19: Placemaking in the ecology of the human habitat
Graham Marshall
Preface: The radical potential of placemaking
Cara Courage
Chapter 20: Displacemaking 2015 and 2020
Catherine Fennell and Daniel Tucker
Chapter 23: Routing out place identity through the vernacular production practices of a community light festival
Gail Skelly and Tim Edensor
Chapter 25: 'If you can make it there, you can make it anywhere…': cultural placemaking at the heart of cities
Sherry Dobbin
Chapter 29: The solution is in the problem; the art of turning a threat into an opportunity by developing resilience using a Creative Placemaking critical praxis
Anita McKeown
Chapter 31: Seven generations: a role for artists in Zuni PlaceKnowing
Theodore S. Jojola and Michaela P. Shirley
Chapter 32: *The Hollywood Forest Story*: Placemaking for the Symbiocene
Cathy Fitzgerald
Chapter 36: Facilitator skills for effective collaborative placemaking
Husam AlWaer and Ian Cooper
Chapter 37: *The Neighbourhood Project*: A case study on community-led placemaking by CoDesign Studio
Lucinda Hartley, Eliza Charley, Sama Choudhury and Harriet McKindlay

40

TRANSFORMING COMMUNITY DEVELOPMENT THROUGH ARTS AND CULTURE

A developmental approach to documentation and research

Victor Rubin

Introduction

The ways in which arts and culture can engage, energize, and motivate residents to work toward ambitious goals to improve their communities are as diverse and colorful as the communities themselves. It recently took the form of a neighborhood-wide treasure hunt for cultural touchstones through Little Tokyo in Los Angeles, following the Japanese tradition of *Takachizu* – touchstones which will prove valuable in the community's battle for control of key development sites. It was found in a homegrown musical theater production in a tiny Minnesota town, bringing together, through surprisingly common myths and traditions, the newcomers from Micronesia and the long-term residents of Scandinavian descent. And, it can be seen in the faces of Zuni children and youth as they explore, and make their own, a new park and community center designed in every respect to embody their tribal history, culture, and artistic heritage. Arts and cultural strategies are proving to be effective and innovative components of the practice of community development. Artists and culture-bearers are now working with community-based organizations in ways that enhance economic, physical, and social outcomes for places and deepen and reinvigorate the voice and agency of residents. In the course of integrating arts and culture into their operations, these community development organizations have, themselves, been fundamentally changed. This activity, and the changes in how community developers and artists work together, represent the latest wave of creative placemaking in the US, one grounded in the principles and practices of equitable development. Previous versions of creative placemaking had emphasized generating pedestrian-friendly environments and 'vibrancy' in underutilized or underappreciated spaces and had sometimes become associated with attracting newcomers rather than enhancing the lives of current residents. As the more equity-centered practices diversify and proliferate, it will be important to document them thoroughly and to understand them within a framework that outlines the process by which arts and culture strategies can make an impact and a difference.

This chapter describes a five-year effort, now in its final stage, to provide this kind of documentation and analysis for one of the most significant philanthropic investments intended to transform community development through arts and culture. While the chapter will provide the context, structure, and evidence of some outcomes of the initiative, the emphasis will be on the goals, terms, parameters, and experience of conducting this research. The initiative was intended to generate learnings that can be more generally applied beyond its grant-supported activities, and the experiences of the grantees and their partners provide guidance for system-level innovation in several fields. The author has been the director of this research and documentation effort as part of a team at PolicyLink, working closely with the staff of the funding consortium ArtPlace America and the leaders of the six grantee organizations. PolicyLink is a national non-profit research, advocacy, and policy change organization and eight staff members participated in the project, as did six local correspondents, one in each of the Community Development Investments (CDI) communities. PolicyLink was chosen in part because of experience in diverse fields including community development, health equity, arts and culture strategies, participatory research methods, and evaluation of philanthropic initiatives. Data collection included site visits, convenings of the grantee organization teams for reflection and dialogue, review of documents, and hundreds of interviews and video recordings of artists, residents, and other stakeholders as well as the lead agency staff.

This approach to researching innovative practices instigated through a one-time grant program implicitly draws from the framework of Developmental Evaluation (DE), which has been 'designed to meet the need for a more expanded view of evidence' than that provided by conventional program evaluations. DE is best suited to inquiry about situations that are continuously evolving. The approach was first conveyed by Michael Quinn Patton (2010) and is enjoying a new wave of attention in recent years. As the research and strategy firm FSG puts it, 'DE can provide stakeholders with a deep understanding of context and real-time insights about how a new initiative, program, or innovation should be adapted in response to changing circumstances and what is being learned along the way' (Parkhurst et al., 2016). PolicyLink did not literally conduct an evaluation, and eschewed the term, for there was neither assessment of a program or its grantees nor expectation that the program would be repeated. Rather, these complex, dynamic experiences offered important lessons for field-building, and the spirit and intention of DE was consistent with our purpose and analytical style in getting at those lessons.

Six well-established organizations, based in diverse urban, rural, and tribal communities, moved deeply into arts and culture practices as part of the CDI Initiative of ArtPlace America. Their plans, struggles, and pathways from ideas to outcomes in housing, local economic development, health, youth development, and parks and recreation were documented and analyzed from their selection in 2015 through early 2020, when they were moving into post-grant sustainability mode. The participating organizations and their partners have taken on and struggled with some of the most pressing and complex issues of our time, including gentrification and displacement, racial health inequities, the isolation of immigrant newcomers, and the historical trauma resulting from racism and oppression (see Figure 40.1). They have combined their expertise and standing with the tools and ways of thinking, imagining, and working of artists. As a result, they have helped residents to own and express the identity of their communities, built cultural resilience, and changed the terms of engagement and the methods of neighborhood planning and placemaking.

The purpose and approach of the CDI initiative

CDI represents a customized approach to linking arts, culture, and community development. Building upon, but significantly altering and expanding what is sometimes known as creative placemaking, ArtPlace, a consortium of foundations and financial institutions, chose for these

Participating organizations and summary of their activities

Cook Inlet Housing Authority, Anchorage, Alaska Guided by Native Alaskan village values, this regional tribally designated housing authority creates housing opportunities to empower people and build community.

Focus: Solving problems in new ways and elevating resident voices.

Key projects: 'Living Big, Living Small,' exploring small space living with set designer Sheila Wyne; '#MIMESPENARD,' mitigating business disruption during a road construction project with performance artists Enzina Marrari and Becky Kendall; the Church of Love, transforming a former church slated for demolition into a community center/art space/performance venue; and embedding story gathering and listening as an organizational practice with Ping Chong + Company.

Fairmount Park Conservancy, Philadelphia, Pennsylvania Urban parks conservancy that leads and supports efforts to improve Fairmount Park's 2,000 acres and 200 other neighborhood parks citywide.

Focus: Working with artists to make city parks relevant for a more diverse population of Philadelphians, and to celebrate the history, culture, and identity of its neighborhoods.

Key projects: A community catalyst residency with the Amber Art & Design collective at the Hatfield House in the Strawberry Mansion neighborhood, including cultural asset mapping, social engagement, and community building; leading a master plan process for the Mander Recreation Center; co-hosting the West Park Arts Fest in East Parkside; and expanding the scope and reach of The Oval, a seasonal pop-up park in downtown Philadelphia.

Jackson Medical Mall Foundation, Jackson, Mississippi Manages a 900,000-square-foot medical and retail facility in central Mississippi with a mission to holistically eliminate health-care disparities through the promotion of creativity and innovation.

Focus: Enhancing their role as a neighborhood anchor by fusing arts and culture with health and economic development goals.

Key projects: Intergenerational programming and festivals linking artistic production and economic development with the delivery of health services; 'Reimagining the Jackson Medical Mall' with Carlton Turner, a leading innovator in Mississippi African American cultural empowerment, to introduce history and storytelling into the design of the space; a new community garden and kitchen; and internal and external creative engagement practices with Significant Developments, LLC.

Little Tokyo Service Center, Los Angeles, California Provides family services, affordable housing and tenant services, and community organizing and planning for the nation's largest Japantown, in downtown Los Angeles.

Focus: Facing increasing pressures of displacement, homelessness, and high costs of living, they launched the +LAB ('Plus Lab') Arts Integration project to test new ways to promote the equitable development of ethnic communities.

Key projects: 'Takachizu' with Rosten Woo and Sustainable Little Tokyo, inviting residents to share treasures from the neighborhood; #MyFSN, which seeks to assert 'moral site control' over the future of the contested First Street North site; 341 FSN, an experimental storefront space designed to explore community control and self-determination; and the +LAB artist residency program.

Southwest Minnesota Housing Partnership, Southwest Region of Minnesota Provides housing development, preservation, rehabilitation, and supportive housing and community development services for a rural 30-county region.

Focus: Partnership Art, which uses arts and cultural strategies to incorporate new voices, including Minnesota's growing immigrant communities, into local planning processes.

Key projects: Milan Listening House, exploring immigration stories and the concept of home to inform the revitalization of public spaces; Healthy Housing Initiative, an outreach and education toolkit reaching new Latinx communities; 'Creative Community Design Build,' where artists engage communities to reimagine underutilized downtown buildings; and hiring Ashley Hanson as an internal artist-in-residence to help sustain their arts and cultural approaches.

Zuni Youth Enrichment Project, Zuni, New Mexico This effort is devoted to enhancing the health and resiliency of youth on the Zuni Pueblo in New Mexico.

Focus: Integrating Zuni arts and culture into planning, design, and construction of a new youth center and park.

Key project: Supporting an ongoing artists' committee and other local artists to co-design and contribute to long-term stewardship, activation, and programming of H'on A:wan ('of the people') Community Park.

Figure 40.1 Participating organizations and summary of their activities (Rubin, 2020).

investments well-established community development organizations which had little or no previous arts and culture experience. The common central priority was to employ arts and culture strategies to advance and strengthen – not to change – the organizations' missions. The criteria and priorities included: selecting organizations rooted in serving low-income communities, with large and medium-sized cities, small towns, and tribal lands among the six, each in a different part of the US; employing an inclusive and broad definition of community development, not only nonprofit housing and neighborhood development organizations, but also community-based health services organizations, and those centrally concerned with youth development and the stewardship of urban parks; encouraging a very broad array of possible arts and cultural contributions, resulting in not only the most commonly expected mode of creative placemaking – public art – but theater, sculpture, printmaking, photography, music, experimental film, storytelling, quilting, beading, interior design, and many other disciplines and crafts, produced and experienced by a very wide range of people. Outcomes as unusual as scores of mimes in the streets and playing cards with pictures and biographies of neighborhood notables were produced, as were some sizable and significant capital projects, from parks to community art centers; seeing the possibilities in certain artists as facilitators, guides, and advisors to the community development organizations who will bring different sensibilities and approaches compared to typical consultants to the nonprofit sector; and cultivating an environment of learning and experimentation, without projects specified in proposals or in the first year of activity, providing plenty of room up front to explore possibilities, map the community's cultural assets, fail at early ideas, adapt, and build the trust and perspective necessary for long-term success. Throughout the initiative, the groups were provided customized technical assistance and peer sharing opportunities, and access to sources of additional capital should they need it.

Defining features of the research and documentation

The CDI research and documentation effort grew up alongside the program itself, and the analytical approach reflects a balance between conveying the unique features of six distinct, experiences and producing lessons of broader relevance to their various fields of practice. Description alone, as useful as it may be, would not be enough; the goal has been to dig deeper into what can be called the 'but for' questions: What difference has the integration of arts and culture strategies made for the practice of community development? What has changed from the agencies' previous way of doing things, and what has been the impact of that shift? With a small sample of six sites and a largely qualitative approach to tracking the planning and implementation that was unfolding throughout the full study period, this did not yield standardized comparative measurements of the relative influence of specific factors. But it has yielded complex accounts of the context, processes, partnerships, strategies, and activities at each site as well as parallel information points about each of them that have fed into addressing three cross-cutting areas of interest and conceptual themes. Those themes, which will be discussed later, became the basis for sets of questions developed through exchanges with the grantees to generate a comparable but distinct 'learning agenda' for each site. The common questions and the site-specific agenda topics were the foundation for the queries used in individual interviews of people in many roles, not only in the CDI grantee agencies but in their numerous partner groups, the artists who worked with them, and the residents and other stakeholders. The common questions were also used to solicit self-generated responses by site leaders via an online interactive audio, video, and text system, VoiceThread, structured in-person dialogues among participants from the six agencies, multi-faceted site visits, and analysis of many types of written materials and budgetary records by and about the sites. Artistic activities and capital projects were documented with still photography and video as well as text, and the myriad creative activities, some of which produced research findings in their own right, became an important part of the documentary record.

Guiding principles for data collection and analysis

This highly interactive and flexible approach to researching an evolving initiative was guided by several priorities which guided who was interviewed or observed in action, when, and for what purposes.

Gather the perspectives of participants at several points in time

Since the program was new, and the grantees began their work period without project plans or specific commitments, virtually every interesting question involved learning about how things were evolving given this open-ended start. That included the work plans and how they were carried out, the relationships with partners, the personal understanding of the role and impact of arts and culture of the organizational leaders, the organizational practices and policies, and much more. It was vital to interview, observe, and otherwise learn from the key players at several points over four years to see how their thinking, and the activities themselves, evolved. It was similarly necessary and rewarding to share drafts of emerging themes and lessons and receive their input on these at every stage.

Encourage and document interaction among site leaders

CDI created a lively community of practice of roughly 20 individuals, including several people from each site, who met roughly monthly in video conference calls and six times in person over four years. They were very insightful about commonalities and differences across their organizations and learned a lot from each other. The monthly online interactions generated by ArtPlace created a regular flow of shared information, and the annual ArtPlace conferences and two additional in-person gatherings organized by PolicyLink produced facilitated dialogues among site team leaders that were recorded and edited for publication and mined extensively for quotes and insights.

Ask about the personal meaning of the work, not only facts, plans, and policies

People who undertake innovative, if not risky, arts and culture activities are motivated by what they hope to achieve, and they draw meaning from the endeavor in personal as well as professional ways. By their nature, creative ventures cannot be reduced to just formal plans, so we added a 'creative documentation' component in the form of extensive video interviewing by photographer and professor of art Chris Johnson. He visited each site twice, almost two years apart, and asked a cross section of 10 participants in each community about the personal meaning and motivation they drew from this work and how it may have changed over time. The video recordings have been edited into a finished production for public use.

Acquire multiple perspectives on the same issue and activity

The staff members of the CDI organizations built new partnerships with artists and/or with community-based groups and fashioned new relationships with residents of their community. It was essential, when documenting these engagements, to learn how this experience looked and felt from the partners' point of view as well as the grantees. These engagements featured false starts and rebooting and faced the need to reconcile different views of how to proceed. The challenging interactions were at least as important to learn from as the ones that went smoothly.

Encourage and record eclectic forms of expression and reflection

Documentation of this kind of initiative calls not only for well-ordered evidence in the conventional social science sense but also for stories, poetry, personal testimonies, photography and videos, including recordings of performances, that are evocative of the spirit of the work and which provide living examples of what can be accomplished. The collective efforts of artists, agency staff members, and community folks in all six places resulted in a plethora of projects and reflections about them which have been captured for future use.

Recognize that social changes will precede, and be the preconditions for, most long-term community outcomes, and that those long-term outcomes will mostly not take place before the study period ends

The activities undertaken through the CDI initiative were creating the conditions for concrete changes to the quality of life: better population health, positive youth development, improved housing, more welcoming public spaces, and so forth. The groups could specify their intended results for the populations they served. However, those ultimate outcomes take years after the initiation of a project to be realized, and with a few exceptions, during the grant period the main impacts were to create changes in the way that people interacted. These preconditions included such phenomena as organizing to build voice, agency, power and collective efficacy, strengthening the social fabric, heightening the sense of community identity, and building bridges across cultures. The research was therefore geared primarily toward documenting and analyzing the role of arts and culture strategies in bringing about these social changes.

Transforming community development through arts and culture: themes, questions, and findings

During the first year of the program, PolicyLink coordinated a series of dialogues among the project coordinators from each CDI site team and the ArtPlace staff to identify common concepts and a framework for the research. The framework that emerged after several rounds of refinement was relevant to all six sites and promised to provide useful comparisons and lessons. It eventually encompassed three categories – Organizational Evolution, Collaborative Practice, and Community Development Outcomes – under which were nested a total of 9 major questions and 25 sub-questions (ArtPlace America, n.d). Organizational Evolution was significant because the chief executives and project directors realized that it would be essential to take advantage of this opportunity to make larger changes in the culture, direction, and internal structure of their agencies in order to better live up to their values and achieve their mission. Collaborative Practice became the category for documenting and comparing the multitude of ways in which community developers and artists or arts organizations became mutually acquainted and more sophisticated about codesigning complex projects. Community Development Outcomes began as an effort to document the tangible changes expected in the health and prosperity of people and their communities, but, as described above in the section on principles of the research, became more focused on the cultural and social changes and new strategies for organizing that were the necessary precursors to better outcomes. For example, a major question from the research framework was: how can community development organizations establish productive relationships with artists and arts organizations, with the follow up questions of: how does a community development organization learn about and build connections within the arts community; what are the approaches to selecting and incorporating artists and arts and cultural organizations as strategic partners in community development work; and to what extent does a community development organization's support of artists or arts and cultural organizations outside of the collaboration (i.e., capacity building, training) strengthen the collaboration?

The most basic task in service of this framework was to track and describe what happened – documentation of the baseline circumstances and then the ideas, plans, deliberative processes, decisions, and actions – as the site teams moved through cultural asset mapping into setting priorities, doing detailed project planning, and then creating numerous partnerships and activities. Beyond the description, however, it was important to discern from the participants why and how the arts and culture strategies may have made a difference to their organizations and communities. The common analytical questions for all the sites were followed by more detailed site-specific questions referred to as learning agendas.

Collaborative practice

The community development organizations needed to figure out how to recruit and partner with artists, and the artists similarly needed to determine how they could make a useful contribution in this different environment. The interviews with all parties in the first two years and reflective dialogues in the third year, firsthand observation of events, and compilation of the materials used to solicit and contract with artists provided a comprehensive picture of how these arrangements were made, complete with the missteps and revisions. When www.communitydevelopment.art, the PolicyLink-managed website on which the CDI research and documentation is being presented, debuted in May 2019, the first brief on the site was 'Working with Artists to Deepen Impact' (Stephens, 2019), 'exploring the theme of collaborative practice, or how these community-based organizations cultivated working relationships with artists, and how they have significantly changed the approaches through which community preservation and revitalization can take place.' Specifically, the brief provides insights on: how the community development organizations matched community development priorities with the expertise and artistic practice of potential collaborators; identifying arts partners and building relationships through cultural asset mapping, calls for artists, collaboration with intermediaries, the compilation of artist rosters and directories, and the formation of arts advisory committees; lessons learned from the process of creating guidelines for this new work, structuring relationships, and establishing roles and responsibilities; overcoming challenges – specifically, how these experiences improved their community development work – and learning to be more transparent, nimble, and reflective; and continuing the work after the program ended (Stephens, 2019). Subsequent conference presentations and publications, including several chapters in the special issue of *Community Development Innovation Review* (Crane et al., 2019), have provided additional stories and reflections from the participants about these techniques, lessons, and challenges.

Organizational evolution

Soon after the initiative was underway it was evident that the ways in which the CDI organizations were changing would be very interesting and important, and that these changes would be ready for documentation and analysis over time well before the community outcomes would be. The CEOs, project managers, and other staff members of the grantee organizations were out-

Incorporating arts and cultural strategies into core work resulted in changes in the overall culture, leadership, and future direction of the organization.

These included:

- Finding new ways to express their core values, leading to greater organizational cohesion and more effective communications
- Identifying different definitions of success based on those values
- Building greater capacity and openness to risk and experimentation
- Crafting more effective interactions with community members and partners

Incorporating arts and cultural strategies deeply in the organization meant actively aligning internal processes and structures.

This included factors such as:

- Assigning and sharing responsibilities throughout the staff for implementing arts and culture
- Diversifying the race and ethnicity of staff and leadership
- Integrating arts into strategic planning processes
- Bringing in artists as advisers and colleagues

Figure 40.2 Incorporating arts and cultural strategies into core work resulted in changes in the overall culture, leadership, and future direction of the organization (Rubin, 2020).

going, candid, and reflective as their efforts unfolded over three years, and their partners shared valuable perspectives about how the community development organizations had changed as a result of the arts and culture strategies. The lessons were pulled together in a brief entitled 'How Organizations Evolve When They Embrace Arts and Culture' (Rubin, 2020). The eight categories of change, four in overall culture, leadership, and future direction and another four covering the active alignment of internal processes and structures, are captured in Figure 40.2. They are illustrated in the brief by the stories and self-assessments of the leaders and other evidence of the new ways of thinking and working.

Community development outcomes

Improvements to the communities and to the health, economic security, and quality of life of the residents are, of course, the ultimate reason for undertaking this initiative. While outlining them was relatively straightforward, it was likely that, in most instances, outcomes of this nature would not be visible or measurable until several years had passed. Most of the projects were, in one way or another, establishing the preconditions for longer-term changes. For example, H'on A:Wan Park, built in the village of the Zuni nation and the largest capital project undertaken with CDI resources, was immediately a significant cultural and physical asset but also the foundation for promoting healthy youth development. While the building and grounds were completed in three years, the changes to the resilience, health, cultural awareness, and future prospects of Zuni children and youth will take longer to become established, let alone measured.

Each site has roughly comparable stories of foundations for change being laid through arts-focused ventures. In Little Tokyo, the projects and activities which promoted the neighborhood's cultural identity reinforced the community's campaign to control future development, an effort which will take years. In Southwest Minnesota, social ties among populations from vastly different cultural backgrounds were established through community theater and other endeavors but would need continual renewal if those ties are to be strengthened and sustained. The trends that could be discerned from the scores of activities and strategies led us to aggregate them into two categories, each of which is the subject of a brief and other articles and presentations. One category includes the ways in which arts and culture strategies led to new forms of community engagement and organizing, for which the Little Tokyo example above is a prime example. This embodied not simply new material to add to the same type of organizing campaign, but also a new way of perceiving the issues, motivating people, and building grassroots voice, agency, and power. The other category includes the strengthening of the community's social fabric, of which the Zuni and Southwest Minnesota examples are emblematic. That stronger social fabric can provide support for both personal growth and collective efficacy.

Audience mapping and its implications for research and writing

The choices of intended readers and listeners for this kind of analysis affect how it is carried out and presented. Researchers were not the principal audience, and few articles or reports were produced directly for fellow analysts, though the products will be hopefully nonetheless useful to them. ArtPlace placed a high priority on reaching people in positions to improve their own practice or alter the prevailing norms and practices of a professional field, and the PolicyLink team worked with ArtPlace in 2017 to identify nine distinct audience segments in order to prepare presentations of findings and stories that would reach each of them effectively. These segments represent sectors in which the CDI grantees work, such as affordable housing, neighborhood planning, community health care, and parks, and media serving demographic groups such as indigenous people (since two of the six sites are Alaska Native or Native American tribal entities). Throughout 2020 there were articles for leaders in nonprofit management and social investment, and a publication or two for practitioners and teachers of the social practice of art. Audiences with intersectoral roles and interests were sought as especially valuable targets for presentations. For example, the findings were brought to a real estate development member organization that was training leaders on health equity, and a national coalition of community development intermediaries that brought together African American, Latino, and Asian American organizations. The most ambitious presentation was the November 2019 issue of *Community Development Innovation Review*, the journal of the Federal Reserve Bank of San Francisco (Crane et al., 2019). The special issue, *Transforming Community Development through Arts and Culture*, includes 27 pieces which not only present this research but provide perspectives on the future direction of the field from leaders in banking, social investment, and philanthropy as well as community development. The intent was for the material to inform and motivate investors as well as practitioners.

Conclusion

The developmental approach taken to documenting and analyzing the ArtPlace Community Development Investments generated a detailed, continually growing body of information about the initiative. The participants co-created the questions that guided the study and their frequent opportunities to revisit and reframe those questions in light of their evolving experience was at the core of the research. The research shows that from a baseline of high accomplishment in their respective fields but little or no familiarity with arts and cultural strategies, the organizations' leaders became, after four years, sophisticated in how to design collaborative ventures with artists that served their communities well. The documentation drew upon not only conventional interviews and reviews of source materials but also explored and illustrated the great breadth and depth of the creative activities undertaken in each place. The research was intended to provide useful feedback to the participants in real time and to generate lessons for practitioners and thought leaders in all the relevant fields. The diversity of audiences and modes of published, online, and in-person presentation of the research have helped to meet that latter goal. Presenting spoken word artists, mimes, and audience dancing at events sponsored by federal reserve banks may have been unconventional, but it proved to be informative, colorful, and engaging, and entirely in the spirit of the initiative and the research. The personal stories and systematic findings have added to the body of evidence about the value of arts and culture for community development, and the leaders in the field will hopefully bring forth many new accounts in the coming years.

References

ArtPlace America (n.d.) *Community Development Investments (CDI) Research & Documentation* [online]. Available at: https://www.artplaceamerica.org/areas-of-work/research/cdi-research (Accessed: 10 June 2020).

Crane, L., Liu, J. and Rubin, V. (eds.) (2019). *Community Development Innovation Review*. San Francisco, CA: Federal Reserve Bank of San Francisco, Volume 14, Number 2.

Parkhurst, M., Preskill, H., Lynn, J. and Moore, M. (2016). 'The case for developmental evaluation', in FSG, 1 March 2016 [online]. Available at: https://www.fsg.org/blog/case-developmental-evaluation (Accessed: 10 June 2020).

PolicyLink. (2019). *Working with Artists to Deepen Impact* [online]. Available at: https://www.policylink.org/resources-tools/working-with-artists (Accessed: 10 June 2020).

Quinn Patton, M. (2010). *Developmental Evaluation: Applying Complexity Concepts to Enhance Innovation and Use*. Guildford: Guilford Press.

Rubin, V. (2020). *How Organizations Evolve When They Embrace Arts and Culture* [online]. Available at: https://communitydevelopment.art (Accessed: 10 June 2020).

Stephens, A. (2019). Working with artists to deepen impact [brief] [online]. Available at: https://www.policylink.org/resources-tools/working-with-artists (Accessed: 10 June 2020).

Further reading in this volume

Chapter 1: Introduction: what really matters – moving placemaking into a new epoch
Cara Courage
Chapter 4: A future of creative placemaking
Sarah Calderon and Erik Takeshita
Chapter 5: Making places for survival: looking to a creative placemaking past for a guide to the future
Jeremy Liu
Chapter 6: Listen, connect, act
Kim Cook
Chapter 9: From the dust of bad stars: disaster, resilience, and placemaking in Little Tokyo
Jonathan Jae-an Crisman
Chapter 10: From moon village to mural village: the consequences of creative placemaking in Ihwa-dong, Seoul
Jason F. Kovacs and Hayun Park
Chapter 11: *Free State Boulevard* and the story of the East 9th Street Placekeepers
Dave Lowenstein
Chapter 12: Public transformation: affect and mobility in rural America
Lyndsey Ogle
Chapter 13: Sensing our streets: involving children in making people-centred smart cities
Sean Peacock, Aare Puussaar and Clara Crivellaro
Chapter 14: Experts in their own tomorrows: placemaking for participatory climate futures
Paul Graham Raven
Chapter 16: More than a mural: participatory placemaking on Gija Country

Samantha Edwards-Vandenhoek
Chapter 17: 'I am not a satnav': Affective placemaking and conflict in 'the ginnel that roared'
Morag Rose
Preface: The radical potential of placemaking
Cara Courage
Chapter 22: Embedded Artist Project: Epistemic Disobedience + Place
Frances Whitehead
Chapter 23: Routing out place identity through the vernacular production practices of a community light festival
Gail Skelly and Tim Edensor
Chapter 24: Artists, creativity, and the heart of city planning
Tom Borrup
Chapter 25: 'If you can make it there, you can make it anywhere…': cultural placemaking at the heart of cities
Sherry Dobbin
Chapter 26: Sculpturing sound in space: on *The Circle and the Square* (2016) by Suzanne Lacy
Trude Schjelderup Iversen
Chapter 31: Seven generations: a role for artists in Zuni PlaceKnowing
Theodore S. Jojola and Michaela P. Shirley
Chapter 32: *The Hollywood Forest Story*: Placemaking for the Symbiocene
Cathy Fitzgerald

41
RITUALS OF REGARD
On festivals, folks, and findings of social impact

Maribel Alvarez

My friend Faeza was a dentist in Iraq until she had to flee her home under the threat of religious persecution. After a long wait, Faeza and her family received asylum in the United States. In 2009, they arrived in Tucson, Arizona – a mid-sized, Sun Belt city, 50 miles north of the US–Mexico border. I met Faeza the next year, when she came to demonstrate Iraqi family dishes at *Tucson Meet Yourself*, the annual traditional arts festival that I lead. Scheduled to participate only on the first day of the event, she came back the next two days eager to soak in the mix of multicultural sounds, tastes, and colors offered in the city's public square. Faeza told me years later that the festival was 'the first time' she 'felt at home in the United States.'

Emotional connections for individuals engaged in large public events are commonly touted as one of the benefits of civic investments in placemaking. Pragmatically, however, these outcomes are remarkably difficult to measure. While large crowds mobilized on behalf of a social cause or a communal ritual can thrive on feelings of shared ownership, festivals are especially challenging because of their reputation for distraction. Despite this potential liability, over the years, I have heard many artists and community members share sentiments of sociability and belonging similar to those expressed by Faeza about the *Tucson Meet Yourself* festival. This was confirmed yet again in 2017 when folklorist Cliff Murphy arrived in Tucson to conduct an evaluation of the event. He was surprised to discover the taxi driver who picked him up at the airport eager to share unsolicited stories about the many ways the annual festival had been part of this man's family history (Murphy, 2017). In this essay, I aim to explore the dynamics that inform a distinctive sense of rootedness *in* and *with* the public in this staged event. The Tucson festival's 'distributed benefits model' of evaluation can offer insights to planners and cultural advocates interested in assessing the impact of placemaking projects through a lens of social equity and inclusion. I hope also to demonstrate the value of methodologies that center self-critique and ethnographic reflection as indispensable practices in the evaluator's toolkit.

In assessing the impact of cultural productions, the most common evaluation methods are able to capture snapshots of activities in delimited times and spaces – numerically, they can account for how many people attend an event or how much is generated through sponsorships and sales. When other social measures are added to the mix, surveys, interviews, and case studies can provide additional insights on how people 'feel' during the event or how they perceive significance in more universal terms (such as 'this festival places our city on the map,' or, 'the event helps increase our community's pride'). Recent interest around the role that racial equity and healing and neighborhood-based wealth creation strategies play in creative placemaking suggest the need to expand the depth and breadth of the questions we ask to assess the impact of public events (Getz et al., 2018; Nowak, 2007). Aligning evaluation efforts in the service of these goals implies paying more attention to the divergent ways that benefits accrue to participants across a range of social positions and levels of investment. In some cases, this will require upending the metrics that conventionally made sense to favor considerations of an 'emotional' nature, such as the sense of belonging (Lee, Lee, & Choi, 2011). As evaluators embrace the challenge to be more responsive to community needs, 'the event' itself may no longer serve as the main unit of analysis. Instead, the radial effects produced cumulatively by the nexus of relationships, hopes, intuitions, affinities, and intentions generated because of the event, and the process of preparing for it, are likely to grow in significance.

In the nonprofit field, the types of supportive actions that enable or increase the ability of people and organizations *to do more for themselves* are called 'capacity-building.' But capacity is a multifarious variable, at times tied very specifically to increased metrics of money or people, and at other types more broadly diffused through enabling relationships and a sense of agency. For *Tucson Meet Yourself*, the task of standing up an infrastructure of production for a *specific event* has always implied sustaining a flexible, aspirational platform that hopes to strengthen the ability of the communities of artists and tradition-bearers served by the event *to do more and better,* with *their own* assets, communities, and aspirations, and in their own terms. Instead of thinking of the event as the final product, the festival inspires a horizon of impact that includes the multiple ways artists and communities might 'use' the festival as fuel for activities of cultural autonomy and renewal. Most of those activities will be after-effects, carried out beyond public view and statistical reach, or outside the curatorial guidance of event organizers.

Most evaluation approaches in the cultural policy arena start by inferring the value of festivals presumptively, through the application of somewhat circular logic: festivals are good for communities because… well, how can they possibly do any harm? (Giorgi et al., 2011). The practices and suggestions for assessment that I discuss here aim to activate different logics of accountability. The experience of *Tucson Meet Yourself* underscores an understanding of *impact* as essentially a measure of 'change' in someone, something, or somewhere. The crucial question is not 'what' gets celebrated (since we all can agree that good music and food is fun for anyone) but for 'whom' does the staging of celebration makes the most ethical and material difference? In this framing, a reimagination of the festival takes place; one less concerned with festivity properly and more oriented to the event as lab and incubator. Attendance goals remain important, but not exclusively. Surely, the festival's credibility in making claims about a horizontal distribution of benefits is in large part connected to its large attendance – approximately 150,000 people over three days. But this numeric accomplishment is always judged in relation to *what else* happens. The festival frequently inspires actions that take place in locations and places where people carry on beyond the spotlight. As a result, the event has gained an identity in the public's perception less as 'party' or 'shopping expedition,' and increasingly more as a restorative ritual of cultural democracy. For many cultural producers, this affirmation will seem too good to be true. The key question I wish to address, therefore, is how does a staged event accomplish this? I believe the answer is embedded in the 'folklife' framing of this event, an architecture of inclusive planning and equitable 'return on investment' that can be adapted to a variety of other placemaking projects.

The grounding: folklife is about folks

Tucson Meet Yourself – the event where Faeza found her bearings as a new American – calls itself a 'folklife festival.' This category of festival represents a specific *type* of multicultural, public, celebratory, and educational gathering, which has synergy with the intentions of creative placemaking (Gillespie, 1996). It resembles other common events – the street fair, craft show, music festival, religious holiday, and ethnic pride day, among several – but it differs from those by prioritizing education as the main rationale for the event. Folklife events grow from ground-level research or 'fieldwork' that lead to collaborations with key cultural authorities and representatives. (Santino, 2016; Bauman and Sawin, 1991). The folklife movement can be understood as one of several antecedents of the 'creative placemaking' field as practiced today. The term folklife evolved from the older form folklore. While folklore referenced primarily the oral transmission of stories and songs, folk-life emerged from the need and desire to emphasize the holistic study of everyday life practices (from food and architecture to body adornment, traditions formed around occupations, and others). Early scholars understood the study of tradition as being imminently attached to geography. Thus, they noted with specificity how the community culture of a maritime area, for example, is different from that of an agricultural setting. As such, the aspects of community associated with folklife are aligned with the premise that good creative placemaking practice begins with acknowledgment of cultural assets that already exist in communities.

Tucson Meet Yourself is one of the largest place-based, heritage arts festivals in the United States. The challenge of producing the event lies in keeping a focus on the educational mission while also trading in the common forms of festival presentations – music and art as entertainment and ethnic 'others' as showmanship (Hutchinson, 2009). Increasingly, the festival has also ventured into practices that are relevant to community development. Given these intersections, the evaluation of the event can go in many directions: how people tend to think about festivals more generally – either too broad, as a free-flowing indulgence, or too narrow, as only an economic booster – further complicates evaluation. The focus on folklife, however, disrupts this simplistic binary. An essential challenge stands at the center of this community-informed framing: how does the project engender value for the people it claims to serve? Cultural strategies for belonging cannot just feel good or make sense to the professional animators, producers, and intermediaries who dream them up (Bissell, 2019). By definition, a festival whose stated mission is to represent the 'life' of the 'folk' carries a special obligation to inclusion and equity. 'Doing right' by the ethnic, cultural, or folk community that has agreed to share in a public setting the ceremonies, objects, stories, or foods that are usually reserved for insiders of the cultural group becomes the evaluative metric by default. The publics served by a folklife event, too, are rarely only tourists, casual browsers, or consumers. Like in other models of participatory action research, the 'folks' in 'folklife' assert their role as self-appointed evaluators with greater 'skin in the game' than mere spectators.

The *Tucson Meet Yourself* festival has been called 'the festival of festivals' because it aggregates under one annual signature City-namesake banner, multiple independent cultural associations, ethnic groups, artists, and small businesses, many of which hold their own smaller events throughout the year. A non-alcohol, family event held every October since 1974, 90 per cent of attendees are local. Sharing a formal affiliation with the University of Arizona through a parent nonprofit organization, the Southwest Folklife Alliance, the festival presents every year close to 400 performing artists, traditional artisans, food vendors, and community exhibitors. All artists receive compensation for their participation. Food vendors – mainly churches, associations, and small ethnic businesses – report sizable earnings during the weekend, monies that in most instances represent their single largest annual fundraising opportunity. Between 45 and 65 different ethnicities or folk-communities are represented at the event in any given year.

Predicaments of evaluation

In the search for revitalization strategies that can help neighborhood and communities, public officials and cultural advocates have turned to festivals as a preferred off-the-shelf tool. Festivals promise to transform local economies, uphold civic pride, attract tourism, and harness a variety of creative placemaking energies that translate to higher and better outcomes of social vitality and community identity (Arcodia and Whitford, 2006; Cibinskiene, 2012; Wood, 2009). In the United States, more people participate in festivals than in any other arts event (NEA, 2010). Festival audiences also tend to more closely resemble the nation's diversity by race, gender, age, region, and ethnicity than any other group of art-goers.

Yet, despite growing enthusiasm, it is common to hear pushback on the value of festivals among certain actors in the community development field. Because of growing skepticism and competition, a resurgence of interest in festival evaluation has taken place recently (Williams and Bowdin, 2007; Wallstam et al., 2020). My own approach to the evaluation of the *Tucson Meet Yourself* festival has undergone a curious evolution. From an initial generic, well-meaning anthropological disposition to 'check in' with participants through informal post-event debriefs, I have reoriented my approach towards an evaluation model that makes *reciprocity* the north star of all we do *for* and *with* festival participants. Reciprocity implies consultation with the communities represented and a fair chance for those ethnic and cultural groups to 'gain something' from being associated with *Tucson Meet Yourself*. Each year we are reminded of this essential principle when festival staff begins anew the meticulous consultation process for festival participation with our two home-based Native Nations – the Pascua Yaqui and Tohono O'odham communities. Participation by these core partners is never assumed *a priori*. Each year, festival curators sit down with tribal representatives to discuss what is working and what needs adjustment. Only then is the decision to participate made. In this light, the arc of festival evaluation shifts from a practice performed *at the conclusion* of the event to a framing that guides the event's planning.

Assessing the success of *Tucson Meet Yourself* started out in conventional terms: gathering reliable data that demonstrated the value of the event to the greater community of sponsors, public officials, and opinion leaders. To that effect, we codified the festival's attendance estimates through a triangulated spreadsheet; surveyed thousands of participants and audience members (over 5,000 live questionnaires taken on site); determined the rate of waste diverted from landfills to assess sustainable practices; and calculated the impact of the festival on the local economy by accounting for cascading related business transactions. In the last few years, we also added the visit of a Guest Folklorist to evaluate the event from an outsider's perspective. Through methods of participant-observation, the Guest Folklorist writes a brief report addressing three questions: what are we getting right; what needs improvement; what practices have you seen or heard about in other festivals that we should consider emulating?

In the process of designing a more grounded evaluation approach, I find particularly useful the distinction between 'indicators' and 'indications' of success articulated by urban planner and scholar Maria Rosario Jackson (2019, p. 10). *Indicators* refer to the regime of recurrent, reliable quantitative data used by professional evaluators and researchers. They represent the menu of metrics that help planners forecast impact and formulate hypotheses. *Indications*, on the other hand, says Jackson, emerge from on-the-ground observations by diverse publics and, most importantly, directly impacted populations, perhaps initially only as 'clues' and forms of 'discernments.' They may not at first glance convey the rigor usually attached to quantitative data, but they are nonetheless in every instance as 'thoughtful and disciplined' observations, signals, and insights as any other form of impact narrative.

Common festival evaluation methods tend to focus too often on narrower transactional outcomes (i.e. hotel bed taxes, sponsorship activations, ticket purchases, brand hegemony, etc.), or in otherwise largely amorphous social outcomes (i.e. social cohesion). The benefits analysis that derives from a folklife perspective defines success through a slightly different lens. Here are three key elements I have learned are foundational to an expanded equitable approach: longitudinal learning; grounded relationships; and equitable returns. Below, I expand on these evaluative lessons as *indications* of impact learned in the trenches of folklife work. My hunch is that organizers and advocates in other sectors of the creative placemaking field have grappled with similar ideas. A shift from a preeminently outcomes-driven approach to evaluation towards more spacious and generous process-evaluation models can reconcile the differences between evaluations deployed as administrative accountability 'from above' and evaluations 'from below' that place greater value in 'the voices, knowledge, expertise, capacity, and experience' of those we are most eager to serve' (Stern et. al., 2019).

Longitudinal learning

Once during a busy *Tucson Meet Yourself* festival a few years back, a woman stopped me dead in my tracks as I was making my way to a stage to introduce two Cuban musicians. She shook her finger at my face and said sternly: 'are you the director of the festival now?' I nodded affirmatively. She retorted, 'So you are the one who brought in "the gays." That is just wrong. This used to be a nice cultural event and now you changed it. Thank you (not) very much!'

The woman was referring to a creative partnership *Tucson Meet Yourself* had established that year with the Southern Arizona AIDS Foundation. On the morning of the last day of our festival, the footprint and infrastructure of the event flips and becomes the site of the annual Tucson AIDSWALK. The partnership is a stroke of genius for efficiency. It benefits both organizations financially, creating substantial cost savings for the AIDSWALK and generating a 'fee for service' revenue stream for *Tucson Meet Yourself*. But the arrangement is not only transactional. We also incorporate the ritual unfolding of the AIDS Names Quilt Project as a public folklife experience featured prominently on the grounds of the festival. In that setting, gay and lesbian couples, transgender individuals, and other relatives of AIDS patients or loved ones co-mingled in public view. In a way, the woman was right. The festival had changed. But the change was not the result of a new ideology, as her comments implied. The change she noticed reflects the normal flow of folklife in context: new communities form and they come closer to each other, or they evolve into something different than we are used to seeing. One year a group from Ghana staged a collective dance on the street. Although I have tried, I have not been able to locate them again to repeat the same performance. The Danish club, a favorite of the public for decades, aged and decided the new scale of the festival imposed an unbearable burden on their single-item *aebleskivers* booth.

Assessing the work that *leads to* inclusive cultural participation demands a different set of tools than have been commonly available in evaluation designs. The temporal frame required to effectively assess how dynamics of public inclusion ebb and flow can exceed the resources of most evaluation budgets. Folklife-informed methods of evaluation consider a festival's publics through patterns of change and adaptation that can escape momentary observations. In *Tucson Meet Yourself*, the immediate absence of the Danish Club – a common negative feedback recorded in the annual surveys – can only make sense in light of the club's very dense history with the festival and their eventual phasing out by necessity of personnel and effort. In the orthodoxies of 'community engagement' we are too often accustomed to deploying the concepts of audience and participant as monolithic wholes. However, more equitable evaluation outcomes could possibly emerge by accounting for the variability of needs and desires of differential actors in the event, many of whom will come in and out of the 'covenant' through zig-zag lines instead of linear progressions. Apprehending the impact of the festival experience as a nexus of feelings, clues, epiphanies, and accruals of value that stretch and change over time can deepen how we understand the fit between what an event offers and what the public wants. Accordingly, city planners, funders, and advocates will have more information to make recommendations for scaling up, downsizing, realigning, or even retiring festivals as these needs and publics change or evolve. Under the current evaluation paradigm, public events either only 'succeed' or 'fail.' Opportunities for festivals to change course, retool, reverse, modify, or refocus are rarely encouraged or rewarded.

Grounded relationships

In 2018, for the first time in 45 years, the *Tucson Meet Yourself* festival was nearly rained out. Previous decades had given us sprinkles or quick showers, but nothing like the two inches of rain that came down that year. The festival went on as scheduled, but the hardship on the staff and logistics was enormous. The public, however, seemed better prepared than the event organizers. Festival photos for 2018 revealed long lines of people at food stations wearing raincoats and ponchos or huddled together under large umbrellas. More aspects of the festival carried on as usual than anyone could have predicted. Technical teams stepped up and not a single act out of 100 performers was cancelled. Some food vendors reported having better sales that year than previously. In hindsight, festival 2018 was a phenomenal testimony to the resiliency of the festival's mission and authentic relationships with community. We also gained insights about how volume and size can cushion downturns. We estimated the rain caused an attendance reduction of approximately 40 per cent; from our usual 150,000 to around 90,000 people. Even then, it felt like business as usual.

One group among all participants was astonishingly cheerful about the rain. The Tohono O'odham, whose tribal name translates in English to 'people of the desert,' took the disruption in stride. As one of the festival staff members hurried among puddles in the street, she ran into Bernard, the elder educational liaison for Tohono O'odham manual artists. Bernard told her: 'we are happy; we pray for rain in our ceremonial songs, and here it is.' A fundamental truth about folklife-informed methods of assessment was revealed in that moment: who gets to define what should count as success? The Tohono O'odham's interpretation of the rain stresses the relational nature of participation as a nexus of interlocking and differential effects for multiple constituencies. Most evaluation methods tend to go in the direction of linear progression towards ascendant certainties, leaving little room for equivocation. I recall once receiving an evaluation questionnaire for a strategic planning process developed by a major consulting firm in which every single question asked, 'on a scale of 1 to 10 *how enthusiastic are you* about X.' This type of phrasing reinforces a 'no way to go, but up' boosterism that is all too frequently the trademark of many festival and creative placemaking efforts. Bernard's comment blunted the commonsense evaluative inclinations of the rain 'crisis,' but not by way of naive optimism, but by centering the climatic event in the larger universe of cultural context and meaning.

As the days passed and both festival and rain stopped, I learned to appreciate the Tohono O'odham's playful sense of reversal. My own evaluation framework relaxed, allowing a more complex picture to emerge of the multiple ways in which 'the rains of 2018' produced both felicitous and dreary outcomes. The best parts of the *Tucson Meet Yourself* festival that year where the improvisations and surprises that mediated between the opposing poles of good and bad, or, sunny and rainy circumstances. For instance, we celebrated the remarkable skills of the electrical crew deployed to protect equipment and power distribution. In his review of strategies for cultural belonging, Evan Bissell notes that when evaluations are based only on 'changes that build towards a pre-determined goal, there is a danger of erasing experiences and processes that are vitally important to people' (Bissell, 2019, p. 43).

From my own vantage point as the person who understood better than most what detrimental effect the rain would cause to festival finances, to the staff member who encountered Bernard in the street, through the loops of informal conversations that circulated competing interpretations of the rain, an essential element of trust was necessary for a critical evaluation of the event to take hold. Trust develops over time as an outgrowth of respect. Trust makes room for dissenting points of view. Although there is a growing body of research on what are recognized to be 'social impacts' of festivals (Robertson et al., 2009), not nearly enough attention has been directed to assessing the role of relationships as the essential anchors that translate the value purported by creative placemaking projects. For instance, what are the mechanisms by which staged events establish different kinds of relationships with various publics? How are different levels and scales of relationships codified as usable data? What effects do relationships have on the accrual of value of an event over time?

For *Tucson Meet Yourself*, the greatest challenge remains, year after year, how to make visible those foundational relationships as the distinguishing mark of our brand of festival. Furthermore, how can the measure of this strategic differential factor be shared with new partners who may or may not have the clarity of mind to understand the unique lens of folklife that frames our work. This challenge caused me to think more carefully about the things people 'see' at the event (the physical built out and visual arrangement of the festival) in relation to the things they 'experience' (the sensory educational offerings). Our grounding in folklife practices advises us to look for points of data not readily apparent. This inclination shows up, for example, in the types of questions we choose to include in the festival's survey. Thus, we decided to be direct and ask pointedly: 'how is *TMY* different from other festivals you've attended'? This data collection exercise ties into a generalized festival culture of reflection, trial-and-error, learning, and adjustment. It contributes to cementing a commitment to an evaluative approach that mirrors the way cultures actually work – never static, monolithic, or simple; instead, always contingent, emergent, and regenerative.

Equitable returns

One afternoon a year ago, I met my friend Danielle for coffee. We had been introduced by a common friend to work on a community social justice project. As we exchanged basic facts about our lives, Danielle told me her mother was 'part of *Tucson Meet Yourself*.' When I asked in what capacity, Danielle said, 'she works at the festival, she is a part of it.' I didn't recognize her mother's name as one of our staff members, so I assumed she was one of the hundreds of volunteers I didn't know personally. But Danielle's repeated assertion that her mother was 'part' of the event caught my attention. After a few more questions, I learned that she volunteers at the food booth of the Thai Buddhist Temple where she also gathers for worship throughout the year. It is undoubtedly reassuring to hear the brand of the festival claimed equitably by community members through multiple points of access – temples, schools, and workplaces. The application of 'equity' in creative placemaking projects implies a recognition that the project must yield benefits that feel tangible and consequential for all partners. In some instances, people will be satisfied to accrue those benefits in the form of social rewards. But no one should be expected to cheer for intangibles if there are other pressing needs or goals at stake. Material benefits, particularly for participants such as artists and food vendors who risk financially the most, must be understood as a valid expectation from the event.

I often joke that at *Tucson Meet Yourself* everyone makes money except *Tucson Meet Yourself*. I am referring, of course, to the administrative costs of putting on the event. Generating surplus revenue from the festival is, objectively, a sound goal that the leadership of *Tucson Meet Yourself* clearly understands as part of a long-term capitalization strategy for the program. However, prioritizing surplus revenue is not where the festival draws its bottom line. This decision is not without risk, but the principle behind it has shaped a number of qualitative measures that have cemented the festival's credibility among the most vulnerable groups the event claims to serve. For example, cash payments to artists, and artist-related expenses like hotel and transportation, account for 30 per cent of the festival's budget. Were *Tucson Meet Yourself* to skip these investments in artists, the potential profit or surplus generated from these savings would be around $65,000 annually. Hypothetically, this is a 'surplus' that the festival could be reinvesting in other programs or set aside in a reserve. The decision to pay the artists a fair level of compensation is inscribed as part of the mission of the event – not just as an optional 'beneficence' the organizers can disavow at will.

The festival's relationship with food vendors is structured slightly differently but illustrates the same principle. Food vendors pay a flat fee for participating in the event; the festival does not ask for a percentage of sales or collects any additional taxes. All 100 per cent of the fees vendors pay go directly to cover costs that the festival must pay to the City, County Health Department, and to subcontractors such as the electrical/power distribution crew that provide the essential infrastructure food vendors need to succeed. In effect, the charge to vendors does not cover, per unit, all the compounded expenses. The festival subsidizes the vendors' expenses through fundraising and sponsorships. In addition, the festival does not ask vendors to provide their own individual liability insurance – a requirement that would represent a hardship and obstacle to participation for many of the small clubs, families, and associations. Since the festival takes on added risk by including food vendors in one comprehensive insurance package, festival staff perform detailed and careful coaching and due diligence working with each individual vendor to mitigate risks. In the 'covenant' of participation between festival and community partners, the festival delivers above the call of duty. This is recognized amply by vendors, who can trace with transparency a direct correlation between the festival's expenditures in marketing, staffing, and infrastructure and their success in sales at the event. The takeaway I want to stress here is not an argument *for* financial risk. Rather, I wish to highlight the reality of how production decisions always pivot on values. Values of social inclusion can orient organizations to take measured risks when greater community goals of equitable participation are at stake. While deficits or break-even budgeting practices should never set the standard for moral high ground in equitable evaluation, I have learned that the finances of an event tell an important story about what and whom is prioritized for investment.

Success as actionable democratic participation

As evaluation practices in creative placemaking continue to evolve, my hope is that the three realms of impact I have described above can offer pause and consideration. While the word 'folklife' may never appear in the vocabulary of the community development field outside those efforts explicitly directed by folklorists, the values and methods of this field of practice can shape the thinking and design of cultural strategy conversations at multiple levels (Zeitlin, 2016). The process of staging cultural events is a call for community-building. Learning to trust the public's response can be painstakingly difficult and simultaneously rewarding. Recently I learned that one festival participant who sells food from his home state of Oaxaca, Mexico, also owns a local pizza business. When he was asked why, if that was his way to make a living, had he not requested to sell pizza at *Tucson Meet Yourself*, he answered unaffectedly: 'the festival is where I share my culture; the pizza is just a business.' In positioning the festival as a platform for democratic participation, there is always an implicit risk that people might interpret the offering in their own terms. What if this man would have insisted in selling pizza? Can an immigrant from

Mexico claim pizza as heritage food? In some ways, the choice of pizza would not comply with the festival's formal requirement for an authentic connection with the culture of the seller. On the other hand, assisting immigrant communities secure a living and earning a 'day in the sun' in the city's civic culture is a core value of the event.

Sometimes the real value of creative placemaking efforts lies not only in leaning forward to uplift opportunities for community celebration, but also in looking side-glance and acknowledging threats against the inclusion of certain actors. Community organizers have always understood the dynamics of social wellness as a paradox of affirmations and inversions (Fox-Piven and Cloward, 1978). Not only is the driving question in the example above one of authentic Mexican food choices, or about giving opportunities to local vendors to make money, but also about protecting the space for agency and dissent. In other words, asking ourselves, what would be lost if we didn't do this work?

Evaluation practices that aim to capture the energy and significance of these tensions are needed now more than ever. Festivals and other staged public events must generate something more than feelings of goodwill. Democracy depends on the circulation of 'oxygen' around more complicated forms of cultural diversity than we have been willing to acknowledge – one in which ethnic and immigrant groups can step outside the coloring lines of stereotypical representation to find common cause with other minoritized folk groupings (Gold, 2005). Evaluations must aim to capture and tell these multilayered and multivocal stories. In city parks, schools, streets, or neighborhood parking lots filled with democratic impulses, evaluators must ask questions that lead in the opposite direction of what is expected.

Acknowledgement: I obtained permission from Faeza (real first name) to share this story. I am grateful to her and her family for their friendship and collaboration.

References

Arcodia, C. and Whitford, M. (2006). 'Festival attendance and the development of social capital', *Journal of Convention and Event Tourism*, 8(2), pp. 1–18.

Bauman, R. and Sawin, P. (1991). 'The politics of participation in folklife festivals', In Karp, I. and Lavine, S.D. (eds) *Exhibiting Cultures: The Poetics and Politics of Museum Display*. Washington, DC: Smithsonian Institution Press, pp. 288–314.

Bissell, E. (2019). *Notes on a Cultural Strategy for Belonging*. Berkeley: Haas Institute for a Fair and Inclusive Society.

Cibinskiene, A. (2012). 'Impact evaluation of events as factors of city tourism competitiveness', *Economics and Management*, 17(4), pp. 1333–1339.

Fox Piven, F. and Cloward, R.A. (1978). *Poor People's Movements: Why They Succeed, How They Fail*. New York: Vintage.

Getz, D., Anderson T.D., Armbrecht, J. and Lundberg, E. (2018). 'The value of festivals', in Mair, J (ed.) *The Routledge Handbook of Festivals*. London: Routledge, pp. 22–30.

Gillespie, A.K. (1996). 'Festival', in Brunvand, J.H. (ed.) *American Folklore: An Encyclopedia*. New York: Garland, pp. 249–251.

Giorgi, L., Sassatelli, M., and Delanty, G. (2011). *Festivals and the Cultural Public Sphere*. London: Routledge.

Gold, S.J. (2005). 'Migrant networks: A summary and critique of relational approaches to international migration', in Romero, M. and Margolis, E. (eds.) *The Blackwell Companion to Social Inequalities*. Malden, MA: Blackwell, pp. 257–275.

Hutchinson, S. (2009). 'The Ballet Folklorico de Mexico and the construction of the Mexican nation through dance', in Najera-Ramirez, O., Cantu, N. and Romero, B. (eds.) *Dancing Across Borders: Danzas y Bailes Mexicanos*. Chicago: University of Illinois Press, pp. 206–225.

Jackson, M.R. (2019). *Creative Placemaking: Rethinking Neighborhood Change and Tracking Progress*. New York: The Kresge Foundation.

Lee, J.-S., Lee, C.-K. and Choi, Y. (2011). 'Examining the role of emotional and functional values in festival evaluation', *Journal of Travel Research*, 50(6), pp. 685–696.

Murphy, C. (2017). *Guest Folklorist Report.* Tucson: Tucson Meet Yourself.
NEA. (2010). *Live from Your Neighborhood: A National Study of Outdoor Art Festivals.* Research Report # 51. Washington, DC: National Endowment for the Arts.
Nowak, J. (2007). *The Power of Placemaking: A Summary of Creativity and Neighborhood Development Strategies for Community Investing.* Philadelphia: The Reinvestment Fund.
Robertson, M., Rogers, P. and Leask, A. (2009). 'Progressing socio-cultural impact evaluation for festivals', *Journal of Policy Research in Tourism, Leisure and Events*, 1(2), pp. 156–169.
Santino, J. (2016). 'What I learned from Ralph Rinzler: The politics and poetics of public presentation', in Cadaval, O., Kim, S. and N'Diaye, D.B. (eds.) *Curatorial Conversations: Cultural Representation and the Smithsonian Folklife Festival.* Jackson: University Press of Mississippi, pp. 33–48.
Stern, A., Guckenburg, S., Persson, H. and Petrosino, A. (2019). *Reflections on Applying Principles of Equitable Evaluation.* Woburn, MA: WestEd Justice and Prevention Research Center.
Wallstam, M., Ioannides, D. and Pettersson, R. (2020). 'Evaluating the social impacts of events: in search of unified indicators for effective policymaking', *Journal of Policy Research in Tourism, Leisure, and Events*, 12(2), pp. 122–141.
Williams, M. and Bowdin, G.A.J. (2007). 'Festival evaluation: An exploration of seven UK arts festivals,' *Managing Leisure*, 12, pp. 187–203.
Wood, E.H. (2009). 'An impact evaluation framework: Local government community festivals', *Event Management*, 12, pp. 171–185.
Zeitlin, S. (2016). 'How folklorists changed the world: The Smithsonian folklife festival as a catalyst for change', in Cadaval, O., Kim, S. and N'Diaye, D.B. (eds.) *Curatorial Conversations: Cultural Representation and the Smithsonian Folklife Festival.* Jackson: University Press of Mississippi, pp. 303–314.

Further reading in this volume

Chapter 6: Listen, connect, act
Kim Cook
Chapter 9: From the dust of bad stars: disaster, resilience, and placemaking in Little Tokyo
Jonathan Jae-an Crisman
Preface: 'Disastrous forces, accidental actions, and grassroots responses'
Tom Borrup
Chapter 11: *Free State Boulevard* and the story of the East 9th Street Placekeepers
Dave Lowenstein
Chapter 12: Public transformation: affect and mobility in rural America
Lyndsey Ogle
Chapter 14: Experts in their own tomorrows: placemaking for participatory climate futures
Paul Graham Raven
Chapter 16: More than a mural: participatory placemaking on Gija Country
Samantha Edwards-Vandenhoek
Chapter 18: '*Homomonument* sounds like a poem': queer placemaking 30 years on: a conversational dialogue with Thijs Bartels, author of *Dancing on the Homomonument* (2003)
Martin Zebracki
Chapter 21: Placemaking through parkour and *Art du Déplacement* (ADD) as a Singaporean applied performance practitioner in London
Adelina Ong
Chapter 22: Embedded Artist Project: Epistemic Disobedience + Place
Frances Whitehead
Chapter 23: Routing out place identity through the vernacular production practices of a community light festival
Gail Skelly and Tim Edensor
Chapter 24: Artists, creativity, and the heart of city planning
Tom Borrup
Chapter 31: Seven generations: a role for artists in Zuni PlaceKnowing
Theodore S. Jojola and Michaela P. Shirley

42

CREATIVE PLACEMAKING AND PLACEKEEPING EVALUATION CHALLENGES FROM THE PRACTITIONER PERSPECTIVE

An interview with Roy Chan

Maria Rosario Jackson

The following is an interview with Roy Chan, Community Planning Manager at Chinatown Community Development Center in San Francisco, California.

Maria Rosario Jackson

Please tell us about Chinatown Community Development Center and the evolution of its creative placemaking/placekeeping practice.

Roy Chan

The Chinatown Community Development Center (CCDC) started in 1977 as the Chinatown Resource Center (CRC), an advocacy organization born from the Civil Rights movement and the movement to create ethnic studies at universities. With new waves of Asian immigration as a result of the Immigration Act enacted in 1968, there was a tremendous need in ethnic neighborhoods across the country to serve new immigrants. CRC was founded by volunteer grassroots organizations that came together to address quality of life issues in San Francisco's Chinatown, which served as a vital immigrant gateway. Many of these volunteer advocates were recent college graduates, who grew up in the neighborhood and understood firsthand the need to address Chinatown's overcrowded living conditions, the need for affordable housing, improved open space, and better access to public transit.

Over time, the Chinatown Resource Center eventually became the Chinatown Community Development Center, which began to develop affordable housing. Unlike other community development corporations, CCDC came out of a grassroots advocacy perspective first and then evolved into realizing the need of building community ownership through the creation of affordable housing. The organization has grown tremendously over the decades and today serves over 3,000 resident tenants in over 30 buildings in not just Chinatown, but throughout the northeast sector of San Francisco and beyond.

CCDC has nurtured the creation of a strong grassroots tenant base in Chinatown. Currently the Community Tenants Association has over 1,000 members, composed of mostly senior citizens. They meet weekly to engage in cultural activities together, support each other's needs, and are active in advocating for policy that supports affordable housing and other neighborhood issues.

Even before the term creative placemaking was coined, CCDC's programs have integrated arts and culture as a strategy for community development. *Arts in the Alley*, for instance, showcased the work of local visual artists in empty or underutilized Chinatown storefronts. *NoodleFest* was a community food competition that creatively highlighted restaurants serving noodle dishes in Chinatown and pasta shops in neighboring North Beach community. *Chinatown Alley Tours* has been an alternative, youth-led walking tour of the neighborhood that showcases the lived experiences of residents and local businesses, many of them located in Chinatown's intricate alleyway system. All of these projects have built intergenerational bridges that also supported the local economy by uplifting the inherent cultural assets and traditions embedded throughout the neighborhood. When I joined the organization, we received funding for our creative placemaking efforts, and the work became more intentional and transparent as a strategy.

What's amazing about CCDC's approach to community development is that it honors lived experiences in the neighborhood in a comprehensive way. CCDC operates from a framework that everything is connected: from land use controls to keep encroachment from the neighboring financial district, to advocating for public transit lines because most residents walk in this dense neighborhood, to advocating for public open space because most residents live in overcrowded conditions with the critical need to socialize. At the heart of CCDC's vision is the belief that Chinatown's cultural identity is only intact when the housing stock is preserved for residents so it can continue to serve as an immigrant gateway and a place for senior citizens to age without having to leave their community. The everyday culture practiced by residents is at the heart of the community's shared identity.

Maria Rosario Jackson
Earlier you mentioned how college courses in Ethnic Studies were a part of the origins of the organization, can you say more about that?

Roy Chan
In 1968, in response to the new waves of immigrant populations, a student movement began at San Francisco State University which advocated for a curriculum that truly spoke to the experiences of Asian Americans and other ethnic groups. Quite frankly, these were perspectives that were left out of the standard US history curriculum. Many of the young Asian American activists, including CCDC's founding director Gordon Chin, saw this movement, known as the Third World Liberation Front, as being vital to the shaping of API (Asian and Pacific Islander) students' understanding of themselves, their families, and their role in the community and city.

The movement started at San Francisco State University in 1968 and then continued at UC Berkeley in 1969. The integration of Ethnic Studies was the birth of a larger movement in the Asian American Pacific Islander population to uplift and preserve places like Chinatown as vital places for generations of immigrants. The movement galvanized a generation of activists coming together to advocate for their communities in the 1970s. This movement led to the creation of several grassroots volunteer organizations in Chinatown that have fought to improve the quality of life and protect residents from displacement in Chinatown.

The story of the International Hotel was a defining moment of this movement. The International Hotel was one of the last standing single-room occupancy hotels in what was

Manilatown, a long-standing neighborhood next to Chinatown. Manilatown was several blocks long and where Filipino men and women lived in Single Room Occupancy (SRO) hotels. During the 1970s, real estate speculation from the nearby financial district and urban renewal pressures led to the demolition of numerous SRO buildings and the displacement of hundreds of residents and neighborhood businesses.

The International Hotel was the last standing building, and for a year, activists came together and fought to protect and preserve the tenants in that building. It wasn't just a residential hotel, but it was also an epicenter of community activism as it housed Everybody's Bookstore and Kearny Street Workshop which created groundbreaking community arts programming. This moment in the movement was an amazing coming together of thousands of activists and residents in Chinatown and Manilatown protesting the eviction of I-Hotel residents in 1977. To this day, if you talk with people from that generation, like our founders former directors Norman Fong and Gordon Chin, many of them would point back to that period as the inspiration to serve the community for so long and to inspire youth leadership.

Maria Rosario Jackson

It sounds like that what was happening back then is what we are now calling 'equitable creative placemaking' and 'creative placekeeping'. The roots of current CCDC 'creative placemaking' had important origins in what you just described.

Roy Chan

Yes.

Maria Rosario Jackson

Please tell us about your role in the organization?

Roy Chan

I was trained as an urban planner at UCLA with an emphasis in affordable housing development. Over time, what really resonated with me was the critical role that culture plays in building wellness for immigrant neighborhoods. After working in the urban planning field for many years, my career path led me to the area of arts administration at the Oakland Asian Cultural Center where I worked directly with artists and cultural keepers to activate public spaces in Oakland's Chinatown. Seven years ago, I returned to community planning work at Chinatown CDC in San Francisco as the practice of creative placemaking began to emerge across the country.

At CCDC, I currently manage our Community Planning department, which is made up of both planners and community organizers to collaboratively advocate for a neighborhood planning approach that responds directly to the needs and assets of residents. Our team's work over the years has focused on a range of quality of life issues in greater Chinatown. A big part of our work is to not only mobilize stakeholders to public meetings, but to walk the neighborhood every day to more deeply understand how residents live, use space, and practice their culture. This then enables our planning work and decisions to be more rooted in lived experiences and cultural rhythms of the neighborhood.

One of our key responsibilities in the Community Planning department has been to uphold the Chinatown zoning regulations adopted in 1987 by CCDC and community leaders who saw the critical need to protect the neighborhood character of Chinatown. Most of the buildings in Chinatown are three to four stories tall with small neighborhood-serving businesses on the ground floor and residential or institutional uses on the upper floors. Chinatown leaders worked with the City to develop land use controls policy to limit building heights to six stories and limit the types of building uses that are allowed, such as restricting office use and formula retail.

During the 1970s and 1980s, there were significant speculation threats to Chinatown from the encroaching financial district nearby. At the time, Chinatown had the same zoning of commercial high rise as the neighboring financial district. Luckily, the City was embarking on a zoning revision of the area, and community leaders argued that Chinatown needed to have its own zoning designation to protect its neighborhood character where so many families and seniors live in a dense urban environment.

Since 1987 when the zoning was adopted, Chinatown CDC's planning department has acted as the gatekeeper for development in the neighborhood. We've built political power and have a pretty close relationship with the planning department in the City. When development proposals that test the limits of the zoning emerge, the city planning development typically consults with us first to ensure that there's community support.

Amazingly, if you walk around the neighborhood, you will immediately notice the edges of the Chinatown zoning where three-story buildings with neighborhood retail and housing stand next to corporate high-rise buildings in the financial district. This community and CCDC effort around adopting and upholding zoning and planning has been essential to keeping Chinatown intact.

Along with that zoning, we work very closely with volunteer advocacy groups made up of residents and leaders in the community. These grassroots organizations preceded CCDC, and today they still serve as the local voice to speak on behalf of the community. One advocacy group is called Chinatown TRIP (Transportation Research and Improvement Project), and their focus has been around improving public transit lines that serve the neighborhood, promoting pedestrian safety, and extending the Central Subway line into the neighborhood, while pushing back on transportation issues that hurt the neighborhood.

We also have an open space advocacy group called the Committee for Better Parks and Recreation in Chinatown (CBP&RC) made up of residents and local community leaders, who advocate for the improvement of local parks and open space and find funding for it. Our team's work has been to provide the technical support for these advocacy groups to function.

Maria Rosario Jackson

What you have described is critical work to sustain the neighborhood and protect it from significant forces. Can you tell us more about why you integrate arts, culture, and community-engaged design into how you do your community development work? What's at the core of it for you today?

Roy Chan

If we look back throughout Chinatown history, it has been a place that has had to constantly remake itself in response to threats over the years. And these strategies always revolved around the preservation of its cultural identity which has been the glue to keep the neighborhood intact. The way we approach arts, culture, and design is to first acknowledge that it is already inherent throughout the community; from our restaurants and the different foods that are created/served, to activities that people do in the park, to people's different cultural practices and traditions during different seasons, such as Lunar New Year and the Moon Festival. These traditions have been going on here for over a century, and they also are at the core of fueling the neighborhood's economy. This deep shared sense of connection through culture continues to bind the neighborhood across generations.

Secondly, our creative placemaking work serves to uplift and strengthen access to the everyday culture that gives residents a sense of belonging. This involves connection to each other and a sense that the neighborhood belongs to them. A big part of this work is removing barriers to cultural practice. One main barrier has been rising commercial and residential rents in the Chinatown that have caused the displacement of local residents and cultural groups (i.e. Chinese opera clubs, lion dance troupes) from gathering places where they have traditionally practiced daily culture together. There is a struggle for local access to public spaces (i.e. recreational clubhouses, dilapidated parks and alleyways) and large community banquet spaces that have historically served as cultural anchors in the neighborhood but have now been privatized by speculators for high-end commercial uses not compatible with the neighborhood. Our creative placekeeping work has centered around creative strategies for community control of these spaces through a reexamination of the local zoning controls and innovative partnerships with city government, community groups, and long-standing property owners. Along with that is the ongoing need to support legacy businesses and cultural institutions that have been operating on thin economic margins while struggling from years of local construction impact in the neighborhood and from the pressures of real estate speculation.

Right now, during our COVID-19 crisis, up to 70 per cent of Chinatown restaurants have closed. If you think about how people experience culture in Chinatown, the restaurants are really at the heart of social gathering through the simple act of sharing meals together. We're realizing that the current pandemic has become a threat to our cultural identity. As part of the creative placemaking work, we are thinking creatively about how to support these restaurants by adapting different business models in a way that restaurants can continue to be a livelihood for immigrant workers. We've been developing a local meal delivery service by creating a community kitchen in our large banquet restaurants and delivering it to SRO residents who have been experiencing food insecurity. This program brings back unemployed restaurant workers, reopens familiar restaurants, addresses food insecurity and responds to the tremendous need for residents to feel connected to their food traditions in a safe way during these isolating times.

We are planning ahead to meet many different kinds of needs. When the Autumn Moon Festival happens in late September, and we're still in this moment of social distancing, we'll have to reconsider how to continue that tradition differently in Chinatown. The Autumn Moon Festival has historically been a time for moon cakes as well as other foods to be made and shared communally throughout the neighborhood as a celebration of harvest festival. As part of this tradition, a street festival brings live lion dance performances and dozens of booths selling different foods and gifts by Chinatown merchants along the Grant Avenue commercial corridor, drawing thousands of people over an entire weekend. This festival along with the Chinese New Year festival and parade in February highlight the annual cultural traditions that are the glue in Chinatown's deep social cohesion for over a century.

A lot of our creative placemaking work has been around the improvement of the public realm with the understanding that it's where everyday arts and culture are practiced. How do we maintain and enhance Chinatown's open space so that daily cultural activities there can thrive? Whether it's working with the City to ensure a truly participatory community design process for our Portsmouth Square master planning or ensuring that new improvements to the Chinese Playground are in line with residents' daily cultural rhythms. A lot of our open space work over the decades has also been about preserving, improving, and cleaning up Chinatown's unique network of over 30 alleyways. Alleyways are informal spaces that function as front porches for many of our residents living in SRO buildings. For instance, our oldest alleyway is Ross Alley. If you walk down that alley, you'll get an amazing snapshot of daily culture happening organically – handmade cookies made for restaurants and visitors at the Golden Gate Fortune Cookie Factory, CCDC's community art gallery 41 Ross that highlights the work of cultural keepers, the lion dance groups that practice at night in the alleyways in anticipation of community events through the year, such as Lunar New Year. Civic engagement of various stakeholders in the neighborhood has been critical to our community design process. It's a collective effort that involved building consensus among residents or family associations who reside upstairs, the store owners on the ground floor, and the daily visitors coming into Chinatown. With culture as a unifying frame, the process involves getting a full sense of the diverse needs and assets of the neighborhood.

What I've learned as a planner in these public workshops is an understanding of how people practice culture by asking the right questions. Oftentimes, as trained planners, we ask the typical questions of 'What are your needs? How would you like to use this space?' But I think what has really opened up this process of getting helpful input is when we asked people how they lived... from when they wake up to what they do and where they go throughout the day. By capturing daily life narratives of residents, we uncovered clues to how use of space fits into people's daily activities and rhythms.

Maria Rosario Jackson

Foundations, government, and others who fund community development work often require proof of impact of creative placemaking? From your perspective, why does the work matter? How does it make a difference?

Roy Chan

Oftentimes, when we're filling out grant report forms, the typical questions that they ask are 'How many people actually came to an event? How many volunteers participated? How much more money has been invested in the neighborhood?' Those are important questions, but we're also discovering new ways of knowing what makes a difference, like just walking the neighborhood each day and seeing firsthand how spaces are transforming over time.

Since we've embarked on a more intentional integration of arts, culture, and design into our community development efforts, we've seen a number of storefronts that, without this intentionality, would have ended up for some other retail use, but were instead transformed into vital spaces for communal storytelling and culture making. Our community gallery space at 41 Ross, a formerly underutilized storefront, is the result of our community engagement work with the property owner, who has enabled CCDC to transform it into a dynamic, multi-purpose space for dialogue, for exhibitions, and communal celebrations. Walking past 41 Ross now is an affirming reminder of the importance in being good stewards of space for art and culture making.

945 Clay is an SRO building that CCDC recently bought from a speculator who tried to evict the tenants. Since we took over ownership, our team has worked closely with our asset management staff to fill the building's storefront space with Asian Improv Arts, a vital performing arts organization. Now the space and block has come alive with dance and music that is inspired by the living history of Chinatown. I don't think there's been a dance or music space in Chinatown for quite some time. It's really an ideal space with tall ceilings and great light that honors the artistic work of dancer Lenora Lee and musician Francis Wong, who are mentors raising up new artists in the community. CCDC has a long history of raising new leaders and now it's exciting to be a part of reviving a storefront to raise new artists in the community.

Another way that we're seeing a difference being made, is in the last two years, an unlikely coalition was formed with diverse community leaders who, in the past, would have never come together. This coalition has been facilitating new conversations around a shared need in Chinatown for more connected cultural spaces that tell our stories in new creative ways.

One coalition of cultural leaders started about a year and a half ago when Chinatown's largest building, the Empress of China building, was sold to a developer who had very different interests than those of the neighborhood. Key leaders made up of property owners, business owners, CCDC, and arts groups such as the Chinese Cultural Center, the Chinese Historical Society, Center for Asian American Media, came together to convene every two weeks. At first, the conversation revolved around strategies to purchase the Empress of China building for community cultural use. Then over time, the coalition became a space to dialogue around a bold new vision of cultural resilience by identifying other key spaces for cultural revitalization as well as a vision to become a designated cultural district in the City of San Francisco.

Another way that we're seeing the work make a difference is the community process for our Portsmouth Square Improvement Project. Portsmouth Square is the largest open space in Chinatown and the most heavily used local park in the City's densest neighborhood. In many ways, it's been referred to as Chinatown's 'living room' where SRO residents are able to socialize and practice their cultural activities together throughout the day. We just finished the conceptual design process through five public workshops that the City led and engaged significant resident participation through the mobilization work of organizers at CCDC. Our planning team partnered with filmmakers and storytellers to capture daily life narratives and living memories about Portsmouth Square as both a community living room and town square. These published stories served as an accessible entry point for over 300 park users to participate in each of the public design workshops, setting new attendance records for public workshops held in the entire City. By reframing these public workshops into storytelling/ sharing events, the design process highlighted the significance of Portsmouth Square as a vital cultural anchor for so many people in the neighborhood.

María Rosario Jackson

You have identified many ways in which the work makes a difference: in determining culturally appropriate uses of space, fostering intergenerational mentoring, leadership development, vitality of street life, new coalitions, identification of new community priorities, increased community engagement in planning and civic activity, among others. How do these kinds of impacts and ways of knowing match up with Community Development standards for evaluation and progress?

Roy Chan

We feel stuck in this loop of having to always answer mostly quantitative questions for Community Development standards. But I think it begins with reframing the questions as a way to rethink how we as practitioners understand and measure impact. Oftentimes, questions are asked about the number of housing units that are built and the income levels that we need to meet for those we serve. What often isn't asked is how many residents in these housing units feel a sense of home; if they feel connected to their place and with their neighbors, if the design/ staffing of the lobby or the community room in our buildings create a shared sense of belonging and community.

We've learned so much in the last five years since we took over management of over 500 units of public housing in Chinatown as part of the Department of Housing and Urban Development's Rental Assistance Demonstration (RAD) program. Previously under the San Francisco Housing Authority's management, these 50-year-old buildings deteriorated over the decades without much thought or understanding of how the physical environment impacted residents' sense of ownership or feeling of home. Then over time, we also started to see the social fabric of these communities deteriorate and racial tensions rise.

And I think what we've been learning since we've taken over management of these projects is how critical the integration of arts, culture, and design are… from the design choices we make in the renovation of the community rooms and lobbies to hiring resident services coordinators who reflect the different ethnicities of the residents and are skilled at building connections/empathy across divides. We started this process as we began relocating tenants out of their units during the renovation, as a way to minimize the trauma of displacement and then their journey home to regain a deeper sense of community and belonging. Bringing in artists and storytellers to work alongside the resident services staff positively impacted critical transitions and home-coming in ways that would not be possible otherwise.

María Rosario Jackson
Do you think that the conventional ways of measuring success are changing?

Roy Chan
In some ways, yes. The transfer of management of public housing to community-based organizations like CCDC through the RAD program has been an acknowledgment that we need to go beyond conventional ways of doing community development and measuring success. When CCDC took over management of 990 Pacific, for example, we had to reimagine public housing and understand what success means from a community wellness perspective. When we began renovation of the building, we had to relocate all the tenants for several months. As a way to process the trauma they would experience, we hired an artist to document the residents' journey of relocation to temporary housing during the renovation and the return home after renovation. The storytelling project was entitled *Coming Home: The 990 Pacific Relocation Story* and became a creative process of resident engagement and an exercise in trust building. The residents' oral histories were presented visually on the walls of the building's community room and during the public reopening ceremony in the community room. The tenants' stories were displayed on the wall through moving photos that dignified their experiences. They revealed the significance of understanding the residents' perspective to better inform how we manage housing and promote the care/investment needed to transform public housing into places of belonging, mutual trust, and connection. And those are the values that CCDC has been trying to uphold during this process of transforming Chinatown's public housing projects.

I would say this is a beginning and an entry point to demonstrate how we can better measure success. I think there's still a lot of work to do but we're starting to see the fruits of the investments we're making in creative placemaking.

María Rosario Jackson
You have alluded to some already, but what kinds of changes in evaluation would you like to see?

Roy Chan
Well for the 990 Pacific Coming Home Project, we had to do the additional work of building social cohesion from a cultural framework. That needs to be acknowledged in evaluation. I think it's this work that moves community development beyond the implementation of basic building improvements to housing units, to a more comprehensive effort that aspires for these units to become places the residents can call home. And I believe there's a quantifiable value to it. The federal government should invest in placemaking as a budget line item when we think of HUD projects like RAD. It's incorporating not just housing developers, but artists and cultural workers to be part of a more comprehensive process of community development.

María Rosario Jackson
The community development field relies greatly on quantification of impacts. Do you think that impacts always have to be quantified?

Roy Chan

I don't think so. I think there's ways that we can capture a sense of belonging in qualitative ways.

I'm a part of the Oakland Cultural Affairs Commission now and they just completed a report about creative placemaking, Roberto Bedoya's report, in which the goal is belonging, culture is the frame, and equity is the driving force. It's interesting because if belonging is the goal, I think the traditional quantitative measures probably won't be able to capture success of that goal. We really need to investigate ways of measuring in more qualitative ways. And I think storytelling and collecting stories is a big part of that work in understanding people's sense of belonging and connection that you might not find in numbers.

María Rosario Jackson

Are the evaluation practices that you hold as an organization different from what is typical in the community development and planning fields – the industry standards?

Roy Chan

It's been an incredible learning process for us as an organization more recently, as we've taken over new management of public housing. Our creative placemaking work has touched different departments across the organization in a way that has brought into conversation new theories of change and ways of measuring success. It hasn't been an easy process. I think for our housing developers and staff in our Community Organizing department, it's always been about how many units we can acquire or how many tenants we can get out to a rally. Sometimes we're stuck as an organization in that conventional pattern of what success looks like because we're constantly reporting to funders with numbers in mind. Because of the creative placemaking funding that we've gotten in the last six years, our planning team has been able to take a step back and try these new methodologies and test new partnerships with folks in the creative field. And these other departments in the organization are taking notice of this work that reminds us all of our mission to build community more deeply and to build not just housing units but homes. And when we tell stories to show the impact in a more qualitative, human way, organizational mindset starts to change. The exhibits we've been showing at 41 Ross Gallery have been a big part of this educational process for our own organization as well as for folks in the community development and planning fields. These past six years have been an incredible learning experience for us and the rest of our organization. We're still constantly filling out grant reports that adhere to a certain industry standard that doesn't really fit with what we know matters. So, the work ahead is to go beyond this mindset and practice, continue to expand our view of what social change looks like, and as practitioners, become evangelists of this expanded view.

María Rosario Jackson

Are you hopeful about things moving in that direction?

Roy Chan

Yes, I am hopeful. I think crises such as the global pandemic we're currently in could help us move in that direction. We're in this moment of isolation where the need for social and cultural cohesion is evident more than ever before. It's particularly evident in a place like Chinatown, where social gathering is a lifeline for our senior tenants living in SROs. They depend so much on the neighborhood for their daily routines and cultural food security is a big part of that. Early on in the pandemic, we realized how important access to local ingredients and foods were for residents sheltered in place. CCDC worked with local restaurants to develop a community kitchen where boxed hot meals were prepared and delivered to the thousands of Chinatown residents unable to safely come out to shop/dine. We've learned that in a state of isolation, ensuring a continued sense of familiarity for residents is so critical to community resilience. City government, funders, and community developers from all over have taken notice. It's these types of creative placekeeping efforts and exposure during a crisis that will push things to move in that direction in which culture keeping is vital to neighborhood sustainability.

María Rosario Jackson
Thank you, Roy, for sharing your experience and your insights.

Acknowledgments

Thanks to Christina Park and Emma Galligan of the Studio for Creativity, Place and Equitable Communities at Arizona State University for editorial assistance with this interview.

Further reading in this volume

Chapter 1: Introduction: what really matters – moving placemaking into a new epoch
Cara Courage
Preface: Placemaking in the age of COVID-19 and protest
Jason Schupbach
Chapter 3: An annotated history of creative placemaking at the federal level
Jen Hughes
Chapter 4: A future of creative placemaking
Sarah Calderon and Erik Takeshita
Chapter 5: Making places for survival: looking to a creative placemaking past for a guide to the future
Jeremy Liu
Chapter 6: Listen, connect, act
Kim Cook
Chapter 7: Conflict and memory: human rights and placemaking in the City of Gwangju
Shin Gyonggu
Chapter 9: From the dust of bad stars: disaster, resilience, and placemaking in Little Tokyo
Jonathan Jae-an Crisman
Chapter 10: From moon village to mural village: the consequences of creative placemaking in Ihwa-dong, Seoul
Jason F. Kovacs and Hayun Park
Chapter 11: *Free State Boulevard* and the story of the East 9th Street Placekeepers
Dave Lowenstein
Chapter 12: Public transformation: affect and mobility in rural America
Lyndsey Ogle
Chapter 14: Experts in their own tomorrows: placemaking for participatory climate futures
Paul Graham Raven
Preface: 'Disastrous forces, accidental actions, and grassroots responses'
Tom Borrup
Chapter 20: Displacemaking 2015 and 2020
Catherine Fennell and Daniel Tucker
Chapter 21: Placemaking through parkour and *Art du Déplacement* (ADD) as a Singaporean applied performance practitioner in London
Adelina Ong
Chapter 23: Routing out place identity through the vernacular production practices of a community light festival
Gail Skelly and Tim Edensor

Preface: The only thing constant is change
Kylie Legge
Chapter 36: Facilitator skills for effective collaborative placemaking
Husam AlWaer and Ian Cooper
Chapter 37: *The Neighbourhood Project*: a case study on community-led placemaking by CoDesign Studio
Lucinda Hartley, Eliza Charley, Sama Choudhury, and Harriet McKindlay

43

A THEORY OF CHANGE FOR CREATIVE PLACEMAKING

The experience of the National Endowment for the Arts' *Our Town* program: An interview with Patricia Moore Shaffer, PhD

Maria Rosario Jackson

Patricia Moore Shaffer, Deputy Director of Research & Analysis at the National Endowment for the Arts, speaks with Maria Rosario Jackson about the Arts Endowment's *Our Town* grant program. The *Our Town* program was launched in 2011.

Maria Rosario Jackson
Why did the National Endowment for the Arts need to establish a theory of change for the *Our Town* program? Why was that necessary?

Patricia Moore Shaffer
We began work on the *Our Town* theory of change in 2016, but our discussions began when the *Our Town* grants program was launched. At that point in time, creative placemaking as a field was just emerging. At the agency, we were reading the White Paper on creative placemaking that Ann Markusen and Anne Gadwa had written and talking about how to measure the impact of *Our Town* grants and document the work of our grantees. When I look back at what we knew then versus what we know now, having gone through the theory of change exercise, we have a far more nuanced understanding of the implementation strategies and outcomes associated with *Our Town* grants. It's due in large part to the *Our Town* theory of change and logic modeling work that we have done, which took us through a deep exploration of our funded projects.

In reflection, developing a theory of change forced the agency to assess its assumptions about the program. We articulated program outcomes as part of that process, and we identified implementation strategies that support those outcomes. When we first started the *Our Town* theory of change work, we had seen it as a critical step toward planning an evaluation study to document outcomes associated with the program. Before I returned to the agency in 2015, I was sitting on the technical advisory group that was planning a summative evaluation of *Our Town*. I was one of the voices at the table that called for a logic model to be in place before we began any evaluation work. When we developed the statement of work for the evaluation contract, we developed it as a two-phase contract. The first phase was to generate the program theory of change and logic model, followed by the actual evaluation study in the second phase.

As sometimes happens during this process of developing a theory of change, the agency developed a much richer understanding of the program and how it worked and, as a result, made changes to the program goal and grant guidelines. What we learned through the process resulted in a revised program model for *Our Town*. While this is a great outcome, it signaled to me that this was not an appropriate time to conduct a summative evaluation. We stopped further planning for a summative evaluation but did additional work to learn more about the program.

Maria Rosario Jackson

The institutional reason for this work, in part, has to do with the fact that this is public money and there has to be accountability. Part of that accountability involves having an evaluation plan, but you recognized that the work was not yet ready for the kind of summative evaluation that would typically be expected. How was this work different from evaluations that the agency has done for other programs?

Patricia Moore Shaffer

At the time of its launch over 10 years ago, *Our Town* was unique among the agency's grant programs for linking the arts with community development and revitalization outcomes. While many *Our Town* grant projects shared a similar intent for attaining these outcomes, they vary considerably in their design. These variations are due in part to differences in local context and needs. While there are some common strategies used across projects and some common outcomes being sought, there are so many unique factors around each project that, from a measurement perspective, it's very difficult to come up with an evaluation design that would yield summative outcomes common across a large body of grants.

By contrast, I'll offer you an example of an evaluation study that was comparatively simple to design. The subject of study was the agency's *Poetry Out Loud* national initiative. It has a strong set of guidelines issued to all states. The project is implemented with a small amount of variation at the school level that is easily documented. We had a strong understanding of anticipated outcomes because there had been some formative evaluation work already completed on this program. Given the combination of a fairly standardized approach to program implementation and previous evaluation work done, the evaluation plan was relatively simple to prepare.

Our Town projects, by contrast, are rich in their differences; we also did not have the benefit of past evaluations to draw upon. We had done some very light portfolio analyses, but we had never looked in depth at any of the narrative data that was part of our grantees' final reports. When we started the work, we knew we needed a working model that allowed us to take into account the very wide variance that we're seeing across projects. That's what started us on this journey and why it differed from other evaluation work that we've done.

Maria Rosario Jackson

What did the work of creating a theory of change entail?

Patricia Moore Shaffer

We decided to use an evidence-based approach to developing the theory of change and the logic model, drawing on evidence associated with grantees' applications, final reports, and other documentation. We also completed a program scan that looked broadly at other creative placemaking and place-based programs to learn about their program models. We consulted with grantees, as well as with subject matter experts through a technical working group.

In terms of grant documentation, we analyzed data and information from grant applications and grantee final reports. We also reviewed existing case studies that are presented in the online

'Exploring *Our Town*' site. During this stage of analysis, we were asking questions about how projects were implemented by grantees. What inputs and strategies were associated with these projects? What outcomes were associated with those projects?

The grant applications and reports offered rich narrative data that helped to explain how *Our Town* projects worked. Looking across hundreds of grant projects, we were able to discern patterns in the essential ingredients of *Our Town* projects. The *Our Town* program leadership at the agency was actively engaged in this analytical process. Sitting in a meeting room for several multi-hour sessions, for example, we worked together identifying themes in a sample of grantee final reports, looking for the strategies and outcomes that were being reported by grantees. I can share that there were a lot of 'Aha!' moments that happened during that process as the patterns of findings began to emerge.

One of those 'Aha!' moments was the realization that several *Our Town* project reports were describing sustained policy or strategy change in local government. There was one project we discussed that involved an artist who, as part of the project, had been hired by municipal government to facilitate community engagement projects related to a planning exercise the city was doing. In the final report, the grantee reported that the city had actually created a permanent job for the artist. This outcome didn't fit neatly in the categories of economic, physical, or social change that we had previously identified. We began to use the term 'systems change' at that point because this was a policy-level change, or even a procedural change, that seemed to reflect a new way of doing business that was sustained after the grant period was over.

In terms of consultation, we engaged with a range of subject matter experts and grantees. We asked very similar questions – How were the *Our Town* projects being implemented? What inputs, strategies, and outcomes were associated with the projects? – but we also inquired about how similar art- and place-based programs were implemented, trying to discern their program theory. We wanted to understand *Our Town* but also situate it in this larger context of creative placemaking and other place-based work. We interviewed key stakeholders and experts recommended to us by the technical working group; we also interviewed other individuals recommended by interviewees using an approach called snowball sampling.

A program scan was another key strategy for our work. We looked at other creative placemaking programs as well as other place-based initiatives and specifically any relevant published studies or program evaluations that had been done. We were especially interested in the underlying program theory as well as evidence of outcomes associated with those programs. We were very interested in understanding the types of measures and measurement approaches other programs used to assess outcomes.

As you might imagine, this process unearthed a lot of information, and we conducted qualitative analysis to identify common project inputs, strategies, tactics, outputs, and outcomes. Following this analysis, our contractor produced a preliminary findings document and facilitated a theory of change workshop with the *Our Town* program staff. This workshop yielded the first theory of change, which we took to the technical working group for constructive feedback. Working with the technical working group and the program staff, we iterated several versions of the theory of change before agreeing upon the version that became the foundation for the revised program guidelines. We also began building a logic model and measurement framework at that point.

The final theory of change highlights the local-level inputs necessary for *Our Town* projects, including local leadership, financial resources, community buy-in, and cross-sector partnerships. The creative placemaking strategies describe broadly the approaches cross-sector partners adopt to address local challenges: imagine/envision new possibilities for a community or place; connect communities, people, places, and economic opportunity via physical spaces or new relationships; inject new or additional energy, resources, activity, people, or enthusiasm into a place, community issue, or local economy; and illuminate or elevate key community assets and issues, voices of residents, local history, or cultural infrastructure. Implementation of these strategies leads to local community changes, including economic, physical, and social change. Both the process of undertaking *Our Town* projects and the positive local changes they engender build community capacity, and ultimately can advance systems change at the local level to sustain the increased utilization of art, culture, and design activities in community development. The theory of change also articulates the vision for *Our Town*: sustained support and recognition of arts, design, and cultural strategies as integral to every phase of community planning and development across the United States. The theory of change also recognizes the importance of external leadership. National leaders in creative placemaking, such as ArtPlace and the Kresge Foundation, definitely influence *Our Town* projects while also contributing toward the vision.

The first phase of the project stopped at this point, and I mentioned before that the original plan had been to move into a summative evaluation. What happened at the end of this first phase was a program goal change that reflected this larger vision for the program. The agency staff also made modifications to the guidelines that embedded the theory of change in the language shared with potential grant applicants. I already mentioned that we made the decision not to pursue a summative evaluation, but we wanted to understand more about longer-term outcomes, particularly systems change. To that end, we designed a small study that employed a grantee survey and a small selection of grant case studies.

The grantee survey was structured to test out every part of the *Our Town* theory of change, asking questions about local inputs, strategies, and outcomes. The value of the survey was that it also allowed us to collect self-reported information about longer-term outcomes, particularly systems change. The agency's final reports must be received 90 days after the close of a grant period, so at best we are collecting information about short-term outcomes. The grantee survey allowed us to ask project directors about outcomes years after their grants had closed. The case studies also focused on longer-term outcomes, examining the mechanisms and indications associated with systems change work.

Using what we learned from the grantee survey and the case studies, we created the final version of the theory of change. In this version, systems change is prominent, weighted equally with local community change. We also developed a logic model based on the theory of change and a measurement model to guide our future performance monitoring and evaluation of *Our Town*.

We're still unpacking a lot of what we learned in Phase Two. In one of our very early conversations, you and I had talked about the importance of looking for those early *indications* of systems change given the challenge of measuring the longer-term outcomes associated with creative placemaking. The grantee survey and case studies helped us identify a set of indications that are correlated with systems change and measurable within or immediately after the grant period. For example, we found that sustained partnerships that continue beyond the immediate grant period is an early indication of systems change. Another early indication is artists trained in creative placemaking, since they can sustain and replicate program activities and perpetuate the successes of projects. Projects that were very active in communicating their project successes also seemed to have a much stronger chance of reporting systems change at a later time. This was particularly true in rural projects, not as much in urban projects, but for rural projects, it seemed to be a very significant factor. We also saw having local leadership, particularly government leadership, that was receptive very early on to arts and creative placemaking, seemed to indicate an early buy-in to creative placemaking that would later result in systems change.

So that's where we are at this point. While we opted not to do a summative evaluation because of the modifications to the program, we feel we're in a much better place now in three to five years to potentially look at an evaluation of this work.

Maria Rosario Jackson
That's super interesting. I'm happy to see that the *indications and indicators distinction* has been useful and has had some traction. Quantitative data to measure or document many of the things you are observing as contributions of creative placemaking work are often not readily available. The notion of 'indications,' which relies on rigorous observation but is not quantitative still allows one to capture the contribution. This distinction between indications (rigorous qualitative observation) vs. indicators (quantitative data) was something I had to come up with as part of my work with the Kresge Foundation. We were at a loss for how to describe what we understood to be movement when the quantitative data that the community development field relies on to track change was a poor fit.

Patricia Moore Shaffer
I can't tell you how happy the researcher was when he shared his report with us. He said, 'We found them! The indications!' And the good news is that they're all measurable. These are indications we can collect at the final report stage and be able to use to tell the story of the grants and that was ultimately what we wanted out of this phase of the work.

Maria Rosario Jackson
What other conceptual or practical challenges or discoveries did you encounter?

Patricia Moore Shaffer
We learned a lot about which methods are most effective in identifying theories of changes. During the first phase of this project, the work relied heavily on analysis of grant documentation. During the second phase, we reached out to our grantees through a survey and case studies. Learning from the field is critical. While we did engage the field during Phase One through interviews and obviously read their words in grant applications and final reports, we would have benefited earlier on by engaging with grantees in a much deeper, more engaged way.

This experience influenced how we worked later with similar projects, prioritizing engagement with grantees and stakeholders.

Maria Rosario Jackson
How has this experience impacted how you commission research going forward?

Patricia Moore Shaffer
It has changed the way we do this work. An example I can offer is the *Creative Forces* program and the work we did establishing a logic model for their community arts engagement portfolio involving military-connected individuals. We made an early decision to hold a series of discussions with a technical working group that included practitioners – from arts organizations, state agencies – who were planning and implementing this type of arts programming. We also facilitated a consultation session with practitioners at a convening to learn more about the range of program models used in the field. We shared with them a very early version of the logic model and used what we learned during the convening to revise the model. We developed a very strong logic model in a much shorter period of time due to the input of practitioners.

Maria Rosario Jackson
Having done this theory of change work, what would you say are the greatest research and evaluation priorities for the creative placemaking field?

Patricia Moore Shaffer

There is a strong interest in the creative placemaking field to collect data to demonstrate outcomes. Documenting how creative placemaking investments change communities is critical for moving the work forward. For me, it's important to support the people working on the ground level to document the outcomes of this important work. The arts sector does not have a strong history of investing in evaluation, but it is so critical to support the evolution of this work.

Another area of research interest for me is in understanding the mechanisms and outcomes associated with systems change at the community level. The arts are not unique in this interest; there are a lot of other sectors, such as education, that are examining systems change and the role of cross-sectoral partnerships. Better documentation of field-level work would improve our understanding of systems change in creative placemaking and, more importantly, improve our ability to facilitate it because we're all interested in seeing this work sustain itself and contribute to change by bringing the arts and culture more deeply into the work of community development.

Maria Rosario Jackson

That makes a lot of sense. One of the things that rings true as I hear you talk about the theory of change process is the necessity to have some descriptive material that can help anchor the questions. So, a lot of this work was about the 'what is it and how does it work' which had to proceed any attempt at summative evaluation. Then there's the question of, what is the timeframe to actually be able to see the impacts? There's a time-related question that perhaps still remains open with some impacts being able to be detected earlier than others. I think that is another priority area.

Particia Moore Shaffer

I agree. The *Our Town* grantee survey confirmed our theories and gave us some self-reported evidence of outcomes. But we need to be doing research that examines longer-term impacts of the work. One tantalizing finding from the survey: our analysis showed that there is some correlation between certain placemaking strategies implemented by *Our Town* grantees and outcomes that they report. This was very exciting to discover. The evidence is not strong enough at this point to say, 'well, if you do this and that, it will lead to X outcomes,' but it does suggest to me a pathway for more research in this area.

If we are able to start generating that type of evidence, it could bolster the type of guidance that we can provide to communities who may be dealing with very specific issues and want to attain certain outcomes.

Maria Rosario Jackson

That does seem valuable. Thank you so much, Patricia, for sharing your work and insights and for making time to talk.

Acknowledgment

Special thanks to Christina Park and Emma Galligan from the Studio for Creativity, Place and Equitable Communities at Arizona State University, for editorial and technical contributions.

Bibliography

Markusen, A. and Gadwa Nicodemus, A. (2010). *Creative Placemaking*. Washington, DC: National Endowment for the Arts [online]. Available at: https://www.arts.gov/sites/default/files/CreativePlacemaking-Paper.pdf (Accessed 12 Oct 2019).

NEA (National Endowment for the Arts). (2019). *Our Town: A Theory of Change and Logic Model for the National Endowment for the Arts' Creative Placemaking Grants Program*, April 2019 [online]. Available at: https://www.arts.gov/sites/default/files/Our-Town-Theory-of-Change.pdf (Accessed: 1 June 2020).

NEA. (n.d.) *Exploring Our Town* [online]. Available at: https://www.arts.gov/exploring-our-town/#:~:text=for%20the%20Arts-,Exploring%20Our%20Town,Explore%20recent%20creative%20placemaking%20projects (Accessed: 1 June 2020).

NEA. (n.d.) *Our Town Grant Program* [online]. Available at: https://www.arts.gov/grants-organizations/our-town/grant-program-description (Accessed: 1 June 2020).

Further reading in this volume

Chapter 1: Introduction: what really matters – moving placemaking into a new epoch
Cara Courage
Preface: Placemaking in the age of COVID-19 and protest
Jason Schupbach
Chapter 2: Placemaking as an economic engine for all
James F. Lima and Andrew J. Jones
Chapter 3: An annotated history of creative placemaking at the federal level
Jen Hughes
Chapter 4: A future of creative placemaking
Sarah Calderon and Erik Takeshita
Chapter 5: Making places for survival: looking to a creative placemaking past for a guide to the future
Jeremy Liu
Chapter 12: Public transformation: affect and mobility in rural America
Lyndsey Ogle
Chapter 14: Experts in their own tomorrows: placemaking for participatory climate futures
Paul Graham Raven
Preface: The radical potential of placemaking
Cara Courage
Chapter 25: 'If you can make it there, you can make it anywhere…': cultural placemaking at the heart of cities
Sherry Dobbin
Preface: The only thing constant is change
Kylie Legge
Chapter 36: Facilitator skills for effective collaborative placemaking
Husam AlWaer and Ian Cooper
Chapter 37: *The Neighbourhood Project*: a case study on community-led placemaking by CoDesign Studio
Lucinda Hartley, Eliza Charley, Sama Choudhury, and Harriet McKindlay

44

CREATIVE PLACEMAKING AND COMPREHENSIVE COMMUNITY DEVELOPMENT

Rethinking neighborhood change and evaluation

Maria Rosario Jackson

Introduction

Over the past several years, there have been important developments relevant to the conception and assessment of equitable Creative Placemaking initiatives and, more generally, the assessment of comprehensive community-development strategies. The observations and insights presented in this chapter draw from an examination of selected grantees of The Kresge Foundation Arts and Culture team's Creative Placemaking initiative who are working primarily in low-income and historically marginalized neighborhoods around the United States. This examination included review of grantee reports and other documents, interviews, and focus group discussions in the first several years of Kresge's Creative Placemaking strategy implementation, starting in 2012. These observations and insights also stem from a still small, but growing, body of research on the roles of arts and culture in communities, some of which precedes the adoption of the term Creative Placemaking, as well as from well-established literature on urban poverty and inequality. As the practice of equitable creative placemaking increasingly aligns with aspirations toward more just communities, we are beginning to recognize that, at its best, the work requires a more nuanced understanding of urban inequality; how arts, culture, and community-engaged design intersect with strategies to expand opportunity and how residents in low-income communities may benefit. The integration of arts, culture, and community-engaged design into community development and planning also challenges us to rethink how we conceive of and track change at the neighborhood level as well as change in the systems that are often part of the root causes of community conditions.

The challenge of describing creative placemaking

The term Creative Placemaking and the related concept of Creative Placekeeping have gained traction in the fields of arts and culture, design, community development, urban planning, public health, and others, and Creative Placemaking is emerging as a nascent field of its own. Creative Placekeeping emerged in response to Creative Placemaking as a way of calling attention to the need for vulnerable, most-often communities of color, to guard against social, cultural, and economic displacement. Displacement of low- and moderate-income communities in some real estate markets has been perceived to be hastened by cultural enhancements to communities, especially when such enhancements are not coupled with strategies to preserve affordability (see Bedoya, 2013). A persistent and critical challenge is that the practice has been difficult to describe. This is true for several reasons. First, the term still has multiple definitions and interpretations, and each sector involved in the work requires a translation of the concept suitable for its audience(s) – often a translation that allows for the bridging of arts, culture, and design to concepts and practices already familiar within a field. Second, the concept continues to be difficult to convey quickly, given that Creative Placemaking activity is often contextual and can manifest in different ways – building on cultural assets specific to communities and in various dimensions of comprehensive community-development processes. Third, for many people, including people in the arts-and-culture sector, definitions of art and the roles of artists in society are frequently narrow and not inclusive of cultural assets in low-income communities, arts-based processes, or the diverse roles of artists, designers, and culture-bearers in planning and community development. All of this often precludes full understanding of Creative Placemaking practices.

A review of grantees of The Kresge Foundation's Creative Placemaking initiative who operate in neighborhoods showed that investments took many forms. What they have in common is the focus on attempting to strengthen comprehensive approaches with arts, culture, and community-engaged design elements. This involves activity such as the inclusion of artists, designers, and culture-bearers in the crafting and implementation of community organizing, empowerment, and visioning efforts; the creation of physical structures and changes in the built environment that are meaningful and beautiful; the delivery of social services that are culturally relevant and appropriate; and the creation of businesses and other enterprises that tap into community imagination, talents, and heritage. As a result, approaches to community development build on the creativity and wisdom of residents, lift up cultural assets, and are, in fact, even more comprehensive.

Consider the following diverse examples of Kresge grantees working at the local level. Surrounded by San Francisco real estate market pressures, the Chinatown Community Development Center strives to preserve and protect its place in the city while recognizing the shifting needs of its community. Inspired and fortified by their cultural heritage, the center has maintained a continuous practice of recognizing and celebrating cultural assets through activities that include: walks in the neighborhood; annual community traditions; art exhibits, films, and events that elevate community history, aesthetics, and style. Residents, artists, culture-bearers, merchants, and community organizers are actively involved in shaping the social character and built environment in the neighborhood. This includes cultural programming intended to increase community pride, connection, and stewardship. Moreover, through culturally relevant community organizing practices, residents stay abreast of critical community issues and contribute to the design of open spaces and transit-oriented developments, helping to maintain and improve a viable, vibrant, and affordable place for longtime Chinatown residents and newcomers alike.

In New Orleans, New Corp, Inc. seeks to revive the historic buildings and craft traditions prevalent in the 7th Ward, while simultaneously addressing employment training and placement needs, blight, and vacancy. New Corp, Inc., along with the New Orleans Master Crafts Guild and other organizations, offers neighborhood residents master-craft apprenticeship training and case management support that result in construction certifications and paths to employment. Additionally, the organization hires graduates of the program to assist with rehabilitation of vacant residential buildings, with a focus on historical design. New Corp, Inc. plans to provide the renovated housing to residents at subsidized prices, buoying a significant, but waning, craft tradition that contributes meaningfully to New Orleans's distinctive architectural character and heritage. That craft tradition has been handed down through generations of Creole craftspeople for more than 200 years.

In Minneapolis, residents in neighborhoods served by Pillsbury United Communities are writing and performing original theater works examining their environment, personal journeys, and the broader human condition. Pillsbury United Communities, a human-services organization made up of four networked neighborhood centers, devises strategies to address intersecting needs and issues. Its creative approaches foster individual and collective resilience and self-sufficiency for participants. With a core value of integrating arts and culture throughout its work, Pillsbury offers a range of programs that tap into the creativity, imagination, experience, and wisdom of the people they serve. Its work impacts program participants and their families and neighbors. Its arts-infused method of working and offering services also contributes an important model and precedent in the human-services field.

In addition to specific, neighborhood-focused initiatives in several cities, sometimes artists are embedded in a range of municipal agencies including planning, transportation, and law enforcement. They help reimagine how such entities, through policies and practices, can better contribute to the creation of healthy, opportunity-rich environments where all people can reach their full potential. Within these systems, artists, designers, and culture-bearers often catalyze different ways of framing issues and new ways of working within bureaucracies, beyond individual policy silos and/or with residents in communities. At their best, these systemic interventions address significant barriers to opportunity and often have the capacity to bolster necessary neighborhood-level work.

One way to get a handle on the range of ways in which arts, culture, and community-engaged design intersect with comprehensive community development and planning is to recognize that essentially activity that falls under the umbrella of equitable creative placemaking and placekeeping serves to help reframe, retool, and repair. *Reframing* has to do with the role that artists, art practices, and art products can play in helping us to see an issue in a different way – by calling a different question, forcing us to shift our gaze away from the consequence of a problem to its root cause. Reframing an issue can help us get unstuck and can lead to fundamentally different approaches to complex and sometimes seemingly intractable circumstances and conditions. *Retooling*, often connected to the concept of reframing, has to do with reimagining the methods by which we seek to address fundamental actions. For example, artists and culture-bearers have made tremendous contributions in providing new ways of encouraging civic engagement through active creative practices – music, theater, film, and other art forms that allow for the interrogation of public concerns and even policy issues. The work of *repairing* through arts and culture, particularly in historically marginalized communities of color, can be deeply profound, meaningful, and necessary. This often requires reckoning with the harm and violence done to root cultures in processes of assimilation and subjugation. The work of repair is often about fighting against cultural erasure and degradation through cultural reclamation and validation. This also often involves helping individuals and communities reclaim agency over their own narrative, history, and physical environment as well as attending to material needs. Recognizing the various roles that arts, culture, and design can play in comprehensive community development has everything to do with our ability to appropriately and usefully assess the specific contributions of creative placemaking as well as our understanding of change in general.

Understanding urban inequality, neighborhood, and systems change and the contributions of creative placemaking

Alignment of Creative Placemaking practices with initiatives to expand opportunity in low-income communities relies on a more nuanced understanding of the root causes and consequences of urban inequality, new ways of thinking about how change happens, and corresponding improved ways of tracking and assessing change.

Urban inequality

To date, most efforts to address urban inequality through Creative Placemaking have embraced the need to be comprehensive or cross-sectoral. To be sure, issues of housing, employment, education, and health, among others, are most often interrelated, and approaches to these issues must account for that. I argue that we must go further to truly uncover the most strategic ways in which the integration of arts, culture, and design in planning and community development can have impact. Drawing from longstanding and extensive research on urban inequality in the United States from sociologists, economists, anthropologists, and scholars in urban planning and public policy, I have posited that poverty and inequality are the result of multiple interrelated factors. These stem from deep flaws in the socioeconomic opportunity structure (racism, sexism, classism, etc.) shortcomings in the institutions that exist to connect people to opportunity, and people's responses to long-term exclusion and disconnection from opportunity. Beyond acknowledging the interrelated nature of socioeconomic conditions and often poorly coordinated diverse areas of public policy, this articulation allows for a more multidimensional way of thinking about comprehensiveness. It affirms the necessity of tackling pressing issues at individual, family, and neighborhood levels, but it also elevates the need to ensure that issues are addressed holistically, inclusive of necessary sustained structural, systemic, and institutional changes.

Mapping the ways in which arts, culture, and design can have plausible impacts at different points of intervention is helpful in both strategy-building and impact assessment. Here, again, the concept of reframing, retooling and repairing apply. Reframing: what role can artists or designers working with residents play in reimagining systems that are exclusionary? Retooling: how might culture-bearers help design programs that are more effective, relevant, and culturally appropriate for residents in historically marginalized communities? Repairing: how can the residents' practice of heritage-based arts-and-culture traditions as part of a comprehensive strategy be impactful? How might changes in the built environment or the creation of businesses that celebrate the cultures of historically denigrated or maligned groups make a difference?

Neighborhood change and barriers to capturing contributions of creative placemaking

As the field struggles to evaluate the impacts of Creative Placemaking, it has become clear that the pace of change is often different from the pace of funding cycles. Foundations and other funders, at their best, know that neighborhood investments may not yield the ultimately desired results during a grant period. Those results may not manifest for years to come. However, research on the role of arts and culture in communities (Jackson et al., 2002; Wali et al., 2002; Walker et al., 2017) and interactions with grantees strongly suggest that many Creative Placemaking efforts can lead to some nearer-term outcomes, including greater social cohesion and sense of agency among residents, increased pride and stewardship of place, physical transformation, and greater control over community narrative. This is particularly true of activities at the neighborhood level that involve celebration of community cultural assets, individual and collective artmaking, interrogation of social issues through artistic media, and physical transformation of previously blighted areas. These are all important contributions in and of themselves. They are also often understood by practitioners on the ground as preconditions for other types of longer-term change. Consistent with previous research, many grantees interpreted these contributions as important steps toward outcomes such as more-equitable economic development, homegrown creative entrepreneurship, and important policy changes including those supporting creativity, innovation, and the preservation, protection, and advancement of community assets.

Recognizing that these preconditions for longer-term change are imperative has implications for how the community-development field initiates strategy development and gauges progress. We are just beginning to realize how this conceptual breakthrough might be embraced and how it might manifest in different field practices and policies. Community development-related theories of change that do not account for the roles of arts and culture, their contributions, and the concept of preconditions for some types of change, are inadequate. They do not represent our best thinking about how change happens and what is required. These need to be updated – modified to reflect our experience and best thinking. Changes in evaluation orthodoxies used by community developers, urban planners, and people from other intersecting fields also will be required. In addition to the need to reassess theories of change and strategies to assess progress inclusive of contributions of arts, culture, and design, other shortcomings must be considered. The following is not an exhaustive discussion of all limitations in these fields. These are selected observations that signal critical areas that warrant attention if Creative Placemaking and better ways of addressing inequality are to be successful.

Recalibrating concepts of neighborhood reinvestment and change in community development and planning fields

In many markets, the traditional focus on attracting reinvestment to the urban core is outdated. The urban core is once again desirable to developers and people who crave more density and related amenities. In those neighborhoods, the challenge is not simply 'more development,' but how to equitably integrate new development while preserving affordability, culture, and community and also creating pathways for existing vulnerable communities to build wealth and benefit from infusions of new resources. For decades, when addressing issues in low-income communities, students of urban planning and community development were trained around the concept of community revitalization and the need to attract investment to neighborhoods hollowed out by urban renewal and white flight. Urban renewal refers to a period in the development of many American cities, during the 1950s and 1960s, in which investments in new highways and the removal of 'urban blight' resulted in the decimation of largely low-income African American and Latino communities and hastened the migration of white people from increasingly racially mixed city centers to more homogenous white communities in the suburbs. No one was sufficiently trained to manage the unbridled reinvestment or return to the urban core and related racialized dynamics that we see in many cities today. The community-development and urban-planning fields have been caught unprepared and must catch up. Similarly, in other related fields, the contemporary experience of practitioners is exposing shortcomings in long-held and outdated assumptions about how things actually work. The observation about the community-development field and dominant thinking about economic revitalization is related to a challenge we encounter with Creative Placemaking: the too-frequent and often overly simplified association of the presence of artists and growing cultural vitality with 'gentrification' or, more specifically, the loss of affordability and the psychological, cultural, and physical displacement of vulnerable populations. In real estate markets where displacement concerns are warranted, it is incumbent upon planners, developers, and community leaders to ensure that Creative Placemaking strategies are integrated with a suite of related interventions that, at minimum, mitigate displacement and, at best, truly expand opportunity.

Recalibrating dominant concepts of impact and excellence in arts and culture

Not unlike the community-development and planning fields, the arts-and-culture field also has shortcomings related to limited ways of understanding impacts, industry standards of excellence, and education and training for artists and designers. In part as a result of the rise of Creative Placemaking, there is growing interest in better understanding and documenting the social impacts of the arts as well as impacts related to health and wellbeing. However, for decades, and even now, the lion's share of research and advocacy focused on arts impacts has concentrated on economic impacts. As such, the body of research available to examine or support Creative Placemaking fully is still emerging. Advocates do not yet have all of the skills and tools required to make the case for the value of the arts in ways that resonate with what we are learning about the various roles of arts and culture in communities.

Another challenge in the arts-and-culture field has to do with existing standards of excellence and corresponding well-developed validation systems that are poor fits for Creative Placemaking. Standards of excellence in the arts field, for the most part, tend to align best with artforms that result in art products for presentation, sale, and consumption in the conventional arts market. Arts-and-culture work that is integrated into community life and is process-heavy does not result in products for conventional presentation or sale, may not even include professional artists, and typically does not aspire to critical acclaim by tastemakers in the art world. The creation of appropriate standards of excellence and validation systems is a crucial piece of work for the ethical advancement of Creative Placemaking and, ultimately, to ensure benefits for already-vulnerable populations. On a related note, the creation of education, training, and professional-development opportunities that can help artists and designers ethically work in communities and collaborate with entities outside of the arts toward equitable outcomes and public good is essential. In recent years, there has been evidence of more academic programs focused specifically on Creative Placemaking as well as growth of programs in public practice, social practice, and similar genres, which are relevant, although not the same. Still, many artists involved in Creative Placemaking acquire their skills on the job, often through baptism-by-fire, working through trial-and-error in and with communities.

Promising developments and trends in creative placemaking evaluation and research

Consistent with earlier observations about standards of excellence and validation systems, there is important work to be done in developing research and evaluation practices aligned with Creative Placemaking, and there are promising signs of progress. Early in the practice of Creative Placemaking, there was premature interest in summative evaluations – a rush to account for causal impact of Creative Placemaking practices in ways that were not appropriate for the work or consistent with the stage of development of the field. Fortunately, there is now recognition that rushing to summative evaluations without better understanding the practices themselves, the contexts in which they are carried out, and the pace of change would be a mistake. Another development is that the national focus on Creative Placemaking has led to a revival of interest in creative ways to enhance and increase resident engagement in civic and community issues, including a revival in community-led and community-engaged research. It has also led to revisiting asset-based community development approaches, evolution in cultural asset mapping tools, and a renewed appreciation for underappreciated features of a community (Kretzmann and McKnight, 1993). To this end, public folklorists and applied ethnographers have surfaced as important collaborators. They bring nuance, rigor, and energy to cultural asset mapping processes, often advancing strategies that include citizen ethnographers and reveal community values, history, and aspirations that otherwise could go unrecognized (see, Alliance for California Traditional Arts and Southwestern Folk Alliance for example.) Artists also have surfaced as resources in this area, bringing arts-based strategies for identifying community cultural assets and assessing neighborhood changes that complement more-conventional research and evaluation, including social science–based methods reliant on secondary and administrative data (for an example of art-based cultural asset mapping see LA County Arts).

Indicators vs. indications

The absence or scarcity of conventional quantitative data serving as clear indicators of progress – recurrent, reliable quantitative data about community characteristics and trends or program performance measures – does not mean that there is no way to measure Creative Placemaking impacts. There are studies of social cohesion, agency, and similar concepts that are important precedents. I have also found that practitioners understand on-the-ground clues that refute, confirm, or expand their hypotheses about the change they expected to see as a result of art-, culture-, and design-based interventions. These observations can be thoughtful and disciplined without leading immediately to rigorous, serial, quantitative data. I call these disciplined observations and discernments 'indications.' In the absence of more quantitative data, indications provide important signals and insights that can inform policy and program development. I am not arguing against the importance of traditional quantitative indicators but offer the concept of indications as an alternative that is useful, particularly when the data infrastructure for a field of practice is in early stages or the subject matter is a poor fit for conventional quantitative methods.

Innovation and measurement

There is increasing evidence of openness to experimenting with new and innovative ways of understanding community conditions and change processes. Examples of this include the Local Initiatives Support Corporation's experimental work with arts-based inquiry into Creative Placemaking initiatives through a collaboration between its research division and artists, as well as the work of PolicyLink with ArtPlace America as they set out to document and assess the process and impacts of the ArtPlace's Community Development Investments program. Also, in the spirit of experimentation, with support from The Kresge Foundation, the Creative Measurement Lab at Arizona State University is a pilot effort involving practitioners, community organizers, and researchers from around the country involved in comprehensive community development and neighborhood change along with graduate students and faculty from fields including the arts, design, architecture, theater, social sciences, sustainability, urban planning, and public policy. They collaborate on questioning existing evaluation orthodoxies, developing new ways of framing community issues, and measuring difficult-to-document community attributes while lifting up the expertise of people who live in the communities in question. Community characteristics explored in the first round of Creative Measurement Lab included agency, social cohesion, stewardship, narrative of place and the systemic, and other factors contributing to the state of those characteristics. Consistent with principles of Equitable Evaluation, ongoing work in Creative Measurement Lab includes modifying existing evaluation methods and tools to account for Creative Placemaking practices in community development, planning and related fields as well as the development of new evaluation approaches (Equitable Evaluation Initiative.) Equitable Evaluation is an emerging practice that calls into question well-established approaches to program assessment and research that may carry inherent biases that preclude a useful and nuanced understanding of conditions, dynamics, and changes in low-income communities. Creative Measurement Lab is also involved in the creation of corresponding pedagogical resources to equip aspiring and existing practitioners in new ways of approaching assessment and research.

Conclusion

To me, the examined experience of Creative Placemaking to date reveals that developing viable, new ways of framing and capturing community and systems change involves taking risks – calling into question our usual ways of working and notions of expertise, creating intentional spaces where people who don't typically come together to work on evaluation do, and where conventional power arrangements are up-ended in an effort to seek truth and proximity to the conditions we wish to see impacted. It also requires calibrating expectations about timelines and management methods inherent in blending different perspectives; attempting to build the structures and validation systems that support new, smarter, and more-holistic, ethical, impactful ways of working. Most importantly, it requires leadership, stewardship, and the resolve to try something different and accept the process of failing, learning, adapting, and trying again.

Acknowledgments

This essay was adapted from a previous publication of the Kresge Foundation: Jackson, M.R. (2019). *Creative Placemaking: Rethinking Neighborhood Change and Tracking Progress*. Troy, MI: The Kresge Foundation.

References

Alliance for California Traditional Arts. [online]. Available at: http://www.actaonline.org/content/building-healthy-communities-cultural-treasures (Accessed: 25 June 2020).

Bedoya, R. (2013). 'Placemaking and the politics of belonging and disbelonging', *Grantmakers in the Arts Reader*, 24(1).

EEI (Equitable Evaluation Initiative). [online]. Available at: https://www.equitableeval.org for more information (Accessed: 25 June 2020).

Jackson, M.R., Herranz, J. and Kabwasa-Green, F. (2002). *Culture Counts in Communities: A Framework for Measurement*. Washington, DC: The Urban Institute.

Kretzmann, J. and McKnight, J. (1993). *Building Communities from the Inside Out: A Path Toward Finding and Mobilizing A Community's Assets*. Chicago: ACTA Publications.

LA County Arts. [online]. Available at: https://www.lacountyarts.org/willowbrook/ (Accessed: 25 June 2020).

Southwestern Folk Alliance. [online]. Available at: https://www.southwestfolklife.org/la-doce-research-findings/ (Accessed: 25 June 2020).

Wali, A., Severson, R. and Longoni, M. (2002). *Informal Arts: Finding Cohesion, Capacity and Other Cultural Benefits in Unexpected Places*. Chicago: Chicago Center for Arts Policy, Columbia College.

Walker, C., Gadwa Nicodemus, A. and Engh, R. (2017). *More Than Storefronts: Insights into Creative Placemaking and Community Economic Development*. New York: Local Initiatives Support Corporation.

Further reading in this volume

Chapter 1: Introduction: What really matters – moving placemaking into a new epoch
Cara Courage
Preface: Placemaking in the age of COVID-19 and protest
Jason Schupbach
Chapter 3: An annotated history of creative placemaking at the federal level
Jen Hughes
Chapter 4: A future of creative placemaking
Sarah Calderon and Erik Takeshita
Chapter 5: Making places for survival: looking to a creative placemaking past for a guide to the future
Jeremy Liu
Chapter 6: Listen, connect, act
Kim Cook
Chapter 10: From moon village to mural village: the consequences of creative placemaking in Ihwa-dong, Seoul
Jason F. Kovacs and Hayun Park
Chapter 11: *Free State Boulevard* and the story of the East 9th Street Placekeepers
Dave Lowenstein
Chapter 12: Public transformation: affect and mobility in rural America
Lyndsey Ogle
Preface: The radical potential of placemaking
Cara Courage
Chapter 20: Displacemaking 2015 and 2020
Catherine Fennell and Daniel Tucker
Chapter 23: Routing out place identity through the vernacular production practices of a community light festival
Gail Skelly and Tim Edensor
Chapter 25: 'If you can make it there, you can make it anywhere…': cultural placemaking at the heart of cities
Sherry Dobbin
Chapter 29: The solution is in the problem: the art of turning a threat into an opportunity by developing resilience using a creative placemaking critical praxis
Anita McKeown
Chapter 31: Seven generations: a role for artists in Zuni PlaceKnowing
Theodore S. Jojola and Michaela P. Shirley
Preface: The only thing constant is change
Kylie Legge
Chapter 36: Facilitator skills for effective collaborative placemaking
Husam AlWaer and Ian Cooper
Chapter 37: *The Neighbourhood Project*: a case study on community-led placemaking by CoDesign Studio
Lucinda Hartley, Eliza Charley, Sama Choudhury, and Harriet McKindlay

CONCLUSION

PREFACE

Cara Courage

Having come through a journey of theory, research, and practice, it is fitting to close this Handbook with a chapter from practitioners, and with the practitioner voice foregrounded, that encompasses all, and, as purposed as the final word in this Handbook, to reorientate the reader to their own place in placemaking.

45

HOW THE CITY SPEAKS TO US AND HOW WE SPEAK BACK

Rewriting the relationship between people and place

Rosanna Vitiello and Marcus Willcocks

Introduction

Let's take a walk. The moment we step out of the door and onto the street we are embraced by place – and from here our relationship with it begins to build. Roland Barthes told us:

> The city is a discourse, and this discourse is truly a language: the city speaks to its inhabitants; we speak our city, the city where we are, simply by living in it, by wandering through it, by looking at it. Still the problem is to bring an expression like 'the language of the city' out of the purely metaphorical stage.
>
> *(Barthes, 1986)*

Certainly, for so many of us these most basic of activities – the wander, the view, or the lived engagements along the street – have once again become precious gems as we reacclimatise to rapidly shifting 'normals' of everyday life. What is different today may surprise Barthes, in that the ways we connect with the places most local to each of us are evolving fast. The languages for connecting have now begun to move beyond the purely metaphorical and towards more symbiotic, responsive, and active discourses. As we continue our stroll, what signs and signals do we now pick up about the street we are on? What prompts us to stop and look again, or acknowledge someone we've not met before?

From our simple walk – and equally thanks to authors including Barthes, and more recently, Deyan Sudjic (2017), Collin Ellard (2015), and Anna Minton (2012) – we learn how the environments we pass speak to us, helping us build perceptions of what lies around us. We have described previously (Vitiello and Willcocks, 2011a, 2011b, and 2006) how we each build our own dialogues with places through urban lexicons – drawing upon the language of the elements that transmit a place's character to us, and defining our own place within it. As we first encounter the setting, we take in cues and clues that build our impressions and actions as we move through the place. As we get to know it, we change what we 'hear.' We also change the ways we want to speak back – through who we connect with and the ways we react to, interpret, or repurpose the street.

Between 2008 and 2019, the *Urban Lexicons* project allowed us to learn from widely diverse neighbours, citizens, placemakers, and urban managers through a series of street workshops (see for example, Vitiello and Willcocks, 2012), in cities including London, Barcelona, New York, Adelaide, and Parramatta. Time and again, we saw that the adjectives, offered by or drawn from a place, can be more important to how a person connects with it than the physical construction or the advertised lifestyle which may be projected. The lexicon of a place is made up from experiences lived through sensorial signs – of love, life, welcome, security, and more – each in interplay with our behaviours, perceptions, and imaginings of that setting. This approach learns from foundations set out between authors including Landry and Murray (2017), Charles Landry (2012b), William H. Whyte (1980), and Kevin Lynch (1960), who each help reflect how people instinctively or unconsciously draw together images of city spots, which frame our impressions and actions in response. And yet, our world and our place within it is being turned upside down.

With the first coronavirus lockdowns of 2020, we were told the world would never be the same again. Almost immediately the streets and neighbourhoods we live between came to occupy so much more of our daily encounters. Traffic reduced dramatically, while the home and all that is most local to us became essential in ways previously taken for granted. People began to revalue and reimagine the role of the street or neighbourhood in their lives. Our efforts to stay connected with others through shifting digital and physical norms became exaggerated, while social distancing became core in our uses of public space. Through all its tragedy, this new era quickly stimulated the start of transformations that many felt were overdue, including evolving physical and social ties with the places most local to us. People spending more time where they live means people noticing things and being part of new shapes of activity in public places. Everyday experiences of streets and the stories that unfold from them are different now. But the changes have been a long time coming.

How the city speaks to us

Before there were streets there was land. Places were made and interpreted, in mind and body. 'The language of landscape is our native language' (Whiston Spirn, 1998, p. 15), even if it's an idiom many urban dwellers have lost. Nature writers George Monbiot and Robert Macfarlane urge us to find this connection again, pointing us to the earliest Aboriginal Australian songlines, which described how 'each significant landform was both a tangible object and an intangible sign' (Macfarlane, 2013, p. 112). Has this human art of symbiotic relations with our setting been forgotten with the world's shifts towards urban life? Neuroscientist Colin Ellard (2015, p. 15) suggests that a defining characteristic of humanity is that we build onto the landscape 'to change perceptions and to influence thoughts and feelings; by these means, we attempt to organize human activity'. This works to impact us so deeply he says (ibid., p. 81), that we 'can grow to love a building or place in much the same way we love a person.' Urban explorer and essayist Walter Benjamin was enthralled by these deeply personal urban encounters:

> To lose oneself in a city – as one loses oneself in a forest – that calls for quite a different schooling. Then signboards and street names, passers-by, roofs, kiosks or bars must speak to the wanderer like a crackling twig under his feet.
>
> *(Tonkiss, 2006, p. 119, quoting Benjamin, 1986)*

For him, relating to the city is through small everyday encounters that require time together to learn its peculiarities and character. Ellard attests that our innate relations with space and our impulse to respond have stayed with us. He points to sensorial and contextual details affecting how our reactions 'can be read easily from our bodies. It is seen in our posture, the patterns of our eye and head movements, and even in our brain activity' which is 'massaged by our experience' wherever we go' (Ellard, 2015,17). Cultural critic Rebecca Solnit explores how the act of walking is the act of getting to know a place.

> A city is a language, a repository of possibilities, and walking is the act of speaking that language, of selecting from those possibilities. Just as language limits what can be said, architecture limits where one can walk, but the walker invents other ways to go.
>
> *(Solnit, 2014. p. 213)*

When cities became vehicle-centric we could argue that we lost that act. But the COVID-19 pandemic slowed patterns of street life, which we had previously assumed were the default. This slowing has gifted us the chance to look and listen more closely, reimagining our localities as we look forward to what's next. Hans Karssenberg, Jeroen Laven, and Mattijs van't Hoff with collaborators, including Jan Gehl, have long-championed the 'city at eye level' and tell how 'our perception of public space depends on viewpoint and distance, the speed at which we move is crucial. Our senses are designed to perceive and process sensory impressions while moving at about 5 km/h: walking pace' (Karssenberg, Laven, Glaser, and van 't Hoff, 2016, p. 318). Through the unusual start to the 2020s, streets and skies quietened, priorities began to change, and so people-first lexicons quickly became more prominent; from 'Heart the NHS' posters and chalk drawings as signs of love, balconies and decorated doorsteps as stages for shared applause or creative expression, and socially distanced parks and wider pavements as signifiers of people-centred freedom. As the meaning of these spaces rapidly shifts, we must remember that the carrier of the message means little without the reader. Just as we each have our own vocabulary borne out of our background and life experiences, we each speak different 'dialects' of place. The perspective of the 'reader' may be of resident, commuter, worker, or tourist. Yet the responsibility of anyone who considers themselves a 'placemaker' is to make this meaning visible and negotiable.

Viewed in this light, the act of placemaking should be a 'revelation' – the art of revealing meanings, connections, and agency in place. Yet, to whom do we turn to reveal the meaning of a street, neighbourhood, or city? Rarely to designers or planners, despite their central role in building. It is the communities who put on the play and make sense of the stage. Architect and author of the pattern-language movement that inspired Wikipedia, Christopher Alexander believes that cities should be shaped by all of us, as they were for millennia. This notion of a shared and open language is essential to its survival, he suggests. 'Architects themselves have lost their intuitions. They no longer have a widely shared language which roots them in the ordinary feelings people have' (Alexander, 1980, p. 233). In response, he offers 256 'patterns for building' as components of collective new lexicons, drawing upon existing understandings: 'To work our way towards a shared and living language once again, we must first learn how to discover patterns which are deep and capable of generating life' (Alexander, 1980, p. xii).

The art of discovery can start close to home – revelations are all around us. Charity Common Ground suggest that however ordinary we think of our streets and neighbourhoods as being, they are the classrooms that offer us the fundamental place-based language to learn from: 'What makes each place unique is the conspiracy of nature and culture, the accumulation of story upon history upon natural history' (Clifford and King, 2006, p. ix). The distinction lies in the detail: 'Small details spark the telling – a line of trees, the shape of a roof, the name of a street. They help us to share knowledge of what makes a place.' They argue that, 'Little things (details) and clues to previous lives and landscapes (patina) may be the very things that breathe significance into streets' (Clifford and King, 2006, p. xi).

As participants in our workshops made evident, every detail plays a role in how it communicates, whether we intend it to or not. For example, so-called security measures were commented upon for their detrimental rather than reassuring contribution to street safety. 'Horrible metal railings,' said Mark, reporting on galvanised palisade fencing, as he exited a station. 'It's just all ugh! … concrete round things for no reason, railings and CCTV,' reports Simon on the same spot. He continues, 'It's bizarre that it's oppressive yet it's really open at the same time, so why should it feel that way?' (Vitiello and Willcocks, 2011, pp. 120–21).

Meaning can be made and stories shared not only through those aspects that persist, but also those that are transient, moving, momentary. Indeed, it's the expression and interpretation of individuality that sets out the changing soul of a place – glimpsed for instance through a hand-painted pattern, or an array of planters by a doorstep. They are signs of identity, often personal, built, carved or painted into a cityscape. Alexander again, calls out that 'the pattern language in your mind is slightly different from the language in the next person's mind; no two are alike; yet many patterns are also shared' (Alexander, 1980, p. 203). The life of a place's language evolves with the people who are attached to it and as external forces trigger change. The pandemic stripped public spaces of their usual function: as places to commute through, work, or socialise. As we rapidly reinvent ways of living, everyday places take on new roles and are animated by different signifiers. Communities are redefining what place can mean through new rituals and reappropriation. At the time of writing, we are still 'feeling it out,' yet it's clear that a stronger connection is unfolding and a sense of specific local character emerging. So, how might we reappropriate our streets and mark meaning through these spaces differently, going forward?

As we rejoin our walk down a street, we find these patterns with rich variation and elements shared. Take Ridley Road, a market street in London's Hackney. Representations of shop owners' identities are woven into the urban fabric, evident when stalls lie still and shopfronts shuttered. Afghan rug sellers, Jamaican grocers, Armenian bakers, Nigerian tailors, Pakistani butchers – all paint clues to a sense of pride in their respective origins, a celebration of idiosyncrasy and personal self-expression. Local entrepreneur Kollier Din-Bangura identifies 'These small business owners …They're landmarks of a space, that's why people come. These people are the currency' (Vitiello, 2020). And yet these well-established signs of Dalston's identity remain under threat. Ubiquitous chains have moved in at the tail of the market, while signs adorn house fronts proclaiming, 'Save Ridley Road.' Sensing this erasure of character, this shared language has been captured over decades – by filmmaker Patrick Kieller in London (1994); The Decorators' Ridley's Restaurant Stall (2006); and by photographer Tamara Stoll's Ridley Road Market (2019). This battle between local distinctiveness versus homogenisation is seen on streets worldwide. It took some 200 voices to make Stoll's record of Ridley Road. To build upon what makes a place special, it will take a multitude of experiences to imagine and act upon its future. Common Ground make a call for 'a radical shift in the way we plan … towards a more responsive, detailed way of changing things' (Clifford and King, 2006, p. xiii). This is about embedding citizens as life-long place makers, leaving frames for meaning and opportunity to be interpreted in step with the flux of communities. It's about revealing how a city speaks to us, but more importantly, allowing us to respond.

How we speak back

Combining collaborative, inclusive design processes with a responsive streetscape can enable natural patterns of call and response. We understand that listening to a place is largely reflective, but how do we speak back? Put simply, this is about how we respond in, and to, a street and its setting. As with any language, these responses can be understood as 'passive' or 'active' voices.

A passive voice reflects those responses prescribed by the environment – passive by design. Take the straight line described by a long narrow pavement and the limited set of experiences we're offered. It encourages us to keep moving along, rather than slow down or engage. For placemakers and managers, these passive forms mean that we *behave* – acting predictably and in line. They also mean that we take our personal experiences with us, with little evidence of our 'language' or dialogue left behind for others to pick up on. In contrast, through speaking back in active voices, we bring our own contributions to a place. We are able to change the experience of that setting for ourselves and others, even if momentarily – usually in balance between forms of affordance and appropriation. If the context 'affords' (Norman, 2013) us the agency to bring something of ourselves to the street, our response may not be predicted but it is supported thanks to the structures in place. Resting against a bollard, sitting on a step, or painting on an open street art wall all make us feel 'heard.' Alternatively, we might speak back by 'appropriating,' to bring a different possibility forward. These alternative and unexpected visions might be seen on a scale between dissent and affirmation. From guerilla gardening to forms of graffiti, from unplanned activities such as kids playing in the fountains in London's Granary Square, or a group of neighbours trying something new for their street. Architect-activist Santiago Cirugeda's (1997) rental of on-street skip spaces supports different kinds of activities for locals, crosses between afforded and appropriated speaking back. His interventions playfully bend the rules to open up possibilities for others.

Whether passive or active, we need to ask 'whose voice?' As host to many walks of life, public realm is, by nature, contested space. It can be dominated by groups who at times have little consciousness of their hold over a site and how their very presence can put off or encourage others. Figure 45.1 offers examples, giving a glimpse of the breadth of voices in play or contestation. Accumulated signs of identity or common interest are of course 'read' by others differently. Yet, in full view of all, public space acts as a soapbox – to welcome or deter the unheard and unseen into the open and reassert ownership. From large-scale protests to small temporary takeovers, actions become gestures that begin to alter the way a place invites to speak back, marking a

EXAMPLES	Unprogrammed	Programmed
Speaking with Passive Voices	Walking along a pavement, to get from A to B; Playing hopscotch on somebody else's chalk drawing on the pavement.	Waiting for the bus at a bus stop; responding to a local authority planning consultation; socially distanced queuing at the post office.
Speaking with Active voices	Hopscotch chalk drawings on the pavement; Skateboarding in the street environment; Graffiti or land art; Protestors tugging down a statue with ropes; Playing music or having coffee from the doorstep with neighbours; leaving plant pots on street near your home	Making new market-stalls for Somers Town at MAKE@Story Garden; Booking to paint at Market Road Gallery; Park(ing) day; Splash Adelaide; Socially distanced street party between neighbours. Santiago Cirugeda's street skips; Weekly doorstep-applause for emergency services.

Figure 45.1 Sample passive and active voices (Vitiello and Willcocks, 2020).

powershift among users. Responsive street design and temporary programming can make places that more people feel fundamental to – breaking down barriers of age, gender, race, class etc. by shifting perceptions of who a space is for. Take gender: public space has classically been a male domain, and 'simply presenting as female in public space increases vulnerability to violence' (Cosgrave, Lam, and Henderson, 2020). Female-led projects such as *Her Barking* by StreetSpace aim to reddress that balance, collaboratively designing and testing low-cost 'experiments' and installations with women to improve perceptions of safety. Or consider children, whose imagination loosens the limits of what a space can be used for. Family-first thinking is seen in projects like urban innovation group 1319's *aUPA* in Argentina, which develops a network of pop-up play spaces that take over abandoned lots, creating a 'structure that can turn any urban void into a child-friendly place' (Trecediecinueve, 2017).

Placemakers and managers urgently need to support richer forms of dialogue through place, to design-in ways for people to speak back in active, not just passive voices. We know these are the relations with a street or city that grow to be resilient and support the ecologies of the place (Dovey et al., 1985). Yet, welcoming active voices can present messy, tricky challenges pressured by long-held approaches to place-making and planning. So often, the citizens' role has been presented as passive consumer, rather than anything more intimate such as an acquaintance, friend, lover, or co-creator. Sociologist Richard Sennett argues for collaboration, suggesting that the worth of collaborative production is that it speaks in the plural, creating different versions of open cities, rather than in the singular. However, current planning systems bias singular definitions. They certainly don't welcome 'messy.' Urbanist Michael Cowdy explained to us that the

> planning standard in many cities around the world has been in existence since the 1920s. This is not responsive to the demands of a twenty-first-century city population. The future city needs to transform from a static planning system to a more adaptable model which is needs based.
>
> *(Cowdy, 2020, personal communication)*

In the name of 'participation,' citizens are asked to attend a meeting or answer online questions, steered by those proposing a development. The 'consultation' box is ticked, and a development process continues out of public access. Such models set out prescriptively passive 'conversations,' in terms of communities' chances to respond to their streetscapes. One example is a recent one-way dialogue at Bedminster Green, Bristol. Here, the lexicon of the neighbourhood is of paint-peeled steel railings, post-war rapid-build buildings, value-engineered infrastructure, and under-maintained but much-loved greenery. Gladly, this is supplemented with a warmth and richness of character thanks to the locals, who carry great soul and welcome throughout the area. As the neighbourhood changes, their voices must be embedded. Thankfully, classic planning consultation systems are evolving. Placemakers JTP pioneered collaborative planning in the UK based on the American Institute of Architects R/UDAT model, whereby charettes with community and urban designers unpack local dreams and fears through rapid feedback sessions and sketches. Platforms such as Neighbourlytics or Neighborland open the conversation up online, enabling citizens to see immediately how design responds – the start of a digital–physical reciprocal relationship.

Increased responsiveness and active voices lend themselves to rapid change and surprise but also to trust and resilience. Yet planning at district or city scale takes a generation. So how do we allow these 'messy' models to become integral in a neighbourhood's evolution? Speaking-back lends itself to the kinds of embedded 'place-prototyping,' modelled by projects such as *MAKE@Story Garden*. Here, community members from Somers Town in London's Kings Cross collaborate both with creative practitioners from neighbouring Central Saint Martins art and design school and with other local organisations and groups. Through hands-on making, they sample and reimagine new designs to respond to how the neighbourhood is changing (Central Saint Martins, 2019). Across the Atlantic, the Brickline Greenway in St Louis, Missouri, bestows a developing green cultural network of public spaces and pathways spanning 15 miles and 20 very different neighbourhoods. Its goal is to connect communities, creating a 'movement to open up St. Louis and its opportunities' (GRG, 2020, p. 3) in a city with extreme social and cultural divides. Susan Trautman, CEO of public development agency Great Rivers Greenway (GRG) explains:

> This isn't just about building a greenway, it's about using the greenway […] to ensure that we look at the entire process with an equity lens […] So, the questions became: How would we make sure that we had voices at the table that had never been at the table?
>
> *(NRPA, 2020)*

GRG gathered working groups on equity, economy, identity, and governance, comprising many people from across the city's neighbourhoods and backgrounds. They opened up a public call to rename the Greenway to the citizens of St Louis. They drew on universal features such as 'the stoop,' celebrated as a symbol of invitation across neighbourhoods, 'where generations have gathered and shared stories, bringing this intimate communal activity out into public space giving people a place to connect' (NRPA, 2020). Landscape architects and local artists worked hand in hand through the design process, including the Artists of Color Council who have been integral to the vision, advocating for voices from the neighborhood; in short, an open and imaginative approach at massive scale.

The call here is for embedding more imagination, and the invaluable roles of collective agency, in the public realm. In *Designing Disorder: Experiments and Disruptions in the City*, Pablo Sendra and Richard Sennett (2020, p. 22) cite the 'overdetermination of both the city's visual form and it's social functions,' with the outcome being, that 'the urban imagination lost its vitality.' As we write this chapter, streets and entire cities are now racing to reinvent themselves and change the way they work. Tragically it has taken a crisis for us to realise it is time to stop and think about what is vital going forward. Now it is time to rewrite our relationships with place.

Revoicing our relationships with place

Through the moment that our cities were forced to close down we have glimpsed our public spaces opening up. Individuals and communities, place-creators and managers are now revisiting the relations we want with our streets and neighbourhoods. This era has fast-forwarded an already-emergent shift, away from place as a commodity, towards place as the living, breathing host of our lives; one where dialogue, interdependence, and people matter more. Richard Sennett has already set out an open city model, which should 'free people from the straitjacket of the fixed and familiar, creating a terrain in which they could experiment and expand their experience' (Sennett, 2019, p. 9). In practice, this hadn't been easy. Yet, to quote Phillipa Bannister, director of Street Space: 'Suddenly… there's a chink in the armour of "normal"' (Street Space 2020). Suddenly the street is reopening as a stage for so much more than driving – long the dominant urban language. With self-assigned agency, people are re-envisioning their streets as more personable places. Different voices are coming forward and rules of dialogue being redefined. In no time, roped-off playgrounds saw pavements turned into children's canvases for self-expression. Social distancing had us reinvent doorsteps, stoops, or parking spaces as gathering spots, and more. Such gestures marked new experiments to reveal future possibilities – as the opening lines to rewriting our relationship with place. Urbanist Stephanie Edwards describes this as 'active use, adaptation and mis-use' (Edwards, 2020, personal communication). None of this is by design, rather personal ingenuity and adaptation in seeing our surroundings through new eyes. Next, we must leave space for this to evolve. The popularity of 'meanwhile' use space (Madanipour, 2017; Bishop and Williams, 2012), has seen fantastic chances to try out and normalise different approaches for using city space. By definition however, the concept hosts limited-term possibilities, always subservient to a 'greater' plan, and normalises inequalities in terms of whose voice can be heard for how long.

How then can placemaking practices establish more equitable, ongoing iterative processes, where conversation remains open, where place is afforded, appropriated, and repeatedly recreated between community members, not just produced between built environment professionals and managers? How can chances to hear diverse active-voices become long-term normals, and different horizons, rather than sporadic opportunities? In our desire to learn from different voices, we shared our questioning with other practitioners and peers. Their responses veer from the technical language often used when speaking about the 'built environment.' Rather, evocative of the street workshops we conducted between 2008 and 2012, they lean upon an emotive vocabulary to talk about place. With thanks to these inputs, what follows are six characteristics that bring together collective ideas and opinions on establishing better dialogue and relations between people and cities.

Trust

If meaning making is about familiarity and growing stronger ties, how do we encourage a life-long relationship, rather than one-night stands with a place? Street Space champions the notion of 'moving at the speed of trust' (Street Space, 2020) as a starting point. In practice this means making changes in iterations with people over the long term, being dependable at every step. Sendra and Sennett (2020, p. 24–25) reflect that 'The bonds of a community cannot be conjured in an instant, with a stroke of a planner's pen; they too require time to develop.' To this end, Shenzhen-based placemakers SANS(三思) trialed a system in Dashilar, Beijing, 'where anthropology students spend extended amounts of time in the neighbourhoods to interview, and provide information to residents regarding interventions within the area' (Gaddes, 2020, personal communication). These students became 'Community Sensors' for deep, qualitative data-gathering, giving locals agency through a safe, unofficial space to learn about projects and give feedback.

Agency and expression

Being yourself and feeling free to share: that is the root of a good relationship. Increased freedoms in the city enable people to express their contribution. As writer Emma Warren asserts, we need to 'make some space' for street-level social energy and creativity, the type that 'make culture – and people – come alive' (Warren, 2019). As architect Ivana Stanisic puts it, to 'assert agency – making individual improvements such as guerrilla gardening or campaigning for improvements locally' (Stanisic, 2020, personal communication). Urban ceramicist, Maria Gasparian says that for people to engage with their environment they must 'be able inhabit it bringing their personal and social activities into it' (Gasparian, 2020, personal communication).

Confidence

Citizen confidence is key in promoting agency in a place. Landscape architect Cannon Ivers, recounts Rab Bennet's assertion that 'The confidence of a city is in the public realm, the insecurities are in the skyline' (Ivers, 2020, personal communication). Participatory City's (2020) research reports growth in personal levels of confidence among participants in the *Everyone Everyday* programme of active community participation (Participatory City, 2020). Scaled up, what could this offer? Urbanist Kate Meyrick talks about 'cities with an irrepressible mojo… exuding a confidence in who they are, warts and all.' Such confidence has to be earned, not prescribed (Meyrick, 2020, personal communication).

Soul

We desperately need urban lexicons which better convey personal character and individuality. Author Barry Lopez articulates this unease: 'We joke that one shopping mall looks just like another, that our housing development on the outskirts of Denver feels no different to us than a housing development outside Kansas City. But we are not always amused by such observations' (Lopez and Gwartney, 2010). Copy-and-paste places are no joke – and come from lack of local knowledge. Vitiello's *Design Like A Storyteller* workshops (2018) train urban designers to seek a deeper understanding of existing identity on the ground, translating symbolism into themes that drive design. Working with photographers, writers, or psychologists as experts in uncovering identity often reveals distinct and unseen insights about a place. Soul is life, defined by many markers. And in place this is fed by people, layering their personalities over time for the collective soul of a place to shine.

Reciprocity

The city is alive, yet often thought of as 'built' rather than 'lived.' Researcher Jessica Riley describes the reciprocal relationship we have with our streets – 'continuously both actors and acted upon.' Riley describes how

> these aesthetics of place shape our behaviours, beliefs and relationships; they are crucial to understanding our position in the world and with each other. Our responses to them offer us a reflection of the society we're embedded in, of others we share it with and of ourselves.
>
> *(Riley, 2020, personal communication)*

Reciprocity also comes in shifting power structures. Sendra and Sennett point out that active 'Communities… hold a powerful knowledge… of great value to urban designers. Partnering with these groups for the process of providing new public spaces and community infrastructure enhances socio-material interactions and starts new ones' (2020, p. 108–109); while SANS(三思) Neill Gaddes advocates for 'creating an exchange between government, stakeholders, users and communities that moves out of a hierarchical relationship based on money or expertise', and in doing so 'share horizontally' (Gaddes, 2020, personal communication). Or, as one of the author's neighbours suggests, seeing 'planet and communities [as] clients of the city' (Day, 2020, personal communication).

Love – and a little magic!

Love, in our relation with the city, is a verb (action), not just a noun (sensation). It can be shown through the intimate details that highlight care – flowers planted by a neighbour around a tree, marks of craftsmanship, or detail that draws us in. They show that the contributors to the place have gone the extra mile, expanding to municipal scale if we move away from classic top-down/bottom-up structures to one of active care. Through passion emerges magic – a word that rarely makes it into an urban design brief. As we spend time with our neighbourhoods, professional and community placemakers need to intertwine excitement into an extraordinary everyday – keeping the love alive over time. Riley reminds us to 'take note of what the materials in your environment are saying, what they're doing to and for you and what it means to you. Question if that is what you want it to be... Talk about it.' Architect Duncan Thomas sees our relationship in those magical places where 'people feel free to have fun and live life' and questions if 'after the pandemic people might feel ready to celebrate life with abandon and not fear the consequences?' (Thomas, 2020, personal communication).

The streets of tomorrow

> It is no exaggeration to say it is work like this which can save the world... it will be through lived experiences of doing the everyday differently.
>
> *(Marc Stears, Sydney Policy Lab, Participatory City, 2019)*

How can we 'do the everyday differently' and turn the process of making places on its head? Sendra and Sennett (2020, p. 109) look towards 'a city that's much more experimental,' involving 'more people messing about in the streets than eyes [on the streets]... The first orders, the second gathers' (ibid.). They point to the importance of learning together by doing together, discussed by Thorpe et al. (2016), and also advocate for 'having facilities in the public realm that allow for a greater diversity of community activities. Addressing such collective and public needs through shared infrastructure.' Willcocks has reported on forums designed as open infrastructures. Here, process-led forms of city-making coalesce via generative acts of discourse and deliberation-through-doing (Willcocks, 2017.p. 832). Now, we see citizen-led communities, public and private organisations push for new interpretations of street forms and processes. So, are we reaching a rare moment to initiate continuums for collaborative experimentation, to build-in place-prototyping, as the norm?

Frameworks that allow urban experiments in the long term support new economic and social infrastructures. In this context, University College London's (UCL) Urban Lab (Campkin and Duijzoings, 2016) discuss an evolving menu of methods for 'engaged urbanism,' and Theatrum Mundi's tools seek to 'expand the crafts of citymaking' (Theatrum Mundi, 2020). Listen deeper to frame a brief that focuses on what we want our streets to say and do for us, then turn to action and speaking back. Platforms, structures, and ongoing programmes for collaboration – such as *Make@Story Garden*, between Central Saint Martins and Somers Town/Camden community members; Splash Adelaide's long-term scheme in South Australia; or Participatory City's *Tomorrow Today Streets* – each give tangible tools and chances to speak with active voice in our neighbourhoods. Communities self-organize and build confidence, with support and knowledge-sharing available through these initiatives. The place becomes the platform, for people to bring different stories, ideas, and actions on the table. Such frameworks are adaptable in physique but also in programme. These approaches remain peripheral to professional practice of built environment design, yet they're often quicker, less expensive, more inclusive, and shape more resonant places. They enable us to prototype, fail safely, reimagine, and can be shared or scaled for broader impact.

The arrival of COVID-19 was a disaster which has also levered-open a door to a new era. The principle urban shifts are not through destruction of the physical city, but through changes to its uses, the agency afforded, and the meanings embodied, in relation to citizens. With budgets cut, the greatest value of placemaking is to enable the relationships that take place within it to flourish. Citizens are critical as lifelong placemakers and must be recognised as such – no place is made until the people bring their mark and their soul. Now is the time to establish practices that enable fluency of place in everyone, through open dialogues of action. Characteristics like trust, confidence, and magic may sound soft or untenable, but these are the powerful building blocks of more equitable, active, and self-sustaining cityscapes. The streets of tomorrow will be made, remade, and re-established through ongoing call and response. Are you ready?

Bibliography

Alexander, C. (1980). *The Timeless Way of Building*. Oxford: Oxford University Press.

Bannister, P. (2020). 'Edit your neighbourhood', in Street Space [online]. Available at: https://medium.com/street-space/edit-your-neighbourhood-826a433c276f (Accessed: 31 May 2020).

Barthes, R. (1967). *Elements of Semiology*. New York: Hill & Wang.

Barthes, R. (1986). 'Semiology and the Urban', in Lagopoulos, A.P. and Gottdiener, M. (eds.) *The City and the Sign: An Introduction to Urban Semiotics*. New York: Columbia University Press, 87–98.

Bishop, P. and Williams, L. (2012). *The Temporary City*. London: Routledge.

Campkin, B. and Duijzings, G. (2016). *Engaged Urbanism*. London, UK: Bloomsbury.

Central Saint Martins. (2019). *MAKE @ Story Garden* [online]. Central Saint Martins. Available at: https://www.arts.ac.uk/colleges/central-saint-martins/csm-public/make-story-garden (Accessed: 31 May 2020).

Chouteau CAC [online]. (2020). *Great Rivers Greenway* [online]. Available at: https://greatriversgreenway.org/chouteau-cac/ (Accessed: 8 May 2020).

Cirugeda, S. (1997). *Skips S.C.: Urban Recipes: Strategies for Subversive Urban Occupation. Steel, Concrete, Plastic.*

City Personality Test [online] (2020). Available at: https://test.urbanpsyche.org (Accessed: 8 May 2020).

Clifford, S. and King, A. (2006). *England In Particular*. London, UK: John Murray Press, Saltyard Books.

Cosgrave, E., Lam, T., and Henderson, Z. (2020). 'Scoping study-London's participation', in *UN Women's Safer Cities and Safe Public Spaces Programme* [online]. Available at: scoping_study-_londons_participation_in_un_womens_safer_cities_and_safe_public_spaces_programme.pdf (Accessed: 12 May 2020).

Destination Brands. (2012). Destination branding *and the* urban Lexicon: London, New York, and Barcelona [online]. Available: https://www.taylorfrancis.com/ (Accessed: 31 May 2020).

Djokovic, B. (2020). *On How to Read the City* [online]. Available at: http://www.disturbis.esteticauab.org/DisturbisII/Branka.html (Accessed 29 May 2020).

Dovey, K., Downton, P. and Missingham, G. (1985). 'Place and placemaking' in *Place and Placemaking*. Presented at the PAPER 85 Conference. Melbourne, Australia: Association for People and Physical Environment Research in association with Faculty of Architecture and Building, Royal Melbourne Institute of Technology.

Ellard, C. (2015). *Places of the Heart: The Psychogeography of Everyday Life*. New York: Bellevue Literary Press.

Fisher, T. and Gamman, L. (2018). *Tricky Design*. Bloomsbury Publishing [online]. Available at: https://www.bloomsbury.com/uk/tricky-design-9781474277181/ Accessed: 14 May 2020).

Gehl, J., Johansen Kaefer, L., and Reigstad, S. (2016). 'Close encounters with building' in Karssenberg, H., Laven, J., Glaser, M. and van 't Hoff, M. (eds.) *The City at Eye Level: Lessons for Street Plinths*. Delft, The Netherlands: Eburon.

GRG (Great Rivers Greenway). (2020). *The Making of a Greenway* [online]. Available at: https://greatriversgreenway.org/making-of-a-greenway/ (Accessed: 15 May 2020).

Happy City. (2020). Inspired by Nature: Superior Creative Placemaking Master Plan [online]. Available at: https://thehappycity.com/project/inspired-by-nature/ (Accessed: 8 May 2020).

Landry, C. (2012a). *The Creative City: A Toolkit for Urban Innovators*. 2nd (Revised) edn. London: Taylor & Francis.

Landry, C. (2012b). *The Sensory Landscape of Cities*. Gloucestershire: Comedia.

Landry, C. and Hyams, J. (2012). *The Creative City Index: Charles Landry*. Gloucestershire: Comedia.

Landry, C. and Murray, C. (2017). *Psychology & the City: The Hidden Dimension*. Illustrated edn. Gloucestershire: Comedia.

Levac, L.R.E. and Wiebe, S.M. (eds.) (2020). *Creating Spaces of Engagement: Policy Justice and the Practical Craft of Deliberative Democracy*. Toronto: University of Toronto Press.

London. (1994). *BFI Production*. London: Koninck Studios.

Lopez, B. and Gwartney, D. (eds.) (2010). *Home Ground: Language for an American Landscape*. San Antonio, TX: Trinity University Press.

Lynch, K. (1960). *The Image of the City*. Cambridge, UK: MIT Press.

Madanipour, A. (2017). *Cities in Time: Temporary Urbanism and the Future of the City*. London: Bloomsbury Academic.

Madanipour, A. (2018). *Temporary Use of SPACE: URBAN PROCESSES Between Flexibility, Opportunity and Precarity*, 55 (5), pp. 1093–1110.

McFarlane, R. (2013). 'Isle of Lewis', in Robson, D. and Evans, G. (eds.) *Towards Re-Enchantment*. London: Artevents.

McFarlane, R. (2016). *Landmarks*. Reprint edn. London: Penguin.

Millthorpe, N. (2019). *The Poetics and Politics of Gardening in Hard Times*. London: Lexington Books.

Minton, A. (2012). *Ground Control: Fear and Happiness in the Twenty First Century City*. 2nd edn. London: Penguin.

Montgomery, C. (2015). *Happy City: Transforming Our Lives Through Urban Design*. London: Penguin.

Nisha, B. (2018). Place and perception: Constructing the urban narrative with spatial schemas. The Centre for Conscious Design [online]. Available at: https://theccd.org/article/place-and-perception-constructing-the-urban-narrative-with-spatial-schemas/ (Accessed: 22 March 2020).

Norman, D. (1988). *The Design of Everyday Things*. New York: Basic Books.

NRPA (2020). *Vitisia Paynich, St. Louis: A Path Toward Equitable Design* [online]. Available at: https://www.nrpa.org/parks-recreation-magazine/2020/march/st-louis-a-path-toward-equitable-design/ (Accessed: 22 March 2020).

Participatory City. (2019). *Y2 Tools to Act* [online]. Available at: https://drive.google.com/file/d/1oHcXJchvC3qPQPazrwa3vDzF8QxSEfJw/view?usp=sharing&usp=embed_facebook (Accessed: 31 May 2020).

Sampson, R.J. (2012). *Great American City: Chicago and the Enduring Neighborhood Effect*. Reprint edn. Chicago: University of Chicago Press.

Sennett, R. (2013). *Together: The Rituals, Pleasures and Politics of Cooperation*. London: Penguin.

Sennett, R. (2019). *Building and Dwelling: Ethics for the City*. London: Penguin.

Sennett, R. and Sendra, P. (2020). *Designing Disorder: Experiments and Disruptions in the City*. London: Verso Books.

Solnit, R. (2014). *Wanderlust: A History of Walking*. London: Penguin.

Stoll, T.R. (2019). *Ridley Road Market*. London: Antenne Books.

Sudjic, D. (2017). *The Language of Cities*. London: Penguin.

The Decorators. (2011). *Ridley's Temporary Restaurant: The Decorators. Wood, Scaffold, Tarpaulin, Paint, Kitchenware*. London: The Decorators.

The New School. (2020). *Civic Imaginaries: Urban Commons*, UTNS5105, Course Catalog [online]. Available at: https://courses.newschool.edu/courses/UTNS5105/ (Accessed: 14 May 2020).

Thorpe, A., Prendeville, A. and Oliver, K. (2016). 'Paper No.114: Learning together by doing together building local government design capacity through collaboration with design education', in *ServDes. Service Design & Innovation Conference 2016 Proceedings*. Presented at the ServDes. Service Design & Innovation Conference. Copenhagen: ServDes, pp. 500–505.

Tonkiss, F. (2006). *Space, the City and Social Theory: Social Relations and Urban Forms*. Cambridge, MA: Polity Press.

Trecediecinueve. (2017). *aUPA Project* [online]. Available at: Trecediecinueve.wixsite.com/1319x/aupa-project (Accessed: 31 May 2020).

Vitiello, R. (2020). 'Local legends an interview series and research project that explores the field of narrative placemaking', in Local Legends [online]. Available at: http://www.local-legends.org (Accessed: 8 May 2020).

Vitiello, R. and Willcocks, M. (2006). 'The difference is in the detail: Its potential as a place branding tool and impact upon perceptions and responses', in Place Branding, pp. 248–262.

Vitiello, R. and Willcocks, M. (2011a). 'Destination branding and the urban Lexicon: London, New York, and Barcelona', in Morgan, N., Pritchard, A. and Pride, R. (eds.) *Destination Brands*. 3rd edn. Oxford: Butterworth-Heinemann.

Vitiello, R. and Willcocks, M. (2011b). *Unravelling the Urban Lexicons of our Everyday Environments*. Morrisville, NC: Lulu.

Warren, E. (2019). *Make Some Space*: Tuning into Total Refreshment Centre. London: Sweet Machine.

Whitson Spur, A. (2020). *Language of Landscape*. Yale University Press [online]. Available at: https://yalebooks.yale.edu/book/9780300077452/language-landscape (Accessed: 31 May 2020).
Whyte, W.H. (1980). *The Social Life of Small Urban Spaces*. New York: Project for Public Spaces Inc.
Zukin, S. (2011). *Naked City: The Death and Life of Authentic Urban Places*. Oxford: Oxford University Press.

Further reading in this volume

Preface: Placemaking in the age of COVID-19 and protest
Jason Schupbach
Preface: 'Disastrous forces, accidental actions, and grassroots responses'
Tom Borrup
Chapter 12: Public transformation: affect and mobility in rural America
Lyndsey Ogle
Chapter 14: Experts in their own tomorrows: placemaking for participatory climate futures
Paul Graham Raven
Chapter 16: More than a mural: participatory placemaking on Gija Country
Samantha Edwards-Vandenhoek
Chapter 18: '*Homomonument* sounds like a poem': queer placemaking 30 years on: a conversational dialogue with Thijs Bartels, author of *Dancing on the Homomonument* (2003)
Martin Zebracki
Chapter 21: Placemaking through parkour and *Art du Déplacement* (ADD) as a Singaporean applied performance practitioner in London
Adelina Ong
Chapter 22: Embedded Artist Project: Epistemic Disobedience + Place
Frances Whitehead
Chapter 24: Artists, creativity, and the heart of city planning
Tom Borrup
Chapter 27: Is 'tactical urbanism' an alternative to neoliberal urbanism?: reflections on an exhibition at the MoMA
Neil Brenner
Chapter 31: Seven generations: a role for artists in Zuni PlaceKnowing
Theodore S. Jojola and Michaela P. Shirley
Chapter 36: Facilitator skills for effective collaborative placemaking
Husam AlWaer and Ian Cooper
Chapter 37: *The Neighbourhood Project*: a case study on community-led placemaking by CoDesign Studio
Lucinda Hartley, Eliza Charley, Sama Choudhury, and Harriet McKindlay
Preface: Evaluating creative placemaking: a collection of observations, reflections, findings, and recommendations
Maria Rosario Jackson
Chapter 39: Translating Outcomes: Laying the groundwork for interdisciplinary evaluation of creative placemaking
Jamie Hand
Chapter 41: Rituals of regard: on festivals, folks, and findings of social impact
Maribel Alvarez
Chapter 42: Creative Placemaking and Placekeeping evaluation challenges from the practitioner perspective: an interview with Roy Chan
Maria Rosario Jackson
Chapter 43: A theory of change for creative placemaking: the experience of the National Endowment for the Arts' *Our Town* program: an interview with Patricia Moore Shaffer, PhD
Maria Rosario Jackson

INDEX

Page numbers in *italics* represent figures or illustrations.

www.ingramcontent.com/pod-product-compliance
Lightning Source LLC
LaVergne TN
LVHW080800011026
849160LV00015B/1494

* 9 7 8 0 3 6 7 6 8 4 0 9 9 *